QUEENSLAND'S

THREATENED ANIMALS

Dedication

To all the individuals and groups who dedicate their lives to the recovery and management of Queensland's threatened animals. We hope this guide will make your lives a bit easier.

To Peter Johnson whose immeasurable contribution to the recovery of Queensland's threatened mammals has not gone unnoticed – by us anyway ...

QUEENSLAND'S THREATENED ANIMALS

Edited by: LEE K. CURTIS, ANDREW J. DENNIS, KEITH R. McDONALD, PETER M. KYNE and STEPHEN J.S. DEBUS

CSIRO PUBLISHING

National Library of Australia Cataloguing-in-Publication entry

Queensland's threatened animals/edited by Lee K Curtis ... [et al.].

9780643096141 (pbk.)
9780643104563 (epdf)
9780643104570 (epub)

Includes bibliographical references and index.

Endangered species – Queensland.
Rare animals – Queensland.

Curtis, Lee K.

591.6809943

Published by

CSIRO PUBLISHING
36 Gardiner Road, Clayton VIC 3168
Private Bag 10, Clayton South VIC 3169
Australia

Telephone: [+613] 9545 8555
Local call: 1300 788 000 (Australia only)
Fax: +61 3 9662 7555
Email: csiropublishing@csiro.au
Web site: www.publishing.csiro.au

Front cover: Gulbaru Gecko *Phyllurus gulbaru* (Eric Vanderduys)
Back cover (clockwise from top): Eclectus Parrot *Eclectus roratus macgillivrayi* (www.ataglance.com.au); Apollo Jewel Butterfly *Hypochrysops apollo Apollo* (MF Braby); map of Queensland; Porcupine Ray *Urogymnus asperrimus* (Susan Theiss)

Set in 10/12 Adobe Minion Pro and Syntax LT Std
Edited by Peter Storer Editorial Services
Cover design by Jenny Cowan
Text design by James Kelly
Typeset by Desktop Concepts Pty Ltd, Melbourne
Index by Russell Brooks
Printed by Ingram Lightning Source

CSIRO PUBLISHING publishes and distributes scientific, technical and health science books and journals from Australia to a worldwide audience and conducts these activities autonomously from the research activities of the Commonwealth Scientific and Industrial Research Organisation (CSIRO). The views expressed in this publication are those of the author(s) and do not necessarily represent those of, and should not be attributed to, the publisher or CSIRO.

Feb26_RP_ILS

Foreword

This book is a sobering reflection of the state of our natural environment.

More than 1300 native Queensland plant and animal species face extinction, with at least 30 already gone forever. We have modified almost the entire landscape, clearing forests and starving entire natural food chains. Not even our reefs or marine life have escaped unscathed from the activities of land-based development, agriculture and over-fishing.

While humans have led the most direct changes to the natural environment, equally profound has been the impact of introduced species, such as livestock, foxes, cats, rabbits and weeds – to name a few.

Yet this book is also a symbol of hope.

Queenslanders have shown overwhelmingly that they want more action to save their native wildlife, with a vast majority in favour of the state government buying up new national parks and identifying and protecting threatened native species.

In recent times, the state government has banned broadscale land-clearing and regulated regrowth clearing, boosted funds for more national parks, expanded marine sanctuaries, reformed pastoral lease conditions, and controlled farm pollution and wetland destruction in Great Barrier Reef catchments. A whole-of-government biodiversity strategy is close to release.

Most Queenslanders want to live in a state where wildlife thrives once more in vast safe havens, where the forests have regained their former glory and productive soils are built up, rather than washed down rivers to smother marine life. They want to enjoy the same quality of life without damaging the planet through use of outdated technologies and an over-reliance on fossil fuels.

WWF believes this vision is achievable given the right political will and community leadership. And so we urge the Queensland Government to be ambitious and to build this dream. The people of Queensland are with you, which is clearly demonstrated by the efforts made to compile this resource guide.

Gilly Llewellyn
Conservation Program Manager
WWF-Australia

Contents

Acknowledgements

Overcoming the enormous challenges inherent to producing a complex publication on a microfibre budget would never have been possible without the invaluable contributions of over 200 volunteers from all over Australia and a few from overseas. These busy professionals wrote and reviewed, in their own time, 205 animal profiles (covering 226 species, subspecies or populations) and made other contributions throughout the book. It has been a true privilege and a pleasure to work with them. The care with which they carried out their work demonstrates their conviction that such a resource guide is very necessary. There were also those who made themselves available as sounding boards, providing wisdom and reassurance at a moment's notice. Their insight is greatly appreciated.

We would also like to gratefully acknowledge and thank the many spouses and families that supported and accommodated the writing and editing of this book, which was done away from salaried office jobs and impinged on family time.

Thanks to the Tropical Rivers and Coastal Knowledge (TRaCK) research hub at Charles Darwin University for supporting Peter Kyne's participation in the project and to TRaCK Director Michael Douglas for facilitating this.

We would like to thank the following people who assisted in various ways, by providing reviews, advice and/or assistance: Ross Alford, Andrew Amey, Angela Arthington, David Baker-Gabb, Derek Ball, John Bannister, Michael Bennett, Walter Boles, Adrian Borsboom, Michael Braby, Scott Burnett, Gavin Butler, Greg Calvert, Ken Chan, Graeme Chapman, Andrew Chin, Howard Choat, Chris Clague, Patrick Couper, Jeanette Covacevich, Ross Daley, Marg Davies, Chris Dickman, Daryl Dickson, Naomi Doak, Sheldon Dudley, David Ebert, Brendan Ebner, Mark Eldridge, Keith Fisher, Jenni Fox, Malcolm Francis, Alastair Freeman, Stephen Garnett, David Geering, Arthur Georges, Peter Gill, Brett Goodman, Greg Gordon, Ken Graham, Ian Gynther, Mark Hamann, Shane Hancock, Graham Harrington, Alastair Harry, Megan Higgie, Harry Hines, Glen Holmes, Tim Holmes, Conrad Hoskin, Deb Hotchkis, Paul Humphries, Bridget Hyder, Dean Ingwersen, Sarah Irvine, Ian Jacobsen, Tina Janssen, Chris Johnson, Jeff Johnson, Rod Kavanagh, Rosie Kerr, Peter Last, Ivan Lawler, Frank Lemckert, Colin Limpus, Lana Little, Susan Logie, Lindy Lumsden, Geoff Lundie-Jenkins, David McFarland, Jenny McLean, Nicola Markus, Wayne Martin, Michael Mathieson, Ed Meyer, Jeff Miller, Lloyd Nielsen, Barry Nolan, Penny Olsen, Lew Oring, John Parmenter, Andrea Phillot, Stirling Peverell, Simon Pierce, Gary Porter, Mark Read, Terry Reardon, Stuart Rowland, Lin Schwarzkopf, Bronwen Scott, Colin Simpfendorfer, Gregory Skomal, Geoff Smith, Susan Smith, David Stewart, Ilona Stobutzki, Stephen Taylor, Bruce Thomson, Michael Todd, Michelle Treloar, Fabian Trinnie, Peter Unmack, Terry Valance, Steve Van Dyck, Melanie Venz, Paul Walbridge, David Westcott, Yvette Williams, John Winter, John Woinarski and other anonymous reviewers. We apologise if we have left anyone out who wished to be listed.

For their insightful and constructive comments on the history chapter, Milena Kim and the editorial team would like to gratefully acknowledge the contributions made by Professor Helene Marsh, Associate Professor Peter Valentine, Lydia Gibsons and Mr Des Boyland.

Our profound thanks also go out to the many who donated photos to the cause. Photographers seldom receive the credit they deserve, but we truly do appreciate your talent and your generosity. Lee would especially like to thank Garrie Douglas, her husband, for instilling in her a profound passion for Australian wildlife, for his painstaking work compiling the maps and working on the photos, and above all, for his undying patience and support during the last 3 years it has taken to compile this resource guide. And, finally, sincere thanks go to the final copy editor, Peter Storer, and John Manger, Tracey Millen and Briana Melideo at CSIRO Publishing for their faith in Lee and in this project as well as for their patience and flexibility.

About the editors

Lee Curtis is the author of the *Green Guide to Kangaroos and Wallabies of Australia* as well as the Whitley award winning *Wallabies, Wombats and Other Mammals of Australia.* She is a freelance journalist and copywriter who is passionate about Australia's natural environment and wildlife. Lee is an active member of the Wildlife Preservation Society of Queensland and has been correspondent for *Wildlife Australia Magazine* since 2002. She writes about wildlife for a variety of publications and does presentations whenever an opportunity to help educate children and/or the public arises. Lee's volunteer work with several threatened species recovery projects in Queensland over the years inspired her to organise the compilation of a resource guide to Queensland's threatened animals in order to facilitate networking and communication among those involved in species management and recovery.

Andrew Dennis has a PhD in tropical ecology. He has conducted research through JCU, QPWS, CSIRO and the EarthWatch Institute on the ecology and management of over a dozen threatened animals in northern Australia, including frogs, mammals and birds. Andrew has also authored or contributed to a range of recovery plans and teams for single and multiple species. In addition, he has investigated the ecological consequences of species loss and its wider ramifications – examining the impacts of changes in species abundance, distribution and behaviour on the process of seed dispersal in tropical rainforest – a process driven primarily by animals, several of which are listed as threatened. Andrew has authored numerous book chapters and journal articles and was lead editor for *Seed Dispersal: Theory and Its Application in a Changing World*, in which several chapters investigate the consequences of species loss to the processes that maintain vegetation communities.

Keith McDonald has worked in Queensland government conservation agencies for the last 40 years. His work has covered most parts of the state and has included threatened species management, monitoring and research with a special emphasis on frogs and reptiles. He has published or co-authored numerous publications on Queensland fauna, coordinated and been involved in recovery teams associated with threatened species and has worked on the management of protected and threatened species on and off the protected area estate. He is the recipient of the Queensland Museum Medal, a Cassowary Award from the Wet Tropics Management Authority and was a member of the international team receiving the CSIRO Medal for studies on the chytrid fungus which decimated frog populations. He is Vice-president (Projects) of the Wildlife Preservation Society Queensland. His involvement in this book is as a private individual.

Peter Kyne has a background in the biology, fisheries and conservation of chondrichthyan fishes (sharks, rays, skates and chimaeras). He completed his PhD on chondrichthyan bycatch in Queensland's trawl fishery, examining the effectiveness of bycatch reduction devices for these fishes. Peter is an active member of the IUCN Shark Specialist Group, having assessed the conservation status of numerous Queensland chondrichthyans. He has also worked on freshwater fishes of northern Australia, and has an interest in bird ecology and conservation. Peter is currently based at Charles Darwin University in Darwin.

Stephen Debus studied biology at Macquarie University and did a graduate thesis on the Little Eagle at University of New England, Armidale, where he has remained. After working as a research assistant in bird ecology at UNE in the 1980s, he did a Masters on owls in the 1990s and a PhD and postdoc on declining woodland birds in the 2000s. Stephen recently reviewed the conservation status of various birds for NSW Department of Environment and Heritage. He now works as an ecological consultant, and is an honorary research associate at UNE, doing projects on various raptors. He co-edits the BOCA journal *Australian Field Ornithology*, edits the Australasian Raptor Association (ARA) journal *Boobook*, and serves on the Australian Bird Study Association committee (handling the raptor special issues of *Corella*). Stephen has authored field guides on raptors and owls, contributed to ARA conference proceedings and participated in several Red Goshawk projects.

Introduction

WHY THIS BOOK?

Lee K Curtis

The seed for this book was planted in my mind as a result of my involvement in a project to recover the endangered Bridled Nailtail Wallaby. My dear friend and continuous source of inspiration, Tina Janssen, has been maintaining a colony of endangered Bridled Nailtail Wallabies rescued from a mine in central Queensland since 2003. While awaiting the go-ahead from then Qld EPA (now DERM) to commence a captive breeding program (which was finally granted in August 2008), Tina's community group, Australian Animals Care and Education (AACE), had to raise funds and recruit people to help with the ongoing maintenance of these animals. This included veterinary expertise, genetic research, food, building materials, and so on. Because Marlborough, located 100 km north of Rockhampton, is not exactly a hub of philanthropy, or teeming with volunteers, funds and helping hands had to be sought far and wide. As a member of AACE with a bit of experience in fundraising, I offered to help.

Tracking down resources for fundraising, recruiting volunteers, PhD students, research scientists and expert consultants – let alone anything else having to do with threatened animal recovery and management in Queensland – proved to be a huge time- and energy-consuming challenge. Slowly but surely, mostly by word of mouth, and endless hours spent surfing the web and scrutinising the phone book, I learned the tricks of the trade and eventually tapped into what resources were available. Around the same time, while researching two books on Australian wildlife, I was surprised at how little networking and information sharing occurred between wildlife experts. I soon discovered that this was not due to an unwillingness to share information but purely to the lack of time available to people working in various wildlife fields. Their workload is enormous and they too appear to suffer from a lack of access to reliable comprehensive resources. After my books were published and didn't make it on to the best seller list, I decided to tackle a more challenging task and began to formulate an idea with the potential for a much broader impact: a comprehensive resource guide that would offer information to guide people in finding out who was doing what, when, where, how and why with threatened animals in Queensland.

The next step was to put together a team of experts who would fill the gaping gaps in my scientific knowledge: a vital component of this guide. Along came Andrew Dennis and Keith McDonald to help design the final structure, to allow me to draw on their extensive network of contacts to provide the best possible sources of up-to-date profiles, to guide the peer review processes and act as scientific and text editors. During the process of pulling it all together, Peter Kyne joined the team to finalise the reviews and editing for the fish and seabirds and Stephen Debus took over the other birds. Needless to say, this book is a gargantuan team effort and each and every person involved in its compilation has reason to be very proud.

INTRODUCTION TO THREATENED SPECIES

Andrew J Dennis and Keith R McDonald

In the last few decades, people have become increasingly concerned by the rate at which species are disappearing. Many plants and animals are already extinct as a result of human actions – direct or indirect – but many more have noticeably declined, and others have had populations disappear. This has led to a worldwide movement towards listing and classifying animals by the level of threat they face. The International Union for the Conservation of Nature (IUCN) Red List of Threatened Species was a result of this concern.

In this section, we broadly define and outline the processes associated with listing species as threatened. Our focus is on legal and government tools and responsibilities because that is the focus of these lists. However, it is not up to governments alone to facilitate species recovery: it is up to each of us, as individuals, communities and organisations, to do our bit and make use of the legal tools, resources and support made available by them; hence this book.

What is a threatened species?

Put simply, a species listed as threatened is one that is at risk of becoming extinct in a short time frame. Different levels of extinction risk are reflected by different categories (see Figure 1.1). Organisms classified as Critically Endangered are at higher risk of extinction and are likely to go extinct faster than those classified as Endangered. Those classified as Vulnerable have a lower risk but are still expected to become extinct over time unless some new circumstance or intervention changes their risk.

Listing species as threatened – the process and reasons

The process of estimating the risk of extinction of a species can be a complicated and exacting task, which is based on measures and criteria that have undergone rigorous assessment and revision over a few decades and continue to evolve and be improved. For example, the IUCN Categories and Criteria are at their seventh iteration. Many systems exist that have been applied to different purposes and at different scales but each measures the risk of extinction. The reason for listing a species is to guide governments, non-government organisations (NGOs) and individuals to reduce threats and enlist broad support to recover the species. Estimating a species' risk of extinction and placing it in one of the categories of risk assists governments and others to allocate limited resources to those species most in need of attention and/or those that can be assisted.

In this book on threatened animals, we draw on three different lists, all of which use similar criteria derived from the IUCN but are directed at different scales and overseen by different authorities. The lists are: globally – the IUCN Red List; for Australia – the *Environment Protection and Biodiversity Act 1999* (EPBC); and for Queensland – the Nature Conservation (Wildlife) Regulation Qld 2006, under the *Nature Conservation Act Qld 1992* (NCA).

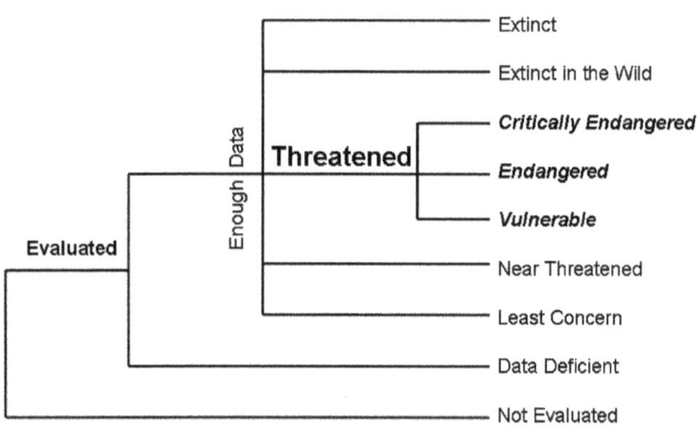

Figure 1.1. The structure of categories into which all species fall based on the IUCN Red List Categories and Criteria Version 3.1. The three threatened categories are in bold. Although lists for each jurisdiction vary to some degree, most reflect this structure.

Different lists have different levels of legal strength. Some provide strong penalties when breached and commit governments to significant investments; others indicate intent, not commitment, but may still provide penalties; others are mostly informative; and all provide guidelines for behaviour towards the species listed. The IUCN Red List is informative. The EPBC Act triggers planning and assessment requirements, possible intervention and/or penalties by the Federal Government, a commitment of resources for the recovery of the species listed and the possibility of a range of other tools such as recovery plans, threat abatement plans, action plans and impact guidelines. The NCA triggers planning and assessment requirements, possible intervention and/or penalties by state agencies and intent to conserve the species. It may also enlist other tools such as conservation plans and management plans for the protection and management of the species.

Below is a simplistic overview of the IUCN Red List criteria (version 3.1), derivatives of which are now widely applied at a range of scales. There are five broad criteria, A–E below, that are used to assess the risk of extinction for a species. Each has detailed definitions and measures and we refer the interested reader to the IUCN Red List website <http://www.iucnredlist.org/documents/redlist_cats_crit_en.pdf> for detail. Broadly speaking, a species must meet one or more of the following criteria to be classified as threatened.

A. Reduction in population size based on an observed, estimated, inferred, projected or suspected population decline.
B. Geographic range in the form of either extent of occurrence OR area of occupancy OR both being small, fragmented, shrinking or fluctuating severely.
C. Population size of mature individuals being small and showing a decline, particularly where subpopulations are very small or most individuals being in only one subpopulation or the number of individuals fluctuating severely.
D. Population size very small or restricted to a small area or few sites and prone to human impacts.
E. Quantitative analysis showing the probability of extinction in the wild is at least 10% within 100 years.

Categories of threat are separated by thresholds such as the speed and magnitude of declines or the size of populations, areas of occurrence or number of populations. For example an Endangered species would have declined faster and by a larger proportion than a Vulnerable species. Alternatively, an Endangered species would have a smaller population in a smaller area with fewer subpopulations than a Vulnerable one.

To further understand the application of measures of extinction risk, we need to understand the geopolitical framework that a species and its listings fit within. A species may be widely or narrowly distributed. For example, the Mountaintop Nurseryfrog *Cophixalus monticola* is found in one small mountainous area in the Wet Tropics of Queensland, whereas the Green Turtle *Chelonia mydas* is found throughout the world's tropics. Both are listed as Endangered on the IUCN Red List. The frog occupies one jurisdiction (Queensland) within one country (Australia), whereas the turtle occupies multiple jurisdictions and countries. Three lists are relevant to the frog and will relate to all conservation measures associated with that frog, but the Queensland list and associated legislation will likely be the most relevant to its management. However, the turtle appears on many more lists because it occurs across so many more jurisdictions. Actions associated with the Queensland legislation may only be relevant to the populations visiting Queensland. Because the species is widely distributed its conservation requires a global outlook and, therefore, the IUCN Red List is most relevant at the global level. Conservation actions will frequently be directed by local legislation and thus, at the national level, the EPBC is the relevant legislation, and at the local (state) level, the NCA is the relevant legislation.

Both the EPBC and the NCA allow the listing of subspecies, evolutionary significant units (ESU) and disjunct populations. An ESU is a genetically distinct subpopulation within a species that does not constitute a subspecies. It is necessary to recognise and manage these subpopulations to maintain the species over time. For example, the Kroombit Tops population of *Litoria pearsoniana* is listed separately to other populations. Queensland legislation enables disjunct populations within a species to be listed if they are threatened. For example, the Southern Cassowary *Casuarius casuarius johnsonii*, Koala *Phascolarctos cinereus* and Yellow-bellied Glider *Petaurus australis* each have separate populations with different conservation status under the NCA.

Why the animals in this book were chosen

We have included species from the three lists mentioned above, despite the book being focused on animals that occur in Queensland. We do this because we feel it best reflects the situation on the ground in Queensland. There are a number of reasons for this. One of the critical ones is illustrated in Figure 1.2. The ovals and circles on the diagrams represent different areas, jurisdictions or legal instruments; in this example we can picture them as the world, Australia and Queensland. The dots represent subpopulations (or individuals) of a species and each instance represents a different time (1–5) when populations have changed. As more populations (dots) disappear, the percentage change in the total population increases and the conservation status of the species reflects this. After around 50% of the populations have disappeared, the species would be accorded a Vulnerable status on a global scale, but, because most of the declines have occurred outside Australia, its status here is not yet in a threatened category. When more than 80% of the world's populations have gone along with 70% of Australia's populations, the species is classified as Critically Endangered and Endangered at those two scales, but still does not rate a mention in Queensland because it has not declined there sufficiently to trigger concern when assessed at the state scale. In some instances, the global status is considered when making an assessment on a local listing. In practice though, many globally listed species are overlooked in such cases and there are a large number of examples in this book where species are threatened globally but not listed locally. In fact, at the time of writing, the Queensland list had only 170 species of animal listed as threatened, whereas the combination of the three lists yielded 226 species, subspecies or populations: 26 invertebrates; 29 cartilaginous fishes; 11 bony fishes; 30 frogs; 36 reptiles; 49 birds and 45 mammals.

Species such as those represented by Figure 1.2, which are locally secure but threatened at broader scales – and there are many examples of them (e.g. see Sharks and Rays) – deserve considerable attention in the one remaining jurisdiction that has the least-threatened population.

Another reason we use the three lists is that the speed of change in animal populations and the regularity of reassessments of lists do not generally match one another. Therefore, the three lists together are more likely to reveal all threatened species. At the time of writing, a significant section of the Queensland list has been reviewed, with the remainder under review, and parts of the Australian list are under review as are some taxonomic groups in the IUCN Red List. Although there are significant reviews happening currently, which serve to underline the dynamic nature of threatened species listings, the Queensland and Australian lists had remained relatively static for over 10 years because legislative inertia results in lists being behind current levels of knowledge. Some taxonomic groups on the IUCN Red List have not been re-assessed for longer periods. Given the rate of decline measured for some species, they can become extinct long before any measures are enacted to save them or their listing is reviewed. Such is the case for a number of frogs in Queensland (see Extinct in Queensland pp. 10–14).

One other compelling reason to look broadly when dealing with threatened species is the sometimes idiosyncratic behaviour of humans. This can affect the listing process in a number of ways. Some researchers, especially those devoted to one group of animals, think on scales that reflect their own personal bent or the distribution or behaviour of the animals concerned. For example, people dealing with very widely distributed species tend to think on global scales because many of the species they deal with operate at those scales. Therefore, they might list species on the IUCN Red List, rather than trying to navigate the numerous local and regional legislative instruments. Some may follow this up with attempts to list in appropriate local legislation, but others leave it at the global scale.

Alternatively, a researcher working on extremely restricted species, might list only within the local jurisdiction. Although listing at the local scale may still afford the species the management actions it might require, the listing at the global scale is less likely to. However, this logic cannot be applied to all listings. For example, a number of restricted crayfish are listed as Critically Endangered on the IUCN Red List, but do not appear in the legislation relevant to their distribution (see Table 1.2).

In addition, some large and charismatic animals engender more sympathy from the public than others. In these cases, the impact of lobbying by special interest groups, and pres-

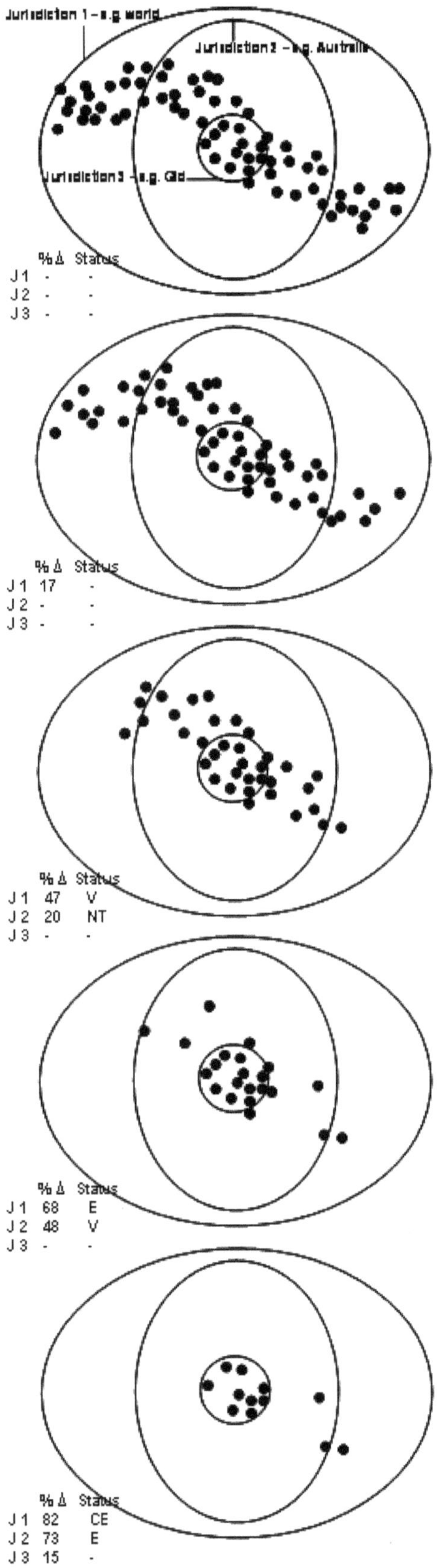

Figure 1.2. A hypothetical outline of a species decline over time showing the corresponding changes in its population size (% Δ) and conservation status (Status) in three nested geopolitical entities such as jurisdictions or legal instruments. The dots represent subpopulations of equal size. This diagram demonstrates that, although an animal can decline by more than 80%, it may not be listed as threatened in the one place that has the least threatened populations, which should be carefully managed and protected.

sure on and from politicians, can distort the listing process and the allocation of resources for recovery. Readers may be surprised to learn that the Koala, Greater Bilby and Southern Cassowary are not the most threatened fauna in Queensland. Instead, sharks, fish, lizards, bats, frogs and even a snail are among our most threatened animals.

Level of risk and knowledge of risk – a dynamic process

There are a number of circumstances where species may be declining and at risk of extinction, but they 'fall through the cracks'. Figure 1.1 outlines the classes into which all animals fall. The lowest two categories in the diagram contain the largest number of species: those that haven't been assessed (Not Evaluated) and those for which too little is known to make an assessment (Data Deficient). Many species that are Data Deficient may be at risk of extinction, but insufficient evidence is available to determine their risk. In many cases, collecting the data required is expensive and there is no compelling reason for governments to fund its collection so the risk to the species remains unknown and declines can continue unmeasured.

Another circumstance feared by many is 'death by 1000 cuts'. For some, this is a very real scenario, particularly where a species is widespread and suffering small losses throughout its range, and the losses are spread over a long time frame. For a species to be listed as threatened, it needs to lose a large proportion of its population or habitat over a short time frame – often 10 years or three generations. Some species, particularly those that occupy habitats on arable soils, coastal lowlands or other areas suitable for urban development have their habitat and populations diminished over a time frame longer than that required to trigger a listing. After the species has disappeared from most of its distribution and its population has diminished to a very low point, a listing will be triggered solely on area of occurrence, number of populations and/or number of individuals. At that point, species are at high risk of extinction through stochastic events such as a single severe storm, fire, drought or bulldozer. Many species that once occurred in the Brigalow Belt could fall into this category.

In some cases, a species might be relatively well known but there can still be a significant lag between a decline being perceived and a listing being finalised. For example, the northern population of the Lemuroid Possum *Hemibelidous lemuroides* disappeared from sites that were regularly monitored by a team from James Cook University. This population is well known for its large number of white individuals compared with the disjunct southern population and worthy of conservation in its own right. Analysis of many years of observational data showed that the species had undergone two distinct declines: one in the late 1980s, possibly associated with an unusually long period of warm dry weather; and the other more recently, and again, possibly associated with a similar pattern of weather. The first decline resulted in an order of magnitude reduction in the rate of sightings and the second in no sightings at all. Since then, just a few individuals have been found. The question now is whether this is a natural population fluctuation or whether it is a significant decline brought about by climate change or dis-

White Lemuroid Possum *Hemibelidous lemuroides*. Photo: Andrew J Dennis

ease. Unfortunately, the available data are insufficient to make a solid case either way. Although the alarm bells have tolled, it will take some years of careful data collection to determine if we have another population threatened with extinction.

Coming off the list

For many species, a significant and rapid decline may occur over a short time, often due to habitat loss, and trigger a threatened status but then populations stabilise at a new, smaller size. For example, the Southern Cassowary population in the Wet Tropics region plummeted rapidly when the coastal lowlands and Atherton Tableland were being cleared during the early 1900s. Since then, they have probably stabilised over most of the Wet tropics, partly because about 90% of their remaining habitat is protected. According to the current criteria used to assess extinction risk, if the Cassowary population remains stable it will begin to be down-listed. This will not mean that it has not undergone a decline in the past, nor deny that its current population size is less than its former numbers. Instead, it demonstrates that its population trend has stabilised and if the species still occurs over sufficient area and in sufficient numbers, its trend towards, and risk of, extinction has lifted. Unfortunately, many species that undergo declines and then stabilise remain listed because their populations are so few and small that they remain at risk. It remains to be seen whether the Cassowary qualifies for down-listing when we can estimate its population size and trend with sufficient rigour.

In some cases, taxonomy is uncertain and a threatened species may be removed from the list if it is demonstrated to belong to a more widespread species that is not under threat. Perhaps more often, additional search effort results in the species being demonstrated to be more widespread and secure. In other cases, better knowledge of the animal's ecology, behaviour and population trend results in a recognition that the species is not at risk. Of course, the opposite can also be true: for example, a disjunct population of a species not at risk may be shown to be a distinct species that is more localised and under threat and therefore becomes listed.

An evolving process

The aim of listing a species is to trigger action to reduce or halt the rate of decline and, ideally, facilitate its recovery. A range of legal tools may assist in this process. The recovery planning process is one that has been instigated by the Commonwealth Government and adopted and supported by state governments and the community and has been one of the primary tools for reducing the risk of extinction of species. We have referred to recovery plans throughout the species profiles and it doesn't take long to notice that only few recovery teams are currently active. As with the listing process, recovery processes are dynamic and the current trend is away from recovery plans for individual species and towards caring for landscapes and threats in the hope that many will look after their own recovery given appropriate conditions and reduction in threat. Some of these more recent approaches are threat abatement plans or multi-species recovery plans. Of course, the success of this approach will depend on the process(es) threatening the species. Some may still require management intervention specifically directed at them (e.g. the Northern Hairy-nosed Wombat p. 354) so that individual recovery plans will retain their currency and application into the future.

LIST OF THREAT CATEGORIES

The following section summarises the categories under the IUCN, EPBC and NCA, which are given under the subheading 'conservation status' in each species profile (where appropriate).

Categories of threatened species under the International Union for the Conservation of Nature (IUCN)

Extinct (EX): A taxon is Extinct when there is no reasonable doubt that the last individual has died.

Extinct in the Wild (EW): A taxon is Extinct in the Wild when it is known only to survive in cultivation, in captivity or as a naturalised population (or populations) well outside the past range.

Critically Endangered (CR): A taxon is Critically Endangered when it is considered to be facing an extremely high risk of extinction in the wild.

Endangered (EN): A taxon is Endangered when it is considered to be facing a very high risk of extinction in the wild.

Vulnerable (VU): A taxon is Vulnerable when it is considered to be facing a high risk of extinction in the wild.

Near Threatened (NT): A taxon is Near Threatened when it does not qualify for Critically Endangered, Endangered or Vulnerable now, but is close to qualifying for or is likely to qualify for a threatened category in the near future.

Least Concern (LC): A taxon is Least Concern when it does not qualify for Critically Endangered, Endangered, Vulnerable or Near Threatened. Widespread and abundant taxa are included in this category.

Data Deficient (DD): A taxon is Data Deficient when there is inadequate information to make a direct, or indirect, assessment of its risk of extinction based on its distribution and/or population status.

Not Evaluated (NE): A taxon is Not Evaluated when it is has not yet been evaluated against the criteria.

Reference: IUCN. (2001). *IUCN Red List Categories and Criteria: Version 3.1.* IUCN Species Survival Commission. IUCN, Gland, Switzerland and Cambridge, UK. <http://www.iucnredlist.org/technical-documents/categories-and-criteria>

Categories of threatened species under the *Environment Protection and Biodiversity Conservation (EPBC) Act*

(1) A native species is eligible to be included in the Extinct **(EX)** category at a particular time if, at that time, there is no reasonable doubt that the last member of the species has died.

(2) A native species is eligible to be included in the Extinct in the Wild **(EW)** category at a particular time if, at that time:
   (a) it is known only to survive in cultivation, in captivity or as a naturalised population well outside its past range; or
   (b) it has not been recorded in its known and/or expected habitat, at appropriate seasons, anywhere in its past range, despite exhaustive surveys over a time frame appropriate to its life cycle and form.

(3) A native species is eligible to be included in the Critically Endangered **(CR)** category at a particular time if, at that time, it is facing an extremely high risk of extinction in the wild in the immediate future, as determined in accordance with the prescribed criteria.

(4) A native species is eligible to be included in the Endangered **(EN)** category at a particular time if, at that time:
   (a) it is not Critically Endangered; and
   (b) it is facing a very high risk of extinction in the wild in the near future, as determined in accordance with the prescribed criteria.

(5) A native species is eligible to be included in the Vulnerable (VU) **category** at a particular time if, at that time:
   (a) it is not Critically Endangered or Endangered; and
   (b) it is facing a high risk of extinction in the wild in the medium term future, as determined in accordance with the prescribed criteria.

(6) A native species is eligible to be included in the Conservation Dependent **(CD)** category at a particular time if, at that time:
   (a) the species is the focus of a specific conservation program the cessation of which would result in the species becoming Vulnerable, Endangered or Critically Endangered; or
   (b) the following subparagraphs are satisfied:
      (i) the species is a species of fish;
      (ii) the species is the focus of a plan of management that provides for management actions necessary to stop the decline of, and support the recovery of, the species so that its chances of long term survival in nature are maximised;
      (iii) the plan of management is in force under a law of the Commonwealth or of a state or territory;
      (iv) cessation of the plan of management would adversely affect the conservation status of the species.

(7) In subsection (6):
   fish includes all species of bony fish, sharks, rays, crustaceans, molluscs and other marine organisms, but does not include marine mammals or marine reptiles.

Species listed as Conservation Dependent and Extinct are not matters of national environmental significance and therefore do not trigger the EPBC Act.

Reference: *Environment Protection and Biodiversity Conservation Act 1999* (EPBC Act). Department of Sustainability, Environment, Water, Population and Communities.

<http://www.environment.gov.au/biodiversity/threatened/species.html#categories>

Categories of threatened species under the Queensland Nature Conservation Act (NCA)

The previous system of categorising species (Presumed Extinct, Endangered, Vulnerable, Rare and Common) did not provide a consistent framework for managing species and were phased out in 2010. The new system is as follows:

Extinct in the Wild (EiW)

A species is Extinct in the Wild if:

- there have been thorough searches conducted for the wildlife; and
- the wildlife has not been seen in the wild over a period that is appropriate for the life cycle or form of the wildlife.

Endangered (E)

A species is Endangered if:

- there have not been thorough searches conducted for the wildlife and the wildlife has not been seen in the wild over a period that is appropriate for the life cycle or form of the wildlife; or
- the habitat or distribution of the wildlife has been reduced to an extent that the wildlife may be in danger of extinction; or
- the population size of the wildlife has declined, or is likely to decline, to an extent that the wildlife may be in danger of extinction; or
- the survival of the wildlife in the wild is unlikely if a threatening process continues.

Vulnerable (V)

A species is Vulnerable if:

- its population is decreasing because of threatening processes; or
- its population has been seriously depleted and its protection is not secured; or
- its population, while abundant, is at risk because of threatening processes; or
- its population is low or localised or depends on limited habitat that is at risk because of threatening processes.

Near Threatened (NT)

An animal is Near Threatened if:

- the population size or distribution of the wildlife is small and may become smaller; or
- the population size of the wildlife has declined, or is likely to decline, at a rate higher than the usual rate for population changes for the wildlife; or
- the survival of the wildlife in the wild is affected to an extent that the wildlife is in danger of becoming vulnerable.

Least Concern (LC)

An animal is Least Concern if:

- the wildlife is common or abundant and is likely to survive in the wild.

Native wildlife may be prescribed as Least Concern wildlife even if:

- the wildlife is the subject of a threatening process; or
- the population size or distribution of the wildlife has declined; or
- there is insufficient information about the wildlife to conclude whether the wildlife is common or abundant or likely to survive in the wild.

Citation: Qld DERM. *Nature Conservation Act 1992.* An amendment to the NCA in May 2010 incorporates new categories for protected species in Queensland. Government of Queensland 2010. <http://www.derm.qld.gov.au/wildlife-ecosystems/wildlife/threatened_plants_and_animals/index.html>

For threatened species listing categories in other states and territories, please refer to the following web sites:

New South Wales: <http://www.environment.nsw.gov.au/threatenedspecies/tscategories.htm>

Victoria: <http://www.dse.vic.gov.au/CA256F310024 B628/0/B10DC61B8DF2A09FCA25760F0010DF1B/$File/Advisory+List+of+Threatened+Vert+Fauna+in+Victoria+-+2007+amended+11+august+2009.pdf>

South Australia: <http://www.environment.sa.gov.au/parks/pdfs/conservation-status-codes.pdf>

Western Australia: <http://www.dec.wa.gov.au/pdf/projects/setting_priorities_wa_threatened_ff.pdf>

Northern Territory: <http://www.nt.gov.au/nreta/wildlife/animals/threatened/classification.html>

Tasmania: <http://www.dpiw.tas.gov.au/inter.nsf/Web Pages/RLIG-5433LB?open>

Australian Capital Territory: <http://www.tams.act.gov.au/__data/assets/pdf_file/0007/154582/crtiteria.pdf>

A HISTORY OF THREATENED SPECIES RECOVERY AND MANAGEMENT IN QUEENSLAND

Milena K Kim

From European settlement to 1945

When Europeans settled in Queensland in the 1820s they found a land of plentiful natural resources. However, this apparently pristine environment had been modified during 60 000 years of human occupation: Indigenous societies manipulated soils with yam sticks and transformed vegetation using fire. Hunting of fauna such as Dugongs and marine turtles remains an important part of Indigenous social and cultural life. Since settlement, the colonial model of economic specialisation resulted in rapid and widespread environmental impacts. As in many other colonies, intensive exploitation meant a sharp decline in biodiversity caused by direct exploitation and habitat destruction.

The commercial exploitation of wildlife was supported by the Queensland Government despite concerns about the viability of such activities. The fur trade, for example, started in the mid 19th century. By 1900, the international demand rapidly increased and, from 1906 to 1936, three million possums and nearly one million Koalas were hunted each year. The historical exploitation of marine wildlife has also been documented:

Dugongs, marine turtles, Bêches-de-mer and whales were intensively harvested on the Queensland coast during the colonial period. Habitat destruction was at least as instrumental as direct exploitation in the decline of biodiversity. After the founding of Moreton Bay in 1824, the rapid expansion to northern and western areas was driven by pastoralism and, by 1920, most of the economically viable land in Queensland was occupied. By the mid 1860s, in addition to grazing, agriculture became an important economic activity, notably sugarcane farming. Between 1865 and 1900, the growth of sugarcane required massive deforestation, both to create farmland and to supply sugar mills with timber as source of fuel. This had a significant impact on flora and fauna.

Australian naturalists in the 1880s were already concerned with the disappearance of native species. Independent bird-watching and scientific societies voiced their desire for the creation of parks and the regulation of species use. By the 1880s, almost every colony had a scientific society. Queensland's Philosophical Society was founded in 1859 and in 1883 it evolved into the Queensland Royal Society. Together with the Queensland branch of the Royal Geographical Society and the Acclimatisation Society, it was instrumental in the first campaigns for national parks. Those efforts resulted in the passing of Queensland's first national parks and state forests legislation in 1906 and the creation of the first national park, Witches Falls, in 1908. Robert Collins, a former president of the Royal Geographical Society and Romeo Lahey, first president of the National Parks Association of Queensland (formed in 1930), were important advocates for protected areas and directly influenced the creation of Lamington National Park in 1915.

Fauna protection campaigns were initially concerned with certain species of birds and mammals; such preoccupation was driven by the increasing social values of these species, simultaneously with the rise in demand for furs and plumage stimulated by European fashion at the end of the 19th century. At the national level, a symptom of the growing scientific concern with species conservation was the creation of a 'Protection of Native Birds and Mammals Committee' at the first general meeting of the Australasian Association for the Advancement of Science (AAAS) in 1888. Another influential group was the Royal Australasian Ornithologists Union (RAOU), which, acting together with other naturalists, bird lovers and journalists, had a meaningful victory when they convinced the federal government to use its customs powers to control the export of Australian fauna. In Queensland these organisations lobbied the state government to enact the two earlier fauna Acts: *The Native Birds Protection Act of 1877* and *The Game and Fishes Acclimatisation Act of* 1898, which offered protection to a limited number of birds and introduced animals, but excluded native mammals. They were followed by *The Native Animals Protection Act of 1906; The Animals and Birds Act of 1921*; *The Native Plant Conservation Act of 1930*; *The Fauna Protection Act of 1937*; *The Fauna Conservation Act of 1952* and the *Fauna Conservation Act 1974.*

However, the enforcement of the acts was difficult and poaching was very common. Moreover, the susceptibility of governments to economic pressures was already evident; a remarkable episode occurred when the Queensland Government withdrew possum and Koala protection in 1927, which led to over one million possums and half a million Koalas being killed for their furs in a single month. The massive killing of Koalas provoked horror in the general public and evidenced the high social value of such species; reactions to the killing of Koalas were (and still are) much stronger when compared with possums.

From the post-war years to the 1990s

Following the Second World War, Australia faced a period of intense economic development, which was mostly driven by resource extraction industries. During this time, the Commonwealth Scientific and Industrial Research Organisation (CSIRO) began to research native species and raise awareness about their value. The Wildlife Survey Section started in 1949 under Francis Ratcliffe, who had undertaken wildlife surveys since 1929. By then, native fauna were considered pests by landowners, but the research carried out by the Section contributed to changing this attitude. The informal work of the Section's scientists also significantly added to the creation of the Australian National Wildlife Collection (ANWC) and the identification of threats to many of these species. During the 1960s and '70s, as well as conducting research, scientists also contributed to raising public awareness about native species, using movies and popular wildlife books.

At that time, lobbying for environmental preservation was not as effective as it had been during the previous decades. Ironically, any protests and arguments favouring conservation were seen as defying progress. Furthermore, many scientists were being incorporated into bureaucracy and industry, and were unwilling to support conservation initiatives. As environmental historians Hutton and Connors noted in their *History of the Australian Environment Movement:*

> *'When a new generation of environmentalists took up the challenge of conservation, they not only had to withstand Cold War attacks from politicians and businesses, and isolation from a timid mainstream Australia: they had to do so with the support of only a handful of principled scientists and the actual hostility of many professional scientific organisations.'*

This new generation of activists played an 'outsider politics' role in this phase, which, as it did in other countries, interfaced with feminist and union movements and somehow merged with pacifists in the anti-nuclear campaigns of the 1970s and 1980s. Environmentalists also supported several indigenous movement causes including the mineral exploitation of Indigenous land. However, species conservation matters such as traditional hunting of threatened species generated controversy among conservationists.

In contrast with the intense timber harvest to support the Second World War, and the incentives provided for freeholding and land clearing by the Queensland Government after 1945, two significant changes, inspired by environmentalists, marked the following decades: the promulgation of the

Forestry Act in 1959, which recognised the need to ensure water quality and soil protection, and the transference of park management to the newly created Queensland National Parks and Wildlife Service (QNPWS) in 1975. The act creating QNPWS also merged the previous *Fauna Conservation Act 1974* and sections of the *Forestry Act 1959* into the one piece of legislation.

During this period, Queensland became the stage for two major wilderness campaigns: the Great Barrier Reef (GBR) and the Wet Tropics. The battle for the protection of the GBR started in 1967 following the proposition by Cairns sugarcane farmers to extract limestone from the Ellison Reef. The campaign against this proposal was led by the Wildlife Preservation Society of Queensland (WPSQ) and people such as Judith Wright, John Busst and Len Webb, among others. WPSQ, together with the Australian Littoral Society and the Save the Reef committee, embraced the movement against the oil and gas exploitation in the Reef, which resulted in the creation of the Great Barrier Reef Marine Park (GBRMP) by the Australian Government.

The movement for the creation of the Wet Tropics World Heritage Area drew wide support after the Douglas Shire Council decided, with state government support, to build a road through a national park from Cape Tribulation to the Aboriginal settlement of Bloomfield. This extremely diverse campaign involved different facets of the movement at that time, evidencing the transition of the conservation movement to mainstream politics: in 1983 and 1984, the local and regional grassroots activism, represented by the blockaders and the battle of Daintree, dominated the scene. In 1985, the campaign was more formalised and state-wide organisations were centred in Brisbane, focused on pressuring the Queensland state government. In the 18 months before the 1987 federal election, the focus was transferred out of Queensland. The campaign was dominated by the professional movement, represented by the Australian Conservation Foundation (ACF) and The Wilderness Society backing the Labor party, which won the election and nominated the World Heritage area.

The nomination of a significant number of World Heritage areas in Queensland to UNESCO by the Australian Government (five of 18 in Australia) represents a great advance in protecting species in their habitats. Nevertheless, the administration of such areas is demanding and contentious due to their size, which impedes the conversion of the entire World Heritage area into a protected area and obligates the management to deal with multiple activities. During the 1990s, the listing of World Heritage areas decreased significantly and the existing ones were subjected to increased pressure from development projects.

Another Queensland feature is the size of marine protected areas compared with terrestrial ones. The relevance of such areas can be traced back to 1929, when Green Island was proclaimed the world's first marine protected area. These areas are concentrated mainly on the eastern coast; marine parks such as Moreton Bay and Hervey Bay, as well as several islands, may have been protected due to the momentum of the creation of the Great Barrier Reef World Heritage Area and the iconic status of this region. In contrast, environmental assets of regions such as Gulf of Carpentaria and Torres Strait are as valuable as the GBRMP, but have not received the same amount of protection and investment. The tradition of marine protected areas also reflects the difficulties of dealing with terrestrial issues in a development model centred on agriculture, grazing and mineral exploitation.

From the 1990s to the present

During the last two decades, demands on biodiversity conservation resulted in the creation of strategies, such as legislation, to protect threatened species. However, the scope of these measures and attempts at habitat maintenance has been strongly hindered by the interests of the production sector.

Although both the federal and state governments have invested effort into species protection, they sometimes played contradicting roles. In Queensland, the *Nature Conservation Act 1992* is the central piece of legislation that deals with threatened species. Among other provisions, it was the first piece of legislation that allowed for listing fauna and flora species under several threat categories. At a national level, the Federal Parliament passed the first legislation designed to reduce species decline after signing the International Convention on Biological Diversity: the *Endangered Species Protection Act 1992*. This was replaced by the *Environment Protection and Biodiversity Conservation Act 1999* (EPBC). An important criticism of the EPBC Act is that it has left too much power to the states. A case in point is the vegetation clearing, which is a major threatening process to many threatened species, especially in Queensland. Historically the Queensland Government favoured expansionist policies providing landowners incentives to clear land. As a signatory to the Convention on Biological Diversity, the federal government has an obligation to ensure that this threat is minimised, but power over land issues remains mainly with states and territories. The EPBC Act considers threatened species as 'matters of national environmental significance', which may trigger Australian Government involvement. The presence of several World Heritage areas in Queensland also amplifies the federal government's powers to intervene, but these mechanisms do not seem to have deterred the state, until broadscale clearing was phased out on 31 December 2006, from having the highest rates of deforestation in Australia.

Key players in species conservation have also changed during the last two decades. In terms of generating new knowledge to support species conservation, the Queensland Government is currently more involved in identifying research needs and funding rather than carrying out the research itself. Stakeholders also determine research priorities: the Commonwealth Environment Research Facilities, for instance, supported investigation with a strong public focus. Consequently, researchers from universities and publicly funded agencies such as CSIRO and AIMS (Australian Institute of Marine Sciences) have lesser autonomy in deciding investigation topics and are required to feed information back into research results. On the other hand, these institutions, as well as NGOs, influence decision making as members of advisory committees.

A process of regionalisation and devolution of state power occurring in the last decades also strengthened the roles of different players in developing conservation action. Natural resource management bodies were responsible for the regional delivery of funds from the Natural Heritage Trust program under the national Liberal Government, and now from the Caring for Our Country Program under the Labor Government. They may develop actions for species conservation in partnership with governments, NGOs and other stakeholders, depending on regional priorities established in their plans and on funding availability. NGOs are also major contributors in terms of lobbying for environmental causes, raising awareness with the public and coordinating a range of on-ground projects with volunteers. They also complement government action by raising funds and acquiring land for conservation. However, these two strands are becoming increasingly professional; the general public is predominantly involved in executing actions on the ground and not necessarily in planning and decision making.

As traditional owners of the land, Indigenous groups receive growing attention regarding environmental matters. The decision Mabo v Queensland (1992) by the High Court determined that Indigenous groups could claim their native title rights over the land. As many Aboriginal-owned lands provide habitat for threatened species, this process triggered the creation of federally supported Indigenous Protected Areas and strengthened the role of jointly managed parks. Considering that Indigenous Australians are a disadvantaged group in society, conservation needs to be associated with benefits and livelihood opportunities. One example of such initiative is the involvement of traditional owners in conservation activities as professional rangers and building capacity for efficient and culturally adequate land management. In terms of species conservation, the traditional use of marine resources agreement developed between traditional owners and the Great Barrier Reef Marine Park Authority (GBRMPA) provide an instance of a jointly conceived agreement regulating traditional hunting in the GBRMP aiming for the sustainable harvest of marine species. In spite of these initiatives, power sharing between government agencies and traditional owners in joint management is still unequal, and tensions between Indigenous aspirations and protected areas have largely not been resolved.

Regarding species conservation via protected areas, there have been important yet scattered advances in the use of ecosystem-based approach. The most relevant example is the GBRMP rezoning; after a long consultation process, from 1998 to 2004, the no-take area of the park went from 5% to 33%. Although the movement for species conservation originates in grassroot movements – and the public affection for charismatic species yields powerful support – the discussion on changes in our development model, which is the major source of threats to species and ecosystems, is still very restricted. Arguably the biggest challenge is to coordinate efforts and integrate species conservation into broader ecosystem-based management. However, this can happen only if all members of society make informed choices on the trade-offs between environmental, social and economic goals, and adjust the current social system to move towards these goals.

The Threatened Species Network

Over its 19 year history, the Threatened Species Network (TSN) evolved from a small, part-time community program to a multi-million dollar national threatened species partnership between WWF-Australia and the Australian Government. The TSN defined its niche as a community-based network, working cooperatively with governments and other stakeholders to empower the community to participate in all aspects of species conservation. In 1999, the introduction of the Threatened Species Network Community Grants Program saw the TSN become the major driver of community-based threatened species recovery in Australia.

The TSN Community Grants ran for 10 years and funded over 400 projects, resulting in over $6 million in grants to help recover nearly 400 threatened plants, animals and ecological communities. This investment generated a further $8.6 million of community investment in threatened species conservation, and established partnerships at the community level that continue to support recovery activities years after the initial projects have concluded.

Funding regimes that support this work concluded on 30 June 2009. The TSN has not, at this stage, been successful in securing sufficient funds to continue its work. However, it remains optimistic that future funding opportunities will allow the work of recovering and conserving Australia's threatened species to continue.

EXTINCT IN QUEENSLAND

Andrew J Dennis and Keith R McDonald

It is clear from the world's somewhat patchy fossil record that species have come and gone over millennia. Extinction is a natural process. For any single species, the road to extinction might be driven by one or more of many potential processes. It might occur over extended geological time frames or relatively rapidly in discrete catastrophic events. In this section, we briefly examine extinctions that stem from a single pervasive source: us. With humans creating the drivers of extinction, species more often decline and disappear more rapidly than they have done in the past, though being sure of this is difficult because fossil and sub fossil evidence rarely measure continuous time. What is more certain is that more species are declining and disappearing from the face of the Earth in a shorter space of time than ever before. As we discussed in a previous section, governments and NGOs throughout the world have created lists that trigger processes to reduce the number of species becoming extinct. But for some it is already too late.

Australia has seen successive waves of extinction over our geologically recent history. In his book on mammal extinctions over the last 50 000 years, Chris Johnson carefully reconstructs all the evidence available as to why numerous

White-footed Rabbit-rat *Conilurus albipes*: an early victim of European settlement. Photo: National Library of Australia

species went extinct. It appears likely that changes brought about by the first waves of humans to spread through the continent altered the dynamics of the landscape, its vegetation and its ecology. Many species of animal were so finely tuned to their environment and the other species that had shared it in isolation for so long, that they had too little resilience to the types and speed of change generated by the newcomers. The declines seem to have occurred in 'waves', each over a few thousand years and some may have been due to mortality only slightly exceeding the species' ability to recruit replacements. Extinctions on this scale also set up cascades of changes that led to further extinctions and changes in vegetation communities. For example, some animals have a strong influence on their ecosystem by being good dispersers of plant propagules, pollen or fungal spores, by being selective browsers or by being effective predators that control the populations of other species. However, more recently, another wave of extinctions has been occurring; one that is happening in a relative 'blink of an eye' over the last 200+ years.

The arrival and spread of Europeans and other peoples, and the animals and plants they brought with them, has caused such rapid and extreme changes in the environment and the pressures facing many species that Australia has earned a reputation as having the worst recent history of mammal extinctions in the world. Other groups of animals have also suffered losses. Many factors have contributed to these extinctions, including direct habitat loss and alteration, introduced

The Mountain Mistfrog *Litoria nyakalensis*, which is now believed to be extinct. Photo: Andrew J Dennis

herbivores, diseases, direct persecution and new fire regimes changing ecosystem dynamics. In some cases, a species may have been driven to extinction by a unique set of circumstances and pressures. However, most – particularly the small- and medium-sized mammals – have been brought to extinction by two: the Red Fox *Vulpes vulpes* and feral Cat *Felis catus*. Although they may work in concert with multiple other new pressures, it is probably these two feral predators that have been responsible for pushing many species over the brink.

Although many of these extinctions have occurred in southern, central and western Australia, Queensland has also lost species completely from within its borders. Some of these have 'earned' the status of Extinct or Presumed Extinct in legislation or lists. Others though, are considered extinct by those holding the most current knowledge but have not been updated in the legislation. Table 1.1 summarises the species and subspecies listed as extinct under various lists and those considered likely to be extinct by current authorities.

It might surprise some readers to know that we have lost 19 species of animal from Queensland. Many of these are mammals of arid and southern Australia and were part of three rapid 'waves' of extinction that occurred from the mid 1800s to the mid 1900s, which were probably driven by the Red Fox and Feral Cat, with the European Rabbit *Orictolagus cuniculus* supporting predator numbers. Others are different stories. For example, the demise of the Paradise Parrot *Psephotus pulcherrimus* after 1928 was probably unrelated to predators; a wave of extinctions in stream breeding frogs in the 1980s and '90s was due to disease; and a flying-fox in Queensland may have been a record keeping error rather than an extinction.

Before European arrival, the Paradise Parrot occurred in rolling downs, open eucalypt woodland and open forest with a grassy understorey in sub-coastal south-east Queensland. It fed on grass seeds that were abundant year round and nested in terrestrial termite mounds. By the late 1880s, it was extremely rare and by 1928 it was sighted for what was probably the last time. The Paradise Parrot probably declined to extinction because its habitat was destroyed or altered to the point that grass seed was no longer available year round. This occurred through the usual set of circumstances that accompanied European colonisation: clearing of vegetation, grazing by sheep and cattle and new fire regimes. In addition to these pervasive pressures that continue to threaten many species today, direct persecution by trappers, the removal of termite mounds (for floors, ovens, tennis courts, etc.), the spread of the Prickly Pear *Opuntia* spp. and a significant drought

Table 1.1. Species or subspecies considered extinct in Queensland; including those listed globally (IUCN – IUCN Red List) (CR – Critically Endangered; EX – Extinct), within Australia (Aust. – *Environment Protection and Biodiversity Conservation Act*, EPBC) (CE – Critically Endangered; Ex – Extinct) and on the Queensland list (Qld – *Nature Conservation Act*, NCA) (E – Endangered; EiW – Extinct in the Wild) and those that are probably extinct, but their status has not yet been reviewed. Some of the mammals have never been recorded live in Queensland, only occurring as bone fragments in owl pellets (*); others probably occurred in Queensland but declined and disappeared so rapidly that there were too few collections to be certain of their total distribution (**).

Species	IUCN	Aust.	Qld
Frogs			
Mountain Mistfrog, *Litoria nyakalensis*	CR	CE	E
Gastric-brooding Frog, *Rheobatrachus silus*	EX	Ex	E
Gastric-brooding Frog, *Rheobatrachus vittelinus*	EX	Ex	E
Sharp-snouted Dayfrog, *Taudactylus acutirostris*	CR	Ex	E
Southern Dayfrog, *Taudactylus diurnus*	EX	Ex	E
Northern Tinkerfrog, *Taudactylus rheophilus*	CR	E	E
Mammals			
Blue-grey Mouse, *Pseudomys glaucus*	EX	Ex	
Gould's Mouse, *Pseudomys gouldii***	EX	Ex	
Short-tailed Hopping-mouse, *Notomys amplus**	EX	Ex	
Long-tailed Hopping-mouse, *Notomys longicaudatus**	EX	Ex	
Darling Downs Hopping-mouse, *Notomys mordax*	EX	Ex	EiW
White-footed Rabbit-rat, *Conilurus albipes*	EX	Ex	EiW
Pig-footed Bandicoot, *Chaeropus ecaudatus***	EX	Ex	
Lesser Bilby, *Macrotis leucura* **	EX	Ex	
Eastern Bettong, *Bettongia gaimardi gaimardi*		Ex	EiW
Desert Rat-kangaroo, *Caloprymnus campestris*	EX	Ex	EiW
Percy Island Flying-fox, *Pteropus brunneus*	EX	Not recognised	EiW
Western Quoll, *Dasyurus geoffroii geoffroii*			EiW
Birds			
Paradise Parrot, *Psophotes pulcherrimus*	EX	Ex	EiW

The Paradise Parrot's extinction is believed to have been mainly caused by the supplanting of its native grass seed by sheep and cattle fodder, the removal of termite mounds and the bird's popularity in the late 19th and early 20th centuries' pet trade. (Artwork by WT Cooper)

probably also combined to contribute to their extinction. The Paradise Parrot is not unique. There are a number of other granivorous animals, both birds and mammals, which occupy similar habitats and are currently listed as threatened (e.g. see finch and mice profiles).

In the 1980s and '90s, a suite of stream-breeding frogs underwent rapid declines to the point that six species in Queensland are now considered extinct and 11 others are extirpated from parts of their previous range. This phenomenon was not restricted to Queensland; throughout the world, frogs were declining and disappearing for a range of reasons. In Australia, as with Central America and the USA, there was epidemiological, pathological and experimental evidence that some species were infected by the fungus *Batrachochytrium dendrobatidis*, and died due to the resultant disease – chytridiomycosis. Although this fungus has all the characteristics that give it the potential to be the sole cause of declines, it is still not certain if other factors predisposed frogs to the disease.

The Percy Island Flying-fox is an unusual case. Its description is based on a single specimen, originally labelled *Pteropus gouldii*, and bought from 'sale rooms'. Although the specimen clearly differs from all other Australian species, its origins seem in doubt. Even if it did come from Percy Island, it may have been a vagrant of another species that happened to have been among a camp of one of the other Australian species. Given that flying-foxes are generally very mobile it is not very likely that one was restricted to a small island. Given the lack of evidence of its validity as an Australian species, or of its decline, we are inclined to think it is a mistake of early collectors and not really an extinction. We hope we are right.

Table 1.2 shows the species, subspecies or populations assessed as more likely to become extinct in the near future than other threatened species. This list highlights a couple of issues. First, not all lists reflect the same level of threat for the same species. Indeed, nearly 80% of these species – those considered most at risk on one list – do not even appear on one or both of the other lists. This is partly skewed by the eight spiny crayfish species that only appear on the IUCN Red List, but also highlights the idiosyncratic behaviour of humans, as discussed in the previous section.

Even the crayfish, though, are examples of a common issue throughout the world. They represent species that have such a naturally restricted distribution that a single event could wipe

Table 1.2. Species (or subspecies or populations) listed as Critically Endangered and therefore, considered under immediate threat of extinction without significant intervention (IUCN: CR – Critically Endangered; EN – Endangered; LC – Least Concern. Aust. – *Environment Protection and Biodiversity Conservation Act*: CE – Critically Endangered; E – Endangered; V – Vulnerable. Qld – *Nature Conservation Act*: E – Endangered; V – Vulnerable – Note, the NCA has no category for Critically Endangered).

Species	IUCN	Aust.	Qld
Invertebrates			
Boggomoss Snail, *Adclarkia dawsonensis*		CE	
Bloomfield Spiny Crayfish, *Euastacus robertsi*	CR		
Currumbin Spiny Crayfish, *Euastacus maidae*	CR		
Eungella Spiny Crayfish, *Euastacus eungella*	CR		
Kroombit Spiny Crayfish, *Euastacus monteithorum*	CR		
Mistake Mountain Spiny Crayfish, *Euastacus jagara*	CR		
Mount Elliot Spiny Crayfish, *Euastacus bindal*	CR		
Mount Glorious Spiny Crayfish, *Euastacus setosus*	CR		
Tully River Spiny Crayfish, *Euastacus yigara*	CR		
Fishes			
Dwarf Sawfish, *Pristis clavata*	CR	V	
Edgabaston Goby, *Chlamydogobius squamigenus*	CR	V	E
Elizabeth Springs Goby, *Chlamydogobius micropterus*	CR	E	E
Freshwater Sawfish, *Pristis microdon*	CR	V	
Green Sawfish, *Pristis zijsron*	CR	V	
Grey Nurse Shark, *Carcharias taurus* (east coast population)	CR	CE	E
Harrisson's Dogfish, *Centrophorus harrissoni*	CR		
Mary River Cod, *Maccullochella peelii mariensis*	CR	E	
Murray Cod, *Maccullochella peelii peelii*	CR	V	
Narrow Sawfish, *Anoxypristis cuspidata*	CR		
Red-finned Blue-eye, *Scaturiginichthys vermeilipinnis*	CR	E	E
Speartooth Shark, *Glyphis glyphis*	EN	CE	
Frogs			
Armoured Mistfrog, *Litoria lorica*	CR	CE	E
Beautiful Nurseryfrog, *Cophixalus concinnus*	CR		V
Eungella Dayfrog, *Taudactylus eungellensis*	CR	E	E
Kroombit Tinkerfrog, *Taudactylus pleione*	CR	V	E
Kuranda Treefrog, *Litoria myola*	CR	E	
Reptiles			
Hawksbill Turtle, *Eretmochelys imbricata*	CR	V	V
Leatherback Turtle, *Dermochelys coriacea*	CR	E	V
Nangur Spiny Skink, *Nangura spinosa*		CE	E
Retro Slider, *Lerista allanae*	CR	E	E
Birds			
Herald Petrel, *Pterodroma heraldica*	LC	CE	E
Night Parrot, *Pezoporus occidentalis*	CR	E	E
Yellow Chat (Dawson subspecies), *Epthianura crocea macgregori*		CE	E
Mammals			
Bare-rumped Sheathtail Bat, *Saccolaimus saccolaimus nudicluniatus*		CE	E
Bramble Cay Melomys, *Melomys rubicola*	CR	E	E
Northern Hairy-nosed Wombat, *Lasiorhinus krefftii*	CR	E	E

them out. Thankfully, most of these crayfish are well protected on mountain tops in national parks in relatively undisturbed habitats that have few weeds or feral animals. If the land tenure was not protected, it could take only a single development or disturbance to the stream systems or a short period of intensive collecting to wipe out one of these species. Even within this tenure, the species are vulnerable to the introduction of a novel parasite or pathogen, as was only too recently witnessed with stream-breeding frogs.

Perhaps most importantly, this Table 1.2 serves as a reminder of the often restricted or overlooked species that face a very uncertain future. Most are unglamorous, some could even be considered a bit dull or even scary and certainly not as cute and cuddly as a Koala. However, do they deserve to be ignored while a few relatively secure species command all the attention and resources? We hope not, and we hope that this book inspires a few people to get more interested in the other, less glamorous species that need our attention.

THREATENED ANIMALS AND REGIONAL ECOSYSTEMS

Ian Fox and Nick Cuff

Although the recovery of a threatened animal frequently requires a focus on a species, the actions that might save that species always incorporate the animals, plants and environmental components in the ecosystem it shares. In Queensland, vegetation communities have been combined with land forms and classified into 'Regional Ecosystems' (see Sattler and Williams 1999) and these have been assigned threat categories similar to those for animals. In this section, we examine regional ecosystems and suggest that their conservation is a fundamental part of recovering threatened animals and frequently, vice versa.

The link between animals and ecosystems

Plants and animals often interact in ways that are ecologically significant. A few of these interactions, or ecological processes, include the pollination of flowers, dispersal of seeds and consumption of plant parts, though there are many more. Animals are often threatened by the loss or decline in the quality of their habitat. Habitat destruction, especially from land clearing, results in the greatest loss of biodiversity of all causes. Equally, ecosystems can be undermined, and eventually threatened, by the loss of species within them, particularly those that are fundamental to key ecological processes.

To maintain viable animal populations frequently requires direct action for the species, maintenance of landscape processes and preservation of habitats and ecosystems. Retention of adequate, well distributed and diverse vegetation types within a landscape is considered appropriate for the protection of ecological processes and ecological health. The Queensland system to classify and map ecosystems has produced data that are fundamental for both landscape and biodiversity planning across the heavily modified landscape. More particularly, these data assist in planning and implementing the recovery of threatened species, ecosystems and landscapes.

Mapping and classifying ecosystems – a brief history

The classification of ecosystems based on vegetation has evolved over the last 100 or more years. Recognition of distinct plant groupings, based not only on floristic composition, but also on life forms and community structure, was the result of the growth in ecological studies in the early 20th century. Perhaps the earliest attempt to classify the vegetation of Australia was a map produced by F.L.E. Diels in 1906. The map contained eight broad vegetation types, including reference to vegetation characterised by rainfall.

Through the 1940s, maps of plant communities were often accompanied by soil and geological maps. However, the classifications and terms used were not uniform across maps. In the 1950s, physiognomy, describing vegetation by its own characteristics and not by its environment, was emphasised. Some of these classifications are still in use today. These differences in approach to classification restricted the compatibility of data and classifications across jurisdictions.

During the period following the Second World War, studies to assess land development opportunities in remote Australia and New Guinea formed the basis for what is now the regional ecosystem framework in Queensland. The CSIRO produced land system maps from the 1940s through to the 1970s based on recurring patterns of topography, soils and vegetation.

As land development expanded in Australia, a corresponding loss of natural systems and species was recognised and the need for a systematic and hierarchical approach to landscape-scale conservation planning grew in importance. In the 1970s, Stanton and Morgan described the bioregions of Queensland using landscape patterns that reflect geology and climate, as well as major changes in plant and animal assemblages. This established a framework for conservation planning in Queensland that coincided with national approaches. In 1990, Morgan and Terrey refined bioregions by describing provinces (now known as subregions) within bioregions, basing them on differences in geology or geomorphology and finer climatic variations.

The Queensland regional ecosystem framework

A hierarchy developed by Sattler in 1993 set up the concept for classifying regional ecosystems in Queensland. It incorporates four levels – landscape, ecosystem, species and genotype. Landscape is represented by bioregions and subregions, which are defined by climate, geology and geomorphology. Ecosystem is represented by regional ecosystems, defined by vegetation, soil and land form. Species and genotype are self explanatory and discussed elsewhere in this book. In Queensland, a regional ecosystem (RE) is a community of vegetation that is consistently associated with a particular combination of geology, land form and soil in a bioregion.

The program to describe and map all Qld REs commenced in the mid-1990s in south-east Queensland and Brigalow Belt bioregions. Maps were required to measure the extent of the remaining vegetation and to provide a basis for planning within bioregions affected by rapid development. From that early mapping, the regional ecosystem framework was refined and the program expanded to include the whole state in what is now regarded as the national benchmark for such work.

The RE framework for each bioregion combines a vegetation community with information about the geology, land form and soil that it sits on.

Most of the state has been mapped at a scale of 1:100 000, with mapped areas as small as 4 hectares. However, in some areas where development pressure is acute, the need for more detailed mapping led to areas being remapped at higher resolution (1:50 000). These maps can be applied to measuring the extent of ecosystems, biodiversity planning, and ecosystem and species recovery.

Application of regional ecosystem mapping

From a national perspective, the Queensland RE mapping has redefined the boundaries of bioregions and subregions for the national Interim Biogeographic Regionalisation of Australia (IBRA). IBRA is used as a tool to monitor progress towards building a comprehensive, adequate and representative national reserve system. It allows the Australian and state and territory governments to focus on gaps in the protected

Table 1.3: Queensland bioregions, approximate area and number of REs in each.

ID	Bioregion name	Area (000's ha)	No of REs
1	Northwest Highlands	7344	43
2	Gulf Plains	21 910	84
3	Cape York Peninsula	12 305	233
4	Mitchell Grass Downs	24 162	60
5	Channel Country	23 217	56
6	Mulga Lands	18 106	65
7	Wet Tropics	1993	185
8	Central Queensland Coast	1484	80
9	Einasleigh Uplands	11 626	145
10	Desert Uplands	6941	76
11	Brigalow Belt	36 528	172
12	South-east Queensland	6248	151
13	New England Tableland	775	25

Source: Queensland Herbarium (November 2009)

area network and prioritise where effort should be applied to ensure the protection of ecosystems that are poorly reserved.

RE mapping underpins the (Queensland) *Vegetation Management Act 1999*. Each RE is assigned a *conservation status* based on its remaining extent. The status classes used are *endangered*, *of concern* or *not of concern*. The Act clearly prescribes activities that are permitted within each. For example, an RE will be mapped as *endangered* if it has less than 10% of its pre-clearing extent across the bioregion, or 10–30% of its pre-clearing extent remains and the remnant vegetation is less than 10 000 hectares. With few exceptions, clearing an *endangered* RE is prohibited by legislation.

However, biodiversity management does not rely solely on the overall extent of ecosystems. Often the condition, geographic distribution and spatial arrangement of an ecosystem are just as important. All REs are also assigned a *biodiversity status* (distinct from *conservation status*): either *endangered*, *of concern* or *not of concern* based on the presence of threatening processes such as weeds, fire or overgrazing, which may be modified after consideration of the extent of a particular RE within the landscape. *Biodiversity status* is not a legislated management class, but provides a guide to the value of ecosystems in management actions such as planning for ecosystem and species recovery.

Information on the spatial extent and conservation status of REs is integral to development of the Biodiversity Assessment and Mapping Methodology (BAMM – a tool) and the implementation of Biodiversity Planning Assessments (BPAs – a process). Such tools and processes allow for the identification and management of biodiversity values to ensure that we have complete and ecologically viable representatives of each RE in each bioregion. Protecting rare and threatened species and ecosystems is particularly important.

RE classifications can have broad correspondence with fauna composition. Consequently they are relied upon as the basis for conservation planning, despite the frequent absence of data that would confirm whether or not an area is actually important for flora and fauna species. However, in areas where RE mapping exists, along with information about individual species or groups of species, and the policy context is conducive, recovery of species described in this book can be effectively planned for and implemented.

Additional information about biodiversity planning, the RE framework as well as datasets and maps for download are available online at the Queensland Government web site at: <http://www.derm.qld.gov.au/wildlife-ecosystems/ecosystems/index.html> and <http://www.derm.qld.gov.au/wildlife-ecosystems/biodiversity/index.html>.

HOW TO USE THIS BOOK

Lee K Curtis

The purpose of this book is to provide a useful resource guide to threatened animals in Queensland. We hope that the information offered will lead you quickly and efficiently to the next step that will help to halt the population decline of threatened animals in our state.

The species are listed under invertebrates, cartilaginous fishes, bony fishes, amphibians, reptiles, birds and mammals and within these groups under their orders or families.

Profiles

Each profile contains the common and scientific names of the species, (including the person(s) who first described it and the year in which it was first described) and family, followed by other common name(s) and Aboriginal names when known or appropriate to report.

Each profile features a photo when available.

Distribution maps: Provide a broad indication of each animal's distribution. They are derived from point records when available, but some are based on catchment boundaries. In cases where a distribution is fragmented, the area in which the fragments are located is boxed. Breeding grounds for sea turtles and birds are in red and their ranges in pink.

Conservation status: Includes the listings as they stood at the time of publication with the International Union for the Conservation of Nature (IUCN), *Environment Protection and Biodiversity Conservation Act* (Aust.), and each state in which the species is listed. See the 'Introduction to Threatened Species' (pp. 1–5) for details.

Description: A brief description is provided, but readers are directed to field guides or other literature appropriate to the animal group for verifying identifications.

Distribution: This section provides information that cannot be gleaned from the map, including international and detailed local information. Population estimates are sometimes included where appropriate.

Habitat: A brief and up-to-date description regarding the habitat preferences and requirements of a species.

Ecology: A succinct summary of the animal's ecology and ecological interactions; this section often provides information that is not readily available in summarised form elsewhere.

Reasons for decline and current threats: Where the data are available, this section provides reasons for listing, including reasons for original decline, its drivers, locations and time frame followed by assessment of continuing population status, current threats and perceived threats, as well as their locations.

Current recovery efforts

Recovery plans: The listing of past and/or current recovery plans.

Other plans: Reference to action, threat abatement, park management, conservation plans, and so on.

Conservation actions: Actions currently being carried out or recently carried out; a brief overview including who is doing them and who is funding them, if room allows.

Future recovery efforts

What you can do: Practical suggestions about what the reader can do as an individual to facilitate the recovery of a threatened species. Contact information for a wide variety of groups can be found in the Appendices.

Conservation needs: Actions that government organisations, NGOs and/or community groups need to carry out to improve species recovery and management.

Research directions: Suggestions for studies and research, unanswered questions and suggestions for the next steps for more effective recovery efforts.

Sources and further reading: The profiles in this book draw mostly from published scientific articles, but also include unpublished information from researchers currently working on the species in question. Because of the format of the book, not all references used can be listed for each profile. Instead we have endeavoured to list some of the recent or more comprehensive references that can lead interested readers deeper into the full literature available for each species.

Glossary and Appendices

Glossary, acronyms and abbreviations: Research terms, uncommon ecological terms, taxon specific terms.

Appendix I: The list of organisations involved in the recovery and management of Queensland's threatened fauna consists of dozens of private and public sector groups that see a strong need for a comprehensive resource guide about Queensland's threatened animals. If any groups have been left out, the oversight was unintentional and omitted groups are encouraged to contact us so that their information can be included in future guides.

Appendix II: How to report a sighting, location record or specimen.

Appendix III: Additional resources to help those involved with threatened species recovery and management access more detailed information.

SPECIES PROFILES

INVERTEBRATES

Many of Australia's invertebrates – 250,000 species described to date, with many more waiting in the wings to be discovered – are endemic to the continent. There are over 1000 species of snails and slugs, 100 crayfish, 300 dragonflies and damselflies, 400 butterflies and moths and, to date, 29,000 different types of beetles. A very small percentage of these creatures are classified as pests, but the majority are vital components of our wildlife's life-sustaining food chain, as well as pollinators and seed dispersers. The few threatened invertebrate species treated in this book are undoubtedly but a tiny fragment of the real picture, which will only be filled in with substantially more research, hopefully not before it is too late.

Insects (Class: Insecta)

Dragonflies (Order: Odonata)

INTRODUCTION TO THREATENED DRAGONFLIES

Gunther Theischinger

Australia is a dry continent, with water being arguably its scarcest essential resource. Generally all dragonfly species depend on some type of freshwater habitat. Conservation of dragonflies, as of other organisms, depends on the conservation of their habitats. Only very few Australian dragonfly species are protected by law. Many more are vulnerable because of their unusual biology and ecology. Australia's dragonfly fauna has a very high percentage of endemism at the family, genus and species level. About 75% of the Australian species are endemic to Australia. Many of them are distributed over very small areas or are even restricted to particular sites, and some represent monotypic genera or are otherwise taxonomically isolated. Very little is known about others.

Examples of threats to dragonfly habitats include deforestation, eutrophication, lowering of the groundwater level, chemical contamination, mining and other land development connected to exploitation or damage to natural freshwater. In the long run, all these and other threats to freshwater habitats can be controlled only by strong political leadership backed up by first-class hydro-biological expertise and competence in hydro-technology.

THYLACINE DARNER

Acanthaeschna victoria Martin, 1901

Photo: Gunther Theischinger

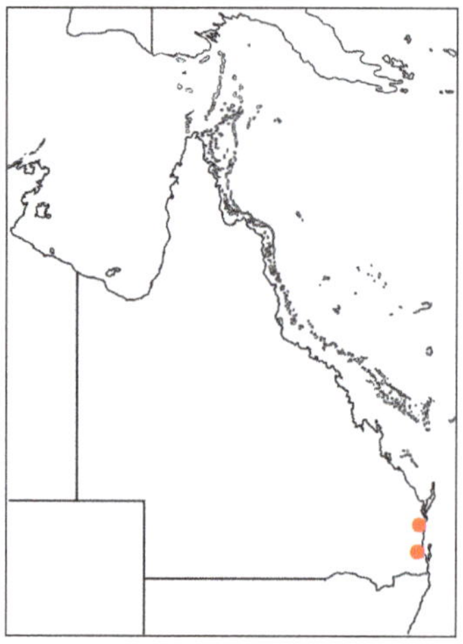

Family: Aeshnidae

Other common name: Blackwater Darner

Conservation status: IUCN: Vulnerable

Description: The Thylacine Darner is a large dragonfly with a wingspan of 80–94 mm and a body length of 57–68 mm. Its large eyes are pale and blue on top and black on the sides. There are black and cream longitudinal markings on the sides of the stout synthorax and the abdomen is club-shaped. Its wings are clear with a slight pale brown darkening in the nodal area. It has two supplementary radial veins in both wings, and there are no brace veins to the long pale brown pterostigma. Both sexes have an unarmed occiput and very short anal appendages. The larva is dark (largely blackish brown) with the base of the antennae markedly lighter. It has a prementum with a single tooth each side of the ligula and relatively small eyes, a moderately armed prothorax, an unarmed synthorax and lateral spines on abdominal segments 6–9. The anal pyramid is short, the epiproct is wedge-shaped, but with a slightly bifid apex.

Distribution: The Thylacine Darner is endemic to eastern Australia. Its range includes southern Qld, NSW and possibly Vic., with all known localities close to the coast. The northern-most locality on record is Elanda Point along Kin Kin Creek and the southernmost is Cockwhy Creek between Ulladulla and Batemans Bay. Populations are probably disjunct according to the remaining suitable habitats.

Habitat: It inhabits slow streams and rivers near the coast; possibly temporary swamps. The low dissolved oxygen content and the low pH of the bodies of water from which larvae have collected are remarkable (black water streams; wallum streams; paper bark swamps).

Ecology: Judging from the available larval material (three fully grown larvae collected from the end of winter to early summer), the Thylacine Darner appears to be univoltine. The emergence of the adults early in the season may indicate a need to be fully developed before the habitat dries up. A copulating pair was collected resting near a recently dried up habitat. From the appearance of the adults and a few observations, possibly even from the poor collecting record, it appears that the species is largely crepuscular and attracted to dark places (malaise traps, space under stools and black cars). Observation and collection is very difficult because of the unusual ecology and behaviour of this species. All information given about ecology, behaviour, life history, etc. is based on observation/collection of around one dozen adults and of three final instar larvae.

Reasons for decline and current threats: For quite some time, the east coast of Australia has been continuously transformed by human activities, such as settlements, pastures and farming. Material was collected and sent to European museums during the 19th century but the species has only rarely been encountered since that time.

CURRENT RECOVERY EFFORTS

Recovery plan: None

Conservation actions: The chain of national parks along the coast of the states in which the Thylacine Darner is found is the most effective means of ensuring their survival. It is important not to reduce the size of their habitat or significantly transform their protected areas. As long as this chain of parks exists, and the status and usage of the national parks is not changed, recovery plans for this species are not necessary.

Sources and further reading

Hawking J and Theischinger G (2004) Critical species of Odonata in Australia. *International Journal of Odonatology* **79**, 113–132.

Peters G and Theischinger G (2007) Die gondwanischen Aeshniden Australiens (Odonata: Telephlebiidae und Brachytronidae). *Denisia* **20**, 517–574.

Theischinger G (2000) The *Acanthaeschna* story. *Linzer Biologische Beitraege* **32**, 235–240.

Theischinger G (2008) Notable range extensions of dragonflies in New South Wales – more species in Victoria? *Victorian Entomologist* **38**, 59–65.

Theischinger G and Endersby ID (2009) *Identification Guide to the Australian Odonata*. Department of Environment, Climate Change and Water NSW, Sydney South, NSW.

Author: Gunther Theischinger

BEAUTIFUL PETALTAIL

Petalura pulcherrima Tillyard, 1913

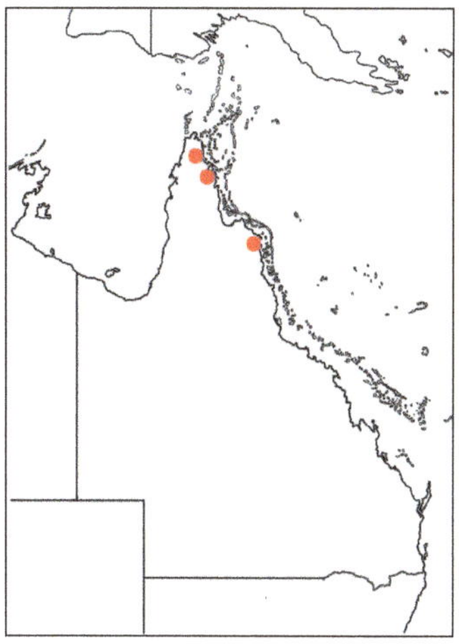

Photo: Gunther Theischinger

Family: Petaluridae

Conservation status: IUCN: Endangered

Description: The Beautiful Petaltail is a very large dragonfly with a wingspan of 118–138 mm and a body length of 90–108 mm. Its eyes are set wide apart and there is a strongly contrasting black and yellow colour pattern on the head, thorax and abdomen. The dark pattern-elements of the head, when compared with its closest ally *P. ingentissima*, are somewhat reduced, leaving a rather small triangular batch on the anterior frons and the postclypeus, distinctly but variably marked with yellow. The costae of the wings are blackish brown and the abdominal tergum 9 is mainly black. The male superior anal appendages are brownish black and of moderate size. The inferior appendage is blackish brown, wide and dentate along the outer portion of the posterior edge. The available final instar exuviae have a prolonged terminal antennal segment (as in *P. ingentissima*). The cerci of the unique available male exuviae are distinctly smaller than those of the available *P. ingentissima* exuviae.

Distribution: The species was described from Cooktown and Kuranda, with Kuranda recognised as doubtful. Material similar to the type material has since been collected mainly from areas north of Cooktown. Material identified from these areas as *P. ingentissima* are possibly Beautiful Petaltail. Individuals of populations on record from near Isabella Creek north-west of Cooktown, Iron Range and Heathlands were identified as this species.

Habitat: Exuviae were found at the confluence of two small creeks. Pairing was observed along margins of small streams and in pandanus swamps; oviposition was observed at the margins of pandanus swamps with hardly any surface water and almost firm ground. Males and females regularly visited these pandanus swamps in rather open monsoon forest.

Ecology: The species is semi-aquatic with the larvae burrowing underground; they probably include aquatic and terrestrial invertebrates in their diet. The life history certainly extends over several years and is probably very flexible. Emergence takes place probably before, certainly during and possibly after the rainy season. The adults are fully diurnal. The males frequently occupy prominent dry branches in order to observe the arrival of females and potential prey, which includes insects as large as cicadas that are seized in flight. Females often appear early in the day at places preferred for oviposition in order to avoid male interference.

Reasons for decline and current threats: An indication of decline is that very few observations were recorded for almost 90 years after the original description. One reason for this was probably that much of the range (Cape York Peninsula) was difficult to access and not commonly visited by dragonfly collectors, particularly near the rainy season. Threats would be road construction and the establishment of extensive facilities for tourists such as camping grounds, sports grounds, etc., in potential habitats.

CURRENT RECOVERY EFFORTS

Recovery plan: None

Conservation actions: DNA studies are currently underway, in cooperation with US-based scientists, to find out about the specific status of the Beautiful Petaltail, particularly the systematic/taxonomic situation in the *P. ingentissima* complex and the phylogeny of the world Petaluridae.

FUTURE RECOVERY EFFORTS

Conservation needs:

- Uphold the currently established use of land and water and the current status of the available national parks and other reserves, with particular considerations for wetlands, to ensure the survival of this species.
- Keep intrusion into the potential wetland habitats to a minimum.
- Make available, publicise and control collecting and export restrictions, as well as implement sensible but

stringent permit requirements, particularly for overseas collectors.

Research directions:

- Find out if there is more than one tropical Australian *Petalura* species.
- Map their distributions.
- Determine if any species occur sympatrically.
- Assess their geographical, altitudinal, ecological or other trends.
- Encourage PhD projects to answer detailed questions about the ecology, ethology, life history, population dynamics and conservation status of this species.
- Clarify the taxonomic status of *P. pulcherrima*. Determine if it is a distinct species, a subspecies of *P. ingentissima* or just a synonym of *P. ingentissima*.

Sources and further reading

Davies DAL (1998) The genus *Petalura*: field observations, habitats and observation status (Anisoptera; Petaluridae). *Odonatologica* **27**, 287–305.

Hawking J and Theischinger G (2004) Critical species of Odonata in Australia. *International Journal of Odonatology* **79**, 113–132.

Theischinger G and Endersby ID (2009) *Identification Guide to the Australian Odonata*. Department of Environment, Climate Change and Water NSW, Sydney South, NSW.

Tillyard RJ (1908) On the genus *Petalura*, with description of a new species. *Proceedings of the Linnean Society of New South Wales* **32**, 708–718.

Tillyard RJ (1913) On some Australian Anisoptera, with description of new species. *Proceedings of the Linnean Society of New South Wales* **37**, 572–584.

Author: Gunther Theischinger

Butterflies and Moths (Order: Lepidoptera)

INTRODUCTION TO THREATENED BUTTERFLIES AND A MOTH

Peter Valentine and Ted Edwards

It is difficult to make comparisons between the conservation status of a butterfly or moth with assessments of vertebrate animals. In common with most insects, butterflies may be ephemeral in the landscape as adults and spend much of their lives as eggs, larvae or pupae, the survey of which can be quite challenging. Many Australian species of butterflies have a single flight period each year; only tropical species fly throughout the year. Even then, their abundance varies greatly throughout the year, being diminished by dry or cold conditions. The task of completing a systematic survey is challenging and very few species have been subjected to any attempt, even in local areas. Much of the current information on butterfly and moth conservation status relies on the casual observations and anecdotal accounts of field scientists and hobbyists.

Substantial knowledge regarding moth distribution and biology would be necessary to be able to determine accurately which species are threatened and which are not. Unfortunately, only a few species have been studied sufficiently. It is highly likely that many moths have become extinct over the years due to human activities, but this happened before they had been described. There is a taxonomy crisis in Australia today: there are very few biodiversity studies being undertaken and currently there is nobody employed to identify most known moth species.

It is usually at the end point of a species conservation decline before surveys are commenced; these may do little more than document extinction. For most species, habitat conservation is the essential priority and conservation agencies can be most effective if they combine excellent knowledge of habitat requirements with programs of habitat protection.

MANGROVE ANT-BLUE BUTTERFLY

Acrodipsas illidgei (Waterhouse & Lyell, 1914)

Photo: Peter Samson

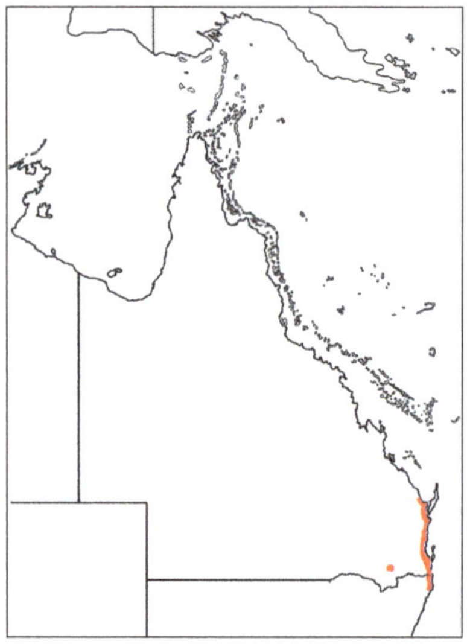

Family: Lycaenidae

Other common name: Illidge's Ant-blue

Conservation status: IUCN: Endangered; Qld: Vulnerable

Description: A small butterfly with a 22 (male) to 24 mm (female) wingspan. The wings are dark brown with a black terminal line on the upper side and faint tornal black spots visible on the hindwing. The underside is pale brown with broad dark brown markings and black tornal spots along the margins. The female is slightly paler and has a large blue area centrally.

Distribution: The species was known from mangroves and adjacent areas from Brisbane to Burleigh in south-east Qld in the 1940s. More recently its range has been extended to Redcliffe and Maryborough north of Brisbane and in extensive mangrove areas east of Beenleigh. Single specimens have also been collected at Toowoomba and Leyburn, both locations far removed from mangrove habitat. Additional coastal locations have been found at Tweed Heads and Brunswick Heads in northern NSW.

Habitat: All breeding populations have been found in mangroves or in ant nests close to mangroves. The habitat requirements appear to be determined by the presence of colonies of the ant *Crematogaster* sp. (*laeviceps* group). These colonies are prevalent in stands of the mangrove *Avicennia marina*, particularly in mature trees. These ant colonies are also present within the bases of mistletoe plants parasitising *Casuarina* trees close to mangroves.

Ecology: The larvae of Mangrove Ant-blue Butterfly prey on the juvenile stages of an ant, *Crematogaster* sp. (*laeviceps* group). This obligate myrmecophilous relationship means that the species survival is strictly linked to the ant colonies on which it depends. Initially, both ant and butterfly were thought to be associated with mangroves, predominantly in mature stands of *Avicennia marina*, but more recently they have been found in adjacent *Casuarina glauca* trees. Adults have been recorded throughout most of the year (apart from May–July), with a seasonal peak in August and September but it is not known how many generations occur each year.

Reasons for decline and current threats: No estimate of their population size is available and only low numbers of individuals have ever been collected, suggesting the species may be naturally rare. The main reason for decline is loss of mangrove habitat throughout its range. Recognition that Mangrove Ant-blue Butterfly was threatened seems to date in the literature from 1988, with further emphasis emerging as part of development conflicts in the Redland Bay area of Brisbane. This contributed to its listing as Permanently Protected Fauna in Qld, under the *Fauna Conservation Act*. Another area of concern is the aerial application of insecticides for midge and mosquito control, usually associated with development adjacent to or within mangrove environments.

CURRENT RECOVERY EFFORTS

Recovery plan: None

Other plans: Sands DPA and New TR (2002) 'The action plan for Australian butterflies'. Environment Australia, Canberra.

Conservation actions: Despite legal actions associated with real estate developments, no on-ground recovery efforts appear to have been taken.

FUTURE RECOVERY EFFORTS

What you can do: Join/donate to Butterfly and Other Invertebrates Club, Inc. (BOIC) and/or other organisations involved in conservation and revegetation projects.

Conservation needs:

- Lobby for improved protection for all mangrove and associated coastal environments and lobby for removal of insecticide use in mangrove and other habitat for this species.

- DERM needs to encourage and enable field research on this species.
- Define the distribution and abundance of the species within suitable habitat.

Research directions:

- Carry out research to establish the taxonomic relationship between Mangrove Ant-blue Butterfly and Black-veined Ant-blue *A. arcana*. Little genetic difference has been found between the two and it has been suggested they could be populations of the same species that are morphologically distinct. However, it is not known if the Mangrove Ant-blue Butterfly uses the same ant host. Resolving these issues will inform management approaches.
- Survey the extensive areas of suitable mangrove habitat that remain unsampled. Only small areas, such as Mary River Heads, have been surveyed intensively. Other areas where the species has been known to occur, including Tallebudgera NP, have not been subject to such intense surveying.
- Conduct further survey work beyond mangrove communities as suggested by the recent occurrence of the species at Toowoomba.
- Gather knowledge about presence and absence and relative abundance.
- Examine the effects that insecticidal sprays used for midge and mosquito control have on the ants, their food and the butterflies.

Sources and further reading

Beale JP (1998) Temporal and spatial distribution of *Acrodipsas illidgei* (Lycaenidae) (Waterhouse and Lyell). *Journal of the Lepidopterists Society* **52**, 139–150.

Beale JP and Zalucki MP (1995) Status and distribution of *Acrodipsas illidgei* (Waterhouse and Lyell) (Lepidoptera: Lycaenidae) at Redland Bay, southeastern Queensland, and a new plant association record. *Journal of the Australian Entomological Society* **34**, 163–168.

Lane DA (1991) A new distribution record for *Acrodipsas illidgei* (Waterhouse and Lyell) (Lepidopters: Lycaenidae). *Australian Entomology Magazine* **18**, 83–84.

Samson PR (1989) Morphology and biology of *Acrodipsas illidgei* (Waterhouse and Lyell), a myrmecophagous lycaenid (Lepidoptera: Lycaenidae: Theclinae). *Journal of the Australian Entomological Society* **28**, 161–168.

Samson PR (1993) Illidge's ant blue, *Acrodipsas illidgei* (Waterhouse and Lyell). In *Conservation Biology of Lycaenidae (butterflies).* (Ed. TR New) pp. 163–165. IUCN, Switzerland.

Authors: Peter Valentine and Steve Johnson

LACED FRITILLARY BUTTERFLY

Argynnis hyperbius inconstans (Butler, 1873)

Photo: Murdoch De Baar

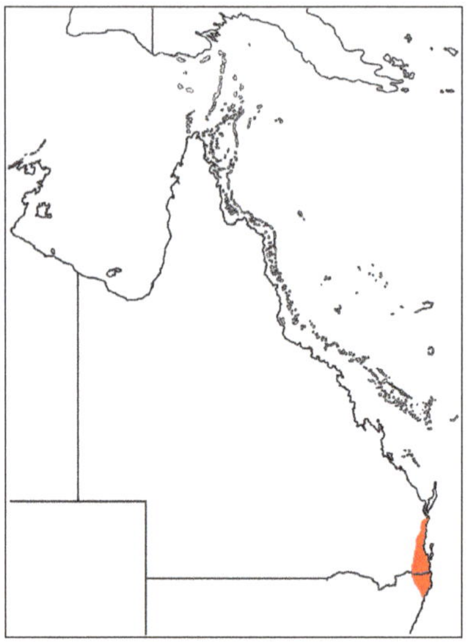

Other common name: Australian Fritillary

Family: Nymphalidae

Conservation status: Qld: Endangered; NSW: Endangered

Description: A medium-sized butterfly from 50 (male) to 65 mm (female) with light orange-brown wings covered with numerous rounded black spots and dark margins with fine lines and spots. The underside is similar but the background colour is more pinkish and some of the spots paler.

Distribution: The species is restricted to scattered locations on the coastal strip from Gympie in Qld to Port Macquarie in NSW. The range appears to have contracted in recent decades due to habitat loss. It may already be extirpated from most of its former range.

Habitat: It occurs in open, coastal, grassy sedgelands and wetlands where its larval food plant, *Viola betonicifolia*, is distributed. However, the larval food plant is much more widespread than the butterfly. It is not known what conditions enable successful colonisation of food plant patches.

Ecology: The life history of the species was described in 1977, but aspects of its ecology remain unclear. For example, the abundance of adults varies over time, but this has not been explained nor is it predictable. It is not known whether any of the juvenile stages undergo diapause. Adults have been observed in most months of the year. Some locations where butterflies were periodically abundant in the past have not experienced rejuvenation recently (e.g. Gympie and Coolum).

Reasons for decline and current threats: Extensive habitat loss or transformation throughout the distribution of the butterfly may be the most important driver of its decline. The loss of *Melaleuca* wetlands and adjacent grasslands has been of greatest concern. The Laced Fritillary appears to have a pronounced irregularity in abundance through time, certainly at any one site. It seems to exhibit strong meta-population characteristics, further complicating any attempt to estimate population numbers and clearly measure decline or recovery.

CURRENT RECOVERY EFFORTS

Recovery plan: Schwencke H and Jordan F (1998) 'Australian Fritillary *Argyreus hyperbius*. Draft interim recovery plan'. Butterfly and Other Invertebrates Club Inc., Brisbane (not readily available).

Other plans: Sands DPA and New TR (2002) 'The action plan for Australian butterflies'. Environment Australia, Canberra.

Conservation actions: Local populations have been artificially established in the past, with some success over a number of years at Mount Tambourine, Birkdale and Wynnum, but persistence has been for a limited period.

FUTURE RECOVERY EFFORTS

What you can do: Join/donate to Butterfly and Other Invertebrates Club, Inc. (BOIC) and/or other organisations involved in conservation and revegetation projects.

Conservation needs:

- Lobby for a recovery plan and encourage and support community involvement.
- A captive breeding program might assist with this butterfly, at least in gaining more information about the operation of diapause in juvenile stages and the likely viability of any reintroduction.

Research directions:

- Conduct extensive field surveys to locate any wild populations within the likely habitat in south-east Qld. A detailed plan for surveys has been proposed, which provides a starting point.
- Study the ecology of the larval food plant.
- Determine the patch-size importance v plant density.
- Study the processes that led to the destruction of the larval food plant communities (fire, weeds, etc).

- Research the reasons for the disjunction between larval food plant and butterfly distribution and abundance.
- Determine whether the larval food plant habitat can be detected using remote sensing techniques.
- Assess the current distribution of suitable habitat for the butterfly and how much of it is in protected areas versus private land.
- Identify mechanisms that enable the butterfly to appear and disappear over time and determine whether or not irregular larval diapause is a factor.
- Genetically assess the relationship between Australian and New Guinea populations to determine if Australian populations are ephemeral ones that periodically recolonise.

Sources and further reading

Braby MF (2000) *Butterflies of Australia: Their Identification, Biology and Distribution*. CSIRO Publishing, Melbourne.

Johnston LM and Johnston DR (1984) Further observations on the life history of *Argyreus hyperbius inconstans* Butler (Lepidoptera: Nymphalidae) in captivity. *Australian Entomological Magazine* **11**, 75–78.

Lambkin TA and Lambkin KJ (1977) Observations of the life history of *Argyreus hyperbius inconstans* Butler (Lepidoptera: Nymphalidae). *Australian Entomological Magazine* **4**, 13–16.

Sands DPA and New TR (2008) Irregular diapause, apparency and evaluating conservation status: anomalies from the Australian butterflies. *Journal of Insect Conservation* **12**, 81–85.

Authors: Peter Valentine and Steve Johnson

APOLLO JEWEL BUTTERFLY

Hypochrysops apollo apollo (Miskin, 1891)

Photo: MF Braby

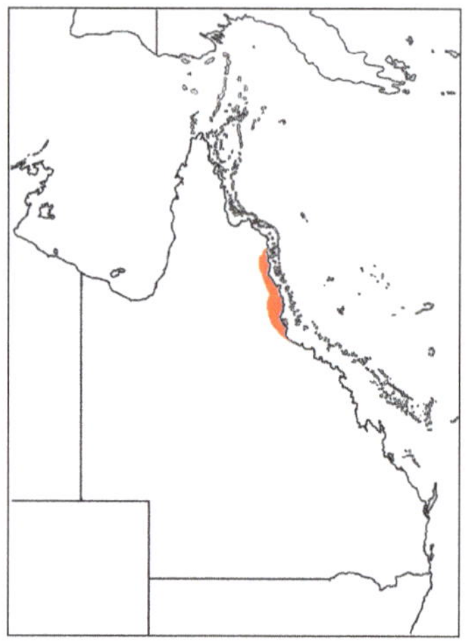

Family: Lycaenidae

Conservation status: Qld: Vulnerable

Description: This is a medium-sized butterfly with a 34 (male) to 36 mm (female) wingspan. The upperside of the wings is bright reddish-orange with a black apex, costa and termen on the forewing, while the hindwing has a black apical patch and at end of veins along termen. The underside is pale orange with brighter orange and brownish patches and a series of pale blue subterminal spots, a post median red-brown line, a series of creamish spots in radial and discal areas and reddish spots in cubital and anal areas. Males and females are similar, though females are paler and have more rounded wings.

Distribution: The southern subspecies occurs along the coast from Cooktown south to Ingham in north Qld. A historical population was known from near Kuranda, but it is not known if it is currently extant as the site was cleared. The remaining populations are highly fragmented. Disjunct colonies are present along the estuaries of the Endeavour and Annan Rivers near Cooktown, in remnant *Melaleuca* swamps north of Cairns, in Trinity inlet at Cairns, in mangroves near Innisfail and further south in *Melaleuca* wetlands extending from Cardwell to Ingham. The larval food plant is known to occur as far south as Saltwater Creek just north of Townsville, but no Apollo Jewel Butterfly has been recorded there.

Habitat: The occurrence of this species is determined by the presence of *Myrmecodia beccarii* host plants (commonly referred to as ant plants). The predominant habitat is coastal *Melaleuca* woodlands, extending into nearby mangrove forests.

Ecology: The life history of this butterfly involves a close association between the larval food plants and the ant species that occupies them (*Philidris cordatus*). Adult butterflies deposit eggs on the outside of the ant plant and larvae penetrate the bulbous base and feed on plant tissue within the galleries, creating more tunnels in which the ants nest. The ants appear to attend the butterfly larvae and the larvae are protected from predators within the chambers of the plant. Resident Golden Ants *Iridomyrmex cordatus* gather a honeydew produced by the larvae. The ants appear to attend the butterfly larvae and the larvae are protected from predators within the chambers of the plant. Eventually, the larva pupates within the bulbous base after creating an exit hole through which the adult butterfly emerges. Adults fly between October and April at Cardwell, though the Cape York Peninsula subspecies may fly throughout the year.

Reasons for decline and current threats: The populations in the Cooktown area appear to be extensive, whereas those near Cairns have been lost due to land clearing for residential development. The populations in southern areas have also been lost or reduced following land clearing for expansion of sugarcane farms and establishment of pine plantations. Frequent burning of *Melaleuca* swamps has resulted in loss of ant plants. An anecdotal account of loss of ant plants due to collection of butterfly larvae has been reported, but no evidence has been produced on the extent of the occurrence or its significance. The granting of permits for further mangrove clearance as part of coastal developments continues to pose a threat. The invasive Big-headed Ant *Pheidole megacephala* may also pose a significant emerging threat, but the impact of its spread is unknown. This ant has displaced the native ant from ant plants in some areas, though a survey in 1999 between Cape Tribulation and Tully found no Big-headed Ants in ant plants.

CURRENT RECOVERY EFFORTS

Recovery plan: None

Other plans: Sands DPA and New TR (2002) 'The action plan for Australian butterflies'. Environment Australia, Canberra.

Conservation actions: No direct efforts to recover this species have been made. However, in the late 1990s, DERM conducted preliminary surveys of the distribution of Big-headed Ants in *Melaleuca* swamps in the Wet Tropics with a view to getting some baseline data to examine the effects of these ants on threatened butterflies and plants in this habitat. Limited, short-term funding meant that this work has not been extended further.

FUTURE RECOVERY EFFORTS

What you can do: Join/donate to Butterfly and Other Invertebrates Club, Inc. (BOIC) and/or other organisations involved in conservation and revegetation projects.

Conservation needs:

- Lobby local and state governments for better planning of development and management in the coastal plain of north Qld, including no further loss of habitat on crown or private land.
- Exercise control over fires in *Melaleuca* wetlands, particularly fires originating from railway property that traverses the wetlands in southern areas.
- Make effective use of firebreaks to contain the regular railway fires.
- Lobby regional councils and Qld Rail to stop railway fires.
- Continue the legal protection of mangrove environments and lobby to reduce the spraying of mangrove habitat by local governments.

Research directions:

- Undertake additional surveys of mangrove communities and other coastal habitats between Cooktown and Ingham for the continuing presence of larval food plants *Myrmecodia* spp. and the butterfly to inform future management and coastal development.
- Clarify the effect of invasive ant species, including the Big-headed Ant, on the native ant *Philidris cordatus*, ant plants and this and other butterfly larvae.
- Conduct a research project to examine the dynamics of vegetation changes in coastal *Melaleuca* swamps, including developing a balanced fire regime to stop rainforest invasion but maximise ant plant abundance to benefit this and other *Melaleuca* swamp specialists.

Sources and further reading

Braby MF (2000) *Butterflies of Australia: Their Identification, Biology and Distribution*. CSIRO Publishing, Melbourne.

Sands DPA (1993) *Hypochrysops* Felder C and R. In *Conservation Biology of Lycaenidae (butterflies)*. (Ed. New TR) pp. 160–162. IUCN, Switzerland.

Authors: Peter Valentine and Steve Johnson

BULLOAK JEWEL BUTTERFLY

Hypochrysops piceata (Kerr, Macqueen & Sands, 1969)

Photo: Peter Samson

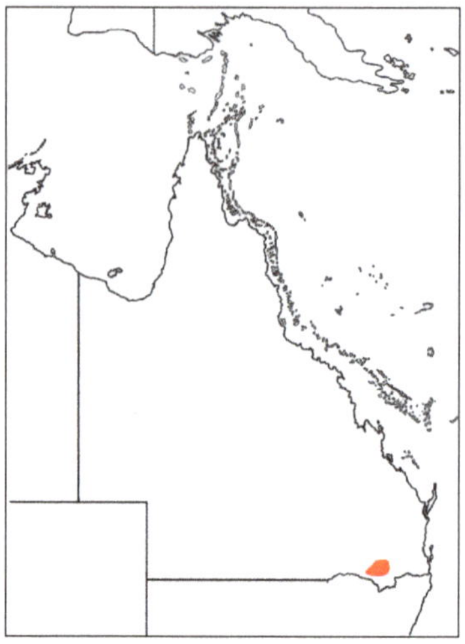

Family: Lycaenidae

Conservation status: Qld: Endangered

Description: A small butterfly with a wingspan of 23 (male) to 25 mm (female). The upperside of both wings is dull purple with broad dark to black margins and orange veins; the underside is pale brownish with lines of orange to reddish patches edged with white, some with black and some with blue or metallic green. The female and male are similar, though the female has more rounded wings and the upperside tends more towards blue.

Distribution: The species is confined to a very small area of the Darling Downs, Qld. Originally discovered near Millmerran and Leyburn, but has not been recorded at the former site for many years. Populations have recently been discovered near Goondiwindi (Bendidee NP). Further populations are now known on the Toowoomba-Karara Road not far from the Ellangowan Nature Reserve.

Habitat: The species is restricted to mature stands of Bulloak *Allocasuarina luehmannii* that are occupied by colonies of the ant *Anonychomyrma* sp. (*itinerans* group). Little of the habitat used by the species occurs in protected areas, with significant areas confined to roadsides or private property.

Ecology: The life history of this species involves an obligate relationship with the ant *Anonychomyrma* sp. (*itinerans* group) and the larval food plant *Allocasuarina luehmannii*. It appears that only old growth stands of the Bulloak are used. There are two generations each year, spring and summer, with adults on the wing for short periods each generation (October, November and January to March). The adults reside close to the breeding area.

Reasons for decline and current threats: The decline of the species has been a direct result of extensive clearing of Bulloak communities on the Darling Downs. The clearing has been for grazing, widening of road corridors and collection of firewood and fencing timber. These activities continue, and little of the small area of suitable habitat is legally protected. A current threat is the ageing and death of Bulloak stands and a lack of recruitment within them. Most host trees are hundreds of years old and run the risk of dying before significant numbers of new trees recruit into the mature phase.

CURRENT RECOVERY EFFORTS

Recovery plan: Lundie-Jenkins G and Payne A (2000) 'Recovery plan for the Bulloak Jewel Butterfly (*Hypochrysops piceatus*) 1999–2003'. Prepared for the Bulloak Jewel Butterfly Recovery Team. Queensland Parks and Wildlife Service.

Other plans: Sands DPA and New TR (2002) 'The action plan for Australian butterflies'. Environment Australia, Canberra.

Conservation actions: Actions commenced include conducting surveys to locate additional populations and model their distribution, and raising public awareness of the species.

FUTURE RECOVERY EFFORTS

What you can do:

- Become involved in the community aspects of the recovery program.
- Join a Landcare group and become involved in establishing succession planning, buffering and expansion of existing stands of Bulloak and replanting of habitat or ensuring recruitment of new trees in existing stands.
- Plant a patch on your own farm if you are in the right area.
- Report any sightings to the Qld Museum (Entomology Department).

Conservation needs:

- Establish plans to manage existing population sites.
- Lobby for updating and revitalising the recovery program to achieve outcomes originally proposed.

- Permanently protect existing known habitat.
- Expand community awareness and engagement program.
- Encourage landholders to help carry out actions.
- Rehabilitate known sites in the Leyburn area.
- Lobby for better protection of habitat and expansion of protected areas to cover known sites.

Research directions:

- Carry out surveys within region to identify additional sites and also further surveys in the Goondiwindi and Inglewood areas to locate new populations among the mature bulloak stands there.
- Monitor the populations to gauge their health and trends.
- Study the butterfly's ecology.
- Develop a predictive model of its distribution, examining its metapopulation dynamics and conducting a population viability analysis.

Sources and further reading

Braby MF (2000) *Butterflies of Australia: Their Identification, Biology and Distribution*. CSIRO Publishing, Melbourne.

Kerr JFR, Macqueen J and Sands DP (1969) A new species of *Hyposchysops* (Lepidoptera: Lycaenidae) from south Queensland. *Journal of the Australian Entomology Society* **8**, 117–120.

Authors: Peter Valentine and Steve Johnson

PALE IMPERIAL HAIRSTREAK BUTTERFLY

Jalmenus eubulus (Miskin, 1876)

Photo: MF Braby

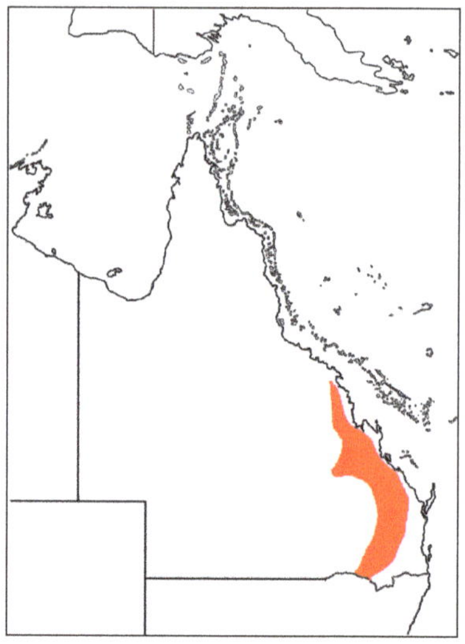

Family: Lycaenidae

Other common name: Common Imperial Blue (pale form)

Conservation status: Qld: Endangered; NSW: Critically Endangered

Description: Once thought to be a subspecies of *Jalmenus evagorus*, it is now recognised as a distinct species. It is a medium-sized butterfly with a wingspan of 32 (male) to 37 mm (female). The upper sides are black with large central areas of pale greenish-white (male) or bluish-white (female). The hindwings also have two orange-red spots at the tornus and a tail at the end of the cubital vein. The undersides are light brown with narrow black bands and orange lines edged with black on the margins. They also have reddish-orange tornal spots reflecting those on the upper side. Apart from the colour and size differences, females have more rounded wings.

Distribution: This butterfly is found in central Qld from near Eungella south to Darling Downs and as far inland as Carnarvon Gorge, and into NSW near Boggabilla.

Habitat: The species is confined to inland mature Brigalow stands *Acacia harpophylla* in central and southern Qld and northern NSW.

Ecology: The butterfly is confined to Brigalow communities, where larvae feed on the leaves of *A. harpophylla* attended by one of several species of ants, generally *Iridomyrmex*. Although the species breeds on small Brigalow trees, it does so only within stands of mature Brigalow. The adults principally fly through February and March, though the exact timing depends on summer rainfall patterns, the peak in flights following the peak in rain.

Reasons for decline and current threats: The decline in populations of the Pale Imperial Hairstreak Butterfly has resulted from clearing of Brigalow areas for grazing and agricultural development. The species still occurs throughout its original range, but populations have become fragmented. Stands of mature Brigalow are now very small and regarded as vulnerable to a range of threats and limited areas remain. Although the *Vegetation Management Act* should ensure protection of remaining stands of mature Brigalow, these are so small that they may not ensure long-term survival of the butterflies and will make genetic continuity difficult. The tiny and fragmented mature Brigalow stands are subject to threats such as edge effects, fire and weed invasion, to name a few.

CURRENT RECOVERY EFFORTS

Recovery plan: None

Other plans: Sands DPA and New TR (2002) 'The action plan for Australian butterflies'. Environment Australia, Canberra.

FUTURE RECOVERY EFFORTS

What you can do:

- Report confirmed sightings to the Qld Museum (Entomology Branch).
- Become involved in establishing succession planning, buffering and expansion of existing stands of Brigalow and replanting of new habitat patches particularly around existing individual trees.

Conservation needs:

- Protect the last remaining stands of mature Brigalow.
- Lobby for the better management of existing stands in protected areas, in particular the protection from unregulated fires.
- Work with private landowners to better protect stands of mature Brigalow on their land.
- Work with landholders to establish connections (corridors or stepping stones) between Brigalow fragments, where feasible.

Research directions: Survey and map all the populations of Pale Imperial Hairstreak Butterfly in Qld and use it as an indicator species for identifying old growth stands of Brigalow of high conservation value.

Sources and further reading

Braby MF (2000) *Butterflies of Australia: Their Identification, Biology and Distribution*. CSIRO Publishing, Melbourne.

Eastwood R and Fraser AM (1999) Associations between lycaenid butterflies and ants in Australia. *Australian Journal of Ecology* **24**, 503–537.

Eastwood R, Braby MF, Schmidt DJ and Hughes JM (2008) Taxonomy, ecology, genetics and conservation status of the Pale Imperial Hairstreak (*Jalmenus eubulus*) (Lepidoptera: Lycaenidae): a threatened butterfly from the Brigalow Belt, Australia. *Invertebrate Systematics* **22**, 407–423.

Eastwood R, Pierce NE, Kitching RL and Hughes JM (2006) Do ants enhance diversification in Lycaenid butterflies? Phylogeographic evidence from a model myrmecophile, *Jalmenus evagoras*. *Evolution* **60**, 315–327.

Authors: Peter Valentine and Steve Johnson

RICHMOND BIRDWING BUTTERFLY

Ornithoptera richmondia (Gray, 1853)

Photo: ScienceImage

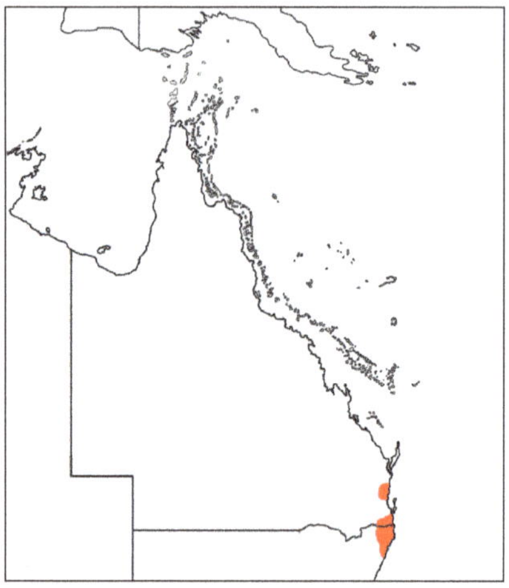

Family: Papilionidae

Conservation status: Qld: Vulnerable

Description: This is a large butterfly with a wingspan of 105 (male) to 115 mm (female). The body is bright yellow and black with patches of red near the legs. Males have a black forewing with a broad bright green subcostal stripe on the upperside, while the underside also includes lines of large greenish spots. The hindwing is mostly shining metallic green with a black edge, a series of round black spots and an occasional yellow spot on the upperside, and a line of golden-yellow spots, green spots and a semicircle of black spots within green patches on the underside. The abdomen is bright yellow and black laterally. Females also have a black forewing, which is dominated by whitish patches and bands on both sides. The hindwing is black with a subterminal row of golden-yellow patches, joined to white patches with a black spot between the two.

Distribution: The Richmond Birdwing Butterfly originally occurred from Maryborough in south-east Qld into northern NSW as far south as Blackwall Range. The species has now contracted to south of Kin Kin and lost about 60% of its previous range.

Habitat: This species is restricted to subtropical rainforest areas containing the larval food plants *Pararistolochia praevenosa* and *P. laheyana*. Suitable areas for *P. praevenosa* are lowland coastal rainforests, but *P. laheyana* occurs in montane rainforests along the Qld/NSW border.

Ecology: Females search for the larval food plants throughout the rainforest patches and lay eggs on mature leaves or on adjacent plants in contact with the vine. Males fly high in the forest at times, often soaring and gliding. At other times, they may search for females in the vicinity of the larval food plant. Larvae pupate on the vine or on adjacent plants that support the vine. There are two or more generations each year in the lowlands, but one in the montane forests. Adults have been reported throughout the year in the lowlands. In the highlands, the populations appear to suffer periodic extirpation due to seasonal conditions and are then recolonised by local migration from adjacent lowland areas.

Reasons for decline and current threats: Although originally widespread and abundant, this species has declined in population and reduced in range since European settlement. The principal reason is the clearing of lowland rainforest for residential and agricultural development. In addition, *Aristolochia elegans*, a decorative vine introduced from Brazil, mimics their native larval food plant but it is toxic to the caterpillars and kills any that hatch on it.

CURRENT RECOVERY EFFORTS

Recovery plan: None

Other plans: Sands DPA and New TR (2002) 'The action plan for Australian butterflies'. Environment Australia, Canberra.

Conservation actions:

- The Richmond Birdwing Conservation Network, first established as the Richmond Birdwing Recovery Network in October 2005, is active in producing and distributing the larval food plants in the lowland environments of south-east Qld and works closely with DERM.
- Significant work is being done with schools and attempts are being made to develop corridors within which the larval food plants are established.

- In the light of existing success in recovery, there are some grounds to reassess the status of, and perhaps de-list, the species in Qld.
- One collaborative project with DERM involves experimenting with captive rearing and out-breeding of butterflies to overcome a loss of genetic diversity in certain wild populations that has led to local extinctions.

FUTURE RECOVERY EFFORTS

What you can do:

- Join/donate to the Richmond Birdwing Conservation Network.
- Plant the larval food plants (*Pararistolochia praevenosa* and *P. laheyana)* in appropriate areas.
- Encourage school-based programs of conservation.

Conservation needs:

- Continue and extend community-based programs for the conservation of the species to assist recovery.
- Improve the management of secure habitats (e.g. removal of *Aristolochia elegans* infestations) and further acquire suitable habitat.
- Lobby for the banning of the nursery sale of *A. elegans*.

Research directions:

- Identify optimal sites for the establishment of corridors to reconnect isolated populations.
- Incorporate this species into landscape planning tools for optimising the value of remnant vegetation, regrowth and re-establishing forests.

Sources and further reading

Braby MF (2000) *Butterflies of Australia: Their Identification, Biology and Distribution*. CSIRO Publishing, Melbourne.

RBRN (2006) Richmond Birdwing Recovery Network Newsletter, Number 5, May. p. 20

Sands D (2008) Conserving the Richmond Birdwing Butterfly over two decades: where to next? *Ecological Management & Restoration* **9**, 4–16.

Authors: Peter Valentine and Steve Johnson

SOUTHERN PINK UNDERWING MOTH

Phyllodes imperialis H. Druce, 1888
(Southern species is undescribed)

Photo: Don Sands (northern ssp.)

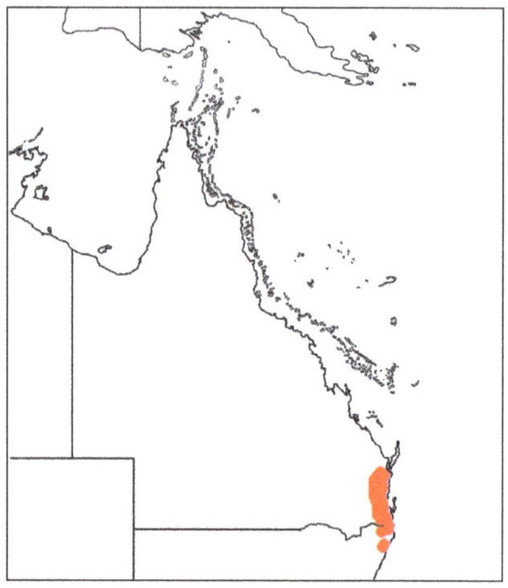

Other common name: Imperial Fruit-piercing Moth

Family: Noctuidae

Conservation status: Aust.: Endangered

Description: This large moth has a wingspan of up to 120 mm. The forewings are leaf-like above, with apex tapered, resembling a leaf petiole, grey-brown in colour with median, leaf-like 'mid-rib' markings, and distinctively-shaped band on the apical cell cross-vein. The hindwings are black with outer margins chequered white at vein ends and a large sub-central pink patch. The proboscis is finely spined but not sclerotised for piercing firm fruit, as in some related species of the tribe Ophiderini.

Distribution: The moth occurs from near Gympie, Qld to the Bellinger River and Dorrigo NSW, where it is associated with tall, coastal and lower montane rainforest. In Qld, the moth occurs near Kin Kin, Conondale, Blackall and D'Aguilar Ranges and McPherson Range at Springbrook. Single, or groups of two to five larvae, were reported from the Blackall Range.

Habitat: The Southern Pink Underwing Moth is only known to breed near stream and river embankments, in steep ravines and in shaded remnants of coastal lower montane rainforest, where larvae feed on the dioecious vine, *Carronia multisepalea*, whereas larvae of northern subspecies feed on a different vine, *Pycnarrhena* spp., restricted to northern Qld and the Torres Strait Islands. The population structure of the moth is unknown but occasional adults have been observed. In daylight hours they rest under leaves in darkened rainforest similar to the northern subspecies and are occasionally attracted to light.

Ecology: There are usually two generations of the Southern Pink Underwing Moth per year in south-east Qld and northern NSW. Adults live for approximately 30 days and imbibe fluid from soft, over-ripe, damaged or decomposing fruit, mainly figs. Adult moths are rarely seen, but are occasionally attracted to light. They may reveal the distinctive hindwing pink patches when disturbed, by arching their leaf-like brown forewings forward. Adult fecundity is unknown but females deposit eggs singly or in groups of three to seven on the underside of sub-terminal foliage of the food plant, the vine, *C. multisepalea* (Menispermaceae). Eggs eclose eight to 10 days after deposition but are often parasitised by a small wasp (*Telenomus* sp.). There are usually five instars (or occasionally six, depending on leaf nutrition).

Reasons for decline and current threats: In the recent past, logging, clearing, burning, farming, housing, weed invasions (especially lantana *Lantana camara*, Madeira vine *Anredera cordifolia*), mining for road-base and grazing (mostly by cattle) and extensive fragmentation, have destroyed or isolated habitats in the moth's known range. In south-east Qld, only about 10% of the pre-clearing suitable habitat remains, mostly on private property. Logging 'plantation' rainforests may also be a threat to the maintenance of vines as well as the need to exclude fire. Few intact habitats remain in national parks and other protected areas in south-east Qld, but insufficient habitats are intact (with both moths and food plant) to ensure the subspecies is not threatened by local disturbance, or in-breeding depression resulting from habitat fragmentation. Fragments of habitats are considered to be too small (e.g. <0.5 ha) by most authorities (but not Sunshine Coast City Council), for designations currently protecting vertebrates and plants on private properties and thus, many habitats remain threatened.

CURRENT RECOVERY EFFORTS

Recovery plan: None

Conservation actions:

- Efforts to encourage landowners to join *Land for Wildlife* programs, enter into Environmental Covenant agreements and Nature Refuge schemes, are productive in the Conondale and Blackall Ranges: major areas of intact habitat for the moth.
- A 2008 Threatened Species Network Project, administered by Barung Landcare, hosted an information workshop for landowners and other interested people. Its aims included rehabilitation of

habitats for the moth; local surveys for habitats and its food plant and to provide landowners with guidance towards rehabilitating the moth's ecosystem. It is not known if progress has been made towards habitat rehabilitation.

FUTURE RECOVERY ACTIONS

Conservation needs:

- Rainforest and its regrowth should not be disturbed after logging (especially hoop pines in 'plantations').
- Mining leases on basalt, andesite, rhyolite and granite, with substantial stands of the food plant vine (*Carronia multisepalea*) should actively avoid disturbing riparian vegetation.
- Exclude horses and cattle from moth habitats, creek beds and embankments and wherever the food plant vine occurs.
- Food plants for the larvae of Southern Pink Underwing Moth included in bush regeneration programs require ongoing threat abatement and management, such as weed removal, shading, support plants, and monitoring of watertables, flooding and soil types.
- A recovery plan must include research on the ecology of the food plant and its propagation, the density of vines required to support moth colonies, mobility of moth between breeding and mating sites and the feasibility of captive breeding.

Research directions:

- Conduct surveys and mapping of breeding subpopulations in each catchment and the tenure of each habitat.
- Carry out research on the moth's natural enemies.
- Determine the genetic variations between isolated populations.
- Gather additional data on the distribution, ecology and pollinators of the food plant vine.

Sources and further reading

Britton D (2006) The hunt for the pink underwing moth, *Phyllodes imperialis* and some associated events, November 2006. *The Society for Insect Studies*, Circular No. 124, pp. 88–90.

Hunter RL (1939) The pink underwing moth, *Phyllodes meyricki* (Ollif). Family Noctuidae, Sub-family Catocalinae. *The North Queensland Naturalist* 7(57), 1–2.

Moss J (2002) Imperial fruit piercing moth (*Phyllodes imperialis*). *Butterfly and Other Invertebrates Club Newsletter* **25**, 4–5.

Sands DPA (1999) Conservation status of Lepidoptera: assessment, threatening processes and recovery actions. In *The Other 99%. The Conservation and Biodiversity of Invertebrates.* (Eds W Ponder and D Lunney) pp. 382–387. Transactions of the Royal Zoological Society of New South Wales, Mosman.

Author: Don Sands

Crustaceans (Class: Crustacea)

Crayfish and Crabs (Order: Decapoda)

INTRODUCTION TO THREATENED DECAPODS

Peter JF Davie

Queensland freshwater decapod crustaceans belong to four main families: the Palaemonidae (long-armed river prawns); the Atyidae (atyid or brush-clawed shrimps); the Parastacidae (crayfish); and the Gecarcinucidae (inland or freshwater crabs). In general, there is still very little knowledge of the exact ranges and biological/ecological requirements for most Queensland decapods. However, perhaps the groups of most immediate concern are the Spiny Crayfish *Euastacus* species and the inland crabs *Austrothelphusa* species.

Although many crustaceans are reasonably resilient to environmental change, there is no doubt that habitat degradation and pollution have had an effect on diversity and abundance, at least on a localised scale. Most freshwater species are more prevalent in upper catchment areas than in lowlands, where excessive growth of algae and siltation are major problems. Coastal wetlands are also disappearing at an alarming rate, and species restricted to such areas appear to be more sensitive to habitat disturbance. When distributions become restricted to isolated patches, their continued survival becomes increasingly vulnerable. Lowering of the watertable, drainage of swamps, and increased levels of pollution, all the result of an ever-increasing human population and urbanisation, are other potential threats.

Another emerging problem is predation and competition from the increasing numbers of introduced species, not only alien fish, but also other translocated native crayfish such as the Redclaw *Cherax tenuimanus* and the Inland Yabby *Cherax destructor*, especially in south-east Queensland. Cane Toads *Rhinella marina* feed on juvenile crayfish and crabs, and although there are no specific data on actual impacts at this time, they share similar habitats, so an impact is almost inevitable. Other exotic species (cats, foxes and pigs) have generally been found to have an impact on crayfish, and thus could be expected to have a similar impact on crabs. Such species could contribute to serious declines in distribution and/or local abundance, especially for Critically Endangered species.

Spiny Crayfish *Euastacus* species

Recent genetic studies suggest that Queensland spiny crayfish evolved from a southern crayfish that expanded its range into Queensland during the Miocene Epoch (5–25 million years ago): a period when eastern Australia was cooler and wetter, and predominantly covered by lush forests. As the climate shifted to hotter and drier conditions during the late Miocene, rainforests began to shrink and became fragmented. So, too, these ancient crayfish had to take refuge in the well-shaded, cool, fast-flowing rainforest streams of mountains and tablelands. Separated on their mountain 'islands', they slowly evolved into the species we recognise today.

A recent assessment of the conservation status of Australian spiny crayfish *Euastacus* spp. under IUCN criteria has provided a sobering classification, with 14 of 15 species listed as threatened (two Vulnerable, four Endangered, eight Critically Endangered) and only one as Least Concern. However, the primary criterion contributing to these assessments is the very restricted ranges occupied by these species (status based on extent of occurrence <10 km^2, severely restricted distribution, and anticipated decline in area, extent and/or quality of habitat and extent of occurrence due to climate change).

Due to the large size and striking colouration of some species, and a common behaviour of leaving the water to forage, spiny crayfish are well known, and some species have had significant fishing and collecting pressure. However, current Queensland legislation protects all spiny crayfish as 'no take' species under the *Fisheries Act 1994*, and they must be released if captured. Nevertheless, there is evidence that illegal poaching does occur. Also in the majority of cases, Queensland spiny crayfish are almost all found within the confines of existing national parks, and thus are also provided a significant level of protection, especially against human-induced habitat destruction.

Freshwater Crabs *Austrothelphusa* species

Freshwater crabs are a key component in many tropical freshwater systems. They are seasonally present, in often large numbers, and are omnivorous scavengers and predators. They are also an important source of food for a range of birds, reptiles and mammals, such as native rats. In 2009, a basic assessment of the 1280 known species of freshwater crab worldwide was done for the IUCN Red List. It was noted that two-thirds of all species of freshwater crab may be at risk of going extinct, with one in six species particularly vulnerable and, of those, 227 species should be considered as Near Threatened, Vulnerable, Endangered or Critically Endangered. For about half (628 species), there is insufficient data to adequately assess their status.

Australia currently has seven species of freshwater crabs in a single genus, *Austrothelphusa*; however, a recent taxonomic and genetic study at the Queensland Museum now suggests that the number of species has been significantly underestimated, and there appears to be an additional 15–20 species across northern Australia. It also appears that many of these have narrow distributions, restricted to particular catchments. At least some of these will almost certainly fall into the Endangered or Critically Endangered categories under the IUCN criteria, based on their narrow known distributional ranges, although many will otherwise be Data Deficient in terms of their biology, life history and population sizes.

In most cases, knowledge is negligible regarding biology and population numbers of individual species. The highly restricted distributional range of many species makes them extremely susceptible to localised impacts, including bushfires, forest management practices, habitat destruction (both from mining and from bush-clearing) and, for some of the more charismatic species, a possible threat from increased wild collecting for the pet-shop trade, particularly if this was to be locally intensive. Climate change could also have a significant impact on the survival rates of tropical species that have evolved under an annual monsoonal regime, if prolonged droughts were to last some years.

SPINY CRAYFISH

Euastacus balanensis Morgan, 1988; Mount Elliot Spiny Crayfish *E. bindal* Morgan, 1989; Eungella Spiny Crayfish *E. eungella* Morgan, 1988; *E. fleckeri* (Watson, 1935); *E. hystricosus* Riek, 1951; Mistake Mountain Spiny Crayfish *E. jagara* Morgan, 1988; Currumbin Spiny Crayfish *E. maidae* (Riek, 1956); Kroombit Spiny Crayfish *E. monteithorum* Morgan, 1989; Bloomfield Spiny Crayfish *E. robertsi* Monroe, 1977; Mount Glorious Spiny Crayfish *E. setosus* (Riek, 1956); Lamington Spiny Crayfish *E. sulcatus* Riek, 1951; *E. suttoni* Clark, 1941; *E. urospinosus* (Riek, 1956); Tully River Spiny Crayfish *E. yigara* Short & Davie, 1993

Lamington Spiny Crayfish. Photo: Jeff Wright, Queensland Museum

Family: Parastacidae

Conservation status: IUCN: Vulnerable: *Euastacus sulcatus*; *E. suttoni*. Endangered: *E. balanensis*; *E. fleckeri*; *E. hystricosus*; *E. urospinosus*. Critically Endangered: *E. bindal*; *E. eungella*; *E. jagara*; *E. maidae*; *E. monteithorum*; *E. robertsi*; *E. setosus*; *E. yigara*

Description: The 15 species found in Qld range in size from less than 90 mm long to an imposing 150 mm. These often heavily built crayfish typically have a shiny carapace and abdomen armed with stout spines, although some of the smaller species are relatively 'hairy' and non-spinous, belying their common name. They can range in colour from drab brown to striking bright blue, bright red or even purple.

Distribution: Spiny crayfish *Euastacus* spp. are found all down the east coast of Australia. However, they are especially interesting because each is confined to separate mountain ranges.

Habitat: Spiny crayfish typically prefer cooler water, so the more tropical the latitude the higher the altitude they occur (only above 250 m in south-east Qld, above 750 m in central Qld and above 900 m in north Qld). They inhabit streams containing abundant instream debris and vegetation, and can even be found away from running water on the rainforest floor during the rainy season. They shelter under rocks and logs in water and may burrow underground during the warmer months.

Ecology: Spiny crayfish feed on rotting debris, bacteria, fungi, and being opportunistic feeders, even on carrion. They breed in late winter to early spring and the female carries her eggs under her tail for up to 5 months. These slow growing crustaceans can take up to 9 years to reach breeding maturity. They have impressive defensive behavioural displays and some large species may even 'vocalise' when approached.

Reasons for decline and current threats: Given highly restricted ranges, these species are extremely susceptible to localised impacts, including bushfires, forest management practices, habitat destruction, and over-exploitation by collectors (mostly through illegal activity).

Climate change, including increasing temperature, alterations to hydrological regimes, severe weather events, loss of suitable cool, forested highland habitat and the increased potential for bushfire may all potentially have an impact upon spiny crayfish. There are potentially significant threats from Cane Toads, although generally there are no specific data on actual impacts. Also, other exotic species (cats, foxes, pigs) have been found to have an impact on crayfish and could contribute to serious declines in distribution and/or local abundance, especially for Critically Endangered species.

CURRENT RECOVERY EFFORTS

Recovery plan: None

Conservation actions:

- All spiny crayfish are 'no take' species in Qld; these species cannot be retained.
- In the majority of cases, Qld spiny crayfish are almost all found within the confines of existing national parks, and thus are also provided a significant level of protection, especially against human-induced habitat destruction.

FUTURE RECOVERY EFFORTS

What you can do:

- Don't catch any spiny crayfish. If you do so by accident, carefully release them back into the stream.
- Keep all streams pollution and rubbish free.

Research directions:

- Assess and monitor the populations.
- Gather biological and life history information.
- Research the habitat requirements of the crayfish.
- Investigate the thermal tolerance of the crayfish.
- Study their resilience to the effects of exotic species.

Sources and further reading

Coughran J and Furse JM (2010) 'An assessment of genus *Euastacus* (49 species) versus IUCN Red List criteria'. A report prepared for the global species conservation assessment of lobsters and freshwater crayfish for the IUCN Red List of Threatened Species. International Association of Astacology: <http://www.iz.carnegiemnh.org/crayfish/iaa/publications.htm>.

Horwitz P (1990) 'The conservation status of Australian freshwater crustacea with a provisional list of threatened species, habitats and potentially threatening processes'. Report Series No. 14. Australian National Parks and Wildlife Service, Canberra.

Morgan GJ (1988) Freshwater crayfish of the genus *Euastacus* Clark (Decapoda: Parastacidae) from Queensland. *Memoirs of Museum of Victoria* **49**, 1–49.

Author: Peter JF Davie

FRESHWATER TIGER CRAB

Austrothelphusa tigrina Short, 1994

Photo: Courtesy Queensland Museum

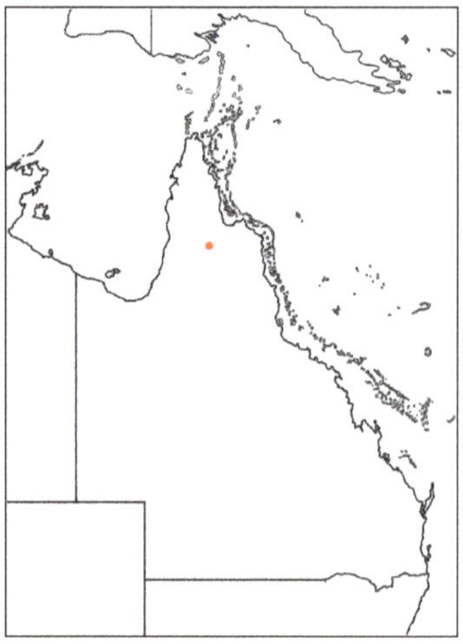

Family: Gecarcinucidae

Conservation status: IUCN: Vulnerable

Description: This is a large species of *Austrothelphusa*, with adults reaching 36 mm carapace breadth, although they become sexually mature at much smaller sizes, with females being reproductive from as little as 13 mm wide. The anterolateral regions of the carapace are generally not swollen, and not distinctly striated, and the anterolateral tooth is small and notch-like, and does not project beyond the lateral border of the carapace. The front of the carapace is slightly concave or straight and its depth is approximately half its greatest width. The length of a male telson is about equal to its basal width. The median suture on the ischium of the third maxilliped is very narrow and sharply defined. The inferior orbital angle is crenulate. The eyestalks are distally expanded. The carapace is tan with conspicuous black markings including one long stripe behind the frontal margin, shorter stripes laterally on the anterior branchial regions, a v-shaped stripe postero-medially, a pair of small posterolateral spots and three large spots across the posterior margin. The claws are cream with a black band distally on the upper palm and the fingers are tinged with purple. Legs with merus yellow and distal segments purple. The eyestalks have a mesial, longitudinal, black stripe.

Distribution: The Freshwater Tiger Crab is known only from One Mile Creek: a seasonally-flowing creek, a tributary of the Alice River, 200 km from the sea on Cape York Peninsula.

Habitat: Crabs occupy sandy burrows among roots of fringing *Melaleuca* trees in riverine habitat and are likely to be nomadic within the catchment area during rain. They are presumed to plug their burrows in the dry season.

Ecology: The species lives in burrows in freshwater habitat. They have an abbreviated larval development, giving birth to juvenile crabs. They are apparently locally abundant during the wet season (November to April), when they are often conspicuous on clean sandy substrates. During the dry season, they occupy sandy burrows among the roots of fringing *Melaleuca* trees.

Reason for decline and current threats: The current status of populations is not known and knowledge is limited regarding its biology and population size. The creek runs through pastoral leasehold and lies largely within the Alice River Mining Field. Gold mining and collection for the pet shop trade are foreseeable threats to the welfare of this crab. Another potential threat would include extended periods of drought as a result of climate change. The survival rates of tropical species that have evolved under a regime of an annual monsoon may be severely affected by prolonged drought.

CURRENT RECOVERY EFFORTS

Recovery plan: None

Recovery actions: At this stage, recovery efforts will be contingent on the outcomes of the required research and assessment outlined below.

FUTURE RECOVERY EFFORTS

Research directions:

- Conduct surveys to ascertain the limits of distribution.
- Assess the magnitude and impact of wild catch for pet shop trade.
- Research the status and requirements of this, and other northern species, to ensure that the population is managed and conserved.

Sources and further reading

Short JW (1994) A new species of freshwater crab (Sundathelphusidae) from Cape York Peninsula. *Memoirs of the Queensland Museum* **35**, 235–240.

Author: Peter JF Davie

MCILWRAITH FRESHWATER CRAB

Austrothelphusa valentula (Riek, 1951)

Photo: Courtesy Queensland Museum

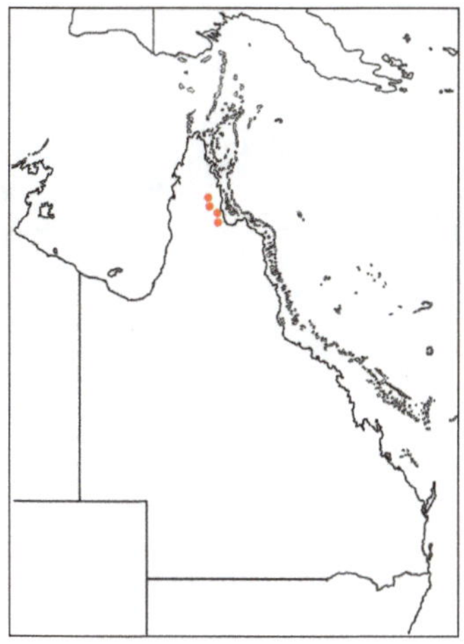

Family: Gecarcinucidae

Conservation status: IUCN: Vulnerable

Description: A large species of *Austrothelphusa* (females reach mature abdomen shape at carapace widths greater than 13 mm). The anterolateral regions of the carapace are markedly swollen, not distinctly striated; the anterolateral tooth is large, projecting beyond lateral carapace border. The carapace is broadly oval and convex; the mid-dorsal surface is finely punctate; the dorsal H-shaped groove system is strongly impressed. The epigastric and postorbital crests are poorly defined but noticeable, particularly in juveniles. The branchial regions more swollen than other *Austrothelphusa*, but not markedly striated. The cervical grooves are incomplete, and poorly defined. The lateral borders are arched, anteriorly keeled; there is a single conspicuous anterolateral tooth on each side projecting beyond lateral margin, and with small, short groove extending medially from base. The orbital spine is blunt. The posterolateral borders are slightly concave, bearing a few oblique grooves. The frontal margin is slightly concave when viewed from above. The male abdomen tapers abruptly to the fifth segment; the sixth segment with proximal and distal borders are subequal in width; the medial breadth is equal to the length. The telson is tongue-shaped and elongate; the length is distinctly greater than basal width. The body, legs and claws of adults are generally fawn brown, perhaps darkening around joints; the middle two-thirds longitudinally are coloured a dark inky purple that is laterally uneven and broken up and indented posterolaterally; there is a large light patch medially and a pair of patches behind the frontal lobes.

Distribution: The McIlwraith Freshwater Crab is found from Coen River southwards, in freshwater along the McIlwraith Ranges to Port Stewart area in Kulla (McIlwraith Range) Cape York Peninsula Aboriginal Land (CYPAL) NP.

Habitat: This is a riverine species (in the Coen River at Coen), which is reported to walk some distance from water across surrounding country during rain.

Ecology: It lives in burrows in freshwater habitats; the burrows are plugged in the dry season, and not easily found. It may be locally abundant during the wet season. It has abbreviated larval development and gives birth to juvenile crabs.

Reason for decline and current threats: The current status of populations is not known; like other *Austrothelphusa*, species, knowledge is negligible regarding biology and population numbers. Most species are geographically confined to relatively narrow ranges. The potential threats include extended periods of drought as a result of climate change (the survival rates of tropical species that have evolved under the regime of an annual wet monsoonal season may be severely affected were this not to occur). Other possible threats result from increased wild collections for pet shop trade, particularly if this was to be locally intensive.

CURRENT RECOVERY EFFORTS

Recovery plan: None

Recovery actions: Not considered necessary at this time.

FUTURE RECOVERY EFFORTS

Research directions:

- Conduct surveys to ascertain the limits of distribution.
- Assess if there is a significant wild catch for pet shop trade.
- Research the status and requirements of this, and other northern species, to ensure that the population is managed and conserved.

Sources and further reading

Bishop JA (1963) The Australian freshwater crabs of the family Potamonidae (Crustacea: Decapoda). *Australian Journal of Marine and Freshwater Research* **14**, 218–238.

Greenaway P (1984) The relative importance of the gills and lungs in the gas exchange of amphibious crabs of the genus *Holthuisana*. *Australian Journal of Zoology* **32**, 1–6.

Greenaway P, Taylor HH and Bonaventura J (1983) Aerial gas exchange in Australian freshwater/land crabs of the genus *Holthuisana*. *Journal of Experimental Biology* **103**, 237–251.

Author: Peter JF Davie

Snails (Class: Gastropoda)

Air-breathing snails (Order: Eupulmonata)

INTRODUCTION TO THREATENED SNAILS

John Stanisic

Currently the Boggomoss Snail *Adclarkia dawsonensis* is the only officially recognised threatened snail species in Queensland. The species is listed as Critically Endangered under the Australian Government's *Environment Protection and Biodiversity Conservation Act 1999*, but has yet to be formally listed under the local *Nature Conservation Act 1992*. A recovery plan for the species was prepared and reviewed in 2011. However, apart from fencing a single habitat in which the snail is known to survive, little has been done to implement the many recovery objectives outlined in the plan. The proposed building of the Nathan Dam on the Dawson River has led to the formulation of a relocation strategy for this snail population. Should this development proceed and elements of the strategy be implemented, a number of key issues surrounding the recovery of the species will be tackled. A number of other Queensland land snails are known to be in the same predicament as the Boggomoss Snail. These species occur mainly in the Brigalow Lands bioregion and, like the Boggomoss Snail, have had their original ranges fragmented and greatly reduced by land clearing, chiefly for farming. They now have to deal with more localised threats such as large-scale mining developments, fire, floods and predation from feral animals. Their situation is perilous and, without active management, they face the very real prospect of extinction. Submissions for their addition to the threatened species list are being prepared at the time of writing.

BOGGOMOSS SNAIL

Adclarkia dawsonensis Stanisic, 1996

Photo: J Stanisic

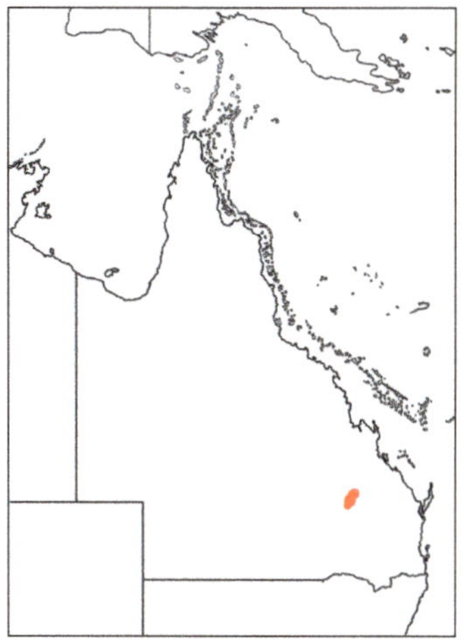

Family: Camaenidae

Other common name: Dawson Valley Snail

Conservation Status: Aust.: Critically Endangered; Qld: under assessment

Description: The Boggomoss Snail is a medium-sized snail with a relatively thin, semi-transparent shell. The shell is subglobose with a flat to slightly elevated spire. Shell diameter ranges from 21–26 mm and shell height ranges from 14–16 mm. The shell colour is light-brown with a slight yellowish-green tinge, occasionally with a reddish-brown band below the suture and a reddish-brown circum-umbilical patch. The aperture is sub-circular and the lip is reflected at adulthood; the umbilicus is small. The shell surface appears smooth, but microscopically is covered by a series of crowded micro-ridgelets that in fresh specimens bear low, elongate periostracal scales; these scales wear off as the snail grows. The animal is grey with an orange-coloured mantle and distinct irregular black blotches on the lung roof that are visible through the shell.

Distribution: The Boggomoss Snail was originally (c. 1980) known from two dead adult shells, which were collected in the vicinity of Theodore, in inland central Qld. Subsequent searches of the area by the Qld Museum failed to locate the snail. Live specimens were discovered as part of broader invertebrate surveys of the Taroom area in the mid-1990s. These were found on a boggomoss on Mount Rose Station, north-east of Taroom and in riparian habitat. Surveys for the snail conducted in 2008–2009 discovered the species on two other boggomosses on the Mount Rose property and also extended the area of riparian corridor occupied by the species both up and down stream of the Isla–Delusion Road crossing.

Habitat: Historically, before clearing, the Boggomoss Snail is presumed to have occupied the gilgaied, closed-canopy Brigalow *Acacia harpophylla* communities on the black soil alluvial plains of the Dawson River between Taroom and Theodore. Although the species is currently largely confined to riparian habitats along the Dawson River, indirect evidence suggests that the adjacent alluvial black soils were the original home of the species. General habitat data from all eastern Australian land snails (1200 odd species) does not identify riparian habitat as the core distribution of any species. However, these communities do act as moist refuges for snails and are important dispersal corridors. The current distribution of the Boggomoss Snail in riparian communities adjacent to cleared tracts of alluvial black soil and the distribution of a sister species in the Dalby-Chinchilla area in remnant patches of gilgaied Brigalow strongly support the former conclusion. The boggomoss habitats are characterised by *Eucalyptus camaldulensis* and *Ficus coronata*/*F. opposita* and the Dawson River riparian habitats by *E. camaldulensis*, *Corymbia tessellaris*, *F. coronata*/*F. opposita* and *Livistona nitida*. Both are marginal, moist refugial habitats in which the snail is able to exist in the absence of core habitat. Their long-term future in these marginal habitats is uncertain.

Ecology: The Boggomoss Snail is a terrestrial nocturnal species that lives among deep, moist accumulated leaf litter and under fallen logs. It is thought to feed on fungi and other biofilm growing among decomposing leaf litter and on fallen timber. The species aestivates by loosely sealing to the underside of logs or to leaves and twigs in the litter. Nothing is known about the general biology, growth rate or population dynamics and genetics of the species.

Reasons for decline and current threats: Land clearing for farming is considered to be the primary cause for the decline in the species. The black soil alluvial flats of the Dawson River between Taroom and Theodore were cleared of Brigalow early last century (c. 1920) for stock grazing and cultivation of wheat and stock feed. Stock grazing also has both direct (trampling) and indirect impacts (altering microhabitat structure) on the species.

Due to the combination of accumulating fuel loads and the high incidence of lightning strikes in the Taroom area, fire is a primary threat either directly through incineration of the snail or indirectly through the destruction of microhabitat (litter and logs). Invasive weeds and grasses (e.g. Buffel *Cenchrus ciliaris*) are a potential secondary threat to the species in that they alter microhabitat structure and provide high fuel loads for fire.

Predation of snails is a natural phenomenon, but in the case of the Boggomoss Snail it appears to have become a primary threat to the species. A large proportion of the dead shells recovered in recent surveys displayed signs of mammal predation. The main culprit is thought to be the common House Mouse *Mus musculus*, which has proliferated in the area in response to availability of grain for food. Some native animals (small mammals, birds and reptiles) will also feed on the snail, but are not present in the numbers to be classed as major threats.

The projected impoundment of the proposed construction of a major dam on the Dawson River in the vicinity of the Nathan Gorge will inundate the Mount Rose boggomosses and subsequently drown the main population of the species. Mitigation measures have been proposed to deal with this threat and are being considered as of the time of writing.

CURRENT RECOVERY EFFORTS

Recovery plan: Stanisic J (2006) 'Recovery plan for the Boggomoss Snail, *Adclarkia dawsonensis* 2007–2011'. Report to the Department of Environment and Heritage (DEH), Canberra. Department of Environment and Resource Management, Brisbane.

Conservation actions: Fencing the Mount Rose boggomoss site to exclude stock.

FUTURE RECOVERY EFFORTS

Conservation needs

- Protect the populations and habitat.
- Implement a public awareness program.
- Map its habitats.
- Reduce the current threats, such as those presented by fire, weed proliferation, grazing and earthworks.

Research directions

- Conduct surveys and monitor the existing populations.
- Thoroughly research the ecology and biology of the snail.

Sources and further reading

Ingram GJ and Stanisic J (1997) 'Dawson River Project. Impact assessment study: Boggomosses (mound springs) and other spring-fed areas'. Queensland Museum, Brisbane.

Stanisic J (1996) New snails from boggomoss environments in the Dawson Valley, southeastern Queensland (Eupulmonata: Charopidae and Camaenidae). *Memoirs of the Queensland Museum* **39**, 343–354.

Author: John Stanisic

CARTILAGINOUS FISHES: Sharks, Rays and Chimaeras

(Class: Chondrichthyes)

INTRODUCTION TO THREATENED CARTILAGINOUS FISHES

Peter M Kyne

The cartilaginous fishes (class Chondrichthyes) are a diverse group of approximately 1150 primarily marine species, comprising the sharks and rays (collectively called the elasmobranchs) and the chimaeras. The diversity of chondrichthyan fishes is high in the Indo-West Pacific, in particular the Indo-Australian region, with high levels of endemism found in shallow, tropical waters. Of the 60 chondrichthyan families worldwide, 45 are represented in Queensland waters, with a total of 177 species recorded from the state (around 15% of the global total).

The conservative life history characteristics (relatively slow growth, late age at maturity, low fecundity, long lifespan, and low natural mortality of all age classes) of chondrichthyan fishes result in an intrinsic vulnerability to overexploitation and population depletion. Their productivity is often similar to those of other long-lived animals, such as many large mammals. Once depleted, the populations of these species may have a limited ability to recover. Productivity among the chondrichthyans varies, however, ranging from relatively short-lived and fast-growing species (e.g. some smooth-hound sharks) to long-lived species with very low fecundity (e.g. the deep-sea gulper sharks). Such variation has significant consequences for fisheries management. Fisheries for some of the more productive species, where careful management is in place, can be sustainable; in contrast, where the capture of chondrichthyans is primarily as bycatch, such management is often lacking and the mortality of bycatch species may exceed sustainable levels.

As a result of the IUCN Shark Specialist Group's Global Shark Assessment program (one of the aims of which is to 'establish a baseline against which to monitor future changes in the global and regional status of chondrichthyan fishes'), the vast majority of chondrichthyan fishes have had their conservation status assessed on the *IUCN Red List of Threatened Species*. As of 2010, there are 1044 species assessed, including 154 species occurring in Queensland (87% of the state's fauna). Assessments for the remainder of the Queensland fauna are underway, and thus the list of species within threatened categories may increase for the state. Readers are urged to consult the IUCN website for updates (<http://www.iucnredlist.org>).

Chondrichthyans on the IUCN Red List are assessed at the global level; that is, the conservation status of a species is evaluated across its entire distribution and encompasses all known populations. At the global level, 27.9% (43 species) of Queensland chondrichthyan species are considered to be threatened with extinction (Critically Endangered: 5 species, 3.2%; Endangered: 5 species, 3.2%; Vulnerable: 33 species, 21.4%), 26.6% (41 species) are Near Threatened, 26.6% (41 species) are Least Concern and 18.8% (29 species) are Data Deficient.

In some cases, in addition to the global assessment, a species has been assessed at the regional level for a particular part of its distribution or for a particular population. These regional assessments highlight when the conservation status of a species for a particular region, country or location differs from the global status. For some species occurring in Queensland, there are relevant regional assessments undertaken for Aus-

Table 4. Chondrichthyan species that, in addition to those treated in this publication, are threatened globally according to the 2010 *IUCN Red List of Threatened Species*, but that are not considered threatened in Australia (and therefore Queensland) due to a lower regional assessment. For more information on the status of these species see Cavanagh *et al.* (2003) and IUCN (2010).

Species	Global IUCN status	Australian regional status
Sharks		
Scalloped Hammerhead *Sphyrna lewini*	Endangered	Data Deficient
Great Hammerhead *Sphyrna mokarran*	Endangered	Data Deficient
Thresher Shark *Alopias vulpinus*	Vulnerable	Data Deficient*
Dusky Shark *Carcharhinus obscurus*	Vulnerable	Near Threatened
Sandbar Shark *Carcharhinus plumbeus*	Vulnerable	Near Threatened
Fossil shark *Hemipristis elongatus*	Vulnerable	Least Concern
Tawny Nurse Shark *Nebrius ferrugineus*	Vulnerable	Least Concern
Sharptooth Lemon Shark *Negaprion acutidens*	Vulnerable	Least Concern
Zebra Shark *Stegostoma fasciatum*	Vulnerable	Least Concern
Rays		
Giant Shovelnose Ray *Glaucostegus typus*	Vulnerable	Near Threatened
Shark Ray *Rhina ancylostoma*	Vulnerable	Near Threatened
Whitespotted Guitarfish *Rhynchobatus australiae*	Vulnerable	Near Threatened
Smoothnose Wedgefish *Rhynchobatus laevis*	Vulnerable	Near Threatened
Blotched Fantail Ray *Taeniurops meyeni*	Vulnerable	Least Concern

*Regional status for the Indo-West Pacific.

tralia, the east coast of Australia or the Indo-West Pacific region. For the purposes of this publication, when there is a regional assessment that encompasses Queensland, that regional assessment is considered to be the relevant conservation status for Queensland. In many cases, regional assessments are not officially published on the IUCN Red List because these are assessments for geographical locations rather than for distinct populations of a species.

For several species threatened globally, their regional threat category for Australia (and hence Queensland) is lower (Near Threatened, Least Concern or Data Deficient) (Table 4). These species are generally those with wide global or Indo-West Pacific distributions, which face considerable pressure from directed fishing and/or bycatch over parts of their distributions, but which do not face comparable pressures in Australia. For example, the Blotched Fantail Ray *Taeniurops meyeni* is exploited as part of directed ray fisheries in Indonesia, while rays are generally not fished across the northern Australian range of the species (where suitable habitat is also protected within marine reserves). Thus, its global IUCN status is Vulnerable, while its Australian regional status is Least Concern. Given the recognised global pressures facing this species, and those listed in Table 1, it should be assured that their populations remain healthy in Queensland. Further information on the conservation status of the species listed in Table 1 can be found in Cavanagh *et al.* (2003) and on the IUCN *Red List of Threatened Species* (see <http://www.iucnredlist.org>).

When adjusted to take into account regional assessments, 19.5% (30 species) of Queensland chondrichthyans are threatened with extinction (Critically Endangered: 6 species, 3.9%; Endangered: 4 species, 2.6%; Vulnerable: 20 species, 13.0%), 26.0% (40 species) are Near Threatened, 35.7% (55 species) are Least Concern and 18.8% (29 species) are Data Deficient.

This section contains profiles for 29 threatened chondrichthyans. An additional threatened species, the School Shark *Galeorhinus galeus*, which is Vulnerable on the IUCN Red List and Conservation Dependent on the EPBC, is not included here. The School Shark occurs only very marginally in southern Queensland waters; this is a species of cooler temperate waters. Its status is the result of a long history of unsustainable exploitation in southern Australia and its conservation is dependent entirely upon actions outside of Queensland. A few of Queensland's other threatened chondrichthyans are also at the northern extent of their range in southern Queensland (e.g. the stingarees *Urolophus* spp.). These are poorly known species in Queensland waters and their threatened listings are also a result of population trends in the southern portions of their ranges. In the case of these stingarees, however, their distribution is much narrower than the school shark – being endemic to eastern Australia – and their inclusion in this book is warranted. For some of the wider ranging Indo-West Pacific species (e.g. Banded Eagle Ray *Aetomylaeus nichofii*, Ornate Eagle Ray *A. vespertilio*) their threatened listings originate from pressures elsewhere, particularly in South-East Asia, and very little information is available on these species in Queensland. A similar situation occurs with the wide-ranging pelagic species (e.g. Oceanic Whitetip Shark *Carcharhinus longimanus*, Porbeagle *Lamna nasus*); their threatened listings are related to global fishing pressure with little information available locally. Indeed, for many of Queensland's threatened chondrichthyans, some of the most basic information on their distribution and occurrence in Queensland is lacking.

Despite the number of species listed as threatened by the IUCN, only the Grey Nurse Shark *Carcharias taurus* is listed on the Queensland NCA. Additionally though, Speartooth Shark *Glyphis glyphis*, White Shark *Carcharodon carcharias* and all sawfish species are protected 'no take' species under Queensland fisheries legislation. At the national level, eight species occurring in Queensland (Dwarf Sawfish *Pristis clavata*, Freshwater Sawfish *P. microdon*, Green Sawfish *P. zijsron*, Grey Nurse Shark, School Shark, Speartooth Shark, Whale Shark *Rhincodon typus* and White Shark) are listed under threatened categories on the EPBC, while proposals are under consideration for two more species (Harrisson's Dogfish *Centrophorus harrissoni* and Endeavour Dogfish *C. moluccensis*). A more comprehensive recognition of the vulnerability of chondrichthyans at both the state and federal level is required to ensure the long-term survival of Queensland's threatened shark and ray fauna.

DOGFISH SHARKS (ORDER: SQUALIFORMES)

PHILIPPINE SPURDOG

Squalus montalbani Whitley, 1931

Photo: ScienceImage

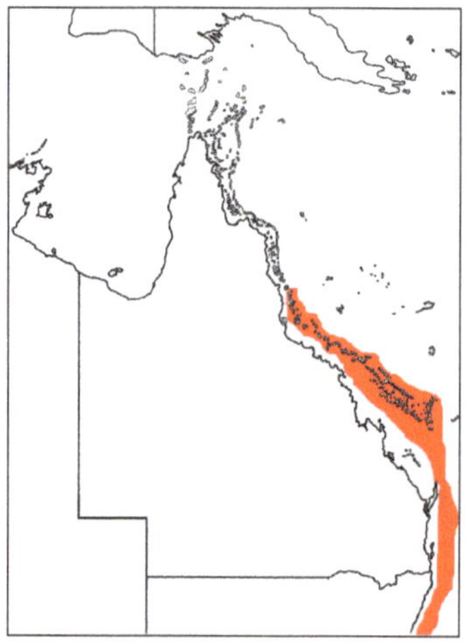

Family: Squalidae

Other common name: Green-eyed Dogfish

Conservation status: IUCN: Vulnerable

Description: The Philippine Spurdog is a medium-sized dogfish, which reaches a maximum total length of about 90 cm. It has a moderately elongated body with a broadly pointed snout. Each dorsal fin possesses a prominent spine and the caudal fin is relatively short. The teeth are blade-like and similar in both jaws. The body is greyish (paler underneath) without any white spots and there is a dark marginal bar above the rear notch of the tail.

Distribution: The Philippine Spurdog has a patchy distribution in the Indo-West Pacific region, being recorded off WA, eastern Australia, eastern Indonesia and the Philippines. Off eastern Australia, its known range is from off Terrigal, NSW to Flinders Reef, Qld.

Habitat: This demersal species inhabits the upper continental slope in a depth range of about 300–700 m. This equates to a narrow band of available habitat because, at these depths, the continental slope is less than 10 nautical miles wide.

Ecology: The biology and ecology of the Philippine Spurdog is very poorly known, in part due to confusion with other similar 'greeneye dogfish' species occurring off eastern Australia. Males mature between 60 and 70 cm total length and females around 80 cm total length. Species of *Squalus* are viviparous, usually having litters of 4–10 young with a continuous reproductive cycle (i.e. females become pregnant again soon after giving birth). The gestation period may be up to 2 years and the size at birth is between 20 and 25 cm total length.

Reasons for decline and current threats: The capture of the species as byproduct and bycatch in commercial trawling activities off NSW has led to a population decline in that area. Fishing pressure may also be high where the species occurs off Indonesia and the Philippines.

Research data from NSW demonstrated the susceptibility of dogfish to population depletion as a result of consistent trawling activities. Trawl surveys conducted on the central and southern NSW upper continental slope (200–650 m depth) showed an overall decline in the catch rate of 'greeneye dogfish' of >97% between 1976–77 and 1996–97. Confusion between 'greeneye dogfish' species precludes species-specific evaluation of declines, but it appears that Philippine Spurdog, Eastern Longnose Spurdog *Squalus grahami* and Greeneye Spurdog *S. chloroculus* have all declined significantly off south-east Australia. The greatest catch rates of 'greeneye dogfish' (including the Philippine Spurdog) during surveys occurred off central NSW (off Sydney), which corresponds to the southern extent of the Philippine Spurdog's distribution off eastern Australia. Current levels of fishing on that part of its NSW range remain high, but fishing pressure over the majority of its distribution off northern NSW and Qld is much lower and any impact on the species is probably small. Species of *Squalus* generally have limiting life history characteristics, which add to their vulnerability.

In Qld, there is no information on the effects of fishing where this species occurs but it is likely that relatively few Philippine Spurdog are caught. Given the extensive areas of its range that are only lightly fished (northern NSW, Qld and WA), the species is probably of lower conservation concern in Australia than some other dogfishes.

CURRENT RECOVERY EFFORTS

Recovery plan: None

Other plans: Australian Fisheries Management Authority (AFMA) (2010) 'Draft Upper-Slope Dogfish Management Strategy. Commonwealth-managed fisheries.' AFMA, Canberra.

Conservation actions: A fishing closure in place off Sydney to mitigate the impacts of Commonwealth fisheries on Endeavour Dogfish *Centrophorus moluccensis* may benefit the Philippine Dogfish.

FUTURE RECOVERY EFFORTS

What you can do:

- Report any fishing catches to the Qld Museum.
- Obtain the Australian Marine Conservation Society's *Australia's Sustainable Seafood Guide* and buy only seafood that is harvested from sustainable sources with limited bycatch (see <http://www.amcs.org.au>).
- Qld commercial fishers should release any captured dogfish safely.

Conservation needs:

- Reduce and limit catches of the species (this is dependent on conservation actions in the southern part of the range of the species where stocks have been severely depleted; currently, there are no catch restrictions in Australia that apply to species of *Squalus*).
- Ensure marine planning processes consider this and other slope elasmobranchs in identifying areas as candidates for marine protected areas.

Research directions:

- Quantify and monitor any catches in Qld fisheries.
- Better define the occurrence and habitat requirements of the species in Qld waters.
- Define the status of the species off eastern Australia, where there may be refuge in the central and northern parts of its range.

Sources and further reading

Graham KJ (2005) Distribution, population structure and biological aspects of *Squalus* spp. (Chondrichthyes: Squaliformes) from New South Wales and adjacent Australian waters. *Marine and Freshwater Research* **56**, 405–416.

Graham KJ, Andrew NL and Hodgson KE (2001) Changes in relative abundance of sharks and rays on Australian South East Fishery trawl grounds after twenty years of fishing. *Marine and Freshwater Research* **52**, 549–561.

Last PR and Stevens JD (2009) *Sharks and Rays of Australia*. 2nd edn. CSIRO Publishing, Melbourne.

Last PR, White WT and Motomura H (2007) Description of *Squalus chloroculus* sp. nov., a new spurdog from southern Australia, and the resurrection of *S. montalbani* Whitley. In *Descriptions of New Dogfishes of the Genus* Squalus *(Squaloidea: Squalidae)*. (Eds PR Last, WT White and JJ Pogonoski) pp. 55–69. CSIRO Marine and Atmospheric Research Paper No. 014. CSIRO Marine and Atmospheric Research, Hobart.

White WT (2008) *Squalus montalbani*. In IUCN 2010. *IUCN Red List of Threatened Species*. Version 2010.4. IUCN, Gland, Switzerland and Cambridge, UK, <http://www.iucnredlist.org>.

Authors: Peter M Kyne and Ken J Graham

HARRISSON'S DOGFISH

Centrophorus harrissoni McCulloch, 1915

Photo: ScienceImage

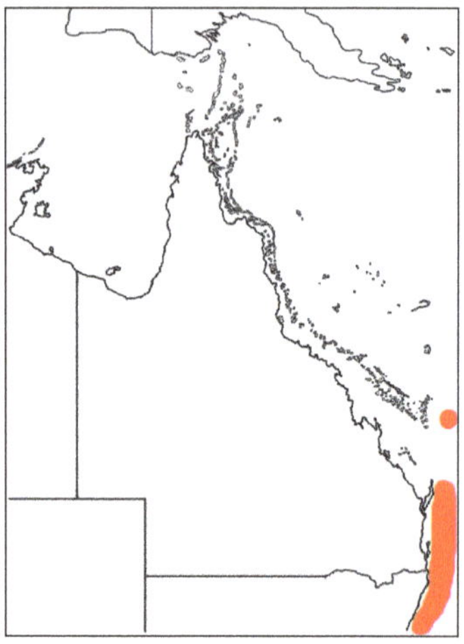

Family: Centrophoridae

Other common names: Dumb Shark, Dumb Gulper Shark, Harrisson's Deepsea Dogfish

Conservation status: IUCN: Critically Endangered; Aust.: under assessment 2011

Description: Harrisson's Dogfish is a small to medium-sized shark, which reaches a maximum total length of 114 cm. The snout is somewhat elongated and dorso-ventrally compressed. The pectoral fins have elongated rear tips. Both dorsal fins have robust spines at the anterior margin and the second dorsal fin is only slightly smaller than the first. The caudal fin has a distinct notch in the posterior margin. Most of the body and fins are greyish apart from the dorsal fins, which have white posterior margins.

Distribution: This species has a disjunct distribution, occurring along Australia's east coast from off Brisbane, Qld to off Hobart, Tas., on the remote Fraser Seamount off Qld and on seamounts and ridges in the Tasman Sea off New Zealand. Gene flow between populations has not been studied, but is assumed to be low due to large expanses of unsuitable habitat between them.

Habitat: Off eastern Australia, Harrisson's Dogfish is constrained to an extremely narrow band of habitat, typically only 3–5 nautical miles wide between the 200 and 700 m depth contours (recorded deeper off New Zealand, to 1050 m). Preferred substrate type has not been studied but can be inferred from commercial fishing records that show significant catches on flat grounds (terraces) as well as canyons and peninsulas. There is little information on the habitat and occurrence of the species off Qld.

Ecology: The productivity of gulper sharks (*Centrophorus* spp.) in general is among the lowest of any elasmobranch. Harrisson's Dogfish is viviparous, late maturing (23–32 years), long lived (42+ years) with limited productivity (1–2 young every 2–3 years). Size at birth is 35 cm total length and maturity is reached in males at about 83 cm total length and in females at about 98 cm total length. The species occupies a high trophic level, and while there is only limited data on diet, it indicates a mid-water fish component.

Reasons for decline and current threats: Overfishing off south-east Australia has caused a major decline in this species.

The narrow restricted habitat and limited productivity of this species makes it particularly vulnerable to human impact. Surveys show this species has been severely depleted on trawl grounds off central and southern NSW. It was later targeted using gill nets off eastern Bass Strait. Output controls have ended targeted fishing, but automatic longline gear now affects this species on rough ground that was previously inaccessible to other fishing methods. Recent surveys by CSIRO show that other areas, including northern NSW and remote offshore seamounts, still harbour populations with higher abundance. Off eastern Bass Strait, the species is still present in good numbers, albeit with a patchy distribution. The greatest potential threat would be harvesting off northern NSW, but management arrangements, complementary to those in place in Commonwealth fisheries, are now being considered by NSW.

Off Qld, the species is unlikely to be affected significantly by fishing activities. Demersal fishing pressure is low or non-existent in the deepwater upper slope environment off Qld. Therefore, the species may have some refuge across its range on the continental slope, as well as poorly explored seamounts off Qld.

CURRENT RECOVERY EFFORTS

Recovery plan: None

Other plans: Australian Fisheries Management Authority (AFMA) (2010) 'Draft Upper-Slope Dogfish Management Strategy. Commonwealth-managed fisheries.' AFMA, Canberra.

Pogonoski JJ, Pollard DA and Paxton JR (2002) 'Conservation overview and action plan for Australian threatened and potentially threatened marine and estuarine fishes'. Environment Australia, Canberra.

Conservation actions:

- Areas closed to fishing have been implemented off Flinders Island (Tas.) and on the Taupo and Barcoo Seamounts off NSW. Other closed areas off south-east Australia may also benefit Harrisson's Dogfish.
- Trip limits exist in Commonwealth and NSW fisheries to prevent further targeting.
- A code of practice for safe handling and release has been developed.

- The CSIRO Wealth from Oceans Flagship is currently involved in collaborative research with AFMA and the Fisheries Research and Development Corporation (FRDC) studying the distribution, movement patterns and home range of the species. This will inform the suitable size and location of fishing closures.

FUTURE RECOVERY EFFORTS

Conservation needs:

- Evaluate options for closed areas in state waters.
- Regulate any fishery expansion into deep (slope) waters (if any fisheries develop, introduce measures to limit catches and collect species-specific data).
- Continue to develop co-management arrangements between the Commonwealth and the states.
- Introduce measures to prevent targeted fishing in state waters.
- Future recovery efforts will depend on the outcome of the EPBC nomination. If the species is listed as threatened, a recovery plan will be needed.

Research directions:

- Collect data on distribution off Qld and remote seamounts.
- Further examine habitat preference, patchiness and usage.
- Examine connectivity between populations off Qld, NSW, Bass Strait, remote seamounts and overseas (focus of a current CSIRO project, but more samples are required from some areas).
- Obtain more information on population demographics to ensure that closures allow populations to mate and pup.

Sources and further reading

Daley RK, Stevens JD and Graham KJ (2002) 'Catch analysis and productivity of the deepwater dogfish resource in southern Australia'. FRDC Final Report 1998/108. Fisheries Research and Development Corporation, Canberra.

Graham KJ, Andrew NL and Hodgson KE (2001) Changes in relative abundance of sharks and rays on Australian South East Fishery trawl grounds after twenty years of fishing. *Marine and Freshwater Research* **52**, 549–561.

Last PR and Stevens JD (2009) *Sharks and Rays of Australia.* 2nd edn. CSIRO Publishing, Melbourne.

Pogonoski J and Pollard D (2003) *Centrophorus harrissoni.* In IUCN 2010. *IUCN Red List of Threatened Species.* Version 2010.4. IUCN, Gland, Switzerland and Cambridge, UK, <http://www.iucnredlist.org>.

Wilson DT, Patterson HM, Summerson R and Hobsbawn PI (2009) 'Information to support management options for upper-slope gulper sharks (including Harrisson's dogfish and southern dogfish)'. Final Report to the Fisheries Research and Development Corporation Project No. 2008/65. Bureau of Rural Sciences, Canberra.

Author: Ross Daley

ENDEAVOUR DOGFISH

Centrophorus moluccensis Bleeker, 1860

Photo: ScienceImage

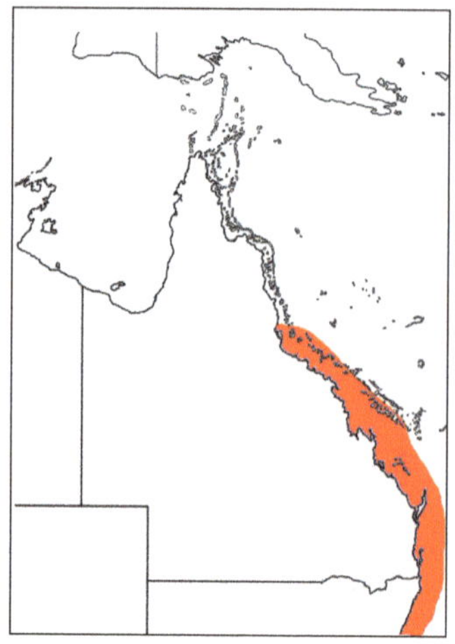

Family: Centrophoridae

Other common names: Smallfin Gulper Shark, Endeavour Shark, Endeavour Dog

Conservation status: IUCN: Endangered (Australian subpopulation), Data Deficient (global); Aust.: under assessment 2011

Description: The Endeavour Dogfish is a small to medium-sized shark, which reaches a maximum total length of 100 cm. The snout is not elongated. The pectoral fins have very elongated rear tips. Both dorsal fins have robust spines at the anterior margin and the second dorsal fin is much smaller than the first. The caudal fin has a distinct notch in the posterior margin. Most of the dorsal surface of the body is greyish. Juveniles have a dark blotch on the first dorsal and caudal fins, but this is paler in adults.

Distribution: This species has a patchy distribution in the Indo-West Pacific, including along Australia's east coast from off Ingham, Qld to off Lakes Entrance, Vic. and on the west coast from the Kimberly to Albany. Despite the wide geographic separation, genetic studies indicate that the eastern and western Australian forms are the same species. However, there is some genetic distance between Australia and elsewhere (i.e. South Africa).

Habitat: This species occurs shallower than other gulper sharks, extending onto the outer continental shelf in waters as shallow as 125 m. It also extends onto the mid-slope to 820 m. However, it occurs primarily in a very narrow band of upper slope habitat in a depth range of 330–550 m.

Ecology: The productivity of gulper sharks (*Centrophorus* spp.) in general is among the lowest of any elasmobranch. These species are viviparous and maturity is late with females only breeding at about 80% of maximum size. Litter size is only one or two (more commonly two). Size at birth of Endeavour Dogfish is 33 cm total length and maturity is reached in males at about 70 cm total length and in females at about 85 cm total length. The sexes are known to segregate; mature females are uncommon in commercial catches and their distribution is not well understood. Diet includes lantern fishes, cephalopods and demersal fish and crustaceans.

Reasons for decline and current threats: Overfishing has caused major declines off southern and central NSW (but most of the range of the species in Australia is off northern NSW and Qld).

The narrow restricted habitat and limited productivity of this species makes it particularly vulnerable to human impacts. Trawl surveys conducted on the central and southern NSW upper continental slope (200–650 m depth) showed declines of >98%. Current levels of fishing pressure on southern NSW trawl grounds remain high. These time series data are considered a robust indicator of decline in this region. However, such declines are not indicated by data from other areas. Recent CSIRO surveys indicate that abundance off northern NSW is much higher in areas that have not been subject to heavy trawl effort, gill netting or automatic longlining.

Off Qld, demersal fishing pressure is low or non-existent in the deepwater upper slope environment. Therefore, the species is likely to have some refuge across its range off Qld. There is a small amateur catch of Endeavour Dogfish off Brisbane.

CURRENT RECOVERY EFFORTS

Recovery plan: None

Other plans: Australian Fisheries Management Authority (AFMA) (2010) 'Draft Upper-Slope Dogfish Management Strategy. Commonwealth-managed fisheries'. AFMA, Canberra.

Conservation actions:

- Although this species is rarely encountered by Qld recreational fishers, there is a recreational take and possession limit of one shark or ray in Qld.
- An area closed to fishing has been implemented off Sydney specifically to mitigate the impacts of Commonwealth fisheries on Endeavour Dogfish.
- Trip limits exist in Commonwealth and NSW fisheries to prevent further targeting.
- A code of practice for safe handling and release has been developed.
- Catch records from off Qld by both amateur and commercial fishers are being examined to better understand distribution.
- Recent CSIRO survey data indicate that where the species occurs off northern NSW it is mainly in state

waters and management measures are being considered by the state.

FUTURE RECOVERY EFFORTS

Conservation needs:

- Evaluate options for closed areas in state waters.
- Identify any fisheries that may be taking minor incidental catches, including off Qld.
- Regulate any fishery expansion into deep (slope) waters (if any fisheries develop, introduce measures to limit catches and collect species-specific data).
- Continue to develop co-management arrangements between the Commonwealth and the states.
- Introduce measures to prevent targeted fishing in state waters.
- Future recovery efforts will depend on the outcome of the EPBC nomination. If the species is listed as threatened, a recovery plan will be needed.

Research directions:

- Better define its occurrence and habitat off Qld.
- Protect available habitat off Qld.
- Acquire ageing data.
- Assess the impact of the recreational catch.

Sources and further reading

Daley RK, Stevens JD and Graham KJ (2002) 'Catch analysis and productivity of the deepwater dogfish resource in southern Australia'. FRDC Final Report 1998/108. Fisheries Research and Development Corporation, Canberra.

Graham KJ, Andrew NL and Hodgson KE (2001) Changes in relative abundance of sharks and rays on Australian South East Fishery trawl grounds after twenty years of fishing. *Marine and Freshwater Research* **52**, 549–561.

Last PR and Stevens JD (2009) *Sharks and Rays of Australia*. 2nd edn. CSIRO Publishing, Melbourne.

Pogonoski J and Pollard D (2003) *Centrophorus moluccensis*. In IUCN 2010. *IUCN Red List of Threatened Species*. Version 2010.4. IUCN, Gland, Switzerland and Cambridge, UK, <http://www.iucnredlist.org>.

Wilson DT, Patterson HM, Summerson R and Hobsbawn PI (2009) 'Information to support management options for upper-slope gulper sharks (including Harrisson's dogfish and southern dogfish)'. Final Report to the Fisheries Research and Development Corporation Project No. 2008/65. Bureau of Rural Sciences, Canberra.

Author: Ross Daley

ANGELSHARKS (ORDER: SQUATINIFORMES)

EASTERN ANGELSHARK

Squatina albipunctata Last & White, 2008

Photo: ScienceImage

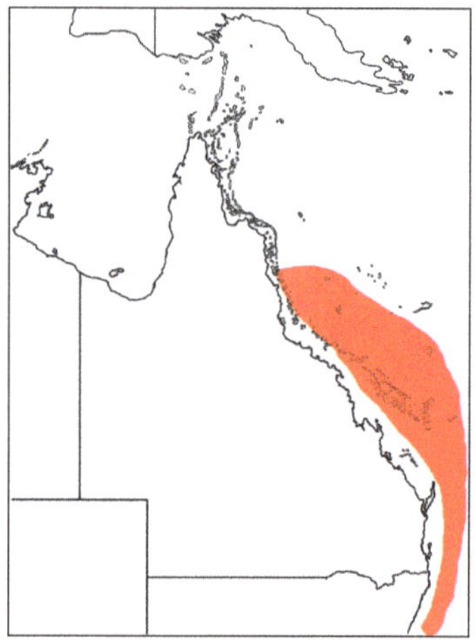

Family: Squatinidae

Conservation status: IUCN: Vulnerable

Description: The Eastern Angelshark is a medium-sized angelshark, which reaches a maximum total length of 132 cm. Angelsharks are characterised by their flattened bodies with extremely broad pectoral fins and small tails. With their eyes and spiracles on the dorsal surface, they resemble rays. The Eastern Angelshark's dorsal surface is covered in a dense pattern of small dark-edged white spots on a brown background.

Distribution: The Eastern Angelshark is endemic to the waters off eastern Australia, where it ranges from off Lakes Entrance, Vic., through NSW to off Cairns in north Qld. It is suspected that the natural abundance of the species in the northern part of its range (e.g. Qld) is lower than that in the southern and central parts of its distribution.

Habitat: This species occurs mainly on the outer continental shelf and the upper slope, but has an overall depth range of 35–415 m. Angelsharks are demersal, but there is very little known of the specific habitat requirements of this species.

Ecology: The biology and ecology of the Eastern Angelshark is not well known. Females mature between 100 and 110 cm total length and males between 80 and 90 cm total length. All angelsharks are viviparous, with litter sizes of 10–13 recorded for the Eastern Angelshark in NSW waters. The species is born at approximately 27–30 cm total length.

Reasons for decline and current threats: Research data from NSW has demonstrated the susceptibility of the species to population depletion as a result of consistent commercial fishing activities; in this case, those primarily targeting bony fish species in deepwater off NSW.

Trawl surveys conducted on the NSW upper continental slope (200–650 m depth) showed an overall decline in the catch rate of Eastern Angelshark of 96% between 1976–77 and 1996–97. If the data are considered by individual survey grounds, declines ranged from 73 to 98%. Although current levels of fishing pressure on southern NSW trawl grounds remain high, landed catches of angelsharks off central NSW have remained relatively stable since the mid 1980s, probably because the species also occupies large areas of the outer shelf (100–200 m depth) that are only lightly trawled. Consistent with this, the species was not assessed higher than Vulnerable by the IUCN because the survey grounds showing the decline of the Eastern Angelshark (central–southern NSW upper slope) represented less than a quarter of its geographic and depth distribution.

Although good time-series data showing a considerable decline was determined for upper slope trawl grounds off central and southern NSW, there is no information on the effects of any fishing activities where this species occurs in Qld. Although the Qld East Coast Trawl Fishery operates seasonally in the shallower parts of its range off southern Qld, the species was not recorded during recent broad-scale surveys of the fishery. However, because these surveys did not sample deeper than 166 m, they did not fully cover the depth range of the species. The compulsory use of turtle exclusion devices (TEDs) in that fishery should assist in limiting any catches. In a large part of the Eastern Angelshark's distribution off Qld, demersal fishing pressure is low or non-existent and therefore some refuge is provided for the species across most of its depth range.

CURRENT RECOVERY EFFORTS

Recovery plan: None

FUTURE RECOVERY EFFORTS

What you can do:

- Report any fishing catches to the Qld Museum.

- Qld commercial fishers should return any captured angelsharks safely to the water.
- Obtain the Australian Marine Conservation Society's *Australia's Sustainable Seafood Guide* and buy only seafood that is harvested from sustainable sources with limited bycatch (see <http://www.amcs.org.au>).

Conservation needs:

- Protect available habitat in Qld, where Eastern Angelshark may have refuge (particularly in deeper waters).
- Reduce and limit bycatch of the species (this is dependent on conservation actions in the southern part of the range of the species where it has declined).

Research directions:

- Quantify and monitor any catches in Qld demersal fisheries.
- Better define the habitat requirements of the species in Qld waters.

Sources and further reading

Graham KJ, Andrew NL and Hodgson KE (2001) Changes in relative abundance of sharks and rays on Australian South East Fishery trawl grounds after twenty years of fishing. *Marine and Freshwater Research* **52**, 549–561.

Last PR and Stevens JD (2009) *Sharks and Rays of Australia*. 2nd edn. CSIRO Publishing, Melbourne.

Pogonoski J and Pollard D (2003) *Squatina albipunctata*. In IUCN 2010. *IUCN Red List of Threatened Species*. Version 2010.4. IUCN, Gland, Switzerland and Cambridge, UK, <http://www.iucnredlist.org>.

Scandol J, Rowling K and Graham K (Eds) (2008) Angel Sharks (*Squatina* spp.) In 'Status of fisheries resources in NSW 2006/07'. pp. 1–3. NSW Department of Primary Industries, Cronulla.

Author: Peter M Kyne

CARPET SHARKS (ORECTOLOBIFORMES)

COLCLOUGH'S SHARK

Brachaelurus colcloughi Ogilby, 1908

Photo: Peter M Kyne

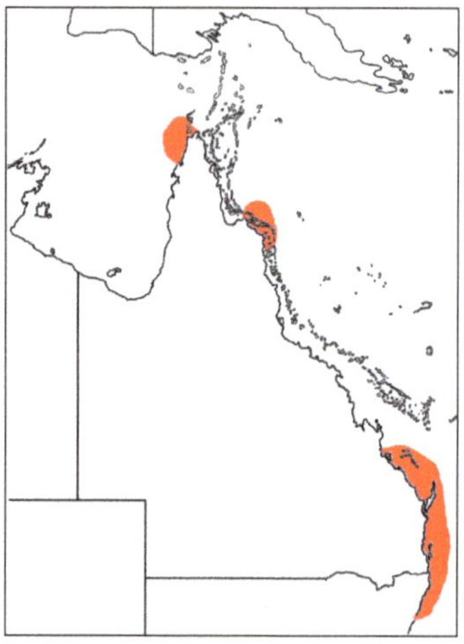

Family: Brachaeluridae

Other common name: Blue-grey Carpetshark

Conservation status: IUCN: Vulnerable

Description: Colclough's Shark is a small, stout shark, which reaches a maximum total length of 85 cm. It has long, weakly bilobed nasal barbels, a narrow mouth situated ahead of the eyes, nostrils situated well behind the snout tip, large spiracles situated close to the eyes, two spineless dorsal fins (close together; the first dorsal fin is slightly larger than the second) and an anal fin with its origin under, or in front of, the second dorsal fin insertion. Juveniles have faint banding on the body and tail that fades with increasing size.

Distribution: Colclough's Shark is endemic to the east coast of Australia from the southern Great Barrier Reef, Qld to Byron Bay, NSW, but most records are from a narrow latitudinal range (<2° of latitude) in south-east Qld and northern NSW waters. The species has also been reported from off Cape York Peninsula, but there are no specimens to support this observation.

Habitat: Colclough's Shark occurs from shallow inshore waters (most records) across the continental shelf with a maximum recorded depth of 217 m. Inner coastal reefs have been identified as critical habitat and inshore seagrass beds appear important for feeding.

Ecology: Relatively little is known of the biology or ecology of this species. It is viviparous, giving birth to six to seven young per litter, which are 17–19 cm total length at birth. Mature females have been observed from 54.5 cm total length and pregnant females from 62 cm total length. Mature males have been observed from 61 cm total length. Blind sharks (family Brachaeluridae) are secretive species that hide in rocky caves and under ledges during the day, becoming active at night to feed around reefs and over adjacent habitats including seagrass beds.

Reasons for decline and current threats: Colclough's Shark's area of occupancy faces intensive human pressure in the form of fisheries, both commercial and recreational, and habitat degradation; its rarity and limited occurrence make it inherently vulnerable to depletion.

The species is taken irregularly as bycatch in the Qld East Coast Trawl Fishery (ECTF), and its area of occupancy receives high effort from the trawl fleet. It was, however, recorded on only a few occasions during a broad-scale study of the elasmobranch bycatch of this fishery, possibly a result of its rarity. The compulsory use of turtle exclusion devices (TEDs) in the ECTF may reduce the bycatch of larger individuals (although individuals up to 67 cm total length have been recorded caught in nets fitted with a TED), but may not be as effective at excluding smaller animals. The species is also taken as bycatch in other inshore fisheries in Qld, particularly in Moreton Bay, but again the catch is low. This shark is also apparently exploited at low levels for the marine aquarium trade.

Many of the environments where the species occurs face considerable pressure as a result of an expanding human populations in coastal regions. Continued loss or alteration of important inshore habitats threatens to reduce the already small area of occupancy of this shark. For example: current pressures on Moreton Bay, a possible stronghold for the species, include large-scale developments that have an impact on near-shore marine habitats (e.g. Port of Brisbane and Brisbane Airport); run-off from terrestrial sources; pollution from marine and terrestrial sources; commercial fishing (trawling, netting and other); and increasing recreational use (particularly recreational fishing using baited hooks and various baited traps). Its rarity (despite decades of survey work and consistent fishing activities across its range, it is not encountered with any regularity), limited productivity and restricted range increase its vulnerability.

CURRENT RECOVERY EFFORTS

Recovery plan: None

Other plans: Pogonoski JJ, Pollard DA and Paxton JR (2002) 'Conservation overview and action plan for Australian threatened and potentially threatened marine and estuarine fishes'. Environment Australia, Canberra.

Conservation actions: Recreational take and possession limit of one shark or ray in Qld.

FUTURE RECOVERY EFFORTS

What you can do:

- Report any live sightings, fishing catches or dead beach-washed specimens to the Qld Museum.
- Both commercial and recreational fishers should release any captured Colclough's Shark safely.
- Join and support the Australian Marine Conservation Society, strong advocates for increasing the protection of the marine environment in Qld (see <http://www.amcs.org.au>).
- Pressure the Qld Government and state agencies (i.e. DERM) on the issue of inadequate protection of marine biodiversity and habitats in Moreton Bay (and elsewhere in Qld).

Conservation needs:

- Afford the species adequate protection at both the federal and state level.
- Add Colclough's Shark to commercial fishing logbooks in order to obtain further information on occurrence and interactions with commercial fisheries.
- Provide advice on the best methods of safe release for the species, together with education on its identification.
- Increase fully protected zones ('marine national park' or 'green' zones) within the Moreton Bay Marine Park (presently these are only 16% of the marine park area).

Research directions:

- Determine the habitat requirements of the species (it may be a microhabitat specialist), including daily and longer-term habitat use and movement patterns.
- Monitor and quantify catches in Qld fisheries.

Sources and further reading

Compagno LJV (2001) Sharks of the world. An annotated and illustrated catalogue of shark species known to date. Volume 2. Bullhead, mackerel and carpet sharks (Heterodontiformes, Lamniformes and Orectolobiformes). *FAO Species Catalogue for Fishery Purposes.* No. 1, Volume 2. FAO, Rome.

Compagno LJV, Last P and Stevens J (2005) *Heteroscyllium colcloughi.* In IUCN 2010. *IUCN Red List of Threatened Species.* Version 2010.4. IUCN, Gland, Switzerland and Cambridge, UK, <http://www.iucnredlist.org>.

Johnson JW (1999) Annotated checklist of the fishes of Moreton Bay, Queensland, Australia. *Memoirs of the Queensland Museum* **43**, 709–762.

Kyne PM, Compagno LJV, Stead J, Jackson MV and Bennett MB (2011) Distribution, habitat and biology of a rare and threatened eastern Australian endemic shark: Colclough's Shark *Brachaelurus colcloughi* Ogilby, 1908. *Marine and Freshwater Research* **62**, 540–547.

Last PR and Stevens JD (2009) *Sharks and Rays of Australia.* 2nd edn. CSIRO Publishing, Melbourne.

Authors: Peter M Kyne and Micha V Jackson

WHALE SHARK

Rhincodon typus Smith, 1828

Photo: Rob Harcourt

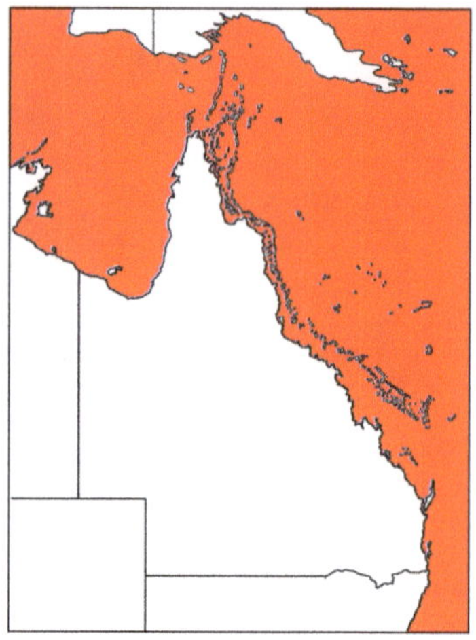

Family: Rhincodontidae

Conservation status: IUCN: Vulnerable; Aust.: Vulnerable

Description: The Whale Shark is the largest fish in the world (>1200 cm total length) and one of only three species of filter-feeding sharks. The dorsal surface is brown/grey, with an intricate pattern of white spots, stripes and swirls; the ventral surface is white.

Distribution: Whale Sharks are distributed globally in tropical and warm-temperate waters. Within Australia, they have been sighted in Qld, WA and the NT, as well as occasional recordings or strandings in southern states. Seasonal aggregations include WA and Christmas Island. This species is also a regular visitor to the Coral Sea in the north of Qld (e.g. Great Barrier Reef and Osprey Reef), and has been associated with spawning aggregations of lanternfish *Diaphus* sp. at Bougainville Reef during the months of October and November. Whale Sharks have been sighted in Qld as far south as the Gold Coast.

Habitat: Whale Sharks inhabit both coastal and pelagic environments and spend much of their time at the surface in the epipelagic layer. Deep vertical movements (>1000 m) and 'yo-yo' style patterns of up and down movements in the water column appear to be common and may be a foraging strategy. Coastal aggregations often occur with predictable timing and form the basis of many ecotourism industries.

Ecology: Whale Sharks are slow growing, late to reach maturity (at total length >700 cm) and have a long lifespan (suggested to be up to 100 years). They are viviparous, with one pregnant female caught off Taiwan carrying more than 300 embryos. A few neonates (size at birth approximately 50–70 cm total length) have been found, but juveniles <250 cm total length are very rarely seen. Whale Sharks can form aggregations from tens to hundreds of animals in some tropical coastal areas; these are thought to form in response to seasonal pulses of planktonic food sources. Sexual segregation is common, with most coastal aggregations consisting of immature males of 300–700 cm total length. Whale Sharks are capable of making large migrations of 1000s of km and appear to have few barriers to movement other than being restricted to warm waters. Evidence from genetic studies indicates there are few subdivisions of populations worldwide, but there appear to be significant differences between Atlantic and Indo-Pacific populations.

Reasons for decline and current threats: Whale Sharks have been targeted by fisheries in South-East Asia (e.g. Taiwan and Philippines) and India for their flesh, fins and liver oil. These fisheries have collapsed and have now been discontinued.

Mortality from sources other than directed fisheries, such as net entanglement and ship strike, appear to be common and many sharks bear propeller scars from encounters with boats. Declines in populations have been recorded in WA, Thailand, South Africa, California, Philippines, Taiwan and India. Little information is available on potential threats to Whale Sharks in Qld waters, though incidental ship strike and entanglement in fishing gear are possibilities.

CURRENT RECOVERY EFFORTS

Recovery plan: Department of the Environment and Heritage (2005) 'Whale Shark (*Rhincodon typus*) Recovery Plan 2005–2010'. Department of the Environment and Heritage, Canberra.

Other plans: Pogonoski JJ, Pollard DA and Paxton JR (2002) 'Conservation overview and action plan for Australian threatened and potentially threatened marine and estuarine fishes'. Environment Australia, Canberra.

Conservation actions:

- Whale Sharks are listed as 'wholly protected' fauna within all state waters of WA, but do not have a similar status in Qld.
- Several scientific studies have been conducted on the Whale Shark population that visits Ningaloo Reef over the past decade by researchers from the Australian Institute of Marine Science (AIMS), CSIRO and Australian and American universities.
- International trade in Whale Sharks is regulated through their listing on Appendix II of CITES.

The current recovery plan aims to:

- increase cooperation among range states within South-East Asia and the Indian Ocean through forums such as CMS in order to reduce fishing pressures and stop the decline of this species in range states (Whale Sharks are listed on Appendix II of CMS);
- monitor Whale Shark numbers in Australia in order to estimate demographics such as population size. This also includes the identification of impacts that will impede species recovery, and assist implementation of appropriate responses.

FUTURE RECOVERY EFFORTS

What you can do:

- Report any observations in Qld to the Qld Museum.
- Vessel operators should maintain acceptable distances from Whale Sharks to avoid altering behavioural patterns. The WA Department of Environment and Conservation suggest boats keep a distance of at least 30 m from Whale Sharks.
- Snorkelers and SCUBA divers should maintain a safe distance (>5 m) if approached by a Whale Shark and avoid touching animals.

Conservation needs:

- Encourage cross-jurisdictional collaboration to provide opportunities for threat abatement and species/habitat protection.
- Increase public awareness of the vulnerability and conservation status of this species through education programs.

Research directions:

- Establish migration pathways between aggregation areas, through long-term tracking programs.
- Identify breeding or pupping areas (such areas need to be conserved to avoid negative impacts on populations).
- Quantify sources of natural and anthropogenic mortality (this would allow more accurate demographic modelling of population status and trends).

Sources and further reading

Castro ALF, Stewart BS, Wilson SG, Hueter RE, Meekan MG, Motta PJ, Bowen BW and Karl SA (2007) Population genetic structure of Earth's largest fish, the whale shark (*Rhincodon typus*). *Molecular Ecology* **16**, 5183–5192.

Last PR and Stevens JD (2009) *Sharks and Rays of Australia*. 2nd edn. CSIRO Publishing, Melbourne.

Meekan MG, Bradshaw CJA, Press M, McLean C, Richards A, Quasnichka S and Taylor JG (2006) Population size and structure of whale sharks *Rhincodon typus* at Ningaloo Reef, Western Australia. *Marine Ecology Progress Series* **319**, 275–285.

Stevens JD (2007) Whale shark (*Rhincodon typus*) biology and ecology: a review of the primary literature. *Fisheries Research* **84**, 4–9.

Wilson SG, Polovina JJ, Stewart BS and Meekan MG (2006) Movements of whale sharks (*Rhincodon typus*) tagged at Ningaloo Reef, Western Australia. *Marine Biology* **148**, 1157–1166.

Authors: Mark G Meekan and Conrad W Speed

MACKEREL SHARKS (ORDER: LAMNIFORMES)

GREY NURSE SHARK (EAST COAST POPULATION)

Carcharias taurus Rafinesque, 1810

Photo: Carley Bansemer

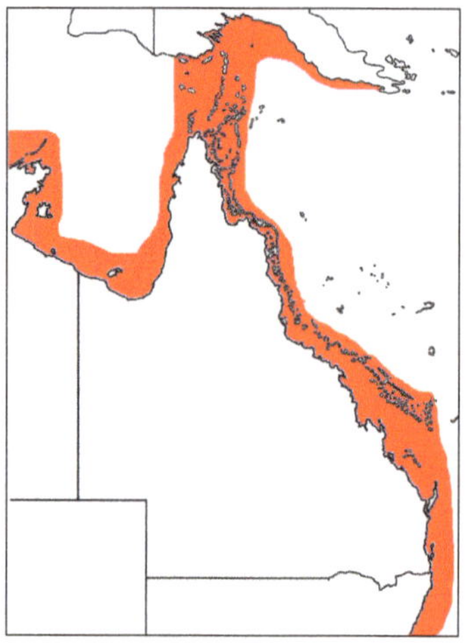

Family: Odontaspididae

Other common names: Sandtiger Shark, Spotted Raggedtooth Shark

Conservation status: IUCN: Critically Endangered (east coast of Australia subpopulation), Vulnerable (global); Aust.: Critically Endangered (east coast of Australia population); Qld: Endangered. NSW: Critically Endangered; Vic.: Data Deficient

Description: The Grey Nurse Shark is a large bulky shark, which reaches a maximum total length of at least 320 cm. The anal and both dorsal fins are equally large and broad-based, and the first dorsal fin is closer to the pelvic fins than to the pectoral fins. The teeth have long central cusps with a relatively short lateral cusplet on either side. Colouration is grey to grey-bronze dorsally, with a paler off-white underbelly and reddish or brownish spots often scattered on the body.

Distribution: The Grey Nurse Shark is wide ranging in warm-temperate and tropical coastal waters of the Atlantic and Indo-West Pacific. In Australia, there are two separate populations: one off the east coast and one off the west coast, with negligible migration between the two. The core distribution of the east coast population is southern Qld and NSW.

Habitat: The prime habitats of Grey Nurse Shark are rocky reefs with caves or large overhangs or gutters, although they have been recorded from a variety of habitats and a depth range of <1–191 m.

Ecology: Maturity is estimated at 220–230 cm total length and 9–10 years for females and 190–195 cm total length and 6–7 years for males. The species has one of the lowest reproductive rates known among elasmobranchs. Reproduction is viviparous and embryos are adelphophagous and oophagous. Gestation is thought to take 9–12 months, with a maximum of two young (about 100 cm total length) born per litter every second or third year. Longevity is ~40 years (females). Grey Nurse Sharks commonly form aggregations, and undergo north–south migrations related to breeding.

Reasons for decline and current threats: Combined with its limited life history characteristics, incidental capture by commercial and recreational fisheries, shark control programs, shark finning and ecotourism are the major threats to the recovery of the Australian east coast population of Grey Nurse Sharks.

Historically, this species was hunted for its oil, flesh, skin and fins and persecuted due to its fierce appearance. As such, large numbers were killed by recreational spear and line fishers and in shark control programs, particularly in eastern Australia. Grey Nurse Sharks are no longer found at several sites along the east coast where they were common during the 1950s and 1960s. The species is susceptible to capture by a wide array of fishing gear and the incidence of hooking and escape or release by fishers remains high; interactions with fishing gear can result in debilitating disease, injury and death.

CURRENT RECOVERY EFFORTS

Recovery plan: Environment Australia (2002) 'Recovery plan for the Grey Nurse Shark (*Carcharias taurus*) in Australia'. Environment Australia, Canberra. This plan is currently under review.

Other plans: Pogonoski JJ, Pollard DA and Paxton JR (2002) 'Conservation overview and action plan for Australian threatened and potentially threatened marine and estuarine fishes'. Environment Australia, Canberra.

Conservation actions:

- Grey Nurse Sharks are protected in Qld and NSW. The species cannot be retained and 14 aggregation sites (four in Qld and 10 in NSW) have controls on fishing and diving activities; additional protection is provided within these states' marine protected area systems.
- Further information about restrictions at Grey Nurse Shark areas in Qld can be obtained from: <http://www.derm.qld.gov.au/wildlife-ecosystems/wildlife/threatened_plants_and_animals/endangered/grey_nurse_shark.html>.
- Research is ongoing into captive breeding, migratory movements, localised movements at critical habitat sites, and the effects of scuba diving.

FUTURE RECOVERY EFFORTS

What you can do:

- Abide by fishing regulations and diving restrictions at known aggregation sites, and follow diving guidelines for aggregation sites if the species is encountered elsewhere. If a Grey Nurse Shark dies after capture, please retain the body and contact the Qld Marine Stranding Hotline (1300 130 372) immediately. If you catch a Grey Nurse Shark that is alive, return it to the water immediately.
- Continue to provide images for the identification of individual sharks by participating in Grey Nurse Shark Watch: http://www.reefcheckaustralia.org/grey-nurse-shark-watch.html (Images of the flank, from the gill slits to the anal fin are most suitable). Further information can be obtained by emailing greynurseshark@derm.qld.gov.au or carley.bansemer@uqconnect.edu.au.

Conservation needs:

- Reduce the impact of commercial and recreational fishing, and shark control activities by protecting Grey Nurse Sharks in the wild at major aggregation sites; captive breeding should not be viewed as the major conservation action of relevance to the recovery of the species.
- Manage the impact of dive ecotourism activities.
- Modify fisheries logbooks to record Grey Nurse Shark catch and biological data.
- Ensure existing fishery observer programs record interactions with the species.

Research directions:

- Monitor all key aggregation sites through photo-identification surveys.
- Establish community-based programs to identify and monitor critical sites.
- Develop population models to assess and monitor recovery of populations.

Sources and further reading

Bansemer CS and Bennett MB (2008) Multi-year validation of photographic identification of grey nurse sharks, *Carcharias taurus*, and applications for non-invasive conservation research. *Marine and Freshwater Research* **59**, 322–331.

Bansemer CS and Bennett MB (2009) Reproductive periodicity, localised movements and behavioural segregation of pregnant *Carcharias taurus* at Wolf rock, southeast Queensland, Australia. *Marine Ecology Progress Series* **374**, 215–227.

Bansemer CS and Bennett MB (2010) Retained fishing gear and associated injuries in the east Australian grey nurse sharks (*Carcharias taurus*): implications for population recovery. *Marine and Freshwater Research* **61**, 97–103.

Goldman KJ, Branstetter S and Musick JA (2006) A re-examination of the age and growth of sand tiger sharks, *Carcharias taurus*, in the western North Atlantic: the importance of ageing protocols and use of multiple back-calculation techniques. *Environmental Biology of Fishes* **77**, 241–252.

Pollard D, Gordon I, Williams S, Flaherty A and McAuley R (2003) *Carcharias taurus* (East coast of Australia subpopulation). In IUCN 2010. *IUCN Red List of Threatened Species*. Version 2010.4. IUCN, Gland, Switzerland and Cambridge, UK, <http://www.iucnredlist.org>.

Authors: Carley Bansemer and Michael B Bennett

BIGEYE THRESHER

Alopias superciliosus (Lowe, 1839)

Photo: ScienceImage

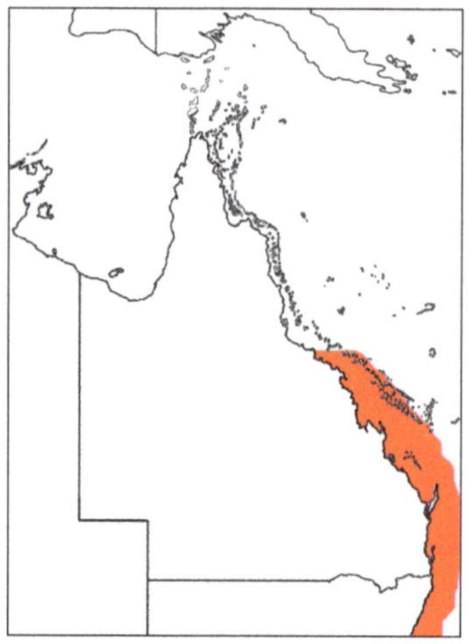

Family: Alopiidae

Conservation status: IUCN: Vulnerable

Description: The Bigeye Thresher is a large oceanic shark, which reaches a maximum total length of 484 cm. This shark has huge eyes that extend onto the dorsal head surface, and pronounced lateral grooves on the top of the head. Like all thresher sharks, it has an extremely long upper caudal fin lobe. The shark's long and bulbous snout sports large, smooth-edged teeth with long cusps. The body is a purple to violet-grey colour dorsally, and creamy white ventrally.

Distribution: This shark is found in all tropical to warm temperate seas. In Australia, it has been recorded on the North-West Shelf, WA, from Townsville to Fraser Island and off Middleton Reef, Qld, off Crowdy Head, NSW, and off SA.

Habitat: This pelagic and oceanic shark occurs from the ocean surface down to a maximum depth of at least 700 m and is found in coastal waters over continental shelves, sometimes close inshore, and on the high seas in the epipelagic zone far from land. It exhibits strong diel movement patterns, with most of the day spent below the thermocline at 300–500 m (12–16°C) and the night between 10–100 m (20–26°C), although it also makes forays into cooler waters (to 6°C).

Ecology: Males mature at 270–300 cm total length, at about 9–10 years old. Females mature at 332–355 cm total length, at about 12–13 years old. Size at birth is 64–140 cm total length. Reproduction is viviparous and embryos are oophagous. There is no apparent reproductive season, with females giving birth to one to four (usually two) young. The diet consists of benthic and pelagic bony fishes and cephalopods, which it stuns with its tail.

Reasons for decline and current threats: The global Vulnerable listing of the Bigeye Thresher on the IUCN Red List is a result of declining population trends due to a combination of slow life history characteristics and high levels of largely unmanaged and unreported mortality in target and bycatch fisheries.

This species has low fecundity and an exceptionally low estimated rate of population increase compared with other species of thresher shark. The epipelagic habitat of this species occurs within the range of many largely unregulated gill net and longline fisheries, in which it is readily caught, and it has been fished throughout its range. Although data are lacking for many parts of its range, significant reductions in thresher catch-per-unit-effort (CPUE) have been reported in pelagic longline fisheries in the Western North Atlantic and the eastern tropical Pacific, and declines are also suspected to have occurred in other areas. Bigeye Thresher possession is now prohibited in US Atlantic fisheries, but as fishing pressure continues in other parts of its range, it is evident that this species, with such low productivity, is highly vulnerable.

In Australian waters, this species is caught incidentally in most oceanic longline fisheries within its distribution, such as the Western Tuna and Billfish Fishery (WTBF) and the Eastern Tuna and Billfish Fishery (ETBF), and is used for meat and fins. The Bigeye Thresher has also been documented in confiscated catches of illegal Taiwanese longline fishing vessels operating in the Torres Strait and off north-east Qld.

CURRENT RECOVERY EFFORTS

Recovery plan: None

Other plans: Australian Fisheries Management Authority (AFMA) (2008) 'Australian Tuna and Billfish Longline Fisheries Bycatch and Discarding Workplan, November 1, 2008 to October 31, 2010'. AFMA, Canberra.

Conservation actions:

- Although this species is rarely encountered by Qld recreational fishers, there is a recreational take and possession limit of one shark or ray in Qld.

Management regulations in the WTBF and ETBF:

- Ban on shark finning (fins can not be landed separate from carcasses).
- Landing limit of 20 sharks per boat per trip (all species) (except for 'high seas only' trips in the WTBF where the limit is 100 sharks per trip).
- Ban on wire-traces from longlines.
- Increased level of fisheries observers monitoring catch.

FUTURE RECOVERY EFFORTS

What you can do: Commercial fishers should release any captured thresher sharks safely.

Conservation needs:

- Prohibit possession of thresher sharks (as has been implemented in US Atlantic fisheries).

- Develop and implement bycatch reduction measures.
- Encourage sustainable fisheries management throughout the range of the species.
- Encourage international cooperation and management action in line with the 1995 UN Agreement on the Conservation and Management of Straddling Fish Stocks and Highly Migratory Fish Stocks (in which thresher sharks are included, and which Australia has ratified) and the International Plan of Action for the Conservation and Management of Sharks.

Research directions:

- Continuously monitor Australian fisheries.
- Investigate the movement patterns and stock structure of the Southern Hemisphere population.

Sources and further reading

Amorim A, Baum J, Cailliet GM, Clò S, Clarke SC, Fergusson I, Gonzalez M, Macias D, Mancini P, Mancusi C, Myers R, Reardon M, Trejo T, Vacchi M and Valenti SV (2007) *Alopias superciliosus*. In IUCN 2010. *IUCN Red List of Threatened Species*. Version 2010.4. IUCN, Gland, Switzerland and Cambridge, UK, <http://www.iucnredlist.org>.

Chen C-T, Liu K-M and Chang Y-C (1997) Reproductive biology of the bigeye thresher shark, *Alopias superciliosus* (Lowe, 1839) (Chondrichthyes: Alopiidae), in the northwestern Pacific. *Ichthyological Research* **44**, 227–235.

Dulvy NK, Baum JK, Clarke S, Compagno LJV, Cortés E, Domingo A, Fordham S, Fowler S, Francis MP, Gibson C, Martínez J, Musick JA, Soldo A, Stevens JD and Valenti S (2008) You can swim but you can't hide: the global status and conservation of oceanic pelagic sharks and rays. *Aquatic Conservation: Marine and Freshwater Ecosystems* **18**, 459–482.

Last PR and Stevens JD (2009) *Sharks and Rays of Australia*. 2nd edn. CSIRO Publishing, Melbourne.

Smith SE, Rasmussen RC, Ramon DA and Cailliet GM (2008) The biology and ecology of thresher sharks (Alopiidae). In *Sharks of the Open Ocean: Biology, Fisheries and Conservation*. (Eds MD Camhi, EK Pikitch and EA Babcock) pp. 60–68. Blackwell Publishing, Oxford.

Author: Lindsay J Marshall

GREAT WHITE SHARK

Carcharodon carcharias Linnaeus, 1758

Photo: ScienceImage

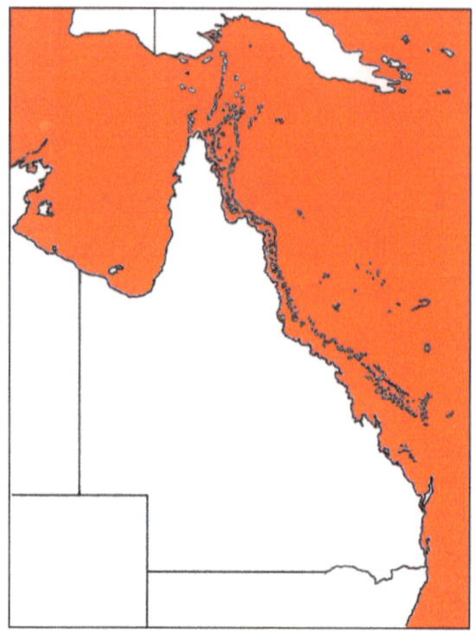

Family: Lamnidae

Other common names: White Shark, White Pointer

Conservation status: IUCN: Vulnerable; Aust.: Vulnerable; NSW: Vulnerable; Tas.: Vulnerable; Vic.: Vulnerable; WA: Vulnerable

Description: The Great White Shark is a large shark, which reaches a maximum total length of at least 600 cm (unverified reports to 700 cm). It has a moderately stout and torpedo shaped body, large serrated triangular teeth and a distinctive lateral keel along the body midline immediately before a crescent shaped tail. The colouration is blue-grey to grey-brown on the upper surface and white below with dusky tips to the under surfaces of its otherwise white pectoral fins.

Distribution: Great White Sharks occur worldwide in coastal temperate and subtropical regions but they can also occur in tropical areas and, in some regions, may spend considerable periods in the open ocean. They are most frequently encountered off South Africa, southern Australia, northern California, the west coast of Mexico and north-east United States. In Australia, Great White Sharks occur from Mackay in central Qld around the south coast to north-west WA, but may occasionally occur further north on both coasts and in the Coral Sea. Northern areas are occupied on a seasonal basis with encounters most frequent in Qld during winter–spring.

Habitat: Juveniles aggregate seasonally at specific nursery area locations in central NSW and south-east Victoria. Satellite tracking also suggests that juveniles may aggregate on the eastern side of Fraser Island, Qld, during late spring. Subadults and adults occur in a wide range of habitats, including around seal colonies and areas where finfish aggregate. All life history stages may move over the outer continental shelf, slope and offshore waters, diving to a depth of up to at least 1000 m. Great White Sharks tagged in SA and New Zealand have been tracked to the region south of the Swains Reef, east of Rockhampton. They are able to tolerate a wide range of temperatures (6–25°C in Australian waters).

Ecology: Great White Sharks are thought to live up to 60 years, although this age has not been validated. Males mature at 360–380 cm (7–9 years) and females at 450–500 cm total length (12–17 years). Reproduction is viviparous and embryos are oophagous. The reproductive rate is low, with a long gestation period, estimated up to 18 months. Development is relatively slow and young measure around 120–150 cm total length at birth (up to 32 kg). However, no pregnant Great White Sharks have yet been reliably examined in Australian waters and so the location of pupping areas remains unknown. This is a versatile predator, feeding primarily on finfish and elasmobranchs as juveniles (<300 cm total length) prior to adding larger prey items to their diet (e.g. marine mammals).

Reasons for decline and current threats: The low abundance and limited productivity of the Great White Shark, together with its capture in targeted fisheries (including persecution) and as bycatch has placed this species at risk.

Great White Sharks are naturally low in abundance with probable low natural mortality. Its biological characteristics imply a low reproductive potential, which has implications for the vulnerability of Great White Sharks to non-natural mortality and means that populations, once depleted, are likely to be very slow to recover. However, their status is poorly known and quantitative population assessments are not currently possible. Most mortality occurs in commercial fisheries. Great White Sharks are taken as bycatch in various Australian commercial fisheries, notably the Southern and Eastern Scalefish and Shark Fishery and shark fisheries in WA. Great White Sharks are also a target species in beach protection programs in NSW and Qld.

CURRENT RECOVERY EFFORTS

Recovery plan: Environment Australia (2002) 'White Shark *Carcharodon carcharias* recovery plan'. Environment Australia, Canberra. This plan is currently under review.

Other plans: Pogonoski JJ, Pollard DA and Paxton JR (2002) 'Conservation overview and action plan for Australian threatened and potentially threatened marine and estuarine fishes'. Environment Australia, Canberra.

Conservation actions:

- Great White Sharks are fully protected in all Australian waters (including as a 'no take' species in Qld); the species cannot be retained by commercial or recreational fishers.

- Research programs have focused on movement patterns and habitat use through a variety of electronic tag technologies.
- International trade in Great White Sharks is regulated through their listing on Appendix II of CITES.
- Great White Sharks are listed on Appendices I + II of CMS.

FUTURE RECOVERY EFFORTS

What you can do:

- Report dead Great White Sharks (fishing catches or beach-washed specimens) to the Qld Marine Stranding Hotline (1300 130 372).
- Fishers should abide by fishing regulations and release accidentally caught Great White Sharks safely.

Conservation needs: Develop methods to assess the population status and long-term trends of Great White Shark populations in Australian waters (in order to establish the efficacy of current recovery plan initiatives and for designing future effective recovery strategies).

Research directions:

- Identify and quantify key habitat regions, including those areas important for pupping and juvenile Great White Shark survival.
- Examine the inter-connectedness of the Australian Great White Shark population within the Exclusive Economic Zone and the overall geographic footprint of the Australasian population.
- Undertake a comprehensive assessment of the estimated take of Great White Sharks from the commercial and recreational fishing sectors.

Sources and further reading

Bonfil R, Meÿer M, Scholl MC, Johnson R, O'Brien S, Oosthuizen H, Swanson S, Kotze D and Paterson M (2005) Transoceanic migration, spatial dynamics, and population linkages of white sharks. *Science* **310**, 100–103.

Bruce BD (2008) The biology and ecology of the white shark, *Carcharodon carcharias*. In *Sharks of the Open Ocean: Biology, Fisheries and Conservation*. (Eds MD Camhi, EK Pikitch and EA Babcock) pp. 69–81. Blackwell Publishing, Oxford.

Bruce BD, Stevens JD and Malcolm H (2006) Movements and swimming behaviour of white sharks (*Carcharodon carcharias*) in Australian waters. *Marine Biology* **150**, 161–172.

Klimley AP and Ainley DG (Eds) (1996) *Great White Sharks: The Biology of* Carcharodon carcharias. Academic Press, San Diego.

Pardini AT, Jones CS, Noble LR, Kreiser B, Malcolm H, Bruce BD, Stevens JD, Cliff G, Scholl MC, Francis M, Duffy CAJ and Martin AP (2001) Sex biased dispersal of great white sharks. *Nature* **412**, 139–140.

Author: Barry D Bruce

LONGFIN MAKO

Isurus paucus Guitart Manday, 1966

Photo: William T White © CMAR

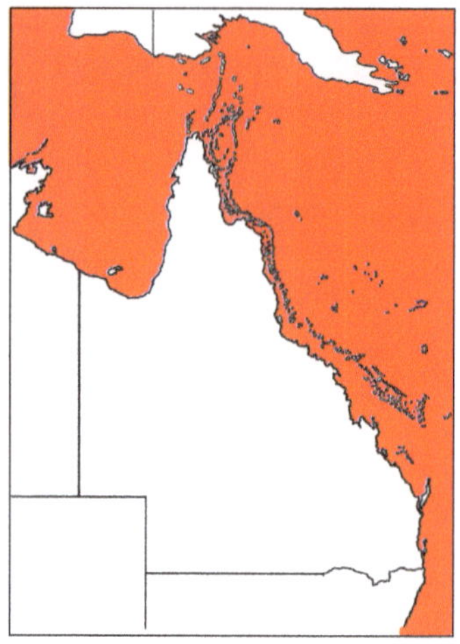

Family: Lamnidae

Conservation status: IUCN: Vulnerable

Description: The Longfin Mako is a large oceanic shark, which reaches a maximum total length of 417 cm. It is distinguished from its close relative the Shortfin Mako *Isurus oxyrinchus* by its more bluntly pointed snout (which is dark underneath) and its long pectoral fins. The second dorsal and anal fins are tiny and the caudal fin is crescent shaped. There is a distinct keel on the tail stock. Its teeth are smooth-edged without cusps on the sides. The colouration is grey-blue on the top and white underneath (except under the snout).

Distribution: This shark is thought to occur circumglobally in tropical and warm temperate waters, although its complete distribution is poorly defined (due in part to confusion with the Shortfin Mako). Within Australian waters, it occurs from Geraldton, WA across northern Australia (including off Qld) to NSW.

Habitat: The Longfin Mako is an epipelagic, oceanic shark, with a usual occurrence in the water column in a depth range of 120–240 m over deeper water (but also has been recorded closer to the surface). It occurs more rarely closer to land.

Ecology: The Longfin Mako is a very poorly known species. Males reach maturity from about 190 cm total length and females at about 245 cm total length. Newborn sharks measure 92–120 cm total length. Reproduction is viviparous and embryos are oophagous with two to eight young per litter. The gestation period is likely to be extended given the large size at birth of the species. This is presumed to be an apex predator.

Reasons for decline and current threats: The global Vulnerable listing of the Longfin Mako on the IUCN Red List is a result of inferred declines from longline fishing activities targeting tuna, billfishes and sharks across its range.

Longline fisheries have resulted in documented declines in Shortfin Mako and, although data collection on Longfin Makos is poor (due to identification issues and its rarity), it is suspected to be similarly affected; it is captured in the same fishing gear and fisheries. The Longfin Mako's apparent rarity and its biological traits, which combine to restrict its ability to sustain exploitation, add to its vulnerability. The meat is of poor quality, but its fins are of high value (thus often resulting in finning and discarding at sea).

Its status in Australia and Qld is poorly known. Before 1995, when it was first documented in Australian waters, all makos were considered to be Shortfin Makos. In Australian waters, this species is a byproduct and bycatch of the Eastern Tuna and Billfish Fishery (ETBF). It is caught infrequently (much less frequently than the Shortfin Mako) and it makes up only a very small portion of the overall shark catch. The general practice in the fishery at present is to release most sharks. Priority species assessments conducted on bycatch species identified the Longfin Mako as a high-risk species. It has been documented in confiscated catches of illegal foreign fishing vessels operating in the Torres Strait and off north-east Qld. Recovery of this species is dependent on reducing catches in longline fisheries globally. Within Australia, the management of Commonwealth fisheries (i.e. ETBF) and illegal fishing activities from foreign vessels are key issues.

CURRENT RECOVERY EFFORTS

Recovery plan: None

Other plans: Australian Fisheries Management Authority (AFMA) (2008) 'Australian Tuna and Billfish Longline Fisheries Bycatch and Discarding Workplan, November 1, 2008 to October 31, 2010'. AFMA, Canberra.

Conservation actions:

- Although this species is rarely encountered by Qld recreational fishers, there is a recreational take and possession limit of one shark or ray in Qld.

Management regulations in the ETBF:

- Ban on shark finning (fins can not be landed separate from carcasses).
- Landing limit of 20 sharks per boat per trip (all species).
- Ban on wire-traces from longlines.
- Increased level of fisheries observers monitoring catch.

FUTURE RECOVERY EFFORTS

What you can do:

- Fishers should release any captured animals safely.
- Lobby the Australian Government to repeal recent legislative changes to the EPBC, which allows recreational fishers and charter boat operators to

continue fishing for this species, despite its listing as Migratory under the EPBC (which should have provided protection for the species in Commonwealth waters). Listing the species as Migratory on the EPBC was required due to its listing on Appendix II of CMS.

Conservation needs:

- Develop and implement bycatch reduction measures.
- Encourage sustainable fisheries management throughout the range of the species.
- Encourage international cooperation and management action in line with the 1995 UN Agreement on the Conservation and Management of Straddling Fish Stocks and Highly Migratory Fish Stocks (in which the Longfin Mako is included, and which Australia has ratified) and the FAO International Plan of Action for the Conservation and Management of Sharks.

Research directions:

- Continuously monitor Australian fisheries.
- Investigate the distribution, habitat and stock structure within Australia and the Indo-Pacific region.

Sources and further reading

Compagno LJV (2001) Sharks of the world. An annotated and illustrated catalogue of shark species known to date. Volume 2. Bullhead, mackerel and carpet sharks (Heterodontiformes, Lamniformes and Orectolobiformes). *FAO Species Catalogue for Fisheries Purposes.* No. 1. Volume. 2. FAO, Rome.

Ebert DA (2003) *Sharks, Rays and Chimaeras of California.* University of California Press, Berkeley.

Gilman E, Clarke S, Brothers N, Alfaro-Shigueto-J, Mandelman J, Mangel J, Petersen S, Piovano S, Thomson N, Dalzell P, Donoso M, Goren M and Werner T (2007) 'Shark depredation and unwanted bycatch in pelagic longline fisheries: industry practices and attitudes, and shark avoidance strategies'. Western Pacific Regional Fishery Management Council, Honolulu, USA.

Last PR and Stevens JD (2009) *Sharks and Rays of Australia.* 2nd edn. CSIRO Publishing, Melbourne.

Reardon MB, Gerber L and Cavanagh RD (2006) *Isurus paucus.* In IUCN 2010. *IUCN Red List of Threatened Species.* Version 2010.4. IUCN, Gland, Switzerland and Cambridge, UK, <http://www.iucnredlist.org>.

Authors: Peter M Kyne and Lee K Curtis

PORBEAGLE

Lamna nasus (Bonnaterre, 1788)

Photo: ScienceImage

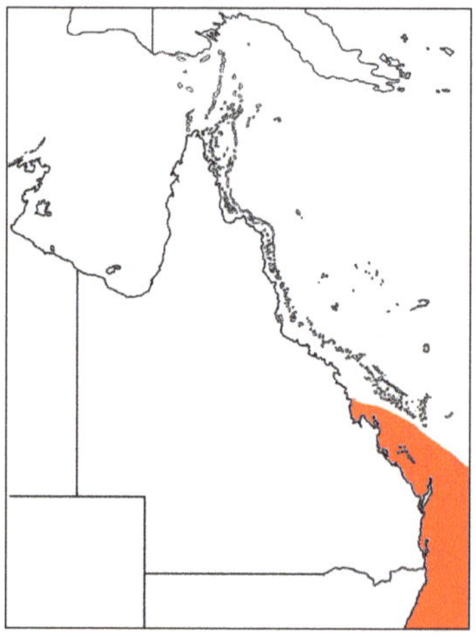

Family: Lamnidae

Other common name: Mackerel Shark

Conservation status: IUCN: Vulnerable

Description: The Porbeagle is a large oceanic shark, which reaches a maximum total length of 336 cm in the North Atlantic (smaller in Australia: 232 cm total length). It has a stout body with a long cone-shaped snout and smooth-edged narrow teeth with one cusp on each side. The trailing free tip of the first dorsal fin is bright white, the second dorsal and anal fins are tiny and the caudal fin is crescent shaped. There is a large keel on the tail stock, with a small secondary keel below it on the lower lobe of the tail. The colouration is dark blue-grey on the top and white underneath.

Distribution: This shark is found in subtropical to temperate waters in the North Atlantic and in a circumglobal band in the Southern Hemisphere. In Australia, it ranges from southern Qld around the south coast to south-west WA. It thus occurs only marginally in Qld waters, being a species of cooler temperate waters, usually less than 18°C.

Habitat: This pelagic and oceanic shark occurs from the ocean surface to a depth of 370 m. It primarily occurs offshore on the continental shelf and in the open ocean, but occasionally enters coastal waters.

Ecology: In the Southern Hemisphere, maturity in males occurs at 8–11 years of age and in females at 15–18 years, with a suggested longevity of 65 years. Newborn sharks measure about 65–80 cm total length. Reproduction is viviparous and embryos are oophagous with one to five (usually four) young born annually after a gestation period of 8–9 months. In Australian waters, males mature at about 165 cm total length and females at about 200 cm total length. This is an apex predator.

Reasons for decline and current threats: The global Vulnerable listing of the Porbeagle on the IUCN Red List is a result of over-exploitation and subsequent population depletion in the North Atlantic.

The high value meat and fins of this species has made it the focus of target fisheries, as well as a retained bycatch in other fisheries. Its biological traits combine to restrict its ability to sustain exploitation and to recover from depletion. Its status in the Southern Hemisphere is poorly known and it has not been as heavily impacted here, although it is a frequent bycatch of longline and some trawl fisheries. This is a very rare species in Qld waters, where it is at the extreme far north of its range. It is, however, included here given the global status of the species, and its inherent vulnerability to population depletion.

In Australian waters, this species is a byproduct and bycatch of the Eastern Tuna and Billfish Fishery (ETBF). Although frequently caught, it makes up only a small portion of the overall shark catch. The general practice in the fishery at present is to release most sharks. Priority species assessments conducted on bycatch species identified the Porbeagle as a high-risk species. Given its marginal occurrence off Qld, recovery of the species is dependent on conservation and management measures elsewhere.

CURRENT RECOVERY EFFORTS

Recovery plan: None

Other plans: Australian Fisheries Management Authority (AFMA) (2008) 'Australian Tuna and Billfish Longline Fisheries Bycatch and Discarding Workplan, November 1, 2008 to October 31, 2010'. AFMA, Canberra.

Conservation actions:

Management regulations in the ETBF:

- Ban on shark finning (fins can not be landed separate from carcasses).
- Landing limit of 20 sharks per boat per trip (all species).
- Ban on wire-traces from longlines.
- Increased level of fisheries observers monitoring catch.

FUTURE RECOVERY EFFORTS

What you can do:

- Lobby the Australian Government to repeal recent legislative changes to the EPBC, which allows recreational fishers and charter boat operators to continue fishing for this species, despite its listing as Migratory under the EPBC (which should have

provided protection for the species in Commonwealth waters). Listing the species as Migratory on the EPBC was required due to its listing on Appendix II of CMS.
- Fishers should release any captured animals safely.

Conservation needs:
- Develop and implement bycatch reduction measures.
- Encourage sustainable fisheries management throughout the range of the species.
- Encourage international cooperation and management action in line with the 1995 UN Agreement on the Conservation and Management of Straddling Fish Stocks and Highly Migratory Fish Stocks (in which the Porbeagle is included, and which Australia has ratified) and the FAO International Plan of Action for the Conservation and Management of Sharks.

Research directions:
- Investigate the movement patterns and stock structure of the Southern Hemisphere population.
- Continuously monitor Australian fisheries.

Sources and further reading

Francis MP, Campana SE and Jones CM (2007) Age underestimation in New Zealand porbeagle sharks (*Lamna nasus*): is there an upper limit to ages that can be determined from shark vertebrae? *Marine and Freshwater Research* **58**, 10–23.

Francis MP, Natanson LJ and Campana SE (2008) The biology and ecology of the porbeagle shark, *Lamna nasus*. In *Sharks of the Open Ocean: Biology, Fisheries and Conservation*. (Eds MD Camhi, EK Pikitch and EA Babcock) pp. 105–113. Blackwell Publishing, Oxford.

Gilman E, Clarke S, Brothers N, Alfaro-Shigueto-J, Mandelman J, Mangel J, Petersen S, Piovano S, Thomson N, Dalzell P, Donoso M, Goren M and Werner T (2007) 'Shark depredation and unwanted bycatch in pelagic longline fisheries: industry practices and attitudes, and shark avoidance strategies'. Western Pacific Regional Fishery Management Council, Honolulu, USA.

Last PR and Stevens JD (2009) *Sharks and Rays of Australia*. 2nd edn. CSIRO Publishing, Melbourne.

Stevens J, Fowler SL, Soldo A, McCord M, Baum J, Acuña E, Domingo A and Francis M (2006) *Lamna nasus*. In IUCN 2010. *IUCN Red List of Threatened Species*. Version 2010.4. IUCN, Gland, Switzerland and Cambridge, UK, <http://www.iucnredlist.org>.

Authors: Peter M Kyne and Lee K Curtis

GROUND SHARKS (ORDER: CARCHARHINIFORMES)

OCEANIC WHITETIP SHARK

Carcharhinus longimanus Poey, 1861

Photo: Ian Gordon-AUSCAPE

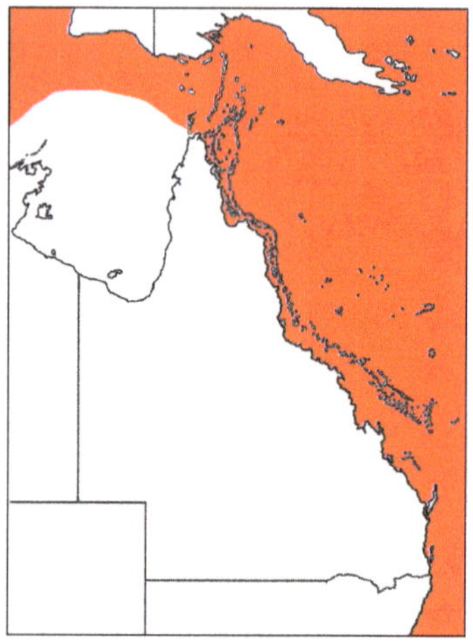

Family: Carcharhinidae

Conservation status: IUCN: Vulnerable

Description: The Oceanic Whitetip Shark is a large, stocky, whaler shark with a blunt snout, which usually reaches a maximum total length of about 300 cm. It has a high first dorsal fin and long, broad, paddle-shaped pectoral fins. Colouration is bronzy-grey dorsally and paler ventrally. Larger animals have distinctive mottled white tips on the pectoral, dorsal, pelvic and caudal fins.

Distribution: The Oceanic Whitetip Shark is distributed worldwide in tropical and warm temperate seas. In Australia, the southern limits of its distribution are Cape Leeuwin in WA and Sydney in NSW, although a single animal was recorded near Port Lincoln in SA. In Qld, it occurs in offshore (and occasionally inshore) waters off the east coast, but is absent from the Gulf of Carpentaria.

Habitat: An oceanic and pelagic species usually found far offshore, although occasionally inshore, in water temperatures of 18–28°C and from the surface to at least 150 m depth. Off Qld, the Oceanic Whitetip Shark is mainly encountered by commercial fishers operating beyond the continental shelf and far offshore. In places such as south-east and north Qld where the continental shelf is relatively narrow, this species may occur closer inshore.

Ecology: Despite being abundant and frequently captured in fisheries, few studies have investigated the ecology of this species. Males mature at 175–195 cm total length and females at 180–200 cm total length. Both sexes mature between 5–7 years of age and the maximum age reported is 22 years. Size at birth is 60–65 cm total length. This species is viviparous and has a gestation period of approximately 12 months. Litter sizes range from one to fifteen. Little study has been undertaken on the Oceanic Whitetip Shark specifically in Australian waters.

Reasons for decline and current threats: The global listing on the IUCN Red List is a result of intense and increasing longline fishing pressure in open ocean waters that targets tuna and billfish throughout virtually all the range of this species.

This species was formerly regarded as one of the more abundant oceanic sharks. Within recent years, however, declines in populations of up to 99% have been suggested in the Northwest and Western Central Atlantic and the Gulf of Mexico. Despite a similar history of exploitation in the Indian and Pacific Oceans, there is a lack of data and population assessments for these regions. Catch of this species is inadequately monitored and effort is likely to persist or increase in the future. The fins are of high value in the international shark fin trade.

At present, fishing pressure on this species in Australian waters is not comparable to that elsewhere. It is a byproduct and bycatch of the Eastern Tuna and Billfish Fishery (ETBF) operating in the waters off Qld. Although frequently caught, it makes up only a small portion of the overall shark catch. The general practice in the fishery at present is to release most sharks. Priority species assessments conducted on bycatch species did not identify the Oceanic Whitetip Shark as a high-risk species. Changing fisheries practices and market prices for shark products could potentially place this species at greater risk in the future.

Individuals occurring in Australian waters are likely part of larger Western and Central Pacific stocks. Therefore, pelagic fisheries in other parts of the Pacific Ocean that retain this species are perhaps the greatest threat to populations in Australian waters, particularly given that the species is presumed to be migratory. The Oceanic Whitetip Shark has also been documented in confiscated catches of illegal Taiwanese longline fishing vessels operating in the Torres Strait and off north-east Qld.

CURRENT RECOVERY EFFORTS

Recovery plan: None

Other plans: Australian Fisheries Management Authority (AFMA) (2008) 'Australian Tuna and Billfish Longline Fisheries Bycatch and Discarding Workplan, November 1, 2008 to October 31, 2010'. AFMA, Canberra.

Conservation actions:

- Although this species is rarely encountered by Qld recreational fishers, there is a recreational take and possession limit of one shark or ray in Qld.

Management regulations in the ETBF:

- Ban on shark finning (fins can not be landed separate from carcasses).
- Landing limit of 20 sharks per boat per trip (all species).
- Ban on wire-traces from longlines.
- Increased level of fisheries observers monitoring catch.

FUTURE RECOVERY EFFORTS

What you can do: Fishers should release any captured Oceanic Whitetip Sharks safely.

Conservation needs:

- Encourage sustainable fisheries management throughout the range of the species.
- Encourage international cooperation and management action in line with the 1995 UN Agreement on the Conservation and Management of Straddling Fish Stocks and Highly Migratory Fish Stocks (in which the Oceanic Whitetip Shark is included, and which Australia has ratified) and the FAO International Plan of Action for the Conservation and Management of Sharks.

Research directions: Investigate the life history, movement patterns and stock structure of the Oceanic Whitetip Shark in Australian waters.

Sources and further reading

Baum J, Medina E, Musick JA and Smale M (2006) *Carcharhinus longimanus*. In IUCN 2010. *IUCN Red List of Threatened Species*. Version 2010.4. IUCN, Gland, Switzerland and Cambridge, UK, <http://www.iucnredlist.org>.

Baum JK, Myers RA, Kehler DG, Worm B, Harley SJ and Doherty PA (2003) Collapse and conservation of shark populations in the Northwest Atlantic. *Science* **299**, 389–392.

Burgess GH, Beerkircher LR, Cailliet GM, Carlson JK, Cortés E, Goldman KJ, Grubbs RD, Musick JA, Musyl MK and Simpfendorfer CA (2005) Is the collapse of shark populations in the Northwest Atlantic Ocean and Gulf of Mexico real? *Fisheries* **30**, 19–26.

Gilman E, Clarke S, Brothers N, Alfaro-Shigueto-J, Mandelman J, Mangel J, Petersen S, Piovano S, Thomson N, Dalzell P, Donoso M, Goren M and Werner T (2007) 'Shark depredation and unwanted bycatch in pelagic longline fisheries: industry practices and attitudes, and shark avoidance strategies'. Western Pacific Regional Fishery Management Council, Honolulu, USA.

Last PR and Stevens JD (2009) *Sharks and Rays of Australia*. 2nd edn. CSIRO Publishing, Melbourne.

Author: Alastair Harry

SPEARTOOTH SHARK

Glyphis glyphis (Müller & Henle, 1839)

Photo: ScienceImage

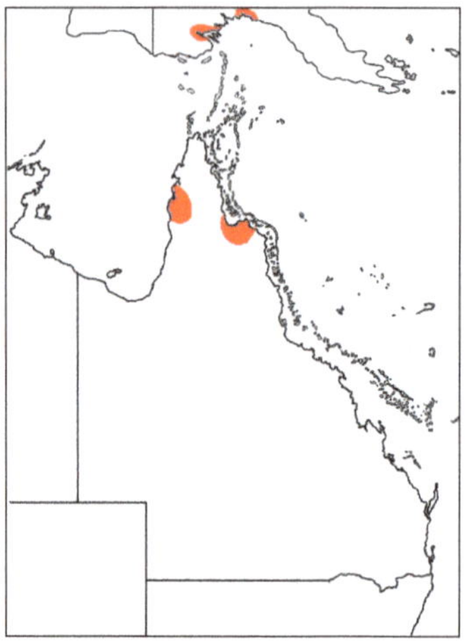

Family: Carcharhinidae

Other common names: Bizant River Shark, Queensland River Shark

Conservation status: IUCN: Endangered; Aust.: Critically Endangered; NT: Vulnerable

Description: The Speartooth Shark is a medium to large sized whaler shark, which is estimated to reach a maximum total length of 300 cm (based on jaws of a specimen of unknown length), although the largest recorded specimen is 175 cm total length. The second dorsal fin is tall, about three quarters of the first dorsal fin height. Its upper teeth are broadly triangular and blade-like, while the lower teeth are narrow, tall and slender with cutting edges confined to spear-like tips on the anterior teeth. The colouration is slate grey above and white below. It has a blackish blotch present on the ventral tips of the pectoral fins. It is often confused with the Bull Shark *Carcharhinus leucas*.

Distribution: The species has been recorded from scattered locations in northern Australia and New Guinea. In Australia, it is known from nine rivers and estuaries including the Port Musgrave system in Qld (Wenlock and Ducie Rivers). There are unconfirmed reports of large (>250 cm) *Glyphis* (thought to be sexually mature Speartooth Sharks) from offshore areas, including off western Cape York. Although first reported in Australia from the Bizant River, Princess Charlotte Bay, Qld, the species has not been recorded there since 1983.

Habitat: Neonate and juvenile Speartooth Sharks use large tropical river systems as their primary habitat and appear to show a distinct preference for highly turbid (muddy), brackish water. This is a euryhaline species capable of living in and moving between freshwater and seawater. There have been no verified captures of Speartooth Sharks outside tidal rivers and estuaries; however, there have been no confirmed records of adults, so their habitat use is unknown. Acoustic tracking of neonate and juvenile sharks has shown the repeated use of a very small proportion of available habitat.

Ecology: The biology and ecology of the species is very poorly known. Size at birth is 50–65 cm total length. The largest male examined (157 cm total length) was not mature. No mature females have been examined. Litter size is unknown. Young are born between November and January, based on the presence of individuals with umbilical scars. Juveniles eat a range of estuarine and freshwater benthic and benthopelagic teleost fishes, as well as freshwater crustaceans.

Reasons for decline and current threats: The conservation status of the species is a result of its narrow geographic distribution, small estimated population size, decline in numbers and/or distribution and overlap with commercial gill netting activities.

Commercial gill net fisheries targeting Barramundi *Lates calcarifer* in rivers, estuaries and coastal foreshores in the Gulf of Carpentaria and Princess Charlotte Bay pose the greatest threat to this species in Qld. Speartooth Sharks caught as bycatch in such fisheries often die due to entanglement. Capture by recreational fishers also poses a threat, as despite being a 'no take' species in Qld, correct identification and safe release are ongoing issues. Any changes to water flow, water quality or habitat modification/alteration due to mining, weirs or barrages are likely to pose a threat to this species given its limited distribution, apparent highly specific habitat requirements and repeated use of available habitat.

CURRENT RECOVERY EFFORTS

Recovery plan: None

Other plans: Pogonoski JJ, Pollard DA and Paxton JR (2002) 'Conservation overview and action plan for Australian threatened and potentially threatened marine and estuarine fishes'. Environment Australia, Canberra.

Conservation actions:

- Speartooth Sharks are 'no take' in Qld; the species cannot be retained by commercial or recreational fishers.
- Education of fishers in the Gulf of Carpentaria to improve identification and release of animals captured as bycatch is ongoing.

FUTURE RECOVERY EFFORTS

What you can do:

- Report any observations to the Qld Museum and DSEWPC. Due to identification issues, take a photograph (preferably from the side to capture important features).

- Fishers should abide by fishing regulations and release any captured Speartooth Sharks safely.

Conservation needs:
- Prepare a recovery plan.
- List the species on the Qld *Nature Conservation Act*.
- Implement protected areas in the habitat of the species.
- Continue the education of fishers on correct identification.

Research directions:
- Establish a long-term monitoring program to examine trends in population status/size to determine the effectiveness of conservation and management initiatives.
- Continue monitoring Qld populations (Wenlock/ Ducie Rivers and Port Musgrave).
- Determine the habitat requirements of adults and increase understanding of the habitat utilisation of juveniles.
- Research interactions with commercial and recreational fisheries, including post-release survival.
- Conduct a comprehensive survey of the Bizant River (and more broadly of north Cape York Peninsula) to establish whether the species has been extirpated on the east coast of Qld.
- Research population genetics to determine the relationship between Qld, NT and WA populations.
- Research the life history of the species.

Sources and further reading

Compagno LJV, White WT and Last PR (2008) *Glyphis garricki* sp. nov., a new species of river shark (Carcharhiniformes: Carcharhinidae) from northern Australia and Papua New Guinea, with a redescription of *Glyphis glyphis* (Müller & Henle, 1839). In *Descriptions of New Australian Chondrichthyans*. (Eds PR Last, WT White and JJ Pogonoski) pp. 203–225. CSIRO Marine and Atmospheric Research Paper No. 022. CSIRO Marine and Atmospheric Research, Hobart.

Last PR and Stevens JD (2009) *Sharks and Rays of Australia*. 2nd edn. CSIRO Publishing, Melbourne.

Peverell SC, McPherson GR, Garrett RN and Gribble NA (2006) New records of the river shark *Glyphis* (Carcharhinidae) reported from Cape York Peninsula, northern Australia. *Zootaxa* **1233**, 53–68.

Pillans RD, Stevens JD, Kyne PM and Salini J (2010) Observations on the distribution, biology, short-term movements and habitat requirements of river sharks *Glyphis* spp. in northern Australia. *Endangered Species Research* **10**, 321–332.

Pillans RD, Stevens JD, Peverell S and Edgar S (2008) 'Spatial distribution and habitat utilisation of the speartooth shark *Glyphis* sp. A in relation to fishing in northern Australia'. A report to the Department of the Environment, Water, Heritage and the Arts, Canberra.

Author: Richard D Pillans

SMOOTH HAMMERHEAD

Sphyrna zygaena (Linnaeus, 1758)

Photo: ScienceImage

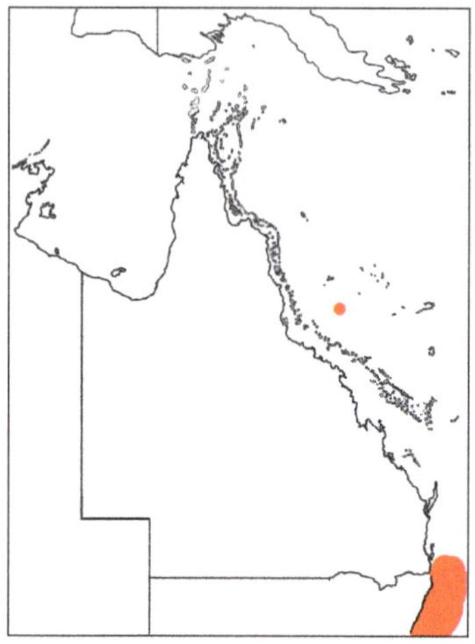

Family: Sphyrnidae

Conservation status: IUCN: Vulnerable

Description: The Smooth Hammerhead is a large shark, which reaches a maximum total length of 370–400 cm. The front border of its head is curved forward anteriorly (but with no median indentation). The head width is less than 40% of total length, and its teeth are finely serrated (occasionally smooth-edged). This species is olive to dark greyish brown above and white below, with the ventral tips of the pectoral fins being dusky.

Distribution: Smooth Hammerheads are widespread in temperate and tropical waters of the Indo-Pacific and Atlantic Oceans. In Australia, the species ranges from the Monte Bello Islands, WA across southern Australia to the NSW/Qld border and in the Coral Sea, off north-east Qld. The full extent of its range in tropical waters may be incompletely known at present, due to probable confusion with the more abundant Scalloped Hammerhead *Sphyrna lewini*.

Habitat: This species is coastal-pelagic and semi-oceanic and occurs over the continental shelf, down to a depth of 200 m. Large individuals are commonly found over deep reefs on the edge of the continental shelf. Off southern WA, juveniles are caught close to the seabed in depths from the shore to at least 60 m.

Ecology: Relative to other hammerhead sharks, the Smooth Hammerhead is poorly known. Maturity occurs at about 250–260 cm and 265 cm total length for males and females, respectively. Females are viviparous giving birth to 49–61 cm total length young between January and March, following a gestation of 10–11 months. Litter size from off eastern Australia ranges 20–49 young. Neonate Smooth Hammerheads are commonly found on smooth sandy substrate in shallow waters, down to depths of 10 m. The species sometimes forms large schools. Squid and teleost fishes are the most common prey, although crustaceans and elasmobranchs have also been reported.

Reasons for decline and current threats: Although the population status of the Smooth Hammerhead is mostly unknown, declines of hammerhead catch rates of up to 99% have been observed from the Northwest and Western Central Atlantic and the Mediterranean Sea, resulting in their global threatened status.

Hammerheads are subject to fishing pressure from both international and domestic fleets and their fins are highly desired for the shark fin trade. In Australia, hammerheads are a byproduct and bycatch of the Eastern Tuna and Billfish Fishery (ETBF). They are caught infrequently and make up only a very small portion of the overall shark catch. The general practice in the fishery at present is to release most sharks; however, hammerheads appear to be vulnerable to post-release mortality. Priority species assessments conducted on bycatch species identified the Smooth Hammerhead as a high-risk species. Hammerheads are also exploited in commercial line and recreational game fisheries off southern Australia. However, fishing pressure is relatively low in Australia and does not appear to have an impact on the Smooth Hammerhead population, in contrast to the global situation.

Interactions with fisheries off Qld are likely to be minimal, although hammerheads are probably caught in longline fisheries off the east coast. The Smooth Hammerhead has also been documented in confiscated catches of illegal foreign fishing vessels operating in the Torres Strait and off north-east Qld. Global recovery of the species is dependent on conservation and management measures elsewhere.

CURRENT RECOVERY EFFORTS

Recovery plan: None

Other plans: Australian Fisheries Management Authority (AFMA) (2008) 'Australian Tuna and Billfish Longline Fisheries Bycatch and Discarding Workplan, November 1, 2008 to October 31, 2010'. AFMA, Canberra.

Conservation actions:

- Although this species is rarely encountered by Qld recreational fishers, there is a recreational take and possession limit of one shark or ray in Qld.

Management regulations in the ETBF:

- Ban on shark finning (fins can not be landed separate from carcasses).
- Landing limit of 20 sharks per boat per trip (all species).
- Ban on wire-traces from longlines.
- Increased level of fisheries observers monitoring catch.

FUTURE RECOVERY EFFORTS

Conservation needs:

- Encourage sustainable fisheries management throughout the range of the species.
- Prohibit landings of this species within Australian fisheries, given demonstrated declines globally and the risk of overexploitation.
- Encourage international cooperation and management action in line with the 1995 UN Agreement on the Conservation and Management of Straddling Fish Stocks and Highly Migratory Fish Stocks (in which hammerhead sharks are included, and which Australia has ratified) and the FAO International Plan of Action for the Conservation and Management of Sharks.
- Encourage nomination of large hammerhead species for listing under Appendix II of CITES. A recent nomination of these species failed at the 15th Conference of Parties to CITES.

Research directions:

- Investigate the distribution, movement patterns and stock structure.
- Continuously monitor Australian fisheries.
- Investigate post-release mortality.
- Improve and sustain data collection and develop stock assessments.

Sources and further reading

Camhi MD, Valenti SV, Fordham SV, Fowler SL and Gibson C (2009) 'The conservation status of pelagic sharks and rays: report of the IUCN Shark Specialist Group Pelagic Shark Red List Workshop'. IUCN Species Survival Commission Shark Specialist Group, Newbury.

Casper BM, Domingo A, Gaibor N, Heupel MR, Kotas E, Lamónaca AF, Pérez-Jimenez JC, Simpfendorfer C, Smith WD, Stevens JD, Soldo A and Vooren CM (2005) *Sphyrna zygaena*. In IUCN 2010. *IUCN Red List of Threatened Species.* Version 2010.4. IUCN, Gland, Switzerland and Cambridge, UK, <http://www.iucnredlist.org>.

Gilman E, Clarke S, Brothers N, Alfaro-Shigueto-J, Mandelman J, Mangel J, Petersen S, Piovano S, Thomson N, Dalzell P, Donoso M, Goren M and Werner T (2007) 'Shark depredation and unwanted bycatch in pelagic longline fisheries: industry practices and attitudes, and shark avoidance strategies'. Western Pacific Regional Fishery Management Council, Honolulu, USA.

Last PR and Stevens JD (2009) *Sharks and Rays of Australia.* 2nd edn. CSIRO Publishing, Melbourne.

Stevens JD (1984) Biological observations on sharks caught by sports fishermen off New South Wales. *Australian Journal of Marine and Freshwater Research* **35**, 573–590.

Author: Charlie Huveneers

BATOIDS (ORDER: RAJIFORMES)

NARROW SAWFISH

Anoxypristis cuspidata (Latham, 1794)

Photo: ScienceImage

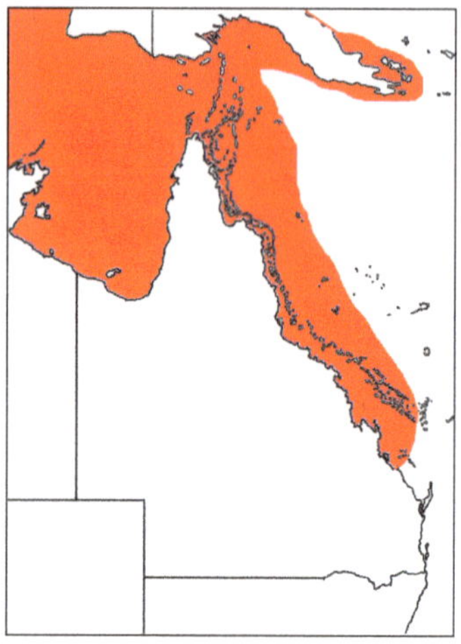

Family: Pristidae

Conservation status: IUCN: Critically Endangered

Description: The Narrow Sawfish is a medium-sized, slender sawfish, which reaches a total length of at least 350 cm. It has a shark-like body, with broadly triangular pectoral fins, which have a slightly concave posterior margin, tall and pointed dorsal fins, with the first dorsal fin origin well behind the pelvic fin origin, and a caudal fin with a very distinct lower lobe (the posterior margin of the caudal fin is deeply concave). The broad snout or saw (rostrum) has 18–22 pairs of rostral teeth; rostral teeth are not present on the basal quarter of the rostrum. The colouration is usually a greyish blue on the upper surface, and pale or white below.

Distribution: The Narrow Sawfish has a wide Indo-West Pacific distribution, including northern Australia from Eighty Mile Beach, WA to central Qld (but is generally rare on the east coast of Qld).

Habitat: The Narrow Sawfish inhabits inshore coastal and estuarine environments, as well as offshore marine environments to a depth of at least 200 m (it has been recorded as far as 200 km offshore). Smaller animals (<200 cm total length) primarily occur in shallow coastal environments, whereas larger animals (>200 cm total length) primarily occur offshore in water deeper than 20 m.

Ecology: The species is born at 70 cm total length, with males mature at about 220 cm total length and females at about 225 cm total length, corresponding to an age at maturity of 4 years for males and 5 years for females. Maximum longevity is estimated at 41 years. Narrow Sawfish are viviparous, with litter size around 15 and a gestation period of 4–5 months. There are no data on the short- or long-term movement patterns of the species. Narrow Sawfish primarily feed on a variety of bony fishes and crustaceans.

Reasons for decline and current threats: Within Australia, commercial and recreational fisheries pose the greatest threat to sawfish populations and the cumulative effects of these have likely caused the decline in abundance in Qld. Along the east coast, populations of all sawfish have declined significantly in the past 30–40 years. The Narrow Sawfish remains more widespread in the Gulf of Carpentaria, but they have undoubtedly suffered considerable declines in the past 40–60 years. However, the reproductive biology of this species (relatively faster growth, earlier maturation and higher litter size) enables it to recover from fishing and sustain higher levels of exploitation than other sawfish species. Despite recent legislation making the Narrow Sawfish a 'no take' species in all commercial and recreational fisheries in Qld, interactions occur with a variety of commercial fisheries (gill net and trawl) and bycatch still results in accidental mortalities due to entanglement. The Narrow Sawfish is the most commonly caught sawfish in the Northern Prawn Fishery, a trawl fishery operating across northern Australia. A study of that fishery in the Gulf of Carpentaria showed that although turtle exclusion devices (TEDs) and bycatch reduction devices (BRDs) did not reduce the total number of sawfish caught, the number of Narrow Sawfish captured in nets with these devices was lower than in nets without. These devices are compulsory in northern Australian trawl fisheries. Narrow Sawfish are also caught by illegal foreign fishing vessels in northern Australian waters.

CURRENT RECOVERY EFFORTS

Recovery plan: None

Other plans: Pogonoski JJ, Pollard DA and Paxton JR (2002) 'Conservation overview and action plan for Australian threatened and potentially threatened marine and estuarine fishes'. Environment Australia, Canberra.

Conservation actions:

- All sawfish species are 'no take' in Qld; the species cannot be retained by commercial or recreational fishers.

- A safe handling and live release document has been released by the Qld Government, which conducts fisher workshops on safe handling practices and tag and release for both recreational and commercial sectors.
- Although not directed specifically at sawfish, spatial and temporal fishing closures will benefit these species.
- JCU is studying movement patterns, habitat utilisation, life history, fishery interactions and genetics of Narrow Sawfish.
- International trade in all sawfishes (with the exception of Freshwater Sawfish *Pristis microdon*) is banned through their listing on Appendix I of CITES.

FUTURE RECOVERY EFFORTS

What you can do:

- Report any live sightings, fishing catches or dead beach-washed specimens to the Qld Museum and DSEWPC.
- Fishers should abide by fishing regulations and release any captured sawfish safely.

Conservation needs:

- List sawfish on the Qld *Nature Conservation Act*.
- Implement protected areas in the habitat of the species, particularly in the Gulf of Carpentaria.
- Continue education of fishers on the correct identification and safe release of the species, including through community-based programs.

Research directions:

- Establish a long-term monitoring program to examine trends in population status and size to determine the effectiveness of conservation and management initiatives.
- Survey the east coast of Qld to identify remaining populations and ensure these are adequately protected.
- Determine the habitat requirements and movement patterns of juveniles and adults.
- Research interactions with commercial and recreational fisheries, including post-release survival.

Sources and further reading

Brewer D, Heales D, Milton D, Dell Q, Fry G, Venables B and Jones P (2006) The impact of turtle excluder devices and bycatch reduction devices on diverse tropical marine communities in Australia's northern prawn trawl fishery. *Fisheries Research* **81**, 176–188.

Compagno LJV, Cook SF, Oetinger MI and Fowler SL (2006) *Anoxypristis cuspidata*. In IUCN 2010. *IUCN Red List of Threatened Species*. Version 2010.4. IUCN, Gland, Switzerland and Cambridge, UK, <http://www.iucnredlist.org>.

Giles J, Pillans R, Miller M and Salini J (2007) 'Sawfish catch data in northern Australia: a desktop study'. Report produced for FRDC Project 2002/064 Northern Australian sharks and rays: the sustainability of target and bycatch fisheries, phase 2.

Last PR and Stevens JD (2009) *Sharks and Rays of Australia*. 2nd edn. CSIRO Publishing, Melbourne.

Peverell SC (2005) Distribution of sawfishes (Pristidae) in the Queensland Gulf of Carpentaria, Australia, with notes on sawfish ecology. *Environmental Biology of Fishes* **73**, 391–402.

Author: Richard D Pillans

DWARF SAWFISH

Pristis clavata Garman, 1906

Photo: ScienceImage

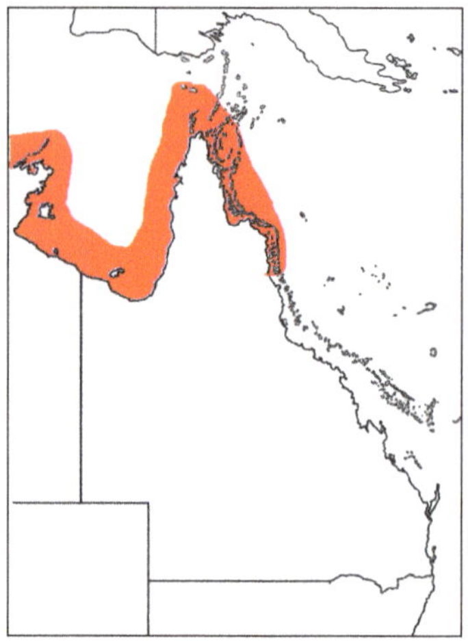

Family: Pristidae

Other common name: Queensland Sawfish

Conservation status: IUCN: Critically Endangered; Aust.: Vulnerable; NT: Vulnerable

Description: The Dwarf Sawfish is a small to medium-sized, slender sawfish, which reaches a maximum total length of 318 cm. It has a shark-like body, with broadly triangular pectoral fins, tall and pointed dorsal fins, with the first dorsal fin origin over or slightly behind the pelvic fin origin, and a caudal fin with a very small lower lobe (posterior margin of caudal fin almost straight). The broad snout or saw (rostrum) has 18–22 pairs of evenly spaced rostral teeth that start near the rostrum base. The colouration is usually a greenish brown or yellow on the upper surface, and pale or white below.

Distribution: The extent of the Dwarf Sawfish's distribution in the Indo-West Pacific is unknown, but it has been recorded from inshore coastal environments, estuaries and rivers across northern Australia from Eighty Mile Beach, WA to Cairns, Qld.

Habitat: The Dwarf Sawfish inhabits shallow (1–10 m) near-shore coastal environments, including estuaries, river mouths, rivers, embayments, and along sandy and muddy beaches. Juveniles have been recorded in rivers, estuaries and inshore coastal environments in salinities between 1 and 41, while adults have been recorded primarily in inshore coastal marine environments. They have not been reported in offshore environments.

Ecology: Young are born at 65 cm total length and males mature at 255–265 cm total length. Maximum longevity is estimated at 48 years. Dwarf Sawfish are viviparous but litter size and breeding frequency are unknown. Short-term tracking of individuals showed that they primarily occupied shallow water. Use of deeper waters occurred mostly when the sawfish were resting at high tides in inundated mangrove areas, or while crossing deeper waters while leaving the mangroves during falling tides. Most movement occurred during falling and rising tides, with little or no movement at high and low tides. Individuals were shown to move distances of 1–10 km during each tidal cycle before returning to the mangrove forest on the next high tide. These movement patterns are most likely related to foraging. Dwarf Sawfish feed on a variety of bony fishes and crustaceans.

Reasons for decline and current threats: Within Australia, commercial and recreational fisheries pose the greatest threat to sawfish populations and the cumulative effects of these have likely caused the decline in abundance in Qld.

Along the east coast, populations of all sawfish have declined significantly in the past 30–40 years. Dwarf Sawfish remain more widespread in the Gulf of Carpentaria, but they have undoubtedly suffered massive declines in the past 40–60 years and abundance is low. Repeated utilisation of small core areas of habitat within inshore environments and high susceptibility to capture in gill net results in high vulnerability and localised extirpation, leading to fragmentation of distribution. Despite recent legislation making the Dwarf Sawfish a 'no take' species in all commercial and recreational fisheries in Qld, interactions occur with a variety of commercial fisheries (primarily inshore gill net fisheries) and bycatch still results in accidental mortalities due to entanglement.

CURRENT RECOVERY EFFORTS

Recovery plan: None

Other plans: Pogonoski JJ, Pollard DA and Paxton JR (2002) 'Conservation overview and action plan for Australian threatened and potentially threatened marine and estuarine fishes'. Environment Australia, Canberra.

Conservation actions:

- All sawfish species are 'no take' in Qld; the species cannot be retained by commercial or recreational fishers.
- A safe handling and live release document has been released by the Qld Government, which conducts fisher workshops on safe handling practices and tag and release for both recreational and commercial sectors.
- Although not directed specifically at sawfish, spatial and temporal fishing closures will benefit these species.
- Murdoch University is studying population genetics of sawfish across northern Australia.
- International trade in all sawfishes (with the exception of Freshwater Sawfish *Pristis microdon*) is banned through their listing on Appendix I of CITES.

FUTURE RECOVERY EFFORTS

What you can do:

- Report any live sightings, fishing catches or dead beach-washed specimens to the Qld Museum and DSEWPC.
- Both commercial and recreational fishers should abide by fishing regulations and release any captured sawfish safely.

Conservation needs:

- Prepare a recovery plan.
- List sawfish on the Qld *Nature Conservation Act.*
- Implement protected areas in the habitat of the species, particularly in the Gulf of Carpentaria.
- Continue education of fishers on the correct identification and safe release of the species through community-based programs.

Research directions:

- Establish a long-term monitoring program to examine trends in population status and size to determine the effectiveness of conservation and management initiatives.
- Survey the east coast of Qld to identify remaining populations and ensure these are adequately protected.
- Determine the habitat requirements and movement patterns of adults and increase understanding of the habitat utilisation of juveniles.
- Research interactions with commercial and recreational fisheries, including post-release survival.

Sources and further reading

Cook SF, Compagno LJV and Last PR (2006) *Pristis clavata.* In IUCN 2010. *IUCN Red List of Threatened Species.* Version 2010.4. IUCN, Gland, Switzerland and Cambridge, UK, <http://www.iucnredlist.org>.

Last PR and Stevens JD (2009) *Sharks and Rays of Australia.* 2nd edn. CSIRO Publishing, Melbourne.

Peverell SC (2005) Distribution of sawfishes (Pristidae) in the Queensland Gulf of Carpentaria, Australia, with notes on sawfish ecology. *Environmental Biology of Fishes* **73**, 391–402.

Stevens JD, McAuley RB, Simpfendorfer CA and Pillans RD (2008) 'Spatial distribution and habitat utilisation of sawfish (*Pristis* spp.) in relation to fishing in northern Australia'. A report to Department of the Environment, Water, Heritage and the Arts. <http://www.environment.gov.au/coasts/publications/pubs/sawfish-report.pdf>.

Thorburn DC, Morgan DL, Rowland AJ, Gill HS and Paling E (2008) Life history notes of the critically endangered dwarf sawfish, *Pristis clavata*, Garman 1906 from the Kimberley region of Western Australia. *Environmental Biology of Fishes* **83**, 139–145.

Author: Richard D Pillans

FRESHWATER SAWFISH

Pristis microdon Latham, 1794

Photo: ScienceImage

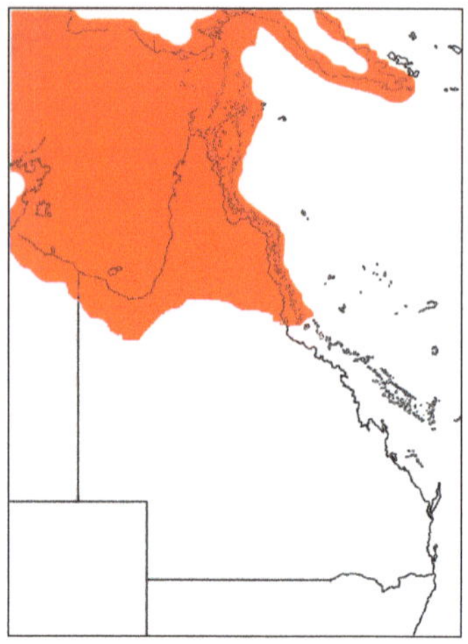

Family: Pristidae

Conservation status: IUCN: Critically Endangered; Aust.: Vulnerable; NT: Vulnerable

Description: The Freshwater Sawfish is a large, slender sawfish, which reaches a maximum total length of at least 600 cm. It has a shark-like body, with broadly triangular pectoral fins, tall and pointed dorsal fins, with the first dorsal fin origin well in advance of the pelvic fin origin, and a caudal fin with a short but distinct lower lobe (much less than half the length of the upper lobe). The broad snout or saw (rostrum) has 17–23 (mainly 20–22) pairs of evenly spaced rostral teeth that start near the rostrum base and are not noticeably closer together at the tip. The colouration is usually a yellow golden-brown on the upper surface and white below.

Distribution: The Freshwater Sawfish has a wide Indo-West Pacific distribution, including rivers, estuaries and marine environments of northern Australia from Shark Bay, WA to Cairns, Qld. Within Qld, the species has been recorded in fresh and saline waters of the Gilbert, Mitchell, Normanby, Wenlock, Mission, Embley and Leichhardt River systems.

Habitat: The Freshwater Sawfish inhabits sandy or muddy bottoms of shallow coastal waters, estuaries, river mouths, freshwater rivers and isolated water holes. Neonate and juvenile animals primarily occur in the freshwater reaches of rivers and estuaries, while adults have been recorded up to 100 km offshore.

Ecology: The species is born at 60–70 cm total length and males and females are thought to mature around 300 cm total length (estimated age at maturity of 8 years). The largest animal examined for age studies (582 cm total length) was estimated to be 28 years old, while maximum longevity is estimated at 80 years. Freshwater Sawfish are viviparous with litter sizes probably around 12; limited evidence suggests animals are capable of breeding once a year. Freshwater Sawfish primarily feed on a variety of bony fishes and crustaceans.

Reasons for decline and current threats: Within Australia, commercial and recreational fisheries pose the greatest threat to sawfish populations, the cumulative effects of which have likely caused the decline in abundance in Qld.

Along the east coast, populations of all sawfish have declined significantly in the past 30–40 years; the Freshwater Sawfish now appears to be restricted to the rivers draining into Princess Charlotte Bay and abundance is extremely low. Freshwater Sawfish remain more widespread in the drainage systems of the Gulf of Carpentaria, but they have undoubtedly suffered massive declines in the past 40–60 years and abundance is low. Despite recent legislation making the Freshwater Sawfish a 'no take' species in all commercial and recreational fisheries in Qld, interactions occur with a variety of commercial fisheries (gill net and trawl) and bycatch still results in accidental mortalities due to entanglement.

Dams, barrages and crossings can prevent the upstream and downstream migration of Freshwater Sawfish, which results in significant reductions in available habitat. Any future proposals affecting flow regimes will further reduce habitat and any extraction of surface and groundwater has the potential to reduce dry season refuge habitats.

CURRENT RECOVERY EFFORTS

Recovery plan: Truelove, K (2003) 'Draft freshwater sawfish (*Pristis microdon*) recovery plan'. Department of the Environment and Heritage, unpublished report.

Other plans: Pogonoski JJ, Pollard DA and Paxton JR (2002) 'Conservation overview and action plan for Australian threatened and potentially threatened marine and estuarine fishes'. Environment Australia, Canberra.

Conservation actions:

- All sawfish species are 'no take' in Qld; the species cannot be retained by commercial or recreational fishers.
- A safe handling and live release document has been released by the Qld Government, which conducts fisher workshops on safe handling practices and tag and release for both recreational and commercial sectors.
- Although not directed specifically at sawfish, spatial and temporal fishing closures will benefit these species.
- Murdoch University is studying movement patterns and habitat utilisation of Freshwater Sawfish in the Kimberley region of WA, and the population genetics of sawfish across northern Australia.
- Although all international trade in all other species of sawfish is banned through their listing on Appendix I

of CITES, Freshwater Sawfish are listed on Appendix II 'for the exclusive purpose of allowing international trade in live animals to appropriate and acceptable aquaria for primarily conservation purposes'.

FUTURE RECOVERY EFFORTS

What you can do:

- Report any live sightings, fishing catches or dead beach-washed specimens to the Qld Museum and DSEWPC.
- Both commercial and recreational fishers should abide by fishing regulations and release any captured sawfish safely.

Conservation needs:

- Update and finalise the draft recovery plan that was initiated in 2003.
- List sawfish on the Qld *Nature Conservation Act.*
- Implement protected areas in the habitat of the species, particularly in the Gulf of Carpentaria.
- Continue education of fishers on the correct identification and safe release of the species through community-based programs.

Research directions:

- Establish a long-term monitoring program to examine trends in population status and size to determine the effectiveness of conservation and management initiatives.
- Survey the east coast of Qld to identify remaining populations and ensure these are adequately protected.
- Determine the habitat requirements and movement patterns (including long-term patterns) of adults and increase understanding of the habitat utilisation of juveniles in rivers and estuaries.
- Research interactions with commercial and recreational fisheries, including post-release survival.
- Examine and manage the impact of water extraction and physical barriers such as weirs, barrages and crossings.

Sources and further reading

Compagno LJV, Cook SF and Fowler SL (2006) *Pristis microdon*. In IUCN 2010. *IUCN Red List of Threatened Species.* Version 2010.4. IUCN, Gland, Switzerland and Cambridge, UK, <http://www.iucnredlist.org>.

Last PR and Stevens JD (2009) *Sharks and Rays of Australia.* 2nd edn. CSIRO Publishing, Melbourne.

Peverell SC (2005) Distribution of sawfishes (Pristidae) in the Queensland Gulf of Carpentaria, Australia, with notes on sawfish ecology. *Environmental Biology of Fishes* **73**, 391–402.

Stevens JD, Pillans RD and Salini J (2005) 'Conservation assessment of *Glyphis* sp. A (speartooth shark), *Glyphis* sp. C (northern river shark), *Pristis microdon* (freshwater sawfish) and *Pristis zijsron* (green sawfish)'. Final report to the Department of the Environment and Heritage. <http://www.environment.gov.au/coasts/publications/pubs/assessment-glyphis.pdf>.

Thorburn DC, Morgan DL, Rowland AJ and Gill HS (2007) Freshwater sawfish *Pristis microdon* Latham, 1794 (Chondrichthyes: Pristidae) in the Kimberley region of Western Australia. *Zootaxa* **1471**, 27–41.

Author: Richard D Pillans

GREEN SAWFISH

Pristis zijsron Bleeker, 1851

Photo: ScienceImage

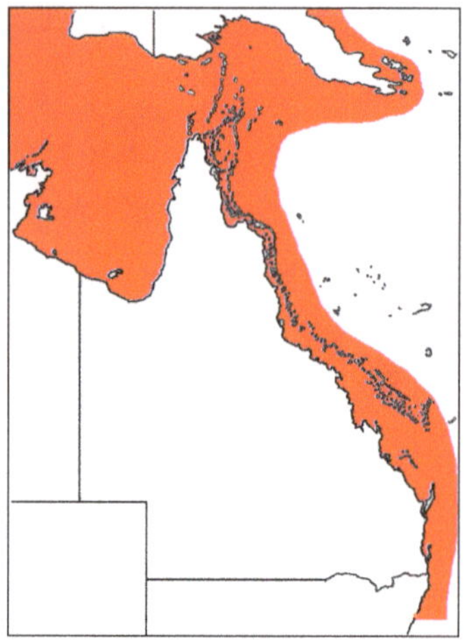

Family: Pristidae

Other common name: Narrowsnout Sawfish

Aboriginal name: Dindagubba

Conservation status: IUCN: Critically Endangered; Aust.: Vulnerable; NSW: Presumed Extinct; NT: Vulnerable; WA: Vulnerable

Description: The Green Sawfish is a very large, slender sawfish, which is reported to reach a maximum total length of 730 cm (largest Australian record: 540 cm total length). It has a shark-like body, with broadly triangular pectoral fins, tall and pointed dorsal fins, with the first dorsal fin origin slightly behind the pelvic fin origin, and a caudal fin with a very small lower lobe. The broad snout or saw (rostrum) has 25–32 (mainly 24–28) pairs of evenly spaced rostral teeth that start near the rostrum base and that are much closer to each other towards the tip of the rostrum. The colouration is usually a greenish brown or olive on the upper surface and pale or white below.

Distribution: The Green Sawfish has a wide Indo-West Pacific distribution, including inshore coastal environments of northern Australia from Shark Bay, WA to Cairns, Qld. Their current distribution is significantly less than historically when they were found in NSW and throughout Qld.

Habitat: The Green Sawfish inhabits near-shore coastal environments, including estuaries, river mouths, embayments and along sandy and muddy beaches (but it does not penetrate into freshwater) in a depth range of <1–70 m. Smaller Green Sawfish (<250 cm total length) are more common in inshore environments, while larger animals (>250 cm total length) are captured both inshore and offshore.

Ecology: The young are born at around 75–80 cm total length and males and females are thought to mature around 300–380 cm total length (estimated age at maturity of 9 years). The largest animal examined for age studies (508 cm total length) was estimated to be 24 years old, while maximum longevity is estimated at 53 years. Green Sawfish are viviparous, with litter sizes probably around 12 but the frequency of breeding is unknown. This species is an active predator, feeding primarily on a variety of bony fishes and crustaceans.

Reasons for decline and current threats: Within Australia, commercial and recreational fisheries pose the greatest threat to sawfish populations, the cumulative effects of which have likely caused the decline in abundance in Qld. Along the east coast, populations of all sawfish, particularly Green Sawfish, have declined significantly in the past 30–40 years and, although they were once recorded as far south as NSW, they are more than likely extinct there. Long-term catch data from 1963–1999 shows the gradual disappearance of Green Sawfish from NSW, south-east Qld, central Qld and north Qld, indicating a significant retraction in its range. Populations of sawfish in the Gulf of Carpentaria have undoubtedly suffered massive declines in the past 40–60 years; high catch rates combined with high fishing effort in the 1970–80s likely contributed to the decline of this species. Despite recent legislation making the Green Sawfish a 'no take' species in all commercial and recreational fisheries in Qld, interactions occur with a variety of commercial fisheries (gill net and trawl) and bycatch still results in accidental mortalities due to entanglement.

CURRENT RECOVERY EFFORTS

Recovery plan: Bunce A (2009) 'Recovery plan for the Green Sawfish (*Pristis zijsron*) in Central Queensland. Conservation overview and status assessment'. Report to the Burnett Mary Regional Group for Natural Resource Management. Centre for Environmental Management, Central Queensland University, Gladstone.

Other plans: Pogonoski JJ, Pollard DA and Paxton JR (2002) 'Conservation overview and action plan for Australian threatened and potentially threatened marine and estuarine fishes'. Environment Australia, Canberra.

Conservation actions:

- All sawfish species are 'no take' in Qld; the species cannot be retained by commercial or recreational fishers.
- A safe handling and live release document has been released by the Qld Government, which conducts fisher workshops on safe handling practices and tag and release for both recreational and commercial sectors.
- Although not directed specifically at sawfish, spatial and temporal fishing closures will benefit these species.

- Murdoch University is studying population genetics of sawfish across northern Australia.
- International trade in all sawfishes (with the exception of Freshwater Sawfish *Pristis microdon*) is banned through their listing on Appendix I of CITES.

FUTURE RECOVERY EFFORTS

What you can do:

- Report any live sightings, fishing catches or dead beach-washed specimens to the Qld Museum and DSEWPC.
- Both commercial and recreational fishers should abide by fishing regulations and release any captured sawfish safely.

Conservation needs:

- Prepare a recovery plan.
- List sawfish on the Qld *Nature Conservation Act*.
- Implement protected areas in the habitat of the species, both in the Gulf of Carpentaria and on the east coast (including if small fragmented populations are found to persist on the east coast south of Cairns).
- Continue education of fishers on the correct identification and safe release of the species, through community-based programs.

Research directions:

- Establish a long-term monitoring program to examine trends in population status and size to determine the effectiveness of conservation and management initiatives.
- Survey the east coast of Qld to identify remaining populations and ensure these are adequately protected.
- Determine the habitat requirements and movement patterns (including long-term patterns) of adults and increase understanding of the habitat utilisation of juveniles.
- Research interactions with commercial and recreational fisheries, including post-release survival.

Sources and further reading

Compagno LJV, Cook SF and Oetinger MI (2006) *Pristis zijsron*. In IUCN 2010. *IUCN Red List of Threatened Species*. Version 2010.4. IUCN, Gland, Switzerland and Cambridge, UK, <http://www.iucnredlist.org>.

Last PR and Stevens JD (2009) *Sharks and Rays of Australia*. 2nd edn. CSIRO Publishing, Melbourne.

Peverell SC (2005) Distribution of sawfishes (Pristidae) in the Queensland Gulf of Carpentaria, Australia, with notes on sawfish ecology. *Environmental Biology of Fishes* **73**, 391–402.

Peverell, SC and Pillans, RD (2004) 'Determining feasibility of acoustic tag attachment and documenting short-term movement in *Pristis zijsron* Bleeker, 1851'. Report for the National Oceans Office. <http://www.environment.gov.au/coasts/mbp/publications/north/pubs/sawfish-tracking.pdf>.

Stevens JD, Pillans RD and Salini J (2005) 'Conservation assessment of *Glyphis* sp. A (speartooth shark), *Glyphis* sp. C (northern river shark), *Pristis microdon* (freshwater sawfish) and *Pristis zijsron* (green sawfish)'. Final report to the Department of the Environment and Heritage. <http://www.environment.gov.au/coasts/publications/pubs/assessment-glyphis.pdf>.

Author: Richard D Pillans

SYDNEY SKATE

Dipturus australis (Macleay, 1884)

Photo: ScienceImage

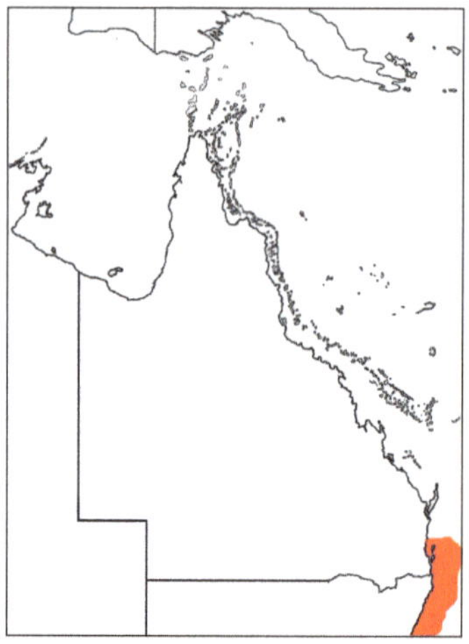

Family: Rajidae

Conservation status: IUCN: Vulnerable

Description: The Sydney Skate is a medium-sized skate, which reaches a maximum total length of about 55 cm. It has a quadrangular shaped disc that is broader than long, with a moderately elongate, pointed snout. In mature males, there are two patches of malar thorns (patches of thorns beside the eyes) on each side, but the species possesses only one small nuchal thorn (a thorn on the nape). Its tail is short, broad and flattened. The colouration on the dorsal surface is yellowish brown with scattered small, pale blotches, while the ventral side is mainly white.

Distribution: The Sydney Skate is restricted to waters off eastern Australia between Moreton Bay in south-east Qld and Tathra on the south coast of NSW. It is most abundant off central and southern NSW, while there are fewer records from south-east Qld and northern NSW.

Habitat: This species occurs on the continental shelf and upper slope in a depth range of 22–325 m. Skates are demersal and usually live on smooth sandy substrates, but there is little known of the specific habitat requirements of the Sydney Skate. One of the few recorded specimens from Qld waters was captured by trawl on the outer continental shelf at a depth of 135 m.

Ecology: The biology and ecology of the Sydney Skate is very poorly known. Males mature at a size of 43–48 cm total length. Like other skates, this species is oviparous.

Reasons for decline and current threats: The Sydney Skate can be threatened by intensive fishing as was demonstrated for deepwater fishing grounds off southern and central NSW, where surveys found a decline in skate numbers caused by continual trawling.

Research data from NSW demonstrated the susceptibility of skates to population depletion through intensive fishing activities. Trawl surveys conducted on the central and southern NSW upper continental slope (200–650 m depth) showed an overall decline in the catch rate of skates (all species combined) of 83% between 1976-77 and 1996-97. Although total skate catches within the upper depth zone surveyed (200–275 m) were relatively small, the Sydney Skate was the main species recorded in that zone. However, the Sydney Skate mainly occupies areas of the outer shelf (100–200 m depth) and, although fishing pressure on established southern NSW trawl grounds remains high, a considerable portion of its range (central and northern NSW) is less intensively fished. Nevertheless, even periodic trawling, such as seasonal prawn trawling on offshore grounds in the northern part of its range, could adversely impact on Sydney Skate numbers.

Although good time-series data showing a considerable decline in skate numbers were documented for upper slope trawl grounds off central and southern NSW, there is no information on any effects of fishing on the Sydney Skate in Qld. Its area of occurrence off southern Qld is fished seasonally by the Qld East Coast Trawl Fishery (ECTF), but during recent broad-scale surveys of the Qld trawl grounds only a single Sydney Skate was recorded. This low number may have been a result of the surveys not sampling deeper than 166 m, thus not covering the full depth range of the species. It is also possible that, because the area south of Moreton Bay is known to be the northern limit of its distribution, Sydney Skate numbers are naturally low off south-east Qld.

CURRENT RECOVERY EFFORTS

Recovery plan: None

FUTURE RECOVERY EFFORTS

What you can do:

- Report any fishing catches to the Qld Museum.
- Commercial fishers should release any captured skates safely.
- Obtain the Australian Marine Conservation Society's *Australia's Sustainable Seafood Guide* and buy only seafood that is harvested from sustainable sources with limited bycatch (see <http://www.amcs.org.au>).

Conservation needs: Reduce and limit bycatch of the species (this is dependent on conservation actions in the southern part of the range of the species where skates have undergone decline and where the Sydney Skate is naturally more common).

Research directions:

- Quantify and monitor any catches in the ECTF (although these are likely to be irregular given the rarity of the species in Qld).
- Undertake research to determine the biological attributes of the species (particularly reproductive biology and age and growth). This will need to occur where the species is more common off southern and central NSW.

Sources and further reading

Graham KJ, Andrew NL and Hodgson KE (2001) Changes in relative abundance of sharks and rays on Australian South East Fishery trawl grounds after twenty years of fishing. *Marine and Freshwater Research* **52**, 549–561.

Kyne PM, Johnson JW, Courtney AJ and Bennett MB (2005) New biogeographical information on Queensland chondrichthyans. *Memoirs of the Queensland Museum* **50**, 321–327.

Last PR and Stevens JD (2009) *Sharks and Rays of Australia.* 2nd edn. CSIRO Publishing, Melbourne.

Stevens JD and Valenti SV (2007) *Dipturus australis.* In IUCN 2010. *IUCN Red List of Threatened Species.* Version 2010.4. IUCN, Gland, Switzerland and Cambridge, UK, <http://www.iucnredlist.org>.

Authors: Peter M Kyne and Ken J Graham

SANDYBACK, YELLOWBACK AND GREENBACK STINGAREES

Urolophus bucculentus Macleay, 1884; *Urolophus sufflavus* Whitley, 1929; *Urolophus viridis* McCulloch, 1916

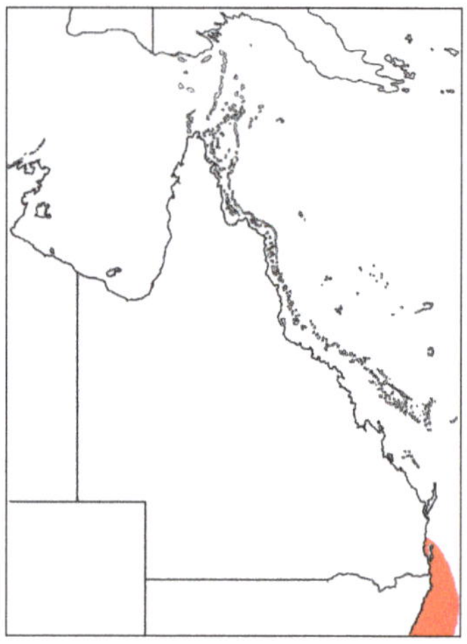

Photo: ScienceImage

Photo: ScienceImage

Photo: ScienceImage

Family: Urolophidae

Common names: Sandyback Stingaree (*U. bucculentus*); Yellowback Stingaree (*U. sufflavus*); Greenback Stingaree (*U. viridis*)

Conservation status: IUCN: Vulnerable

Description: These three species of medium to large sized stingarees reach maximum total lengths of 89 cm (Sandyback Stingaree), 42 cm (Yellowback Stingaree) and 51 cm (Greenback Stingaree). Stingarees are small rays with an oval shaped disc, a short tail (usually with stinging spine) and a well-developed caudal fin. The tail is about equal or shorter in length than the disc; this feature, combined with the presence of a caudal fin, easily separates them from the true stingrays (family Dasyatidae). All three species treated here lack broad fleshy lobes on the lateral margins of the nostrils (a character of the genus *Urolophus*). The three species vary in their dorsal surface colour pattern: the Sandyback Stingaree is yellowish to brownish, usually with a complex pattern of light spots and reticulations; the Yellowback Stingaree is lighter uniform yellowish brown; and the Greenback Stingaree is uniform light green with paler edges. The Sandyback Stingaree has a small but prominent dorsal fin just anterior to the stinging spine; the Yellowback and Greenback Stingarees both lack a dorsal fin. The tail is short in the Sandyback and Yellowback Stingarees (generally <75% of the disc length), and moderately long in the Greenback Stingaree (>75% of the disc length).

Distribution: All three species are endemic to south-east Australia and range only marginally into southern Qld waters to about North Stradbroke Island; these are primarily temperate species. The Sandyback Stingaree ranges from eastern SA through Vic., Tas. and NSW and into Qld (although its occurrence appears to be patchy across this distribution). The Yellowback Stingaree is restricted to the east coast from Green Cape, NSW to North Stradbroke Island. The Greenback Stingaree ranges from western Vic. through Tas. and NSW and into Qld.

Habitat: These three stingarees generally occupy deeper water on the outer continental shelf and upper slope than many of their relatives, which have a preference for inshore shallow water environments. The Sandyback Stingaree occurs in a depth range of 65–265 m, the Yellowback Stingaree 45–320 m (mainly 100–160 m), and the Greenback Stingaree 20–300 m (mainly 80–180 m, rarely <60 m). All stingarees are demersal and these three species probably occur on soft (including sandy) substrates. There is no specific information on habitat in Qld for the species treated here.

Ecology: Estimates of size at the onset of maturity for the three species treated here are as follows: Sandyback Stingaree, males ~41 cm total length, females ~50 cm total length; Yellowback Stingaree, males ~23 cm total length, female size at maturity unknown; Greenback Stingaree, males ~28 cm total length, females 26–31 cm total length. Stingarees are viviparous, with mostly annual reproduction and low fecundity. The Sandyback Stingaree, however, reproduces biennially with a gestation period of 14–19 months, the longest reported for a stingaree species. Litter size is one to five young. The Greenback Stingaree has a gestation period of 10–12 months and litter size of one to three young. A small litter size has also been confirmed for the Yellowback Stingaree, with a female caught off southern Qld containing two embryos. Preliminary investigation of age for the Greenback Stingaree suggest a maximum age of ~10 years and for the Sandyback Stingaree of ~17 years; no information is known about maximum age of the Yellowback Stingaree, but, based on relative size, ~10 years is likely. The larger relative size of the Sandyback Stingaree suggests a slower growth rate for that species.

Reasons for decline and current threats: Intensive commercial fishing pressure principally targeting bony fish species (mainly off NSW) is responsible for declines in these stingarees.

Trawl surveys conducted on the NSW upper continental slope demonstrated an overall decline in the catch rate of stingarees of 66% between 1976–77 and 1996–97. If the data are considered by individual survey grounds, declines were as high as 91%. The Sandyback and Greenback Stingarees represented the largest part of these catches, while the Yellowback Stingaree was caught in smaller numbers. Historical and current levels of fishing pressure on these trawl grounds is high. Fishing also occurs on the continental shelf where these species occur, but not as intensively as on the slope.

Although good time-series data showing a considerable decline was determined for upper slope trawl grounds off central and southern NSW, there is no information on the effects of any fishing activities where these species occur in Qld. The Qld East Coast Trawl Fishery (ECTF) operates seasonally in their Qld range (which is small, extending only to about North Stradbroke Island in southern Qld). During recent broad-scale surveys of the ECTF, only a single Yellowback Stingaree was recorded. However, these surveys did not sample deeper than 166 m, and thus did not fully cover the depth ranges of the species. At the northern extent of their geographic distribution in Qld, it is possible that these species may have a preference for deeper waters, which is often the case for more temperate fishes. Alternatively, they may be naturally rare here, given that these are primarily of temperate occurrence. There may be some refuge at depth for the species in southern Qld, as fishing only occurs seasonally there.

There is no evidence that the use of turtle exclusion devices (TEDs) on Qld trawlers reduces the catch of stingarees (the Common Stingaree *Trygonoptera testacea* and Kapala Stingaree *U. kapalensis* are regular components of the bycatch), although they may be effective for adult Sandyback Stingaree, given the large size of the species.

Data from NSW demonstrate the susceptibility of stingarees to population depletion as a result of consistent fishing activities. This susceptibility, combined with their low fecundity and common propensity to abort embryos upon capture and handling (thus reducing reproductive output), suggests that stingarees are an intrinsically vulnerable group of species. However, it should be noted that populations of some other stingaree species remain high where trawl fisheries regularly operate (i.e. south-west Australia), suggesting that not all species share the vulnerability of the deeper water species. Such differences may relate to discarding practices and the subsequent survivability of animals returned to the water (survivability is likely to be lower for stingarees trawled from deep water). In the case of the three outer shelf and upper slope species treated here, their relatively restricted depth range and geographic distributions (particularly the Yellowback Stingaree), much of which is commercially fished, adds to their vulnerability. Both the Sandyback and Greenback Stingarees receive some refuge in their range off western Tas. and in Bass Strait where there is reduced trawling pressure.

CURRENT RECOVERY EFFORTS

Recovery plan: None

Conservation actions: Deakin University and the Victorian Marine and Freshwater Fisheries Research Institute are conducting research into the biological productivity of these species for an ecological risk assessment to understand the impacts of fishing.

FUTURE RECOVERY EFFORTS

What you can do:

- Report any fishing catches to the Qld Museum.
- Commercial fishers should return any captured stingarees safely to the water.
- Obtain the Australian Marine Conservation Society's *Australia's Sustainable Seafood Guide* and buy only seafood that is harvested from sustainable sources with limited bycatch (see <http://www.amcs.org.au>).

Conservation needs:

- Reduce fishing effort and bycatch levels to ensure the recovery of these species; this is dependent on reducing interactions with fisheries in southern states.
- Protect available habitat in Qld, where stingarees may have refuge (particularly in deeper waters).

Research directions:

- Quantify and monitor any catches in the Qld ECTF.
- Better determine the occurrence of the three species in Qld waters.

Sources and further reading

Graham KJ, Andrew NL and Hodgson KE (2001) Changes in relative abundance of sharks and rays on Australian South East Fishery trawl grounds after twenty years of fishing. *Marine and Freshwater Research* **52**, 549–561.

Kyne PM, Johnson JW, Courtney AJ and Bennett MB (2005) New biogeographical information on Queensland chondrichthyans. *Memoirs of the Queensland Museum* **50**, 321–327.

Last PR and Compagno LJV (1999) Urolophidae. Stingarees. In *FAO Species Identification Guide for Fishery Purposes. The Living Marine Resources of the Western Central Pacific. Volume 3: Batoid Fishes, Chimaeras and Bony Fishes Part 1 (Elopidae to Linophrynidae).* (Eds KE Carpenter and VH Niem) pp. 1469–1476. FAO, Rome.

Last PR and Marshall LJ (2006) *Urolophus bucculentus.* In IUCN 2010. *IUCN Red List of Threatened Species.* Version 2010.4. IUCN, Gland, Switzerland and Cambridge, UK, <http://www.iucnredlist.org>.

Last PR and Marshall LJ (2006) *Urolophus sufflavus.* In IUCN 2010. *IUCN Red List of Threatened Species.* Version 2010.4. IUCN, Gland, Switzerland and Cambridge, UK, <http://www.iucnredlist.org>.

Last PR and Marshall LJ (2006) *Urolophus viridis.* In IUCN 2010. *IUCN Red List of Threatened Species.* Version 2010.4. IUCN, Gland, Switzerland and Cambridge, UK, <http://www.iucnredlist.org>.

Last PR and Stevens JD (2009) *Sharks and Rays of Australia.* 2nd edn. CSIRO Publishing, Melbourne.

Trinnie FI (2010) The reproductive biology of five species of stingarees from the genera *Urolophus* and *Trygonoptera* found within south-eastern Australia. PhD thesis. Deakin University, Australia.

Author: Peter M Kyne

ESTUARY STINGRAY

Dasyatis fluviorum Ogilby, 1908

Photo: ScienceImage

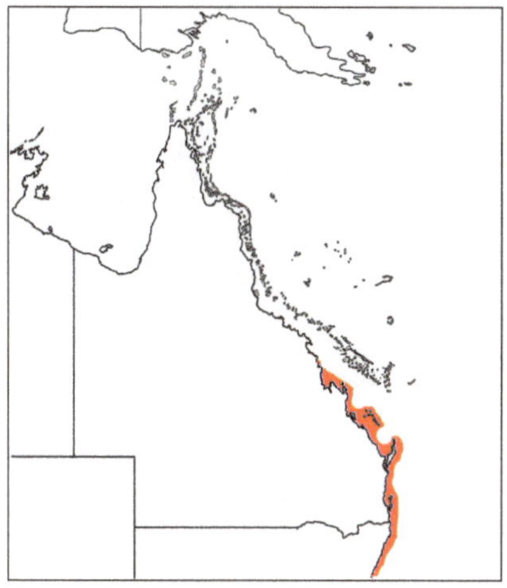

Family: Dasyatidae

Other common names: Brown Stingray

Aboriginal name: Banko (Jandai language; North Stradbroke Island (Minjerribah), south-east Qld)

Conservation status: IUCN: Vulnerable

Description: The Estuary Stingray is a medium-sized stingray, which reaches a maximum disc width of about 100 cm. Colouration is dark brown to light tan on the dorsal surface with a cream ventral surface. There is a row of enlarged thorns along the midline of the dorsal surface extending onto the tail. The tail is black or dark past the stinging spine.

Distribution: The Estuary Stingray is endemic to eastern Australia, from Repulse Bay in central Qld to Wagonga Inlet, NSW. Being somewhat habitat specific, their occurrence may not be continuous across this range and it is unclear how much interchange occurs between sites. Although the species had been reported from the NT and New Guinea, these appear to have been misidentifications of a similar, undescribed stingray that is restricted to tropical waters.

Habitat: The Estuary Stingray is found predominantly in rivers, estuaries and sheltered bays, particularly over mangrove-fringed intertidal flats. The maximum depth record is 28 m and the species can penetrate rivers to pure freshwater, though it is thought to prefer salinities of 14–35.

Ecology: Disc width at birth is 11 cm, with maturity reached at around 41 cm disc width in males and 63 cm disc width in females, corresponding to estimated ages of around 7 years and 13 years, respectively. Reproduction is viviparous. This ray is the most common large (>5 kg) predator found over intertidal sand flats in Moreton Bay and may be important to ecosystem function. Populations of Estuary Stingrays are noticeably segregated within Moreton Bay, with small juveniles absent from areas where larger individuals are regularly caught. The species has a slow growth rate, late age of maturity and a relatively low rate of population increase under natural conditions.

Reasons for decline and current threats: The Estuary Stingray is endemic to the highly urbanised east coast, where a suite of anthropogenic threats affect its favoured habitats. Anecdotal and observational reports suggest that populations of the species have been depleted in areas where they were historically abundant.

A variety of activities and processes, including commercial and recreational fisheries, habitat modification and possibly pollution, may have contributed to reports of declines of the species. There remains some uncertainty regarding the extent of declines, which requires a future research focus. Most known Estuary Stingray habitats are currently affected by urban expansion and development. Aside from direct impacts on the stingrays, these processes can also cause long-term declines in the abundance and availability of prey species. Limited survey data from south-east Qld suggests that Estuary Stingrays do not use canal estate habitat, even when the species is present in adjacent unmodified areas. However, the species has also persisted in some highly modified environments (such as Sydney Harbour) over the medium to long term.

Fishing, including recreational catches, and bycatch in the Qld trawl (i.e. the East Coast Trawl Fishery; ECTF) and inshore net (i.e. the East Coast Inshore Fin Fish Fishery; ECIFFF) fisheries, also affect this species. A large survey of Estuary Stingrays in Moreton Bay found ~11% had either amputated tails (tails are often severed by fishers to remove the venomous stinging spine) or internally retained hooks as evidence of past interactions with fisheries. Deep-hooking can result in significant pathology, including peritonitis and hepatitis that may result in elevated post-release mortality.

The impacts of extrinsic threats are likely to be exacerbated by the Estuary Stingray's low biological productivity (it has life history characteristics that are often associated with a low ability to recover from population decline). Modelling studies have demonstrated that even low levels of additive human-induced mortality can result in long-term decline in the species.

CURRENT RECOVERY EFFORTS

Recovery plan: None

Other plans: Pogonoski JJ, Pollard DA and Paxton JR (2002) 'Conservation overview and action plan for Australian threatened and potentially threatened marine and estuarine fishes'. Environment Australia, Canberra.

Conservation actions:

- Recreational take and possession limit of one shark or ray in Qld.
- Qld DEEDI is conducting research on recreational fishing catches of sharks and rays, including Estuary Stingrays.
- INFOFISH and Qld DEEDI are conducting education campaigns with anglers to emphasise best-practice handling and release techniques for accidental catches of the species.
- UQ is conducting research on the biology, population status and threatening processes affecting the species (including the development of basic population models).

FUTURE RECOVERY EFFORTS

What you can do:

- Use barbless, non-stainless steel hooks and try to avoid fishing in areas where Estuary Stingrays are known to be present.
- Release stingrays carefully if they are accidentally caught (unhook the fish if possible and be careful of the sting, but do not cut off the tail).

Conservation needs:

- Protect known and potential key habitats from further modification or significant disturbance. Restore modified habitats where feasible by allowing mangroves to recolonise and anthropogenic inputs to be minimised.
- Restrict fishing in critical habitats to minimise mortality.

Research directions:

- Conduct field surveys to clarify the current distribution and population size of the species.
- Conduct tagging/tracking studies to characterise preferred habitat and to examine interchange between potential subpopulations.
- Undertake quantitative analyses on survivorship and population trajectories.
- Monitor and quantify catches in the ECTF and ECIFFF.

Sources and further reading

Kyne PM, Pollard DA and Bennett MB (2003) *Dasyatis fluviorum*. In IUCN 2010. *IUCN Red List of Threatened Species*. Version 2010.4. IUCN, Gland, Switzerland and Cambridge, UK, <http://www.iucnredlist.org>.

Pierce SJ (2009) Biology, demography and conservation of rays in Moreton Bay, Queensland, Australia. PhD thesis. The University of Queensland, Australia.

Pierce SJ and Bennett MB (2010) Destined to decline? Intrinsic susceptibility of the threatened estuary stingray to anthropogenic impacts. *Marine and Freshwater Research* **61**, 1468–1481.

Pierce SJ and Bennett MB (2010) Distribution of the estuary stingray (*Dasyatis fluviorum*) in Australia. *Memoirs of the Queensland Museum - Nature* **55**, 9–17.

Authors: Simon J Pierce and Michael B Bennett

RETICULATE WHIPRAY

Himantura uarnak (Forsskål, 1775)

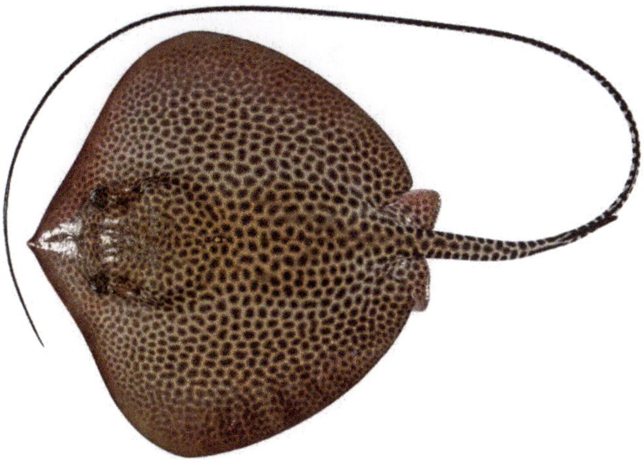

Photo: ScienceImage

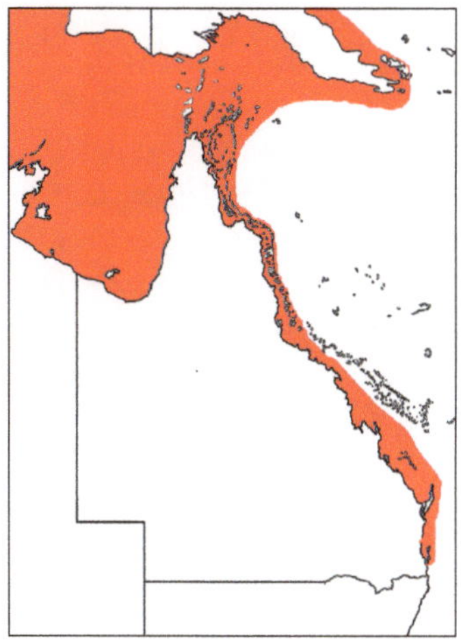

Family: Dasyatidae

Other common names: Honeycomb Ray, Coachwhip Ray, Longtail Ray

Conservation status: IUCN: Vulnerable

Description: The Reticulate Whipray is a large stingray, which reaches a maximum disc width of about 160 cm. Adults have a series of fine, dark reticulations over a yellowish brown skin on the dorsal surface. Juveniles exhibit a different colouration, displaying an extremely dense pattern of small dark brown spots. The Reticulate Whipray shows some resemblance to other whipray species, in particular the recently described Leopard Whipray *Himantura leoparda*. As a consequence, the species has been misidentified on occasion.

Distribution: The Reticulate Whipray has a widespread distribution encompassing the Indo-Pacific region, from along the eastern African coastline to the Red Sea, the eastern Mediterranean and east to South-East Asia, New Guinea and northern Australia. In Australia, it is found in northern coastal and continental shelf regions from Shark Bay, WA to Moreton Bay, south-east Qld.

Habitat: The Reticulate Whipray inhabits intertidal and shallow subtidal waters, including estuaries, coastal bays, sand flats, mangroves and seagrass beds. It occurs in subtropical and tropical waters to 50 m depth, although it is commonly observed in waters less than 25 m. Although there is limited data on Qld populations, a recent faunal survey of Hervey Bay reported this species from a range of habitats including mangroves, seagrass beds and intertidal/subtidal sand flats in 2–25 m of water.

Ecology: The biology and ecology of the Reticulate Whipray remains poorly documented and poorly understood. Previous reports suggest males attain maturity at 82–84 cm disc width, although the onset of maturity may occur at a smaller size. Like other whiprays, the Reticulate Whipray is viviparous with low fecundity. Although females are known to have up to four young, female size at maturity, size at birth and gestation periods are not adequately defined for this species. In areas of low disturbance (e.g. sand flats), this species is thought to be an important benthic predator in near-shore environments, due to its residency throughout the year.

Reasons for decline and current threats: The Reticulate Whipray is listed globally as Vulnerable on the IUCN Red List as a result of intense and increasing fishing pressures outside of Australia, particularly in South-East Asia.

The low reproductive output of the species, its large size and its occurrence in easily accessible fishing areas makes it particularly susceptible to inshore fishing pressures, including demersal trawl, gill net, longline and seine net fishing. When targeted by commercial fisheries, it is retained primarily for its meat, skin and cartilage. On a broader scale, its prevalence in shallow coastal regions, such as mangroves and seagrass beds, makes the species susceptible to distribution retraction due to habitat loss and degradation.

When compared with other areas of its range, fishing pressures on the Reticulate Whipray in Australia are considered to be low. It is caught as bycatch in prawn trawl fisheries of northern and eastern Australia, although the number and frequency of these interactions has probably reduced with the introduction of turtle exclusion devices (TEDs). Qld net fisheries targeting shark species (i.e. the East Coast Inshore Fin Fish Fishery; ECIFFF) also retain various ray species on occasion. Although specific data regarding interactions with the Reticulate Whipray are not available, this species may form a component of retained stingray catches.

CURRENT RECOVERY EFFORTS

Recovery plan: None

Conservation actions: Recreational take and possession limit of one shark or ray in Qld.

FUTURE RECOVERY EFFORTS

What you can do:

- Report any live sightings, fishing catches or dead beach-washed specimens to the Qld Museum.
- Both commercial and recreational fishers should release any captured whiprays safely.

Conservation needs:

- Encourage sustainable fisheries management and widespread habitat protection throughout the range of the species. Fisheries that target and retain the Reticulate Whipray should be monitored and regulated to ensure the ongoing exploitation of this species is limited.
- Afford national protection of key habitats, and restrict fishing in these to ensure the ongoing health of the Australian population.

Research directions:

- Monitor and quantify interactions with the ECIFFF and identify the fate of captured individuals.
- Encourage and continue faunal assessments within known and unknown habitats in Australian and Qld waters.
- Undertake genetic examination to identify the population structure.

Sources and further reading

Last PR and Stevens JD (2009) *Sharks and Rays of Australia*. 2nd edn. CSIRO Publishing, Melbourne.

Manjaji BM and White WT (2004) *Himantura uarnak*. In IUCN 2010. *IUCN Red List of Threatened Species*. Version 2010.4. IUCN, Gland, Switzerland and Cambridge, UK, <http://www.iucnredlist.org>.

Manjaji-Matsumoto BM and Last PR (2008) *Himantura leoparda* sp. nov., a new whipray (Myliobatoidei: Dasyatidae) from the Indo-Pacific. In *Descriptions of New Australian Chondrichthyans*. (Eds PR Last, WT White and JJ Pogonoski) pp. 293–301. CSIRO Marine and Atmospheric Research Paper No. 022. CSIRO Marine and Atmospheric Research, Hobart.

Vaudo JJ and Heithaus MR (2009) Spatiotemporal variability in a sandflat elasmobranch fauna in Shark Bay, Australia. *Marine Biology* **156**, 2579–2590.

Author: Adrian N Gutteridge

PORCUPINE RAY

Urogymnus asperrimus Bloch & Schneider, 1801

Photo: Susan Theiss

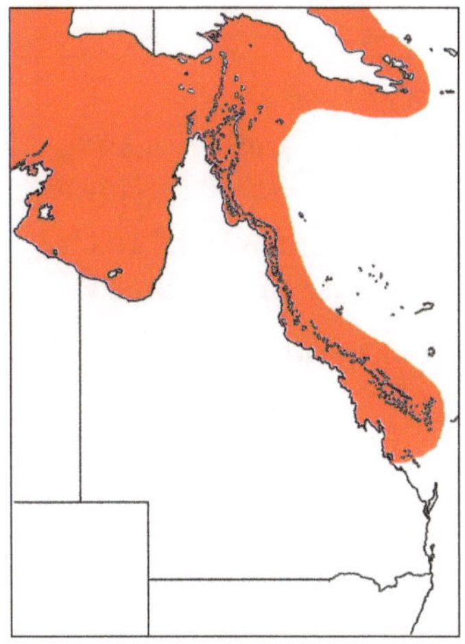

Family: Dasyatidae

Other common names: Solander's Ray, Roughskin Stingaree, Thorny Ray

Aboriginal name: Daringyan ('stingray')

Conservation status: IUCN: Vulnerable

Description: The Porcupine Ray is a large, robust ray with an oval shaped disc, which reaches a maximum disc width of at least 115 cm (possibly to 147 cm). It is dark brown to grey on the dorsal surface, and lighter on the ventral surface. The disc is covered with flat, plate-like denticles and sharp thorns on the dorsal side. The short tail lacks stinging spines and skin folds. If caught, individuals should be handled carefully because the thorns are sharp and can inflict injury.

Distribution: Porcupine Rays are widespread throughout the Indo-West Pacific from the east coast of Africa to the Philippines, and south to Fiji and the northern coast of Australia. In Qld, they occur from Heron Island on the southern Great Barrier Reef to the Gulf of Carpentaria. Although widespread, they appear to be rare compared with other stingray species, and informative records are lacking.

Habitat: Porcupine Rays are found inshore on sand near coral reefs or among coral rubble, mainly in shallow water, though there is some indication that larger individuals may be found at greater depths.

Ecology: Males mature at ~90 cm disc width while females mature at ~100 cm disc width. The Porcupine Ray is viviparous, but reproductive traits such as fecundity and gestation period are unknown. Population size and structure in Qld is unknown, although individuals from Heron Island appeared to be solitary. Healthy populations have recently been observed in Ningaloo Marine Park, WA, where these rays have been found to aggregate. Their diet is presumed to consist primarily of bivalves and crustaceans. Contrary to the feeding strategies of other stingray species, the Porcupine Ray has been observed to plough through the substrate in search of food, expelling large plumes of sand from its spiracles.

Reasons for decline and current threats: Overfishing is the main threat to the Porcupine Ray (primarily in its range outside of Australia).

Although this species is difficult to handle, the meat, cartilage and skin fetch a high market price. Porcupine Rays are commonly caught in several South-East Asian countries, where intense inshore demersal fisheries operate (trawls, gill nets, trammel nets), and fishing pressure is high on the inshore coastal environment. Over the last few decades, the Porcupine Ray appears to have become very rare, if not disappeared altogether, from catches landed in Bangkok (from the Gulf of Thailand) and the Bay of Bengal.

The decrease in global abundance and lack of regular records has led to the Porcupine Ray being listed as Vulnerable on the IUCN Red List. Given the rarity of this species in Australia, it is unlikely to be regularly caught and is most likely not retained if taken as bycatch. There is little information on interactions with commercial fisheries operating in northern Australian waters, though the Porcupine Ray is susceptible to some trawl, gill net and possibly line fisheries. Australian trawl fisheries are likely to have less impact than those from South-East Asia due to the use of turtle exclusion devices (TEDs) in trawl nets.

To a lesser extent, human impacts such as coastal environmental modification and degradation may be having a negative impact on the habitat of the Porcupine Ray, particularly in highly populated and/or polluted areas.

CURRENT RECOVERY EFFORTS

Recovery plan: None

Other plans: Pogonoski JJ, Pollard DA and Paxton JR (2002) 'Conservation overview and action plan for Australian threatened and potentially threatened marine and estuarine fishes'. Environment Australia, Canberra.

Conservation actions: Although this species is rarely encountered by Qld recreational fishers, there is a recreational take and possession limit of one shark or ray in Qld.

FUTURE RECOVERY EFFORTS

What you can do:

- Report any live sightings, fishing catches or dead beach-washed specimens to the Qld Museum.
- Both commercial and recreational fishers should release any captured Porcupine Rays safely.

Conservation needs:

- Protect populations in Qld and Australian waters (Australia must act as a secure refuge for the species as its prognosis elsewhere is bleak).
- Encourage sustainable fisheries management throughout the range of the species. Fisheries taking the Porcupine Ray in South-East Asia are generally unmanaged; monitoring and regulating these fisheries would be an important step towards improving the conservation status of this and other elasmobranchs.

Research directions:

- Better determine its occurrence and distribution in Australian waters.
- Determine its population size in Australian and Qld waters.
- Determine its life history characteristics.

Sources and further reading

Compagno LJV (2005) *Urogymnus asperrimus*. In IUCN 2010. *IUCN Red List of Threatened Species*. Version 2010.4. IUCN, Gland, Switzerland and Cambridge, UK. <http://www.iucnredlist.org>.

Last PR and Stevens JD (2009) *Sharks and Rays of Australia*. 2nd edn. CSIRO Publishing, Melbourne.

Theiss SM, Kyne PM and Chisholm LA (2010) Distribution of the porcupine ray *Urogymnus asperrimus* (Bloch & Schneider, 1801) in Australian waters, with new records from Queensland. *Memoirs of the Queensland Museum - Nature* **55**, 101–105.

White WT, Last PR, Stevens JD, Yearsley GK, Fahmi and Dharmadi (2006) *Economically Important Sharks and Rays of Indonesia*. Australian Centre for International Agricultural Research, Canberra.

Author: Susan M Theiss

BANDED EAGLE RAY

Aetomylaeus nichofii Bloch & Schneider, 1801

Photo: ScienceImage

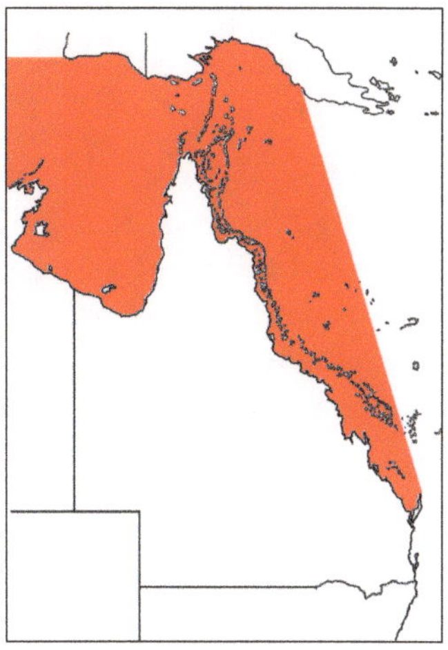

Family: Myliobatidae

Conservation status: IUCN: Vulnerable

Description: The Banded Eagle Ray is a small eagle ray, which reaches a maximum disc width of 72 cm. It has a very broad disc, which is greyish brown above in adults; juveniles possess about five to eight bluish bands on a paler disc. The tail is long when undamaged (1.5–2 times the width of the disc), with no stinging spine.

Distribution: The Banded Eagle Ray has a widespread distribution in the Indo-West Pacific from at least India (possibly further west) to South-East Asia, north to Japan and south to northern Australia (WA from Doubtful Bay, through the NT to the east coast of Qld). In Qld waters, it is reported from the Gulf of Carpentaria, Torres Strait and down the east coast to Hervey Bay.

Habitat: This species occurs in both inshore (including the intertidal zone) and offshore waters on the continental shelf to a depth of 115 m. Specimens held in museums that were collected from Qld waters were caught in depths of 8–43 m. No other specific information is available in relation to habitat in Qld, but the species probably occurs on sandy substrates, and is possibly associated with coral reefs.

Ecology: The ecology of the Banded Eagle Ray is poorly known and there is little specific information on its reproductive biology. Males mature at 39–42 cm disc width, but size at maturity is not known for females. Size at birth is ~17 cm disc width. Eagle rays are viviparous and reproductive output is low in this species, with litter sizes of up to four young.

Reasons for decline and current threats: The global Vulnerable listing on the IUCN Red List is a result of intense and increasing fishing pressure throughout a large part of the known distribution of the species outside of Australia, in particular South-East Asia, where the exploitation of batoids is the highest in the world.

Intense inshore demersal fisheries (trawls, gill nets, trammel nets) operate in South-East Asia, and fishing pressure is high on the inshore coastal environment. Eagle rays are landed and marketed for their meat and cartilage, often in considerable numbers; the Banded Eagle Ray is one of the main commercial species. This has resulted in observed and suspected declines of some elasmobranch species. For example, eagle rays are now rare in the Gulf of Thailand, despite being once common there; the Banded Eagle Ray was previously more common in Thai fish markets. Combined with this heavy fishing pressure, the limiting life history parameters of eagle rays, such as low reproductive output, make them a highly vulnerable group.

Fishing pressure on this species in Australian waters is not comparable with that elsewhere; it is not targeted and probably not retained when taken as bycatch in Australia. However, there is little information on interactions with commercial fisheries operating in northern Australian waters. This appears to be an uncommon species in Australian waters, and it is unlikely to be caught with any regularity here. It would be susceptible to some trawl (less so now with the introduction of turtle exclusion devices (TEDs) to trawl nets, although smaller individuals could still be caught) and gill net fisheries. The species has been reported as bycatch in the Northern Prawn Fishery, which operates partially off Qld in the Gulf of Carpentaria, but only in ~1% of trawls. In a broad-scale survey of elasmobranch bycatch in the Qld East Coast Trawl Fishery, only a single individual was recorded. There is no other information available on any interactions with Qld commercial fisheries.

CURRENT RECOVERY EFFORTS

Recovery plan: None

Conservation actions: Although this species is rarely encountered by Qld recreational fishers, there is a recreational take and possession limit of one shark or ray in Qld.

FUTURE RECOVERY EFFORTS

What you can do:

- Report any live sightings, fishing catches or dead beach-washed specimens to the Qld Museum.

- Both commercial and recreational fishers should release any captured eagle rays safely.

Conservation needs:

- Afford the species adequate protection at both the federal and state level (given its rarity and global threatened status it would benefit from protection in Qld waters).
- Encourage sustainable fisheries management throughout the range of the species. Fisheries taking the Banded Eagle Ray in South-East Asia are generally unmanaged; monitoring and regulating these fisheries would be an important step towards improving the conservation status of this and other elasmobranchs.

Research directions: Quantify any catches in Qld inshore gill net fisheries.

Sources and further reading

Kyne PM, Compagno LJV and Bennett MB (2003) *Aetomylaeus nichofii*. In IUCN 2010. *IUCN Red List of Threatened Species*. Version 2010.4. IUCN, Gland, Switzerland and Cambridge, UK, <http://www.iucnredlist.org>.

Kyne PM, Johnson JW, Courtney AJ and Bennett MB (2005) New biogeographical information on Queensland chondrichthyans. *Memoirs of the Queensland Museum* **50**, 321–327.

Last PR and Stevens JD (2009) *Sharks and Rays of Australia*. 2nd edn. CSIRO Publishing, Melbourne.

Stobutzki IC, Miller MJ, Heales DS and Brewer DT (2002) Sustainability of elasmobranchs caught as bycatch in a tropical prawn (shrimp) trawl fishery. *Fishery Bulletin* **100**, 800–821.

White WT, Last PR, Stevens JD, Yearsley GK, Fahmi and Dharmadi (2006) *Economically Important Sharks and Rays of Indonesia*. Australian Centre for International Agricultural Research, Canberra.

Author: Peter M Kyne

ORNATE EAGLE RAY

Aetomylaeus vespertilio Bleeker, 1852

Photo: ScienceImage

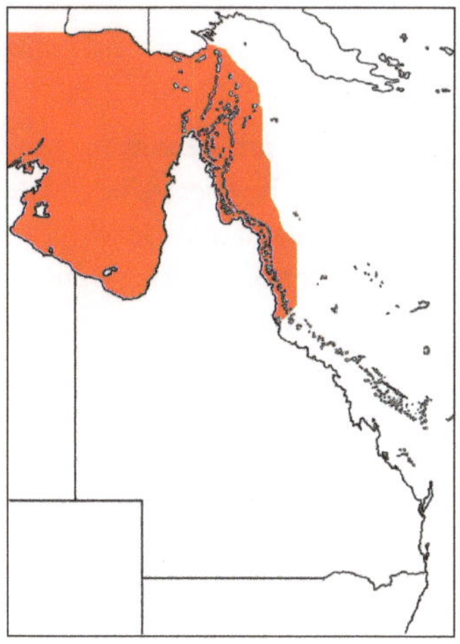

Family: Myliobatidae

Conservation status: IUCN: Endangered

Description: The Ornate Eagle Ray is a large eagle ray, which reaches a maximum disc width of at least 300 cm (possibly to 350 cm). It has a very broad disc with a networked pattern of dark lines on the dorsal surface. The tail is extremely long when undamaged (nearly three times the width of the disc), with no stinging spine.

Distribution: The global distribution of this uncommon species is poorly known. It has a widespread but patchy occurrence in the Indo-West Pacific from east Africa, the Red Sea, Maldives, South-East and East Asia, and northern Australia (WA from Shark Bay, through the NT to the east coast of Qld). In Qld waters, it is reported from the Gulf of Carpentaria, Torres Strait and down the east coast to Cardwell. However, this is a rare and poorly known species in Qld (and indeed Australia), with few records and no detailed information on its distribution or occurrence in these waters.

Habitat: Occurs in both inshore (including muddy bays and around coral reefs) and offshore waters on the continental shelf to a depth of 110 m. No specific information is available in relation to habitat in Qld.

Ecology: Very poorly known: among the most poorly known of the eagle rays. Males reach maturity at about 170 cm disc width, but size at maturity is not known for females. Details of the reproductive biology of the species are unknown. Eagle rays are viviparous and have limiting life history characteristics. Fecundity is expected to be low in the Ornate Eagle Ray as is the case with other species, such as the Spotted Eagle Ray *Aetobatus narinari* and the Banded Eagle Ray *Aetomylaeus nichofii*, which have litter sizes of up to four young. Given its large size, the Ornate Eagle Ray is likely to be a moderately long-lived species.

Reasons for decline and current threats: The global Endangered listing on the IUCN Red List is a result of intense and increasing fishing pressure throughout a large part of the known distribution of the species outside of Australia, in particular South-East Asia, where the exploitation of batoids is the highest in the world.

Intense inshore demersal fisheries (trawls, gill nets, trammel nets) operate in South-East Asia, and fishing pressure is high on the inshore coastal environment; eagle rays are landed and marketed for their meat and cartilage, often in considerable numbers. This has resulted in observed and suspected declines of some elasmobranch species. For example, eagle rays are now rare in the Gulf of Thailand, despite once being common there. Combined with this heavy fishing pressure, the limiting life history parameters of eagle rays, such as low reproductive output, make them a highly vulnerable group. In the case of the Ornate Eagle Ray, its apparent natural rarity (nowhere has it been observed in any considerable numbers) makes it even more susceptible to population depletion.

Fishing pressure on this species in Australian waters is not comparable to that elsewhere; it is not targeted and probably not retained when taken as bycatch in Australia. However, there is little information on interactions with commercial fisheries operating in northern Australian waters. Given that it is very rarely observed in Australia, it is unlikely to be caught with any regularity, although it would be susceptible to some trawl (less so now with the introduction of turtle exclusion devices (TEDs) to trawl nets) and gill net fisheries. The species has been reported as bycatch in the Northern Prawn Fishery, which operates partially off Qld in the Gulf of Carpentaria, but was not recorded in surveys of elasmobranch bycatch in the Qld East Coast Trawl Fishery. There is no other information available on any interactions with Qld commercial fisheries.

CURRENT RECOVERY EFFORTS

Recovery plan: None

Conservation actions: Although this species is rarely encountered by Qld recreational fishers, there is a recreational take and possession limit of one shark or ray in Qld.

FUTURE RECOVERY EFFORTS

What you can do:

- Report any live sightings, fishing catches or dead beach-washed specimens to the Qld Museum.

- Both commercial and recreational fishers should release any captured eagle rays safely.

Conservation needs:

- Afford the species adequate protection at both the federal and state level (given its rarity and global threatened status it would benefit from protection in Qld waters).
- Encourage sustainable fisheries management throughout the range of the species. Fisheries taking the Ornate Eagle Ray in South-East Asia are generally unmanaged; monitoring and regulating these fisheries would be an important step towards improving the conservation status of this and other elasmobranchs.

Sources and further reading

Compagno LJV and Last PR (1999) Myliobatidae. Eagle rays. In *FAO Species Identification Guide for Fishery Purposes. The Living Marine Resources of the Western Central Pacific. Volume 3. Batoid Fishes, Chimaeras and Bony Fishes Part 1 (Elopidae to Linophrynidae).* (Eds KE Carpenter and VH Niem) pp. 1511–1519. FAO, Rome.

Last PR and Stevens JD (2009) *Sharks and Rays of Australia.* 2nd edn. CSIRO Publishing, Melbourne.

Stobutzki IC, Miller MJ, Heales DS and Brewer DT (2002) Sustainability of elasmobranchs caught as bycatch in a tropical prawn (shrimp) trawl fishery. *Fishery Bulletin* **100**, 800–821.

White WT (2006) *Aetomylaeus vespertilio.* In IUCN 2010. *IUCN Red List of Threatened Species.* Version 2010.4. IUCN, Gland, Switzerland and Cambridge, UK, <http://www.iucnredlist.org>.

White WT and Dharmadi (2007) Species and size compositions and reproductive biology of rays (Chondrichthyes, Batoidea) caught in target and non-target fisheries in eastern Indonesia. *Journal of Fish Biology* **70**, 1809–1837.

Author: Peter M Kyne

PURPLE EAGLE RAY

Myliobatis hamlyni Ogilby, 1911

Photo: Jeff Johnson-Queensland Museum

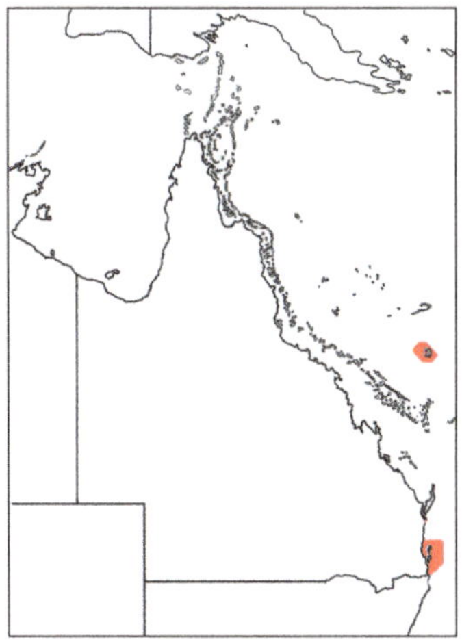

Family: Myliobatidae

Conservation status: IUCN: Endangered

Description: The Purple Eagle Ray is a small eagle ray, which reaches a maximum disc width of at least 59 cm. It has a very broad disc, which is plain on the dorsal surface (purplish to greenish grey centrally, more olive brown laterally). When undamaged, the tail is long (about 1.5 times the width of the disc) with a prominent stinging spine.

Distribution: The Purple Eagle Ray is endemic to Australian waters, with patchy records from both the east and west coasts. This species has been recorded in Qld from only two individuals, one from off Cape Moreton and one from off Swain Reefs. More records are known from NSW and a very small number are known from WA.

Habitat: The Purple Eagle Ray occurs on the outer continental shelf and the upper slope in a depth range of 117–350 m. It is benthopelagic (occurring near or just above the ocean floor). The known area of occurrence of the species appears to be very small, because the outer continental shelf and upper slope where it has been recorded represents a relatively narrow band of available habitat.

Ecology: This species differs from other eagle rays, which are aggregating species; the Purple Eagle Ray appears to be individually dispersed. There are no specific details relating to the biology of this species. Eagle rays are viviparous and have limiting life history characteristics; fecundity is expected to be low in the Purple Eagle Ray as is the case with other species, such as the Spotted Eagle Ray *Aetobatus narinari* and the Banded Eagle Ray *Aetomylaeus nichofii*, which have litter sizes of up to four young.

Reasons for decline and current threats: The IUCN Red List assessment for the Purple Eagle Ray states: '... it appears to be an extremely rare species and as a precautionary measure (even though comprehensive population data do not exist) the species is assessed as Endangered given its inferred very low population size, fragmented distribution, limited available habitat (habitat at depth range is not broad) and ongoing fishing pressure where [most] specimens were recorded.'

Across a large part of the outer shelf and upper slope of NSW, where the vast majority of known records of this species are from, consistent commercial fishing has resulted in the collapse of continental slope shark and ray populations. It is possible that the Purple Eagle Ray has been depleted there. Off southern Qld (where one of the two known Qld specimens was collected), the outer shelf is fished seasonally by the Qld East Coast Trawl Fishery, but there may be some refuge at depth for the species, because this fishery does not extend to the depths of trawl fisheries in NSW. The Swain Reefs specimen was caught as champagne lobster trawl bycatch, but this area is subject to only low levels of fishing effort. In the northern part of the range of the species, areas of the outer continental shelf and upper slope off Qld may represent important habitat for the species, because fishing pressure there has been historically low.

Only 17 records of this species exist, despite decades of survey work and consistent fishing activities across its range, and thus its population size is suspected to be very low. The threatened species listing by the IUCN is based partially on its suspected very low population size (estimated to be <2500 mature individuals). This is a rare species, which likely has life history characteristics restricting its ability to rebound after population depletion. Even though it is very poorly known, a lack of knowledge should not be used as a case to delay conservation measures.

CURRENT RECOVERY EFFORTS

Recovery plan: None

FUTURE RECOVERY EFFORTS

What you can do:

- Report any fishing catches or dead beach-washed specimens to the Qld Museum.
- Commercial fishers should release any captured eagle rays safely.
- Obtain the Australian Marine Conservation Society's *Australia's Sustainable Seafood Guide* and buy only

seafood harvested from sustainable sources with limited bycatch (see <http://www.amcs.org.au>).

Conservation needs:

- Afford the species adequate protection at both the federal and state level.
- Protect available habitat in Qld (where the species may have refuge at depth as fishing pressure has not been as high historically as that off NSW).
- Add the Purple Eagle Ray to commercial fishing logbooks in order to obtain further information on occurrence and interactions with commercial fisheries.
- Provide advice to fishers on the best methods of safe release for the species.

Research directions: Quantify any catches in Qld fisheries.

Sources and further reading

Compagno LJV and Last PR (1999) Myliobatidae. Eagle rays. In *FAO Species Identification Guide for Fishery Purposes. The Living Marine Resources of the Western Central Pacific. Volume 3: Batoid Fishes, Chimaeras and Bony Fishes Part 1 (Elopidae to Linophrynidae).* (Eds KE Carpenter and VH Niem) pp. 1511–1519. FAO, Rome.

Graham KJ, Andrew NL and Hodgson KE (2001) Changes in relative abundance of sharks and rays on Australian South East Fishery trawl grounds after twenty years of fishing. *Marine and Freshwater Research* **52**, 549–561.

Kyne PM and Last PR (2006) *Myliobatis hamlyni*. In IUCN 2010. *IUCN Red List of Threatened Species*. Version 2010.4. IUCN, Gland, Switzerland and Cambridge, UK, <http://www.iucnredlist.org>.

Kyne PM, Johnson JW, Courtney AJ and Bennett MB (2005) New biogeographical information on Queensland chondrichthyans. *Memoirs of the Queensland Museum* **50**, 321–327.

Last PR and Stevens JD (2009) *Sharks and Rays of Australia*. 2nd edn. CSIRO Publishing, Melbourne.

Author: Peter M Kyne

JAVANESE COWNOSE RAY

Rhinoptera javanica Müller & Henle, 1841

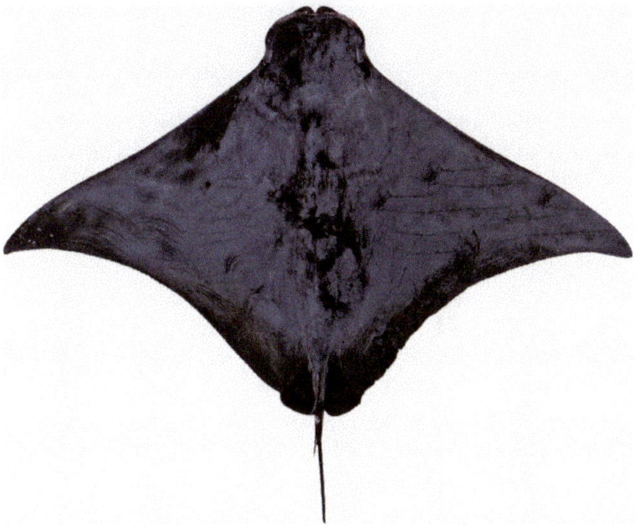

Photo: ScienceImage

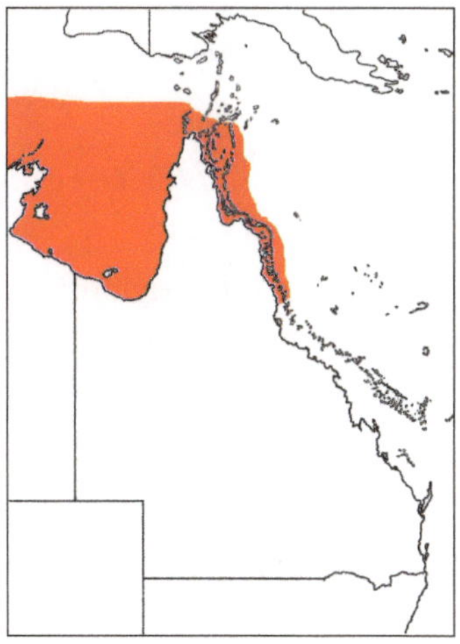

Family: Rhinopteridae

Conservation status: IUCN: Vulnerable

Description: The Javanese Cownose Ray is a large cownose ray, which reaches a maximum disc width of about 150 cm. This family of rays has a distinctive bilobed snout that easily separates them from the eagle rays (family Myliobatidae). This species is brown on the dorsal surface and has a relatively long whip-like tail with a small stinging spine. It can be distinguished from the very similar, but smaller, Australian Cownose Ray *Rhinoptera neglecta* (which also occurs in Qld waters south of Cairns), by its dorsal fin origin being well behind the pelvic fin insertion and by thinner teeth in the middle row of the upper jaw.

Distribution: The global and Australian distribution of this species is not completely known due to the poorly resolved taxonomy and systematics of the group. The Javanese Cownose Ray is possibly widespread in the Indo-West Pacific, including northern Australia where it is currently known from only a few records from the NT to Innisfail in north Qld.

Habitat: Restricted to coastal inshore habitats of the continental shelf to a depth of at least 20 m, it occurs in shallow estuaries, coastal bays, mangroves and around coral reefs. No specific information is available in relation to its habitat in Qld.

Ecology: There is little specific information on the biology and ecology of the species. It is a schooling species, sometimes occurring in schools of several hundred individuals. Some cownose ray species are highly migratory (e.g. Cownose Ray *Rhinoptera bonasus* in the Atlantic Ocean), but details of movement patterns are not available for the Javanese Cownose Ray. Males mature at a size greater than 100 cm disc width. Cownose rays are viviparous and have limiting life history characteristics; a pregnant female observed in Indonesia carried only a single late-term embryo. Some cownose rays are reported to have litter sizes of up to six young, but it has been postulated that the Javanese Cownose Ray is not likely to have more than two young per litter given its apparent large size at birth (a late-term embryo has been observed at 61 cm disc width). The gestation period of this species is unknown, as is the reproductive periodicity; the gestation period of the closely-related *R. bonasus* has been reported as 11–12 months, with an annual reproductive cycle. Given the large size of the Javanese Cownose Ray, it is likely to be a moderately long-lived species (the smaller *R. bonasus* has been aged to 18 years).

Reasons for decline and current threats: The global Vulnerable listing on the IUCN Red List is a result of intense and increasing fishing pressure throughout a large part of the known distribution of the species outside of Australia, in particular South-East Asia, where the exploitation of batoids is the highest in the world.

The Javanese Cownose Ray is widely landed and used for its meat across its Asian range. Its schooling behaviour and inshore occurrence make it very susceptible to capture in a variety of demersal fisheries, including by trawling, gill netting and tangle netting. Throughout much of the region, these fisheries are unregulated. Combined with this heavy fishing pressure, the limiting life history parameters of cownose rays (i.e. low reproductive output), make them a highly vulnerable group.

Fishing pressure on this species in Australian waters is not comparable to that elsewhere; it is not targeted and probably not retained when taken as bycatch in Australia. However, there is little information on interactions with commercial fisheries operating in northern Australian waters, although the species would be susceptible to some trawl (less so now with the introduction of turtle exclusion devices (TEDs) to trawl nets), gill net and possibly line fisheries. Information on the species in Qld waters is scant and it is not known to what level it interacts with commercial fisheries in the state.

CURRENT RECOVERY EFFORTS

Recovery plan: None

Conservation actions: Although this species is rarely encountered by Qld recreational fishers, there is a recreational take and possession limit of one shark or ray in Qld.

FUTURE RECOVERY EFFORTS

What you can do:

- Report any live sightings, fishing catches or dead beach-washed specimens to the Qld Museum.
- Both commercial and recreational fishers should release any captured cownose rays safely.

Conservation needs:

- Encourage sustainable fisheries management throughout the range of the species. Fisheries taking the Javanese Cownose Ray in South-East Asia are generally unmanaged; monitoring and regulating these fisheries would be an important step towards improving the conservation status of this and other elasmobranchs.

Research directions:

- Quantify any catches in Qld inshore gill net fisheries.
- Better determine its actual occurrence and distribution in Australian waters.
- Determine its population size in Australian and Qld waters.
- Research movement patterns in northern Australia to identify if there is a separate Australian stock or if there is considerable exchange between Australian and South-East Asian waters.

Sources and further reading

Compagno LJV and Last PR (1999) Rhinopteridae. Cownose rays. In *FAO Species Identification Guide for Fishery Purposes. The Living Marine Resources of the Western Central Pacific. Volume 3: Batoid Fishes, Chimaeras and Bony Fishes Part 1 (Elopidae to Linophrynidae).* (Eds KE Carpenter and VH Niem) pp. 1520–1523. FAO, Rome.

Dudley SFJ, Kyne PM and White WT (2006) *Rhinoptera javanica.* In IUCN 2010. *IUCN Red List of Threatened Species.* Version 2010.4. IUCN, Gland, Switzerland and Cambridge, UK, <http://www.iucnredlist.org>.

Last PR and Stevens JD (2009) *Sharks and Rays of Australia.* 2nd edn. CSIRO Publishing, Melbourne.

Neer JA and Thompson BA (2005) Life history of the cownose ray, *Rhinoptera bonasus*, in the northern Gulf of Mexico, with comments on geographic variability in life history traits. *Environmental Biology of Fishes* **73**, 321–331.

White WT, Last PR, Stevens JD, Yearsley GK, Fahmi and Dharmadi (2006) *Economically Important Sharks and Rays of Indonesia.* Australian Centre for International Agricultural Research, Canberra.

Author: Peter M Kyne

Bony Fishes (Class: Osteichthyes)

INTRODUCTION TO THREATENED BONY FISHES

Peter M Kyne

Queensland's marine and fresh waters are home to over 2600 species of bony fishes. The vast majority of these are marine and estuarine, with Queensland's freshwater fishes represented by about 130 native taxa. Although marine diversity is high, the diversity of freshwater species is very low, when compared with other continents, largely a result of Australia's dry climate and inconsistent rainfall. Bony fishes are of high recreational, commercial and social importance in Queensland, but, surprisingly, as a group they have not received the focus needed to assess and understand their conservation status. This relates particularly to Queensland's marine fishes. A lack of knowledge of basic ecological information for many species, together with an incomplete understanding of threats facing individual species, results in a data poor situation. Considerable future work needs to be done to accurately assess the conservation status of the diverse assemblage of Queensland bony fishes.

Not a single marine bony fish occurring in Queensland is listed at the federal (EPBC) or state level (NCA). At the state level, however, some species are designated 'no take' under the *Fisheries Act 1994*, thus affording them a high level of protection within Queensland. However, relative to the diversity of species, very few species are afforded this protection. The IUCN Red List includes 10 threatened marine bony fishes that occur in Queensland waters. One of these, the Humphead Maori Wrasse *Cheilinus undulatus* is considered Endangered and is treated in this book. The other nine species are listed as Vulnerable: Bumphead Parrotfish *Bolbometopon muricatum*, Barramundi Cod *Cromileptes altivelis*, Queensland Grouper *Epinephelus lanceolatus*, Common Seahorse *Hippocampus kuda*, Hedgehog Seahorse *H. spinosissimus*, Low-crowned Seahorse *H. trimaculatus*, Squaretail Coral Grouper *Plectropomus areolatus*, Blacksaddled Coral Grouper *P. laevis* and Bigeye Tuna *Thunnus obesus* and profiles are not presented for these species. The reef fishes listed above (parrotfish, Barramundi Cod, Queensland Grouper and coral groupers) are all wide-ranging Indo-West Pacific species and their conservation status is largely a result of exploitation pressure elsewhere across their range. Generally, there is no evidence for declines in Queensland and, indeed, many are afforded some protection in the Great Barrier Reef Marine Park and some are 'no take' species. In the case of the seahorses, their threatened listings are also a result of targeted catch and trade elsewhere in their range.

In contrast to the state's marine fauna, a larger proportion of freshwater fishes have had their conservation status assessed. Profiles follow for 10 taxa, with all species listed on the EPBC and five on the NCA. The majority of these are also considered to be threatened by the IUCN. Additionally, two freshwater fishes for which profiles have not been provided are also considered Vulnerable by the IUCN: Silver Perch *Bidyanus bidyanus* and Cairns Rainbowfish *Cairnsichthys rhombosomoides*. The IUCN assessments for these and the other IUCN listed freshwater fish (undertaken in 1996) require updating and this should be a priority. Some threatened classifications may no longer be valid, such as the Lake Eacham Rainbowfish *Melanotaenia eachamensis*, which has a broader occurrence than originally thought and whose habitat is well protected in the Wet Tropics region. Similarly, many populations of the Cairns Rainbowfish are within national parks. Most of those species appearing in this book, however, do require ongoing conservation and management to ensure viable future populations. Furthermore, concern has been expressed for some other Queensland freshwater fishes, such as the restricted range Bloomfield River Cod *Guyu wujalwujalensis*, and a broader assessment of the conservation status of Queensland's freshwater and marine bony fishes is needed to update conservation priorities and recovery plans.

AUSTRALIAN LUNGFISHES (ORDER: CERATODONTIFORMES)

AUSTRALIAN LUNGFISH

Neoceratodus forsteri Krefft, 1870

Photo: Gunther Schmida

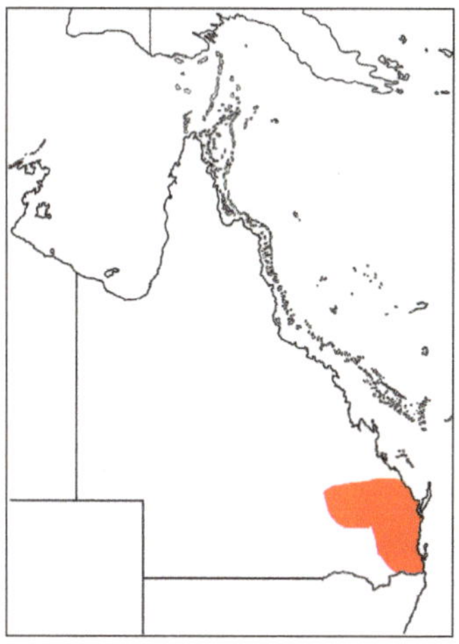

Family: Ceratodontidae

Other common name: Queensland Lungfish

Aboriginal name: Dala

Conservation status: Aust.: Vulnerable

Description: The Australian Lungfish is a long, heavy-bodied freshwater fish, which reaches a maximum total length of 150 cm (adult weight 6–40 kg). It has large, overlapping, heavy scales, a large tail, which incorporates the dorsal and anal fins, and flipper-like paired pectoral and pelvic fins, which are often used alternately, like limbs, to lift the body above the substrate. The small mouth is terminal. Many of these features resemble their now extinct dipnoan relatives from 300 mya and are the same as a species found in fossil deposits in Australia up to 150 mya, giving Australian Lungfish the status of 'living fossils'. Adult Australian Lungfish are glossy brown viewed from above with cream, yellow or pink to orange bellies.

Distribution: Australian Lungfish are restricted to freshwaters of south-east Qld. It currently occurs naturally only in the Burnett and Mary Rivers and their tributaries. Australian Lungfish have been translocated to several other river systems and persist in the Brisbane (including Lake Wivenhoe and Enoggera Reservoir) and North Pine Rivers (including Lake Samsonvale). There is little evidence that other translocations have persisted. Australian Lungfish occur up to 310 km inland from the mouth of the Burnett River, but they are much more common downstream.

Habitat: Australian Lungfish require still or slow flowing, well vegetated, 0.5–1 m deep pools with clean, oxygenated water in which to spawn. The species is restricted to areas of permanent water. It does not aestivate like other living lungfish. In the Mary River, it occupies sandy bottomed pools with overhanging riverside vegetation and woody debris, which is particularly attractive to lungfish during the day. Eggs, larval fish and juveniles are always found in dense cover such as beds of macrophytes in water ≤0.5 m deep. Juveniles probably remain in this habitat for many months, or even years, after hatching. Large juveniles can also be found in deeper pools with macrophytes not usually used for spawning.

Ecology: In the wild, this species spawns from August through to December and, very occasionally, January. Australian Lungfish congregate in particular locations to spawn for several nights, rest, and then spawn again. Bouts of spawning activity may be coordinated by a lead male for each group. Time between bouts of spawning are spent in deeper, shaded pools adjoining vegetated riffles to the side of the main river flow. Before spawning, Australian Lungfish swim from the deeper pools into the shallow rifles and begin to pair up and court. After spawning they disperse up and down the river, feeding in small groups. They may forage at almost any time of the day, though they are generally more active at night. In dark tannin stained or turbid waters, such as Tinana Creek off the Mary River, Australian Lungfish can be quite active during the day. This species locates food items (tadpoles, fish, invertebrates and plants) by olfaction, sight and electro-location as they swim in slow undulating movements just above the substrate.

Reasons for decline and current threats: The greatest threat to the Australian Lungfish is human interference in their waterways, particularly in the form of impoundments.

The construction of dams on rivers can prevent movement past the dam wall, either up or down the river, which is a requirement for moving to and from feeding and spawning grounds. The impoundment itself 'drowns' spawning sites and larval habitat, and intermittent release of water from the impoundment downstream can gouge out the shallow weed beds required for spawning and for the first few years of the hatched Australian Lungfish's life. Withholding of water flowing downstream can dry out those same spawning sites. The Australian Lungfish grows very slowly and can live to a great age (80–100 years or more). To maintain their populations, Australian Lungfish do not need to recruit regularly, but they do need to do so occasionally and, if all spawning/larval rearing sites in a river have been severely affected by dams, lack of recruitment may not become apparent until the adult population has become senescent.

CURRENT RECOVERY EFFORTS

Recovery plan: A recovery plan is currently being prepared.

Conservation actions:

- Protected species in Qld waters.
- Qld DEEDI monitoring program, which is collecting the population data required to form an action plan.
- International trade in Australian Lungfish is regulated through their listing on Appendix II of CITES.

FUTURE RECOVERY EFFORTS

What you can do:

- Approach your water use thoughtfully; use less, use efficiently and explore new ways of obtaining and saving water. By so doing, the need for new dams on rivers inhabited by Australian Lungfish should not arise.
- Urge the government to invest in smart new water supply/water-saving technologies.
- Remind the government that new, large-scale water impoundment projects are not the answer to water supply issues. Instead, encourage integrated, sustainable water supply planning.

Conservation needs: Maintain, and where necessary restore, natural flow regimes.

Sources and further reading

Arthington AH (2009) Australian lungfish, *Neoceratodus forsteri*, threatened by a new dam. *Environmental Biology of Fishes* **84**, 211–221.

Brooks SG and Kind PK (2002) 'Ecology and demography of the Queensland lungfish (*Neoceratodus forsteri*) in the Burnett River, Queensland with reference to the impacts of Walla Weir and future water infrastructure development'. Final Report. Prepared for the Queensland Department of Natural Resources and Mines. Department of Primary Industries, Brisbane.

Jorgensen JM and Joss J (Eds) (2010) *The Biology of Lungfishes.* Science Publishers, New Hampshire.

Kemp A (1986) The biology of the Australian lungfish, *Neoceratodus forsteri* (Krefft 1870). *Journal of Morphology* **190** (Suppl. 1), 181–198.

Pusey B, Kennard M and Arthington A (2004) *Freshwater Fishes of North-Eastern Australia.* CSIRO Publishing, Melbourne.

Author: Jean Joss

SILVERSIDES (ORDER: ATHERINIFORMES)

LAKE EACHAM RAINBOWFISH

Melanotaenia eachamensis Allen & Cross, 1982

Photo: Gunther Schmida

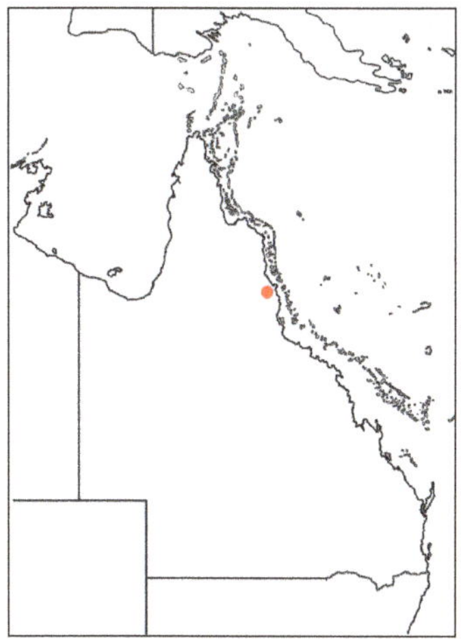

Family: Melanotaeniidae

Conservation status: IUCN: Vulnerable; Aust.: Endangered

Description: The Lake Eacham Rainbowfish is a small laterally compressed fish, which reaches a maximum total length of 70 mm. The species possesses first dorsal spines V–VI, second dorsal rays I,9–13, anal rays I,15–21 and pectoral rays 11–14. Colour varies between localities, but overall the body tends to be silvery or bluish with a darker midlateral stripe, and two to three thinner dark bands present ventrally of the midlateral stripe, which extend almost to the pectoral fins. Fins tend to be uniform bright red with no or little yellow blotches. Fin margins of breeding males tend almost towards black. Specimens from the type locality at Lake Eacham tended to be less brightly coloured.

Distribution: The species was originally described from Lake Eacham on the Atherton Tablelands, north Qld, where the population became extinct by 1987 after the translocation of predatory native fish. The species was considered to be extinct in the wild until genetic studies discovered extant populations in other locations. More recent studies have determined that this species is restricted to the headwaters of the Johnstone, Tully and Barron Rivers in the Wet Tropics of north Qld above an altitude of 500 m.

Habitat: This small schooling species prefers shallow waters among, or directly adjacent to, aquatic vegetation, submerged terrestrial vegetation and root masses. The species occurs both in small streams and in lakes, with populations in crater lakes Euramoo and Bromfield Swamp and the artificial Lake Tinaroo and Koombooloomba Dam (and historically Lake Eacham).

Ecology: Spawning occurs from August to April when water temperatures exceed 17°C. The eggs are demersal and adhesive, and are most commonly attached to submerged root masses. In captivity, females produce batches of 40–50 small eggs (1.2 mm diameter). The fry are also small (4.0 mm) and lack a well developed yolk sac; feeding commences shortly after hatching. The Lake Eacham Rainbowfish is omnivorous, but more than 50% of the diet is composed of aquatic invertebrates.

Reasons for decline and current threats: The extinction of this species at the type locality, Lake Eacham, is directly attributed to predation pressure by translocated native fishes, principally Mouth Almighty *Glossamia aprion*.

Experiments with captive specimens from the Lake Eacham population subsequent to their extinction showed that they were predator naive. Potential threats to remaining populations include destruction of habitat, flow regulation, hybridisation with other rainbowfish species and interactions with introduced aquatic species (including larger predatory fishes). The removal of riparian vegetation and subsequent changes to instream channel morphology and the presence of invasive weeds are also likely to result in reductions in abundance.

The validity of listing this species as threatened, however, needs to be reconsidered, given the much broader distribution than originally considered. Furthermore, despite its type locality and occurrence in lake systems, several researchers believe this species is primarily stream dwelling and any future conservation actions should take this into account.

CURRENT RECOVERY EFFORTS

Recovery plan: None

Other plans: Wagner R and Jackson P (1993) 'The action plan for Australian freshwater fishes'. Australian Nature Conservation Agency, Canberra.

Conservation actions:

- A captive population of the extinct Lake Eacham population is maintained by members of the Australia New Guinea Fishes Association.
- A considerable proportion of the distribution of this species is currently protected within the Wet Tropics region.

FUTURE RECOVERY EFFORTS

Research directions:

- Undertake a more extensive inventory of the distribution of the Lake Eacham Rainbowfish on the Atherton Tablelands (to assess the current listing of this species).
- Determine the conditions that favour Eastern Rainbowfish *Melanotaenia splendida splendida* and whether this is to the detriment of Lake Eacham Rainbowfish given the threat of hybridisation between these two species.

Sources and further reading

Allen GR and Cross NJ (1982) *Rainbow Fishes of Australia and Papua New Guinea*. T.F.H. Publications, New Jersey.

McGuigan KL (2001) An addition to the rainbowfish (Melanotaeniidae) fauna of north Queensland. *Memoirs of the Queensland Museum* **46**, 647–655.

Pusey B, Kennard M and Arthington A (2004) *Freshwater Fishes of North-Eastern Australia*. CSIRO Publishing, Melbourne.

Pusey BJ, Bird J, Kennard MJ and Arthington AH (1997) The distribution of the Lake Eacham rainbowfish in the Wet Tropics region, north Queensland. *Australian Journal of Zoology* **45**, 75–84.

Zhu D, Degnan S and Moritz C (1998) Evolutionary distinctiveness and status of the endangered Lake Eacham rainbowfish (*Melanotaenia eachamensis*). *Conservation Biology* **12**, 80–93.

Author: Steven Brooks

HONEY BLUE-EYE

Pseudomugil mellis Allen & Ivantsoff, 1982

Photo: Gunther Schmida

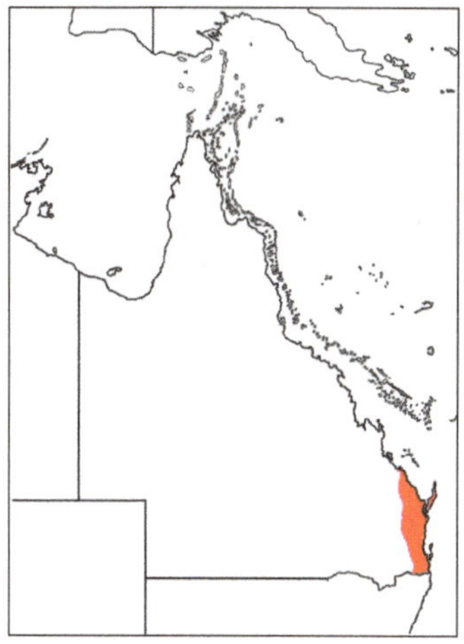

Family: Pseudomugilidae

Conservation status: IUCN: Endangered; Aust.: Vulnerable; Qld: Vulnerable

Description: The Honey Blue-eye is a small fish, which commonly reaches 25–30 mm total length (males may reach 38 mm total length). It possesses first dorsal spines IV–VI, second dorsal rays 6–9, anal rays 10–13 and pectoral rays 12–13. The mouth of the species is small and subvertical with small villiform teeth. The eyes and scales are relatively large, their margins outlined with black spots. The first dorsal fin originates before the pectoral fin tip about mid way along the body; the second dorsal fin originates well behind the anal fin origin. The caudal fin is forked, with rounded tips. There is a single dark midlateral line extending along the midlateral scale row from the pectoral tip to the base of the caudal fin. Breeding males are honey to copper coloured, with a dusky head and white edges and black submarginal bands on the fins; breeding females are plain amber with clear fins. Non-breeding fish of both sexes are greyish. The eyes are blue and the cheeks iridescent blue.

Distribution: The Qld endemic Honey Blue-eye has a very restricted and patchy distribution in south-east Qld, extending from Tibrogargan Creek on the mainland in the Glass House Mountains region, northwards to the Bundaberg area, Shoalwater Bay and Fraser Island. This species remains relatively abundant in the Noosa River and Fraser Island localities.

Habitat: The Honey Blue-eye is confined mainly to dystrophic, acidic (pH 4.4–6.8), freshwater systems draining through sandy coastal lowlands and *Banksia* dominated heath ecosystems (the 'wallum'). It lives in slow-flowing pools and along the margins of rivers/creeks in swampy drainages, perched dune lakes, ponds and a few brackish creeks within wallum ecosystems close to the coast. This species prefers sheltered habitats with dense, aquatic vegetation and undercut banks. It can tolerate wide fluctuations in stream temperature (14°C in winter to 29°C in summer), but prefers well-oxygenated water.

Ecology: The Honey Blue-eye completes its entire life cycle in fresh water and is easy to breed in captivity. Fish mature at a relatively small size (16.6–21.0 mm total length), corresponding to 10–12 months of age. Breeding in spring and summer coincides with increasing water temperatures (26.5–28.0°C), increasing day length and low likelihood of elevated discharges in south-east Qld streams. Males defend territories and exhibit courting behaviour within beds of emergent sedges in water depths of around 15 cm. The Honey Blue-eye is a batch spawner and, in aquaria, pairs spawn several times a day, with females producing 1–15 eggs over 24 hours. Eggs are large (1.3–1.6 mm) relative to body size and are covered by multiple filaments that vary in number, placement and size, and form rope-like attachments to aquatic vegetation. Incubation periods range from 4.5 to 8 days at 25–27°C. The lifespan of the Honey Blue-eye in the wild is unknown, but may be 1–2 years. This species is a microphagic omnivore, foraging mostly on algae, aquatic insects, microcrustaceans and other small invertebrates. In aquaria, juveniles and adults eat the eggs of conspecifics.

Reasons for decline and current threats: Contemporary distribution and abundance patterns of the Honey Blue-eye reflect landscape evolution, climatic conditions and recent human influences, which have reduced or heavily degraded available habitat.

Large expanses of habitat suitable for this species have been destroyed, fragmented or degraded by road construction, agriculture, forestry, residential and resort development, and water pollution. Small weirs and other obstructions may interrupt movements by blocking fish passage, and should be removed. The Honey Blue-eye may be threatened by large populations of the introduced Mosquitofish *Gambusia holbrooki* in several localities. These threatening processes frequently coincide and interact in disturbed coastal catchments and may lead to the extirpation of entire populations of this threatened species.

CURRENT RECOVERY EFFORTS

Recovery plan: None

Other plans: Arthington AH and Marshall CJ (1993) 'Distribution, ecology and conservation of the honey blue-eye, *Pseudomugil mellis*, in south-eastern Queensland'. Final Report to the Australian Nature Conservation Agency. Volume 1. Australian Nature Conservation Agency, Canberra.

Wagner R and Jackson P (1993) 'The action plan for Australian freshwater fishes'. Australian Nature Conservation Agency, Canberra.

Conservation actions: Honey Blue-eye are relatively secure in national parks and other protected areas (Cooloola NP, Noosa River, Fraser Island World Heritage Area and the Forestry Scientific Reserve near Beerwah).

FUTURE RECOVERY EFFORTS

What you can do:

- Don't collect this species from its wild habitats or disturb those habitats.
- Don't release or move Mosquitofish and other exotic fishes into natural wallum streams and wetlands.

Conservation needs:

- Protect undisturbed wallum stream and wetland habitats and maintain water quality.
- Rehabilitate degraded creeks and restore channel and habitat structure (including planting of riparian vegetation and the elimination of invasive species such as Mosquitofish and Para Grass *Urochloa mutica*).
- Ensure any stocking activities to replenish populations at risk use stock reared from system specific genotypes.

Research directions: Investigate the mechanisms by which degraded habitat, degraded water quality, and Mosquitofish, have an impact on populations.

Sources and further reading

Arthington AH and Marshall CJ (1995) Threatened fishes of the world: *Pseudomugil mellis* Allen & Ivantsoff, 1982 (Pseudomugilidae). *Environmental Biology of Fishes* **43**, 268.

Howe E (1987) Breeding behaviour, egg surface morphology and embryonic development in four Australian species of the genus *Pseudomugil* (Pisces: Melanotaeniidae). *Australian Journal of Marine and Freshwater Research* **38**, 885–895.

Pusey B, Kennard M and Arthington A (2004) *Freshwater Fishes of North-Eastern Australia*. CSIRO Publishing, Melbourne.

Semple GP (1991) Reproductive behaviour and early development of the honey blue-eye, *Pseudomugil mellis* Allen and Ivantsoff 1982 (Pisces: Pseudomugilidae), from the North-east Coast Division, south-eastern Queensland, Australia. *Australian Journal of Marine and Freshwater Research* **42**, 277–286.

Author: Angela H Arthington

RED-FINNED BLUE-EYE

Scaturiginichthys vermeilipinnis Ivantsoff, Unmack, Saeed & Crowley, 1991

Photo: Gunther Schmida

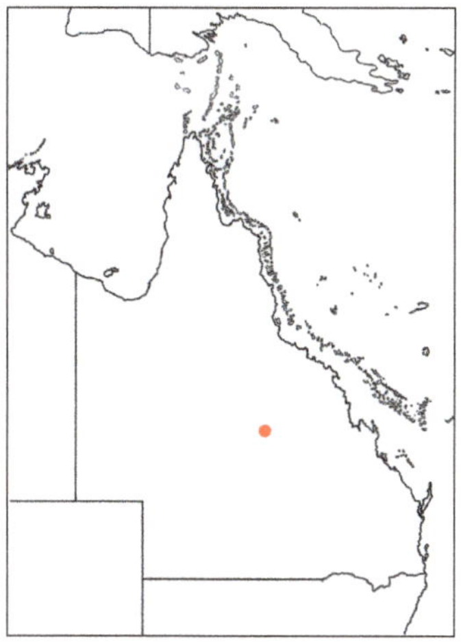

Family: Pseudomugilidae

Conservation status: IUCN: Critically Endangered; Aust.: Endangered; Qld: Endangered

Description: The Red-finned Blue-eye is a small fish, which reaches a maximum total length of 28 mm. It possesses first dorsal spines I,3-4, second dorsal rays I,6-7 and anal rays I,6-10. This fish is translucent silver to golden, darkening dorsally with iridescent spangles forming an anterior mid-lateral band. The posterior half of the body is golden-yellow in juveniles. Fins in juveniles and females are clear to faintly yellow, and the dorsal and anal fins of mature males are larger and edged in red. The eye is silvery-blue with a vertical stripe through the orbit.

Distribution: This species is only found in Edgbaston Springs on Edgbaston Station, 80 km north of Aramac in central western Qld (22°44'S, 145°25'E). When first discovered in 1990, Red-finned Blue-eye were found in eight springs; five populations have since been lost and subsequent colonisation of two further springs has occurred.

Habitat: Edgbaston springs are located on a spinifex plain within Mitchell grass country. The complex contains approximately 51 active springs within an area of approximately 50 km^2. The combined total wetland is approximately 2.4 ha, though only 0.4 to 0.6 ha of this is usable habitat for Red-finned Blue-eye. Maximum depth varies between springs and may be as shallow as 5 mm. Adults generally occur in depths of 10–40 mm, while juveniles prefer depths of 5–10 mm. Grasses and other emergent aquatic vegetation grow throughout the springs. Maximum and minimum recorded water temperatures range from 37°C to 3°C, with as much as 21°C variation over several hours. Red-finned Blue-eye are usually only found in areas within the springs that exceed 16°C (and this influences their seasonal distribution within the springs).

Ecology: Similar to other pseudomugilid fishes, they form loose schools up to several hundred individuals when disturbed, breaking into smaller schools when undisturbed. Red-finned Blue-eye mature at 15 mm total length. Males defend spatially and temporally variable territories against other males, usually among emergent vegetation. Eggs are laid on algae or emergent vegetation. Displaying males, eggs and juveniles have been found throughout the year, but are more prevalent during the warmer months. Spawning commences at 20°C in captive populations, but is more productive at 27°C. Eggs are spherical, opaque, 1.2 mm in diameter, with filaments that attach to the vegetation and substrate and take 7–10 days to hatch at 28°C. Fry are 4.5 mm long on hatching and begin feeding one day after hatching.

Reasons for decline and current threats: A combination of the effects of exotic Mosquitofish *Gambusia holbrooki*, habitat loss and alteration, and the impacts of domestic animals, have all caused detrimental effects on Red-finned Blue-eye.

Mosquitofish compete directly with Red-finned Blue-eye for habitat and resources and are implicated in the recent extinction of Red-finned Blue-eye in two springs and a reduction in the population at three others. Mosquitofish are thought to have invaded the springs during extensive flooding of the region and may invade other springs during future flooding. Trampling of the remaining habitat by domestic stock (now removed from Edgbaston Station) and feral pigs and eutrophication of water due to the faeces and decomposing carcases of these species have also impacted on Red-finned Blue-eye. Flow reduction and the alteration of the springs to provide watering opportunities for domestic stock have reduced available habitat for the species.

CURRENT RECOVERY EFFORTS

Recovery plan: An initial recovery plan was completed in 1995 (Wager R (1995) 'Recovery plan for Queensland artesian spring fishes'. Report for the Australian Nature Conservation Agency, Canberra) and recently updated in 2010 (Fensham RJ, Ponder WF and Fairfax RJ (2010) 'Recovery plan for the community of native species dependent on natural discharge of groundwater from the Great Artesian Basin'. Report to the Department of

Environment, Water, Heritage and the Arts, Canberra. Qld Department of Environment and Resource Management, Brisbane).

Other plans: Wagner R and Jackson P (1993) 'The action plan for Australian freshwater fishes'. Australian Nature Conservation Agency, Canberra.

Conservation actions:
- Edgbaston Station was recently purchased by Bush Heritage Australia.
- Eradication methods for Mosquitofish are presently being investigated by Bush Heritage Australia.
- Two springs on Edgbaston Station have been fenced off (Red-finned Blue-eye is now extinct in one of these).
- All domestic stock has been removed from Edgbaston Station.

FUTURE RECOVERY EFFORTS

What you can do: Those interested in helping manage Edgbaston Springs should contact Bush Heritage Australia (<http://www.bushheritage.org.au>).

Conservation needs:

Recommendations from the recovery plan:
- Mitigate the impacts of feral animals, possibly through further fencing.
- Investigate possibilities for the removal of Mosquitofish and methods to prevent further invasions.
- Investigate possible reintroduction/translocation sites.
- Control introduced flora.
- Ensure that the flow rate of the springs does not decrease further. The Great Artesian Basin Sustainability Initiative (responsible for regulating the flow of artesian bores and capping free flowing bores) should target bores having an impact on Edgbaston Springs.
- Investigate the potential impacts of mineral exploration/mining within the recharge area.

Research directions: Investigate the interaction between Red-finned Blue-eye and Mosquitofish.

Sources and further reading

Allen GR, Midgley SH and Allen M (2002) *Field Guide to the Freshwater Fishes of Australia.* Western Australian Museum, Perth.

Fairfax R, Fensham R, Wager R, Brooks S, Webb A and Unmack P (2007) Recovery of the red-finned blue-eye: an endangered fish from springs of the Great Artesian Basin. *Wildlife Research* **34**, 156–166.

Fensham RJ and Fairfax RJ (2003) Spring wetlands of the Great Artesian Basin, Queensland, Australia. *Wetlands Ecology and Management* **11**, 343–362.

Ivantsoff W, Unmack P, Saeed B and Crowley L (1991) A redfinned blue-eye, a new species and genus of the family Pseudomugilidae from central western Queensland. *Fishes of Sahul* **6**, 277–282.

Wager RNE and Unmack PJ (2004) Threatened fishes of the world: *Scaturiginichthys vermeilipinnis* (Ivantsoff, Unmack, Saeed & Crowley 1991) (Pseudomugilidae). *Environmental Biology of Fishes* **70**, 330.

Author: Steven Brooks

PERCH-LIKE FISHES (ORDER: PERCIFORMES)

EASTERN FRESHWATER COD

Maccullochella ikei Rowland, 1985

Photo: © Gary Bell/OceanwideImages.com

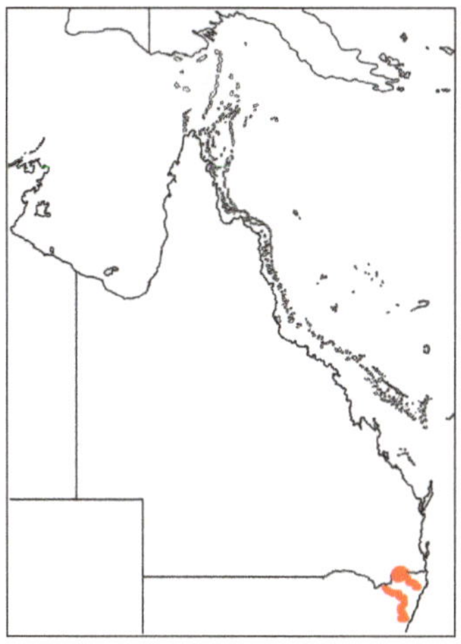

Family: Percichthyidae

Aboriginal name: Diinbam

Conservation status: IUCN: Endangered; Aust.: Endangered; NSW: Endangered

Description: The Eastern Freshwater Cod is a large, elongate, slightly compressed, deep-bodied fish, which reaches a maximum total length of >100 cm (>40 kg). It has relatively small eyes and a short rounded jaw. The species has dorsal spines X–XII and dorsal rays 13–16. The caudal fin is convex and large in comparison to the overall size of the fish. The pelvic fins have white tapering filaments that extend beyond the lower margins of the fin by 1–2 cm. Most specimens are yellow-green to golden in colour, with dark green to black reticulated mottling extending both dorsally and laterally over the upper body to the base of the fins. The dorsal, pectoral, caudal and anal fins are dark grey-green to semi-clear with pale grey and white margins.

Distribution: Historically, large populations of Eastern Freshwater Cod were present in the Clarence and Richmond River systems of northern NSW, and possibly the Brisbane River, Qld. Little is known about cod in the Brisbane River system, which may in fact have been Mary River Cod *Maccullochella peelii mariensis*, but there is no historical genetic material available to verify this. By the 1970s, cod were considered extirpated in the Richmond and Brisbane systems, and only small populations remained in isolated sub-catchments of the Clarence system. Recovery actions implemented post-1970s have resulted in a general increase in the distribution and abundance in the Clarence and Richmond systems.

Habitat: Remnant populations inhabit deep pools within clear, slow-flowing rivers and creeks, and individuals use boulders, woody debris, undercut banks and grassy tussocks as cover as well as inundated riparian vegetation during flooding.

Ecology: The Eastern Freshwater Cod is a relatively long-lived species (beyond 15 years in the wild). Maturity occurs at 4–5 years when around 40 cm total length and 1 kg. Breeding is annual, but the season is relatively short, commencing in early spring and lasting around 8–10 weeks. A nesting, guarding spawner, males select and prepare nesting sites 1–2 days pre-spawning, and remain at the site for up to 24 days post-spawning to care for eggs and larvae. Larvae hatch 8–10 days after fertilisation and commence feeding 12–14 days later when 10.5-12 mm total length. Little is known of the habitats of post-dispersal juveniles. Considered a generalist predator, it consumes a wide variety of food items including fish, freshwater shrimp and prawns, aquatic insects, and terrestrial animals such as birds, rats and mice. Genetically, there is distinct structuring and extremely low diversity between and within populations.

Reasons for decline and current status: Fish kills, environmental conditions, overfishing and habitat degradation caused historical declines in the Eastern Freshwater Cod.

The major decline in the Clarence system is attributed to large fish kills during the 1920s and 1930s, resulting from prolonged drought conditions, large bushfires and finally flooding. In the Orara River and much of the Richmond system, major declines occurred during construction of the North-Coast railway system in the early 1900s. Overfishing and habitat degradation continued to place pressure on stocks in all rivers (including the Brisbane River, Qld) throughout the 20th century. There has been some recovery in NSW rivers post-1970s, particularly within remnant populations.

CURRENT RECOVERY EFFORTS

Recovery Plan: NSW Fisheries Threatened Species Recovery Planning Program (2004) 'Eastern (Freshwater) Cod *(Maccullochella ikei)* recovery plan'. NSW Fisheries, Nelson Bay.

Other plans: Wagner R and Jackson P (1993) 'The action plan for Australian freshwater fishes'. Australian Nature Conservation Agency, Canberra.

Conservation actions:

- Year-round prohibition on targeting and/or retaining Eastern Freshwater Cod throughout its original range in NSW.
- A total fishing exclusion from August to October each year to protect breeding individuals within the main remnant population in Mann and Nymboida Rivers.
- Restocking program undertaken in the Clarence and Richmond, but not in the Brisbane system due to cross-state jurisdictional issues and uncertainty over the original genetic structure of stocks.
- Habitat restoration programs (e.g. stock-exclusion fencing and off-stream watering, riparian re-vegetation, removal of weirs and road crossings and re-snagging).
- Demonstration reach established in 2005-06 within the Richmond system to demonstrate cost-effective ways of improving habitat quality for Eastern Freshwater Cod.
- Ongoing monitoring and research within remnant and restocked populations as funding permits.

FUTURE RECOVERY EFFORTS

What you can do: Anglers should abide by fishing regulations.

Conservation needs:

- Increase community awareness of the imperilled state of the species.
- Review recovery plan.
- Review conservation status and raise level to Critically Endangered if required.
- Establish long-term monitoring and research program.
- Re-establish the recovery team.
- Restock fry and fingerlings in areas where species is extirpated or where populations are not self-maintaining.
- Protect and restore habitat.

Research directions:

- Determine the effects of recreational angling, particularly on breeding individuals.
- Determine the dispersal patterns and habitat selection of larvae and juveniles.
- Trial a range of artificial and natural structures to use as spawning sites.
- Determine reasons for the limited success of restocking programs.

Sources and further reading

Butler GL and Rowland SJ (2008) Using traditional age and growth techniques in endangered species management: eastern freshwater cod, *Maccullochella ikei*. *Marine and Freshwater Research* **59**, 684–693.

Butler GL and Rowland SJ (2009) Using underwater cameras to describe the reproductive behaviour of the endangered eastern freshwater cod *Maccullochella ikei*. *Ecology of Freshwater Fish* **18**, 337–349.

Nock CJ, Ovenden JR, Butler GL, Wooden I, Moore AS and Baverstock PR (2011) Population structure, effective population size and adverse effects of stocking in the endangered Australian eastern freshwater cod, *Maccullochella ikei*. *Journal of Fish Biology* **78**, 303-321.

Rowland SJ (1993) *Maccullochella ikei*, an endangered species of freshwater cod (Pisces: Percichthyidae) from the Clarence River system, NSW and *M. peeli mariensis*, a new subspecies from the Mary River system, Qld. *Records of the Australian Museum* **45**, 121–145.

Rowland SJ (1996) Threatened fishes of the world: *Maccullochella ikei* Rowland, 1985 (Percichthyidae). *Environmental Biology of Fishes* **46**, 350.

Authors: Gavin L Butler and Stuart J Rowland

MARY RIVER COD

Maccullochella peelii mariensis Rowland, 1993

Photo: Gunther Schmida

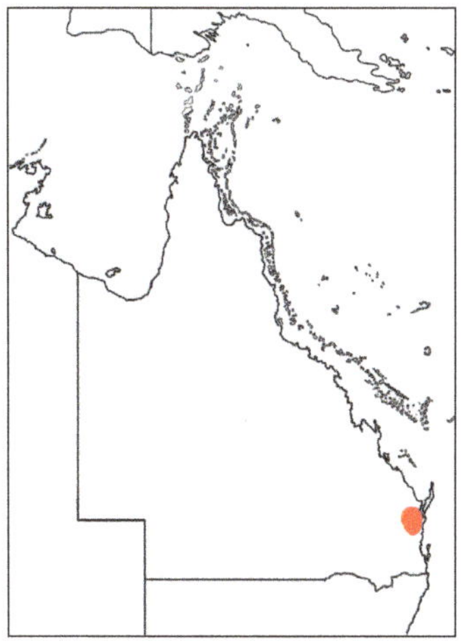

Family: Percichthyidae

Conservation status: IUCN: species (*M. peelii*) is Critically Endangered; Aust.: Endangered

Description: The Mary River Cod is a heavy bodied, elongate fish, which reaches a maximum total length of 180 cm (but uncommon >70 cm). It has a concave head profile above the eyes, and a large mouth with jaws about equal or the lower protruding in larger animals. It has a single dorsal fin with dorsal spines X–XII and dorsal rays 13–16 (usually XI, 15) and a large, rounded caudal fin. Colour varies, but includes dark mottling over a green, dark brown or golden-yellow back and sides and a grey or whitish belly. The soft dorsal, anal and caudal fins have thin, whitish margins, and the whitish pelvic fin has elongated white filaments.

Distribution: Freshwater cod were reportedly abundant in several south-east Qld river systems at the time of European settlement. Within the Mary River system, remnant populations of cod could only be confirmed in a few locations by the 1970s (it is not known if cod that disappeared from other river systems were Mary River Cod, or other taxa such as Eastern Freshwater Cod *Maccullochella ikei*). The Mary River Cod now only occurs in three natural subpopulations, occupying <30% of its former range in the Mary River system.

Habitat: Mary River Cod have a strong preference for riverine habitats that provide abundant cover, particularly large woody debris and boulders. This is usually combined with a preference for larger pool habitats, although smaller individuals can often be found in shallower, faster flowing areas. In large impoundments, cod frequent the edges of weed beds, drop-offs and rock faces.

Ecology: Mary River Cod reach maturity at 4–5 years of age (~1 kg). Spawning occurs in spring and early summer (water temperature 20–21°C) and involves a male selecting a nest site, into which a female deposits several hundred to tens of thousands of adhesive eggs (3.0–3.5 mm diameter). The male guards both the eggs (until hatching; 4–7 days at 21°C) and fingerlings (until they disperse; 7–9 days). Adult cod move long distances during flow events, returning to their previous location sometimes after many months absence. During more sedentary periods, Mary River Cod establish a well defined home range up to 1 km in length containing core occupancy areas (often a large sunken log). This is an opportunistic ambush predator feeding mainly on fish and crustaceans.

Reasons for decline and current threats: The decline of Mary River Cod appears to relate to large-scale modification and general degradation of riverine habitats.

Widespread clearing of catchment and riparian vegetation for agriculture and forestry in the 1800s and 1900s led to siltation and loss of many of the deeper, shaded pool habitats that likely underpinned the historical abundance of the Mary River Cod. In many areas, grazing and disturbance by cattle inhibits the regeneration of riparian vegetation, and contributes to ongoing stream bank erosion. These processes, in conjunction with the construction of dams, weirs, culverts and low level crossings, led to fragmentation of suitable habitat (and thus Mary River Cod populations). Overfishing has also likely played a role, with anecdotal reports of large numbers of cod being taken historically using set lines, gill nets and gelignite fishing.

CURRENT RECOVERY EFFORTS

Recovery plan: Simpson RR and Jackson PD (1996) 'The Mary River Cod research and recovery plan'. Australian Nature Conservation Agency, Canberra. It was updated in 1999 and 2003 and is currently overdue for review.

Other plans: Wagner R and Jackson P (1993) 'The action plan for Australian freshwater fishes'. Australian Nature Conservation Agency, Canberra.

Conservation actions:

- Protected species in Qld flowing waters.
- Minimum size limit of 50 cm and an in possession bag limit of one per person in stocked impoundments outside the Mary River catchment.
- Closure to all angling in sections of Obi Obi, Coondoo and Tinana Creeks (Mary River) to protect wild populations.
- Captive breeding and restocking into the Mary River and other south-east Qld rivers and streams since

1997 (little monitoring overall to gauge the success of restocking).

- Community awareness commenced in 1996 (including through a nationally funded Mary River Cod Community Liaison Officer).
- Landcare Australia and the Mary River Catchment Coordinating Committee continue community engagement, but without dedicated funds.
- Instream and riparian habitat restoration have been undertaken to improve cod habitat in many areas across its original range.

FUTURE RECOVERY EFFORTS

What you can do: Anglers should abide by fishing regulations.

Conservation needs:

- Reinstate the recovery team to review and update the recovery plan, and direct research, restocking and habitat rehabilitation.
- Reinstate a fully funded Community Liaison Officer to assist with community involvement.
- Minimise the impact of any barriers to fish movement.
- Preserve large woody debris within the range of the species.

Research directions:

- Investigate in more detail the movements and habitat use by Mary River Cod.
- Collect baseline data on existing stocked and remnant populations, followed by regular monitoring.
- Develop a predictive model to describe population dynamics.
- Investigate breeding behaviour and larval recruitment.
- Assess the genetic structure of Mary River Cod to support and direct the restocking program.

Sources and further reading

Pusey B, Kennard M and Arthington A (2004) *Freshwater Fishes of North-Eastern Australia*. CSIRO Publishing, Melbourne.

Rowland SJ (1993) *Maccullochella ikei*, an endangered species of freshwater cod (Pisces: Percichthyidae) from the Clarence River system, NSW and *M. peelii mariensis*, a new subspecies from the Mary River system, Qld. *Records of the Australian Museum* **45**, 121–145.

Simpson RR (1994) 'An investigation into the habitat preferences and population status of the endangered Mary River cod (*Maccullochella peelii mariensis*) in the Mary River system, south-eastern Queensland'. Queensland Department of Primary Industries Information Series QI94011, Brisbane.

Simpson RR (2003) Mary River cod. *Nature Australia* **27**, 26–27.

Simpson RR and Mapleston AJ (2002) Movements and habitat use by the endangered Australian freshwater Mary River cod, *Maccullochella peelii mariensis*. *Environmental Biology of Fishes* **65**, 401–410.

Author: Bob Simpson

MURRAY COD

Maccullochella peelii peelii (Mitchell, 1838)

Photo: Gunther Schmida

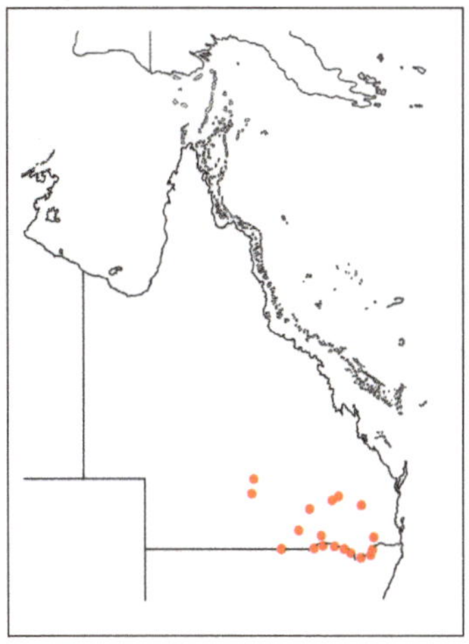

Family: Percichthyidae

Aboriginal names: Goodoo, Ponde, Mewuruk, Bewuk, Mungee

Conservation status: IUCN: species (*M. peelii*) is Critically Endangered; Aust.: Vulnerable; Vic.: Endangered

Description: Murray Cod is Australia's largest freshwater bony fish, which reaches a maximum length of 180 cm (113 kg) and is an important recreational and cultural species. It is a large, elongate, groper-like fish with a broad head, rounded snout and large mouth. The jaws are equal or the lower jaw is overhanging. The head is sloped and concave above the eye. Its body is covered by small scales. It has a single dorsal fin with dorsal spines X–XII and dorsal rays 13–16 (usually XI, 15) and a large, rounded caudal fin. Colour varies but is typically olive green to yellow dorsally with black, blue-black, brown to pale green mottling. The underbelly is creamy white.

Distribution: Murray Cod is widely distributed throughout the rivers of the Murray-Darling Basin, being recorded in Qld, NSW, Vic., SA and the ACT. Important areas for Murray Cod in Qld include the Dumaresq, MacIntyre, Condamine and Warrego Rivers.

Habitat: Murray Cod occupy a wide range of habitats in lower to mid river reaches, although they are also abundant in some head waters such as rivers on the NSW Northern Tablelands. Murray Cod are demersal, river channel specialists that also inhabit impoundments. Adult and young-of-the-year Murray Cod use habitats that are largely influenced by high amounts of instream wood, overhanging vegetation, are close to the bank and have slower water velocities.

Ecology: Murray Cod generally mature at 5 years of age (about 50–60 cm total length) and lay 3–3.5 mm diameter adhesive eggs (fecundity up to 110 000 for large females) on hard substrata (logs, rocks, or hard clay banks). Spawning occurs in spring at water temperatures greater than 15°C, with the eggs then guarded by the male. Hatching occurs after 5–13 days, depending on water temperature, and the larvae (5–8 mm long) then drift for about 5–7 days. Although they spend relatively sedentary periods with limited home ranges and high site fidelity, Murray Cod also undertake larger pre-spawning movements (up to 130 km). Murray Cod are a long-lived (to 47 years), apex predator that feed on fish and a range of larger macroinvertebrates (e.g. crayfish) and terrestrial food items.

Reasons for decline and current threats: Murray Cod were once widespread and abundant but now have a more patchy distribution, with abundances greatly reduced as a result of a number of threatening processes.

Murray Cod are threatened by: changes to flows (in particular reduced flooding); reduced water quality (increased suspended sediment, low oxygen levels, herbicides, extreme temperatures); cold-water releases from large dams (estimated to affect >2800 km of Murray-Darling Basin rivers); barriers to movement (prevention of upstream movement, mortality to larvae flowing over weirs); larval and juvenile fish loss into irrigation off-takes; removal of instream wood; past commercial catches (these fisheries are now closed); unsustainable recreational catches; and the potential impacts of stocking of hatchery-reared Murray Cod on the genetics and fitness of wild populations.

CURRENT RECOVERY EFFORTS

Recovery plan: National Murray Cod Recovery Team (2010) 'National recovery plan for the Murray Cod *Maccullochella peelii peelii*'. Victorian Government Department of Sustainability and Environment, Melbourne.

Other plans: Murray-Darling Basin Commission (2004) *Native Fish Strategy for the Murray-Darling Basin 2003–2013.* Murray-Darling Basin Commission, Canberra. <http://www.mdbc.gov.au>.

Conservation actions:

- Qld recreational fishing rules and regulations: closed season from 1 September until 30 November each year; overall daily bag limit of two per person; minimum size limit of 60 cm; maximum size limit of 110 cm.
- Captive breeding programs: government and/or commercial hatcheries in Qld, NSW and Vic.

produce Murray Cod fingerlings for stock enhancement and conservation. Murray Cod are regularly stocked into a range of Qld waters for recreational angling purposes.

- Instream, riparian and catchment habitat rehabilitation efforts provide benefits to Murray Cod.

FUTURE RECOVERY EFFORTS

What you can do: Anglers that catch a Murray Cod should abide by fishing regulations and consider just taking a photo and returning it to the water unharmed.

Conservation needs:

Recommendations from the recovery plan:

- Manage river flows to enhance recruitment of Murray Cod populations.
- Evaluate the risks of threats and benefits of recovery options on Murray Cod populations for each management unit.
- Manage the recreational fishery for Murray Cod in a sustainable manner while recognising the social, economic and recreational value of the fishery.
- Encourage community ownership for Murray Cod conservation.

Research directions:

- Determine the distribution, structure and dynamics of Murray Cod populations across the Murray-Darling Basin.
- Determine the habitat requirements of Murray Cod life stages and populations.

Sources and further reading

Humphries P (2005) Spawning time and early life history of Murray cod, *Maccullochella peelii peelii* (Mitchell) in an Australian river. *Environmental Biology of Fishes* **72**, 393–407.

Koehn JD (2009) Multi-scale habitat selection by Murray cod *Maccullochella peelii peelii* in two lowland rivers. *Journal of Fish Biology* **75**, 113–129.

Koehn JD, McKenzie JA, O'Mahony DJ, Nicol SJ, O'Connor JP and O'Connor WG (2009) Movements of Murray cod (*Maccullochella peelii peelii*) in a large Australian lowland river. *Ecology of Freshwater Fish* **18**, 594–602.

Lintermans M and Phillips B (Eds) (2005) 'Management of Murray cod in the Murray-Darling Basin: statement, recommendations and supporting papers. Proceedings of a workshop held in Canberra, ACT, 3–4 June 2004'. Murray-Darling Basin Commission and Cooperative Research Centre for Freshwater Ecology, Canberra.

Rowland SJ (1989) Aspects of the history and fishery of the Murray cod, *Maccullochella peeli* (Mitchell) (Percichthyidae). *Proceedings of the Linnean Society of New South Wales* **111**, 201–213.

Author: John D Koehn

OXLEYAN PYGMY PERCH

Nannoperca oxleyana Whitley, 1940

Photo: Gunther Schmida

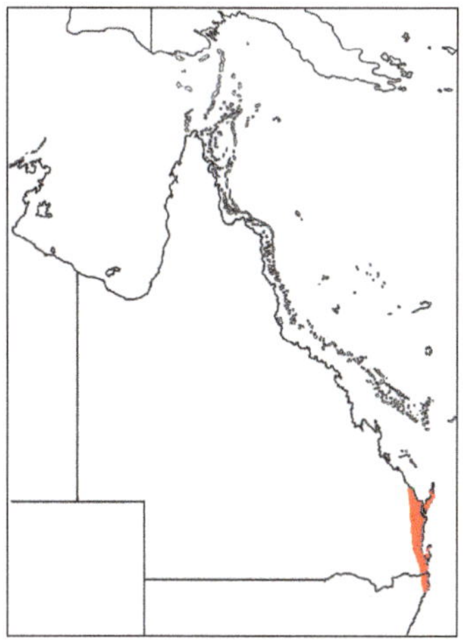

Family: Percichthyidae

Conservation status: IUCN: Endangered; Aust.: Endangered; Qld: Vulnerable; NSW: Endangered

Description: The Oxleyan Pygmy Perch is a small fish, which reaches a maximum total length of about 60 mm (more commonly 45 mm). It possesses first dorsal spines VI–VIII, second dorsal rays 7–9, anal rays III,7–9 and pectoral rays 11–13. This species has a moderately compressed body covered by ctenoid scales without a lateral line. The mouth is relatively small, with jaws reaching to below the anterior part of the pupil; teeth in front of the lower jaw are enlarged. It has a truncate caudal fin. The body is light brown to olive, darker on the back; the sides are paler and mottled with three to four patchy dark brown bars extending from head to tail. The opercular flap has a blue iridescence, and scales have dusky outlines. There is a prominent round black spot with an orange margin at the base of the caudal fin, with fins mainly clear. During breeding, males develop more intense red and brown fin and body colouration than females, and have jet black pelvic fins.

Distribution: The Oxleyan Pygmy Perch has been recorded in 106 coastal water bodies extending along 534 km of coastline from Coongul Creek on Fraser Island, Qld (25°16'S, 153°09'E) south to a tributary of the Wooli River, NSW (29°56'S, 153°09'E). It also occurs on Moreton and North Stradbroke Islands.

Habitat: This species lives in dystrophic, acidic (pH range 3.32–6.90), freshwater systems draining sandy coastal lowlands and *Banksia* dominated heath (the 'wallum'). It is found in slow-flowing pools and backwaters of rivers, in swampy drainages, perched dune lakes and ponds, where plants, undercuts banks, rootlets, snags and litter provide shelter. These habitats are typically well-oxygenated.

Ecology: The Oxleyan Pygmy Perch spawns and completes its entire life cycle in freshwater and is easy to breed in captivity. Maturation commences at relatively small sizes (females 20–30 mm total length, males 23–27 mm total length) with maturity in aquaria bred fish reached in 4–5 months. Peak spawning occurs in spring and early summer, coinciding with increasing water temperatures (26.5–28.0°C), increasing day length and low flows in coastal streams. Eggs are deposited onto substrates or aquatic vegetation. In aquaria, females lay a few eggs per day (mean 7.8) and a mean total of 1323 eggs per fish. Eggs held at 25°C hatch at 50 hours and larvae commence exogenous feeding 5 days after hatching. Wild individuals may live up to 3–6 years. This species consumes crustaceans, aquatic insects and small amounts of terrestrial invertebrates.

Reasons for decline and current threats: Contemporary distribution and abundance patterns of the Oxleyan Pygmy Perch reflect landscape evolution, climatic conditions and recent human influences.

Large expanses of habitat suitable for this species, from the Glasshouse Mountains in Qld to the township of Broadwater in NSW, have been destroyed, fragmented or degraded by residential and resort development, road construction, agriculture, forestry, sandmining and water pollution. Small weirs and other obstructions may interrupt fish movements. This species may be threatened by large populations of the introduced Mosquitofish *Gambusia holbrooki*.

CURRENT RECOVERY EFFORTS

Recovery plans: Qld: Arthington AH (Ed.) (1996) 'Recovery plan for the Oxleyan Pygmy Perch *Nannoperca oxleyana*'. Final Report to the Australian Nature Conservation Agency. Centre for Catchment and In-Stream Research, Griffith University, Queensland.

National: NSW Department of Primary Industries (2005) 'Oxleyan Pygmy Perch: recovery plan and background paper'. New South Wales Department of Primary Industries, Nelson Bay.

Other plans: NSW Threat Abatement Plan: NSW National Parks and Wildlife Service (NPWS) (2003) 'Predation by *Gambusia holbrooki* – the Plague Minnow'. NSW NPWS, Hurstville.

Wagner R and Jackson P (1993) 'The action plan for Australian freshwater fishes'. Australian Nature Conservation Agency, Canberra.

Conservation actions:

- Oxleyan Pygmy Perch are relatively secure in national parks and other protected areas in Qld.
- Declaration of Critical Habitat under the NSW *Fisheries Management Act 1994*, the Qld *Nature Conservation Act 1992* and the Commonwealth *Environment Protection and Biodiversity Conservation Act 1999* affords protection of all water bodies supporting this species both inside and outside protected areas.
- Consolidated Rutile Ltd established a captive breeding program to guard against possible population extinction due to sandmining on North Stradbroke Island.

FUTURE RECOVERY EFFORTS

What you can do:

- Don't collect this species from its wild habitats or disturb those habitats.
- Don't release or move Mosquitofish *Gambusia holbrooki* and other exotic fishes into natural wallum streams and wetlands.

Conservation needs:

- Protect undisturbed wallum stream and wetland habitats and maintain water quality.
- Rehabilitate degraded creeks, including planting riparian vegetation and eliminating invasive species (Mosquitofish and Para Grass *Urochloa mutica*).
- Improve management of lands held by public authorities to account for the species presence (e.g. amendment of fire control plans and fish management activities and the development of low-impact drain management practices in NSW).
- Ensure any stocking activities use stock reared from system specific genotypes.

Research directions: Investigate the mechanisms by which degraded habitat, degraded water quality, and Mosquitofish, have an impact on populations.

Sources and further reading

Knight JT and Arthington AH (2008) Distribution and habitat associations of the endangered Oxleyan pygmy perch, *Nannoperca oxleyana* Whitley, in eastern Australia. *Aquatic Conservation: Marine and Freshwater Ecosystems* **18**, 1240–1254.

Knight JT, Butler GL, Smith PS and Wager RNE (2007) Reproductive biology of the endangered Oxleyan pygmy perch *Nannoperca oxleyana* Whitley. *Journal of Fish Biology* **71**, 1494–1511.

Knight JT, Nock CJ, Elphinstone MS and Baverstock PR (2009) Conservation implications of distinct genetic structuring in the endangered freshwater fish *Nannoperca oxleyana* (Percichthyidae). *Marine and Freshwater Research* **60**, 34–44.

Pusey B, Kennard M and Arthington A (2004) *Freshwater Fishes of North-Eastern Australia*. CSIRO Publishing, Melbourne.

Authors: Angela H Arthington and James T Knight

HUMPHEAD MAORI WRASSE

Cheilinus undulatus Rüppell, 1835

Photo: Anthony Coward

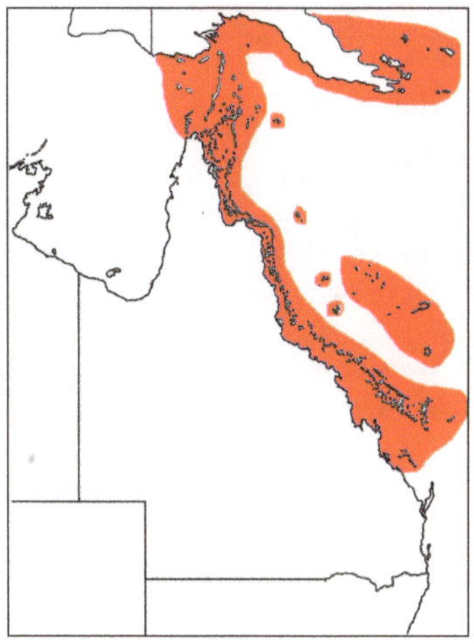

Family: Labridae

Other common names: Maori Wrasse, Napolean Wrasse, Napoleon Fish, Humphead Wrasse, Giant Wrasse

Conservation status: IUCN: Endangered

Description: The Humphead Maori Wrasse is a large, deep-bodied fish, which reaches a normal maximum total length for males of 100–150 cm. It has a hump on its head above its eyes, which grows bigger with age and is completely absent in small juveniles. It also has distinctive fleshy lips and three red-brown lines radiating from the eye, one running down to the back of the mouth and two back towards the top of the gill opening. The name Maori refers to the pattern of thin brown lines on the head, which are similar to facial tattoos. Juveniles are almost white with a pattern of dark brown spots and lines. Young individuals and females are pale with red-brown vertical scale markings, but large males have green scale markings and a blue-tinged head.

Distribution: The Humphead Maori Wrasse has a wide Indo-West Pacific distribution, from the Red Sea and South Africa east to the Tuamotu Archipelago (French Polynesia), north to southern Japan and south to New Caledonia. Within Australia, they inhabit the waters of the Great Barrier Reef, the Coral Sea and oceanic reefs off northern WA.

Habitat: This wrasse usually lives near the bottom in areas of coral reef habitat, but large adults may swim some distance between reefs and are occasionally seen out in open water. They range from the surface down to a depth of 100 m. Juveniles are found in sheltered lagoonal areas among stands of staghorn *Acropora* corals, or sometimes in algal and seagrass beds.

Ecology: The Humphead Maori Wrasse is a slow growing, protogynous hermaphrodite, with all young fish maturing first as females when they are about 8 years old. At about 15 years of age, some Humphead Maori Wrasse change sex to males and begin to grow very rapidly. In spring and summer, groups of up to 30 males and females gather on steep reef front walls where they court and spawn. Males and females can both reach a maximum age of about 30–35 years. Outside the breeding season, this species is usually solitary but may also move around in loose groups of two to five individuals. They are active during the day and usually shelter in a cave or crevice at night. This fish is a top predator and eats a wide range of invertebrates, including molluscs, echinoderms, crustaceans and worms, but larger individuals become active piscivores, feeding on a wide variety of reef fishes.

Reasons for decline and current threats: Humphead Maori Wrasse were extensively targeted (including by spearfishing) from the earliest days of fishing in Qld and, combined with considerable fishing pressure across large parts of its range, particularly in South-East Asia, pressure has been placed on their populations.

Humphead Maori Wrasse are considered a prime food fish and within Qld waters, there is anecdotal evidence that numbers were formerly much higher than they are presently, although accurate data are lacking. Commercial fishers verbally reported that catches were high when 'new' areas were first exploited 50 years ago. Numbers of this fish have only been recorded in scattered Great Barrier Reef locations over the past decade, but there is not much information available on current numbers or changes in numbers. Measured densities are usually around one to two fish per hectare, but in preferred areas such as the reefs off Mackay densities may reach four per hectare. This equates to between 200 and 500 individuals on an average sized reef of the Great Barrier Reef. As far as is possible to determine, given the limited data available, Humphead Maori Wrasse numbers in Qld appear have been stable over the past 5–10 years. Great Barrier Reef densities are on average higher than in many other parts of the Indo-West Pacific.

Outside of Australia, there is considerable pressure from fishing activities across large parts of its range, particularly in South-East Asia, where they are much in demand in the Asian live fish trade. This is a large, slow growing fish that is susceptible to overfishing.

CURRENT RECOVERY EFFORTS

Recovery plan: None

Other plans: Pogonoski JJ, Pollard DA and Paxton JR (2002) 'Conservation overview and action plan for Australian threatened and potentially threatened marine and estuarine fishes'. Environment Australia, Canberra.

Conservation actions:

- Humphead Maori Wrasse are totally protected in all Qld waters and in the Great Barrier Reef and must be returned to the water if caught.
- Considerable habitat protection occurs in the Great Barrier Reef Marine Park, with almost a third of the Park area closed to fishing.
- Considerable research into the diet, growth and reproduction of Humphead Maori Wrasse has been undertaken in the Great Barrier Reef region.
- International trade in Humphead Maori Wrasse is regulated through their listing on Appendix II of CITES.

FUTURE RECOVERY EFFORTS

What you can do:

- Fishers should abide by fishing regulations and return any captured Humphead Maori Wrasse safely to the water.
- Divers should not harass any encountered Humphead Maori Wrasse.

Research directions: Monitor population sizes and status in order to develop reliable population trends.

Sources and further reading

Chateau O and Wantiez L (2007) Site fidelity and activity patterns of a humphead wrasse, *Cheilinus undulatus* (Labridae), as determined by acoustic telemetry. *Environmental Biology of Fishes* **80**, 503–508.

Choat JH, Davies CR, Ackerman JL and Mapstone BD (2006) Age structure and growth in a large teleost, *Cheilinus undulatus*, with a review of size distribution in labrid fishes. *Marine Ecology Progress Series* **318**, 237–246.

Colin PL (2010) Aggregation and spawning of the humphead wrasse *Cheilinus undulatus* (Pisces: Labridae): general aspects of spawning behaviour. *Journal of Fish Biology* **76**, 987–1007.

Russell B (2004) *Cheilinus undulatus*. In IUCN 2010. *IUCN Red List of Threatened Species*. Version 2010.4. IUCN, Gland, Switzerland and Cambridge, UK, <http://www.iucnredlist.org>.

Sadovy Y, Kulbicki M, Labrosse P, Letourneur Y, Lokani P and Donaldson TJ (2003) The humphead wrasse, *Cheilinus undulatus*: synopsis of a threatened and poorly known giant coral reef fish. *Reviews in Fish Biology and Fisheries* **13**, 327–364.

Author: Tony Ayling

ELIZABETH SPRINGS GOBY

Chlamydogobius micropterus Larson, 1995

Photo: Peter Unmack

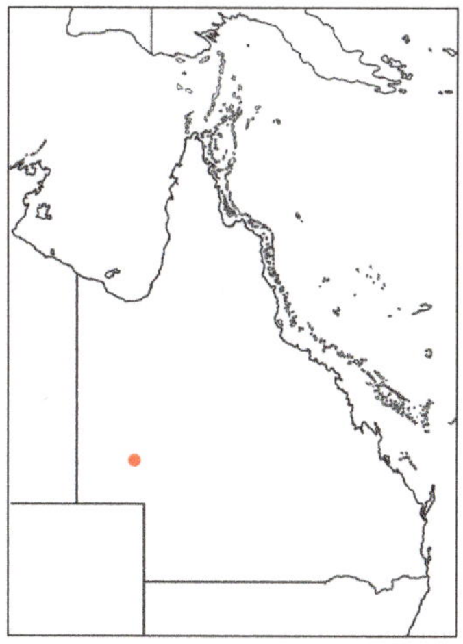

Family: Gobiidae

Conservation status: IUCN: Critically Endangered; Aust.: Endangered; Qld: Endangered

Description: The Elizabeth Springs Goby is a small goby, which reaches a maximum total length of 62 mm. It possesses first dorsal spines III–VI; second dorsal rays I,6–8, anal rays I,5–7 and pectoral rays 11–13. It differs from other species by the reduced sizes of all the fins, in particular the first dorsal and pelvic fins, and less intense colouration. Dorsal colouration is greenish brown, lighter on the sides with a cream to white belly. Males become golden brown when mature. The first dorsal fin is grey with a pale yellow distal margin and a pale blue spot on the posterior edge. The second dorsal, anal and caudal fins are blue grey with a whitish blue margin.

Distribution: This species is restricted to Elizabeth Springs on Springvale Station, 80 km south-east of Boulia in western Qld (23°30′S, 140°35′E).

Habitat: Elizabeth Springs is a group of approximately 40 low mound springs associated with the Great Artesian Basin. The springs cover an area of some 60 ha on the banks of Spring Creek. The springs range from dry/non-flowing mounds with little or no vegetation, vegetated marshy soaks with no surface water, to the largest, which has well defined outflows and vegetated areas. These larger springs have wetted areas up to 140 m in diameter, but a maximum depth of only 0.1 m. The discharge of these springs is now reduced to only a few litres per second. Elizabeth Springs gobies are predominantly found in the largest of the springs; monitoring between 1994 and 2002 found gobies in as few as five, up to a maximum of 14, springs.

Ecology: During daylight the gobies shelter within emergent vegetation and have been observed apparently foraging in the open areas of the springs at night. Reproduction within the species is only known from captive populations where spawning commenced at temperatures above 26°C. The male selects a section of substrate (often a rock in captivity) and attracts a female by extending his fins and swimming in jerky movements around the site. After spawning is complete, the male is left to guard the 40–100 eggs, which are attached to the substrate. The water hardened eggs are between 2.5 and 3.0 mm long and hatch after 10 days. Newly hatched larvae are 5–6 mm total length. Spawning sites are yet to be located in the wild. The diet of this species is considered to include invertebrates and algae.

Reasons for decline and current threats: Decline of this species is directly attributed to a decline in the rate of flow of the springs and the associated loss of habitat.

The flow rate at Elizabeth Springs has reduced some 70% in the last 100 years due to water extraction from the Great Artesian Basin. Trampling of the remaining habitat by domestic stock and feral pigs and eutrophication of water due to the faeces and decomposing carcases of these species have also had an impact on the Elizabeth Springs Goby. Although not currently present at this site, Mosquitofish *Gambusia holbrooki*, will become a problem if they become established (their establishment at other springs has been highly detrimental to native spring-dependent species).

CURRENT RECOVERY EFFORTS

Recovery plan: An initial recovery plan was completed in 1995 (Wager R (1995) 'Recovery plan for Queensland artesian spring fishes'. Report for the Australian Nature Conservation Agency, Canberra) and recently updated in 2010 (Fensham RJ, Ponder WF and Fairfax RJ (2010) 'Recovery plan for the community of native species dependent on natural discharge of groundwater from the Great Artesian Basin'. Report to the Department of the Environment, Water, Heritage and the Arts, Canberra. Queensland Department of Environment and Resource Management, Brisbane).

Other plans: Wagner R and Jackson P (1993) 'The action plan for Australian freshwater fishes'. Australian Nature Conservation Agency, Canberra.

Conservation actions:

- The Elizabeth Springs Conservation Park was established in 2000.
- All but one spring was fenced to exclude cattle in 1993–94.

FUTURE RECOVERY EFFORTS

Conservation needs: Consider the establishment of another population (captive or translocated) given the restricted range and area of this spring complex.

Recommendations from the recovery plan:

- Allocate water from the Great Artesian Basin for Elizabeth Springs (a priority).
- Maintain feral animal control and fencing.
- Regularly monitor the Spring Creek catchment for the presence of Mosquitofish.
- Cap and pipe flowing bores to reduce suitable habitat for Mosquitofish within the catchment.
- Identify and control bores that may influence spring discharge.
- Ensure that the flow rate of the springs does not decrease further. The Great Artesian Basin Sustainability Initiative (responsible for regulating the flow of artesian bores and capping free flowing bores) should target bores that have an impact on Elizabeth Springs.

Research directions: Research the basic biology and ecology of the species, which is little known.

Sources and further reading

Allen GR, Midgley SH and Allen M (2002) *Field Guide to the Freshwater Fishes of Australia*. Western Australian Museum, Perth.

Fensham RJ and Fairfax RJ (2003) Spring wetlands of the Great Artesian Basin, Queensland, Australia. *Wetlands Ecology and Management* **11**, 343–362.

Larson HK (1995) A review of the Australian endemic gobiid fish genus *Chlamydogobius*, with description of five new species. *The Beagle, Records of the Museums and Art Galleries of the Northern Territory* **12**, 19–51.

Unmack PJ and Wager RNE (2007) Threatened fishes of the world: *Chlamydogobius micropterus* Larson, 1995 (Gobiidae). *Environmental Biology of Fishes* **80**, 489–490.

Author: Steven Brooks

EDGBASTON GOBY

Chlamydogobius squamigenus Larson, 1995

Photo: Gunther Schmida

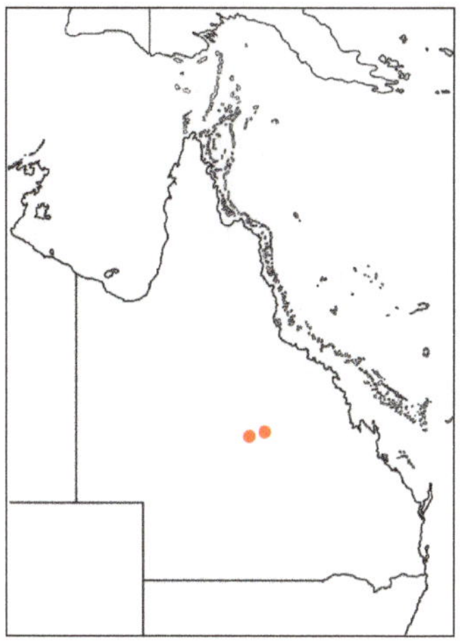

Family: Gobiidae

Conservation status: IUCN: Critically Endangered; Aust.: Vulnerable; Qld: Endangered

Description: The Edgbaston Goby is a small goby, which reaches a maximum total length of about 60 mm. It possesses first dorsal spines V–VII, second dorsal rays I,6–8, anal rays I,5–7 and pectoral rays 13–14. It differs from other species of the genus by the presence of cycloid scales in the preopercular region. Males in breeding condition become quite colourful; the body colour grades from olive on the back to golden yellow on the belly, with a chequered pattern of dark blotches on the back and sides. The fins are varying shades and patterns of blue to black and have a greyish white margin. The top of the first dorsal fin has a yellow flash. Females are mostly mottled drab with uncoloured fins.

Distribution: This species is found in Edgbaston Springs on Edgbaston Station, 80 km north of Aramac (22°44'S, 145°25'E) and bores and drains on Crossmoor Station (22°54'S, 144°35'E) between Longreach and Muttaburra in central western Qld. Although the species may once have been more widespread, Edgbaston Springs appears to be its stronghold. The Edgbaston Goby has been found in 19 springs, but they have only been recorded in four springs consistently over several years of monitoring. During monitoring in May 1997, the species was recorded in 11 springs, six of these springs represented new distribution records indicating dispersal among the springs during flooding.

Habitat: Edgbaston springs are located on a spinifex plain within Mitchell grass country. The complex contains approximately 51 active springs within an area of approximately 50 km^2. The combined total wetland is approximately 2.4 ha, though only 0.4 to 0.6 ha of this is usable habitat for Edgbaston Goby. Maximum depth varies between springs and may be as shallow as 5 mm. Grasses and other emergent aquatic vegetation grow throughout the springs. In deeper areas around the outflow, the habitat comprises a patchwork of small, open water areas and emergent clumps of substrate/vegetation. Narrower channels connect the small, open water areas. This appears to be the preferred habitat of the Edgbaston Goby. Maximum and minimum recorded water temperatures range from 39°C to 3°C, with as much as 21°C variation over several hours. Gobies display a preference for the warmer head areas of the springs. The presence of Edgbaston gobies at Crossmoor indicates that flowing bores may also provide suitable habitat for the species.

Ecology: During daylight, adult gobies shelter near or among the emergent vegetation. Juveniles may be found throughout the springs, but are usually found in shallower waters than the adults. Reproduction within the species is only known from captive populations where spawning commenced at temperatures above 20°C; limited data suggest that other details are similar to the Elizabeth Springs Goby *Chlamydogobius micropterus*. Spawning sites are yet to be located in the wild, though males in breeding colouration have often been observed near and defending the burrows of Freshwater Crayfish *Cherax destructor* and undercut tussocks. It is likely that this species is an omnivore and an opportunistic feeder.

Reasons for decline and current threats: A combination of the effects of exotic Mosquitofish *Gambusia holbrooki*, habitat loss and alteration, and the impacts of domestic animals have all caused detrimental effects on the Edgbaston Goby.

Gobies have only rarely been recorded in springs containing Mosquitofish; where this has occurred, the goby populations do not persist. Mosquitofish are thought to have invaded the springs during extensive flooding of the region and may invade other springs during future flooding. Trampling of the remaining habitat by domestic stock (now removed from Edgbaston Station) and feral pigs and eutrophication of water due to the faeces and decomposing carcases of these species have also impacted on the gobies. Flow reduction, the introduction of ponded pasture species and the alteration of the springs to provide watering opportunities for domestic stock have all reduced the available habitat for the species.

CURRENT RECOVERY EFFORTS

Recovery plan: An initial recovery plan was completed in 1995: Wager R (1995) 'Recovery plan for Queensland artesian spring fishes'. Report for the Australian Nature

Conservation Agency, Canberra. It was recently updated in 2010: Fensham RJ, Ponder WF and Fairfax RJ (2010) 'Recovery plan for the community of native species dependent on natural discharge of groundwater from the Great Artesian Basin'. Report to the Department of the Environment, Water, Heritage and the Arts, Canberra. Queensland Department of Environment and Resource Management, Brisbane.

Conservation actions:

- Edgbaston Station was recently purchased by Bush Heritage Australia.
- Eradication methods for Mosquitofish are presently being investigated by Bush Heritage Australia.
- Two springs on Edgbaston Station have been fenced off.
- All domestic stock has been removed from Edgbaston Station.

FUTURE RECOVERY EFFORTS

What you can do: Those interested in helping manage Edgbaston Springs should contact Bush Heritage Australia (<http://www.bushheritage.org.au>).

Conservation needs:

Recommendations from the recovery plan:

- Mitigate the impacts of domestic and feral animals, possibly through further fencing.
- Investigate possibilities for the removal of Mosquitofish and methods to prevent further invasions.
- Investigate possible reintroduction/translocation sites.
- Control introduced flora.
- Ensure that the flow rate of the springs does not decrease further. The Great Artesian Basin Sustainability Initiative (responsible for regulating the flow of artesian bores and capping free flowing bores) should target bores that have an impact on Edgbaston Springs.
- Investigate the potential impacts of mineral exploration/mining within the recharge area.

Research directions: Investigate the interaction between gobies and Mosquitofish.

Sources and further reading

Allen GR, Midgley SH and Allen M (2002) *Field Guide to the Freshwater Fishes of Australia*. Western Australian Museum, Perth.

Fensham RJ and Fairfax RJ (2003) Spring wetlands of the Great Artesian Basin, Queensland, Australia. *Wetlands Ecology and Management* **11**, 343–362.

Larson HK (1995) A review of the Australian endemic gobiid fish genus *Chlamydogobius*, with description of five new species. *The Beagle, Records of the Museums and Art Galleries of the Northern Territory* **12**, 19–51.

Author: Steven Brooks

Amphibians (Class: Amphibia)

Frogs (Order: Anura)

INTRODUCTION TO THREATENED FROGS

Keith R McDonald

Frogs worldwide have been experiencing declines and extinctions since the 1990s, which is a cause for concern because many species have quickly disappeared or become threatened over a short time frame. A range of impacts have contributed to different declines. These include loss of habitat, especially loss of wetlands, expanding urbanisation, disease such as ranavirus and the chytrid fungus, and overexploitation for human use. However, the occurrence of the chytrid fungus has had the largest worldwide impact, even in protected areas. Despite this fungus being implicated in declines in the Americas, much of the initial research on it, documenting declines and identifying the cause, was undertaken in Australia, especially on the stream-dwelling frogs of eastern Queensland.

There are currently 127 frog species from five native families in Queensland, with a few other species waiting formal scientific description. Thirty-one species (24%) in four of the families are listed as threatened (14 Endangered and 17 Vulnerable) in Queensland's NCA regulations. Twenty-three of these are found only in Queensland, with the others also occurring in northern New South Wales. One species, *Litoria pearsoniana,* has a population at Kroombit Tops that is listed as Endangered. In this book, two additional species are incorporated because they are listed in the *IUCN Red List of Threatened Species*, although they are not recognised as threatened in any Australian legislation. The EPBC Act lists four species of the NCA listed Endangered species as Presumed Extinct (see Extinct in Queensland pp. 10–14).

Seventy-eight per cent of the threatened frogs are found in the rainforest and wet forest of eastern Queensland: areas of high plant and animal diversity. All have been recorded from protected areas, but, despite the protection of habitat, declines from chytrid fungus disease have occurred in stream-dwelling species. Chytrid fungus is known, or believed to have, affected 18 of the threatened species, mostly stream-inhabiting species. Five other declining species are found on the coastal lowlands and are threatened by development for urbanisation (especially south-east Queensland), pine plantations and agriculture.

Many of the species listed as Vulnerable occur only in small areas and, although currently not experiencing declines in populations or habitat, have populations so small and isolated that they can be impacted by stochastic events in the future. Several of the nurseryfrogs are in this category and are restricted to mountain tops, which could be affected by climate change. Additional studies and population monitoring will need to be undertaken over time to confirm this is occurring. Rock-dwelling species are restricted to small areas, especially in the northern Wet Tropics and Cape York Peninsula. Although currently secure, stochastic events could have significant impacts, even within protected areas.

Although the chytrid fungus appears to have reached its limit of range in Queensland because of environmental restraints such as temperature and moisture patterns, it continues to affect populations where it is found by preventing or slowing their recovery to levels prior to chytrid introduction, in addition to other ongoing impacts such as loss of wetlands and adjacent habitat, habitat splitting and loss, and expanding urbanisation. With the globalisation of trade, potential introduction of disease and predators through hitch-hiker species on international produce and cargo containers are a cause for concern.

The following profiles provide the best current information on the species of frogs that are threatened. It is expected additions will happen as research results become available and reassessment of species is incorporated into legislation and lists.

TUSKED FROG

Adelotus brevis (Günther, 1863)

Photo: Harry B Hines-DERM

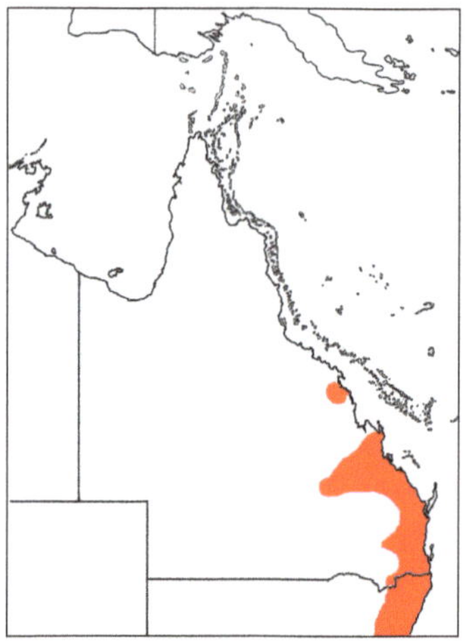

Family: Limnodynastidae

Conservation status: IUCN: Near Threatened; Qld: Vulnerable; NSW: Endangered (Nandewar and New England Tablelands bioregions)

Description: This robust medium-sized frog measures up to 50 mm in length. Both sexes have paired tusks at the front of the lower jaw, visible only when the mouth is open. Adult males are larger and more robust than females, with a broader head and larger tusks. The dorsal surface is rough and grey, beige or brown in colour, with irregular darker markings. A dark butterfly shaped marking on the head commences between the eyes. The ventral surface is smooth with black/grey and white marbling or mottling. There is black and red marbling in the groin and on the hind sides of the calf. Fingers and toes have rudimentary webbing with no discs at the tips. The second finger of the female is flanged. The tympanum is hidden. Tadpoles are dark brown to black above and dark grey below and reach a maximum length of about 35 mm. They are deep bodied, ovoid, with a rounded snout, terminal mouth and dorsolateral eyes. The fins are low, translucent and suffused with black. The tail musculature is grey brown in colour and the tail tip rounded. The spiracle is sinistral and the vent is dextral. The oral disc is surrounded by marginal papillae except for an anterior gap. Labial tooth row formula: 4-5(3-5)/3(1).

Distribution: Historically widely distributed in temperate and subtropical forests along the east coast, from the Clarke Range in mid-east Qld to mid-east NSW. There are disjunct populations at Carnarvon Gorge and Blackdown Tableland. In Qld, it has disappeared from the New England Tableland and Bunya Mountains and is very scarce or patchy in upland areas of the Scenic Rim. It remains widespread in coastal lowlands and foothills in the south-east.

Habitat: Breeds in ponds and slow moving sections of streams in rainforest, wet sclerophyll forest and, less commonly, dry open forest and is also known to breed in dams and ornamental ponds in urban and peri-urban areas.

Ecology: The Tusked Frog is typically encountered in close proximity to ponds and slow moving sections of streams, but is occasionally seen well away from water under leaf litter, logs or in loose soil. Males typically call from concealed locations within or beside water, use their tusks to attack rivals by biting vigorously and can be heard calling night and day, mostly in spring and early summer. Amplexus is inguinal. Females lay unpigmented eggs (between 148 and 638) in a floating foamy egg mass, often concealed under a log or rock. Tadpoles may over-winter in upland sites. Diet is varied and includes arthropods, molluscs, crustaceans, annelids, onycophorans and small vertebrates (other frogs). The male advertisement call is a short soft 'pluck', 'p-tuk' or 'p-t-t-tuk'.

Reasons for decline and current threats: The decline and disappearance of the Tusked Frog, and a number of other species, from the New England Tableland and upland areas of south-east Qld probably occurred in the late 1970s to mid 1980s.

Although the cause of these population declines is not known, the amphibian chytrid fungus disease is strongly suspected. Since the mid 1990s, a number of moribund or dead Tusked Frogs have been diagnosed with this disease. It is one of a handful of frog species that survives in highly modified and human-made habitats (e.g. ornamental ponds in suburban Brisbane); although at the local level there is ongoing loss and fragmentation of natural habitat through increased urbanisation.

CURRENT RECOVERY EFFORTS

Recovery plan: None

Other plans: Department of the Environment and Heritage (DEH) (2006) 'Threat abatement plan: infection of amphibians with chytrid fungus resulting in chytridiomycosis'. Department of Environment and Heritage, Canberra.

Conservation actions:

- Ongoing population surveys and monitoring and investigations into basic life history are being carried out.

- Research is being conducted on the epidemiology of the amphibian chytrid fungus.
- DERM manages parks and state forests where this species occurs, with particular emphasis on control of feral pigs; undertakes surveys and monitoring; liaises with universities and other research institutions. Activities are severely limited by funding availability.
- Mary River Catchment Coordinating Committee conducts surveys and population monitoring.
- UQ, Griffith University and JCU provide funding, research expertise and student volunteers to assist with research and recovery actions.
- JCU, CSIRO and UQ support amphibian disease research.
- Volunteers assist with surveys, monitoring, data management and logistical support.

FUTURE RECOVERY EFFORTS

What you can do:

- Donate to the Qld Frog Society Trust Fund.
- Write letters to state and federal environment ministers urging much greater funding allocations to implement recovery and research recommendations.
- Clean and disinfect soiled clothes/equipment when moving between streams and wetland areas to reduce the risk of spreading disease among frog populations.

Conservation needs:

- Prepare a recovery plan.
- Protect and restore habitat.
- Implement the threat abatement plan for amphibian chytrid fungus.
- Develop ecologically sensitive weed management in riparian areas.

Research directions:

- Develop a better understanding of the threatening processes, especially the role of amphibian chytrid fungus.
- Conduct population monitoring.
- Determine the genetic structure of the population.

Sources and further reading

Gillespie G and Hines H (1999) Status of temperate riverine frogs in south-eastern Australia. In *Declines and Disappearances of Australian Frogs*. (Ed. A Campbell) pp. 109–130. Environment Australia, Canberra.

Hines HB, Mahony M and McDonald K (1999) An assessment of frog declines in wet subtropical Australia. In *Declines and Disappearances of Australian Frogs*. (Ed. A Campbell) pp. 44–63. Environment Australia, Canberra.

Ingram GJ and McDonald KR (1993) An update on the decline of Queensland's frogs. In *Herpetology in Australia*. (Eds D Lunney and D Ayers) pp. 297–303. Surrey Beatty & Sons, Sydney.

Katsikaros K and Shine R (1997) Sexual dimorphism in the tusked frog, *Adelotus brevis* (Anura, Myobatrachidae): the roles of natural and sexual selection. *Biological Journal of the Linnean Society* **60**, 39–51.

Meyer E, Hines H and Hero J-M (2001) *Wet Forest Frogs of South-east Queensland*. Griffith University, Brisbane.

Author: Harry B Hines

BEAUTIFUL NURSERYFROG

Cophixalus concinnus Tyler, 1979

Photo: Conrad J Hoskin

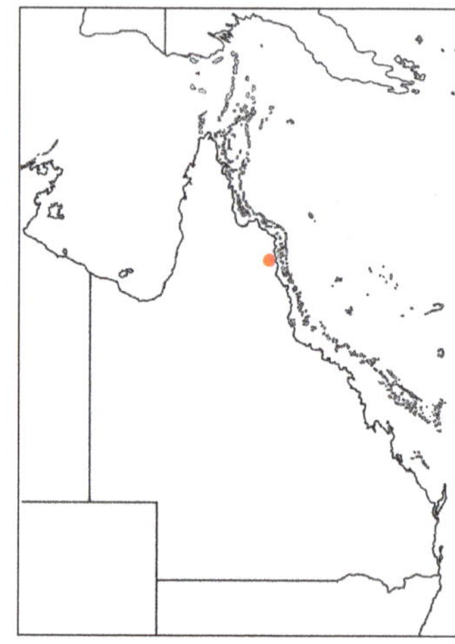

Family: Microhylidae

Conservation status: IUCN: Critically Endangered; Qld: Vulnerable

Description: This small frog measures 18–26 mm in length, with a squat body shape, a blunt face and short hindlimbs. The skin is smooth or slightly bumpy. The dorsal surface colour is variable, but is generally black, grey or brown in males, and paler in females. The ventral surface is distinctive, with red or orange blotches and dark stippled areas and blotches on a white background. This red/orange is particularly bright on the throat/chest of males. The finger discs are large and truncated. The toe discs are expanded and round. There is no webbing between the fingers or toes.

Distribution: Very restricted distribution to higher altitudes (>1100 m) of Thornton Peak, north-east Qld. Distribution is wholly within Daintree NP, in the Wet Tropics World Heritage Area. No estimate of population size, but occurs at reasonable high abundance (although patchily) within distribution.

Habitat: Montane rainforest and boulder fields. Particularly common where rainforest grows on or among boulders.

Ecology: Following rain or heavy mist, males call at night from elevated positions among plants and rocks or from within boulder piles. The call is a short creaking sound. Females rarely encountered, but have been seen foraging among shrubs at night. Frogs sometimes found during the day sheltering beneath logs and rocks on the ground. Eggs are laid on the ground, with development of froglets occurring within the eggs (no free swimming tadpole). A clutch of 17 eggs has been found beneath a rock and attended by an adult. Diet consists of invertebrates.

Reasons for decline and current threats: Listed as Critically Endangered on the IUCN Red List due to extremely small distribution, all individuals being in a single location/population, and predicted declines due to global climate change in the number of individuals, and extent and quality of habitat. Environmental modelling of future climate change predicts this species to be particularly threatened. No population decline has been recorded to date, but monitoring has been limited.

Microhylid frog populations are not affected by chytrid fungus disease in the wild, but the sudden impact of this disease on other frog species in apparently pristine rainforest in the Wet Tropics highlights the potential threat of emerging diseases.

CURRENT RECOVERY EFFORTS

Recovery plan: None

Conservation actions: No recovery efforts are in place, but surveys and monitoring have been conducted over recent years.

FUTURE RECOVERY EFFORTS

What you can do:

- Act at a local level to minimise your negative impacts towards global climate change.
- Support state and federal governments that strive to act effectively on global climate change, and pressure those that don't.
- Write letters to state and federal environment ministers to urge much greater funding allocations to implement recovery and research recommendations.

Research directions:

- Conduct continued surveys and long-term monitoring.
- Research the ecology and physiology of the montane microhylid frogs of the Wet Tropics, particularly regarding the likely impacts of climate change and the potential for these species to adapt to it.

Sources and further reading

Hoskin CJ (2004) Australian microhylid frogs (*Cophixalus* and *Austrochaperina*): phylogeny, taxonomy, calls, distributions and breeding biology. *Australian Journal of Zoology* **52**, 237–269.

Hoskin CJ and Hero J-M (2008) *Rainforest Frogs of the Wet Tropics, North-east Australia.* Griffith University, Gold Coast.

Hoskin CJ and Higgie M (2005) Minimum calling altitude of *Cophixalus* frogs on Thornton Peak, northeastern Queensland. *Memoirs of the Queensland Museum* **51**, 572.

Williams SE, Bolitho EE and Fox S (2003) Climate change in Australian tropical rainforests: an impending environmental catastrophe. *Proceedings of the Royal Society B* **270**, 1887–1892.

Williams YM, Williams SE, Alford RA, Waycott M and Johnson C (2006) Niche breadth and geographical range: ecological compensation for geographical rarity in rainforest frogs. *Biology Letters* **2**, 532-535.

Author: Conrad J Hoskin

NORTHERN NURSERYFROG

Cophixalus crepitans Zweifel, 1985

Photo: Keith R McDonald-DERM

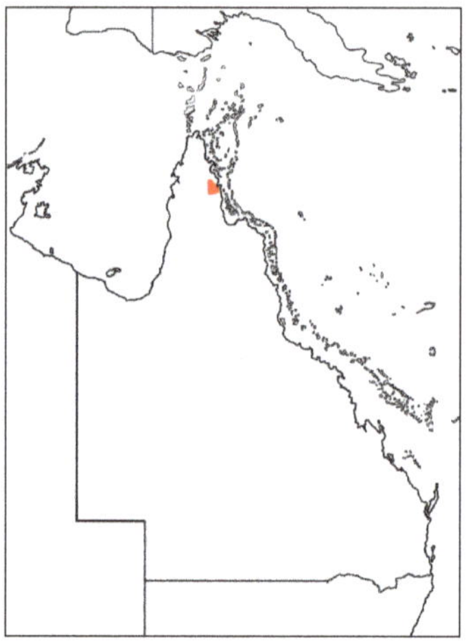

Family: Microhylidae

Conservation status: IUCN: Near Threatened; Qld: Vulnerable

Description: This very small frog measures 11.6–14.1 mm in length and has relatively long legs, about half its total length. The dorsal colours are brown to rusty-brown with darker mottling and blotches, including a W-shaped mark between the shoulders and obscure dark spots in the lumbar region. The limbs have indistinct darker cross bands. The ventral surface is brown with paler spots and mottling. The skin is smooth, with a weak fold above the tympanum. The finger and toe discs are moderately developed, with that of the 4th toe wider than the 3rd. The digits lack webbing. The tympanum is small and barely visible.

Distribution: Restricted to McIlwraith Range, Cape York Peninsula.

Habitat: The Northern Nurseryfrog occurs in rainforest and monsoon vine forest between 400 and 540 m altitude on granite hills.

Ecology: This frog is found in leaf litter and under logs, rocks and branches on the forest floor. Males call during spring and summer on moist nights or overcast days from debris on the forest floor. The call is a rattling sound comprised of 31–44 double pulsed clicks and lasting ~2 seconds. No egg clutches have been located, but the species most likely reproduces similar to other nurseryfrogs with eggs being laid in moist sheltered locations on the ground and development occurring within the egg (no free swimming tadpole).

Reasons for decline and current threats: The small distribution and population size of this species makes it vulnerable to stochastic events threatening it with extinction. Currently, it appears secure, with most of its population occurring in protected area with little activity or impacts. No declines or threats are known.

CURRENT RECOVERY EFFORTS

Recovery plan: None

Conservation actions: The Northern Nurseryfrog has most of its habitat in Kulla (McIlwraith Range) Cape York Peninsula Aboriginal Land (CYPAL) NP and therefore not at any immediate risk.

FUTURE RECOVERY EFFORTS

Research directions:

- Conduct surveys to determine any additional populations within the area.
- Because it is restricted to altitude in its range, research into its physiological limits of tolerance of drying and temperature and variation in microhabitat conditions within its habitat may help predict its susceptibility to climate change.
- Undertake periodic monitoring to ensure the species remains secure.

Sources and further reading

Hoskin CJ (2004) Australian microhylid frogs (*Cophixalus* and *Austrochaperina*): phylogeny, taxonomy, calls, distributions and breeding biology. *Australian Journal of Zoology* **52**, 237–269.

Hoskin CJ (2008) A key to the microhylid frogs of Australia, and new distributional data. *Memoirs of the Queensland Museum* **52**, 233–237.

Zweifel RG (1985) Australian frogs of the family Microhylidae. *Bulletin American Museum of Natural History* **182**, 265–388.

Author: Keith R McDonald

DAINTY NURSERYFROG

Cophixalus exiguus Zweifel & Parker, 1969

Photo: Keith R McDonald-DERM

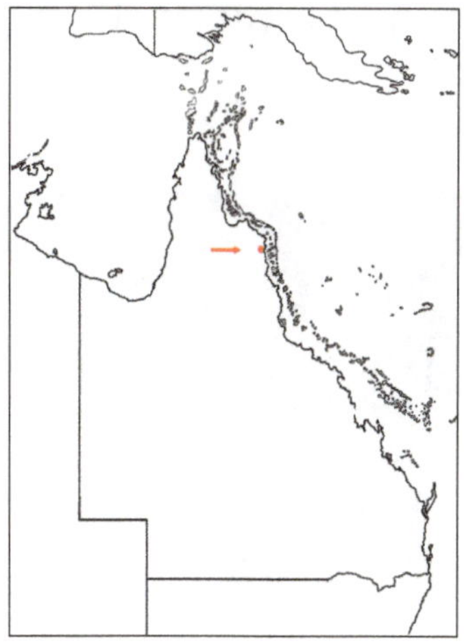

Family: Microhylidae

Other common name: Northern Tapping Nursery-Frog

Conservation status: IUCN: Near Threatened; Qld: Vulnerable

Description: The males of this small squat frog measure 14–16 mm in length and the females 16–19 mm. They have unwebbed fingers and small discs on the fingers and toes. Dorsal colours are brown, tan or grey-brown, sometimes with a vertebral stripe, and a faint ocellate marking above the groin. There is a dark stripe above ear to forearm and a mid body dorso-lateral blotch. The ventral surface is pale, sometimes with brown markings, especially around the throat and chest.

Distribution: Coastal ranges of the northern Wet Tropics bioregion. The species is known from three disjunct populations covering an area of <20 km^2 above 180 m altitude within Timber Reserve 165 Monkhouse, some areas of which will become national park.

Habitat: Wet rainforest on granite hills immediately adjacent to the coast.

Ecology: The frog is found in leaf litter and under logs, rocks and branches on the forest floor. Males call during spring and summer on moist nights or overcast days from forest floor debris. The call is a tapping sound, like a marble bouncing on tile. Eggs are laid on the ground and there is direct development within the egg (no free swimming tadpole). The only nest located had 11 eggs with a male in attendance.

Reasons for decline and current threats: A very restricted species. No declines or threats currently known.

CURRENT RECOVERY EFFORTS

Recovery plan: None

Conservation actions: No recovery efforts have been undertaken to date, but the species distribution is partly in national park.

FUTURE RECOVERY EFFORTS

Conservation actions: Transfer Timber Reserve 165 Monkhouse to protected area.

Research directions:

- Conduct surveys to determine any additional populations within the area.
- Research the physiological limits of the species and variation in microhabitat within its broader habitat. This will be a good baseline for future climate change impacts.

Sources and further reading

Hoskin CJ (2004) Australian microhylid frogs (*Cophixalus* and *Austrochaperina*): phylogeny, taxonomy, calls, distributions and breeding biology. *Australian Journal of Zoology* **52**, 237–269.

Hoskin CJ (2008) A key to the microhylid frogs of Australia, and new distributional data. *Memoirs of the Queensland Museum* **52**, 233–237.

Hoskin CJ and Hero J-M (2008) *Rainforest Frogs of the Wet Tropics, North-east Australia.* Griffith University, Gold Coast.

Zweifel RG (1985) Australian frogs of the family Microhylidae. *Bulletin American Museum of Natural History* **182**, 265–388.

Author: Keith R McDonald

RATTLING NURSERYFROG

Cophixalus hosmeri Zweifel, 1985

Photo: Andrew J Dennis

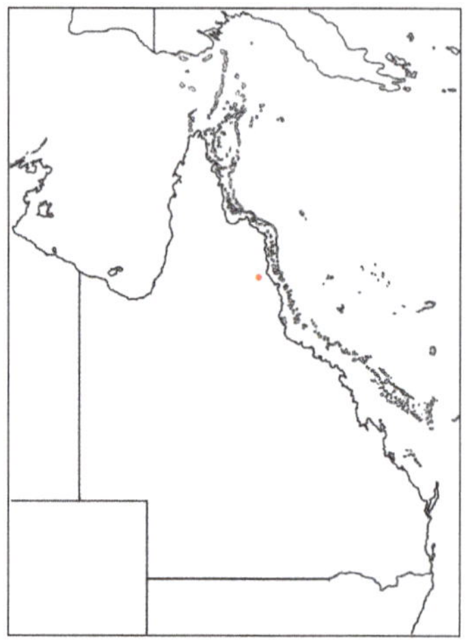

Family: Microhylidae

Conservation status: IUCN: Least Concern; Qld: Near Threatened

Description: This very small frog measures 11–17 mm in length (males average 13 mm, females slightly larger) and has a squat body shape, blunt face and small hind limbs. The skin texture ranges from smooth to bumpy. The dorsal colour and pattern is highly variable, with generally dark blotches or mottling on a grey, brown and orange background. The ventral surface is smooth and pale, grey or yellowish. Males and females are of similar appearance. First finger ('thumb') is very short. There is no webbing between the fingers or toes. The finger and toe discs are small and rounded.

Distribution: Restricted to higher altitudes (>800 m) of the Carbine Tableland, north-east Qld. Distribution is within the Daintree NP, Mount Lewis NP and Brooklyn Nature Refuge in the Wet Tropics World Heritage Area. There are currently no estimates of population size, but this species generally occurs at reasonably high abundance within its distribution.

Habitat: Montane rainforest on granitic soils.

Ecology: Males call at night following rain or heavy mist, from elevated positions (generally <30 cm) such as palm axils, leaf litter caught in vegetation, logs, rocks and embankments, sometimes also from leaf litter on the ground. Call is highly variable: typically it is a fast-paced high-pitched tapping, but also a slower tapping or a buzz. Genetic and morphological variation suggests this call variation represents a single species, but the reason for such call variation remains unresolved. Females are rarely encountered. Both sexes can be found during the day sheltering beneath logs and rocks on the ground. Eggs are laid on the ground, with development of froglets occurring within the eggs (no free swimming tadpole). Males have been found with small egg clutches, and a male has been found among leaf litter on Mt Lewis with seven hatchlings. Diet consists of invertebrates.

Reasons for decline and current threats: Listed as Vulnerable on the IUCN Red List due to small distribution, known from fewer than five locations, and predicted declines in the number of individuals due to global climate change, and extent and quality of habitat. No population decline has been recorded to date, but monitoring has been limited.

Microhylid frog populations are not affected by chytrid fungus disease in the wild, but the sudden impact of this disease on other frog species in apparently pristine rainforest in the Wet Tropics highlights the potential threat of emerging diseases.

CURRENT RECOVERY EFFORTS

Recovery plan: None

Conservation actions: No recovery efforts are in place, but surveys and monitoring have been conducted over recent years.

FUTURE RECOVERY EFFORTS

What you can do:

- Act at a local level to minimise your negative impacts towards global climate change.
- Support state and federal governments that strive to act effectively on global climate change, and pressure those that don't.
- Pressure state and federal governments to increase funding for research and management of threatened species and habitats.

Research directions:

- Conduct continued surveys and long-term monitoring.
- Research the ecology and physiology of the montane microhylid frogs of the Wet Tropics, particularly regarding the likely impacts of climate change and the potential for these species to adapt to it.

Sources and further reading

Hoskin CJ (2004) Australian microhylid frogs (*Cophixalus* and *Austrochaperina*): phylogeny, taxonomy, calls, distributions and breeding biology. *Australian Journal of Zoology* **52**, 237–269.

Hoskin CJ and Hero J-M (2008) *Rainforest Frogs of the Wet Tropics, North-east Australia*. Griffith University, Gold Coast.

Shoo LP and Williams YM (2004) Altitudinal distribution and abundance of microhylid frogs (*Cophixalus* and *Austrochaperina*) of north-eastern Australia: baseline data for detecting biological responses to future climate change. *Australian Journal of Zoology* **52**, 667–676.

Williams SE, Bolitho EE and Fox S (2003) Climate change in Australian tropical rainforests: an impending environmental catastrophe. *Proceedings of the Royal Society B* **270**, 1887–1892.

Williams YM, Williams SE, Alford RA, Waycott M and Johnson C (2006) Niche breadth and geographical range: ecological compensation for geographical rarity in rainforest frogs. *Biology Letters* **2**, 532-535.

Author: Conrad J Hoskin

MOUNT ELLIOT NURSERYFROG

Cophixalus mcdonaldi Zweifel, 1985

Photo: Conrad J Hoskin

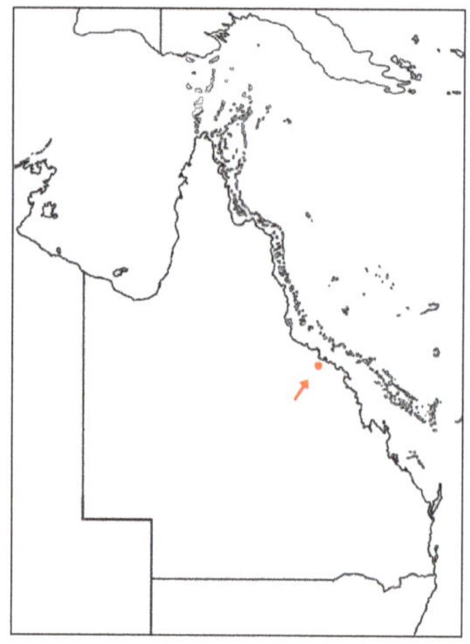

Family: Microhylidae

Other common name: McDonald's Nurseryfrog

Conservation status: IUCN: Endangered; Qld: Vulnerable

Description: The males of this small squat frog measure 16–23 mm in length and the females 21–26 mm. The finger and toe discs are expanded and similar in size. The first finger is very short. The dorsal surface is light to dark brown with slightly darker markings, a darker bar between the eyes, and sometimes a faint ocellate marking above the groin. There is a head stripe above ear to face and mottling on flanks. The ventral surface is pale sometimes with brown mottling.

Distribution: Only known from above 900 m on Mount Elliot in Bowling Green Bay NP, south of Townsville with an area of occurrence of approximately 10 km^2.

Habitat: Restricted montane rainforest on granitic soils. This frog is found in leaf litter and forest floor debris.

Ecology: Males call at night during moist spring and summer nights from leaf litter or raised sites on forest floor debris. The call is a short creak. Eggs are laid on the ground, with development to froglet occurring within the egg (no free swimming tadpole). Seventeen eggs have been found in a dissected female.

Reasons for decline and current threats: There are no threats or declines currently known. This species could potentially be affected by climate change, but all predicted impacts are based on models at a landscape level.

CURRENT RECOVERY EFFORTS

Recovery plan: None

Conservation actions: No recovery efforts have been undertaken to date, but the species is totally in national park.

Other plans: Bowling Green Bay National Park Management Plan 2000.

FUTURE RECOVERY EFFORTS

Research directions: Research the physiological limits of the species and variation in microhabitat conditions within its habitat. This will be a good baseline for future climate change impacts.

Sources and further reading

Hoskin CJ (2004) Australian microhylid frogs (*Cophixalus* and *Austrochaperina*): phylogeny, taxonomy, calls, distributions and breeding biology. *Australian Journal of Zoology* **52**, 237–269.

Hoskin CJ (2008) A key to the microhylid frogs of Australia, and new distributional data. *Memoirs of the Queensland Museum* **52**, 233–237.

Hoskin CJ and Hero J-M (2008) *Rainforest Frogs of the Wet Tropics, North-east Australia.* Griffith University, Gold Coast.

Zweifel RG (1985) Australian frogs of the family Microhylidae. *Bulletin American Museum of Natural History* **182**, 265–388.

Author: Keith R McDonald

MOUNTAIN-TOP NURSERYFROG

Cophixalus monticola Richards, Dennis, Trenerry & Werren, 1994

Photo: Andrew J Dennis

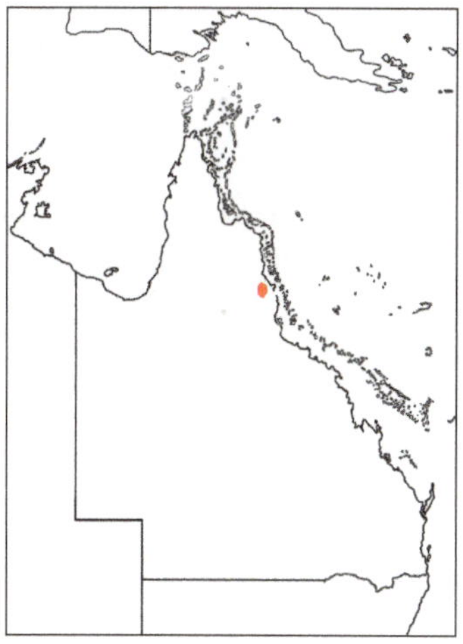

Family: Microhylidae

Conservation status: IUCN: Endangered; Qld: Vulnerable

Description: This very small, squat frog measures 17–20 mm in length and has highly variable colours and patterns, ranging from pale to dark brown and even bright orange patches. Most patterns involve lateral or vertebral stripes of varying size, but it usually has two obvious pale circular markings above the groin. When undisturbed it has scattered tubercles, although when handled its skin can become quite smooth. Its limbs and toes are short with round pads, and the first finger is half the length of the next.

Distribution: The Mountain-top Nurseryfrog is found in only a few areas of the Carbine Tablelands rainforests above 1100 m. It occurs in small patches within each area, but may occur in other high altitude patches in other parts of the tableland.

Habitat: The Mountain-top Nurseryfrog occurs in montane rainforest with high rainfall and relatively cool temperatures for the tropics. Within these forests, males frequent patches of *Linospadix* palms, a small understorey palm that can grow in large aggregations in moist areas. Being such small frogs, they are unlikely to travel large distances, so females and immature individuals presumably also reside in the vicinity of male groups, but may be more terrestrial.

Ecology: Males of the Mountain-top Nurseryfrog call a short trill, generally from leaf litter trapped in the axils of *Linospadix* palms where they are known to remain there day and night. They may also call from similar patches of elevated leaf litter or moist crevices on tree trunks, fallen branches and occasionally exposed on low vegetation. Calling is contingent on storm season rains and is generally at its most intense after heavy rain has ceased. Clutches of 8–13 eggs have been recorded at male calling sites in *Linospadix*, nestled into the moist litter. When not calling, the male attends the eggs and will arch over them to protect them if disturbed. As with other microhylids, the tadpoles remain within the egg and young frogs hatch directly from eggs. They likely consume tiny invertebrates and may, like many other microhylids, show some degree of preference for ants.

Reasons for decline and current threats: The species is listed as threatened on the basis of its restricted distribution and small and patchy populations. There is no indication of a decline and no disturbance threatening its well-protected habitat. Like other species listed on the basis of a restricted distribution or small population size, it is vulnerable to stochastic events such as climatic anomalies or the invasion of a novel disease. In addition, being adapted to a cool, moist, high altitude environment, climate change, which is predicted to include warming and drying, could be extremely detrimental. This is possibly of greater concern for this species than many others due to its habit of sheltering and breeding in elevated patches of leaf litter in the understorey: a microhabitat that is likely to dry faster than those associated with the forest floor or boulder piles.

CURRENT RECOVERY EFFORTS

Recovery plan: None

Conservation actions: The habitat of the species is protected within World Heritage listed national park throughout its entire range. In addition, vehicle access is only available along a single road and camping and other facilities are not available, which reduces human impacts.

FUTURE RECOVERY EFFORTS

Conservation needs: The species is currently well protected and its habitat secure. Future actions for this species should be general ones to reduce the severity of climate change and reduce the possibility of novel diseases or pests entering the country.

Research directions: The Mountain-top Nurseryfrog is an excellent candidate for ongoing monitoring of the impacts of climate change.

Sources and further reading

Hoskin CJ (2004) Australian microhylid frogs (*Cophixalus* and *Austrochaperina*): phylogeny, taxonomy, calls, distributions and breeding biology. *Australian Journal of Zoology* **52**, 237–269.

Hoskin CJ (2008) A key to the microhylid frogs of Australia, and new distributional data. *Memoirs of the Queensland Museum* **52**, 233–237.

Hoskin CJ and Hero J-M (2008) *Rainforest Frogs of the Wet Tropics, North-east Australia*. Griffith University, Gold Coast.

Richards SJ, Dennis AJ, Trenerry MP and Werren GL (1994) A new species of *Cophixalus* (Anura: Microhylidae) from northern Queensland. *Memoirs of the Queensland Museum* **37**, 307–310.

Shoo LP and Williams Y (2004) Altitudinal distribution and abundance of microhylid frogs (*Cophixalus* and *Austrochaperina*) of north-eastern Australia: baseline data for detecting biological responses to future climate change. *Australian Journal of Zoology* **52**, 667–676.

Author: Andrew J Dennis

BELLENDEN KER NURSERYFROG

Cophixalus neglectus Zweifel, 1962

Photo: Andrew J Dennis

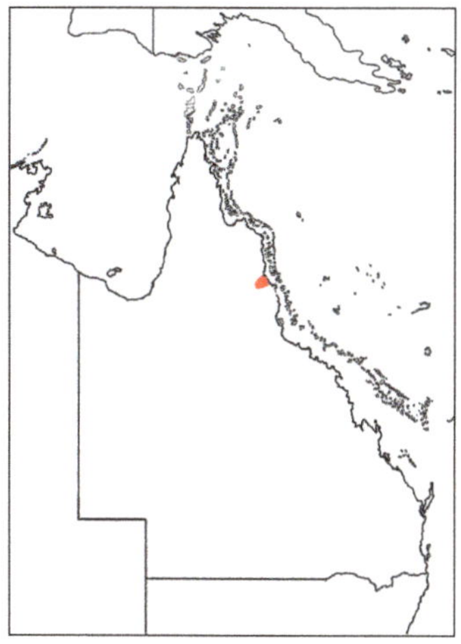

Family: Microhylidae

Conservation status: IUCN: Endangered; Qld: Vulnerable

Description: The males of this small squat frog measure 20–23 mm in length and the females 21–28 mm. They have poorly developed finger and toe discs and the first finger is very short and with no disc. The dorsal surface is brown to orange-brown, sometimes with a vertebral stripe, and occasionally a faint ocellate marking above the groin. There is a head stripe above the ear to face. The ventral surface is pale, sometimes with brown flecks.

Distribution: Restricted to two disjunct populations between 850 and 1600 m on the Bellenden Ker Range in Wooroonooran NP.

Habitat: Montane notophyll and microphyll vine forest on granitic soils. This frog is found in leaf litter and under logs, rocks and branches on the forest floor.

Ecology: Males call during spring and summer on moist nights or overcast days from forest floor debris. The call is a buzz or squelch. Egg numbers vary from 10–32 eggs in a 'necklace' string and are laid on the ground under cover. There is no free swimming tadpole stage and the young emerge as fully developed froglets.

Reasons for decline and current threats: Climate change is the most likely threat because recent records are primarily 1200 m and above.

CURRENT RECOVERY EFFORTS

Recovery plan: None

Conservation actions: No recovery efforts have been undertaken to date, but the species is totally in national park.

FUTURE RECOVERY EFFORTS

Research directions:

- Research the natural history of the species, especially the breeding biology, the physiological limits of temperature and moisture and variation in the microhabitat conditions within its habitat. This will be a good baseline for future climate change impacts.
- This pivotal species must be monitored over a range of altitudes for indications of population contraction associated with climate change in upland areas of the Wet Tropics.

Sources and further reading

Hoskin CJ (2004) Australian microhylid frogs (*Cophixalus* and *Austrochaperina*): phylogeny, taxonomy, calls, distributions and breeding biology. *Australian Journal of Zoology* **52**, 237–269.

Hoskin CJ (2008) A key to the microhylid frogs of Australia, and new distributional data. *Memoirs of the Queensland Museum* **52**, 233–237.

Hoskin CJ and Hero J-M (2008) *Rainforest Frogs of the Wet Tropics, North-east Australia*. Griffith University, Gold Coast.

Zweifel RG (1985) Australian frogs of the family Microhylidae. *Bulletin American Museum of Natural History* **182**, 265–388.

Author: Keith R McDonald

CAPE YORK NURSERYFROG

Cophixalus peninsularis Zweifel, 1985

Never been photographed

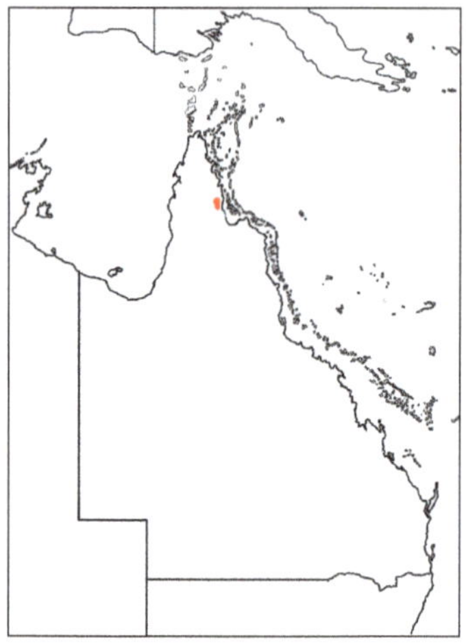

Family: Microhylidae

Conservation status: IUCN: Data Deficient; Qld: Vulnerable

Description: This small frog measures 17–18 mm in length and has widely separated nostrils. Its dorsal surface is brown above with darker markings on the snout, face, shoulders and flanks, and faint dark spots in the lumbar region. The limbs have indistinct darker spots and crossbars. The ventral surface is brown with paler spots and mottling. The skin is smooth except for faint skin folds behind the eye and on the shoulder. The finger and toe discs are well developed and the digits lack webbing.

Distribution: This frog is restricted to McIlwraith Range, Cape York Peninsula and known only from the two type specimens.

Habitat: It occurs in rainforest at 500 m altitude on granite hills.

Ecology: The only frogs seen were found calling in leaf litter during spring. The call is a rattling sound comprised of a series of clicks

Reasons for decline and current threats: A very restricted species, known only from two specimens. No declines or threats are currently known.

CURRENT RECOVERY EFFORTS

Recovery plan: None

Conservation actions:

- No recovery efforts have been undertaken to date, but the distribution of the species is in national park.
- The Cape York Nurseryfrog has its habitat in Kulla (McIlwraith Range) CYPAL NP and therefore, many risks are minimised.

FUTURE RECOVERY EFFORTS

Research directions: Conduct surveys to collect basic information on its extent of occurrence, population size and find additional populations on which to conduct basic ecology research. However, because it occurs in a remote location that is relatively inaccessible during its season of activity, this is likely a major undertaking requiring helicopter support.

Sources and further reading

Hoskin CJ (2004) Australian microhylid frogs (*Cophixalus* and *Austrochaperina*): phylogeny, taxonomy, calls, distributions and breeding biology. *Australian Journal of Zoology* **52**, 237–269.

Hoskin CJ (2008) A key to the microhylid frogs of Australia, and new distributional data. *Memoirs of the Queensland Museum* **52**, 233–237.

Zweifel RG (1985) Australian frogs of the family Microhylidae. *Bulletin American Museum of Natural History* **182**, 265–388.

Author: Keith R McDonald

BLACK MOUNTAIN BOULDERFROG

Cophixalus saxatilis Zweifel & Parker, 1977

Photo: Michael Trenerry

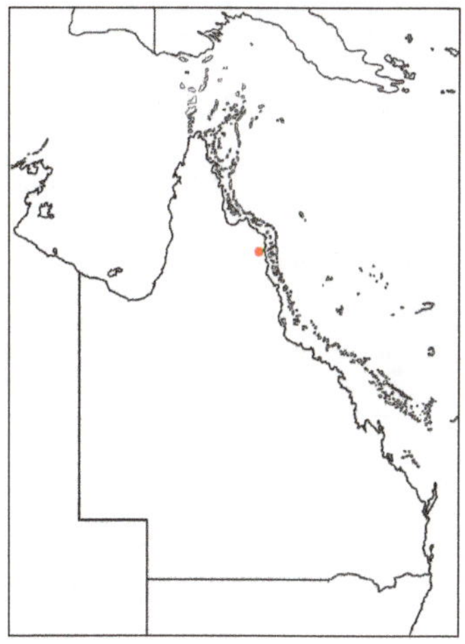

Family: Microhylidae

Conservation status: IUCN: Vulnerable; Qld: Vulnerable

Description: A large microhylid frog, with males measuring 29–35 mm in length and females 39–46 mm, with well developed finger and toe discs. The dorsal surface on females is vivid bright yellow at night and fades to a tan colour during the day. Males are mottled brown or grey, sometimes with a W-shaped marking on the shoulders. Some individuals have ocellate markings above the groin. The ventral surface is whitish with peppering of dark brown.

Distribution: Northern edge of the Wet Tropics bioregion, south of Cooktown. Restricted to 6 km^2 in the Black Mountain NP and adjacent Water Reserve, between 100 and 300 m altitude. The Black Mountain Boulderfrog is currently secure in a small distribution within protected habitat, with no indication of population declines.

Habitat: This frog is found in the Black Trevethan Range, within the granite boulder field with limited rainforest vegetation in gullies.

Ecology: This frog lives in the humid boulder piles, coming to the surface during moist conditions in the wet season. Males call from crevices between boulders, with the call a series of taps over 2–3 seconds. Females lay terrestrial eggs, with one clutch of 13 recorded.

Reasons for decline and current threats: No known decline or threat exists, but the species has a very restricted distribution in specialised boulder habitat, which may be affected by stochastic events such as prolonged drought.

CURRENT RECOVERY EFFORTS

Recovery plan: None

Conservation actions: No recovery efforts have been undertaken to date, but the distribution of the species is mainly in national park.

FUTURE RECOVERY EFFORTS

Conservation needs: Add the area of boulder field in the Water Reserve to the adjacent national park.

Research directions: Research the natural history of the species, especially breeding biology, the physiological limits of temperature and moisture, and variation in microhabitat conditions within its habitat. This will be a good baseline for future climate change impacts.

Sources and further reading

Hoskin CJ (2004) Australian microhylid frogs (*Cophixalus* and *Austrochaperina*): phylogeny, taxonomy, calls, distributions and breeding biology. *Australian Journal of Zoology* **52**, 237–269.

Hoskin CJ (2008) A key to the microhylid frogs of Australia, and new distributional data. *Memoirs of the Queensland Museum* **52**, 233–237.

Hoskin CJ and Hero J-M. (2008) *Rainforest Frogs of the Wet Tropics, North-east Australia*. Griffith University, Gold Coast.

Zweifel RG (1985) Australian frogs of the family Microhylidae. *Bulletin American Museum of Natural History* **182**, 265–388.

Author: Keith R McDonald

MELVILLE RANGE BOULDERFROG

Cophixalus zweifeli Davies & McDonald, 1998

Photo: Keith R McDonald-DERM

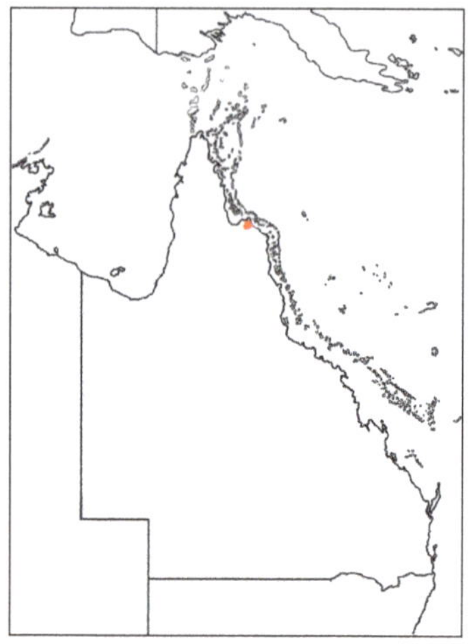

Family: Microhylidae

Conservation status: IUCN: Data Deficient; Qld: Vulnerable

Description: This is a large microhylid frog. Females measure 40–45 mm in length, with large finger and toe discs. Male measurements are unknown. It has no webbing on fingers and toes. The back is beige to tan with scattered, irregular brown speckles and the flanks have dark brown mottling extending to form a dark brown streak from the ear above the eye to the snout. The hidden parts of the limbs are ruby. The ventral surface is pale with a yellow wash on the lower abdomen and upper legs. The throat is mottled light purple.

Distribution: Only known from one stream catchment within the Melville Range, Cape Melville NP. The species is narrowly restricted in national park, with no obvious impacts on the limited populations. It is currently secure.

Habitat: Granite boulder field along a stream from 20–240 m altitude. Females were located after rain near the base of boulders and a juvenile near water flowing underneath boulders.

Ecology: No calls have been heard and nothing is known of the breeding biology.

Reasons for decline and current threats: No known decline or threat, but the species has a very restricted distribution in specialised boulder habitat, which may be affected by stochastic events such as prolonged drought.

CURRENT RECOVERY EFFORTS

Recovery plan: None

Conservation actions: No recovery efforts have been undertaken to date but the species is entirely in national park.

FUTURE RECOVERY EFFORTS

What you can do:

- Visitors to the national park need to observe management guidelines and good environmental hygiene.
- Report confirmed sightings to DERM.

Research directions:

- Monitor the populations for potential disease or threats from invasive species such as weeds and feral animals.
- Conduct surveys for additional populations within the Melville Range.
- Research the natural history of the species, especially breeding biology and microhabitat conditions. This is essential for management and will be a good baseline for future climate change impacts.

Sources and further reading

Davies M and McDonald KR (1998) A new species of frog (Anura: Microhylidae) from Cape Melville, Queensland. *Transactions of the Royal Society of South Australia* **122**, 159–165.

Hoskin CJ (2004) Australian microhylid frogs (*Cophixalus* and *Austrochaperina*): phylogeny, taxonomy, calls, distributions and breeding biology. *Australian Journal of Zoology* **52**, 237–269.

Hoskin CJ (2008) A key to the microhylid frogs of Australia, and new distributional data. *Memoirs of the Queensland Museum* **52**, 233–237.

Author: Keith R McDonald

WALLUM FROGLET

Crinia tinnula (Straughan & Main, 1966)

Photo: Ed Meyer

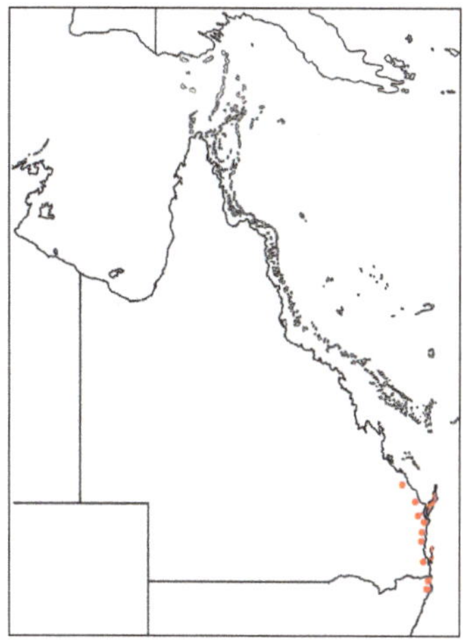

Family: Myobatrachidae

Other common name: Tinkling Froglet

Conservation status: IUCN: Vulnerable; Qld: Vulnerable; NSW: Vulnerable

Description: A small ground-dwelling frog, less than 22 mm in length. The dorsal body surface is grey, beige, red-brown or dark brown in colour, with irregular darker markings or longitudinal stripes. The dorsal skin of this species may be smooth, bumpy or ridged. The ventral body surface is coarsely granular and off-white, grey or dark grey in colour with dark and/or light flecking or stippling. Usually, there is a narrow pale stripe running down the middle of the throat, which, in some animals, extends back as far as the belly. The toes and fingers have no discs or pads and are free of webbing. Tadpoles are oval-bodied and grey or brown above with irregular darker markings. The snout is rounded in profile and eyes are situated dorsolaterally. The tail musculature is brown and fins semi-transparent, with dark spots, flecking and/or or fine reticulations. Silver or gold flecking may also be present on the tail fins and musculature and the underside of the body. The spiracle is sinistral, opening lateroventrally, while the vent tube is dextral opening halfway up the ventral fin. The labial tooth row formula for Wallum Froglet tadpoles is: 2(1)/3(1). Tadpoles of the Wallum Froglet reach a maximum size of 37 mm.

Distribution: Scattered populations from Littabella NP (near Bundaberg), south to around Kurnell, mid-eastern NSW. It is known from Fraser, Bribie, Moreton and North Stradbroke Islands, as well as adjacent mainland areas on the Fraser, Gold and Sunshine Coasts. Scattered populations also known from Caboolture Shire, Redland Shire, and Brisbane and Logan City areas.

Habitat: Largely restricted to coastal wallum (sandy lowland) environs, it is associated with wet heath, acid paperbark (*Melaleuca*) swamps, wallum lakes and sedge swamps. Occurs in disturbed habitat including recently burnt heath and 4WD-affected sites; also recorded in eucalypt woodland/forest and dry heath some distance from water.

Ecology: It may breed any time of year following rain. Males call from wet ground or sedges near water. Amplexus is inguinal and eggs are laid singly or in small clumps attached to submerged vegetation, usually in shallow ephemeral water. Tadpoles are rarely found sympatric with fish. Spawn and tadpoles tolerate acidic (low pH) waters extremely well. Non-breeding habitat usage is poorly understood. Adult animals subsist on a diet of small arthropods. Tadpoles feed on sediment, detritus and algae. Often common in suitable habitat, the Wallum Froglet sometimes occur at densities of hundreds of animals per hectare. The male advertisement call is a short high-pitched, staccato 'tcheh' or 'tchich.'

Reasons for decline and current threats: The population has declined due to habitat loss, fragmentation and degradation resulting from urban development, establishment of pine plantations, agriculture and sandmining. Sandmining and urban development remain a threat to the species, as does the introduction and spread of Mosquitofish *Gambusia holbrooki*. Competition with the related Beeping Froglet *Crinia parinsignifera* in disturbed areas may also be of concern. It breeds mainly in oligotrophic waters, so nutrient enrichment (in particular increased nitrate levels) may also pose a threat. This is of particular concern where breeding habitat adjoins urban or rural land, where there is use of biocides to control weeds and insects. The vulnerability of the Wallum Froglet to amphibian chytrid fungus disease is unknown, but declines attributable to the diseases have not been documented in this species.

CURRENT RECOVERY EFFORTS

Recovery plan: Meyer E, Hero J-M, Shoo L and Lewis B (2006) 'National recovery plan for the wallum sedgefrog and other wallum-dependent frog species'. Report to Department of the Environment and Water Resources, Canberra. Queensland Parks and Wildlife Service, Brisbane.

Other plans: Department of Environment and Heritage (2006) 'Threat abatement plan for infection of amphibians with chytrid fungus resulting in chytridiomycosis'. Commonwealth Department of Environment and Heritage, Canberra.

Conservation actions:

- Monitoring of 'acid' frogs species (including Wallum Froglet) on Bribie Island by DERM and Bribie Island Environmental Protection Association (BIEPA).
- Occasional surveys on Fraser Coast by Mary River Catchment Coordinating Committee (MRCCC) and Wildlife Preservation Society of Queensland (WPSQ).
- Acquisition of 'acid' frog habitat by local and government agencies for conservation.
- Limited monitoring of Wallum Froglet and other 'acid' frog species on North Stradbroke Island by the sandmining industry.
- Rehabilitation of affected 'acid' frog habitat by sandmining companies.

FUTURE RECOVERY EFFORTS

What you can do:

- Volunteer to assist MRCCC, BIEPA, DERM and WPSQ with monitoring and surveys.
- Avoid the use of garden fertilisers and ensure detergent used to wash cars is kept out of stormwater drains.
- Clean and disinfect soiled clothes/equipment when moving between streams and wetland areas to reduce the risk of spreading disease among frog populations.

Conservation needs:

- Carry out additional research into non-breeding habitat usage; improved protection of important habitat areas on freehold and leasehold land.
- Implement better management of potential threats.
- Lobby local and state governments to ensure adequate protection of remaining habitat areas on leasehold and freehold land.

Research directions:

- Conduct further surveys and monitoring.
- Further research the threat posed by chytrid fungus disease.
- Gain a better understanding of the competition between wallum-restricted 'acid' frog species (e.g. the Wallum Froglet) and ecologically similar sibling species (e.g. the Beeping Froglet).

Sources and further reading

Anstis M (2002) *Tadpoles of South-eastern Australia*. Reed New Holland, Sydney.

Hines HB, Mahony M and McDonald K (1999) An assessment of frog declines in wet subtropical Australia. In *Declines and Disappearances of Australian Frogs*. (Ed. A Campbell) pp. 44–63. Environment Australia, Canberra.

Ingram GJ and Corben CJ (1975) The frog fauna of North Stradbroke Island with comments on the 'acid' frogs of the wallum. *Proceedings of the Royal Society of Queensland* **86**, 49–54.

Straughan I and Main AR (1966) Speciation and polymorphism in the genus *Crinia* Tschudi (Anura: Leptodactylidae) in Queensland. *Proceedings of the Royal Society of Queensland* **78**, 11–28.

Author: Ed Meyer

MELVILLE RANGE TREEFROG

Litoria andiirrmalin McDonald, 1997

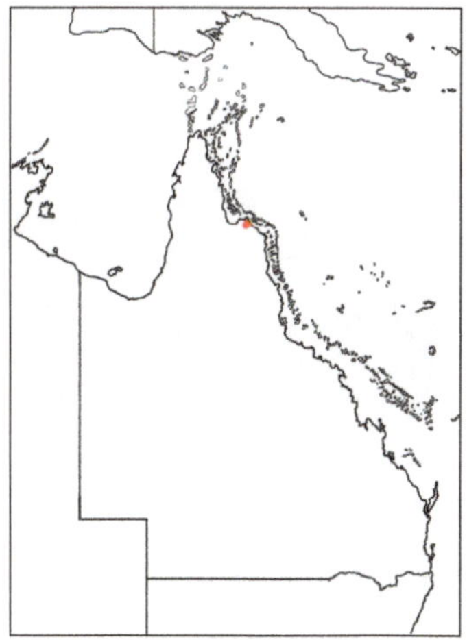

Photo not published at the request of, and out of respect for, the local Indigenous Australians.

Family: Hylidae

Aboriginal name: Andiirrmalin

Conservation status: IUCN: Vulnerable; Qld: Vulnerable

Description: The males of this large treefrog measure 60–76 mm in length, and the females 92–110 mm. They have large finger and toe discs, no webbing on the fingers but extensive webbing on the toes. The dorsal surface is mottled brown with irregular cream markings and an indistinct pale dorsolateral stripe from eye bulge to groin. The ventral surface is cream with throat faintly greyish to arms. Juveniles have a distinct pale dorsolateral stripe from eye to groin and a black throat, which fades with increasing size.

Distribution: Only known from four perennial streams in the boulder fields of the Melville Range, Cape Melville NP from 20–400 m altitude. The species is narrowly restricted in national park with no obvious impacts on the limited populations. It is currently secure.

Habitat: Restricted to perennial streams within the granite boulder field with gallery rainforest sometimes with *Melaleuca* emergents. This nocturnal species is observed on granite boulders, vines, branches in or near perennial streams where there are riffles and cascades.

Ecology: Faecal pellets and gut contents contained a range of terrestrial and aquatic invertebrates, freshwater prawns, and small frogs and lizards. Females carry pigmented eggs from May until the onset of the monsoon season. Males have been heard calling in November.

Reasons for decline and current threats: There are no known reasons for decline or current threats but this frog has a very restricted distribution, which may be affected by stochastic events such as prolonged drought.

CURRENT RECOVERY EFFORTS

Recovery plan: None

Conservation actions: No recovery efforts have been undertaken to date, but the species is located in a national park.

FUTURE RECOVERY EFFORTS

What you can do:

- Clean and disinfect soiled clothes/equipment when moving between streams and wetland areas to reduce the risk of spreading disease among frog populations.
- Do not remove water at base flow during drier months or during drought. Excessive water collection by campers may affect part of one stream population.

Research directions:

- Monitor the populations for potential disease or threats from invasive species such as weeds and feral animals.
- Conduct surveys for additional populations within the Melville Range to provide necessary baseline information for management.
- Research the natural history of the species, especially breeding biology, the physiological limits of temperature and moisture, and variation in microhabitat conditions within its habitat. This will be a good baseline for climate change impacts.

Sources and further reading

McDonald KR (1997) A new stream-dwelling *Litoria* from the Melville Range, Australia. *Memoirs of the Queensland Museum* **42**, 307–309.

Author: Keith R McDonald

COOLOOLA SEDGEFROG

Litoria cooloolensis (Liem, 1974)

Photo: Ed Meyer

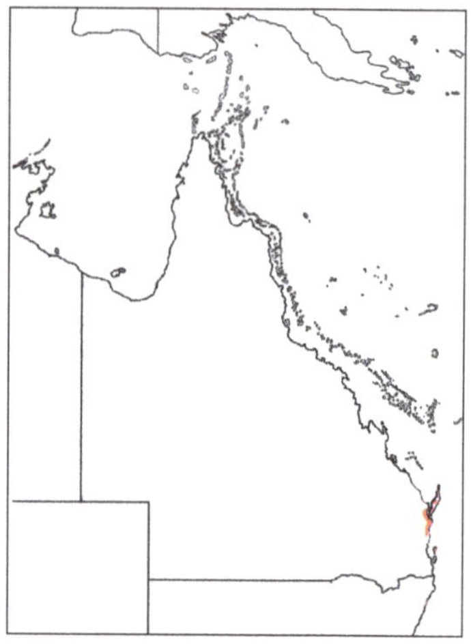

Family: Hylidae

Conservation status: IUCN: Endangered; Qld: Near Threatened

Description: This small narrow-bodied frog measures up to 30 mm in length. The dorsal surface is smooth, green, beige, or yellow in colour, with or without dark stippling. The ventral body surface is granular and mostly white, except in breeding males where the throat is yellow or orange-yellow in colour. The finger discs and toe pads are readily discernible and the toes are fully webbed. The posterior thigh is mostly orange in colour, with a purple-brown streak uppermost. Considered diagnostic for the species, this streak may be absent or difficult to discern in some individuals, making positive identification difficult. Tadpoles are deep-bodied and high-finned with eyes laterodorsal in position. Tadpoles may be uniformly light in colour or boldly marked with dark spots, splotches and reticulations. The belly is silver-white in colour with the coiled intestinal mass fully obscured. The labial tooth row formula of Cooloola Sedgefrog tadpoles is 2(2)/3. The vent tube is dextral, opening half way up the tail fin, while the spiracle is sinistral and ventrolateral in position. Tadpoles of the Cooloola Sedgefrog reach a maximum size of 45 mm.

Distribution: Endemic to Qld. Restricted to coastal wallum (sandy lowland) environs. Scattered populations within Great Sandy NP (i.e. the sand-masses of Fraser Island and Cooloola), south to the Noosa River. Also known from lakes and larger swamps on North Stradbroke Island. Not known from Moreton, Bribie or South Stradbroke Islands. North Stradbroke Cooloola Sedgefrogs represent a distinct evolutionary lineage warranting special consideration with regards to conservation/management actions.

Habitat: Associated with oligotrophic (low nutrient), acidic freshwater lakes (pH <5.5) in coastal wallum. May also be found in larger wallum swamps. Usually found among reeds near water. However, the species has also been recorded from rainforest and dry sclerophyll forest some distance (many hundreds of metres) from water.

Ecology: Breeding occurs in spring, summer and autumn, usually after rain. Male frogs generally call from reeds or trees, near water. Amplexus is axillary and eggs are laid in clumps attached to reeds. Tadpoles may take 8 weeks or more to complete their development with metamorphs emerging in spring and summer. Rarely is the Cooloola Sedgefrog found sympatric with the Eastern Sedgefrog *Litoria fallax*, a potential competitor. It occurs in sympatry with native fish, but rarely with the introduced Mosquitofish *Gambusia holbrooki*. Non-breeding habitat usage is poorly understood. The male advertisement call is a loud, high-pitched staccato 'rik-kik', 'kik-kik' or simple 'kik', similar to that of the Eastern Sedgefrog. Adult frogs feed on small arthropods, including mosquitoes and damselflies. Tadpoles appear to feed on detritus and algae enveloping reeds.

Reasons for decline and current threats: Localised population declines are likely primarily attributable to habitat loss (due to sandmining) and the spread of Mosquitofish (an introduced species that preys upon spawn and tadpoles of native frog species). Sandmining and predation by Mosquitofish remain a threat, particularly on North Stradbroke Island. Additional threats include trampling of reed beds around freshwater lakes (particularly in areas subject to heavy human visitation), establishment of, and subsequent competition with, the Eastern Sedgefrog, and increased frequency or intensity of fire around wallum swamps and lakes. The threat posed by amphibian chytrid fungus disease is unknown.

CURRENT RECOVERY EFFORTS

Recovery plan: Meyer E, Hero J-M, Shoo L and Lewis B (2006) 'National recovery plan for the wallum sedgefrog and other wallum-dependent frog species'. Report to Department of the Environment and Water Resources, Canberra. Queensland Parks and Wildlife Service, Brisbane.

Other plans: Department of Environment and Heritage (2006) 'Threat abatement plan for infection of amphibians with chytrid fungus resulting in chytridiomycosis'.

Commonwealth Department of Environment and Heritage, Canberra.

Conservation actions:

- Limited monitoring of Cooloola Sedgefrog and other 'acid' frog species on North Stradbroke Island by the sandmining industry.
- Rehabilitation of affected 'acid' frog habitat by sandmining companies.

FUTURE RECOVERY EFFORTS

What you can do:

- Clean and disinfect soiled clothes/equipment when moving between wetland areas to reduce the risk of spreading disease among frog populations.
- Tread lightly in areas inhabited by the Cooloola Sedgefrog (i.e. avoid trampling reed beds).
- Report the occurrence of Eastern Sedgefrogs, a potential competitor species, on Fraser Island to DERM.

Conservation needs:

- Discourage the spread of Mosquitofish and other exotic fish in wallum areas.
- Investigate/assess non-breeding habitat usage, population size and habitat condition (particularly on North Stradbroke Island where habitat is threatened by sandmining, frequent fires and/or water extraction).
- Manage the impact of visitors to freshwater lakes (e.g. trampling of reed beds).
- Protect existing habitat and rehabilitate degraded habitat.

Research directions:

- Carry out studies on the threat posed by chytrid fungus disease to wallum frog species.
- Attain a better understanding of competition between wallum-restricted 'acid' frog species, such as the Cooloola Sedgefrog, and ecologically similar sibling species (e.g. the Eastern Sedgefrog).

Sources and further reading

Hines HB, Mahony M and McDonald K (1999) An assessment of frog declines in wet subtropical Australia. In *Declines and Disappearances of Australian Frogs.* (Ed. Campbell A) pp. 44–63. Environment Australia, Canberra.

Ingram GJ and Corben CJ (1975) The frog fauna of North Stradbroke Island with comments on the 'acid' frogs of the wallum. *Proceedings of the Royal Society of Queensland* **86**, 49–54.

James C (1996) 'Conservation genetics of island and mainland populations of the sedge frogs *Litoria cooloolensis* and *Litoria olongburensis*'. Unpublished report for the Queensland Department of Environment and Heritage.

Liem DS (1974) A new species of the *Litoria bicolor* species group from southeast Queensland, Australia (Anura: Hylidae). *Memoirs of the Queensland Museum* **17**, 169–174.

Author: Ed Meyer

WALLUM ROCKETFROG

Litoria freycineti (Tschudi, 1838)

Photo: Harry B Hines-DERM

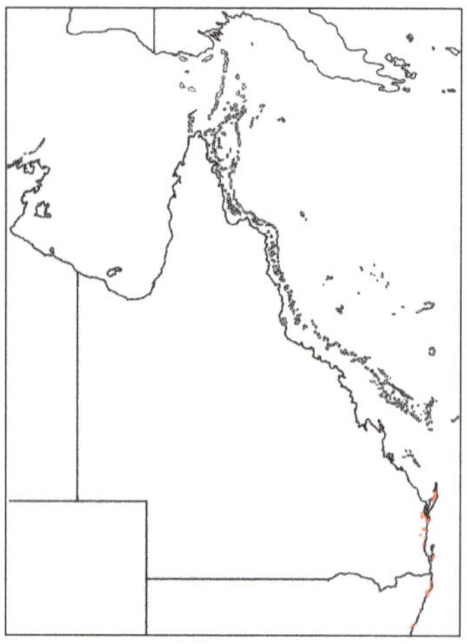

Family: Hylidae

Other common name: Freycinet's Frog

Conservation status: IUCN: Vulnerable; Qld: Vulnerable

Description: This moderate-sized frog measures up to 42 mm in length. Like other rocketfrog species, it has long legs and a pointed snout. The dorsal surface is grey, grey-brown or beige in colour with irregular darker markings, often with reddish-brown spots and streaking as well. A dark lateral stripe runs back from the snout, through the eye to the level of the shoulder. This stripe is broken by a white streak in front of the eye which extends back along the upper lip to the shoulder. The dorsal surface is typically warty or rugose, while the ventral surface is coarsely granular and mostly white. The posterior thigh is dark brown, olive-brown or grey-brown in colour with fawn, yellow or pinkish-yellow mottling. The toes are partly webbed, while the fingers are free of webbing. Tadpoles are deep-bodied with moderately high fins and reach a maximum size of 53 mm. The upper body of tadpoles is dark grey or grey-brown in colour, sometimes with irregular darker markings, while the underside is silver white with a faint copper-pink sheen. Both the tail musculature and fins are heavily suffused with black or dark brown. The eyes are laterodorsal in position and the vent tube dextral, opening half way up the ventral fin. The spiracle is sinistral and opens ventrolaterally. The labial tooth row formula of tadpoles is 2(1)/3(1).

Distribution: Coastal sandy lowlands of south-east Qld and eastern NSW, from Fraser Island south to Jervis Bay, mid-eastern NSW. Scattered populations on Fraser, Moreton, Bribie and North Stradbroke Islands, as well as adjacent mainland areas on the Fraser and Sunshine Coasts.

Habitat: In Qld, the Wallum Rocketfrog is restricted to coastal sandy lowlands in the south-east of the state (coastal wallum). It is found mainly around sedge swamps, drainage lines and perched lakes. It may also be found some distance from water in eucalypt forest near areas of wet heath.

Ecology: Breeding takes place in spring and summer after rain. Males call from wet ground near water among sedges. Amplexus is axillary and spawning usually occurs in shallow ephemeral water. Larval development is rapid, with tadpoles completing metamorphosis in less than 60 days. Non-breeding habitat usage is unknown. The species has, however, been found some distance from water in eucalypt forest, near areas of wet heath. Adult frogs feed on small arthropods; tadpoles feed on detritus. The male advertisement call is similar to that of the Striped Rocketfrog *Litoria nasuta*, comprising occasional wavering 'yehrps' and shorter oft repeated 'yips'.

Reasons for decline and current threats: Has declined in abundance due to habitat loss, fragmentation and degradation resulting from urban development, establishment of pine plantations, agriculture and sandmining. Sandmining and urban development remain a threat, as may the spread of Mosquitofish *Gambusia holbrooki* (an introduced species that preys upon spawn and tadpoles of native frogs). Competition with the related Striped Rocketfrog in disturbed areas may also be of concern. It breeds in acidic, oligotrophic waters, so nutrient enrichment (in particular increased nitrate levels) may also pose a threat to the species. This is of particular concern where breeding habitat adjoins urban or rural land, where the use of biocides to control weeds and insect pests may also pose a threat. Rising sea levels as a result of global warming may pose a threat to populations at or near sea level. The threat posed by amphibian chytrid fungus disease is unknown.

CURRENT RECOVERY EFFORTS

Recovery plan: Meyer E, Hero J-M, Shoo L and Lewis B (2006) 'National recovery plan for the wallum sedgefrog and other wallum-dependent frog species'. Report to Department of the Environment and Water Resources, Canberra. Queensland Parks and Wildlife Service, Brisbane.

Other plans: Department of Environment and Heritage (2006) 'Threat abatement plan for infection of amphibians with chytrid fungus resulting in chytridiomycosis'. Commonwealth Department of Environment and Heritage, Canberra.

Conservation actions:

- Monitoring of 'acid' frogs species (including Wallum Rocketfrog) on Bribie Island by DERM and Bribie

Island Environmental Protection Association (BIEPA).

- Occasional surveys on Fraser Coast and Sunshine Coast by Mary River Catchment Coordinating Committee (MRCCC) and the Wildlife Preservation Society of Queensland (WPSQ).
- Acquisition of 'acid' frog habitat by local and government agencies for conservation.
- Limited monitoring of Wallum Rocketfrog and other 'acid' frog species on North Stradbroke Island by the sandmining industry.
- Rehabilitation of affected 'acid' frog habitat by sandmining companies.
- Research investigating genetic diversity involving Griffith University staff/students.

FUTURE RECOVERY EFFORTS

What you can do:

- Avoid the use of garden fertilisers and ensure that detergent used to wash cars is kept out of stormwater drains.
- Clean and disinfect soiled clothes/equipment when moving between streams and wetland areas to reduce the risk of spreading disease among frog populations.
- Volunteer with MRCCC, BIEPA, DERM and WPSQ to assist with monitoring and surveys.

Conservation needs:

- Additional research into non-breeding habitat usage.
- Improved protection of important habitat areas on freehold and leasehold land.
- Better management of potential threats.
- Lobby local and state governments to ensure adequate protection of remaining habitat areas on leasehold and freehold land.

Research directions:

- Conduct further surveys and monitoring.
- Studies need to be done to better understand non-breeding habitat usage and the threat posed by chytrid fungus disease to wallum frog species and to better understand the competition between wallum-restricted 'acid' frog species (e.g. the Wallum Rocketfrog) and ecologically similar sibling species (e.g. the Striped Rocketfrog).
- Research into the risk posed by the chytrid fungus.

Sources and further reading

Anstis M (2002) *Tadpoles of South-eastern Australia.* Reed New Holland, Sydney.

Hines HB, Mahony M and McDonald K (1999) An assessment of frog declines in wet subtropical Australia. In *Declines and Disappearances of Australian Frogs.* (Ed. A Campbell) pp. 44–63. Environment Australia, Canberra.

Ingram GJ and Corben CJ (1975) The frog fauna of North Stradbroke Island with comments on the 'acid' frogs of the wallum. *Proceedings of the Royal Society of Queensland* **86**, 49–54.

Author: Ed Meyer

ARMOURED MISTFROG

Litoria lorica Davies & McDonald, 1979

Photo: Andrew J Dennis

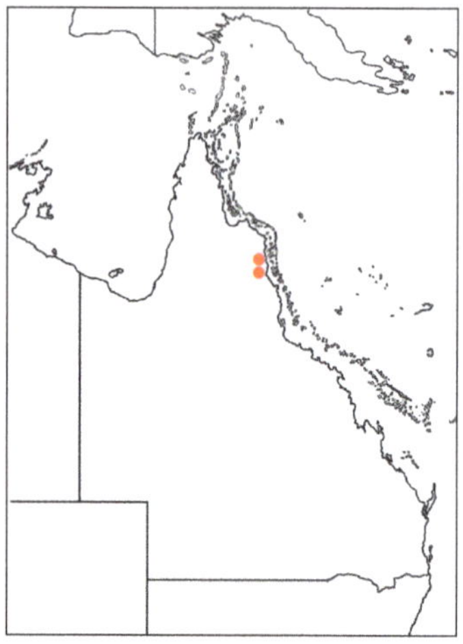

Family: Hylidae

Other common name: Little Waterfall Frog

Conservation status: IUCN: Critically Endangered; Aust.: Critically Endangered; Qld: Endangered

Description: This small to medium-size frog measures 30–42 mm in length. The dorsal surface is smooth, with scattered small tubercules and a pale brown, sandy or grey surface with dark blotches or mottling. The ventral surfaces are pale and smooth. Forearms are robust and fingers and toes have large, rounded discs. Males have a large prepollex; black spinulated nuptial pads are on the base of the thumbs; in some individuals small dark spinules also extend to upper chest and chin. Females can have a large prepollex, but with sparser dark spinules than males; its function in females is unknown.

Distribution: Before severe population crashes in the early 1990s, the Armoured Mistfrog was known from four localities in the northern Wet Tropics of north-east Qld. A small and localised population was rediscovered in 2008 at a remote location on the Carbine Tableland.

Habitat: Inhabits waterfalls and cascades along permanent streams and rivers. Thought to be a rainforest specialist, but rediscovered population occurs on a river in dry sclerophyll habitat downstream from rainforest, along 2 km of waterfalls and riffles. Habitat requirements and ecology appear very similar to those of the closely related Waterfall Frog *Litoria nannotis*. The Armoured Mistfrog has only been found in, or immediately adjacent to, the splash zone of waterfalls and cascades, on rocks and low vegetation. It may also use surrounding terrestrial habitats, often used by the Waterfall Frog during nocturnal foraging.

Ecology: Males, females and juveniles occur in the same local areas, and are often found with Waterfall Frogs. The tadpole is in the process of being described and is most similar to the tadpole of the Waterfall Frog. Not much is known about the reproductive biology of the species; gravid females are found year round in the dry forest population.

Reasons for decline and current threats: The disappearance of the Armoured Mistfrog coincided with the declines of sympatric species, such as Waterfall Frog and the Sharp-snouted Dayfrog *Taudactylus acutirostris*, in pristine and protected areas above 400 m altitude throughout their rainforest range. Populations declined dramatically from these sites, probably due to outbreaks of chytrid fungus disease. The only currently known population of the Armoured Mistfrog exists in dry sclerophyll habitat. Sampling during winter showed that the majority of individuals were infected by the chytrid fungus, but repeated visits showed that the population was stable; at the new site, the species is presently coexisting with the pathogen. The prevalence and intensity of chytrid infections responds strongly to environmental conditions; present coexistence may be tenuous in the face of future fluctuations in weather or climate. With only a single known population, the species may be highly vulnerable to disease outbreaks, feral pigs and cattle, and human activities that affect the stream-side habitat as well as water quality.

CURRENT RECOVERY EFFORTS

Recovery plan: Northern Queensland Threatened Frogs Recovery Team (2001) 'Recovery plan for the stream-dwelling rainforest frogs of the Wet Tropics biogeographic region of north-east Queensland 2000–2004'. Report to Environment Australia, Canberra. Queensland Parks and Wildlife Service, Brisbane.

Other plans: Department of Environment and Heritage (2006) 'Threat abatement plan: infection of amphibians with chytrid fungus resulting in chytridiomycosis'. Commonwealth Department of Environment and Heritage, Canberra.

Conservation actions: A working group has been established between QPWS and amphibian disease specialists from JCU.

FUTURE RECOVERY EFFORTS

What you can do: Clean and disinfect soiled clothes/equipment when moving between streams and wetland areas to reduce the risk of spreading disease among frog populations.

Conservation needs:

Captive breeding is recommended for this species by the IUCN Red List assessment by specialists who assessed its conservation status. Several factors need to be achieved before this can be considered:

- Investigate the mechanism of the population's resistance to chytrid fungus. This is crucial because if the resistance is due to environmental conditions, it will be necessary to replicate these during captive maintenance and to ensure that any reintroductions occur in suitable environments.
- Determine if enough individuals can be removed from the wild for a viable *ex situ* program to be established without significantly affecting the wild population.
- Obtain basic knowledge about how the peculiar biology of this species enables it to live and breed in torrent environments.
- Develop captive breeding techniques for the two closely related, ecologically sympatric and more common species, Waterfall Frog and Common Mistfrog *L. rheocola.*

Research directions:

- Continue to monitor the newly discovered population.
- Data on the biology and ecology of the species is required.
- Continue to conduct surveys. Preliminary disease surveys have determined that the rediscovered Armoured Mistfrog population is infected with the chytrid fungus at high prevalences, but no dead have been found so far. The real impact of chytrid fungus on this species requires further evaluation.

Sources and further reading

Berger L, Speare R, Daszak P, Green DE, Cunningham AA, Goggin CL, Slocombe R, Ragan MA, Hyatt AD, McDonald KR, Hines HB, Lips KR, Marantelli G and Parkes H (1998) Chytridiomycosis causes amphibian mortality associated with population declines in the rainforests of Australia and Central America. *Proceedings of the National Academy of Science* USA **95**, 9031–9036.

Berger L, Speare R and Hyatt A (1999) Chytrid fungi and amphibian declines: overview, implications and future directions. In *Declines and Disappearances of Australian Frogs.* (Ed. A Campbell) pp. 23–33. Environment Australia, Canberra.

Cunningham M (2002) Identification and evolution of Australian torrent treefrogs (Anura: Hylidae: *Litoria nannotis* group). *Memoirs of the Queensland Museum* **48**, 93–102.

Davies M and McDonald KR (1979) A new species of stream-dwelling hylid frog from Northern Queensland. *Transactions of the Royal Society of South Australia* **103**, 169–176.

Hoskin CJ and Hero J-M. (2008) *Rainforest Frogs of the Wet Tropics, North-east Australia.* Griffith University, Gold Coast.

Authors: Robert Puschendorf, Ross A Alford and Conrad J Hoskin

KURANDA TREEFROG

Litoria myola Hoskin, 2007

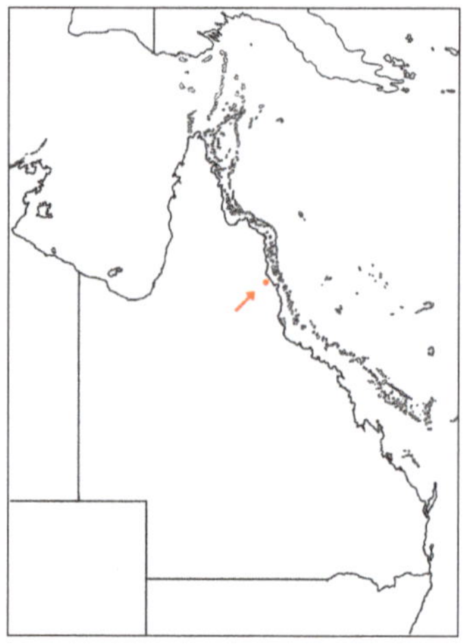

Photo: Conrad J Hoskin

Family: Hylidae

Conservation status: IUCN: Critically Endangered; Aust.: Endangered

Description: This is a recently recognised species, formerly considered a population of the tapping Green-eyed Treefrog *Litoria serrata* (which itself was formerly included in *L. genimaculata*). The males of this medium-sized frog measure 36–44 mm in length (average 40 mm) and the females 57–69 mm. Males are slender and females considerably more robust. The skin is smooth or with scattered small lumps. The dorsal pattern is a highly variable tan or brown, with faint or distinct darker brown, orange or green blotches or mottling. The ventral surface is smooth and pale, sometimes with grey on the chin and throat. There is a pale, serrated ridge of skin along the outer edge of the foot and similar, but less obvious, serrated ridge along the hind edge of the forearm. The fingers are half webbed, toes nearly fully webbed with large, rounded finger and toe discs. The frog has a green crescent across the upper iris.

Distribution: The very small distribution is restricted to the Kuranda–Myola–Kowrowa region, in north-east Qld. The 'extent of occurrence' is approximately 13.5 km^2, and the 'area of occupancy' has been estimated to be 3.5 km^2. The species is found in Jumrun Creek Conservation Park. Monitoring at several sites since 2000 has revealed a stable population at one site, but declines at others.

Habitat: Mature and regenerating rainforest along permanent and ephemeral streams. Breeding occurs in streams, but the frogs also use rainforest away from the streams.

Ecology: The Kuranda Treefrog inhabits rainforest and breeds in streams. Males call along streams through the spring/summer months, primarily near riffles and small cascades. The call is a soft tapping. The Kuranda Treefrog breeds in both permanent streams and small drainage lines that only flow during and following heavy wet season rainfall. Males spend much of their time near streams during the breeding season, but females, juveniles and non-breeding males also use rainforest away from the streams. Females visit the streams to breed, but forage in rainforest trees both near and considerable distances from the streams. The frogs are nocturnal. The eggs are laid as a cohesive clump (clutch size approximately 500 eggs) in flowing water and the tadpoles take about 2 months until metamorphosis. The total population is estimated to be <1000 adults. Thirteen breeding sites are known: all short sections of adjacent streams flowing into the Barron River. Most of the known breeding sites are reasonably close to each other. There is potentially movement of frogs between some sites, although movements between others may be limited by intervening areas of unsuitable habitat.

Reasons for decline and current threats: Very restricted geographic range, consisting of small and fragmented breeding populations. There have been observed declines in abundance at some sites, and projected declines at some sites due to potential residential development. The species occurs in an area under intense pressure for residential development. Threats include: the clearing of rainforest vegetation (including regrowth); impacts to the streams in terms of water flow, water quality and sedimentation – initiated both at the sites where the species occurs as well as upstream in the catchments – breeding sites are in the downstream sections of streams so directly vulnerable to impacts upstream in the catchment; fragmentation of habitat and breeding populations; stochastic events, due to the very small distribution and population size; frog chytrid fungus and other unforeseen frog diseases. Chytrid fungus disease is probably a minor threat because it is known to be present in the region and also because the closely related Green-eyed Treefrog *L. serrata* appears to be able to tolerate chytrid fungus better than many other Wet Tropics stream-breeding hylids.

Automobile traffic and cats are also a threat to the Kuranda Treefrog given its restricted distribution and population size in an area of moderate (and increasing) human usage.

CURRENT RECOVERY EFFORTS

Recovery plan: None

Conservation actions:

- No official recovery efforts are in place.
- Surveys and monitoring have been conducted.
- Actions of community groups (Kuranda EnviroCare and Friends of the Earth Kuranda) are directly benefiting the species through protecting and replanting habitat.

FUTURE RECOVERY EFFORTS

What you can do:

- Become involved in local community groups, such as Kuranda EnviroCare.
- Protect and rehabilitate rainforest and stream habitat, and ensure local streams and gullies are not affected by changes in water flow, water quality and sedimentation. This includes urban runoff such as fertilisers, detergents, oil, emptying pools, erosion, etc.
- Don't collect or move tadpoles or frogs between sites, and don't attempt to rear tadpoles. These activities may affect natural levels of interaction between the Kuranda Treefrog and Green-eyed Treefrog, and may increase the potential for disease transmission between sites.
- Clean and disinfect soiled or wet clothes/equipment when moving between streams and wetland areas to reduce the risk of spreading disease among frog populations.
- When doing rehabilitation or other activities in the area, take all measures to reduce the potential for disease transmission between sites, including washing/disinfecting shoes and other gear that comes in contact with water or wet earth.
- Pressure the local, state and federal governments to protect habitat and limit development in the Kuranda–Myola–Kowrowa area.

Conservation needs:

- The species requires *in situ* conservation efforts associated with habitat retention and rehabilitation and replanting of rainforest at breeding sites and in the areas in between these.
- Minimising impacts to breeding sites regarding water flow, water quality and sedimentation is dependent on activities both in the vicinity of the breeding sites and in upstream areas of the catchments.

Research directions:

- Long-term transects need to be established to continue monitoring key sites in a systematic manner.
- The degree of connectivity/gene flow between the breeding sites needs to be determined to identify which should be considered subpopulations.
- Research is required on the ecology of this species.

Sources and further reading

Hoskin CJ (2007) Description, biology and conservation of a new species of Australian Tree Frog (Hylidae: *Litoria*) and an assessment of the remaining populations of *Litoria genimaculata* Horst, 1883: systematic and conservation implications of an unusual speciation event. *Biological Journal of the Linnean Society* **91**, 549–563.

Hoskin CJ and Hero J-M (2008) *Rainforest Frogs of the Wet Tropics, North-east Australia*. Griffith University, Gold Coast.

Hoskin CJ, Higgie MA, McDonald KR and Moritz C (2005) Reinforcement drives rapid allopatric speciation. *Nature* **437**, 1353–1356.

Author: Conrad J Hoskin

WATERFALL FROG

Litoria nannotis (Andersson, 1916)

Photo: Harry B Hines

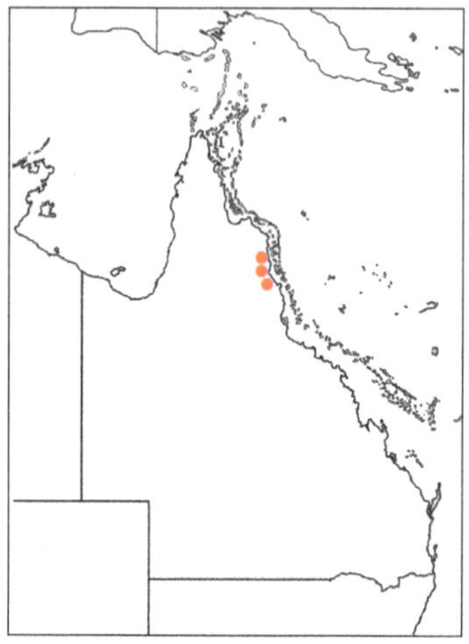

Family: Hylidae

Other common name: Torrent Treefrog

Conservation status: IUCN: Endangered; Aust.: Endangered, Qld: Endangered

Description: Adult males measure 31–52 mm in length and females 48–59 mm. The dorsal surface is smooth, with scattered small tubercules; the dorsal colour varies from green, brown, dull brown, pale brown, slaty grey, all with dark mottling. The ventral surface is smooth and pale. The limbs are robust, especially in males with particularly well developed forearms. The fingers have basal webbing and the toes are almost fully webbed. Males have a large black spinulated nuptial pad on the base of the thumb and accessory black tipped spinules on upper chest and occasionally on the lower jaw.

Distribution: Formerly abundant between 180 and 1300 m altitude throughout the Wet Tropics, between Paluma and Cooktown. It suffered strong population declines, primarily due to outbreaks of chytrid fungus disease. After the initial outbreaks, the Waterfall Frog declined dramatically and became locally extinct at altitudes above 400 m. In recent years, some individuals have reappeared at higher altitude sites, but these are small populations in which diseased and dying animals are commonly found through winter months.

Habitat: It inhabits waterfalls and cascades along permanent streams and rivers. It was thought to be a rainforest specialist, but it is also on some nearby streams in dry sclerophyll areas such as Blencoe Falls and Davies Creek.

Ecology: This frog is restricted to waterfalls and cascades, around which individuals sit on rocks and low vegetation. It is usually found in the splash-zone or immediately next to it. It is common to find several individuals clustered in a small area of suitable habitat. Males, females and juveniles occur in the same microhabitat. The call is described as a repeated 'crawk, crawk, crawk...' or a gentle, popping, slow growl-like sound difficult to hear above that of flowing water. This species does not seem to vocalise often. Breeding occurs in streams throughout the year. The eggs are unpigmented and laid under rocks within the water. The tadpoles are well-adapted for torrent environments and characterised by a hydrodynamic, depressed body shape, large ventral oral disc, and muscular tail with low fins. The large suctorial mouthparts are used to adhere to, and feed on, the microbial community on rocks in fast flowing water. The body colour is variable, ranging from a uniform grey to pale yellow/olive green with irregular dark blotches. Most individuals show dark blotches of pigment on the tail musculature and smaller dark spots and striations on the fin. The labial tooth row formula is 2/3. The development time is typically between 9 and 12 months and tadpoles metamorphose at a body length of between 19 and 27 mm. Tadpoles experience high prevalence of infection with the chytrid fungus. Habitat requirements and ecology of the Waterfall Frog appear very similar to those of the Armoured Mistfrog *Litoria lorica*.

Reasons for decline and current threats: Chytrid fungus disease remains the most serious threat and highlights the threat posed by future unforseen diseases. Lowland populations are persisting well, but there have been only limited signs of recovery at higher altitude rainforest sites, and these populations still show obvious mortality in winter months. Populations of the Waterfall Frog crashed at the same time as individuals of other rainforest stream dwellers, such as the Armoured Mistfrog *Litoria lorica*, the Sharp-snouted Dayfrog *Taudactlus acutirostris*, the Common Mistfrog *L. rheocola* and the Lace-eyed Treefrog *Nyctimystes dayi*.

Other threats include disturbances caused by human activities, which potentially affect the streamside habitat as well as water quality.

CURRENT RECOVERY EFFORTS

Recovery plan: Northern Queensland Threatened Frogs Recovery Team (2001) 'Recovery plan for the stream-dwelling rainforest frogs of the Wet Tropics biogeographic region of north-east Queensland 2000–2004.' Report to Environment Australia, Canberra. Queensland Parks and Wildlife Service, Brisbane.

This recovery plan was not completely implemented. Most work on this species has been research-based, investigating aspects of its biology and interaction with chytrid fungus disease.

Conservation actions: At present, most aspects on the conservation of this species are dealt with through collaborations between DERM, NGOs and university-based researchers.

FUTURE RECOVERY EFFORTS

What you can do:

- Clean and disinfect soiled clothes/equipment when moving between streams and wetland areas to reduce the risk of spreading disease among frog populations.
- Pressure local, state and federal governments to protect threatened species.

Research directions:

- Captive breeding has been suggested, but never attempted. Currently there is no urgent need for it because lowland populations are stable and higher altitude populations seem to be recovering. However, it would be very useful to develop captive techniques for this species due to its unusual biology (adapted to live and breed in torrent environments), and these could be used as a model in the development of such techniques for more threatened species of similar ecology (e.g. the Armoured Mistfrog).
- Further research is required to understand how, and from where, this species is recolonising higher altitude sites, and the origin and mechanism of resistance of some populations to chytrid fungus disease.

Sources and further reading

Cunningham M (2002) Identification and evolution of Australian torrent treefrogs (Anura: Hylidae: *Litoria nannotis* group). *Memoirs of the Queensland Museum* **48**, 93–102.

Hoskin CJ and Hero J-M (2008) *Rainforest Frogs of the Wet Tropics, North-east Australia.* Griffith University, Gold Coast.

McDonald K and Alford R (1999) A review of declining frogs in Northern Queensland. In *Declines and Disappearances of Australian Frogs.* (Ed. A Campbell) pp. 14–22. Environment Australia, Canberra.

Richards SJ, McDonald KR and Alford RA (1993) Declines in populations of Australia's endemic tropical rainforest frogs. *Pacific Conservation Biology* **1**, 66–77.

Rowley JJL and Alford RA (2007) Movement patterns and habitat use of rainforest stream frogs in northern Queensland, Australia: implications for extinction vulnerability. *Wildlife Research* **34**, 371–378.

Authors: Robert Puschendorf, Ross A Alford, Conrad J Hoskin and Scott Cashins

WALLUM SEDGEFROG

Litoria olongburensis (Liem & Ingram, 1977)

Photo: Ed Meyer

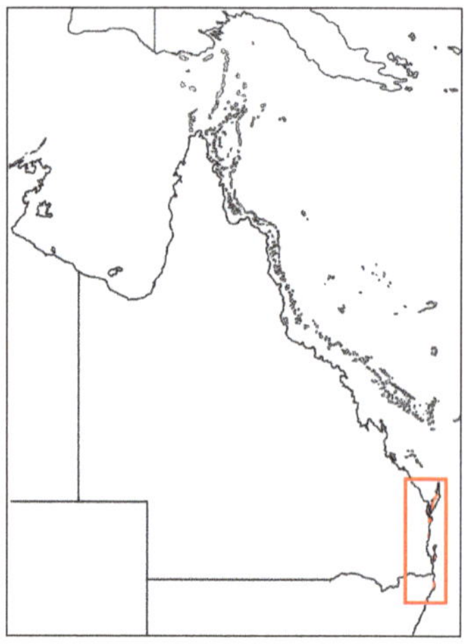

Family: Hylidae

Other common names: Olongburra Frog, Olongburra Treefrog, Sharp-snouted Treed Frog, Wallum Treefrog

Conservation status: IUCN: Vulnerable; Aust.: Vulnerable; Qld: Vulnerable; NSW: Vulnerable

Description: The Wallum Sedgefrog is a small narrow-bodied frog that measures up to 30 mm in length with a sharply pointed snout. Its dorsal surface is smooth and either green, grey-brown or beige in colour, sometimes with dark flecking and spotting. The ventral body surface is mostly granular and white, except in breeding males where the throat is typically brown to greyish-brown in colour. In females, the throat may also be coloured green. The finger discs and toe pads are readily discernible in this species. The toes are partly webbed while the fingers have only a small amount of webbing. A dark brown stripe runs from the snout back through the eye and tympanum. A prominent white lateral streak, starting below the eye, runs back over the shoulder onto the flanks, breaking up into a series of raised glandular spots. In females, the posterior thigh is typically blue or purple-blue in colour, with a little orange ventrally. Usually, there is some blue colouration in the groin as well. In males, this blue colouration is often absent or indistinct. The male advertisement call is a high-pitched wavering, insect-like 'wzzzzrt'. Tadpoles are deep-bodied with moderately high fins and reach a maximum size of 40 mm.

Distribution: Scattered populations along the south-east Qld coast south to northern NSW. Known from Fraser, Moreton, Bribie and North Stradbroke Islands, and adjoining mainland areas on the Fraser and Sunshine Coasts and the far southern Gold Coast.

Habitat: Restricted to coastal wallum (sandy lowland) environs, this frog prefers acidic freshwater swamps (pH <5.5) with emergent reeds, ferns or sedges. The water at breeding sites is usually clear and heavily tannin-stained. Adults are typically found among reeds and sedges near water, less commonly around creeks and large freshwater lakes. The Wallum Sedgefrog is found in association with sedges all year round, even when swamps are dry.

Ecology: Breeding occurs in spring, summer and autumn, usually after rain. Males call from emergent reeds near water after rain. Amplexus is axillary and eggs are attached to grasses and sedges. Tadpoles may take 8 weeks or more to complete their development. Rarely is the Wallum Sedgefrog found sympatric with the Eastern Sedgefrog *Litoria fallax*, a potential competitor. The Wallum Sedgefrog is generally associated with fish-free environs, though it may co-occur with native fish (where these occur at low densities). Rarely does the Wallum Sedgefrog co-occur with introduced Mosquitofish *Gambusia holbrooki*. Adult frogs feed on small arthropods, including mosquitoes. Tadpoles feed on detritus and algae enveloping sedges.

Reasons for decline and current threats: Habitat loss, fragmentation and degradation resulting from urban development, the establishment of pine plantations, agriculture and sandmining. Sandmining and urban development remain a threat, with the latter a major threat to mainland populations. Additional threats include the spread of Mosquitofish; and establishment of, and subsequent competition with, the Eastern Sedgefrog, particularly in disturbed wallum habitat. The Wallum Sedgefrog breeds in oligotrophic waters, so nutrient enrichment (in particular increased nitrate levels) may also pose a threat. This is of particular concern where breeding habitat adjoins urban or rural land, where biocides are used. Rising sea levels as a result of global warming may pose a threat to populations at or near sea level. Excessive extraction of ground water from dune islands may also pose a threat to this species. The threat posed by amphibian chytrid fungus disease is unknown.

CURRENT RECOVERY EFFORTS

Recovery plan: Meyer E, Hero J-M, Shoo L and Lewis B (2006) 'National recovery plan for the wallum sedgefrog and other wallum-dependent frog species'. Report to Department of the Environment and Water Resources, Canberra. Queensland Parks and Wildlife Service, Brisbane.

Other plans: Department of Environment and Heritage (2006) 'Threat abatement plan for infection of amphibians with chytrid fungus resulting in chytridiomycosis'. Commonwealth Department of Environment and Heritage, Canberra.

Conservation actions:

- Monitoring of 'acid' frog species (including Wallum Sedgefrog) on Bribie Island by DERM and Bribie Island Environmental Protection Association (BIEPA).
- Occasional surveys on Fraser Coast by Mary River Catchment Coordinating Committee (MRCCC) and the Wildlife Preservation Society of Queensland (WPSQ).
- Acquisition of 'acid' frog habitat by local and government agencies for conservation.
- Limited monitoring of Wallum Sedgefrog and other 'acid' frog species on North Stradbroke Island by the sandmining industry.
- Rehabilitation of affected 'acid' frog habitat by sandmining companies.
- Research into habitat usage and population dynamics by researchers at Griffith University.

FUTURE RECOVERY EFFORTS

What you can do:

- Volunteer with MRCCC, BIEPA, DERM and WPSQ to do monitoring and surveys.
- Avoid the use of garden fertilisers and ensure detergent used to wash cars is kept out of stormwater drains.
- Clean and disinfect soiled clothes/equipment when moving between streams and wetland areas to reduce the risk of spreading disease among frog populations.

Conservation needs:

- Lobby local and state governments to encourage adequate protection of remaining habitat areas on leasehold and freehold land.
- Additional research into non-breeding habitat usage.
- Improve protection of important habitat areas on freehold and leasehold land.
- Implement better management of potential threats.

Research directions:

- Conduct further surveys and monitoring.
- Further study the threat posed by chytrid fungus disease to wallum frog species.
- Gain a better understanding of competition between wallum restricted 'acid' frog species (e.g. Wallum Sedgefrog) and ecologically similar sibling species (e.g. the Eastern Sedgefrog).

Sources and further reading

Anstis M (2002) *Tadpoles of South-eastern Australia.* Reed New Holland, Sydney.

Ingram GJ and Corben CJ (1975) The frog fauna of North Stradbroke Island with comments on the 'acid' frogs of the wallum. *Proceedings of the Royal Society of Queensland* **86**, 49–54.

James C (1996) 'Conservation genetics of island and mainland populations of the sedge frogs *Litoria cooloolensis* and *Litoria olongburensis*'. Unpublished report for the Queensland Department of Environment and Heritage.

Lewis BD and Goldingay RL (2005) Population monitoring of the vulnerable wallum sedge frog (*Litoria olongburensis)* in north-eastern New South Wales. *Australian Journal of Zoology* **53**, 185–194.

Liem DS and Ingram GJ (1977) Two new species of frogs (Anura: Myobatrachidae, Pelodryadidae) from Queensland and New South Wales. *Victorian Naturalist* **94**, 255–262.

Author: Ed Meyer

CASCADE TREEFROG (KROOMBIT TOPS POPULATION)

Litoria pearsoniana (Copland, 1961)

Photo: Harry B Hines-DERM

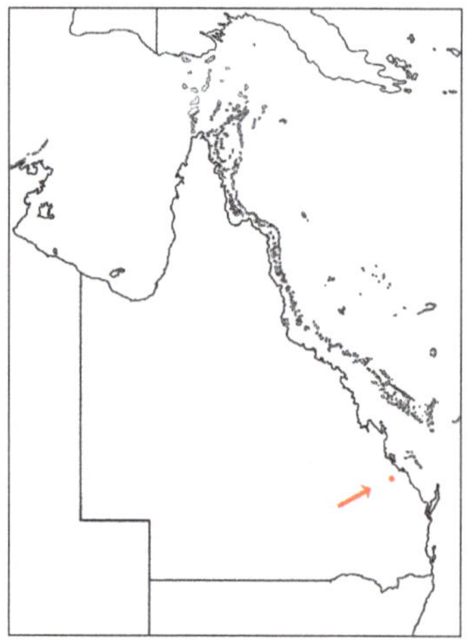

Family: Hylidae

Conservation status: Qld: Endangered

Description: The Cascade Treefrog measures up to 40 mm in length. Adults weigh 2–5 g, with males smaller than females. These frogs are variable in colour: the dorsal surface is smooth, bright green and/or beige, often peppered or mottled with black or brown. A dark lateral stripe (bordered above by a thin gold stripe) runs back from the nostril, through the eye and onto the flanks. The heel and outside of the foot are white-edged. The upper lip is usually pale. The ventral surface of the body is white and granular. The tips of fingers and toes possess disc-like pads. The snout is blunt in profile. Tadpoles: see Cascade Treefrog profile (p. 164)

Distribution: The distinct and very divergent population of cascade treefrogs occurs at Kroombit Tops, south-west of Gladstone. It is an evolutionarily significant unit (ESU), with a mitochondrial lineage of considerable antiquity.

Habitat: Mid-to-high altitude streams in temperate and subtropical rainforest and tall open eucalypt forest at Kroombit Tops.

Ecology: Males mostly call at dusk and into the night in spring and summer, but have been heard during the day when conditions are overcast, warm and humid. They call from debris, rocks and low vegetation in or at the edge of streams. Amplexus is axillary and eggs are laid in water as a single clump or several small clumps on the underside of rocks or attached to vegetation or debris in streams. Tadpoles probably take at least 8 weeks to complete their development, with metamorphs emerging in late spring to autumn. Non-breeding habitat requirements at Kroombit Tops are not known. Adult frogs feed on small arthropods, while tadpoles feed on bottom sediment and algae. The male advertisement call is a crisp 'wreek-kik-kik'.

Reasons for decline and current threats: There are no population estimates, but populations are thought to have declined. The reason for the decline is not known, but amphibian chytrid fungus disease is strongly suspected. In Qld, a number of stream-associated species have declined or disappeared due to chytrid fungus disease. At Kroombit Tops, the amphibian chytrid fungus has been detected in a number of amphibian species and was determined to be the cause of recent die-offs of Cascade Treefrogs in this area.

Current threats include the chytrid fungus, climate change, fire regimes and invasion of the weed Lantana *Lantana camara*. Trampling of stream banks and removal of soil-stabilising vegetation by pigs, cattle and horses increases turbidity of the streams causing a build up of silt which is likely to affect tadpole development. Pigs are also likely to prey upon adult frogs and disturb winter shelter sites. The Kroombit Tops population of Cascade Treefrogs is dependent upon the relatively cool and humid environments provided by rainforest and wet sclerophyll communities at mid-high altitudes for its survival. Projected increases in temperature and a decline in rainfall at Kroombit Tops may reduce the extent of suitable habitat for this species and lead to an increase in the frequency of severe fire events, resulting in significant habitat loss.

CURRENT RECOVERY EFFORTS

***Recovery plan*:** Hines HB and the South-east Queensland Threatened Frogs Recovery Team (2002) 'Recovery plan for stream frogs of south-east Queensland 2001–2005'. Report to Environment Australia, Canberra. Queensland Parks and Wildlife Service, Brisbane.

***Other plans*:** Department of the Environment and Heritage (2005) 'Predation, habitat degradation, competition and disease transmission by feral pigs'. Department of the Environment and Heritage, Canberra.

Department of the Environment and Heritage (2006) 'Threat abatement plan: infection of amphibians with chytrid fungus resulting in chytridiomycosis'. Department of the Environment and Heritage, Canberra.

Conservation actions:

- Ongoing population surveys and monitoring and investigations into basic life history are being carried out.
- Research is being conducted into the epidemiology of the amphibian chytrid fungus.
- DERM coordinates recovery plan implementation; manages Kroombit Tops NP which holds all known populations, with particular emphasis on control of introduced mammals and fire; undertakes surveys and monitoring; liaises with universities and other research institutions. These activities are severely limited by availability of funding.
- Various NRMs (particularly Fitzroy Basin Association and Burnett Mary Regional Group), NGOs, Qld Frog Society Inc., universities and government research institutes carry out and/or fund monitoring, surveys, feral animal control and scientific research.
- Volunteers assist with surveys, monitoring, data management and logistical support.

FUTURE RECOVERY EFFORTS

What you can do:

- Volunteer with DERM.
- Donate to the Qld Frog Society Trust Fund.
- Write letters to state and federal environment ministers to urge much greater funding allocations to implement recovery and research recommendations.
- Clean and disinfect soiled clothes and equipment between sites to reduce the risk of spreading disease among frog populations.

Conservation needs:

- Review recovery plan.
- Control introduced animals, especially feral pigs.
- Implement threat abatement plans for feral pigs and amphibian chytrid fungus.
- Ongoing fire planning and management targeting protection of rainforest patches.

Research directions:

- Study the threatening processes, especially the role of amphibian chytrid fungus.
- Clarify the taxonomic status of Kroombit Tops animals.
- Determine the genetic structure of the population.
- Monitor the populations.

Sources and further reading

Anstis M (2002) *Tadpoles of South-eastern Australia: A Guide with Keys*. Reed New Holland, Sydney.

Donnellan SC, McGuigan K, Knowles R, Mahony M and Moritz C (1999) Genetic evidence for species boundaries in frogs of the *Litoria citropa* species-group (Anura: Hylidae). *Australian Journal of Zoology* **47**, 275–293.

Hines HB (2008) Presidential address 2008. Frogs of the Southeast Queensland Bioregion. *The Queensland Naturalist* **46**, 29–43.

McDonald KR and Davies M (1990) Morphology and biology of the Australian treefrog *Litoria pearsoniana* (Copland) (Anura: Hylidae). *Transactions of the Royal Society of South Australia* **114**, 145–156.

Author: Harry B Hines

CASCADE TREEFROG

Litoria pearsoniana (Copland, 1961)

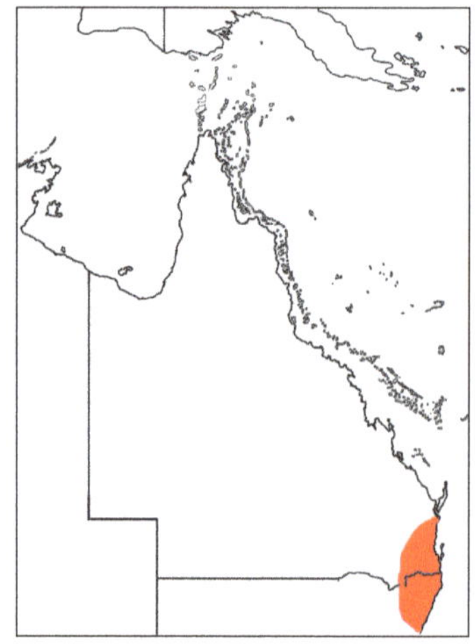

Photo: Harry B Hines-DERM

Family: Hylidae

Other common names: Mountain Stream Treefrog, Pearson's Frog, Pearson's Green Treefrog, Pearson's Treefrog

Conservation status: IUCN: Near Threatened; Qld: Vulnerable

Description: The Cascade Treefrog measures up to 40 mm in length. Adults females and males weigh 2 and 5 g, respectively. They are variable in colour: the dorsal surface is smooth, bright green and or beige, often peppered or mottled with black or brown. A dark lateral stripe (bordered above by a thin gold stripe) runs back from the nostril, through the eye and onto the flanks. The heel and outside of the foot are white-edged. The upper lip is usually pale. The ventral surface is white and granular. The tips of fingers and toes have disc-like pads. The snout is blunt in profile.

Tadpoles grow to about 40 mm, have a globose body, and are dark-grey to brown above, with silver stippling below. The intestinal mass, heart and gills are visible through the ventral body wall when tadpoles are viewed from below. The eyes are dorsolateral in position. The tail is low-finned and 1.5 times as long as the body. The fins are translucent with dark stippling, and the tail musculature is light brown to grey in colour. The spiracle is sinistral and the vent tube dextral. The labial tooth row formula of the tadpoles is 2(2)/3.

Distribution: The Cascade Treefrog occurs throughout the wet forests of the Scenic Rim, D'Aguilar, Conondale and Blackall Ranges. A taxonomic uncertainty exists about certain populations. This profile only considers populations occurring in far southern Qld and excludes the Kroombit Tops Cascade Treefrog (see profile p. 162).

Habitat: Along streams in temperate and subtropical rainforest and wet sclerophyll forest from near sea level to over 1000 m. Highest densities are seen in streams at mid-altitudes.

Ecology: Males mostly call at dusk and into the night in spring and summer, but have been heard calling during the day when conditions are overcast, warm and humid. They call from debris, rocks and low vegetation in, or at the edge of, streams. Amplexus is axillary and eggs are laid in water as a single clump, or several small clumps, on the underside of rocks or attached to vegetation or debris in streams. Clutch size is about 500 eggs. Tadpoles probably take at least 8 weeks to complete their development, with metamorphs emerging in late spring to autumn. In upland streams, tadpoles may over-winter. Non-breeding habitat requirements are poorly known, but large winter aggregations of adults under rocks, in rock crevices and cracks in bridge girders have been documented. Adult frogs feed on small arthropods, while tadpoles feed on bottom sediment and algae. The male advertisement call is a crisp 'wreek-kik-kik'.

Reasons for decline and current threats: No population estimates exist for the species, but populations in southern Qld suffered serious declines during the late 1970s to early 1980s and the population from Girraween NP appears to be extinct. Amphibian chytrid fungus disease is strongly suspected. Over the past 10–15 years, other populations have shown signs of recovery. In Qld, a number of stream-associated species have declined or disappeared due to chytrid fungus disease. The chytrid fungus has been detected in a number of amphibian species in south-east Qld and was determined to be the cause of death of the Cascade Treefrog at several sites.

Current threats include chytrid fungus, feral pigs and weed invasion. Feral pig activity increases the turbidity of streams causing a build-up of silt likely to affect tadpole development. Pigs also prey upon adult frogs and disturb winter shelter sites. A number of weeds invade the riparian habitat of this species but their impact is not well understood.

CURRENT RECOVERY EFFORTS

***Recovery plan*:** Hines HB and the South-east Queensland Threatened Frogs Recovery Team (2002) 'Recovery plan for stream frogs of south-east Queensland 2001–2005'. Report to Environment Australia, Canberra. Queensland Parks and Wildlife Service, Brisbane.

Other plans: Department of the Environment and Heritage (2005) 'Predation, habitat degradation, competition and disease transmission by feral pigs'. Department of the Environment and Heritage, Canberra.

Department of the Environment and Heritage (DEH) (2006) 'Threat abatement plan: infection of amphibians with chytrid fungus resulting in chytridiomycosis'. Department of the Environment and Heritage, Canberra, Canberra.

Conservation actions:

- Ongoing population surveys and monitoring and investigations into basic life history are being carried out.
- Research is being conducted on the epidemiology of the amphibian chytrid fungus.
- NRMs (particularly Burnett Mary Regional Group), NGOs, universities (Griffith University Gold Coast), Gold Coast City Council, Mary River Catchment Coordinating Committee, Qld Frog Society Inc., and government research institutes carry out and/or fund monitoring, surveys, feral animal control and scientific research.
- DERM coordinates recovery plan implementation; manages Kroombit Tops NP which holds all known populations, with particular emphasis on control of introduced mammals and fire; undertakes surveys and monitoring; and liaises with universities and other research institutions.

FUTURE RECOVERY EFFORTS

What you can do:

- Volunteer with DERM.
- Donate to the Qld Frog Society Trust Fund.
- Write letters to state and federal environment ministers urging greater funding allocations to implement recovery and research recommendations.
- Clean and disinfect soiled clothes and equipment between sites to reduce the risk of spreading disease among frog populations.

Conservation needs:

- Review recovery plan.
- Control feral pigs.
- Implement threat abatement plans for feral pigs and amphibian chytrid fungus.

Research directions:

- Monitor the populations.
- Study the threatening processes, especially the role of amphibian chytrid fungus.
- Clarify the taxonomic status of the Kroombit Tops population.
- Develop ecologically sensitive weed management in riparian areas.

Sources and further reading

McDonald KR and Davies M (1990) Morphology and biology of the Australian treefrog *Litoria pearsoniana* (Copland) (Anura: Hylidae). *Transactions of the Royal Society of South Australia* **114**, 145–156.

McGuigan K, McDonald K, Parris K and Moritz C (1998) Mitochondrial DNA diversity and historical biogeography of a wet forest-restricted frog (*Litoria pearsoniana*) from mid-east Australia. *Molecular Ecology* 7, 175–186.

Parris KM (2001) Distribution, habitat requirements and conservation of the cascade treefrog (*Litoria pearsoniana*, Anura: Hylidae). *Biological Conservation* **99**, 285–292.

Author: Harry B Hines

COMMON MISTFROG

Litoria rheocola Liem, 1974

Photo: Harry B Hines-DERM

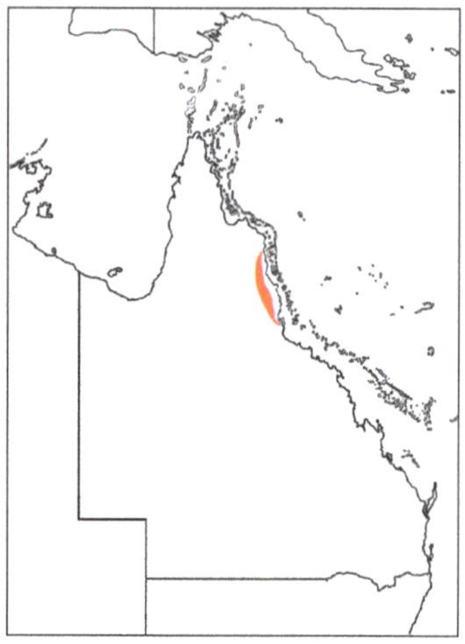

Family: Hylidae

Conservation status: IUCN: Endangered; Aust.: Endangered; Qld: Endangered

Description: This medium-sized frog measures 27–41 mm in length and weighs 1.2–4.5 g, with females being larger than males. Its dorsal surface is grey or brown, mostly smooth but with a few tubercles and often with irregular dark markings or mottling. It has granular white skin below. There is frequently a distinct line between the eyes, paler in front and darker behind, fading towards the rump. The fingers and toes have large discs and, while the toes are fully webbed, the fingers are only partly webbed. The tympanum is covered by skin, though is often distinct.

Distribution: Occurs throughout the Wet Tropics from north of Ingham to the Big Tableland south of Cooktown. It once occurred from sea level to mountain tops at 1200 m, but no longer occurs above about 600 m and only rarely above 400 m.

Habitat: Primarily along rainforest streams, particularly in or near riffles, cascades and waterfalls where they perch on streamside vegetation, rocks, roots and logs. During the day they can often be flushed from among leaf litter trapped besides flowing water. They also inhabit similar streamside microhabitats in a few streams outside rainforest. Tadpoles inhabit riffles and have large mouthparts for clinging to rocks and muscular tails for swimming in a current.

Ecology: Common Mistfrog males have a repetitive rising 'reeek' call, which can be heard at all times of year, though can be quite suppressed when it is cool. Males also use foot-flagging, possibly to communicate when territorial boundaries have been breached by a rival. The large, unpigmented eggs can be laid throughout the year, usually in a gelatinous clump beneath a rock in fast flowing water. Tadpoles hatch to live among the rocks in a riffle and feed on algae growing on the rocks. Adults feed on a wide range of invertebrates associated with their streamside habitat.

Reasons for decline and current threats: Populations of this species collapsed to extinction in the areas above 600 m within their range in the late 1980s and early 1990s. The cause is thought to be the chytrid fungus, which appears to have swept through the area decimating this and six other species, and causing a reduction in abundance of an eighth species. The disease is now firmly established in extant populations and is likely to be suppressing the frog's ability to recolonise high-altitude streams. This is possibly because cooler temperatures at altitude may make the fungus more pathogenic and may suppress the immune response of frogs. Although its impacts have apparently stabilised, this pathogen may still pose a significant threat by suppressing reproductive output and increasing mortality in the species, leading to populations with low resilience in the face of future perturbations. Future perturbations are likely to come in the form of continuing coastal development, slowly encroaching into the lower hill slopes.

CURRENT RECOVERY EFFORTS

Recovery plan: Northern Queensland Threatened Frogs Recovery Team (2001) 'Recovery plan for the stream-dwelling rainforest frogs of the Wet Tropics biogeographic region of north-east Queensland 2000–2004'. Report to Environment Australia, Canberra. Queensland Parks and Wildlife Service, Brisbane.

Other plans: Australian Government Department of Environment and Heritage (2006) 'Threat abatement plan for infection of amphibians with chytrid fungus resulting in chytridiomycosis'. Department of the Environment and Heritage, Canberra.

Conservation actions:

- This species, and other stream-dwelling frogs in the Wet Tropics region, has been the subject of some of the most intensive and long-term monitoring and research of any frogs. Thus, their population trends and disease profiles have been well documented and significant discoveries, relevant to frog declines worldwide, have been made about the disease process. Much of this work was done as a consequence of the recovery plan and process initiated after the declines occurred.

- Much of the species range is protected within World Heritage-listed national park, although a significant proportion is also subject to coastal development pressures.
- A set of protocols designed to minimise the spread of disease has been established for researchers and enthusiasts frequenting rainforest streams (see <http://www.jcu.edu.au/school/phtm/PHTM/frogs/adms/scope.htm>).
- In addition, the movement of tadpoles has been restricted, except under permit. Unfortunately, it appears all populations already have chytrid fungus, so these measures are of little value in controlling the spread of this pathogen but could be of benefit should another emerge (see Australian Lacelid p. 174).

FUTURE RECOVERY EFFORTS

Conservation needs:

- Investigate the possibility of reintroducing the species to areas from which it has disappeared. This needs to be done with careful consideration to the movement of the chytrid fungus with the frog and our ability to treat the disease in wild populations (see Australian Lacelid Frog).
- Consider the needs of this and other species when making land management decisions, particularly those involving stream habitats.
- Take all measures available to reduce the possibility of novel diseases or pests entering the country.

Research directions:

- Continue to monitor the populations periodically, particularly at the 400–600 m range to see if they are recovering and recolonising parts of their previous range.
- Further investigate the role of disease as a continuing threatening factor and how to treat it in wild populations, as well as its impact on the resilience of the species modelled.
- Continue study of the autecology of this species to assist management of the species in the coastal landscape.

Sources and further reading

Berger L, Speare R and Hyatt A (1999) Chytrid fungi and amphibian declines: overview, implications and future directions. In *Declines and Disappearances of Australian Frogs*. (Ed. A Campbell) pp. 23–33. Environment Australia, Canberra.

Hodgkison S and Hero J-M (2003) Seasonal, sexual and ontogenetic variations in the diet of the 'declining' frogs *Litoria nannotis, Litoria rheocola* and *Nyctimystes dayi*. *Wildlife Research* **30**, 345–354.

Hoskin CJ and Goosem M (2010) Road impacts on abundance, call traits, and body size of rainforest frogs in northeast Australia. *Ecology and Society* **15(3)**, 15 (online).

Hoskin CJ and Hero J-M (2008) *Rainforest Frogs of the Wet Tropics, North-east Australia*. Griffith University, Gold Coast.

McDonald KR and Alford RA (1999) A review of declining frogs in northern Queensland. In *Declines and Disappearances of Australian Frogs*. (Ed. A Campbell) pp. 14–22. Environment Australia, Canberra.

Author: Andrew J Dennis

NEW ENGLAND TREEFROG

Litoria subglandulosa Tyler & Anstis, 1983

Photo: Marion Anstis

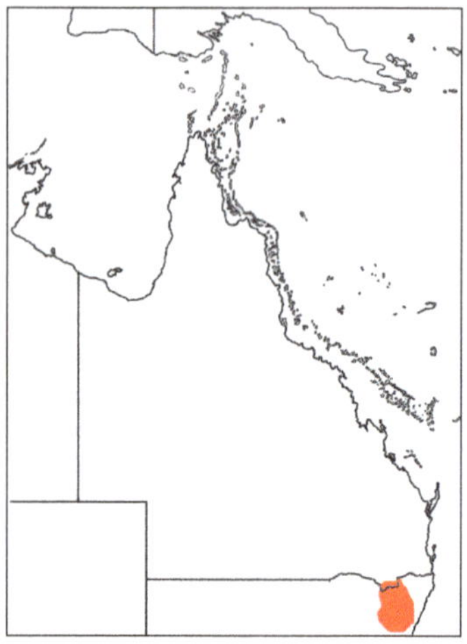

Family: Hylidae

Other common name: Glandular Treefrog

Conservation status: IUCN: Vulnerable; Qld: Vulnerable; NSW: Vulnerable

Description: Males measure 34.5–40.3 mm in length and females 45.8–50.4 mm. The head is broader than long and the snout smoothly contoured. The tympanum is indistinct and bordered above by a thick glandular ridge. The eyes are prominent with a golden iris. The dorsal surface of active frogs is predominantly bright green, but some are golden brown with green patches on the back, sides and limbs. Inactive or cold frogs are often very dark green. A dark brown lateral stripe, bordered above by a prominent pale gold stripe, extends from tip of the snout to the eye and breaks into broad gold patches and spots on a darker background over the sides of body and flanks. A narrow white stripe runs along the edge of the upper lip to the arm, with green above. The dorsal skin is smooth and the ventral surface is slightly granular and white, with a smooth gland around the margin of lower jaw. The groin, under surfaces of the hind limbs and armpits are translucent yellow. The fingers are unwebbed and toes about two-thirds webbed; finger discs are prominent and toe discs slightly less.

Distribution: This frog has a restricted distribution along the eastern escarpment of the Great Dividing Range above 600 m altitude, from near Walcha in central NSW to Girraween NP 250 km further north in south-east Qld. Populations within this range are disjunct and many in the southern parts of have dwindled since 1978. A low density patchy population persists in Girraween NP and appears to be the only area where this frog is known historically in Qld.

Habitat: The New England Treefrog is associated with permanent streams and rivers in rainforest, montane areas or wet sclerophyll forest. Tadpoles are most commonly found in the shallower, less heavily flowing sections at the sides of the stream, often among rocks or on the substrate. Groundcover in the habitat beside streams varies from dry leaf litter in sandy granite country to moist leaf litter, rotting logs and fertile top soil in rainforests.

Ecology: This frog breeds in spring and summer. Males usually call in small groups spaced out along the stream bank, where they mostly sit on ferns, sedges or low shrubs. The typical call is a sequence of about 5–7 repeated, slightly drawn out 'or-āk' notes. Calling increases in intensity during and after light rain. During the breeding season, gravid females may be found in the vicinity of calling male territories. Eggs are laid in a single clump of two to three layers of adhering capsules and, from observations of the closely related Davies's Treefrog *L. daviesae*, are likely to be attached to a submerged rock face, or partly submerged vegetation in stream pools. Hatchlings are dark grey with a grey yolk sac and average 6.2 mm in length. The tadpoles are stream-adapted and grow to about 37 mm in length, and have unique mouthparts with numerous prominent pointed papillae, but no black keratinised jaws or tooth rows like most other species. They appear to feed on fine silt as they graze over the bottom and on rocks, but their unique mouthparts indicate they may be highly specialist feeders. Adult frogs are known to hibernate communally from early winter under rotting logs within the vicinity of the stream.

Reasons for decline and current threats: Chytrid fungus disease has been strongly implicated in declines of many stream species. Other possible threats include excess logging closer than about 80 m to stream banks, alterations in stream flow or water quality, loss of streamside vegetation, introduced fish and general habitat pollution from any human-induced source. Lack of management and monitoring is perhaps an even greater threat, because, without specific evidence of threats and disease occurrence throughout its range, its fate is uncertain at best.

CURRENT RECOVERY EFFORTS

Recovery plan: None

Other plans: Department of the Environment and Heritage (2006) 'Threat abatement plan: infection of amphibians with chytrid fungus resulting in chytridiomycosis'. Department of the Environment and Heritage, Canberra.

Conservation actions: Qld populations are in protected areas.

FUTURE RECOVERY EFFORTS

What you can do:

- Clean and disinfect soiled clothes/equipment when moving between streams and wetland areas to reduce the risk of spreading disease among frog populations.
- Write letters to state and federal environment ministers to urge much greater funding allocations as well as the implementation of research recommendations and development of a management plan.

Conservation needs:

- Implement the threat abatement plan for the amphibian chytrid fungus.
- Philanthropic and private industry funding is essential to support recovery of this species, as government funding is inadequate.

Research directions:

- Continue surveys and monitoring to further clarify the species current distribution in south-east Qld. The Qld population is of special interest because it is only known from such a small range.
- Conduct additional studies of the frog's microhabitat to assess the potential impact of any changes to their habitat, such as excessive human recreational activity in the national park.
- Develop a better understanding of the threatening processes, especially the role of amphibian chytrid fungus on the Qld population.

Sources and further reading

Anstis M (1997) Glandular Treefrog. In *Threatened Frogs of New South Wales: Habitats, Status and Conservation*. (Ed. Ehmann H) pp. 214–221. Frog and Tadpole Study Group of NSW Inc., Sydney.

Anstis M (2002) *Tadpoles of South-eastern Australia: A Guide with Keys*. Reed New Holland, Sydney

Anstis M and Littlejohn MJ (1996) The breeding biology of *Litoria subglandulosa* and *L. citropa* (Anura: Hylidae), and a re-evaluation of their geographic distribution. *Transactions of the Royal Society of South Australia* **120**, 83–99.

Mahony M, Knowles R, Foster R and Donnellan S (2001) Systematics of the *Litoria citropa* complex in northern New South Wales and southern Queensland, Australia, with the description of a new species. *Records of the Australia Museum* **53**, 37–48.

Author: Marion Anstis

FLEAY'S BARRED FROG

Mixophyes fleayi Corben & Ingram, 1987

Photo: Harry B Hines-DERM

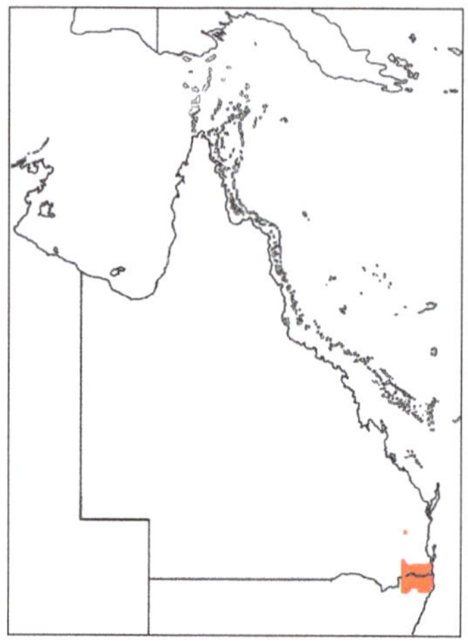

Family: Myobatrachidae

Other common names: Fleay's Frog, Silverblue-eyed Barred Frog

Conservation status: IUCN: Endangered; Aust.: Endangered; Qld: Endangered; NSW: Endangered

Description: The Fleay's Barred Frog measures up to 90 mm in length and adult females and males weigh about 30 and 100 g, respectively. The dorsal surface in adults is finely granular, beige or brown with an irregular vertebral band extending back from a line between the eyes. The limbs have a series of dark cross-bars. A dark stripe running back from the tip of the snout through the nostril and eye extends over the tympanum before stopping abruptly at above the level of the forelimb. Well-defined spots and splotches are often found on flanks. The upper lip is usually with some small irregular darker markings. The ventral surface is typically yellow and smooth. The tips of fingers and toes do not have disc-like pads; the fingers are not webbed, while the feet are about one-third webbed. The pupil forms a vertical slit when constricted. The iris colour is highly variable: bright flame orange in juveniles; in adults, palest at the top (pale-yellow brown, through silver, to blue), darker below. Tadpoles grow to about 100 mm (typically 60–70 mm) and have a fusiform body, with a tail twice as long as the body.

Distribution: Fleay's Barred Frog occurs throughout the wet forests of the Scenic Rim, with an outlying, genetically distinct, population in the Conondale Range. Populations have previously been recorded at Mount Tamborine and the Bunya Mountains. It also occurs in far north-eastern NSW. There are no population estimates, but populations probably suffered serious declines during the late 1970s to early 1980s and the populations from Mount Tamborine and the Bunya Mountains appear to be extinct.

Habitat: Occurs in temperate and subtropical rainforest and tall open forests, mostly within national parks.

Ecology: Males call from the stream bed or banks and from under forest litter or loose soil at dusk and into the night from spring through autumn, with calls being most intense when the stream levels fall back towards basal flow following rain. Breeding occurs sporadically during this period. Amplexus is axillary. The eggs are laid within streams in gently flowing water. The clutch size is up to 1000 eggs. Tadpoles probably take 6 months to 2 years to complete their development. Metamorphs usually emerge from late spring to summer. Non-breeding habitat requirements are little known. Juveniles and females are rarely encountered along streams and can be found many hundreds of metres from breeding sites, sometimes on ridges. Adult frogs feed on arthropods, while tadpoles feed on bottom sediment, algae, detritus, fallen fruit and carrion. Highest densities are seen along streams at mid to high altitudes. The male advertisement call is a throaty 'ok-ok-ok-ok-ok-ok'. Tadpoles are long lived and usually over-winter in streams.

Reasons for decline and current threats: Amphibian chytrid fungus disease is a strongly suspected cause for the decline of this species. Chytrid fungus has been detected and was determined to be the cause of this frog's decline at a number of sites. Tadpoles carry the infection in their mouthparts and a current study at Main Range NP shows that the prevalence in the tadpole population is close to 100%. Over the past 10 to 15 years populations appear to be recovering.

Current threats include the amphibian chytrid fungus, feral pigs, domestic stock trampling and fouling streams, invasion of weeds and road deaths. A number of weeds invade the riparian habitat of this species, but their impact is not well understood. A study into the population genetics of this species showed limited gene flow between catchments. At Green Mountains, female Fleay's Barred Frogs have been found killed along catchment boundary roads, possibly affecting female population size and further restricting gene flow between catchments.

CURRENT RECOVERY EFFORTS

Recovery plan: Hines HB and the South-east Queensland Threatened Frogs Recovery Team (2002) 'Recovery plan for stream frogs of south-east Queensland 2001–

2005'. Report to Environment Australia, Canberra. Queensland Parks and Wildlife Service, Brisbane.

Other plans: Department of the Environment and Heritage (2005) 'Predation, habitat degradation, competition and disease transmission by feral pigs'. Department of the Environment and Heritage, Canberra.

Department of the Environment and Heritage (DEH) (2006) 'Threat abatement plan: infection of amphibians with chytrid fungus resulting in chytridiomycosis'. Department of the Environment and Heritage, Canberra, Canberra.

Conservation actions:

- In Qld, ecological studies, population surveys and monitoring have recently ceased and a project studying the epidemiology of the amphibian chytrid fungus in tadpoles of this species is near its completion.
- There is a captive holding facility at Lone Pine Koala Sanctuary.
- DERM coordinates recovery plan implementation; manages national parks where this species occurs, with particular emphasis on control of feral pigs; undertakes surveys and monitoring; liaises with universities and other research institutions. These activities are severely limited by availability of funding and other priorities.
- NRMs (particularly SEQ Catchments), NGOs, universities (Griffith University Gold Coast), Qld Frog Society Inc., Lone Pine Koala Sanctuary, volunteers and government research institutes carry out and/or fund monitoring, surveys, feral animal control and scientific research.

FUTURE RECOVERY EFFORTS

What you can do:

- Donate to the Qld Frog Society Trust Fund.
- Write letters to state and federal environment ministers urging greater funding allocations to implement recovery and research recommendations.
- Clean and disinfect soiled clothes and equipment between sites to reduce the risk of spreading disease among frog populations.

Conservation needs:

- Review recovery plan.
- Implement threat abatement plans.
- Control riparian weeds.
- Monitor the populations.
- Develop ecologically sensitive weed management in riparian areas.

Research directions: Study threatening processes, especially the role of amphibian chytrid fungus.

Sources and further reading

Corben CJ and Ingram GJ (1987) A new barred river frog (Myobatrachidae: *Mixophyes*). *Memoirs of the Queensland Museum* **25**, 233–237.

Hines HB, Mahony M and McDonald K (1999) An assessment of frog declines in wet subtropical Australia. In *Declines and Disappearances of Australian Frogs*. (Ed. Campbell A) pp. 44–63. Environment Australia, Canberra.

Symonds EP, Hines HB, Bird PS, Morton JM and Mills PC (2007) Surveillance for *Batrachochytrium dendrobatidis* using *Mixophyes* (Anura: Myobatrachidae) larvae. *Journal of Wildlife Diseases* **43**, 48–60.

Author: Harry B Hines

GIANT BARRED FROG

Mixophyes iteratus Straughan, 1968

Photo: Harry B Hines-DERM

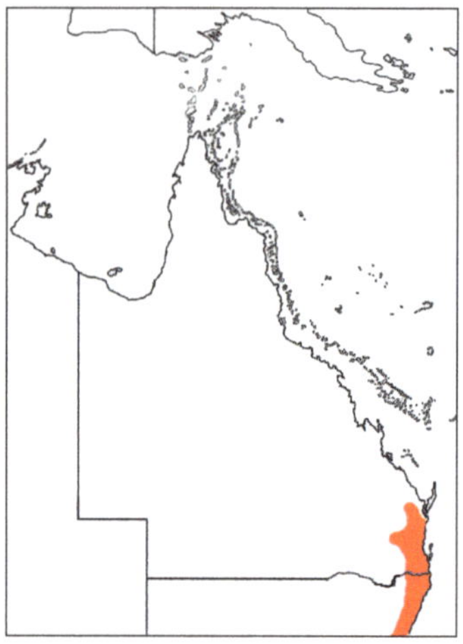

Family: Myobatrachidae

Other common names: Giant Barred River Frog, Gold-eyed Barred Frog, Great Barred Frog, Southern Barred-frog

Conservation status: IUCN: Endangered; Aust.: Endangered; Qld: Endangered; NSW: Endangered

Description: The Giant Barred Frog measures up to 120 mm in length and the adult females and males weigh 80 and 190 g, respectively. The dorsal surface of adults is finely granular, beige to dark-brown, with an irregular vertebral band extending back from a line between the eyes. The limbs have a series of dark cross-bars and there is often rufous colouration, especially along the outer toe or finger. A dark stripe runs back from the tip of the snout through the nostril and eye, extending over the tympanum before stopping abruptly at above the level of the forelimb. The flanks are mottled. The upper lip has irregular darker markings. The ventral surface is yellow to white, smooth. The tips of fingers and toes do not have disc-like pads; the fingers are not webbed, while the feet are about three quarters webbed. The pupil forms a vertical slit when constricted. The upper half of the iris is yellow to gold, and golden brown below. Tadpoles grow to about 100 mm, and have a deep ovoid body, with a tail twice as long as the body.

Distribution: The Giant Barred Frog occurs from mid-eastern NSW to south-east Qld. In Qld, there is a small population in the foothills of McPherson Range, then populations occur in coastal streams north from Burpengary to Doongul Creek near Maryborough. The largest populations are in the upper Stanley River catchment and along the larger streams flowing into the Mary River, where riparian vegetation has been retained. The populations from Main Range and the Bunya Mountains appear to be extinct, while the population in the Conondale Range appears to be slowly recovering.

Habitat: Currently this species occurs along larger streams in the hills and coastal valleys and plains. Highest densities are seen along well-vegetated (typically rainforest) lower order streams at low altitudes. Most of the currently occupied habitat of this species falls on private land.

Ecology: Males call from under soil or forest litter or from exposed positions along stream banks. They are most likely to call at dusk and through the night on warm to hot nights in late-spring and summer. Amplexus is axillary. Eggs are fertilised in water then lifted out of water and stuck to the underside of a bank or to a rock overhanging a pool. The average clutch size is around 2000 eggs. Tadpoles probably take at least 6 months to complete their development, with metamorphs emerging in late summer and autumn. Non-breeding habitat requirements are poorly known, but the species is usually encountered in close proximity to the stream (typically less than 20 m from the stream). Adult frogs feed mostly on arthropods, while tadpoles feed on bottom sediment, algae, detritus, fallen fruit and probably carrion. The male advertisement call is a deep guttural 'woh'.

Reasons for decline and current threats: There are no population estimates, but populations probably suffered serious declines during the late 1970s to early 1980s. Amphibian chytrid fungus disease is strongly suspected and the fungus has been detected in a number of amphibian species in south-east Qld, including the Giant Barred Frog. Fifteen per cent of juveniles sampled at a site on the Stanley River had chytrid fungus disease. Current threats include the chytrid fungus, loss and fragmentation of habitat, changes in water quality and flow regimes, introduced fish, feral pigs, domestic stock trampling and fouling streams, invasion of weeds and road deaths.

CURRENT RECOVERY EFFORTS

***Recovery plan*:** Hines HB and the South-east Queensland Threatened Frogs Recovery Team (2002) 'Recovery plan for stream frogs of south-east Queensland 2001–2005'. Report to Environment Australia, Canberra. Queensland Parks and Wildlife Service, Brisbane.

Other plans: Department of the Environment and Heritage (2005) 'Predation, habitat degradation, competition and disease transmission by feral pigs'. Department of the Environment and Heritage, Canberra.

Department of the Environment and Heritage (DEH) (2006) 'Threat abatement plan: infection of amphibians with chytrid fungus resulting in chytridiomycosis'.

Department of the Environment and Heritage, Canberra, Canberra.

Conservation actions:

- In Qld, ecological studies, population surveys and monitoring undertaken by DERM have recently ceased.
- DERM coordinates recovery plan implementation; manages national parks where this species occurs, with particular emphasis on control of feral pigs; undertakes surveys and monitoring; liaises with universities and other research institutions. These activities are severely limited by availability of funding and other priorities.
- NRMs (Particularly the Burnett Mary Regional Group and SEQ Catchments), Qld Frog Society Inc., NGOs (Mary River Catchment Coordinating Committee), universities (Griffith University Gold Coast), volunteers and government research institutes carry out and/or fund monitoring, surveys, feral animal control and scientific research.

FUTURE RECOVERY EFFORTS

What you can do:

- Donate to the Qld Frog Society Trust Fund.
- Write letters to state and federal environment ministers to urge much greater funding allocations to implement recovery and research recommendations.
- Clean and disinfect soiled clothes and equipment between sites to reduce the risk of spreading disease among frog populations.

Conservation needs:

- Protect and restore riparian habitat to increase area, quality and connectivity.
- Implement threat abatement plans for feral pigs and amphibian chytrid fungus.
- Review recovery plan.
- Develop ecologically sensitive weed management in riparian areas.

Research directions:

- Monitor the population.
- Determine the genetic structure of the species.
- Develop a better understanding of the threatening processes, especially the role of amphibian chytrid fungus.

Sources and further reading

Hines HB, Mahony M and McDonald K (1999) An assessment of frog declines in wet subtropical Australia. In *Declines and Disappearances of Australian Frogs*. (Ed. Campbell A) pp. 44–63. Environment Australia, Canberra.

Ingram GJ and McDonald KR (1993) An update on the decline of Queensland's frogs. In *Herpetology in Australia*. (Eds D Lunney and D Ayers) pp. 297–303. Surrey Beatty & Sons, Sydney.

Lemckert F and Brassil T (2000) Movements and habitat use of the endangered giant barred river frog (*Mixophyes iteratus*) and the implications for its conservation in timber production forests. *Biological Conservation* **96**, 177–184.

Straughan IR (1968) A taxonomic review of the genus *Mixophyes*, (Anura, Leptodactylidae). *Proceedings of the Linnean Society of New South Wales* **93**, 52–59.

Author: Harry B Hines

AUSTRALIAN LACELID

Nyctimystes dayi (Günther, 1897)

Photo: Andrew J Dennis

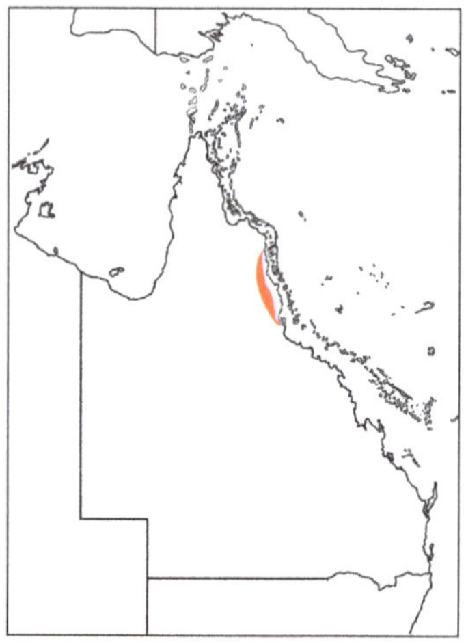

Family: Hylidae

Conservation status: IUCN: Endangered; Aust.: Endangered; Qld: Endangered

Description: Males of this medium-sized frog measure 30–42 mm in length of and weigh 1.75–4 g, while females are 40–60 mm and 2.6–10 g. The Australian Lacelid Frog ranges from orange through brown to grey above, sometimes with indistinct mottling and occasionally with cream to white, lichen like patches of variable size. The skin on the dorsal surface is smooth or slightly granular and cream and granular on the ventral surface. The eye is large and dark and, when closed, the fine golden 'lace' is visible on the lower lid. The fingers and toes have large discs and while the toes are near fully webbed; the fingers are only half webbed. The tympanum is visible but indistinct. The tadpole has large mouthparts for clinging to rocks and a muscular tail for swimming in a current.

Distribution: The Australian Lacelid Frog occurred throughout the Wet Tropics from Paluma, north of Townsville, to the Big Tableland, south of Cooktown. It once occurred from near sea level to mountain tops at 1200 m altitude but no longer occurs above about 400 m and has not been recorded from Paluma since its population and range has declined.

Habitat: The Australian Lacelid Frog occurs in rainforests and their margins. Males frequent streams where they occupy perches and rocks throughout all microhabitats. Females come to streams when gravid, but otherwise may be found well away from streams and may occupy the canopy. Its tadpole inhabits riffles and other fast flowing sections of streams.

Ecology: Males of the Australian Lacelid Frog have a soft growl-like call, which can only rarely be heard above the sound of running water. It breeds from October to April; with females laying clumps of >100 large, unpigmented eggs on or under rocks near the water's surface. When the tadpoles hatch, they remain together for a short time before dispersing to live among rocks in fast flowing water, feeding on algae. Adults feed on a range of invertebrates.

Reasons for decline and current threats: Populations of this species collapsed to extinction in the areas above 400 m in the late 1980s and early 1990s. As with other hylids, the cause of decline is thought to be chytrid fungus disease. The disease is now throughout all populations and is likely to be suppressing the frog's ability to recolonise high altitude streams (see Common Mistfrog p. 166). Future threats are likely to stem from continuing coastal development encroaching on their habitat, which incorporates significant areas away from streams.

CURRENT RECOVERY EFFORTS

Recovery plan: North Queensland Threatened Frogs Recovery Team (2001) Recovery plan for the stream-dwelling rainforest frogs of the Wet Tropics biogeographic region of north-east Queensland 2000–2004. Report to Environment Australia, Canberra. QPWS, Brisbane.

Other plans: Australian Government Department of Environment and Heritage (2006) 'Threat abatement plan for infection of amphibians with chytrid fungus resulting in chytridiomycosis'. Department of the Environment and Heritage, Canberra.

Conservation actions: See Common Mistfrog. Much research and action has centred on understanding the chytrid fungal disease, detecting it, measuring its occurrence and prevalence, controlling its spread, reducing its impact and how to recover frogs after their populations have been infected. Much headway has been made into these aims, although treating and recovering wild populations with the disease is still not realised.

FUTURE RECOVERY EFFORTS

Conservation needs:

- Focus future actions on keeping impacts on remaining populations low, so as not to threaten a species with low resilience and high disease burden. This can be done by managing coastal and lower montane development with stream-dwelling frogs and their habitat requirements in mind and using existing planning act triggers for managing developments that may impact this species.

- Develop chytrid fungus treatment methods able to be applied in the field, selecting suitable sites to try to control and in places 'eradicate' chytrid fungus and implementing a treatment program.

Research directions:

- Continue to survey and monitor the populations.
- Apply acquired knowledge of their environmental requirements in future planning and development along the tropical coast.
- Research the control, or perhaps local eradication, of the chytrid fungus disease in the wild. This will include understanding how it is spread so that 'isolated' populations might be created that are disease free.
- Breed captive populations of this and other species to assist in research into disease and its spread and its occurrence in other hosts.

Sources and further reading

Davies M and SJ Richards (1990) Developmental biology of the Australian hylid frog *Nyctimystes dayi* (Gunther). *Transactions of the Royal Society of South Australia* **114**, 207–211.

Hoskin CJ and Hero J-M (2008) *Rainforest Frogs of the Wet Tropics, North-east Australia.* Griffith University, Gold Coast.

Richards SJ, McDonald K and Alford RA (1993) Declines in populations of Australia's endemic tropical rainforest frogs. *Pacific Conservation Biology* **1**, 66–77.

Woodhams DC, Rollins-Smith LA, Carey C, Reinert L, Tyler MJ and Alford RA (2006) Population trends associated with skin peptide defenses against chytridiomycosis in Australian frogs. *Oecologia* **146**, 531–540.

Author: Andrew J Dennis

MAGNIFICENT BROODFROG

Pseudophryne covacevichae Ingram & Corban, 1994

Photo: Andrew J Dennis

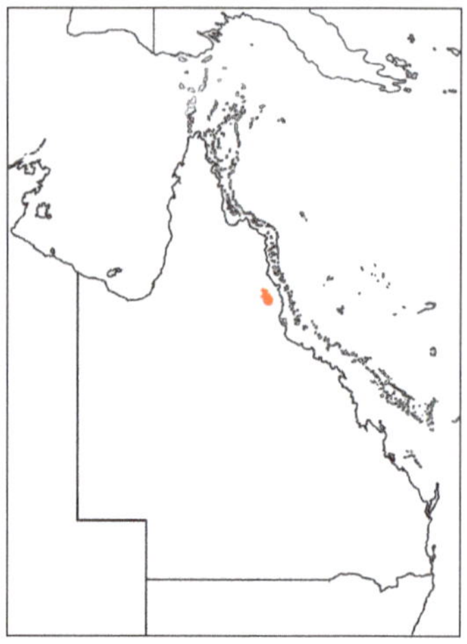

Family: Myobatrachidae

Conservation status: IUCN: Endangered; Aust.: Vulnerable; Qld: Vulnerable

Description: The Magnificent Broodfrog measures 24–28 mm in length. The dorsal surface is rufous brown on the head and back bounded by a dorso-lateral dark region. It has a distinctive bright yellow patch above the cloaca and upper forearm. The ventral surface is marbled black and white. The metamorphosing tadpole can have the distinctive yellow markings, as well as a pale yellow area on the head.

Distribution: Found in five disjunct populations above 800 m altitude in the vicinity of Ravenshoe and Herberton. Each population consists of a set of discrete breeding sites with the population centred on Ravenshoe having the largest number of sites.

Habitat: Dry season habitats are unknown, but during the wet season populations aggregate for breeding in seepage areas and along first-order streams. These occur in open eucalypt woodlands generally with grassy understoreys on acid volcanic and granitic substrates.

Ecology: Frogs use grass tussocks or leaf litter in seepages for calling and egg-laying. These seepages are dry for most of the year, only flowing after heavy rainfall associated with the monsoonal wet season. The frogs are primarily nocturnal, but may call on overcast days. Most calling occurs on warm, wet nights from October to March. Eggs, laid in clutches from 6–82, are found under vegetation in moist soil. Larger clutches consist of eggs at different stages, suggesting multiple laying times. Tadpoles move into small pools to develop. Their diet is unknown, but likely to consist of ants or other small invertebrates. The call is a short 'ark'.

Reasons for decline and current threats: Known populations were threatened by cattle grazing (trampling in seepages and reduction of cover), road development and logging activities, which altered water quality and flow patterns affecting small isolated breeding sites. Clearing of vegetation for road works and urban infrastructure was also a significant threat causing the loss of an entire breeding site when a land fill site was buried.

Currently there are few threats other than those associated with ongoing road maintenance in areas where the frog occurs. Impacts from road works may include erosion or silting and run off from road works and dropping watertables. The continued survival of the species relies on ensuring that land management practices reflect its needs (e.g. grazing intensity is well managed). Impacts from grazing may include trampling by horses and cattle, and invasion by dense stands of weeds.

The species outlook has improved since its recovery plan was partially instigated. However, the species is very restricted, with multiple small isolated breeding sites and is therefore likely to remain vulnerable. Of most concern is the continued expansion of urban developments and infrastructure and the demand for recreation areas for horse riding. Because each breeding site is in a small area, this species could suffer 'death by a thousand cuts', with the loss of any one breeding site not seeming significant but the loss of many sites over time would have a negative impact on its conservation status.

CURRENT RECOVERY EFFORTS

Recovery plan: McDonald KR, Bolitho E, Dennis A, Simpson N and Winter JW (2002) 'Recovery plan for the magnificent broodfrog *Pseudophryne covacevichae* 2000–2004'. Environmental Protection Agency, Brisbane.

Conservation actions:

- Data compiled in 2000 indicated 36 breeding sites spread across four more or less discrete 'populations'.
- Subsequent surveys, conducted as part of the recovery process and assisted by predictive models, located an additional 62 breeding sites and one additional 'population' by 2005.
- As a result of other recovery actions carried out (see recovery plan), most threats have been reduced. This reduction has occurred through changes in land management practices (reduction of grazing and cessation of logging), which were assisted by strong

interest and involvement from the local community and management agencies, combined with an education campaign.

FUTURE RECOVERY EFFORTS

What you can do: Avoid riding horses off existing trails or through gully heads.

Conservation needs:

- All management and infrastructure agencies and departments conducting works within the species range are required to refer to DERM through the *Sustainable Planning Act 2009*. Land management practices (including infrastructure development and maintenance) should be monitored and controlled to ensure they do not affect the habitat of the species. All development activities within the range of the species should be subject to an environmental impact study.
- Carry out the rapid transfer of state forests to national parks within the distribution of the species.
- Exclude horses and cattle from breeding sites.
- Cater fire management plans to the needs of the species.

Research directions:

- Carry out an analysis of the genetic structure across the metapopulation and data on habitat requirements and movements outside the breeding season to guide appropriate management of the species.
- Research the natural history of the species, especially breeding biology, the physiological limits of temperature and moisture, and variation in microhabitat conditions within its habitat to provide baseline information to manage the impacts of climate change.

Sources and further reading

Ingram G and Corban CJ (1994) Two new species of broodfrog (*Pseudophryne*) from Queensland. *Memoirs of the Queensland Museum* **37**, 267–272.

Authors: Andrew J Dennis and Keith R McDonald

EUNGELLA DAYFROG

Taudactylus eungellensis Liem & Hosmer, 1973

Photo: Harry B Hines-DERM

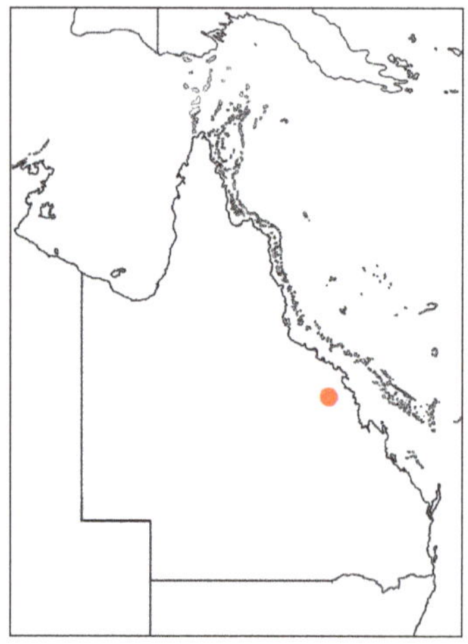

Family: Myobatrachidae

Other common name: Eungella Torrent Frog

Conservation status: IUCN: Critically Endangered; Aust.: Endangered; Qld: Endangered

Description: The Eungella Dayfrog males and females measure 28–36 mm in length. They have a somewhat pointed snout and a body that is yellowish tan, grey-brown or dark brown above with irregular dark brown markings. There is a dark 'X' or butterfly-shaped marking across the back and dark crossbands across the limbs. Subadult animals may also be marked with green on their dorsal surface. The throat and abdomen are cream coloured, while the limbs are a pale yellowish-grey below, with or without dark brown spots. The skin is somewhat rough above and mostly smooth on their ventral surface. Fingers and toes have expanded tips, and are broadly fringed but lack webbing. Males have greyish rounded nuptial pads at the base of the second and third fingers. Males do not possess a vocal sac and the tympanum is hidden. Tadpoles are ovoid and deep-bodied with a shortish muscular tail and dorsolateral eyes. The tail is low-finned and terminates in a blunt rounded tip. The tail musculature is cream with brown and gold splotches, while the fins are mostly clear with a little dark flecking. Tadpoles are light brown to orange-brown above, with a prominent dark V-shaped collar. The coiled intestinal mass is visible through the ventral body wall, although partly obscured by gold pigment at the sides. The spiracle is sinistral, opening laterally. The vent is dextral. The oral disc is umbrella shaped and surrounded by a single row of papillae. Tooth rows are absent, although there are distinct papillar ridges on the lower labium. The horny beak is nearly all white, with only trace amounts of dark pigment around the edge of the jaw sheaths.

Distribution: The Eungella Dayfrog is restricted to the Clarke Range in mid-east Qld. Currently, it is known only from Finch Hatton Creek, Boulder Creek, Urannah Creek and Cattle Creek catchments. Previously it was more widespread in the area, particularly at higher altitudes. There are no population estimates for this species. Where present, the species generally occurs at low or moderate densities.

Habitat: Streams in rainforests in Eungella NP and adjoining state forests at altitudes of 200–1000 m. The Eungella Dayfrog inhabits rocky sections in the splash zones of riffles, cascades and waterfalls of permanent streams. Frogs may be found under large rocks and boulders, in crevices or in more exposed situations.

Ecology: Animals basking on exposed rocks will dive into water when disturbed, remaining submerged for some time. Though largely diurnal, The Eungella Dayfrog can often be seen at night sitting out on vegetation or wet boulders along streams. Male frogs have been observed communicating by visual cues, including flicking and waving legs, head bobbing and distinctive hops. The Eungella Dayfrog is a late summer to autumn breeder, with calling occurring January to May. Females lay 30–50 large (2.2–2.6 mm) eggs under rocks in the water. Tadpoles of the Eungella Dayfrog are bottom-dwellers, feeding on algae and silt in slower-moving sections of stream. Male frogs make a soft chirping call, which is barely audible above the sound of running water.

Reasons for decline and current threats: The Eungella Dayfrog underwent a rapid range contraction in the mid 1980s, and after a period of apparent absence, has now been located at a small number of sites within its former distribution.

The cause of the population decline remains unknown, but the disease amphibian chytrid fungus disease is strongly suspected.

CURRENT RECOVERY EFFORTS

Recovery plan: Northern Queensland Threatened Frogs Recovery Team (2001) 'Recovery plan for the stream-dwelling rainforest frogs of the Wet Tropics biogeographic region of north-east Queensland 2000–2004'. Report to Environment Australia, Canberra. Queensland Parks and Wildlife Service, Brisbane.

***Other plans*:** Department of the Environment and Heritage (DEH) (2006) 'Threat abatement plan: infection of amphibians with chytrid fungus resulting in chytridiomycosis'. Department of the Environment and Heritage, Canberra, Canberra.

Conservation actions:

- DERM coordinates recovery plan implementation; manages national parks where this species occurs, with particular emphasis on control of feral pigs; undertakes surveys and monitoring; liaises with universities and other research institutions. These activities are severely limited by availability of funding and other priorities.
- NRMs, NGOs, universities, volunteers and government research institutes carry out and/or fund monitoring, surveys, feral animal control and scientific research.

FUTURE RECOVERY EFFORTS

What you can do:

- Donate to the Qld Frog Society Trust Fund.
- Write letters to state and federal environment ministers to urge much greater funding allocations to implement recovery and threat abatement plans and research/management recommendations.
- Clean and disinfect soiled clothes and equipment between sites to reduce the risk of spreading disease among frog populations.

Conservation needs:

- Review the recovery plan.
- Implement a threat abatement plan for amphibian chytrid fungus.

Research directions:

- Develop a better understanding of the threatening processes, especially the role of amphibian chytrid fungus.
- Conduct population monitoring.
- Determine the genetic structure of the population.
- Develop captive husbandry skills.

Sources and further reading

Liem DS and Hosmer W (1973) Frogs of the genus *Taudactylus* with descriptions of two new species (Anura: Leptodactylidae). *Memoirs of the Queensland Museum* **16**, 435–457.

McDonald KR (1990) *Rheobatrachus* Liem and *Taudactylus* Straughan and Lee (Anura: Leptodactylidae) in Eungella NP, Queensland: distribution and decline. *Transactions of the Royal Society of South Australia* **114**, 187–194.

Retallick RWR and Hero J-M (1998) The tadpoles of *Taudactylus eungellensis* and *T. liemi* and a key to the stream-dwelling tadpoles of the Eungella rainforest in east-central Queensland, Australia. *Journal of Herpetology* **32**, 304–309.

Retallick RWR, McCallum H and Speare R. (2004) Endemic infection of the amphibian chytrid fungus in a frog community post-decline. *PLoS Biology* **2**, 1965–1971.

Authors: Harry B Hines, Keith R McDonald and John M Clarke

KROOMBIT TINKERFROG

Taudactylus pleione Czechura, 1986

Photo: Harry B Hines-DERM

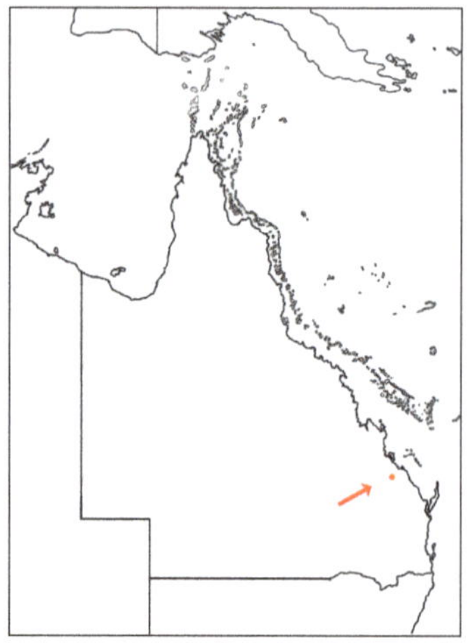

Family: Myobatrachidae

Other common name: Pleione's Torrent Frog

Conservation status: IUCN: Critically Endangered; Aust.: Vulnerable; Qld: Endangered

Description: The Kroombit Tinkerfrog measures 35 mm in length and weighs about 3 g. Adults are grey to reddish-brown on their dorsal surface with darker markings, most notably a prominent 'butterfly' or X-shaped mark on the back and a bar between the eyes. In some individuals, however, these darker markings may be absent or indistinct. The legs are prominently marked with transverse bars. The sides are grey to grey-brown and the ventral surface is mottled or speckled with cream and grey. The toe-discs are small and toes have indistinct flanging. Tadpoles of the Kroombit Tinkerfrog have never been found and have yet to be described.

Distribution: Kroombit Tops, south-west of Gladstone. Only recorded in the headwaters of Degalgil, Diglum, Kroombit, Madsen and Munholme Creeks. The current total population size is unlikely to exceed 1000 adults. Extremely limited distribution: restricted to 12 discrete rainforest patches totalling 596 ha within a total extent of occurrence of 2960 ha.

Habitat: First-order streams at Kroombit Tops originating above 825 m altitude within patches of simple notophyll gully vine forest that span more than 168 m of elevation, are generally greater than 20 ha in area with a high area–perimeter ratio and a catchment of over 56 ha and are no more than 207 m from the nearest occupied patch.

Ecology: Knowledge of the ecology of this species is based on the males calling from crevices, typically from within rock piles. Temperature, atmospheric moisture and rainfall are important determinants of male calling activity. Adults have been occasionally observed eating small arthropods. The male advertisement call consists of a series of 3–17 high-pitched metallic 'tinks'. Males are most likely to call at dusk and into the night in spring and summer.

Reasons for decline and current threats: The population in the headwaters of Kroombit Creek has declined and is possibly extinct. Elsewhere on the plateau, at Kroombit Tops, there is unoccupied suitable habitat, suggesting declines prior to the discovery of the species, with amphibian chytrid fungus disease strongly suspected as the cause.

Current threats include the amphibian chytrid fungus, introduced pigs, horses and cattle, climate change, fire regimes, small/fragmented population size, and invasion of the weed Lantana *Lantana camara*. Grazing and browsing by pigs, cattle and horses may prevent the recovery of canopy vegetation following natural perturbations such as storm or fire damage. Trampling of stream banks and removal of soil-stabilising vegetation also increases turbidity of the streams causing a build up of silt. The weed Lantana invades the disturbed areas, seriously affecting the rate and ability of rainforest patches to recover from fire and increasing subsequent fire intensity. Rainfall and humidity are particularly important for breeding, so predicted changes in temperature and rainfall could also have an impact upon breeding success in this species. The implications of projected climate change on the Kroombit Tinkerfrog could be profound.

CURRENT RECOVERY EFFORTS

Recovery plan: Hines HB and the South-east Queensland Threatened Frogs Recovery Team (2002) 'Recovery plan for stream frogs of south-east Queensland 2001–2005'. Report to Environment Australia, Canberra. Queensland Parks and Wildlife Service, Brisbane.

Other plans: A captive husbandry plan has been prepared. It has not been formally adopted by government agencies and has not been implemented.

Meyer E (2003) 'Captive husbandry plan for the endangered Kroombit tinkerfrog *Taudactylus pleione*'. Unpublished report to Queensland Parks and Wildlife Service and the Southeast Queensland Threatened Frogs Recovery Team, Griffith University, Gold Coast Campus.

Other plans: Department of the Environment and Heritage (2005) 'Predation, habitat degradation, competition and disease transmission by feral pigs'. Department of the Environment and Heritage, Canberra.

Department of the Environment and Heritage (DEH) (2006) 'Threat abatement plan: infection of amphibians with chytrid fungus resulting in chytridiomycosis'. Department of the Environment and Heritage, Canberra.

Conservation actions:

- Ongoing population surveys and monitoring and investigations into basic life history are being carried out. Research informing possible future captive husbandry of the Kroombit Tinkerfrog is being undertaken on a related analogues species, the Eungella Tinkerfrog *Taudactylus liemi*.
- Research is being conducted into the epidemiology of the amphibian chytrid fungus.
- There are captive holding facilities at Currumbin Wildlife Sanctuary.
- DERM coordinates recovery plan implementation; manages national parks where this species occurs, with particular emphasis on control of feral pigs; undertakes surveys and monitoring; liaises with universities and other research institutions. These activities are severely limited by availability of funding and other priorities.
- NRMs (particularly Fitzroy Basin Association and Burnett Mary Regional Group), NGOs, Currumbin Sanctuary; Qld Frog Society Inc., universities, volunteers and government research institutes carry out and/or fund monitoring, surveys, feral animal control and scientific research.

FUTURE RECOVERY EFFORTS

What you can do:

- Donate to the Qld Frog Society Trust Fund.
- Write letters to state and federal environment ministers to urge much greater funding allocations to implement recovery and research recommendations.
- Clean and disinfect soiled clothes and equipment between sites to reduce the risk of spreading disease among frog populations.

Conservation needs:

- Review the recovery plan.
- Control introduced animals, especially feral pigs.
- Implement threat abatement plans for feral pigs and amphibian chytrid fungus.
- Carry out ongoing fire planning and management targeting protection of rainforest patches.
- Implement the captive husbandry plan and establish a captive population.

Research directions:

- Monitor the population.
- Develop a better understanding of the threats, especially chytrid fungus.
- Determine the genetic structure of the population.
- Determine the reproductive biology and captive husbandry requirements.

Sources and further reading

Clarke JM, Borsboom AC, Cunningham M and Hines HB (1999) The recovery process for the Kroombit tinkerfrog, *Taudactylus pleione*. In *Rainforest Recovery for the New Millennium.* (Ed. Boyes BR) pp. 109–123. *Proceedings of the World Wide Fund For Nature 1998 South-east Queensland Rainforest Recovery Conference*. WWF-Australia, Sydney.

Czechura GV (1986) A new species of *Taudactylus* (Myobatrachidae) from southeastern Queensland, Australia. *Memoirs of the Queensland Museum* **22**, 299–307.

Authors: Harry B Hines and John M Clarke

NORTHERN TINKERFROG

Taudactylus rheophilus Liem & Hosmer, 1973

Photo: Andrew J Dennis

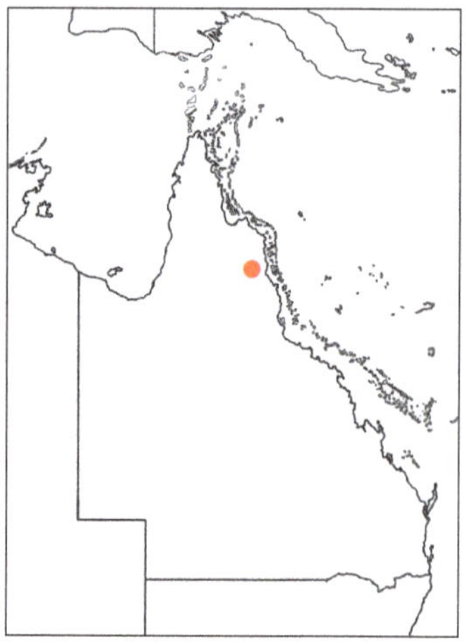

Family: Myobatrachidae

Conservation status: IUCN: Critically Endangered; Aust.: Endangered; Qld: Endangered

Description: This small frog measures 24–31 mm in length. Its colouration and pattern is variable. The dorsal surface may be brown, grey, olive and occasionally orange or bright green with a pattern that that can range from uniform to intricately mottled. The ventral surface is white with brownish marks. A pale grey streak runs from the head to the groin. The fingers and toes are without webbing; the pads are small and distinct. The snout is blunt and rounded.

Distribution: The Northern Tinkerfrog was once found in four separate localities in the Wet Tropics World Heritage Area: Thornton Peak, Carbine Tableland, Lamb Range and Bellenden Ker Range. Dramatic population declines in the late 1980s and early 1990s were thought to have resulted in their extinction by 1991. Between 1996 and 2000, spasmodic records from Mount Lewis and Mount Bellenden Ker confirmed their persistence at these sites. However, since a record of calling males on Mount Bellenden Ker in 2000, no confirmed observations have been made and the continued existence of this species is questionable.

Habitat: This frog is confined to upland, rocky, perennial streams in rainforest above 700 m altitude. Northern Tinkerfrogs hide and call from beneath rocks, ferns and streamside debris immediately adjacent to flowing water. When active, they emerge to nearby mossy rocks and climb ferns to call and presumably feed. They have also been recorded in seepage and trickle areas adjacent to fast flowing streams.

Ecology: A secretive species, it is mainly active at night, although males have been observed calling on overcast days. On Bellenden Ker, males were heard calling from among boulders covered by dense *Helmholtzia acorifolia*. The call is a gentle 'tink tink tink', reminiscent of a teaspoon hitting a glass bottle. A gentle rattling sound has also been recorded. Eggs have not been found in the field and the tadpoles have yet to be described, although 35–50 eggs have been found in specimens in museums. Juveniles have been observed in the months of December and May. Its diet is unknown.

Reasons for decline and current threats: Chytrid fungus disease is the most likely cause of the dramatic population declines that occurred in the late 1990s. The Northern Tinkerfrog has also been identified as one of eight species of frog that are most likely to be affected by climate change in the Wet Tropics because of its restricted, high altitude distribution.

CURRENT RECOVERY EFFORTS

Recovery plan: Northern Queensland Threatened Frogs Recovery Team (2001) 'Recovery plan for the stream-dwelling rainforest frogs of the Wet Tropics biogeographic region of north-east Queensland 2000–2004'. Report to Environment Australia, Canberra. Queensland Parks and Wildlife Service, Brisbane.

Other plans: Department of the Environment and Heritage (DEH) (2006) 'Threat abatement plan: infection of amphibians with chytrid fungus resulting in chytridiomycosis'. Department of the Environment and Heritage, Canberra, Canberra.

FUTURE RECOVERY EFFORTS

Conservation needs: Develop a conservation plan for the Northern Tinkerfrog that details the management needs of the species over the medium to long term. The actions in this plan should then be implemented as a matter of urgency.

Research directions:

- Survey for Northern Tinkerfrogs at sites they were most recently recorded to ascertain the population status of the species.
- Conduct more generalised surveys of historic localities. If they are found, extant populations should be monitored to ascertain the current status and population trends.
- Avoid all invasive research until stable populations are known.

- Study existing museum specimens, which may be of value in examining the history of chytrid fungus infection, the patterns of environmental stresses over time using studies of symmetry, and some aspects of their ecology, such as diet.

Sources and further reading

Dennis A and Trenerry M (1984) Observations on species diversity and habitat compartmentalisation of the frogs of Mt Lewis. *North Queensland Naturalist* **184**, 2–9.

Freeman AB (2003) An observation of calling northern tinker frogs (*Taudactylus rheophilus*) on Mount Bellenden Ker. *Memoirs of the Queensland Museum* **1**, 295–297.

Hoskin CJ and Hero J-M (2008) *Rainforest Frogs of the Wet Tropics, North-east Australia.* Griffith University, Gold Coast.

Marshall C (1998) The reappearance of *Taudactylus* (Anura: Myobatrachidae) in north Queensland streams. *Pacific Conservation Biology* **4**, 39–41.

McDonald KR (1992) Distribution patterns and conservation status of north Queensland rainforest frogs. *Queensland Department of Environment and Heritage Conservation Technical Report* **1**, 1–51.

Author: Alastair Freeman

Reptiles (Class: Reptilia)

INTRODUCTION TO THREATENED REPTILES

Keith R McDonald

In Queensland, there are currently 485 known species of reptile in 14 families, with other species awaiting formal scientific description. Many are poorly known, with limited knowledge of their distribution and habitat preferences and use. However, there are exceptions, such as the freshwater turtles, sea turtles and the reptiles of north-east Queensland wet tropical rainforests.

Thirty-five species (<1%) are listed as threatened (11 Endangered and 24 Vulnerable) in the Queensland NCA Wildlife regulations. In this book, an additional species (*Myuchelys bellii*) is incorporated because it is listed in the *IUCN Red List of Threatened Species*, although currently not recognised as threatened in any Australian legislation. Several species urgently require review because much more is now known since the initial listing in the early 1990s; for example, the Estuarine Crocodile *Crocodylus porosus* is now stable and widespread.

Although a smaller proportion of reptiles are declining than for other groups, those that are doing so are for a diverse range of reasons. For example, most frogs are declining due to the chytrid fungus, and many mammals due to habitat modification and loss as well as introduced pests such as foxes and cats, but reptiles are declining for a range of reasons, such as habitat modification or the loss of woodland, dry vine forest and riparian environments, and harvesting.

Twenty-five threatened species of reptile are found only in Queensland, three occur in other Australian states and seven species (the sea turtles and Estuarine Crocodile) are also found in international waters and other countries.

The family with most threatened species is the skinks. Several of these are threatened for a range of reasons, including loss of habitat, while others are geographically localised and known from limited habitat, which could be affected by stochastic events. The geckoes and marine turtles each have six species listed as threatened. All of the geckoes are listed because they are very localised in small areas of special habitat. Other families have three or fewer species threatened.

Of the threatened reptiles, nine (25.5%) are found in rainforest and wet forest of eastern Queensland, three (8.5%) in wetlands and streams, six (17%) in marine habitat, two (6%) in rock outcrops, and the remainder (14; 40%) in forests and woodlands. The Estuarine Crocodile is found in both marine and freshwater environments.

Many (11) of the Vulnerable (category) listed species are geographically localised in small areas, especially rainforest- and rock-dwelling reptiles and, although currently not experiencing declines in populations or habitat, may well be affected by stochastic events in the future.

The most threatened species are found in forests and wetlands and are at risk from loss of habitat through clearing and habitat modification and changes in wetlands such as rivers, streams and riparian vegetation. Many are poorly represented in protected areas, although many protected areas have not been surveyed for reptiles. Ongoing loss and fragmentation of habitat is of concern, especially in the Brigalow Belt and south-east Queensland bioregions.

The following profiles provide the best current information on the threatened reptiles. It is expected that more species will be added as research results become available and re-assessment of species is incorporated into legislation and lists.

CROCODILES (ORDER: CROCODILIA)

ESTUARINE CROCODILE

Crocodylus porosus (Schneider, 1801)

Photo: www.ataglance.com.au

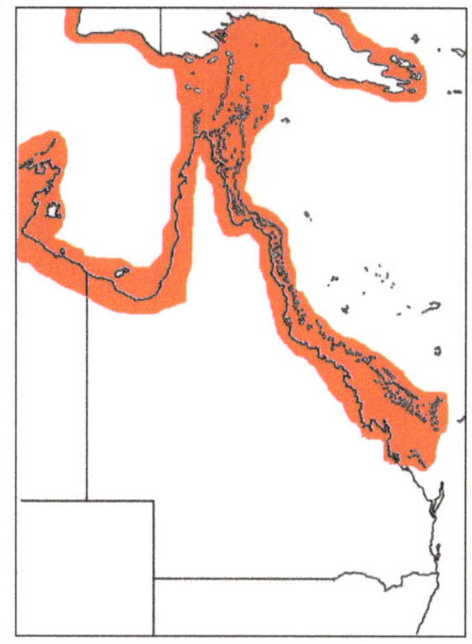

Family: Crocodylidae

Other common names: Saltwater Crocodile, Saltie

Aboriginal name: Pukpuk

Conservation status: IUCN: Least Concern; Qld: Vulnerable

Description: Normal maximum length of the female crocodile is 2.7–3.4 m and weight 120–>200 kg; the male is between 4.2 and >7 m and weighs 350–>1000 kg. Dorsal colouration of hatchlings ranges from yellow with dark spots, which often form dark bands, to almost black. Mature animals may be black, grey-black, tan or even yellow. The ventral surface is uniformly cream to pale yellow. Distinctive features are a large, skull with enlarged body ridges, raised scutes of dermal bone across the dorsal surface and enlarged raised scutes on the tail.

Distribution: It is found from Gladstone in Qld across northern Australia to Derby in WA. Before European settlement, the population of Estuarine Crocodiles in Qld was probably pristine and undisturbed with isolated disturbances by Aborigines, but any impact was probably minimal.

Habitat: It is found on reefs and in coastal and inland waterways throughout its range in Qld. The inland distribution seems to be constrained by the presence of major geographical features that inhibit individuals travelling upstream. Where there is no barrier, inland distribution may extend to over 100 km. Animals have been recorded approximately 130 km upstream in the Fitzroy River and in the Gulf of Carpentaria as far inland as Georgetown. They have also been recorded at almost every island and cay within the northern Great Barrier Reef and at its outer edge.

Ecology: Females lay a single clutch of eggs, between 40 and 70, usually at the beginning of the wet season (late November–January), with a nesting peak in December. The female builds a mound nest 0.5 m to 1.0 m high and between 1.5 m and 2.4 m in diameter from grass or other vegetation. The eggs are laid in the top of the mound, covered and the female usually protects them for the duration of the incubation period (between 60 and 95 days). The sex of the offspring is determined primarily by incubation temperature. Low temperatures (<31°C) produce predominantly females; high temperatures (between 31.8°C and 33°C) produce predominantly males and very high temperatures (>33°C) produce predominantly females. This opportunistic feeder hunts primarily at the water's edge, although larger individuals will feed in deep water and on land. They feed mainly during the night, on fresh food or occasionally carrion. Smaller animals eat invertebrates, small crustaceans and fish, and larger animals include a larger proportion of vertebrates in their diet. Mature animals feed on birds, mud crabs, mammals, fish, turtles and sea snakes. Research has revealed that animals maintain distinctive home ranges, with males occupying larger home ranges than females.

Reasons for decline and current threats: The species was hunted intensively in Qld from the mid-1940s to mid-1970s when it was fully protected under the Qld *Fauna Conservation Act 1974*. Populations are undergoing a limited recovery for most waterways of the southern Gulf of Carpentaria and the populated east coast, with waterways on Cape York supporting higher numbers. The skewed population structure suggests that the population may still be recovering from hunting or that larger animals are still being removed from the population by human interference.

Slow recovery of Estuarine Crocodile populations in Qld (estimated at approximately 30 000 animals) is due to a lack of suitable nesting habitat. The majority of 'good' nesting habitat is located on Cape York Peninsula, with that available in coastal regions south of 13°S described as 'marginal' or 'poor'. There is also considerable netting for commercially important finfish in northern Qld, and, although formal reports are rarely lodged, anecdotal and fisheries observer data indicate that a number of Estuarine Crocodiles >1m long are caught and drowned every year. The recent urban development of the Qld coast south of Cooktown, together with the clearing of riparian vegetation, reclamation of wetlands, commercial gill netting in estuaries and removal of crocodiles identified as 'problem crocodiles', appears to be keeping crocodile numbers low in this area.

CURRENT RECOVERY EFFORTS

Recovery plan: None

Conservation actions:

- DERM's conservation and management plan objectives are to maintain viable wild populations across its range in Qld, increase public awareness and safety in crocodile habitat, educate the public to appreciate the ecological significance of crocodiles and the need for conservation, and to permit and control legitimate commercial enterprises that use crocodiles.
- DERM conducts limited vessel-based spotlight surveys in selected 'index' rivers to provide contemporary information on the current distribution and abundance of the species in Qld waters. This is also combined with the ongoing identification and removal of selected animals ('problem crocodiles') to reduce the risk to people and livestock from crocodiles.
- Researchers from UQ in collaboration with Australia Zoo are looking at the movement patterns and diving behaviour of Estuarine Crocodiles.

FUTURE DIRECTIONS

What you can do:

- Volunteer with DERM to conduct crocodile surveys.
- Volunteer to work with crocodile researchers from The University of Qld and/or make donations to support their research.

Research directions:

- Research the population trends, ecology, movement patterns, breeding biology and endocrinology.
- Research behavioural information on how the species interacts with and maintains sympatric populations with its smaller relative, the Freshwater Crocodile *Crocodylus johnstoni*.

Sources and further reading

Brien MJ, Read MA, McCallum HI and Grigg GC (2008) Home range and movements of radio-tracked estuarine crocodiles (*Crocodylus porosus*) within a non-tidal waterhole. *Wildlife Research* **35**, 140–149.

Campbell HA, Watts ME, Sullivan S, Read MA, Choukroun S, Irwin SR and Franklin CE (2010) Estuarine crocodiles ride surface currents to facilitate long-distance travel. *Journal of Animal Ecology* **79**, 955–964.

Read M, Wright B and Enoch C (2004) Crocodile management in Queensland: an overview. pp. 13–27. In *Crocodiles: Proceedings of the 17th Working Meeting of the IUCN-SSC Crocodile Specialist Group.* 24–29 May, 2004, Darwin, NT, Australia. IUCN, Gland, Switzerland.

Read MA, Grigg GC, Irwin SR, Shanahan D and Franklin CE (2007) Satellite tracking reveals long distance coastal travel and homing by translocated estuarine crocodiles. *PLos One* **2**, e949.

Read MA, Miller JD, Bell IP and Felton A (2004) The distribution and abundance of the estuarine crocodile, *Crocodylus porosus*, in Queensland. *Wildlife Research* **31**, 527–534.

Author: Mark Read

TURTLES (ORDER: TESTUDINES) – Sea Turtles (Chelononiodea)

LOGGERHEAD TURTLE

Caretta caretta (Linnaeus, 1758)

Photo: ©Jean-Paul Ferrero-AUSCAPE

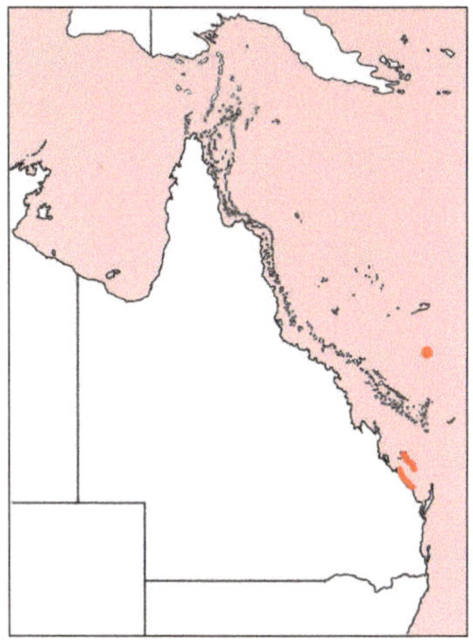

Family: Cheloniidae

Conservation status: IUCN: Endangered; Aust.: Endangered; Qld: Endangered; NSW: Endangered; Tas.: Endangered; NT: Endangered; SA: Vulnerable; WA: Endangered

Description: The smooth-domed carapace of Loggerhead Turtles is brown with light brown, reddish-brown and black variegated areas; the carapace scutes are non-overlapping. There are typically five pairs of costal scutes and three or four enlarged inframarginal scutes on the bridge between the carapace and the plastron. The large inframarginal scales do not have pores. Breeding females are similar in carapace size and colour to the males: around 96 cm. When on the nesting beach, they crawl with an alternating gait, leaving tracks up to ~1 m wide. The eggs measure about 4 cm in diameter and weigh about 37 g; the clutches rarely contain yolkless eggs. Hatchlings are dark brown dorsally and lighter brown ventrally. The straight carapace length of an average hatchling is 4.3 cm; they weigh about 20 g.

Distribution: The eastern Australian population contains three breeding aggregations. One nesting aggregation occurs on the mainland coast of south-east Qld (including Mon Repos, Woongarra Coast beaches and Wreck Rock Beach). A second aggregation, nests in the Capricorn-Bunker Group in the southern Great Barrier Reef. The third, nests on the islands of the Swain Reefs and at Bushy Island off Mackay. Their foraging range encompasses the eastern Arafura Sea, Gulf of Carpentaria, Torres Strait, Gulf of Papua, Coral Sea and western Tasman Sea to southern NSW, including the Great Barrier Reef, Hervey Bay and Moreton Bay.

Habitat: Large immature and adult-sized Loggerhead Turtles from eastern Australia feed in a wide range of tidal and subtidal habitats, including coral and rocky reefs, sea grass meadows, and soft-bottomed sand or mud areas. Nesting beaches occur on the south Qld mainland coast in brown sand and on offshore islands of the Great Barrier Reef in white coralline sand.

Ecology: In the coastal waters of eastern Australian, the Loggerhead Turtle feeds principally on gastropod and bivalve molluscs, portunid crabs and hermit crabs; its diet includes about 100 taxa, occasionally including jellyfish, anemones, holothurians, sea urchins and fish. Turtles nesting at one rookery have migrated from numerous foraging areas; most females return to the same small beach in successive nesting seasons. Females do not breed every year; the average interval between reproductive seasons is 3.82 years. The nesting season in south Qld extends from late October and to early March, peaking in late December. During a breeding season, a female will produce three or more clutches of eggs, each containing about 125 eggs at about 2-weekly intervals. Duration of incubation (8–9 weeks) varies with the temperature. The temperature during the middle third of incubation determines the sex of the hatchling (males at cooler and females at warmer temperatures). After emerging onto the beach surface, hatchlings disperse to an oceanic life for a decade or more. There is high mortality among small turtles as they grow to adulthood; it takes two or more decades before they reach sexual maturity.

Reasons for decline and current threats: Loggerhead Turtles have been subjected to a range of human-induced disturbances. From the late 1960s until the mid 1980s, clutches laid on the mainland coast beaches experienced high levels of fox and low levels of dingo, dog and goanna predation. These beaches predominantly produce female hatchlings, so a reduced number of female hatchlings are now maturing.

Other causes of turtle mortality include boat strike, entanglement in ropes and fishing line, dredging of channels and ingestion of synthetic material. In addition, the combined impacts of other human activities that occur outside Australian territory (e.g. boat strike, entanglement in fishing line and ropes, entanglement in crab-pot float lines, ingestion of synthetic debris, oceanic long-line fishing) have the potential to compromise or delay the recovery of this severely depleted stock. The long period to reach maturity and the lengthy intervals between breeding seasons put Loggerhead Turtle populations at risk from even modest increases in mortality at all stages in their life.

CURRENT RECOVERY EFFORTS

Recovery plan: Environment Australia (2003) 'Recovery plan for marine turtles in Australia'. Environment Australia, Canberra.

Conservation actions:

- Order in Council under the *Qld Fisheries Act* (18 July 1968), declared a year-round closed season for the harvest of all species of marine turtles and their eggs in Qld.
- The introduction of turtle exclusion devices to the Northern Prawn Fishery and the Qld East Coast Trawl Fishery resulted in a significant reduction in the number of trawl turtle captures.
- In an attempt to reduce hatchling disorientation, low pressure sodium vapour lights have been installed in street lights at some beaches. At others, trained personnel relocate eggs to artificial nests on nearby dark beaches.
- Greater than 80% of all Loggerhead Turtles nesting in Qld occurs within the protected habitat of conservation parks and national parks. The majority of internesting habitat is protected by the Great Barrier Reef (GBR), Coast Marine Park (Qld) and the adjacent Great Barrier Reef Marine Park (Australian) zoning regulations. Loggerhead Turtles in foraging areas within the GBR are protected both through habitat protection and through trawling regulations.
- Since late 1990s, the extensive areas of shallow seagrass habitat that support a high density of foraging Loggerhead Turtles have been managed as 'go-slow' areas within the Moreton Bay Marine Park; as a result, the number of boat strikes has been reduced.

FUTURE RECOVERY EFFORTS

What you can do: Report sick, injured, dead and/or tagged sea turtles to DERM (1300 130 372).

Research directions:

Over the next few years, a better understanding of the population dynamics, including links between foraging areas and nesting areas, should emerge from a combination of genetic research and population modelling. In the meantime, studies need to be carried out on:

- the condition of turtle habitats, both in the general distribution of the habitat and the specific habitat used by the turtles;
- the factors affecting the survival of egg clutches and hatchlings (especially factors associated with people, such as presence of feral animals);
- the loss and degradation of coastal habitat due to poorly planned or unplanned coastal development and tourism;
- the impact of changed light horizons on adult female selection of nesting beaches at Australian rookeries.

Sources and further reading

Dodd CKJ (1988) Synopsis of the biological data on the loggerhead sea turtle *Caretta caretta* Linnaeus 1758. *US Fish and Wildlife Service Biological Report* **88**, I–VIII, 1–110.

James R (compiler) (1994) *Proceedings of the Australian Marine Turtle Conservation Workshop.* Queensland Department of Environment and Heritage and Australian Nature Conservation Agency.

Limpus CJ (2009) *A Biological Review of Australian Marine Turtles. 1. Loggerhead Turtle* Caretta caretta *Linnaeus, 1758.* Queensland Environmental Protection Agency, Brisbane.

Limpus CJ and Miller JD (1993) *Family Cheloniidae. Fauna of Australia. Volume 2A. Amphibia & Reptilia.* (Eds CJ Glasby, GJB Ross and PL Beesley) pp. 133–138. Australian Government Publishing Service, Canberra.

Author: Jeff Miller and Colin Limpus

GREEN TURTLE

Chelonia mydas (Linnaeus, 1758)

Photo: Greg Calvert

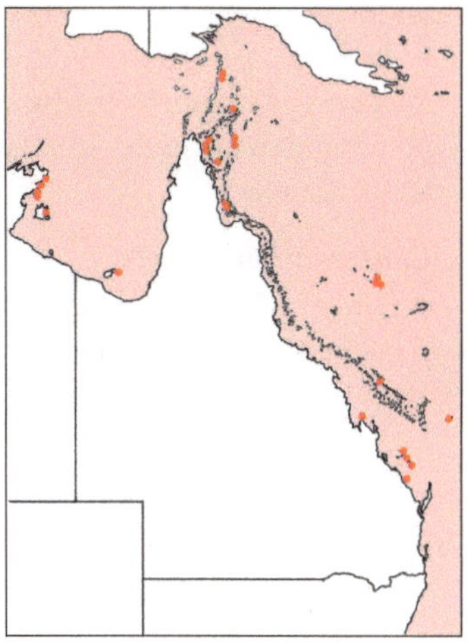

Family: Cheloniidae

Conservation status: IUCN: Endangered; Aust.: Vulnerable; Qld: Vulnerable; NSW: Vulnerable; SA Vulnerable; Tas.: Vulnerable; WA: Vulnerable

Description: Adult Green Turtles have a smooth, high-domed carapace with non-overlapping scutes and usually four costal scutes. Adults are olive-green above with brown, reddish-brown and/or black variegated markings; the ventral surface is whitish or cream coloured. The head has one pair of prefrontal scutes and no preoccular scutes. Adult males can be distinguished by their much longer tails (<20 cm long) and longer curved claws. The curved carapace length of adult Green Turtles ranges between ~90 cm and <120 cm; large adults can weigh >200 kg. Green Turtle eggs are white and spherical; they average about 4.5 cm in diameter and about 50 g in weight. The carapace of hatchlings is dark brown to black; the margin of the carapace, the trailing edge of the flippers and the plastron are white. The average straight carapace length of hatchlings is approximately 5 cm and they weigh approximately 25 g.

Distribution: Green Turtles are found globally in all tropical and subtropical oceans. Genetic analysis has identified seven separate genetic groups of Green Turtles in northern Australia: four of these occur in Qld (southern and northern Great Barrier Reef (GBR), Coral Sea and Gulf of Carpentaria). Turtles from these different breeding aggregations share feeding areas with other Australian and international stocks. The southern GBR breeding unit nests mainly on the islands of the Capricorn Bunker Group; the northern GBR breeding unit nests mainly on the islands of the outer edge of the northern GBR (e.g. Raine Island, Moulter Cay, No.7 and No.8 Sandbanks). The Coral Sea stock nests on many, but not all, of the cays (e.g. North-East Herald and Coringa) in the region. In the Gulf of Carpentaria, the majority of nesting occurs near Mornington Island (e.g. Bountiful, Pisonia and Rocky Islands).

Habitat: Green Turtle foraging habitat is distributed along the shallow coastal areas of Qld, among the reefs of the Great Barrier Reef, and in the Gulf of Carpentaria. Mating areas used by Green Turtles tend to be in the vicinity of the nesting beaches. Nesting beaches are distributed on the offshore islands, cays and, to a lesser degree, along the mainland coast. Usually, internesting habitat is located within 30 km of the nesting beach.

Ecology: Green Turtles feed primarily on seagrass and algae, but occasionally consume non-plant material (jellyfish, dead fish and small crustaceans). Breeding is correlated with the Southern Oscillation Index (SOI) 2 years before the breeding season. The SOI presumably affects the quantity and/or quality of the forage available; as a result, the number of breeding Green Turtles fluctuates widely from year to year. The majority of turtles migrate less than 1000 km. Although migrating turtles do not follow a single path or travel in groups, most females return to the same nesting area between seasons and to the same (or adjacent) beach within a nesting season. Female Green Turtles produce about five clutches (range one to nine) at about 14 day intervals. The eggs require temperatures in the range of 23–33°C throughout incubation. The temperature during the middle third of incubation determines the sex of the hatchling (males at cooler and females at warmer temperatures). Hatchlings emerge from the nests after about 8–9 weeks of incubation. Hatchlings disperse to an oceanic life for an undetermined period.

Reasons for decline and current threats: The main reasons for decline and the current threats include: fisheries bycatch, unquantified levels of harvest by Indigenous Australians and by people in neighbouring countries; predation of eggs by native and introduced animals; coastal development; deteriorating water quality; ingestion of marine debris; and loss of habitat and disease. Although the contribution of each type of mortality may not be large, when considered together, human activities are having a negative impact on the populations of Green Turtles in Qld. The long period required to reach maturity, coupled with the long intervals between breeding seasons, suggest that Green Turtle populations are at risk from small increases in mortality that act over an extended time.

CURRENT RECOVERY EFFORTS

Recovery plan: Environment Australia (2003) 'Recovery plan for marine turtles in Australia'. Environment Australia, Canberra.

Conservation actions:

- Order in Council under the *Queensland Fisheries Act* (18 July 1968), declared a year-round closed season for the harvest of all species of marine turtles and their eggs in Qld.
- Threat abatement, population assessment and habitat monitoring are the major issues dealt with in the recovery plan. Some threatening processes have been tackled through management, but the populations are still at risk.
- Most of the research issues require long-term studies. However, by participating in local planning groups to affect the decisions made about coastal development, especially the lighted areas behind nesting areas, individuals contribute to the conservation management of marine turtles.
- The introduction of turtle exclusion devices (TEDs) in the Northern Prawn Fishery and the Qld East Coast Trawl Fishery minimises captures in trawl fisheries off Qld.

FUTURE RECOVERY EFFORTS

What you can do:

- Report sick, injured, dead and/or tagged sea turtles to DERM (1300 130 372).
- Get involved with the Northern Gulf Resource Management Group and Carpentaria Ghost Nets Programme projects.
- Join/donate/volunteer: the Sea Turtle Foundation, Australian Marine Conservation Society, Bribie Island Environmental Protection Association, Cape York Sustainable Futures, Cape York Turtle Rescue, DERM.

Research directions:

- Assess the condition of the coastal nesting and foraging habitat used by Green Turtles.
- Identify and rectify the factors affecting the survival of eggs and hatchlings (especially factors associated with people and feral animals).
- Conduct genetic research and population modelling over the next few years to develop a better understanding of the population dynamics.

Sources and further reading

Hirth HF (1997) Synopsis of biological data on the Green Turtle *Chelonia mydas* Linnaeus, 1758. U.S. Dept. of the Interior, Fish and Wildlife Service, *Biological Report* **1**, 1–120.

James R (compiler) (1994) *Proceedings of the Australian Marine Turtle Conservation Workshop*. Queensland Department of Environment and Heritage and Australian Nature Conservation Agency.

Limpus CJ (2009) *A Biological Review of Australian Marine Turtles. 2. Green Turtle* Chelonia mydas *(Linnaeus)*. Queensland Environmental Protection Agency, Brisbane.

Limpus CJ and Miller JD (1993) *Family Cheloniidae. Volume 2A. Amphibia & Reptilia*. (Eds CJ Glasby, GJB Ross and PL Beesley) pp. 133–138. Australian Government Publishing Service, Canberra.

Authors: Jeff Miller and Colin Limpus

LEATHERBACK TURTLE

Dermochelys coriacea (Vandelli, 1761)

Photo: T Montford-Biosphoto-AUSCAPE

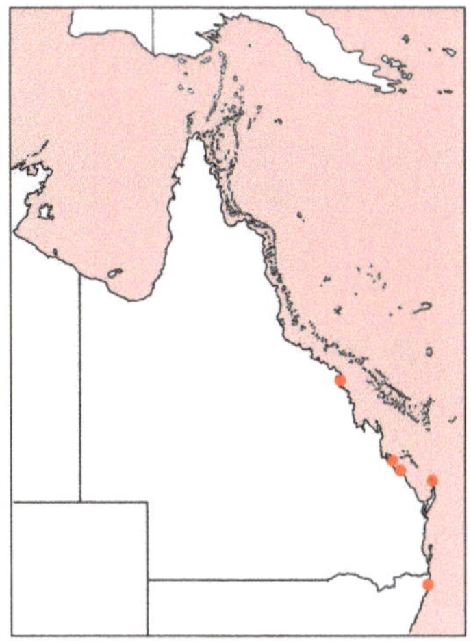

Family: Dermochelyidae

Other common name: Luth

Conservation status: IUCN: Critically Endangered; Aust.: Endangered; Qld: Vulnerable; NSW: Vulnerable; Tas.: Vulnerable; NT: Vulnerable; Vic.: Critically Endangered

Description: The Leatherback Turtle is the largest marine turtle. Adults are black with pale spots and have a very pointed posterior carapace. The beak has a distinct notch on each side. Pink blotches highlight the head. The carapace has five distinct longitudinal ridges, not including the carapace margin, and lacks the large keratinised scutes found on the carapace of other marine turtles. The flippers do not have claws. Breeding males and females are similar in appearance, colour and carapace size (mean curved carapace length = 162 cm), but males have a much longer tail. Leatherback Turtle eggs are billiard-ball sized (average egg diameter = 5.33 cm); the clutches contain about 85 eggs and about 45 yolkless eggs. Hatchlings are black (dark grey when dry) and have five distinct ridges on the carapace, not including the carapace margins. Each ridge on the carapace is highlighted by a row of white scutes. The carapace of the hatchling is covered with small scutes that are lost in adults. The mean straight carapace length of hatchling Leatherback Turtles is about 5.8 cm and they weigh 46 g.

Distribution: The Leatherback Turtle has a circumglobal distribution. It forages over the continental shelf around Australia, including the Great Barrier Reef and the Gulf of Carpentaria in Qld. Leatherback Turtles nest in low density (0–3 females per year) at two sites in southern Qld: Wreck Rock Beaches and Rules Beach. A total of 28 nesting attempts occurred between northern Hervey Bay (Bundaberg) and Roundhill Head from 1968 to the 1980s.

Habitat: The species has been reported from coastal waters in central eastern Australia and in ~3 m deep intertidal waters of the south-eastern Gulf of Carpentaria. Unfortunately, there have been no quantified surveys of habitat use in Qld coastal waters. Because Leatherback hatching and emergence success are low at Qld beaches, they may not be prime nesting habitat. Nothing is known about internesting habitat used by Leatherback Turtles in Australia.

Ecology: Leatherback Turtles are migratory (i.e. their resident foraging areas are removed from nesting sites). Because of the paucity of recapture data, it is presumed that the majority of Leatherback Turtles encountered in Qld waters are part of the larger nesting populations in Indonesia, Papua New Guinea and the Solomon Islands. Mating has not been recorded for this species in Australian waters. During a breeding season, a female Leatherback Turtle will produce up to four clutches of eggs at less than 2-weekly intervals (mean re-nesting interval = 9.17 days). Leatherback Turtle eggs require temperatures in the range of 23–33°C throughout incubation. The temperature during the middle third of incubation determines the sex of the hatchling (males at cooler and females at warmer temperatures). Hatchlings emerge from nests after about 8–9 weeks of incubation. After emerging onto the beach surface, hatchlings disperse to an oceanic life for an undetermined period. The Leatherback Turtle is carnivorous and feeds throughout the water column. In Australian waters its diet is not well documented, but includes colonial tunicates, jellyfish and other soft-bodied invertebrates. Although little is known about Leatherback populations in Qld and Australia, the population seems to be declining based on the lack of nesting at traditional sites. The low density nesting begins in mid December, reaches a peak in January, and ends in about mid February.

Reasons for decline and current threats: Unquantified mortality in neighbouring countries from longline and gill net fisheries has reduced the regional population. Commercial harvest of this species, or its eggs, is not permitted under any state or federal legislation in Australia. In Qld, infrequent mortality occurs from anthropogenic causes. There are no records of disease-related mortality of Leatherback Turtles in Australia.

CURRENT RECOVERY EFFORTS

Recovery plan: Environment Australia (2003) 'Recovery plan for marine turtles in Australia'. Environment Australia, Canberra.

Conservation actions:

- Order in Council under the *Queensland Fisheries Act* (18 July 1968), declared a year-round closed season for the harvest of all species of marine turtles and their eggs in Qld.
- The introduction of turtle exclusion devices (TEDs) in the Northern Prawn Fishery and the Qld East Coast Trawl Fishery minimises captures in trawl fisheries off Qld.

FUTURE RECOVERY EFFORTS

What you can do:

- Report sick, injured, dead and/or tagged sea turtles to DERM (1300 130 372).
- Participate in local planning groups that influence decisions made about coastal development, especially the lighted areas behind the nesting area, to contribute to the conservation management of marine turtles.
- Volunteer with DERM.

Conservation needs:

- Identify the threatening processes.
- Coordinate abatement efforts across the entire range of the species.

Research directions:

Over the next few years a better understanding of the population dynamics should emerge from a combination of genetic research and population modelling.

- Survey the population and their habitats.
- Monitor the populations.
- Study how the various populations (foraging area and nesting area) are interrelated to support the development of more specific modelling.
- Assess the distribution of suitable habitat and specific sites used by turtles.
- Identify and rectify the factors affecting the survival of eggs and hatchlings (especially those associated with people) and degradation of coastal habitat (due to poorly or unplanned coastal development and tourism, and through lack of coordination and collaboration between sectors).

Sources and further reading

James R (compiler) (1994) *Proceedings of the Australian Marine Turtle Conservation Workshop.* Queensland Department of Environment and Heritage and Australian Nature Conservation Agency.

Limpus CJ (2009) *A Biological Review of Australian Marine Turtles. 6. Leatherback Turtle,* Dermochelys coriacea *(Vandelli).* Department of Environment and Resource Management, Brisbane.

Limpus CJ and Miller JD (1993) Family Cheloniidae. *Fauna of Australia. Volume 2A. Amphibia & Reptilia.* (Eds CJ Glasby, GJB Ross and PL Beesley) pp. 133–138. Australian Government Publishing Service, Canberra.

Authors: Jeff Miller and Colin Limpus

HAWKSBILL TURTLE

Eretmochelys imbricata (Linnaeus, 1758)

Photo: A Coward

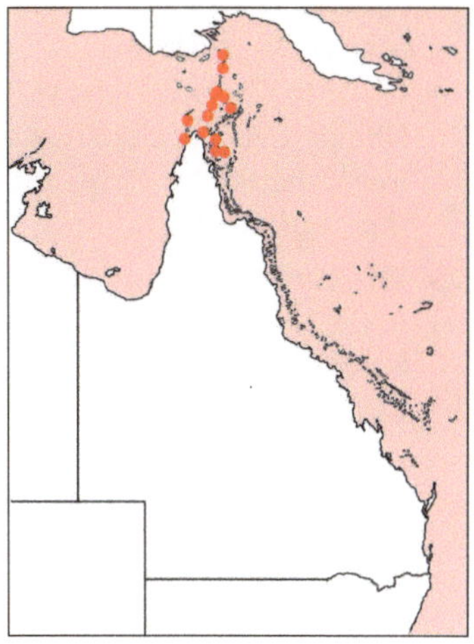

Family: Cheloniidae

Conservation status: IUCN: Critically Endangered; Aust.: Vulnerable; Qld: Vulnerable; WA: Vulnerable; NT: Data Deficient

Description: The carapace is high domed and covered with imbricate (overlapping at the posterior edge) scutes. The species has four pairs of costal scutes, two pairs of prefrontal scutes and no preoccular scutes. The carapace is brown and variegated with dark brown to black; the plastron is cream to yellowish. Breeding females are similar in curved carapace length (about 80 cm) and colour to the males and weigh about 50 kg. Hawksbill Turtle clutches contain about 120 white and spherical eggs (average diameter 3.5 cm, mass 26 g). The carapace of hatchlings is dark brown, occasionally with darker spotting; the ventral surface is light brown. The straight carapace length of hatchling Hawksbill Turtles is about 4 cm and they weigh about 14 g.

Distribution: Hawksbill Turtles have a circumglobal, tropical distribution that includes the islands and reefs of northern Australia and the Great Barrier Reef (GBR). Hawksbill Turtles migrate between foraging and nesting areas located in the Torres Strait–northern GBR, as well as in neighbouring countries. In Qld, most nesting areas of the northern GBR are within national parks. However, almost all nesting areas in Torres Strait and along western Cape York Peninsula are on Aboriginal or Torres Strait Islander land and as a result, are outside of protected habitat.

Habitat: Foraging Hawksbill Turtles live in tidal and sub-tidal coral and rocky reef habitats in warm temperate areas. At least 72 Hawksbill Turtle nesting locations are known from Qld, including the islands of the GBR and Torres Strait, and along the mainland coast of Western Cape York Peninsula north of Cotterell River. Only three islands in Torres Strait host more than >500 nesting females per year; the majority of the remainder support fewer than 100 nesting females per year.

Ecology: Hawksbill Turtles are omnivorous; although the diet has not been studied in Qld, they feed primarily on sponges, but also consume algae and seagrass. The majority of Hawksbill Turtles migrate less than 1000 km from their foraging area to their nesting site. Hawksbill Turtles nesting at the one location (rookery) have migrated from numerous foraging areas. Within a single breeding season, a female will spend the interval between laying successive clutches in an internesting area located within a few tens of kilometres of the nesting beach. In northern Qld, breeding occurs year round, with a peak in January and February. However, individual females have an interval of 2 or more years between breeding seasons; females produce multiple clutches (mean = 2.4) at about 2-week intervals. The sex of the hatchling turtles is determined by incubation temperature (males at cooler and females at warmer temperatures). There is no parental care of the eggs or hatchlings. Hatchlings emerge after 55–60 days of incubation.

Reasons for decline and current threats: The main reasons for decline and the current threats include fisheries bycatch, unquantified harvest by Indigenous Australians and by people in neighbouring countries, predation of eggs by native and introduced animals, coastal development, deteriorating water quality, marine debris and loss of habitat.

If the current rate of decline (as measured at the Milman Island index beach) continues, the Torres Strait–northern GBR Hawksbill Turtle stock may decline by >80% in less than one generation. In the Gulf of Carpentaria, ghost nets entangle an unquantified number (possibly many hundreds) of turtles annually. An unquantified proportion of the turtles killed in these nets are immature Hawksbill Turtles. Other accidental mortality in Qld occurs at low levels. Diseases in wild Hawksbill Turtles in Australia are poorly studied. Although many threatening processes have been tackled through management, the populations are still at risk. The long period to maturity and the long intervals between breeding seasons put Hawksbill Turtle populations at risk from even modest increases in mortality, particularly if the impact occurs over an extended period of time.

CURRENT RECOVERY EFFORTS

Recovery plan: Environment Australia (2003) 'Recovery plan for marine turtles in Australia'. Environment Australia, Canberra.

Conservation actions:

- Order in Council under the *Queensland Fisheries Act* (18 July 1968), declared a year-round closed season for the harvest of all species of marine turtles and their eggs in Qld.
- The introduction of turtle exclusion devices (TEDs) in the Northern Prawn Fishery and the Qld East Coast Trawl Fishery minimises captures in trawl fisheries off Qld.

FUTURE RECOVERY EFFORTS

What you can do:

- Report sick, injured, dead and/or tagged sea turtles to DERM (1300 130 372).
- Participate in local planning groups that influence decisions made about coastal development, especially the lighted areas behind the nesting area, to contribute to the conservation management of marine turtles.
- Get involved with the Northern Gulf Resource Management Group and Carpentaria Ghost Nets Programme projects.
- Join/donate/volunteer: the Sea Turtle Foundation, Australian Marine Conservation Society, Bribie Island Environmental Protection Association, Cape York Sustainable Futures, Cape York Turtle Rescue, DERM.

Conservation needs:

- Threat abatement, scientific research, population and habitat monitoring.
- Identify the threatening processes.
- Coordinate the abatement efforts across the entire range of the species, not just in Qld.

Research directions:

- Study the condition of turtle habitats, both in the general distribution of the habitat and the specific habitat used by the turtles.
- Research the factors affecting the survival of eggs and hatchlings (especially factors associated with people and feral animals) and degradation of the coastal habitat.
- Conduct studies on how the various populations (foraging area and nesting area) are interrelated via genetics which will support the development of population models and more conservation action.

Sources and further reading

James R (compiler) (1994) *Proceedings of the Australian Marine Turtle Conservation Workshop.* Queensland Department of Environment and Heritage and Australian Nature Conservation Agency.

Limpus CJ (2009) *A Biological Review of Australian Marine Turtles: 3. Hawksbill Turtle* Eretmochelys imbricata *(Linnaeus).* Department of Environment and Resource Management, Brisbane.

Witzell WN (1983) Synopsis of biological data on the Hawksbill Turtle, *Eretmochelys imbricata* Linnaeus, 1766. *FAO Fisheries Synopsis* No. 137. Food and Agriculture Organization of the United Nations, Rome.

Authors: Jeff Miller and Colin Limpus

OLIVE RIDLEY TURTLE

Lepidochelys olivacea (Eschscholtz, 1829)

Photo: Frans Lanting-AUSCAPE

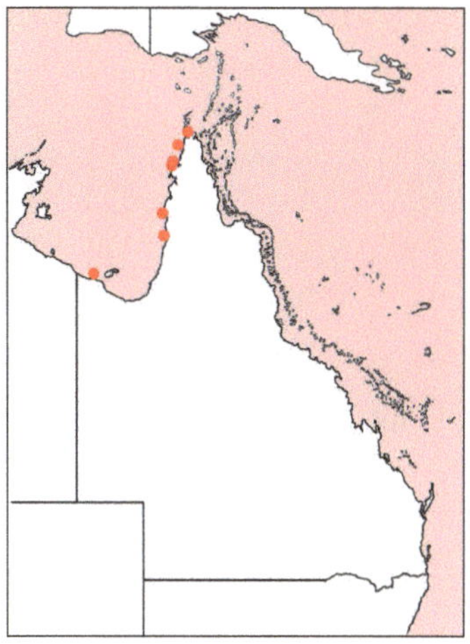

Family: Cheloniidae

Other common name: Pacific Ridley Turtle

Conservation status: IUCN: Vulnerable; Aust.: Endangered; Qld: Endangered; NT: Data Deficient; WA: Endangered

Description: The carapace of the Olive Ridley Turtle is roughly heart shaped and covered by large keratinised scales. The presence of more than five costal scutes and inframarginal pores is diagnostic for the species. The carapace colour of adults and large immature turtles is uniform grey to olive-grey; the colour of the plastron is whitish. The curved carapace length of nesting females Olive Ridley Turtles is approximately 70 cm and they weigh about 41 kg. Nesting females move with a narrow (>0.7 m), alternating gait. Eggs are small (average egg diameter = 3.7 cm; weight 30–32 g). The carapace of hatchling Olive Ridley Turtles is black-brown with slight ridges on the costal and vertebral scutes; the ventral surface is light brown; they have an average straight carapace length of 4.12 cm.

Distribution: The Olive Ridley Turtle is widespread and regularly encountered over soft-bottom habitats inside the Great Barrier Reef from south Qld northward through Torres Strait, Gulf of Papua, Gulf of Carpentaria, Arafura Sea to at least offshore of Exmouth Gulf in WA.

Habitat: Much of the foraging habitat used by the Olive Ridley Turtle along the eastern Australia coast is in the Great Barrier Reef World Heritage Area and the associated Great Barrier Reef Marine Park and adjacent Qld State Marine Parks. The shallow waters of the Gulf of Carpentaria are also foraging habitat. Low-density nesting occurs between Weipa and Bamaga on the north-western coast of Cape York Peninsula. Although females preparing to breed have been captured in foraging areas within the Great Barrier Reef near Townsville, there is no record of Olive Ridley Turtles nesting along the east coast of Qld.

Ecology: There is a paucity of data concerning all aspects of the life history the Olive Ridley Turtle in Australian waters because no species-specific studies have been conducted. Adult and large immature Olive Ridley Turtles are carnivorous and feed on gastropod molluscs and small crabs. The interval between reproductive seasons for female Olive Ridley Turtles is 1–3 years (no Australian data); males probably breed more frequently than females, possibly annually. Olive Ridley Turtles may migrate long distances from foraging areas to their rookeries. It is presumed that the Olive Ridley Turtles will show fidelity to traditional nesting beaches. Nesting occurs year-round in northern Qld, with a peak during the early part of the dry season (April to June). Within a nesting season, the Olive Ridley Turtle lays one or two clutches at about 14-day intervals (but may produce more). Each clutch contains about 100–107 eggs; the clutches rarely contain numerous yolkless eggs. The nest depth to the top of the eggs is about 30 cm; the bottom of the nest chamber is about 45– 50 cm deep. The sex of Olive Ridley Turtle hatchlings, as with other species of marine turtle, is a function of the nest temperature during the middle third of incubation (males result from cooler and females from warmer temperatures). Hatchlings emerge from nests after slightly less than 2 months of incubation.

Reasons for decline and current threats: Based on available data, the magnitude of cumulative mortality from the wide array of anthropogenic sources affecting Olive Ridley Turtles within Australia cannot be assessed. Known sources of mortality include entanglement in drift nets and gill net fisheries. In the Gulf of Carpentaria, the number of adult and large immature Olive Ridley Turtles killed by the ghost net fishery is unquantified, but may exceed hundreds of turtles annually. The nests are subject to unknown levels predation by pigs, dingoes and goannas along the northern Qld coast.

Coastal Indigenous people collect Olive Ridley Turtle eggs along the beaches in western Cape York Peninsula, Qld, but the egg and/or hatchling and adult turtle harvest for human consumption has not been measured.

CURRENT RECOVERY EFFORTS

Recovery plan: Environment Australia (2003) 'Recovery plan for marine turtles in Australia'. Environment Australia, Canberra.

Conservation actions:

- Order in Council under the *Queensland Fisheries Act* (18 July 1968), declared a year-round closed season for the harvest of all species of marine turtles and their eggs in Qld.

- The introduction of turtle exclusion devices (TEDs) in the Northern Prawn Fishery and the Qld East Coast Trawl Fishery minimises captures in trawl fisheries off Qld.

FUTURE RECOVERY EFFORTS

What you can do:

- Report sick, injured, dead and/or tagged sea turtles to DERM (1300 130 372).
- Participate in local planning groups that influence decisions made about coastal development, especially the lighted areas behind the nesting area, to contribute to the conservation management of marine turtles.
- Get involved with the Northern Gulf Resource Management Group and Carpentaria Ghost Nets Programme projects.
- Join/donate/volunteer: the Sea Turtle Foundation, Australian Marine Conservation Society, Bribie Island Environmental Protection Association, Cape York Sustainable Futures, Cape York Turtle Rescue, DERM.

Conservation needs:

- Abatement of threats.
- Coordinate the abatement efforts across the entire range of the species, not just in Qld.

Research directions:

- Monitor the populations.
- Survey nesting and foraging populations and their habitat.
- Conduct species-specific studies.
- Identify the threatening processes.

Sources and further reading

James R (compiler) (1994) *Proceedings of the Australian Marine Turtle Conservation Workshop.* Queensland Department of Environment and Heritage and Australian Nature Conservation Agency.

Limpus CJ (2009) *A Biological Review of Australian Marine Turtles. 4. Olive Ridley Turtle* Lepidochelys olivacea *(Eschscholtz).* Department of Environment and Resource Management, Brisbane.

Authors: Jeff Miller and Colin Limpus

FLATBACK TURTLE

Natator depressus (Garman, 1880)

Photo: John Parmenter

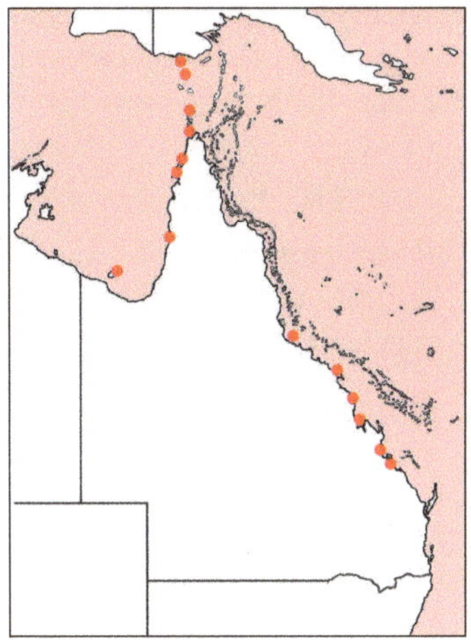

Family: Cheloniidae

Conservation status: IUCN: Data Deficient; Aust.: Vulnerable; Qld: Vulnerable; WA: Vulnerable

Description: True to its name, the Flatback Turtle has a flat low-domed oval-shaped olive-grey carapace with turned up outer edges. The carapacial scutes do not overlap and are smooth and waxy, a result of poor keratinisation. The plastron is white. Carapace length of adult females is 93 cm and they weigh 86 kg. Olive-green hatchlings are the largest of all marine turtles with a carapace length of approximately 6 cm and weighing about 45 g. The hatchling's ventral surface is bright white and the iris of the eye is a brilliant blue.

Distribution: The Flatback Turtle is found only in the northern tropical waters of the Australian continental shelf. They feed in the Gulf of Papua New Guinea, Indonesian waters and within the Great Barrier Reef (GBR). The only reports of nesting sites are in Australian waters, where four management units have been identified; two of these are in Qld. Major rookeries are located mostly on inshore continental islands. Current population estimates are scarce. The distribution of the Flatback Turtle population is atypical, probably because it is missing an oceanic pelagic hatchling phase in its life cycle.

Habitat: This turtle stays away from reefs and intertidal seagrass beds, due to the lack of keratinisation of its skin and scutes. The post-hatchling lives inside the GBR lagoon and, as it reaches adulthood, it moves among various habitats eventually ending up in sub-tidal soft bottomed habitats located closer to shore, possibly within the same general geographical distribution where it matured. It is known to nest on several offshore islands of the GBR, and Gulf of Carpentaria; secondary nesting beaches occur along the mainland coast of Queensland.

Ecology: Large Flatback Turtles are carnivorous, and feed on soft-bodied invertebrates such as soft corals, sea pens, holothurians and jellyfish. Post-hatchlings feed on plankton. The eastern Australian Flatback Turtles nest from early October to early January with the Gulf of Carpentaria turtles nesting year round, peaking during the winter months. Unlike other sea turtles that will migrate thousands of miles to their nesting sites, the female Flatback Turtle will travel only up to 1300 km from her foraging grounds to reach her same nesting site – a single beach, or a few adjacent beaches – every 1 to 5 years. The hatching success is a high 70–89%. The survival of hatchlings crossing the beach once they leave the nest has been measured as >96% at Mon Repos, 97% at Crab Island in December and 62% in April–May. The Flatback Turtle takes decades to reach maturity. Information regarding the various stages of any marine turtle's life cycle, other than that obtain from beached nesting females, is difficult to come by due to the logistics involved. Because Flatback Turtles stay away from clear shallow waters, it has not been possible to conduct any feeding ground studies.

Reasons for decline and current threats: The main reasons for the decline and current threats include fishery bycatch, predation of eggs by introduced species, unknown levels of harvest by Indigenous Australians, coastal development, a decline in water quality, marine debris and loss of habitat.

Indigenous Australians harvest Flatback Turtles and their eggs for communal, non-commercial purposes under the provisions of the *Native Title Act*. This activity may have increased over the years, due to growing population numbers and mechanised transport. Trawling capture rates were reduced more than 95% by the compulsory use of turtle exclusion devices (TEDs) in the Torres Strait, Northern Prawn Fishery and the East Coast Trawl Fishery. However, inshore Gulf of Carpentaria and southern West Papua gill net fisheries and Gulf ghost nets possibly kill many hundreds of adult and large immature Flatback Turtles annually.

Native animals, including goannas, prey heavily on eggs and hatchlings on the mainland but not as much on the major island rookeries. Introduced animals such as pigs, foxes and feral dogs destroy nests, with rates reaching a high of 90% on mainland Cape York Peninsula and 95% at Curtis Island and Wreck Rock Beach. Mainland coastal development including development of port infrastructure and shipping may deplete Flatback Turtle populations. These activities intensify with increasing regional human population. Development near beaches may alter light horizons causing the females to choose less favourable nesting sites and increasing hatchling mortality brought on by faulty dispersal – they get tangled up in vegetation, become road kill, suffer heat exhaustion and are more likely to be preyed on by birds and fish.

CURRENT RECOVERY EFFORTS

Recovery plan: Environment Australia (2003) 'Recovery plan for marine turtles in Australia'. Environment Australia, Canberra.

Conservation actions:

- Order in Council under the *Queensland Fisheries Act* (18 July 1968), declared a year-round closed season for the harvest of all species of marine turtles and their eggs in Qld.
- The majority of the foraging distribution of the eastern Qld population is within the GBR World Heritage Area and GBR Marine Park and approximately 70% of the nesting of this population distribution is contained within national parks and conservation parks. The Gulf of Carpentaria population is not as protected.
- The introduction of turtle exclusion devices (TEDs) in the Northern Prawn Fishery and the Qld East Coast Trawl Fishery minimises captures in trawl fisheries off Qld.

FUTURE RECOVERY EFFORTS

What you can do:

- Report sick, injured, dead and/or tagged sea turtles to DERM (1300 130 372).
- Participate in local planning groups that influence decisions made about coastal development, especially the lighted areas behind the nesting areas, to contribute to the conservation management of marine turtles.
- Get involved with the Northern Gulf Resource Management Group and Carpentaria Ghost Nets Programme projects.
- Join/donate/volunteer: the Sea Turtle Foundation, Australian Marine Conservation Society, Bribie Island Environmental Protection Association, Cape York Sustainable Futures, Cape York Turtle Rescue, DERM.

Conservation needs:

- Identify the threatening processes.
- Coordinate the abatement efforts across the entire range of the species, not just in Qld.

Research directions:

- Monitor populations.
- Study how the various populations (foraging area and nesting area) are interrelated to support the development of more specific modelling.
- Study the condition of turtle habitats, both in the general distribution of the habitat and the specific habitat used by the turtles.
- Identify and rectify the factors affecting the survival of egg clutches and hatchlings loss and degradation of coastal habitat due to poorly planned or unplanned coastal development and tourism, and through lack of coordination and collaboration between sectors.
- Survey the nesting and foraging populations and their habitat.

Sources and further reading

Hamann M, Limpus CJ and Read MA (2007) Chapter 15. Vulnerability of marine reptiles in the Great Barrier Reef to climate change. In *Climate Change and the Great Barrier Reef: A Vulnerability Assessment.* (Eds JE Johnson and PA Marshall) pp. 465–496. Great Barrier Reef Marine Park Authority, Townsville and Australian Greenhouse Office, Canberra.

Limpus CJ (2007) *A Biological Review of Australian Marine Turtle Species. 5. Flatback Turtle,* Natator depressus *(Garman).* Queensland Environmental Protection Agency, Brisbane.

Theissinger K, FitzSimmons NN, Limpus CJ, Parmenter CJ and Phillott AD (2008) Mating system, multiple paternity and effective population size of the endemic flatback turtle (*Natator depressus*) in Australia. *Conservation Genetics* **10**, 329–346.

Walker TA and Parmenter CJ (1990) Absence of a pelagic phase in the life cycle of the flatback turtle, *Natator depressa* (Garman). *Journal of Biogeography* **17**, 275–278.

Author: Lee K Curtis

Freshwater Turtles (Chelidae)

GULF SNAPPING TURTLE

Elseya lavarackorum White & Archer, 1994

Photo: Arthur White

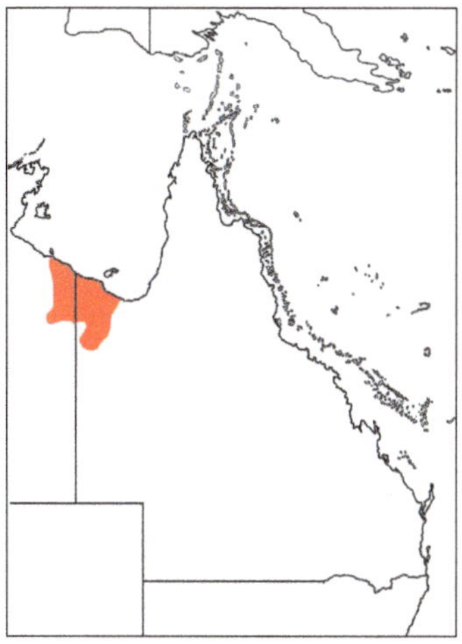

Family: Chelidae

Conservation status: Aust.: Endangered; Qld: Vulnerable

Description: The Gulf Snapping Turtle is a large, short-necked freshwater turtle with a carapace length of 39 cm (recent fossil material has carapace lengths up to 46 cm). The carapace is broadly oval in shape with minor flaring of the marginal scutes. It is very dark in older animals and is often covered with clinging algae. The head is also dark, but blanching of the head occurs in older animals. There are two prominent chin barbels. The Gulf Snapping Turtle has the unusual distinction of having been described from recent fossil material before living animals were discovered. Juvenile turtles have rounded carapaces, with saw-toothing of the posterior peripheral bones. The carapace has a central crest that becomes less pronounced with age, eventually disappearing in mature animals. The carapace and head is brown but becomes increasingly darker with age. There are no chin or head markings.

Distribution: The distribution is inadequately known. It is currently only known from rivers and creeks draining into the Gulf of Carpentaria, including the upper reaches of the Gregory River, Lawn Hill Creek, and the O'Shanassy, Calvert and Nicholson Rivers

Habitat: It is found in slow flowing freshwater rivers and streams of the Gulf of Carpentaria, often lined with mature fig trees (such as Cluster Fig *Ficus glomerata* or Sandpaper Fig *F. coronata*), and may occur in alluvial as well as in rocky areas.

Ecology: The ecology is poorly known. Gulf Snapping Turtles eat a range of plant food including leaves, flowers and fruits, as well as pandanus root, but they appear to be mainly frugivorous; they have been observed feeding on figs floating in the water and jumping up at figs on low, overhanging branches near the water's edge. They have also been observed feeding on the excreta of flying-foxes camped in trees overhanging a stream. There appears to be a marked difference in the site distribution of age classes, with older, more mature animals often concentrating around cascades, tufa dams and other river obstructions, whereas smaller and younger animals are more frequently found in the open rivers away from these obstructions. Mature turtles have been observed repelling smaller turtles from sites near tufa dams at Lawn Hill Creek. Egg-laying sites have not been found to date, but juvenile turtles have been located in smaller watercourses off a main river or stream, suggesting that egg-laying may occur on sandy banks associated with these smaller tributaries. In 2010, gravid females were captured in Boodjamulla NP. Recently hatched turtles were found in April, May and early June of 2010. Nothing is known about longevity, predation or seasonal movements.

Reasons for decline and current threats: The largest observed mortality of Gulf Snapping Turtles has been through food harvesting by the local Indigenous community. Turtles are either caught on hand lines or using nets. In large communal gatherings of Indigenous people, large numbers of turtles (of various species) are collected and eaten; small turtles are usually returned to the water.

Pigs are common along the rivers where the Gulf Snapping Turtle occurs, so it is likely that pig predation of turtle nests would also account for the loss of many animals. In some areas, cattle are also common around the riparian areas and so trampling of nests, as well as the general degradation of riparian habitat with subsequent negative impacts on food availability, is also likely.

Mortality associated with drowning in traps set by recreational fishers to catch crustaceans is also a possible threat.

CURRENT RECOVERY EFFORTS

Recovery plan: None

Other plans: Department of the Environment and Heritage (2005) 'Predation, habitat degradation, competition and disease transmission by feral pigs'. Department of the Environment and Heritage, Canberra.

FUTURE RECOVERY EFFORTS

What you can do:

- Join/donate to a local Landcare group and/or The Australian Freshwater Turtle Conservation and Research Association.
- Report observations of turtles to DERM.
- Identify and manage threats such as feral animal impacts on nesting sites.

Conservation needs:

- Prepare a joint Qld/NT recovery plan.
- State agencies need to liaise with local Indigenous communities with respect to harvesting of these turtles.

Research directions:

- Obtain ecological and biological information about the Gulf Snapping Turtle.
- Conduct surveys to better determine the distribution and abundance of the species, especially in the downstream sections of rivers leading into the Gulf of Carpentaria.
- Design and implement a monitoring program for the species, especially in areas where Indigenous harvesting regularly occurs.
- Identify the key threats to the species.
- Initiate a captive breeding program.
- Survey the wild populations: the only population that has been surveyed was in the Gregory River at Riversleigh in 1998.

Sources and further reading

Georges A and Adams M (1996) Electrophoretic delineation of species boundaries within the short-necked freshwater turtles of Australia (Testudines: Chelidae). *Zoological Journal of the Linnean Society* **118**, 241–260.

Georges A and Thompson S (2010) Diversity of Australasian freshwater turtles, with an annotated synonymy and key to species. *Zootaxa* **2496**, 1–37.

Thomson S, White AW and Georges A (1997) Re-evaluation of *Emydura lavarackorum*: identification of a living fossil. *Memoirs of the Queensland Museum* **42**, 327–336.

White AW (1999) Initial observations and survey results of Freshwater turtles in the Gregory River and Lawn Hill Creek, northwestern Queensland. *Herpetofauna* **29**, 37–48.

White AW and Archer M (1994) *Emydura lavarckorum*, a new Pleistocene turtle (Pleurodira: Chelidae) from fluviatile deposits at Riversleigh, northwestern Queensland. *Records of the South Australian Museum* **27,** 159–167.

Author: Arthur White

MARY RIVER TURTLE

Elusor macrurus Cann & Legler, 1994

Photo: Michael Cermak

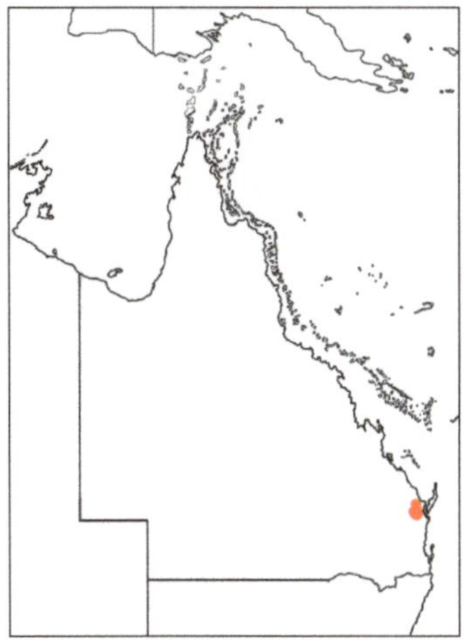

Family: Chelidae

Conservation status: IUCN: Endangered; Aust.: Endangered; Qld: Endangered

Description: This turtle is characterised by a reversed sexual dimorphism, with the adult male weighing 5500 g, with a carapace measuring 420 mm, being significantly larger than the female that weighs 3500 g and measures 340 mm. This turtle possesses a distinctive massive tail that is more than half the length of the carapace in adult males. In all sexes and ages, the tail has a large precloacal portion and is laterally compressed. The head is relatively small for its carapace size, displaying a prominent headshield that becomes more distinct from the surrounding skin colouring with age.

Distribution: The Mary River Turtle is endemic to a single river system: the Mary River of south-east Qld. It was originally seen in pet shops throughout eastern Australia during the 1960s and only discovered to occur in the Mary River in 1990. It occurs primarily in the mainstream of the Mary River and major tributaries, including Tinnana Creek, Yabba Creek and Obi Obi Creek.

Habitat: It is restricted to the permanent flowing stream and large pool habitats of the Mary River catchment. Adults are usually found in a diversity of instream habitats including under overhanging banks and vegetation, macrophyte beds and around submerged logs and rock crevices. They can be seen basking on logs and rocks. Hatchlings and immature turtles are thought to occur in rocky areas with sand or gravel on the river bed, in a variety of water depths.

Ecology: The Mary River Turtle relies on bimodal respiration via cloacal ventilation and can extract up to 50% of its oxygen requirements from the water, which it obtains through well developed gill systems lining their cloacal bursae. It is slow growing, with delayed maturity taking more than 15 years before either sex commences breeding. The turtle aggregates for nesting on traditional nesting sand banks that are revisited across decades showing strong nesting site fidelity. Successive studies have recorded nesting occurring from late October to December each year, with the average clutch size being 14.7 eggs. Adult turtles are mostly herbivorous and feed on a range of macrophytes (filamentous algae, *Vallisneria*, *Myriophyllum*) as well as the fallen fruits of riparian trees including figs, lillypillies and black beans. They also eat aquatic insects and occasionally molluscs. There has been an estimated reduction in the population of breeding female Mary River Turtles of around 95% between 1970 and 2000. This equates to a reduction from many hundreds of females nesting near Tiaro in the 1960s and 1970s to only tens of individuals nesting on the same sand banks in 1998 and 1999. The population size is unknown.

Reasons for decline and current threats: This species was subject to intense egg harvest to procure turtle hatchlings for the pet trade in the 1960s through to 1974. This loss of significant numbers of egg production, and subsequent lack of recruitment of hatchlings back into the Mary River, is reflected in the population decline resulting in mainly ageing animals. Monitored nesting success rates during 1998–2002 identified an almost 100% loss of nests at some monitored sites on sand banks of the Mary River from feral animals (dogs and foxes), native predators (goannas *Varanus* sp. and water rats *Hydromis chrysogaster*) and illegal collection of eggs. In the absence of nest protection actions, the continuing lack of recruitment through egg predation is now the critical threat to the survival of this species.

Cloacal ventilating turtle species, such as the Mary River Turtle, are among those most susceptible to the impacts of habitat modification from climate change, and from human impacts within the river (e.g. dams) and adjoining catchment land uses. On a broader landscape scale, the Mary River catchment has only 10% remnant vegetation and has increasing land use intensity, which negatively affect the riparian and instream conditions.

CURRENT RECOVERY EFFORTS

Recovery plan: None

Conservation actions:

- Protective measures taken resulted in approximately 75% hatchling success from sand banks into the river. These measures included caging of natural nests to protect them from predators and monitoring from illegal harvest.

- Since 2001, the Tiaro and District Landcare Group, in partnership with the Queensland DERM, has been protecting nests in order to increase the survival of the Mary River Turtle hatchlings. Other groups that have been involved in recovery efforts through a number of government grants include the Burnett-Mary Regional Body and Greening Australia.

FUTURE RECOVERY EFFORTS

What you can do:

- Join/donate to a local Landcare group and/or The Australian Freshwater Turtle Conservation and Research Association.
- Landholders are encouraged to restore and protect riparian vegetation, including reducing the impacts of weeds and cattle access to stream sections.

Conservation needs:

- Develop a whole-of-catchment management plan.
- Public awareness and education of the Mary River Turtle is essential to secure its long-term conservation.
- Carry out sensitive land management of sandbanks along riparian areas.

Research directions:

- Obtain baseline data on biology and ecological requirements (including immature turtles' habitat requirements).
- Survey the upper reaches to determine distributional limits.
- Investigate the optimal nesting bank conditions and characteristics.
- Achieve a better understanding of how to restore nesting banks after flood events.
- Achieve a better understanding of the aquatic respiration capabilities.
- Investigate the occurrence and persistence of populations in some existing instream impoundments.

Sources and further reading

Cann J and Legler JM (1994) The Mary River tortoise: a new genus and species of short-necked Chelid from Queensland, Australia (Testudines: Pleurodira). *Chelonian Conservation and Biology* **1**, 81–96.

Flakus S (2002) The ecology of the Mary River turtle, *Elusor macrurus*. MSc thesis. Department of Zoology and Entomology, The University of Queensland, Brisbane.

Limpus CJ (2008) *Freshwater Turtles in the Mary River, Queensland*. 2nd edn. Queensland Environmental Protection Agency, Brisbane.

Mathie NJ (2007) New dams may flush bottom-breathers out. *Australasian Science 2007* (June), 36–38.

van Kampen T, Emerick S and Parkes D (2006) Increasing the survivorship of the Mary River turtle: Tiaro District of Southeast Queensland, October 2002–March 2003. Tiaro and District Landcare Group Greening Australia, Tiaro.

Author: Darren Fielder and Arthur Georges

WESTERN SAWSHELLED TURTLE

Myuchelys bellii (Gray 1844)

Photo: Robert Ashdown

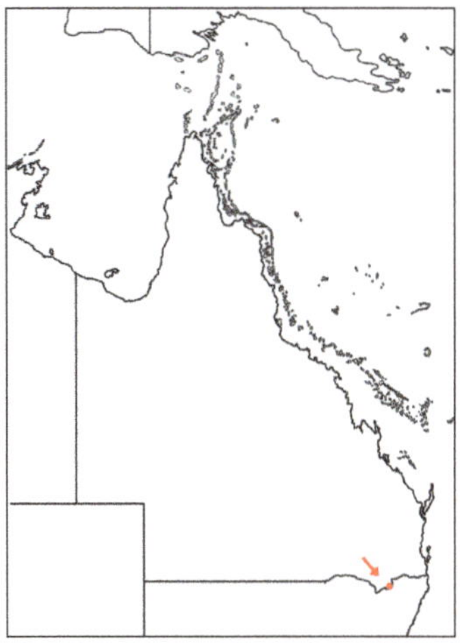

Family: Chelidae

Aboriginal name: The genus name *Myuchelys* is a combination of an Aboriginal word for clear water, Myuna, and the Greek word for tortoises, *chelys*.

Conservation status: IUCN: Endangered; Aust.: Vulnerable; Qld: Least Concern; NSW: Vulnerable

Description: This turtle weighs 1200 g and male carapaces measure up to 220 mm, while females grow up to 300 mm and weigh 3200 g. The Western Sawshelled Turtle has a robust head with a distinctive head shield, extending down the parietal ridge towards the tympanum. There are numerous pronounced pointed tubercles along the dorsal surface of the neck. The iris is a distinctive olive-grey, lacking leading and trailing black spots. Yellow or salmon stripe extends along the side of the head and neck from the symphysis of the jaws to the plastron, decreasing in intensity with age. The cervical scute is present (except as rare variant). The carapace is dark grey or brown, dorso-ventrally flattened, oval, with deep serrations along the posterior margins of the carapace, from which the common name 'sawshelled turtle' is derived. The lobe of the anterior plastron is broad; the plastron narrows through the bridge and tapers posteriorly. The plastron and ventral surfaces of the carapace are predominantly black in adults with areas of light brown and cream rayed colouration on their edges.

Distribution: It is endemic to a single river system, the Darling, which drains the New England Tablelands bioregion of the Murray–Darling Basin. Three disjoint populations inhabit headwater streams of the Darling: the Namoi River and Gwydir River, and Bald Rock Creek of the Border Rivers region. The Qld population from Bald Rock Creek is restricted to less than 10 km of stream habitat and is estimated to have fewer than 400 individuals. Most individuals occur in two separate permanent waterholes (approximately 3 km in combined length) at either end of the stream section; the intermediate reach is less favourable habitat. The Western Sawshelled Turtle is the most restricted freshwater turtle species in Qld. Recent examination of genetic divergence between the Bald Rock Creek populations and those of the adjacent Namoi and Gwydir Rivers found only minor differences.

Habitat: The habitat of Western Sawshelled Turtles is permanent cold flowing upland streams above about 700 m altitude. Granite boulders and bedrock are common features of these streams. The instream habitat is complex, with underwater caverns formed by granite boulders. Large woody debris and overhanging banks are also common features. Coarse granite sand deposits occur in lower velocity areas and are usually covered in fine silt, green algal growth and macrophytes.

Ecology: The species tends to be crepuscular in habit, being most active in the early morning and evenings and can be seen basking on protruding boulders or logs throughout the warmer months. Adults are omnivorous. Nesting of Western Sawshelled Turtles occurs during the months of October and mid-January where a single clutch of between eight and 23 eggs will be laid in loamy soil adjoining the watercourses.

Reasons for decline and current threats: The Qld population has incurred significant habitat modification and loss through construction of a weir in 1987, which completely impounded the downstream permanent natural waterhole. In conjunction with this weir, there was the complete removal of the riparian zone and instream habitats. There are now a limited number of turtles present in this highly modified pool. Recreational fishing is a known cause of adult turtle deaths and there is concern about fox predation of nests and loss of immature turtles from stocked native fish (Murray Cod *Maccullochella peelii peelii* and Golden Perch *Macquaria ambigua*). Approximately 10% of the Namoi and Gwydir turtle populations are suffering serious eye damage and blindness from eye cataracts. The population is also vulnerable to catastrophic events (climate change, severe drought) that could lead to its complete extirpation in Bald Rock Creek through loss of permanent waterholes and instream food resources.

CURRENT RECOVERY EFFORTS

Recovery plan: None

Conservation actions:

- Current research into Western Sawshelled Turtle populations is focused on genetic and taxonomic investigations, breeding biology, ecology, diving ecophysiology and population estimates. The upstream extent of its range is within Girraween NP where active management is being undertaken to protect its habitat including exotic goldfish and fox control.
- DERM has an active role through its freshwater turtle research project, which provides support to population dynamics research at Bald Rock Creek and implements a number of conservation measures to sustain the population in Girraween NP. These measures include monitoring water levels during drought conditions and applying restrictions on recreational use for campers in the national park. The University of New England and the University of Canberra provide funds and laboratory space for ongoing research.

FUTURE RECOVERY EFFORTS

What you can do:

- Join/donate to a local Landcare group and/or The Australian Freshwater Turtle Conservation and Research Association.
- Volunteer with DERM.

Conservation needs:

- Landholders need to adopt targeted grazing management of riparian zones, including installing off-stream watering points and fencing off river banks.
- Restore degraded riparian zones and instream habitats. Protect the remaining native vegetation and waterholes in areas outside protected estate.
- Actively control feral animals (foxes, introduced fish) to ensure recruitment of turtles into adult populations.
- Conduct education and awareness campaigns of the habitat requirements and restricted distribution for landholders and recreational fishers.
- Consider contribution of private industry funds from water managers and agriculture for conservation throughout the catchments.
- Erect interpretive signs at public locations.

Research directions:

- Study the potential effects of inbreeding in the Qld population.
- Research the effects of nest predation and introduced fish on recruitment.
- Study habitat restoration and range extension.
- Assess the potential impacts of climate change.

Sources and further reading

Cann J (1998) *Australian Freshwater Turtles.* Beaumont Publishing Pty Ltd, Singapore.

Fielder DP (2011) Seasonal and diel dive performance and behavioral ecology of the bimodally respiring freshwater turtle *Myuchelys bellii* of eastern Australia. *Journal of Comparative Physiology A*, in press.

Fielder DP, Vernes K, Alacs EA and Georges A (2011) Mitochondrial gene variation among Australian freshwater turtles in the genus *Myuchelys*, with special reference to the endangered *Myuchelys bellii. Endangered Species Research*, in press.

Georges A and Thomson S (2010) Diversity of Australasian freshwater turtles, with an annotated synonymy and key to species. *Zootaxa* **2496**, 1–37.

Thomson S and Georges A (2009) *Myuchelys* gen. nov. – a new genus for *Elseya latisternum* and related forms of Australian freshwater turtle (Testudines: Pleurodira: Chelidae) *Zootaxa* **2053**, 32–42.

Author: Darren Fielder

FITZROY RIVER TURTLE

Rheodytes leukops Legler & Cann, 1980

Photo: Matthew Gordos

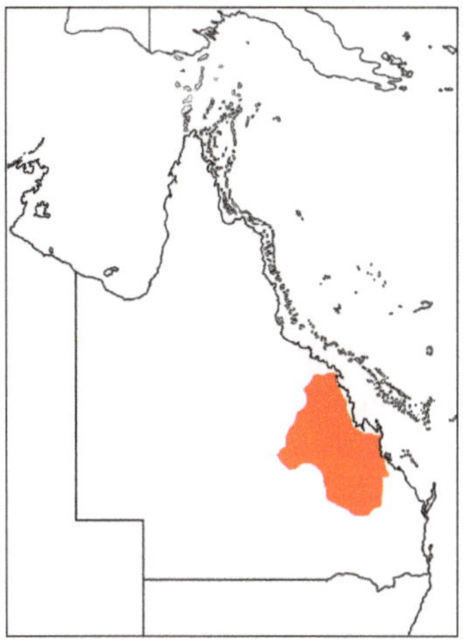

Family: Chelidae

Conservation status: IUCN: Vulnerable; Aust.: Vulnerable; Qld: Vulnerable

Description: The carapace measures up to 260 mm in males and up to 300 mm in females. Males and females weigh up to 2000 and 2500 g, respectively. The carapace is medium to dark brown with a tapered ellipse shape and furrowed surface. The carapace margins are smooth in adults, but serrated in hatchlings. A cervical scute is generally present. The carapace interlateral seams contact the posterior parts of the sixth and eighth marginal scutes. The plastron is smooth, and cream to pale yellow in colour. The carapace cross section is relatively low. The skin is olive-grey, with the upper eyelids, ear and throat of older males becoming pinkish-orange. Adults display pronounced conical tubercles on the dorsal surface of the neck; these are compressed in juveniles. The head is a similar width to the neck. A single pair of pale barbells occurs under the chin. The iris is a conspicuous white in adults, but silvery in hatchlings. The appendages are webbed, with all sizes displaying a gaping cloaca.

Distribution: Distribution is limited to the Fitzroy River catchment in central Qld.

Habitat: This turtle was originally described as a habitat specialist, preferring shallow (<2.0 m) riffle zones characterised by high water velocities. However, individuals have since been recorded from large pools (>1 km long; >2 m deep) and impoundments. Nesting is associated with alluvial sand-loam riverbanks.

Ecology: Known as the 'bum-breathing turtle', it possesses highly modified cloacal bursae that ventilate water for aquatic respiration. Under favourable conditions, it can meet its entire oxygen demand from the water, allowing it to undertake extended dives of several days to weeks in duration. It is primarily carnivorous, feeding on aquatic invertebrates when in riffle zones and on freshwater sponges when residing in pools. The Fitzroy River Turtle can take up to 20 years to reach reproductive maturity. Nesting occurs in spring, with females laying one to two clutches of 18 small eggs in sand and loam alluvial deposits within 20 m of the river margin. Incubation takes 46 days.

Reasons for decline and current threats: The lack of understanding of the turtle's ecology and population dynamics limits effective management. Conjecture still occurs as to whether the species is a habitat specialist preferring fast-flowing riffle zones, given the capture of individuals from slow-flowing pools and impoundments. Additionally, the population density and distribution across the catchment are poorly understood, with overall numbers assumed to be on the decrease.

The turtle has a limited distribution to the Fitzroy catchment, where riverine habitat has been highly modified through deleterious water and land management activities, such as water impoundment, water extraction and riparian clearing. Weirs and dams inundate upstream lotic habitat, changing the fluvial and physico-chemical properties of the water. Water extraction further exacerbates the habitat impacts associated with water impoundment by altering hydrological patterns within remaining lotic sections.

Concerns have also been raised about population recruitment, with up to 100% nest mortality rates being recorded due to predation (e.g. foxes, pigs and goannas) and nest trampling (livestock). Additionally, few immature turtles have been recorded in the field, suggesting low recruitment into the breeding population.

CURRENT RECOVERY EFFORTS

Recovery plan: None

Other plans: Cogger HG, Cameron EE, Sadlier RA and Eggler P (1993) 'The action plan for Australian reptiles'. Australian Nature Conservation Agency, Canberra.

Conservation actions:

- UQ previously conducted research into the turtle's diving physiological ecology.
- Currently, DERM undertakes annual investigations into population dynamics and nesting ecology at select sites

- Greening Australia (Rockhampton) has been active in nest protection and increasing hatchling recruitment within the Fitzroy Barrage. Greening Australia also promotes education about habitat protection and management.

FUTURE RECOVERY EFFORTS

What you can do:

- Join/donate to a local Landcare group and/or The Australian Freshwater Turtle Conservation and Research Association.
- Landholders are encouraged to enact riparian management plans that consider bank protection and revegetation as well as stock exclusion, fencing and off-stream watering.

Conservation needs:

- Identify and protect nesting banks to enhance hatchling recruitment.
- Reassess the current status.
- Environmental flows should be considered to enhance lotic zones that remain between impoundments. Future water development should also be limited until adequate data is available to guide species conservation and water management.
- Protect and rehabilitate riparian zones, especially sandy nesting banks.
- Undertake education and public awareness, particularly for landholders and local government, to promote guidelines to improve land use practices and water quality.

Research directions:

- Study population dynamics and distribution within the basin to inform and direct conservation actions.
- Investigate dietary and habitat preferences relative to growth and reproductive rates, nesting success, and juvenile recruitment and survivorship.
- Assess the hydrological flows, habitat association and dietary preference for turtles occurring within riffle zones versus still water habitats to determine unambiguous guidance for water managers.

Sources and further reading

Gordos MA, Franklin CE and Limpus CJ (2003) Seasonal changes in the diving performance of the bimodally respiring freshwater turtles, *Rheodytes leukops*, in a natural setting. *Canadian Journal of Zoology* **81**, 617–625.

Gordos MA, Franklin CE, Limpus CJ and Wilson G (2004) Blood respiratory and acid-base changes during extended diving in the bimodally respiring freshwater turtle *Rheodytes leukops*. *Journal Comparative Physiology B* **174**, 347–354.

Legler JM and Cann J (1980) A new genus and species of chelid turtle from Queensland, Australia. *Contributions in Science – Natural History Museum of Los Angeles County* **324**, 1–18.

Limpus CJ, Limpus CJ, Parmenter CJ, Hodge J, Forrest MJ and McLachlan J (2007) 'An assessment of the potential implications and management measures for Fitzroy Turtles (*Elseya albagula* and *Rheodytes leukops*)'. Environmental Protection Agency, Brisbane.

Priest T and Franklin CE (2002) Effect of water temperature and oxygen levels on the diving behaviour of two freshwater turtles: *Rheodytes leukops* and *Emydura macquarii*. *Journal of Herpetology* **36**, 555–561.

Author: Matthew Gordos

SNAKES AND LIZARDS (ORDER: SQUAMATA)

BLACK MOUNTAIN GECKO

Nactus galgajuga (Ingram, 1978)

Photo: Eric Vanderduys

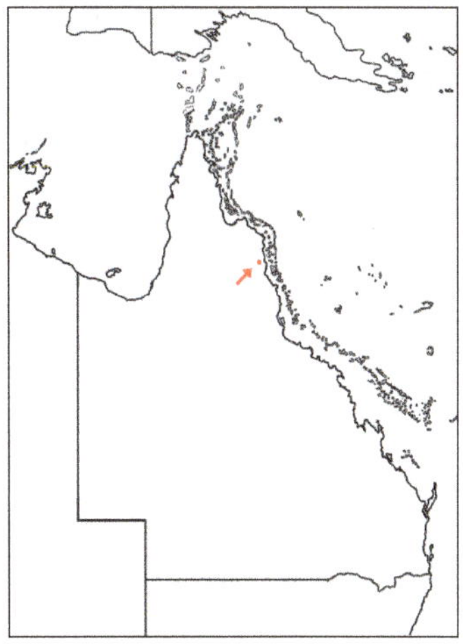

Family: Gekkonidae

Conservation status: Qld: Vulnerable

Description: This small gecko has a snout–vent length to 50 mm and distinctive white bands on a dark grey to black dorsal surface and head, and a light brown ventral surface. The body is slender with a flattened head designed for living in boulder crevices. The legs are long and slender with clawed digits.

Distribution: The Black Mountain Gecko is only found on the northern edge of the Wet Tropics bioregion, south of Cooktown. This gecko is restricted to approximately 6 km^2 in the Black Mountain NP and adjacent Water Reserve of the Black Trevethan Range, at an altitude of between 100 and 300 m. It is currently secure in a small distribution within protected habitat, with no indication of population declines.

Habitat: The Black Mountain Gecko is found in the granite boulder field with scattered vegetation, mainly of figs.

Ecology: This agile gecko lives in the boulder piles and is frequently seen at night on boulders, retreating to the lower side of the boulder when disturbed. This gecko typically shelters during the day deep within rock piles and in rock crevices, but may be active in the late afternoon following rain in the wet season. It produces two eggs.

Reasons for decline and current threats: There is no known decline or threat, but the species has a very restricted distribution in a specialised boulder habitat which may be affected by stochastic events such as prolonged drought.

CURRENT RECOVERY EFFORTS

Recovery plan: None

Conservation actions: No recovery efforts have been undertaken to date, but the species distribution is mainly in national park.

FUTURE RECOVERY EFFORTS

Conservation needs: The area of boulder field in the Water Reserve should be added to the adjacent national park.

Research directions: Research the natural history of the species, especially breeding biology, the physiological limits of temperature and moisture, and variation in microhabitat conditions within its habitat. This will be a good baseline for future climate change impacts.

Sources and further reading

Covacevich J, Ingram GJ and Czechura GV (1982) Rare frogs and reptiles of Cape York Peninsula, Australia. *Biological Conservation* **22**, 283–294.

Ingram GJ (1978) A new species of gecko, genus *Cyrtodactylus*, from Cape York Peninsula, Queensland, Australia. *Victorian Naturalist* **95**, 142–146.

Author: Brett A Goodman

McILWRAITH RANGE LEAF-TAILED GECKO

Orraya occultus (Couper, Covacevich & Moritz, 1993)

Photo: Eric Vanderduys

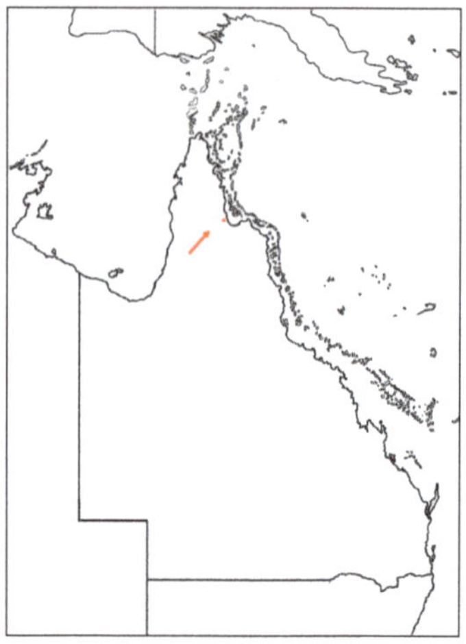

Family: Carphodactylidae

Conservation status: IUCN: Data Deficient; Qld: Vulnerable

Description: The McIlwraith Range Leaf-tailed Gecko is a medium-sized gecko with a snout–vent length to 100 mm. The dorsal surface is moderately prickly (covered in small tubercles) and grey with darker blotches. The ventral surface is smooth and cream with brown mottling. The body and head are flattened, the neck is proportionally long and the limbs are long and very slender. The digits are clawed. The tail is a rounded, broad leaf-like shape. The margin of the tail is spiny, particularly on original tails.

Distribution: The McIlwraith Range Leaf-tailed Gecko is known only from a very small area of McIlwraith Range, in Kulla NP (CYPAL). It is currently secure in a small distribution within protected habitat with no indication of population declines.

Habitat: It is found in jumbled granite boulders along creeks in rainforest (microphyll/notophyll vine forests) at high altitude.

Ecology: The gecko shelters in crevices among boulders during the day and emerges at night to forage among the rocks for spiders and insects. Females found in winter had enlarged egg follicles, suggesting that eggs are laid in spring or summer. A clutch consists of two eggs.

Reasons for decline and current threats: There is no known decline or threat, but the species has a very restricted distribution in a specialised habitat.

CURRENT RECOVERY EFFORTS

Recovery plan: None

Conservation actions: The species is protected within the recently declared Kulla NP (CYPAL).

FUTURE RECOVERY EFFORTS

Research directions:

- Monitor the species to assess trends in abundance.
- Conduct further surveys to determine the full extent of the distribution and estimate population size.

Sources and further reading

Couper PJ, Covacevich JA and Moritz C (1993) A review of the leaf-tailed geckos endemic to eastern Australia: a new genus, four new species, and other new data. *Memoirs of the Queensland Museum* **34**, 95–124.

Couper PJ, Schneider CJ, Hoskin CJ and Covacevich JA (2000) Australian leaf-tailed geckos: phylogeny, a new genus, two new species and other new data. *Memoirs of the Queensland Museum* **45**, 253–265.

Hoskin CJ, Couper PJ and Schneider CJ (2003) A new species of *Phyllurus* (Lacertilia: Gekkonidae) and a revised phylogeny and key for the Australian leaf-tailed geckos. *Australian Journal of Zoology* **51**, 153–164.

Author: Conrad J Hoskin

BULBURIN LEAF-TAILED GECKO

Phyllurus caudiannulatus Covacevich, 1975

Photo: www.ecosmartecology.com.au

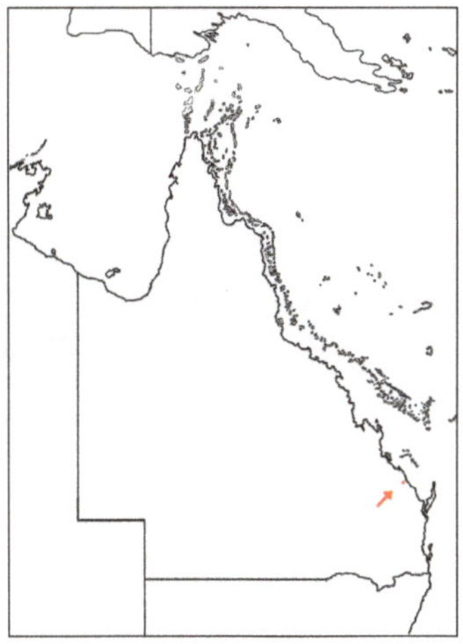

Family: Carphodactylidae

Conservation status: Qld: Vulnerable

Description: The Bulburin Leaf-tailed Gecko has a snout–vent length to approximately 100 mm. The body and head are flattened, the limbs are long and slender, and the digits are clawed. The tail is long and cylindrical, tapering to a fine tip. It is a prickly gecko with large, conical tubercles on the dorsal surface of the body, tail, head and limbs. The dorsal surface is brown with darker blotching, and the original tail is brown with dark blotching and 5–6 cream cross bands. The regenerated tail is smooth and lacks pale banding. The ventral surface of the body, head, limbs and tail is cream with brown 'peppering'.

Distribution: The Bulburin Leaf-tailed Gecko is restricted to Bulburin NP, in the Many Peaks and Dawes Ranges of south-east Qld. The species is currently secure in a protected area, with no known threats.

Habitat: This gecko is restricted to rainforest. It can be particularly common on the trunks of strangling fig trees.

Ecology: This species is found foraging at night on boulders or tree trunks. The diet consists of invertebrates such as beetles. During the day, the gecko hides in secluded spots such as rock crevices, inside hollow fig trees, or behind bark. The clutch size consists of two eggs laid in spring or summer.

Reasons for decline and current threats: The Bulburin Leaf-tailed Gecko has a very restricted distribution. No known decline has been detected and no threatening processes have been identified.

CURRENT RECOVERY EFFORTS

Recovery plan: None

Conservation actions: The habitat is protected in Bulburin NP.

FUTURE RECOVERY EFFORTS

Research directions:

- Monitor the population to assess trends in abundance through time.
- Conduct further surveys to determine the fine-scale distribution and approximate population size of this gecko.

Sources and further reading

Couper PJ, Covacevich JA and Moritz C (1993) A review of the leaf-tailed geckos endemic to eastern Australia: a new genus, four new species, and other new data. *Memoirs of the Queensland Museum* **34**, 95–124.

Couper PJ, Hamley B and Hoskin CJ (2008) A new species of *Phyllurus* (Lacertilia: Gekkonidae) from the Kilkivan district of south-eastern Queensland. *Memoirs of the Queensland Museum* **52**, 139–147.

Couper PJ, Schneider CJ, Hoskin CJ and Covacevich JA (2000) Australian leaf-tailed geckos: phylogeny, a new genus, two new species and other new data. *Memoirs of the Queensland Museum* **45**, 253–265.

Covacevich JA (1975) A review of the genus *Phyllurus* (Lacertilia: Gekkonidae). *Memoirs of the Queensland Museum* **17**, 293–303.

Hoskin CJ, Couper PJ and Schneider CJ (2003) A new species of *Phyllurus* (Lacertilia: Gekkonidae) and a revised phylogeny and key for the Australian leaf-tailed geckos. *Australian Journal of Zoology* **51**, 153–164.

Author: Conrad J Hoskin

GULBARU GECKO

Phyllurus gulbaru Hoskin, Couper & Schneider, 2003

Photo: Angus McNab

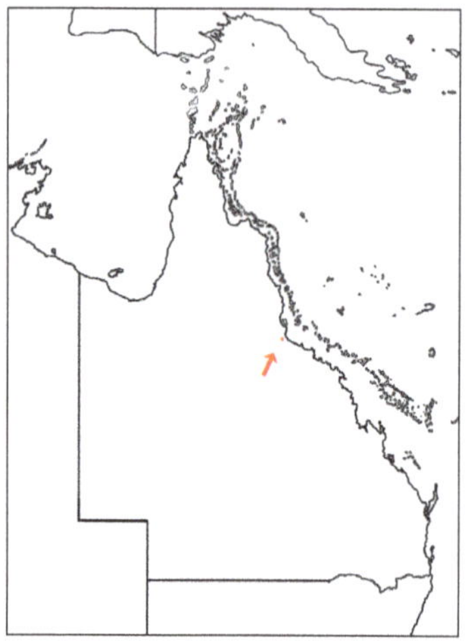

Family: Carphodactylidae

Conservation status: IUCN: Critically Endangered; Qld: Endangered

Description: The Gulbaru Gecko has a snout–vent length to approximately 85 mm and a tail of about the same length. The body and head are flattened and the limbs are long and slender. The digits are clawed. The tail is a cylindrical shape tapering to a point. The dorsal surface is moderately prickly, with small tubercles, and is grey with darker mottling. The original tail is mottled grey with distinct white cross bands. The regenerated tail is smooth and lacks white banding. The ventral surface is smooth and cream peppered with brown specks.

Distribution: This is a narrowly distributed species, restricted to the Pattersons Gorge area west of Townsville, including Mount Cataract Forest Reserve.

Habitat: The Gulbaru Gecko occurs in rocky gullies within microphyll-notophyll vine forest dominated by Hoop Pine *Araucaria cunninghamii*.

Ecology: Found at night among boulders and rock crevices, foraging for invertebrates. Shelters in crevices during the day. It is slow moving and very cryptic. Gravid females, each carrying two eggs, have been found in summer.

Reasons for decline and current threats: One of Qld's most narrowly distributed reptile species. Part of the distribution is protected in Mount Cataract Forest Reserve; the remainder is unprotected. Given such a small distribution, any reduction or degradation of rainforest habitat could pose a threat. Potential threats include fire, particularly late dry season burning in adjacent open forest and pastoral land.

CURRENT RECOVERY EFFORTS

Recovery plan: None

Conservation actions: The recently declared Mount Cataract Forest Reserve protects part of the habitat.

FUTURE RECOVERY EFFORTS

Conservation needs:

- Conversion of Mount Cataract Forest Reserve and adjacent state lands to national park will provide maximum protection for the species.
- Implement an active program to reduce late dry season hot fires.

Research directions:

- Monitor the species and its habitat to detect changes in abundance and distribution through time.
- Conduct further surveys to determine the fine-scale distribution of the Gulbaru Gecko and to estimate population size.
- Assess the impact of late dry season fires on vine-thicket habitat in this area.

Sources and further reading

Hoskin CJ, Couper PJ and Schneider CJ (2003) A new species of *Phyllurus* (Lacertilia: Gekkonidae) and a revised phylogeny and key for the Australian leaf-tailed geckos. *Australian Journal of Zoology* **51**, 153–164.

Author: Conrad J Hoskin

MOUNT BLACKWOOD LEAF-TAILED GECKO

Phyllurus isis Couper, Covacevich & Moritz, 1993

Photo: Angus McNab

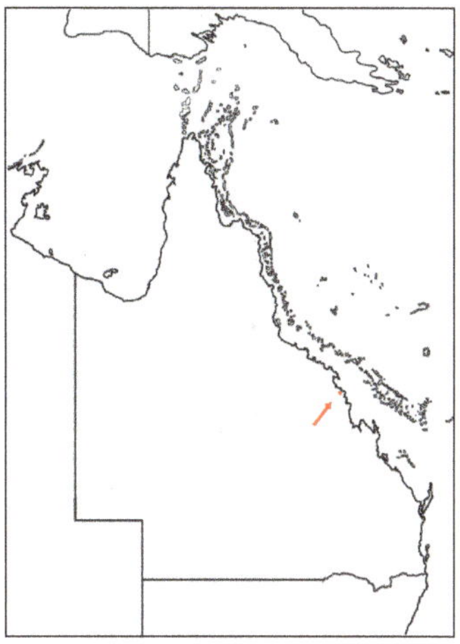

Family: Carphodactylidae

Conservation status: Qld: Vulnerable

Description: The Mount Blackwood Leaf-tailed Gecko is a small gecko with a snout–vent length to approximately 70 mm. The body and head are flattened and the limbs are long and slender. The digits are clawed. The tail is flared at the base and then tapers to a pointed tip, forming a broad leaf-like shape. The dorsal surface is covered in fine tubercles, but less prickly than most *Phyllurus* geckos, and greyish brown with darker mottling. The original tail is dark brown to black with two white broken blotches on the flared portion and three white bands on slender tip. The regenerated tail is smooth and lacks white markings. The ventral surface is smooth and cream.

Distribution: The Mount Blackwood Leaf-tailed Gecko is restricted to Mount Jukes and Mount Blackwood, north of Mackay.

Habitat: It is found in complex notophyll vine forests on rocky substrates, including granite. The Mount Blackwood Leaf-tailed Gecko has primarily been found on rock outcrops.

Ecology: It is nocturnal and its diet includes invertebrates such as moths, cockroaches and crickets. It shelters during the day in crevices. Gravid females, carrying two eggs, have been found in early October.

Reasons for decline and current threats: The Mount Blackwood Leaf-tailed Gecko has a very restricted distribution in a specialised habitat, but there is no known decline or identified threat. The impact of fire on its habitat warrants consideration. As with other species that have very localised distributions, this species could be affected greatly by stochastic events.

CURRENT RECOVERY EFFORTS

Recovery plan: None

Conservation actions: The species habitat is protected in Pioneer Peaks NP.

FUTURE RECOVERY EFFORTS

Conservation needs: Assess the threat late dry season fires pose to habitat requirements.

Research directions:

- Monitor the species to assess any abundance trends through time.
- Conduct further surveys to determine the fine-scale distribution of this gecko and to estimate the population size.

Sources and further reading

Couper PJ, Covacevich JA and Moritz C (1993) A review of the leaf-tailed geckos endemic to eastern Australia: a new genus, four new species, and other new data. *Memoirs of the Queensland Museum* **34**, 95–124.

Couper PJ, Schneider CJ, Hoskin CJ and Covacevich JA (2000) Australian leaf-tailed geckos: phylogeny, a new genus, two new species and other new data. *Memoirs of the Queensland Museum* **45**, 253–265.

Hoskin CJ, Couper PJ and Schneider CJ (2003) A new species of *Phyllurus* (Lacertilia: Gekkonidae) and a revised phylogeny and key for the Australian leaf-tailed geckos. *Australian Journal of Zoology* **51**, 153–164.

Author: Conrad J Hoskin

OAKVIEW LEAF-TAILED GECKO

Phyllurus kabikabi Couper, Hamley & Hoskin 2008

Family: Carphodactylidae

Conservation status: Qld: Endangered

Description: The Oakview Leaf-tailed Gecko has a snout-vent length to approximately 70 mm. The body and head are flattened, the limbs are long and slender, and the digits are clawed. The tail is long and cylindrical, tapering to a point. It is a prickly gecko with conical tubercles, which are largest on the flanks and on the edge of the tail. The dorsal surface is brown with darker blotching, and the original tail is brown with dark blotching and about five cream cross bands. The regenerated tail is smooth and lacks pale banding. The ventral surface is cream with brown 'peppering'.

Distribution: The Oakview Leaf-tailed Gecko is restricted to Oakview NP, near Kilkivan in south-east Qld. Records come from a very small area, at an altitude of 540 m.

Habitat: The Oakview Leaf-tailed Gecko occurs in moist semi-evergreen vine forest growing on a deeply layered rocky substrate. Suitable habitat is highly localised and patchy.

Ecology: It shelters in crevices during the day and emerges at night to forage for invertebrates on rocks and tree growing among rocks. It is slow moving and very cryptic. One individual was found in September with two eggs visible through the body wall, suggesting a clutch of two eggs is laid in late spring.

Reasons for decline and current threats: One of Qld's most narrowly distributed reptile species, its entire known distribution is protected in national park. Potential threats include habitat degradation from forest die-back caused by *Phytophthora* fungus, fire, weeds and feral animals.

CURRENT RECOVERY EFFORTS

Recovery plan: None

Conservation actions: The species' habitat is protected in Oakview NP.

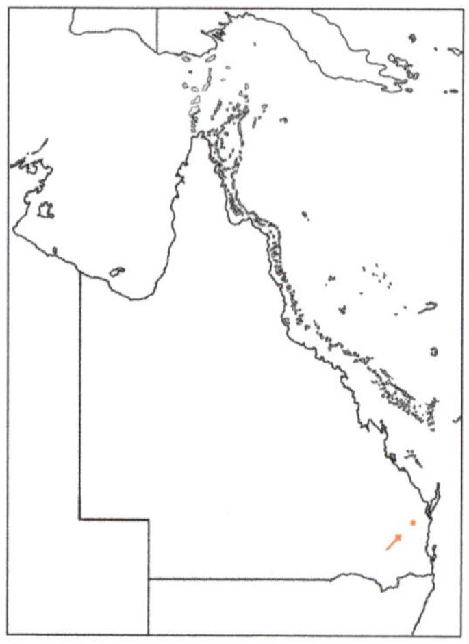

FUTURE RECOVERY EFFORTS

Research directions:

- Monitor the population to assess if the species is affected by environmental factors such as prolonged drought.
- Conduct surveys of other areas of potentially suitable habitat in Oakview NP in order to determine the full extent of the distribution and to estimate population size.

Sources and further reading

Couper PJ, Hamley B and Hoskin CJ (2008) A new species of *Phyllurus* (Lacertilia: Gekkonidae) from the Kilkivan district of south-eastern Queensland. *Memoirs of the Queensland Museum* **52**, 139–147.

Couper PJ, Schneider CJ, Hoskin CJ and Covacevich JA (2000) Australian leaf-tailed geckos: phylogeny, a new genus, two new species and other new data. *Memoirs of the Queensland Museum* **45**, 253–265.

Hoskin CJ, Couper PJ and Schneider CJ (2003) A new species of *Phyllurus* (Lacertilia: Gekkonidae) and a revised phylogeny and key for the Australian leaf-tailed geckos. *Australian Journal of Zoology* **51**, 153–164.

Author: Conrad J Hoskin

BORDER THICK-TAILED GECKO

Uvidicolus sphyrurus (Ogilby, 1892)

Photo: Harry B Hines

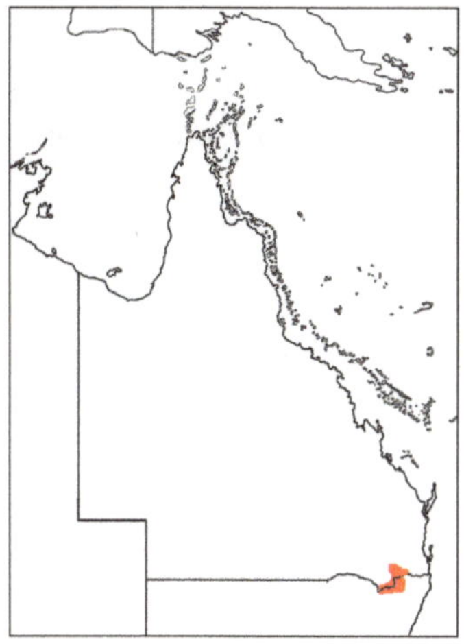

Family: Gekkonidae

Common names: Nandewar Thick-tailed Gecko, Granite Thick-tailed Gecko, Granite Belt Thick-tailed Gecko

Conservation status: IUCN: Near Threatened; Aust.: Vulnerable; Qld: Near Threatened; NSW: Vulnerable

Description: The snout–vent length is up to 70 mm, with a total length to 105 mm. The dorsal surface is brownish with faint darker flecks and reticulations, and numerous transversely aligned white spots centred on very small conical tubercles. An original tail is squarish, tapering rapidly to a fine rounded terminal point, dark grey-brown to blackish, with up to four pale cross bands formed by rows of enlarged tubercles. The ventral surface is pale. The feet are bird-like, without enlarged toe pads. When threatened, individuals raise themselves high on extended legs and wave the tail slowly from side to side, sometimes rushing forward with mouth open emitting an audible barking noise.

Distribution: Most of the range of this species falls in the Interim Biogeographic Regionalisation of Australia Nandewar bioregion, with the southern most record from near Tamworth, NSW. In Qld, the species has been recorded from Girraween and Sundown NPs, north to Durikai State Forest and Cherrabah.

Habitat: This is poorly documented. In Qld, the gecko occurs in temperate dry open eucalypt or *Callitris* woodland, principally on the northern and western slopes of the New England Tableland. It appears to favour rocky to stony hills with boulders, rock slabs, stones, fallen timber and deep leaf litter.

Ecology: It is active at night, foraging on the ground. By day it has been found sheltering under logs, stones and rock slabs. Otherwise, it is very poorly known. General information published for the genus shows a diet of a broad range of arthropods and clutches comprising two parchment-shelled eggs. There are no population estimates. In Qld, this species is rarely encountered and appears to occur at low density.

Reasons for decline and current threats: It is not known to have declined, but occurs in an area with high human impact, with increasing loss and fragmentation of natural vegetation cover and heavy grazing by introduced herbivores. The main identified threats include: vegetation clearing for agriculture and development; the collection of bush rock for landscaping and fallen timber for firewood; altered fire regimes; grazing and trampling of habitat by domestic stock and feral ungulates; timber harvesting activities; predation by cats, foxes and pigs; and disturbance to habitat by pigs. Illegal collection for pets is also a possible threat to individuals and key microhabitats.

CURRENT RECOVERY EFFORTS

Recovery plan: None

Other plans: Cogger HG, Cameron EE, Sadlier RA and Eggler P (1993) 'The action plan for Australian reptiles'. Australian Nature Conservation Agency, Canberra.

Department of the Environment and Heritage (2005) 'Threat abatement plan for predation, habitat degradation, competition and disease transmission by feral pigs'. Department of Environment and Heritage, Canberra.

Department of the Environment, Water, Heritage and the Arts (2008) 'Threat abatement plan for competition and land degradation by unmanaged goats'. Department of the Environment, Water, Heritage and the Arts, Canberra.

Department of the Environment, Water, Heritage and the Arts (2008) 'Threat abatement plan for predation by feral cats'. Department of the Environment, Water, Heritage and the Arts, Canberra.

Department of the Environment, Water, Heritage and the Arts (2008) 'Threat abatement plan for predation by the European red fox'. Department of the Environment, Water, Heritage and the Arts, Canberra.

Conservation actions: The NRM groups Condamine Alliance and Queensland Murray Darling Committee provide advice and funding for biodiversity conservation and management to landholders in the area occupied by this species.

FUTURE RECOVERY EFFORTS

What you can do: Write letters to state and federal environment ministers to urge funding allocation to develop and implement a recovery plan for this species.

Conservation needs:

- Implement threat abatement plans for feral cats, feral pigs, goats and foxes.
- Philanthropic and private industry funding is essential to the recovery of this species because government funding is inadequate.

Research directions:

- Conduct a population survey and monitoring.
- Investigate the basic ecology of the species, including habitat requirements and threats.
- Determine the genetic structure of the population.
- Determine the current distribution, population size and population trend, habitat requirements and evaluate the role of potential threats.
- Develop site-specific management actions to protect and enhance important populations.
- DERM encourages students to develop research programs that relate directly to future management of the species.

Sources and further reading

Department of Conservation and Climate Change (DECC) (2009) *Border Thick-tailed Gecko Profile.* <http://www.threatenedspecies.environment.nsw.gov.au/tsprofile/profile.aspx?id=10823>.

Department of the Environment, Water, Heritage and the Arts (DEWHA) (2009) *Approved Conservation Advice for* Underwoodisaurus sphyrurus *(Border Thick-tailed Gecko).* <http://www.environment.gov.au/biodiversity/threatened/species/pubs/1660-conservation-advice.pdf>.

Hines HB (2008) Some observations on the vertebrate fauna of Durikai State Forest, southeast Queensland, 2000-2008. *The Queensland Naturalist* **46**, 44–69.

Author: Harry B Hines

STRIPED-TAILED DELMA

Delma labialis Shea, 1987

Photo: Greg Calvert

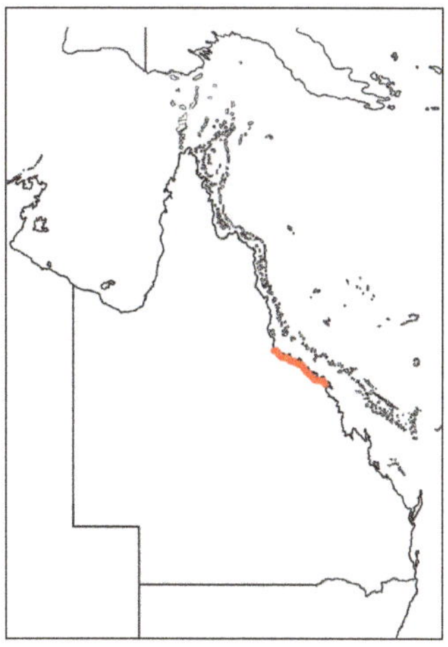

Family: Pygopodidae

Other common name: Single-striped Delma

Conservation status: IUCN: Vulnerable; Aust.: Vulnerable; Qld: Vulnerable

Description: This large legless lizard has a snout–vent length to 126 mm (usually less) and a 490 mm tail (usually less). It is dull tan to silvery brown, with the head usually yellow tinted with obscure to distinct pale bands from the throat and belly onto the sides – these bands are often just restricted to the head and neck, but may extend to first half of the body. The tail colour is usually copper, darkening to reddish brown towards the tip. A single, thin dark stripe is present on either side of the body extending from about half to a quarter of the rear of the body and onto the sides of the tail – this stripe often fades towards the tail tip. Midbody scales are in 16 rows.

Distribution: On the mainland, this species is known from coastal ranges and flats in several areas in north-east Qld, from Seaview Range near Paluma south to Mount Abbott and Horseshoe Bay in Bowen and Conway State Forest near Proserpine. It is also known from several rocky islands – Magnetic Island (Townsville), South Molle Island (Whitsundays), Shaw Island (Whitsundays) and Keswick Island (Cumberland group, north-east of Mackay). It is not known whether mainland populations are contiguous or disjunct, especially the existence of populations between Townsville and Bowen. One population in Townsville (Castle Hill) is almost certainly isolated from nearest known populations in outer Townsville suburbs (Douglas/Wulguru). It is a known inhabitant of conservation areas at Magnetic Island NP, Townsville Town Common Conservation Park, Keswick Island NP and Whitsunday Islands NP.

Habitat: The habitat includes two very dissimilar land forms: coastal sandy dunes and rocky hill slopes. Vegetation type is a poor predictor of occurrence, because this species has been located in open eucalypt woodland, *Melaleuca viridiflora* woodland, low open forest, wet sclerophyll, rainforest (simple notophyll vine forest), vine thicket on hillside gullies, beach scrub and moist gallery forest with dense leaf litter, but, in the absence of sand dunes, nearly always in areas with abundant boulders or exposed rock outcrops and platforms. It apparently prefers areas with thick ground cover, especially Spinifex *Triodia* spp., Spear Grass *Heteropogon contortus*, Kangaroo Grass *Themeda triandra,* Razor grasses *Scleria* spp. and Saw Sedges *Gahnia* sp. It may also occupy areas of introduced grasses such as Guinea Grass *Megathyrsus maximus* where these are adjacent to large tracts of country with thick ground cover. It ranges in altitude from sea level to 750 m.

Ecology: The Striped-tailed Delma probably lays two eggs in a well sheltered site, such as under logs or rocks, but particulars of breeding are unknown. It feeds on insects and spiders, but, like most *Delma* spp., the particulars of diet are unknown. The species has generally been recorded as diurnal, particularly midday to late afternoon, but individuals have been encountered at night as late as 9 pm. In the cooler months, they have been recorded under ground cover such as rocks, logs and artificial rubbish. This may indicate a period of winter inactivity. Such records are rare for warmer months, except when an individual is about to slough its skin, which is a time of increased vulnerability.

Reasons for decline and current threats: There is no direct evidence of declines in this species because there are too few records. There are no pre-clearing records from the large (now cleared) areas of agricultural and urban land from Proserpine to Ingham. However, it can be reasonably inferred that clearing and modification for agricultural, urban and rural residential development in these areas has led to loss of populations, declines in numbers and loss and fragmentation of habitat. Increased road networks into new coastal developments probably render this skink increasingly susceptible to significant levels of road-kill. It is likely that any loss of the complex structure of natural ground vegetation and leaf litter will lead to loss or reduction in suitable habitat for this species. In the vicinity of urban areas, intense predation by household cats may be a threat, especially when combined with loss of habitat structural complexity. Fire regimes that result in broad scale loss of ground vegetation are also likely to render areas unsuitable, or at least suboptimal for this species. Frequently, in the vicinity of urban areas, bushland fires are an all too common occurrence as a result of arson. Mortality from a late season intense wildfire has been recorded, but the subsequent loss of cover is likely to be an additional threat by rendering them vulnerable to predation.

CURRENT RECOVERY EFFORTS

Recovery plan: None

Other plans: Cogger HG, Cameron EE, Sadlier RA and Eggler P (1993) 'The action plan for Australian reptiles'. Australian Nature Conservation Agency, Canberra.

FUTURE CONSERVATION EFFORTS

Conservation needs:

- Areas of suitable habitat need to be managed.
- Connectivity should be maintained between areas of known or likely habitat to avoid fragmentation of coastal populations.
- Disseminate information regarding best-practice survey techniques for Striped-tailed Delmas to fauna surveying companies and sub-contracted consultants conducting Environmental Impact Assessments in potential Striped-tailed Delma habitat.

Research directions:

- Conduct surveys of potential habitat inside reserves such as Bowling Green Bay NP, Cape Upstart NP and in suitable habitat in military areas to the west of Townsville.
- Research the basic biology and ecology and assessment of threat impacts on populations.

Sources and further reading

Lloyd R (2005) A high altitude observation of the North Queensland pygopod *Delma labialis* (Sauria: Pygopodidae). *Herpetofauna* **35**, 40–41.

Low T (1978) The reptiles of Magnetic Island, North Queensland. *Herpetofauna* **9**, 10–14.

Shea GM (1987) Two new species of *Delma* (Lacertilia: Pygopodidae) from North-eastern Queensland and a note on the status of the genus *Aclys*. *Proceedings of the Linnean Society of NSW* **109**, 203–212.

Wilson S (2005) *A Field Guide to Reptiles of Queensland*. Reed New Holland, Australia.

Author: Eric Vanderduys

ATHERTON DELMA

Delma mitella Shea, 1987

Photo: Greg Calvert

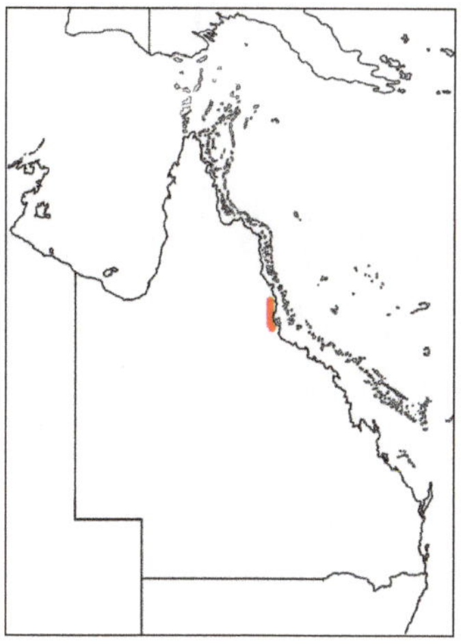

Family: Pygopodidae

Other common name: Atherton Legless Lizard

Conservation status: Aust.: Vulnerable; Qld: Least Concern

Description: This is a very large legless lizard with a snout–vent length to 200 mm and a tail of 660 mm (usually less). The dorsal surface is coppery brown with ventral and lower lateral surfaces cream to lime green. The head markings are usually darker copper with three to four narrow pale bands. The mid-body scales are in 16 rows.

Distribution: Tolga to Paluma, north-east Qld. It has been recorded in Kirrama and Paluma Range NPs.

Habitat: The Atherton Delma inhabits the wetter open forests of the Wet Tropics bioregion above 600 m altitude. It is mostly known from areas with thick ground cover and leaf litter. Common grasses forming much of the cover in areas from which it is known are Kangaroo Grass *Themeda triandra*, Blady Grass *Imperata cylindrica* and *Mnesithea rottboellioides*.

Ecology: The lizard feeds on insects and spiders, but, like most *Delma* spp., the particulars of diet are unknown. It probably lays two eggs in a well sheltered site, such as under logs or rocks embedded in the soil, but particulars of breeding are unknown.

Reasons for decline and current threats: Declines and threats to the Atherton Delma can only be inferred: the species is not known to have declined in range or abundance. It is a poorly known species that may be affected by loss of habitat through clearing and habitat modification, urbanisation and rural residential development in areas around the Atherton Tablelands. It appears to be secure in other locations where it has been recorded on the Cardwell, Kirrama and Paluma Ranges. It is likely that loss of the complex structure of natural ground vegetation and leaf litter will lead to loss or reduction in suitable habitat for this species. Where grazing results in a loss of ground cover and structural complexity, it is likely to cause a reduction in suitable habitat. Similarly, a reduction in suitable habitat for this species is likely if broad-scale wildfires cause a loss of cover over large areas. In the vicinity of urban areas, intense predation by household cats may be a threat, especially when combined with loss of habitat structural complexity.

CURRENT RECOVERY EFFORTS

Recovery plan: None

Other plans: Cogger HG, Cameron EE, Sadlier RA and Eggler P (1993) 'The action plan for Australian reptiles'. Australian Nature Conservation Agency, Canberra.

FUTURE RECOVERY EFFORTS

Conservation needs: Identify and manage areas of suitable habitat.

Research directions:

- Conduct surveys of potential habitat to gain an understanding of geographic range and threats.
- Research the basic biology and ecology and assessment of threat impacts on populations.

Sources and further reading

Shea GM (1987) Two new species of *Delma* (Lacertilia: Pygopodidae) from North-eastern Queensland and a note on the status of the genus *Aclys*. *Proceedings of the Linnean Society of NSW* **109**, 203–212.

Van Barneveld S (2006) Two new records of *Delma mitella* Shea 1987 from the Wet Tropics. *Herpetofauna* **36**, 66–67.

A statement for the purposes of approved conservation advice (s266B of the *Environment Protection and Biodiversity Conservation Act 1999*) Approved Conservation Advice for *Delma mitella* (Atherton Delma). <http://www.environment.gov.au/biodiversity/threatened/species/pubs/25931-conservation-advice.pdf>.

Author: Eric Vanderduys

COLLARED DELMA

Delma torquata Kluge, 1974

Photo: Eric Vanderduys

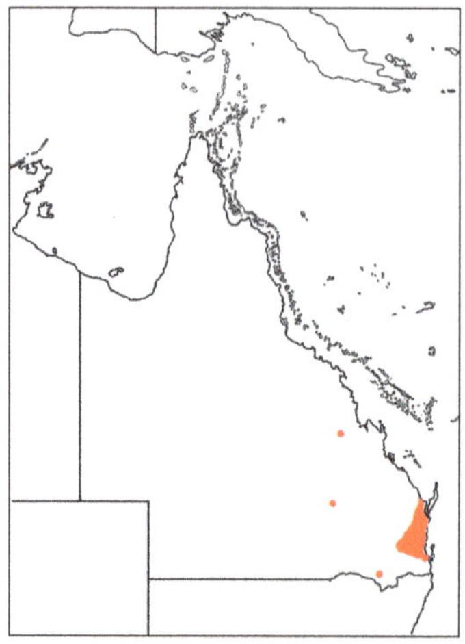

Family: Pygopodidae

Other common name: Adorned Delma

Conservation status: IUCN: Vulnerable; Aust.: Vulnerable; Qld: Vulnerable

Description: Collared Delma adults have a snout–vent length of 50–70 mm (average: 56 mm), a tail length of 103–126 mm (average 110.5 mm) and an average weight of 1.36 g. The body is brown, the tail has a bluish/grey tinge, and the head and neck have bold black bands with cream/yellow interspaces. The chin and neck are white with bold grey/black marbling and the remainder of the ventral surface is white. The mid-body scales are in 16 rows. There are two supranasal scales and two pre-anal scales.

Distribution: The majority of the Collared Delma records are from south-east Qld around the western suburbs of Brisbane and the Toowoomba Ranges. Recent field work has significantly increased the known distribution from Blackdown Tablelands NP in the north and the Roma area to the west. The western suburbs of Brisbane have the highest density of active sites recorded for the Collared Delma. Even within this area, populations are highly fragmented, with some populations of relatively high density separated by tens of kilometres. Recent field work has recorded a significant number of populations from the Toowoomba Ranges area, which may prove to be equally as, or even more, important than the western suburbs of Brisbane because there is less habitat development pressure. The total number of known localities for this species has increased with surveys, but population sizes are unknown.

Habitat: The Collared Delma is typically associated with westerly facing ridgelines supporting dry open eucalypt/acacia woodlands with an open midstorey and a ground cover consisting of native grasses, thick leaf litter, with an abundance of loose surface rocks. It has also been recorded from Red Gum *Eucalyptus tereticornis* woodlands and Brigalow *Acacia harpophylla* communities on cracking clay soils in low lying areas. Although habitat disturbance has been identified as a threatening process for the Collared Delma, populations have been recorded from highly disturbed areas. Within these areas, the native tree cover has usually been maintained but the midstorey has been removed and the ground cover has been replaced by introduced pasture species. An important feature of these disturbed sites is the retention of the surface rocks.

Ecology: The Collared Delma appears to be mostly diurnal. In captivity, it has been observed foraging during the day in the leaf litter among rocks and in low grass tussocks. It is insectivorous, feeding predominantly on small cockroaches. The Collared Delma appears to be solitary in habit because only single specimens have been recorded from each rock shelter. In captivity, they have been recorded squeaking while foraging. The microhabitat of the Collared Delma is loose sandstone, basalt or quartzite surface rocks or leaf mats on stony lithosols or basalt derived podzolic soils. Surface rocks have been identified as a significant habitat component. Two relatively large eggs (14.3 × 5.2 mm, 0.27 g) are produced in December and hatch in February/March.

Reasons for decline and current threats: The Collared Delma has always been considered an uncommon species and there has not been a documented dramatic decline in abundance or in its distribution that can be attributed to any one significant threatening process. However, considering the naturally fragmented nature of its distribution, the species is susceptible to relatively local pressures, including land clearing for pastoralism and urban development in the south-eastern area of its distribution.

Other threats include unsuitable fire regimes and invasive weeds. Hot uncontrolled fires can result in undesirable habitat modification or the direct loss of populations. However, the total removal of fire in these fire-tolerant ecosystems can lead to both native and introduced vegetation thickening, which renders the habitat unsuitable.

CURRENT RECOVERY EFFORTS

Recovery plan: Davidson C (1993) 'Recovery plan for the Collared Legless Lizard (*Delma torquata*). ANCA Endangered Species Program'. Unpublished report to Queensland Department of Environment and Heritage.

Other plans: Brisbane City Council (2004) 'Collared Delma (*Delma torquata*) conservation action statement'.

Current conservation actions:

- DERM manages the state-wide recovery plan for the Collared Delma and the species has been nominated as a 'critical' priority species within its threatened species management program 'Back on Track'. DERM funds and encourages voluntary conservation agreements on private land and manages protected areas with known populations.
- Brisbane City Council acquires significant habitat through its Bushland Acquisition Program, encourages voluntary conservation partnerships with private landholders, promotes public awareness of the species through its conservation action statements and supports and encourages local catchment groups with the on-ground management of core habitat.

FUTURE RECOVERY EFFORTS

What you can do: Join a local catchment or bushcare group and help restore bushland in your local area.

Conservation needs:

- Develop and implement species-specific land management guidelines for known habitat.
- Encourage conservation agreements for known populations on private land.
- Maintain and increase public awareness of the species.

Research directions:

- Conduct ongoing monitoring of known populations.
- Survey existing and potential new habitats to identify additional populations.

Sources and further reading

Department of the Environment, Water, Heritage and the Arts (2009) *Delma torquata*. In *Species Profile and Threats Database*. Department of the Environment, Water, Heritage and the Arts, Canberra. Available at <http://www.environment.gov.au/sprat>.

Environmental Protection Agency (2006) Conservation Management Profile – Collared delma, *Delma torquata*. Environmental Protection Agency, Brisbane. Available at <http://www.environment.gov.au/biodiversity/threatened/species/pubs/1656-conservation-advice.pdf>.

Hines BM, Hannah D, Venz M and Eyre T (2000) New distribution and habitat data for the vulnerable Pygopodid, *Delma torquata* (Kluge 1974). *Memoirs of the Queensland Museum* **45**, 391–393.

Kluge AG (1974) 'A taxonomic revision of the lizard family Pygopodidae'. Miscellaneous publications Museum of Zoology, University of Michigan, No. 147. Ann Arbor Museum of Zoology, University of Michigan.

Porter R (1998) A preliminary field investigation of the Collared Delma, *Delma torquata*. *Memoirs of the Queensland Museum* **42**, 565–572.

Author: Stephen Peck

BRIGALOW SCALY-FOOT

Paradelma orientalis (Gunther, 1876)

Photo: www.ecosmartecology.com.au

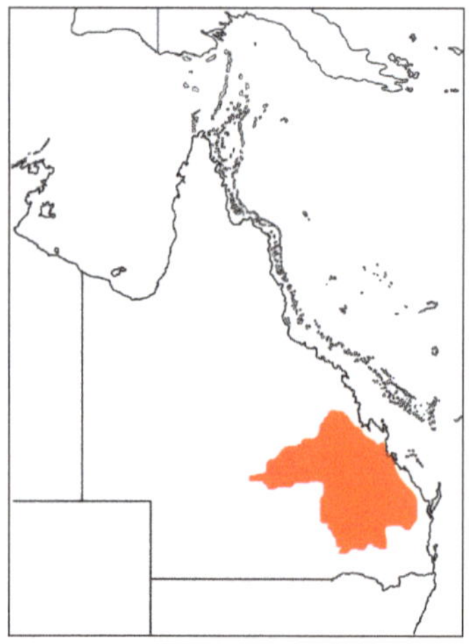

Family: Pygopodidae

Conservation status: IUCN: Vulnerable; Aust.: Vulnerable; Qld: Vulnerable

Description: This is a robust lizard with a rounded snout. Adults have a snout–vent length to 197 mm and a tail similar in length. The dorsal surface is dull grey to brown with a distinctive opaque sheen. The head is a pale yellow-brown, being lighter posteriorly, almost forming a pale band. There is a dark brown to black bar on the nape; the ventral surface is white. The mid-body scales range from 18 to 20 and the ventral scales are larger than the adjacent body scales. There are four preanal scales. The juveniles have similar colouration to the adults, except the pale nape band is much lighter, being almost white.

Distribution: The Brigalow Scaly-foot is primarily restricted to the Brigalow Belt bioregion from Dipperu NP in the north, south to the inland central NSW border region, east to the Windera areas and west to Idalia NP on the eastern edge of the Mulga Lands bioregion. It has long been considered as being endemic to Qld, but has recently been recorded in central inland NSW. It has been recorded from 13 national parks, one conservation park and 10 state forests in Qld. The Brigalow Scaly-foot is normally considered uncommon, but it can be locally abundant. Nineteen individuals were recorded over a 12 day period from a 1.3 ha study site on Boyne Island. Multiple specimens were also recorded in pitfall trap lines during fauna surveys at Chesterton Range NP, where the species is considered common. Although there is no current population estimate, it is regularly recorded during fauna surveys within its distribution. The total number of known localities for this species has increased, but there is no knowledge of population size.

Habitat: It is recorded from a variety of open woodland communities and soil types, which includes Brigalow *Acacia harpophylla* on grey cracking soils, Narrow-leaved Ironbark *Eucalyptus crebra* woodlands with a dense mid-storey layer of Cypress Pine *Callitris columellaris* and Bulloak *Allocasuarina luehmannii* on sandy clays, Spotted Gum *Corymbia citriodora* woodland on shallow stony soils, mixed species woodlands with a Spinifex *Triodia mitchelli* dominated ground cover on sand and ooline, and Ooline *Cadellia pentastylis*/Brigalow *A. harpophylla* closed forest with a thick mixed species ground cover. Its microhabitat includes sandstone slabs, logs, fallen bark, leaf litter and grass tussocks.

Ecology: The Brigalow Scaly-foot is nocturnal and mostly terrestrial. However, in the Boyne Island population, an element of arboreal behaviour has been observed: climbing Rough Barked Acacias *Acacia falciformis* where it feeds on sap. Prey items include spiders and crickets. In captivity, it will readily feed on crickets and cockroaches. The Brigalow Scaly-foot has been recorded laying two elongate eggs (35.5 × 11.5 mm, 3.9 g) in November, hatching 82 days later. The neonates had an average snout–vent length of 70.5 mm (range 69–72 mm) and a tail length of 74.5 mm (71–78 mm), and an average weight of 1.6 g (1.5–1.7 g). When disturbed, the Brigalow Scaly-foot will raise its head and neck off the ground and rapidly flicks its tongue in what is considered to be a snake-mimicking posture.

Reasons for decline and current threats: The Brigalow Scaly-foot is typical of a number of Brigalow Belt endemic reptiles, which are considered as at risk due to extensive clearing and development of this bioregion. Other threats include clearing of roadside vegetation, predation from feral animals, habitat disturbance from feral animals and livestock, and fire regimes, which alter the vegetation structure.

CURRENT RECOVERY EFFORTS

Recovery plan: Richardson R (2006) 'Queensland Brigalow Belt reptile recovery plan 2008–2012'. Report to the Department of the Environment, Water, Heritage and the Arts, Canberra. WWF-Australia, Brisbane.

Conservation actions: DERM manages recovery plans and has nominated the species as a 'critical' priority species within its threatened species management program 'Back on Track'. It funds and encourages voluntary conservation agreements on private land and manages protected areas with known populations.

FUTURE RECOVERY EFFORTS

What you can do:

- Join a local catchment or bushcare group and restore bushland in your local area.
- Consider entering into a voluntary conservation agreement to help protect threatened reptiles.

Conservation needs:

- Develop and implement species-specific land management guidelines for its known habitat.
- Encourage conservation agreements for the known populations on private land.
- Maintain and increase public awareness of the species.

Research directions:

- Survey the known and potential new habitat to identify additional populations.
- Monitor the known populations, especially those subject to habitat modification or loss.

Sources and further reading

Covacevich JA, Couper PJ and McDonald KR (1998) Reptile diversity at risk in the Brigalow Belt, Queensland. *Memoirs of the Queensland Museum* **42**, 475–486.

Kutt AS, Hannah DS and Thurgate NY (2003) Distribution, habitat and conservation status of *Paradelma orientalis* (Gunther 1876) (Lacertilia: Pygopodidae). *Australian Zoologist* **32**, 261–264.

Schulz M and Eyre TJ (1997) New distribution and habitat data for the pygopodid, *Paradelma orientalis* (Gunther, 1876) *Memoirs of the Queensland Museum* **42**, 212.

Shea GM (1987) Notes on the biology of *Paradelma orientalis*. *Herpetofauna* **17**, 5–6.

Tremul PR (2000) Breeding, feeding and arboreality in *Paradelma orientalis*: a poorly known, vulnerable pygopodid from Queensland, Australia. *Memoirs of the Queensland Museum* **45**, 599–609.

WWF and Queensland Murray-Darling Committee (2008) *Reptiles of the Brigalow Belt: An Information Kit*. WWF-Australia, Australia.

Author: Stephen Peck

LONG-LEGGED WORM-SKINK

Anomalopus mackayi Greer & Cogger, 1985

Photo: Eric Vanderduys

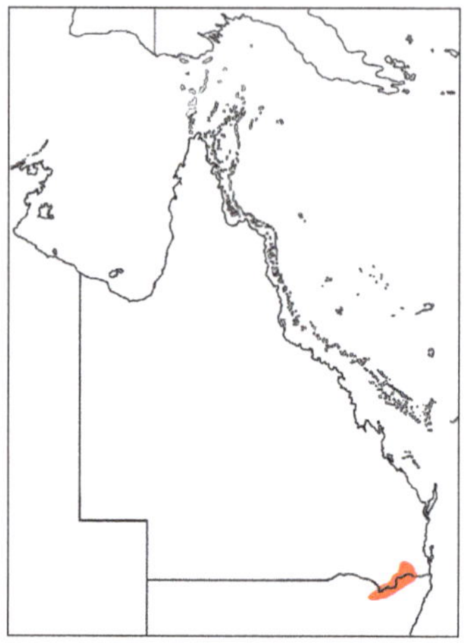

Family: Scincidae

Other common name: Five-clawed Worm-skink

Conservation status: IUCN: Vulnerable; Aust.: Vulnerable; NSW: Endangered; Qld: Endangered

Description: The Long-legged Worm-skink grows to a snout–vent length to 27 cm and is easily distinguished by having reduced limbs with three fingers and two toes. Its scales are smooth. The dorsal surface is grey-brown with rows of dark dots on each scale, giving a striped appearance. Some populations have reduced scale markings. The belly is yellow-green with dark flecking. The species may be confused with other reduced limb skinks found within its distribution.

Distribution: Found in the eastern Darling Downs area of the southern Brigalow Belt in Qld and the western slopes of the Dividing Range in NSW. There are no records from protected areas.

Habitat: The worm-skink has been found in eucalypt open woodland, *Callitris* woodland with a grass understorey and grassland on cracking black clay loam or sandy soils.

Ecology: This skink burrows and shelters under logs, rocks and loose soil. It lays one to three eggs in a clutch in spring. The diet is believed to be arthropods.

Reasons for decline and current threats: The skink has declined in range and threatened through destruction, fragmentation and modification of its habitat from agriculture and intensive irrigation, soil compaction due to overgrazing, loss of shelter habitat, such as fallen timber and logs, and the loss of ground litter from fires.

CURRENT RECOVERY EFFORTS

Recovery plan: Richardson R (2006) 'Queensland Brigalow Belt reptile recovery plan 2008–2012'. Report to the Department of the Environment, Water, Heritage and the Arts, Canberra. WWF-Australia, Brisbane.

Other plans: Cogger HG, Cameron EE, Sadlier RA and Eggler P (1993) 'The action plan for Australian reptiles'. Australian Nature Conservation Agency, Canberra.

FUTURE RECOVERY EFFORTS

Conservation needs:

- Implement land management practices that promotes the maintenance of secure, viable populations outside of reserve systems.
- Defer licenses to clear remnant woodland within the known range of this species.
- Use buffers to ensure no clearing of native vegetation in areas in which the species occurs.
- Develop and promote guidelines for landowners to help reduce the impact of current land use.
- Establish appropriate reserves for the species.
- Develop community awareness of the species.
- Ensure fire regimes maintain the species' habitat.
- Reduce the impact of grazing.
- Contain or eradicate weeds.
- Ensure that habitat and shelter exists to protect the species from prey.

Research directions:

- Undertake ground surveys to determine the full geographic range and habitat requirements of the species
- Research the basic biology and ecology of the species.
- Research the species decline and major factors behind the decline.
- Survey the known habitat in existing reserves on the eastern Darling Downs to determine if the Long-legged Worm-skink is present.

Sources and further reading

Greer AE and Cogger HG (1985) Systematics of the reduced-limbed and limbless skinks currently assigned to the genus *Anomalopus* (Lacertilia: Scincidae). *Records of the Australian Museum* **37**, 11–54.

Queensland Environment Protection Agency (Qld EPA) (2007). Long-legged worm-skink *Anomalopus mackayi* conservation management profile. <http://www.derm.qld.gov.au/register/p02321aa.pdf>.

Sadlier RA and Pressey RL (1994) Reptiles and amphibians of particular conservation concern in the western division of New South Wales: a preliminary review. *Biological Conservation* **69**, 41–54.

Shea GM, Millgate M and Peck S (1987) A range extension for the rare skink *Anomalopus mackayi*. *Herpetofauna* **17**, 16–19.

Author: Rod Hobson

YAKKA SKINK

Egernia rugosa De Vis, 1888

Photo: C Eddie

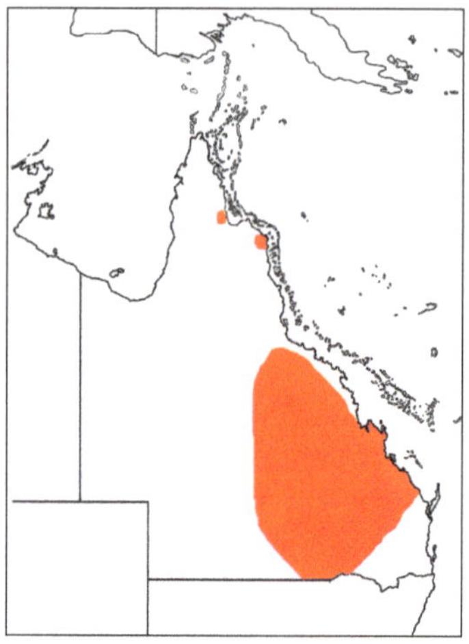

Family: Scincidae

Conservation status: IUCN: Least Concern; Aust.: Vulnerable; Qld: Vulnerable

Description: The Yakka Skink has a snout–vent length to 158–243 mm, with a tail length of 165–248 mm, and weighs 0.1–0.41 kg. It is a large brown, robust skink with a broad dark brown stripe down the back, bordered on each side by a straw-coloured line extending from the nape to the base of the tail. The underside is cream to yellow, with dark flecks, lines or spots on the throat and chin. The tail is thick, tapered and usually a little longer than the body. The iris is orange.

Distribution: It is patchily distributed from the McIlwraith Range south to the Qld/NSW border. Its core range is the Brigalow Belt and Mulga Lands bioregions.

Habitat: Yakka Skinks inhabit dry eucalypt and acacia woodlands and open woodlands. typically dominated by Narrow-leaved Ironbark *Eucalyptus crebra*, Dusky-leaved Ironbark *E. fibrosa*, Silver-leaved Ironbark *E. melanophloia*, Poplar Box *E. populnea*, Bulloak *Allocasuarina luehmannii*, Belah *Casuarina cristata*, White Cypress Pine *Callitris glaucophylla*, Mulga *Acacia aneura*, Bendee *A. catenulata*, Lancewood *A. shirleyi* or Brigalow *A. harpophylla*. Substrates include rock, loam, sand and clay.

Ecology: Yakka Skinks are terrestrial lizards that shelter in logs, abandoned rabbit warrens, rock crevices, roots of tree stumps and sink holes. They construct a burrow where there is suitable substrate. They bask close to their shelter site and are most active during morning and afternoon. Yakka Skinks bear live young but little is known about their breeding biology. Their diet includes invertebrates and fruit. They often use a regular defecation site. Yakka Skinks live in colonies that comprise adults and juveniles. They are capable of a quick retreat to their shelter site and are easily overlooked.

Reasons for decline and current threats: Habitat destruction (vegetation clearing) has contributed to the loss and fragmentation of many populations and localised contractions of range, particularly in the Brigalow Belt.

Yakka Skinks are predated by wild dogs, cats, foxes and pigs. In some parts of their range, particularly the Mulga Lands bioregion, feral goats and rabbits degrade Yakka Skink habitat.

Mining activities, especially open-cut coal mines and coal seam gas exploration, have expanded throughout the range of the Yakka Skink. Direct impacts include clearing of habitat, vehicle strike and accidental deaths when caught in pipeline trenches.

Populations occurring in roadside vegetation strips are susceptible to fire and potential disturbance from roadwork.

CURRENT RECOVERY EFFORTS

***Recovery plan*:** Richardson R (2006) 'Queensland Brigalow Belt reptile recovery plan 2008–2012'. Report to the Department of the Environment, Water, Heritage and the Arts, Canberra. WWF-Australia, Brisbane.

Other plans: Cogger HG, Cameron EE, Sadlier RA and Eggler P (1993) 'The action plan for Australian reptiles'. Australian Nature Conservation Agency, Canberra.

Conservation actions:

- Habitat containing Yakka Skinks has been purchased by non-government organisations (e.g. Australian Bush Heritage Fund, Indigenous Land Corporation and Australia Zoo) in the Brigalow Belt and Mulga Lands bioregions.
- Australia Zoo, Sunshine Coast, Qld has captive holding facilities.
- Brochures, booklets and species profiles have been produced by WWF-Australia, Threatened Species Network and Queensland Murray-Darling Committee (QMDC).
- Preparation of the recovery plan and co-ordination of recovery team meetings has been conducted by WWF-Australia in partnership with QMDC.
- Fine-scale habitat mapping is being undertaken at the Australian Centre for Sustainable Catchments (University of Southern Queensland).

- The Qld Museum has a public education display concerning threatened Brigalow Belt reptiles.
- The distribution and population dynamics of Yakka Skinks in the Mulga Lands is being investigated by DERM.

FUTURE RECOVERY EFFORTS

What you can do:

- Secure habitat, particularly in the Brigalow Belt and Mulga Lands, by supporting the purchase of properties for conservation purposes, particularly by NGOs.

Landholders:

- Negotiate a management agreement with DERM to protect key populations in vegetation remnants outside of reserves.
- Control pigs, cats, goats and foxes and supporting integrated pest-management activities.
- Leave timber piles unburnt and avoid removing logs, tree stumps, dead trees and rock piles from paddocks.
- Graze paddocks conservatively, maintaining ground cover, which ensures a source of invertebrate prey.
- Encourage corridors of native vegetation to regrow between remnant patches, allowing reptiles to disperse across the landscape.
- Implementing mosaic burns, which retain unburnt areas and a proportion of woody debris, leaf litter and ground layer vegetation.
- Inspect abandoned rabbit warrens for yakka skinks or their signs (e.g. tracks, scat piles, sloughed skin) prior to ripping.

Conservation needs:

- Conduct regular media campaigns and further disseminate educational material to raise public awareness of the plight of Yakka Skinks, particularly in the Brigalow Belt bioregion.
- Companies with mining tenements over areas containing Yakka Skinks or their habitat need to ensure that searches for skinks are routinely conducted by specialist observers as part of environmental assessments.
- Yakka Skink populations are to be avoided and protected where they are found; open pipeline trenches should be regularly checked and trapped animals removed by fauna specialists.
- Minimise habitat fragmentation by using existing clearings.

Research directions:

- Study all aspects of the biology and ecology of Yakka Skinks.
- Conduct targeted surveys throughout its range.

Sources and further reading

Cogger HG (2000) *Reptiles and Amphibians of Australia*. 6th edn. Reed New Holland, Sydney.

Covacevich JA (1996). Realities in the biodiversity holy grail: prospects for reptiles of Queensland's Brigalow biogeographic region. *Proceedings of the Royal Society of Queensland* **106**, 1–9.

Covacevich JA, Couper, PJ and McDonald, KR (1998). Reptile diversity at risk in the Brigalow Belt, Queensland. *Memoirs of the Queensland Museum* **42**, 475–486.

Department of the Environment and Heritage (2009). *Egernia rugosa*. In *Species Profile and Threats Database*. Department of the Environment and Heritage, Canberra. <http://www.deh.gov.au/sprat>.

Author: Craig Eddie

BARTLE FRERE BARSIDED SKINK

Eulamprus frerei Greer, 1992

Photo: Luke Shoo

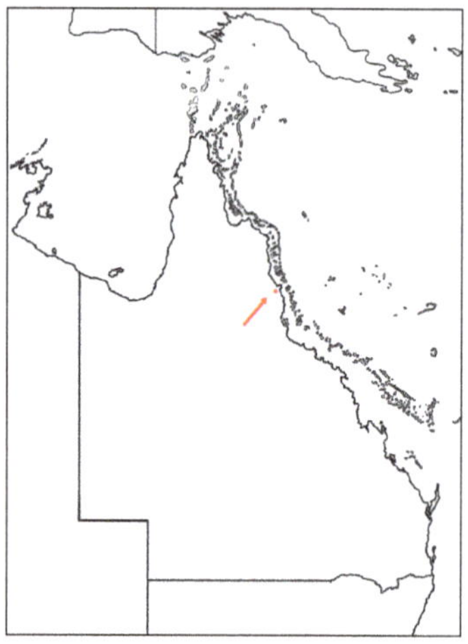

Family: Scincidae

Conservation status: Qld: Vulnerable

Description: The Bartle Frere Barsided Skink has a robust, medium-sized body, with a snout–vent length to 70 mm. The dorsal colouration is grey-brown, with a series of irregular, narrow, dark brown to black cross-bands or spots, without a dark vertebral stripe on the neck region. A dark brown upper lateral zone commencing as a dark streak below the canthus extends to the base of the tail. This zone has ragged edges and may be continuous or broken up into a series of large, black blotches on the hind portion of the body. The lower flanks are paler brown, cream, or white, liberally spotted and flecked with dark brown. The lips are barred with dark brown, while the ventral surface is whitish to yellowish. The chin shields are edged with dark brown and there is a series of dark brown flecks on the throat, which sometimes extend to the chest. The limbs are well developed and overlap when adpressed.

Distribution: The Bartle Frere Barsided Skink is believed to be restricted to the boulder-strewn heath habitats and rocky outcrops and peaks of the Mount Bartle Frere summit above 1400 m altitude: an area of approximately 400 ha.

Habitat: The Bartle Frere Barsided Skink is a restricted, Wet Tropics endemic that appears to be heavily reliant on rocks and boulders, but has also been observed using tree trunks as basking sites.

Ecology: The Bartle Frere Barsided Skink is a diurnal posturing/basking species. As for other members of the genus *Eulamprus*, this skink is presumably viviparous; gravid females have been observed in the field from December–January. There is currently no data on clutch or hatchling size. The Bartle Frere Barsided Skink is also presumed to have a diet that ranges from insectivorous to omnivorous.

Reasons for decline and current threats: The Bartle Frere Barsided Skink is a geographically localised species restricted within Wooroonooran NP, with no known current threats. Currently, the greatest potential threat to this species is climate change, which is expected to cause drier winter conditions in the Wet Tropics region. Drying may, in turn, result in an increase in wildfires, which will have a negative effect on populations in forested areas. Climate change may threaten the viability of hatchlings by exposing gravid females to extreme temperatures and drought conditions, in association with changes in rainfall patterns. This species is also adapted to a cool, stable climate and thus may be at risk from an increase in temperature extremes, particularly if individuals are unable to find thermally suitable refugia.

In addition, habitat alteration, in the form of die-back of rainforest trees, has been recorded from several sections of the summit of Mount Bartle Frere and may present a further threat under a climate change scenario by exposing individuals (e.g. new-born offspring) and populations to extremes in temperature.

CURRENT RECOVERY EFFORTS

Recovery plan: None

Conservation actions: The Bartle Frere Barsided Skink is in a national park managed by DERM.

FUTURE RECOVERY EFFORTS

Conservation needs: The impacts of camping sites and walking tracks in the area should be monitored and regular assessment made for feral animal and weed presence.

Research directions:

- Conduct surveys to determine accurate estimates of population size and distribution range and commence ongoing monitoring of this species.
- A recent review forecast that increases in temperature associated with global climate change is likely to impact most strongly on tropical reptiles from forested and desert regions of the world. To determine how susceptible mountain-top restricted tropical reptile species, including the Bartle Frere Barsided Skink, are to increases in temperature, it is imperative to establish key aspects of thermal biology. Pertinent data must include the

measurement of voluntary thermal maximum (VTMax), voluntary thermal minimum (VTMin) and the preferred body temperature range of the species. Other pertinent data for quantifying the susceptibility of this species to probable risks would be to determine the ability of the species to find and use suitable thermal refugia (e.g. rock crevices, burrows).

Sources and further reading

Goodman BA, Krockenberger AA and Schwarzkopf L (2007) Master of them all: performance specialization does not result in trade-offs in tropical lizards. *Evolutionary Ecology Research* **9**, 527–546.

O'Connor D and Moritz C (2003) A molecular phylogeny of the Australian skink genera *Eulamprus*, *Gnypetoscincus* and *Nangura*. *Australian Journal of Zoology* **51**, 317–330.

Turner G (2007) Notes on the habits of three skinks from the Mt Bartle Frere summit, north Queensland. *Herpetofauna* **37**, 2–7.

Williams SE (2006) *Vertebrates of the Wet Tropics Rainforests of Australia: Species Distributions and Biodiversity.* Cooperative Research Council for Tropical Rainforest Ecology and Management, Cairns.

Worboys S and Gadek PA (2004) 'Rainforest dieback: risks associated with roads and walking track access in the Wet Tropics World Heritage Area'. Report to the Wet Tropics Management Authority, Rainforest CRC, Cairns.

Author: Brett A Goodman

ORANGE-SIDED SKINK

Eulamprus luteilateralis (Covacevich & McDonald 1980)

Photo: www.ecosmartecology.com.au

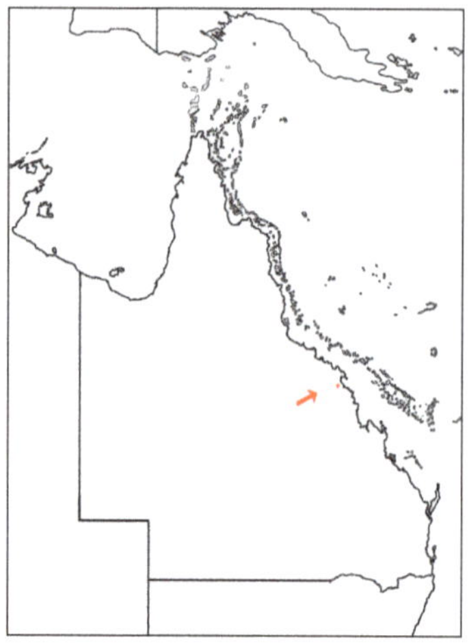

Family: Scincidae

Conservation status: IUCN: Least Concern; Qld: Vulnerable

Description: The Orange-sided Skink is a robust lizard with a snout–vent length to 92 mm. The dorsal surface is rich brown to bronze with darker flecks and there is a prominent black blotch above the forelimbs, and occasionally darker mottling on the neck and above ear. The flanks between the limbs are burnt orange with fine white spots. The belly is white. Juveniles are similar in colouration to adults, although the flanks are not as intensely orange.

Distribution: Known from above 900 m altitude in the Clarke Range west of Mackay, where currently it appears restricted to an area centred on Mount William, Mount David and Mount Dalrymple. The known extent of occurrence is only 19 km^2, but likely and possible unsurveyed habitat above 900 m suggests the skink's extent could potentially be up to around 85 km^2. Most of its extent is in Eungella NP.

Habitat: It is found in notophyll vine forest (rainforest) with palms on granitic ranges. The Orange-sided Skink is a terrestrial lizard of the rainforest floor where it takes refuge in rotting logs. It has been observed out during the day on palm trunks, palm fronds and among forest litter.

Ecology: The species is territorial, using tail wagging and fighting to ward off other Orange-sided Skinks. Juveniles encountered by adults are not attacked. The skink is a live bearer, usually giving birth to five young, which have a snout–vent length ranging from 38.8 to 45.3 mm when born (total length around 7 cm).

Reasons for decline and current threats: There is no known decline or threat, but the species has a very restricted distribution in upland rainforest habitat, which may be affected by stochastic events such as prolonged drought associated with climate change. Climate change modelling predicts a reduction in rainfall for this general area of Qld. Although Cane Toads, foxes, cats and pigs occur in the general region in which Eungella NP lies, there is currently no evidence suggesting these feral pests are affecting the skink.

CURRENT RECOVERY EFFORTS

Recovery plan: None

Conservation actions: No recovery efforts have been undertaken to date, but the skink's distribution is largely in Eungella NP, where it is protected from broad-scale clearing and other activities associated with private land use and management.

FUTURE RECOVERY EFFORTS

Conservation needs:

- Exclude or minimise development of facilities within Eungella NP in areas the skink is known or highly likely to occur.
- If at any stage feral pests are shown to be having an adverse impact on the skink, then management strategies to reduce these impacts will need to be developed, instigated and evaluated.

Research directions:

- Conduct regular monitoring of Cane Toads, foxes, cats and pigs.
- Determine the full distribution and abundance in the Clarke Range.
- Research the skink's natural history, in particular breeding biology, population structure and numbers, diet, movement patterns and physiological temperature and moisture limits. This information may be important as baseline data if managing climate change impacts becomes necessary.

Sources and further reading

Covacevich JA and McDonald KR (1980) Two new species of skink from mid-eastern Queensland rainforest. *Memoirs of the Queensland Museum* **20**, 95–101.

Winter JW and McDonald KR (1986) Eungella – the land of cloud. *Australian Natural History* **22**, 39–43.

Author: Adrian Borsboom

RETRO SLIDER

Lerista allanae (Longman, 1937)

Photo: Eric Vanderduys

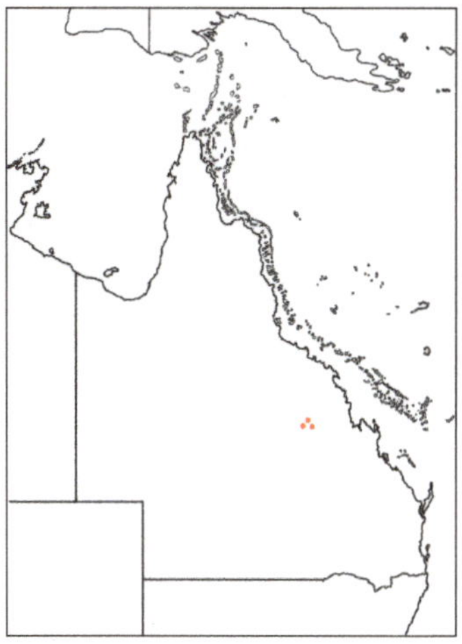

Family: Scincidae

Other common name: Allan's Lerista

Conservation status: IUCN: Critically Endangered; Aust.: Endangered; Qld: Endangered

Description: This reduced-limb lizard has a snout–vent length to 88 mm with a tail of the same length, if not longer. It weighs up to 5.7 g. The body is robust, the foreleg is absent, the hind leg has one clawed digit, the eyelid is movable and the postmental scale usually contacts two infralabial scales on each side. The dorsal surface and flanks are grey-brown with longitudinal lines of dark spots. The ventral surface is pale with dark edging to the scales on the chest, belly and tail, most prominent on the tail. The dorsal head scales are dark-edged and obscurely mottled. The face has a dark streak through the loreal and temporal scales and barring on the lips.

Distribution: Until recently the total knowledge base for the Retro Slider consisted of 13 museum specimens collected from three geographically proximate localities, Retro Station, Logan Downs and Clermont, on the black soil downs of Qld's central Brigalow Belt North bioregion. The known distribution, including recent records south-east of Clermont, is approximately 895 km^2. The area of occupancy (based on amount of suitable habitat) is expected to be significantly less. In 1996, evidence was provided to suggest this skink may be extinct, but a small population was recently found.

Habitat: Specimens have been collected from plains formed on Tertiary basalt, Permian shales, sandstone and unconsolidated sediments. When the Retro Slider specimens were collected in 1927 and 1936, this habitat supported open grasslands with scattered gum trees, moderately heavy groves of tea trees and an occasional bottle tree. Correspondence in the Qld Museum archives states that the skinks were found a few inches down in black soil under grass tussocks. However, this information may be erroneous because recent records are from friable, rich brown, surface soils and associated leaf litter.

Ecology: Almost nothing is known of the ecology of this species. However, based on the habits of morphologically similar congeners (the reduced-limb *Lerista* species), it is probably crepuscular to nocturnal and forages in the surface, humus-rich litter layer for small arthropods.

Reasons for decline and current threats: The probable decline of the Retro Slider has been attributed to habitat loss through broad-scale changes in land use practices.

CURRENT RECOVERY EFFORTS

Recovery plans: Richardson R (2006) 'Queensland Brigalow Belt reptile recovery plan 2008–2012'. Report to the Department of the Environment, Water, Heritage and the Arts, Canberra. WWF-Australia, Brisbane.

Other plans: Cogger HG, Cameron E, Sadlier R and Eggler P (1993) 'The action plan for Australian reptiles'. Australian Nature Conservation Agency, Canberra.

Conservation actions: The rediscovery of this species initiated a collaborative survey by staff from DERM and the Qld Museum. Survey results are detailed in a report provided to the Australian Government in 2010. Further survey work requires funding.

FUTURE RECOVERY EFFORTS

Conservation needs:

- Develop guidelines and incentives for landowners to reduce negative impacts.
- Establish appropriate reserves.
- Develop local community awareness and support.
- Encourage landholders to enter into voluntary conservation agreements with the Fitzroy Basin Association.

Research directions: Conduct surveys and monitoring to determine the current population extent and viability.

Sources and further reading

Borsboom AC, Couper PJ, Amey A, Hobson R and Wilson SK (2010) 'Rediscovery of the endangered Retro Slider (*Lerista allanae*) in the Clermont region of central Queensland'. Report. Queensland Department of Environment and Resource Management and the Queensland Museum, Brisbane.

Couper PJ and Ingram GJ (1992) A new species of skink of *Lerista* from Queensland and a re-appraisal of *L. allanae* (Longman). *Memoirs of the Queensland Museum* **16**, 49–67.

Covacevich JA, Couper PJ and McDonald KR (1996) *Lerista allanae* (Scincidae: Lygosominae): 60 years from exhibition to extinction. *Memoirs of the Queensland Museum* **39**, 247–256.

Greer AE and Couper PJ (1998) A further diagnostic character and some basic biological information for *Lerista colliveri*. *Memoirs of the Queensland Museum* **42**, 474.

Longman HA (1937) Herpetological notes. *Memoirs of the Queensland Museum* **11**, 165–168.

Slevin JR (1955) Notes on Australian amphibians. *Proceedings of the California Academy of Sciences* **4**, 355–392.

Authors: Patrick Couper, Andrew Amey and Adrian Borsboom

LEGLESS SLIDER

Lerista ameles Greer, 1979

Photo: Mark Hutchinson

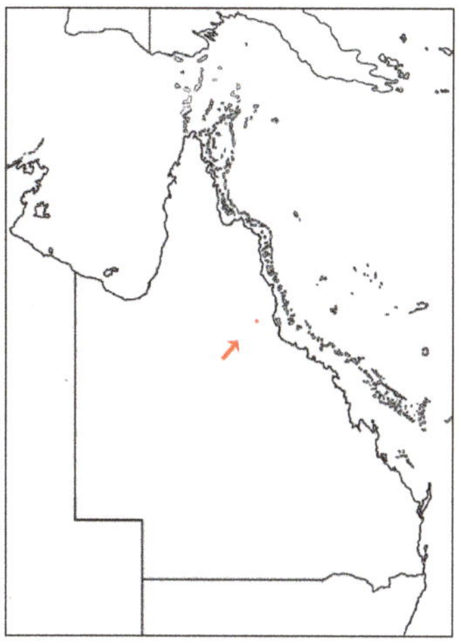

Family: Scincidae

Conservation status: Qld: Vulnerable

Description: A small, worm-like burrowing skink growing to around 110 mm with a snout–vent length to 58 mm, no limbs and a movable eyelid with a transparent disc. The dorsal surface is dark, silvery-grey to grey-brown with a series of dotted lines from neck to tail. The tail is grey, the flanks and belly grey-brown or grey.

Distribution: It is found in the Mount Surprise district of north Qld. The current best estimate of extent of occurrence is around 426 km^2, but with few records and little survey work for the skink it is possible its extent may be larger. Most records and two of three known sites for the skink are within the northern of four separate basalt areas that form the Undara-Toomba Basalts Province of the Einsleigh Uplands bioregion. The basalt section in which the skink occurs has a total estimated area of 6143 km^2. The skink's known habitat preferences would suggest its area of occupancy within its extent could be very limited.

Habitat: The Legless Slider has been found under a rock on a sparsely vegetated granite outcrop with low trees, shrubs and vines, under a basalt rock in open, grassy, eucalypt woodland and under a log on a sandstone escarpment.

Ecology: Little is known about the skink's ecology and general biology. *Lerista* species are considered to be burrowers of leaf litter and soft soils, although habitat where it occurs accumulates limited leaf litter. Its mode of reproduction is unknown, but is probably an egg-layer based on reproduction in other *Lerista* species. The role of fire in its ecology is unknown.

Reasons for decline and current threats: The level of decline in this species is unknown, although the type locality has been quarried and habitat where the skink has been recorded can be subject to heavy grazing pressure. Feral pigs may also be a threat because they are known to predate on small reptiles and, in areas where the skink occurs, can turn over large numbers of rocks in their search for food. Feral cats are mapped as widespread and common where the skink occurs, but their level of predation on the skink is unknown. However, predation may not be significant because of this burrowing skink's likely cryptic behaviour.

CURRENT RECOVERY EFFORTS

Recovery plan: None

Conservation actions: This skink has been recorded once from Undara NP, but is currently not specifically managed within the park.

FUTURE RECOVERY EFFORTS

Conservation needs: Exclude facility development within Undara NP from areas the skink is known, or highly likely, to occur. Manage these areas appropriately for fire, stray stock, cats and pigs if research shows them to be having an adverse impact on the skink.

Research directions:

- Conduct detailed surveys to determine the skink's distribution and abundance, especially in habitat similar to the few sites in which it has been recorded.
- Research the skink's natural history, in particular its breeding biology, diet, movement patterns and microhabitat use and preferences.
- Evaluate the role of fire in its ecology and, if required, instigate and evaluate appropriate fire management.
- Assess the impact of feral pigs and cats and, if required, implement management strategies to reduce any adverse impact.

Sources and further reading

Greer AE (1979) A new species of *Lerista* (Lacertilia: Scincidae) from northern Queensland, with remarks on the origin of the species. *Records of the Australian Museum* **32**, 383–388.

Author: Adrian Borsboom

MOUNT COOPER STRIPED SKINK

Lerista vittata Greer, McDonald & Lawrie, 1983

Photo: David Knowles

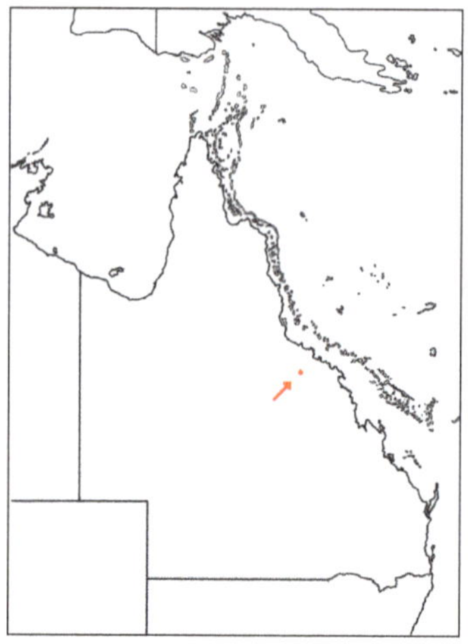

Family: Scincidae

Conservation status: IUCN: Endangered; Aust.: Vulnerable; Qld: Vulnerable

Description: A small, elongate skink lacking front limbs, with a single clawed digit on its short hindlimbs. The snout-vent length is up to 70 mm and the total length up to 120 mm. The dorsal surface is brown with four narrow black to dark brown stripes or lines of dots longitudinally, while each side has a broad black to dark brown band from snout to tail. Juveniles have a red flush to their tail. This species superficially resembles several other closely related species, from which it can be diagnosed by its head scale combination.

Distribution: The Mount Cooper Striped Skink was originally described from Mount Cooper Station, 90 km south-east of Charters Towers, north-east Qld and, until recently, was known only from there. A population of very similar skinks has recently been discovered at Blackbraes NP, approximately 250 km west of Charters Towers. Specimens from this population are indistinguishable in scalation from those from Mount Cooper, and seem to differ only in colour pattern. The taxonomic status of the Blackbraes population has not been evaluated. The Mount Cooper population occurs in an area of approximately 2000 ha, but the population size has not been estimated. The size and area of occupancy of the Blackbraes population are unknown.

Habitat: The Mount Cooper population is known only from dry rainforest, living in the leaf litter or under cover such as logs. This habitat is included in the 'Semi-evergreen vine thickets of the Brigalow Belt (North and South) and Nandewar bioregions', which is an EPBC Act-listed threatened ecological community. The literature and museum records suggest that several other species of *Lerista* may be heavily dependent on dry rainforest patches of north Queensland (*L. ameles*, *L. cinerea*, *L. rochfordensis* and *Lerista storri*) and there are indications that more species remain to be discovered and described. The Mount Cooper Striped Skink may not be the only *Lerista* isolated in a small patch of vulnerable habitat.

Ecology: Very little research on this species has been undertaken, but it is presumably similar to its congeners in being nocturnal or crepuscular and active in the leaf litter, feeding on invertebrates.

Reasons for decline and current threats: Habitat loss is the major threatening process for this species. Between collection of the original specimens and the description of the species, the only known habitat, a dry rainforest patch, was reduced from 6000 ha to 2000 ha to make way for Buffel Grass *Cenchrus ciliaris* pasture. The Mount Cooper Striped Skink is not known to occur in any protected areas, except for the western population at Blackbraes NP (should that population prove to be this species). Additional threats are any deterioration in habitat quality such as inappropriate fire regime, weeds or feral animals.

CURRENT RECOVERY EFFORTS

Recovery plan: Richardson R (2006) 'Queensland Brigalow Belt reptile recovery plan 2008–2012'. Report to the Department of the Environment, Water, Heritage and the Arts, Canberra. WWF-Australia, Brisbane.

Other plans: Cogger HG, Cameron EE, Sadlier RA and Eggler P (1993) 'The action plan for Australian reptiles'. Australia Nature Conservation Agency, Canberra.

FUTURE RECOVERY EFFORTS

Conservation needs:

- Implement the Brigalow Belt reptile recovery plan actions.
- Prevent clearing and protect dry rainforest, the only known habitat of the Mount Cooper Striped Skink, and a much-assailed habitat type.
- Develop community awareness.
- Establish reserves.
- Encourage land-owners to manage their land appropriately.

Research directions:

- A larger knowledge base is essential before the Mount Cooper Striped Skink can be properly managed.

- Conduct surveys to ascertain the current health of the Mount Cooper population and whether other populations exist.
- Survey known and potential habitat.
- Ecological studies are needed to determine the needs of this species with respect to micro-climate, fire regimes, habitat size, etc.
- Taxonomic work is needed to clarify the status of the Blackbraes population.

Sources and further reading

Amey AP and Couper P (2009) A new limb-reduced skink (Scincidae: *Lerista*) from the dry rainforest of north Queensland, Australia. *Zootaxa* **2173**, 19–30.

Department of the Environment, Water, Heritage and the Arts (2008a) *Lerista vittata* – Mount Cooper Striped Lerista. <http://www.environment.gov.au/cgi-bin/sprat/public/publicspecies.pl?taxon_id=1308>.

Department of the Environment, Water, Heritage and the Arts (2008b) Approved conservation advice for *Lerista vittata* (Mount Cooper Striped Lerista). <http://www.environment.gov.au/biodiversity/threatened/species/pubs/1308-conservation-advice.pdf>.

Environmental Protection Agency (2008) Mount Cooper striped lerista. <http://www.epa.qld.gov.au/nature_conservation/wildlife/az_of_animals/mount_cooper_striped_lerista/>.

Greer AE, McDonald KR and Lawrie BC (1983) Three new species of *Lerista* (Scincidae) from northern Queensland with a diagnosis of the *wilkinsi* species group. *Journal of Herpetology* **17**, 247–255.

Author: Andrew Amey

BLACK MOUNTAIN SKINK

Liburnascincus scirtetis (Ingram & Covacevich, 1980)

Photo: Eric Vanderduys

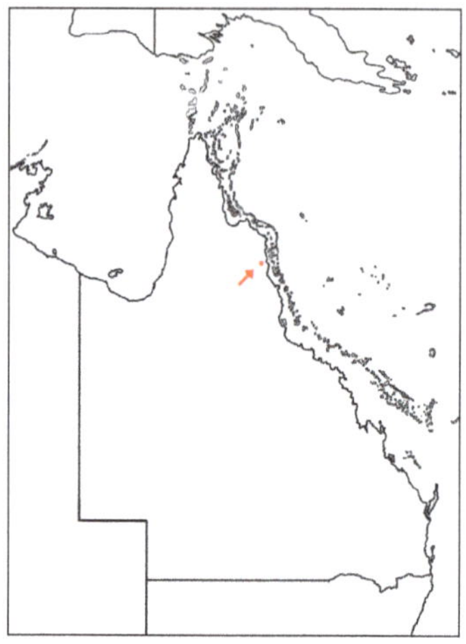

Family: Scincidae

Conservation status: IUCN: Least Concern; Qld: Vulnerable

Description: This small skink has a snout–vent length to 70 mm and weighs 4–6 g. The dorsal colour is uniform dark grey-black, often with distinctive cream-golden brown ocelli on a dark grey to black background, sometimes aligned to form a stripe-like pattern. The belly is white cream to dark grey. Dorsal scales are posteriorly rounded and have three low keels. The body shape is designed for living in crevices and on boulders with the head flattish, the snout slightly upturned and the limbs long, slender and well developed. It has four fingers and five toes. The ear opening is surrounded by sharply pointed lobules.

Distribution: The Black Mountain Skink is only found on the northern edge of the Wet Tropics bioregion, south of Cooktown. This skink is restricted to approximately 6 km^2 in the Black Mountain NP and adjacent Water Reserve of the Black Trevethan Range, between 100 and 300 m elevation. The Black Mountain Skink is currently secure in a small distribution within protected habitat, with no indication of population declines.

Habitat: The Black Mountain Skink is found only in the boulder field with scattered vegetation, mainly composed of figs.

Ecology: This agile skink lives in the boulder piles and is frequently seen in the early morning on the surface of granite boulders retreating to the shade on the lower sides of the boulders and crevices during the heat of the day. It can also be found in leaf litter and roots associated with figs. The diet of the skink consists of figs, small insects and spiders. One to two eggs are laid during the late spring and summer, with possibly more than one clutch per season.

Reasons for decline and current threats: No known decline or threat, but the species has a very restricted distribution in a specialised boulder habitat, which may be affected by stochastic events such as prolonged drought.

CURRENT RECOVERY EFFORTS

Recovery plan: None

Conservation actions: No recovery efforts have been undertaken to date, but the species distribution is mainly in national park.

FUTURE RECOVERY EFFORTS

Conservation needs: Add the area of boulder field in the Water Reserve to the adjacent national park.

Research directions: Research the natural history of the species, especially breeding biology, the physiological limits of temperature and moisture, and variation in microhabitat conditions within its habitat. This will be a good baseline for future climate change impacts.

Sources and further reading

Covacevich J, Ingram GJ and Czechura GV (1982) Rare frogs and reptiles of Cape York Peninsula, Australia. *Biological Conservation* **22**, 283–294.

Goodman BA (2004) Frugivory in the Black Mountain Rainbow skink, *Carlia scirtetis* Ingram and Covacevich, 1980. *Memoirs of the Queensland Museum* **49**, 700.

Ingram GJ and Covacevich J (1980) Two new lygosomine skinks endemic to Cape York Peninsula. In *Contemporary Cape York*. (Eds NC Stevens and A Bailey) pp. 45–48. Royal Society of Queensland, Brisbane.

Ingram GJ and Covacevich J (1989) Revision of the genus *Carlia* (Reptilia, Scincidae) in Australia with comments on *Carlia bicarinata* of New Guinea. *Memoirs of the Queensland Museum* **27**, 443–490.

Author: Brett A Goodman

MAGNETIC ISLAND SKINK

Menetia sadlieri Greer, 1991

Photo: Angus McNab

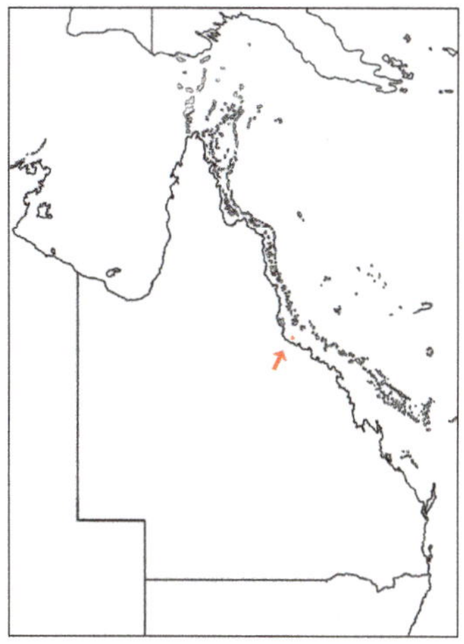

Family: Scincidae

Conservation status: Qld: Vulnerable

Description: The Magnetic Island Skink is a small, fossorial skink with a snout–vent length to 27.9 mm and total length to around 63 mm. It has four fingers and five toes. The eye is covered by a fixed spectacle and is therefore unable to blink. The dorsal surface is light brown and the sides are dark chocolate brown. The ventral surface is cream to white. It is very secretive and rarely seen above leaf litter. It is distinct from most common skinks on Magnetic Island by its lack of demonstrative behaviours; the Magnetic Island Skink does not wave its tail or sun itself obviously, as most *Carlia* spp. do.

Distribution: Known only from Magnetic Island off Townsville, north Queensland. All except two records, discussed below, were obtained from the lowlands of the island. Records are known from West Point, Bolger Bay Conservation Park, near Balding Bay, Unallocated State Land (USL) (near the upper reaches of Gustav Creek), and in suburban gardens in Nelly Bay and Arcadia.

Habitat: It is known primarily from the lowlands of Magnetic Island. All records are from areas with thick ground cover and/or leaf litter. Most records, including one specimen lodged with the Qld Museum, were obtained from one suburban backyard in Nelly Bay, bordering Gustav Creek. The yard contained extensive leaf litter and was well vegetated. The Magnetic Island Skink has also been recorded from mixed woodlands dominated by Poplar Box *Eucalyptus platyphylla*, Pink Bloodwood *Corymbia clarksoniana*, Weeping Tea Tree *Melaleuca leucadendron* and Pandanus *Pandanus whitei*, always with a thick ground cover of various grasses, typically including Spear Grass *Heteropogon contortus* and Kangaroo Grass *Themeda triandra* and herbaceous plants such as Flax Lilies *Dianella* spp., Native Wandering Jew *Commelina* spp., Polynesian Arrowroot *Tacca leontopetaloides* and Razor grasses *Scleria* spp. and Saw-sedges *Gahnia* spp. and often with leaf litter. The Magnetic Island Skink has also been recorded in deep bamboo litter in an ephemeral creek bed and in deep mango and Weeping Tea Tree litter in a low lying dry swamp. Two records away from the lowlands are known, as follows. The holotype was collected under a rock in dry rainforest growing in a granite boulder gully above Balding Bay (north-east side of the island). In addition to the granite boulders, this gully contained thick leaf litter and a ground layer of Saw-sedges and Spinifex *Triodia stenostachya*. The record from USL was from the access track that services the Nelly Bay header tank. The habitat contained granite boulders, thick leaf litter, Saw-sedges and Hoop Pine *Araucaria cunninghamii*.

Ecology: Very little is known of this species beyond its litter-dwelling habits. Small, litter-dwelling arthropods such as termites, cockroaches and small spiders are likely to be important prey. In captivity, these skinks feed readily on termites. The only reproductive data available is from a single female, captured gravid, which laid two eggs in early March. Both failed to hatch. Very small juveniles (snout–vent length of 14 mm) have been observed in November–December, suggesting birth in spring. The gravid female was captured in February, indicating that this species has either a protracted breeding season or the ability to multi-clutch, or both.

Reasons for decline and current threats: No known threats have been documented, but feral and domestic cats may pose a threat in and around urban parts of Magnetic Island, especially when combined with loss of habitat structural complexity. Nine of 14 individuals measured had lost and regrown part of their tail, suggesting predation pressure is high, but the cause is not known. Urban development, where older established yards are replaced by more manicured lawns and 'tidied up', with a consequent loss of ground cover vegetation and leaf litter, may pose a threat to the species. However, this is only supposition based on limited known locations for the species.

CURRENT RECOVERY EFFORTS

Recovery plan: None

Conservation actions: The Magnetic Island Skink has been recorded from protected areas and appears to be able to cope with some disturbance.

FUTURE RECOVERY EFFORTS

Conservation needs: Infrastructure management within the national park should be directed away from the known habitat of this species. However, in light of the current poor knowledge of the range and habitat requirements of this species on the island, the conservative approach of adopting a minimum footprint should be adopted in any inferred habitat.

Research directions:

- Conduct detailed surveys to determine the distribution of the species. Dominant regional ecosystems present on Magnetic Island should be surveyed, especially those occupying large portions of the island. Select areas of the nearby mainland, including coastal woodlands of Cape Cleveland, should also be surveyed for the species to determine if this skink is restricted to Magnetic Island.
- Research the ecology of the Magnetic Island Skink, especially breeding biology, food and microhabitat preferences within its habitat which will provide a good baseline for future impacts.

Sources and further reading

Borsboom A (2005) Nomination to re-classify the rare *Menetia sadlieri* to vulnerable under the *Nature Conservation Act 1992*. Biodiversity Sciences, Environmental Sciences Division. Environmental Protection Agency, Queensland Government.

Greer AE (1991) Two new species of *Menetia* from northeastern Queensland, with comments on the generic diagnosis of *Lygisaurus* and *Menetia*. *Journal of Herpetology* **25**, 268–272.

Vanderduys E (2005) Additional information on *Menetia sadlieri*, a poorly known skink from Magnetic Island, north Queensland. *Herpetofauna* **35**, 54–60.

Author: Eric Vanderduys

NANGUR SPINY SKINK

Nangura spinosa Couper, Covacevich & James 1993

Photo: Stewart Macdonald-UgMedia.com.au

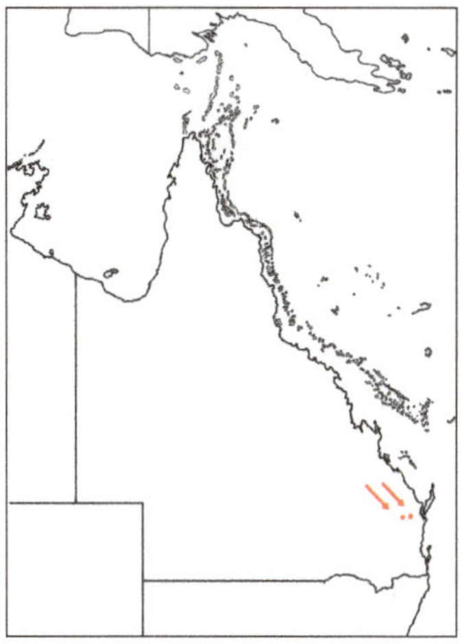

Family: Scinidae

Conservation status: Aust.: Critically Endangered; Qld: Endangered

Description: This skink has a snout–vent length of 15–102 mm, with a total length to about 190 mm, and weighs 1.5–30 g. The head scales are rugose with most other dorsal and lateral scales keeled; the keels most pronounced on the tail with those dorsally forming longitudinal lines. It has a brown dorsal surface with a series of weak, irregular, black bands from the nape to the tail base. The ventral surface is cream. The ear openings are large and the eyes dark.

Distribution: It is only known from two locations 36 km apart, which are situated west of Gympie in south-east Qld. Recent surveys have failed to find new populations at a number of sites with similar habitat near the known locations. The estimated extent in 2010 at one location is 868 ha and the second is only just over 7 ha. Based on limited density data, the estimated total adult population in 2005 was 183 (43 and 140, respectively, at the two known sites).

Habitat: It is recorded in semi-evergreen vine thicket on dark basaltic soils, and araucarian notophyll vine forest on Quaternary alluvials. It is also recorded in Hoop Pine plantations. The skink prefers a closed canopy with little vegetation at ground level.

Ecology: The skink uses single entrance burrows up to 60 cm long for refuge. Burrows are usually occupied by a one adult, but can be shared by up to five individuals, that can include a combination of young, sub-adults and/or an additional adult. Burrow entrances are normally associated with rocks, tree roots and bases, or open ground (usually on a slope or embankment). Outside the burrow entrance is often a smooth, bare earth resting platform regularly used by the skink. Some burrows have been in use for over 10 years, but it is unknown if use is by the same individual. In warmer weather the skink can be active diurnally and in the early evening at or near its burrow, but appears inactive inside its burrow in cooler months. Limited survey work suggests the skink forms small colonies. Little is known about the skink's breeding biology. Its reproductive mode is unknown, but it is suspected of giving birth to live young. Juveniles believed to be newly born have been found in late February to early March. Faecal pellets contain arthropods remains, mainly beetles and spiders.

Reasons for decline and current threats: Although there is no pre-clearing information on the skink's distribution, clearing has reduced and fragmented available habitat and has probably caused a major decline in the skink's numbers. At one site, around 89% of the original semi-evergreen vine thicket has been cleared and the remainder fragmented into 19 small blocks. The skink occurs in the largest remaining fragment (750 ha), which has also been selectively logged.

Hoop Pine plantation harvesting and replanting is scheduled to start in the next few years in plantation compartments where the skink occurs, and will be reliant on heavy machinery, which would kill the skink and destroy burrows. Where the skink is found on road and track-side banks and edges, inappropriate road maintenance by machinery could potentially kill a significant number.

Feral pigs, cats and foxes are present where the skink occurs. All will prey on small reptiles and have the potential to have a significant impact. Unless a regular and systematic program to control pigs is maintained, they will continue to degrade and disturb habitat, especially along drainage lines. Cane Toads are a suspected significant threat. The toad is widespread where the skink occurs and is known to use the skink's burrows as retreats. Cane Toads are voracious feeders of arthropods and therefore are potential competitors for food. The toad will take vertebrate prey, including lizards, so larger toads are a likely predator of the skink.

The shrubby weed Lantana *Lantana camara* is a suspected threat currently under investigation. It can form dense cover at ground level, which the skink may avoid. The flammability of lantana facilitates fire spreading into vine thickets, leading to death of fire-intolerant species and further Lantana invasion. The weed can form a monodominant shrub layer, suppressing forest succession processes.

CURRENT RECOVERY EFFORTS

Recovery plan: The draft national recovery plan was at the public consultation stage in 2010.

Conservation actions:

- Lantana control was carried out at one site for the skink in 2006 and was initiated at the second known site in August 2009.
- Monitoring of the skink's burrows commenced in 2005 and was funded to June 2010.
- Locked gates were installed in 2009 to prevent easy access to one site for the skink The gates will reduce the risk of illegal collecting and unnecessary disturbance of the skink.
- Because of legal requirements under the EPBC Act to protect this Critically Endangered skink, negotiations are underway to prevent scheduled harvesting and replanting of Hoop Pine plantations where it occurs, or to ensure these activities will have no impact.
- Recent research has found the two known populations to be genetically distinct and this information will be considered in the management of both populations.

FUTURE RECOVERY EFFORTS

Conservation needs: Lobby government for funding to initiate conservation measures and related research outlined in the draft national recovery plan, once the plan has been endorsed..

Research directions:

- Survey known habitat for potential new populations.
- Determine the population size and structure, and all aspects of reproduction, growth, maturation and diet.
- Determine the burrow spacing and use, and home ranges, movement and dispersal patterns.
- Detail the impact of Cane Toads other feral animals and weeds such as Lantana, and monitor the effectiveness of ameliorative measures.

Sources and further reading

Borsboom AC, Couper PJ, Amey A and Hoskin CJ (2010) Distribution and population genetic structure of the critically endangered skink *Nangura spinosa*, and the implications for management. *Australian Journal of Zoology* **58**, 369–375.

Covacevich JA, Couper PJ and James C (1993) A new skink, *Nangura spinosa* gen. et sp. nov., from a dry rainforest of southeastern Queensland. *Memoirs of the Queensland Museum* **34**, 159–167.

Hannah D, Agnew G, Hamley B and Hogan L (1997) New information on the narrowly restricted skink, *Nangura spinosa*. *Memoirs of the Queensland Museum* **42**, 90.

Author: Adrian Borsboom

BARTLE FRERE SKINK

Techmarscincus jigurru (Covacevich, 1984)

Photo: Luke Shoo

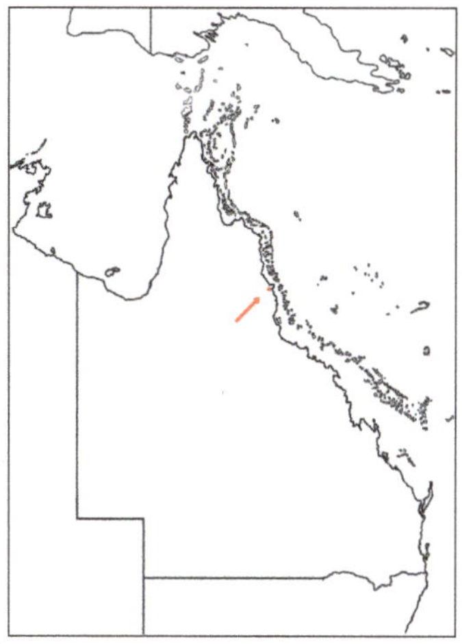

Family: Scincidae

Conservation status: Qld: Vulnerable

Description: The Bartle Frere Skink is a medium-sized lizard, with a snout–vent length from 35 to 80 mm, tail 36 to 172 mm and weighing 1.8 to 8.9 g. The dorsal surface is typically a rich metallic brown above, with scattered, dark brown flecks and streaks on the head. The back and tail have numerous scattered, pale brown and dark brown flecks and spots, with the dark spots often aligned longitudinally and the pale spots arranged transversely, except on the tail, where the dark spots predominate, often arranged transversely. A narrow, blackish streak runs from the eye to the snout, which continues behind the eye as a broad lateral line extending to the hind limb, and encloses a series of pale spots or blotches along its lower half. The ventral surface is metallic-cream or pale green. The adpressed limbs are overlapping.

Distribution: The Bartle Frere Skink is endemic to north-east Qld and restricted to altitudes above 1400 m on Mount Bartle Frere in an area of about 700 ha.

Habitat: The high altitude habitat on Mount Bartle Frere is a mosaic of three habitat types: low dense rainforest, rocky outcrops and boulder-strewn heath, with the Bartle Frere Skink most frequently observed on the boulder-strewn heath and on rocky outcrops.

Ecology: The Bartle Frere Skink is a diurnally active, basking species that possesses a number of morphological traits common to other rock-using species of skink. These include a flattened head and body, which are adaptations for rocky habitats. As a consequence, this species may experience a reduction in survivorship and fitness if forced to occupy other types of habitat. Although this species also uses open areas of bare ground and will bask on logs and tree trunks some distance from rocks and rocky outcrops, it appears less common in heavily forested areas, where rocks and boulders are absent. Recent work demonstrated that this species uses rocks in preference to other available microhabitats. The Bartle Frere Skink is oviparous and produces one to four eggs in November–December. Hatchlings emerge in January–February and have a snout–vent length to 27 ± 0.5 mm. Casual observations and scat analysis suggests that, like its closest relatives, the Bartle Frere Skink is primarily insectivorous.

Reasons for decline and current threats: The Bartle Frere Skink is a geographically localised species restricted to within Wooroonooran NP with no known current threats. The biggest potential threat to this species is climate change. Climate change is expected to cause drier winter conditions in the Wet Tropics region, which may result in an increase in wildfires, which will, in turn, have a negative effect on populations in forested areas. In addition, climate change may threaten the viability of hatchlings, by exposing nest sites to extreme temperatures and drought conditions, in association with changes in rainfall patterns. This species is also adapted to a cool, stable climate and thus may be at risk from an increase in temperature extremes, particularly if individuals are unable to find thermally suitable refugia.

In addition, habitat alteration, in the form of die-back of rainforest trees has been recorded from several sections of the summit of Mount Bartle Frere and may present a further threat under a climate change scenario by exposing both individuals (e.g. hatchlings) and incubating eggs in nests to extremes in temperature.

A recent review forecast that increases in temperature associated with global climate change are likely to have the highest impact on reptiles from warm environments, such as the tropics and desert regions of the world.

CURRENT RECOVERY EFFORTS

Recovery plan: None

Conservation actions: The Bartle Frere Skink is in a protected area managed by DERM.

FUTURE RECOVERY EFFORTS

Research directions:

- Monitor the impacts of camping sites and walking tracks in the area.
- Monitor and regularly assess the presence of feral animals and weeds.

- Study the population size and distribution range and commence ongoing monitoring of this species.
- Quantify key aspects of their thermal biology to determine how susceptible mountain-top restricted tropical reptile species, including the Bartle Frere Skink, are to increases in temperature. Pertinent data to include: the measurement of critical thermal maximum (CTMax), critical thermal minimum (CTMin), and the preferred body temperature range of several populations of this species. Other pertinent data for quantifying the susceptibility of this species to probable risks would be to determine its ability to find and use suitable thermal refugia (e.g. rock crevices and burrows).

Sources and further reading

Deutsch CA, Tewksbury JJ, Huey RB, Sheldon KS, Ghalambor CK, Haak DC and Martin PR (2008) Impacts of climate warming on terrestrial ectotherms across latitude. *Proceedings of the National Academy of Science USA* **105**, 6668–6672.

Goodman BA (2007) Microhabitat separation and niche overlap among five assemblages of tropical skinks. *Australian Journal of Zoology* **55**, 15–27.

Goodman BA (2009) Nowhere to run: the role of habitat openness and refuge use in defining patterns of morphological and performance evolution in tropical lizards. *Journal of Evolutionary Biology* **22**, 1535–1544.

Goodman BA, Miles DB and Schwarzkopf L (2008) Life on the rocks: habitat use drives morphological and performance evolution in lizards. *Ecology* **89**, 3462–3471.

Turner G (2007) Notes on the habits of three skinks from the Mt Bartle Frere summit, north Queensland. *Herpetofauna* **37**, 2–7.

Author: Brett A Goodman

ORNAMENTAL SNAKE

Denisonia maculata (Steindachner, 1867)

Photo: Mark Sanders

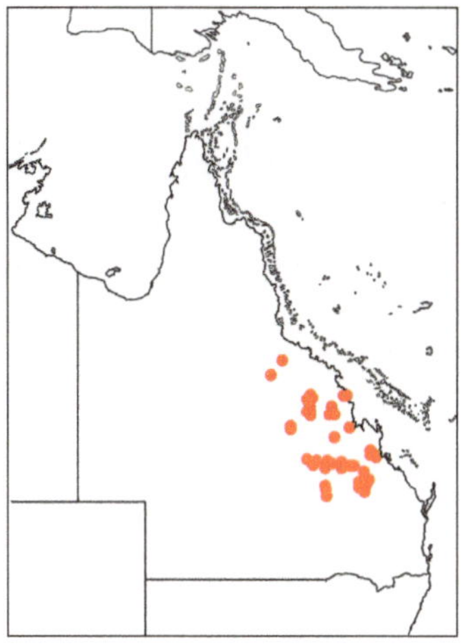

Family: Elapidae

Conservation status: IUCN: Vulnerable; Aust.: Vulnerable; Qld: Vulnerable

Description: The Ornamental Snake is a small, robust snake with a snout–vent length from 247 mm in females, and from 230 mm in males, to 500 mm. The dorsal colouration is grey-brown, brown or almost black. The lateral colours are lighter with darker streaks or flecks, sometimes with black hind edges. The ventral colours are white to cream with dark flecks on scale margins. The head is grey to dark grey with a narrow pale collar. The lips are barred. The body scales are smooth. This is a venomous snake and should be treated with care.

Distribution: The Ornamental Snake occurs south from about Charters Towers to the Dawson River valley and is most frequently encountered in seasonally flooded wetlands with deep cracking clay soils and gilgai, with or without Brigalow *Acacia harpophylla*, in the Fitzroy and Burdekin River basins. It is also recorded from the neighbouring Pioneer and Styx River basins. This distribution includes the eastern north and central Brigalow Belt bioregion and small parts of the Desert Uplands and Central Coast bioregions. Highest populations densities are from the Isaac River (Dysart-Moranbah-Nebo) and Dawson River catchments (Moura). The species is poorly represented in conservation reserves (Nuga Nuga and Dipperu NPs), and is typically found on rural lands during resource development projects.

Habitat: The Ornamental Snake is associated with moist or seasonally moist areas (flood plains, clay pans, water bodies) with appropriate shelter (deep cracking soils, fallen timber) and aquatic or fringing vegetation with seasonally abundant frogs. Records also occur in adjoining, slightly elevated lands and from cleared land. Snakes are found in a diverse range of natural cover (coarse litter, rocks) and human-made litter (galvanised iron sheets) as well as in fissures in cracking clay soils and, rarely, in homestead buildings. The species tolerates relatively simple habitat structures (grasslands, cleared paddocks, woody regrowth) where shelter and frogs are available.

Ecology: The Ornamental Snake is a live-bearing species, producing 3–11 offspring during the pre-wet season breeding season (September to November). This nocturnal snake requires daytime shelter. It feeds almost exclusively on frogs and congregates around ponds where frogs are abundant. Although snakes are encountered in a variety of seasons, they are mostly located while active in summer. Activity peaks within days of heavy summer rain, as frogs congregate to breed, and later as the young frogs emerge. The diet consists of many frog species.

Reasons for decline and current threats: A decline in this species is implied by the widespread destruction of suitable habitat – especially Brigalow dominated ecosystems supporting gilgai.

They are not usually protected beyond national park wetlands. Feral pig wallows and rooting, as well as trampling by cattle, and overgrazing degrade cover and disrupt the habitat for frogs.

Burning of grazing lands and post-clearing landscapes eliminates groundcover and simplifies the habitat structure. Open-cut coal mining and associated infrastructure eliminate local snake populations when within mine sites.

The Ornamental Snake is often absent or rare where Cane Toads are abundant, implying mortality from consuming toads. Although toad consumption has not been observed, the snake does strike adult toads, so it's likely that small toads will be consumed and the snake will die. Even if the snake is resistant to toad toxins, effects on its physiology will increase vulnerability to predation or environmental challenges. Degraded wetlands and clearing favour reproduction by toads over native species, increasing the likelihood of the Ornamental Snake encountering toads.

CURRENT RECOVERY EFFORTS

Recovery plan: Richardson R (2006) 'Queensland Brigalow Belt reptile recovery plan 2008–2012'. Report to the Department of the Environment, Water, Heritage and the Arts, Canberra. WWF-Australia, Brisbane.

Other plans: Cogger HG, Cameron EE, Sadlier RA and Eggler P (1993) 'The action plan for Australian reptiles'. Australian Nature Conservation Agency, Canberra.

Conservation actions: Habitat use and home range studies were completed in 2009 and 2010 respectively.

FUTURE RECOVERY EFFORTS

Conservation needs:

- Increase protected habitat within national park and covenanted private lands – especially in industrial buffer zones.
- Recover and protect degraded habitat.
- Encourage managers (via incentives) to implement land-use practices to protect wetlands and surrounding habitat.
- Promote public awareness of the needs of the species through industry sectors, the community and local government.

Research directions:

- The extent, general ecology and habitat characteristics of the Ornamental Snake are known, augmented by a study of habitat use and ranging behaviour in 2010. Many observations come from discrete monitoring for resource projects and few from within conservation tenure – reflecting the paucity of habitat protection. Consolidate and publish current knowledge, which is scattered among technical reports and unpublished observations.
- Conduct studies on the current status and conservation biology of this species.
- Monitor the current status throughout its range, including long-term monitoring sites, to ascertain population fluctuations and changes in distribution.
- Research the impacts of climate change (especially increased drought duration and severity). Conduct trials of habitat and population restoration and develop management guidelines for rural lands. Ornamental Snakes probably have limited ranging behaviour and, given their extensive distribution across multiple catchments, discrete genetic subpopulations are likely and require study before translocation is contemplated.

Sources and further reading

Department of the Environment, Water, Heritage and the Arts (2010) *Denisonia maculata*. In *Species Profile and Threats Database*. Department of the Environment and Heritage, Canberra. Available from: <http://www.environment.gov.au/sprat>.

Mackay R (1956) The status of *Denisonia maculata* Steindachner and related species (Ophida: Elapidae). *Proceedings of the Royal Society of New South Wales* **1954–1955**, 29–33.

Queensland Murray-Darling Committee (2008) *Reptiles of the Brigalow Belt: An Information Kit. Ornamental Snake*. WWF-Australia and Queensland Murray-Darling Committee. [Online]. Available from: <http://www.qmdc.org.au/publications/download/39/fact-sheets-case-studies/reptile-recovery/ornamental-snake.pdf> (Note: does not cover the full range of the species).

Shine R (1983) Food habits and reproductive biology of the Australian elapid snakes of the genus *Denisonia*. *Journal of Herpetology* **17**, 171–175.

Author: Alastair Melzer

DUNMALL'S SNAKE

Furina dunmalli (Worrell, 1955)

Photo: www.ecosmartecology.com.au

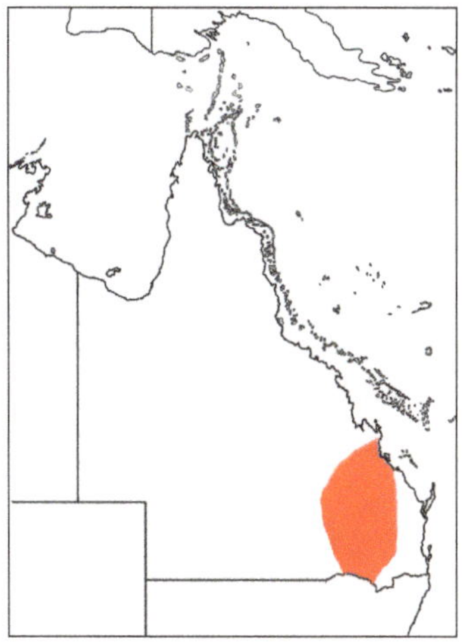

Family: Elapidae

Conservation status: IUCN: Vulnerable; Aust.: Vulnerable; Qld: Vulnerable

Description: A small snake with a total length to 750 mm. The dorsal surface is dark brown, grey or immaculate black with pale blotching on upper lips. The belly is white. The head is moderately depressed with a distinct neck and small black eyes. The scales are smooth and glossy. The mid-body scales number 21 and the ventral scales 175–190. The tail subcaudal scales are divided and range from 35–50. The anal scale is divided. The rostral scale is broader than deep and the frontal scale broadly contacting upper labial scales. The single preocular scale is widely separated from the nasal scale, which is divided by a large nostril.

Distribution: This species is confined to the Brigalow Belt bioregion of south-eastern Qld and north-eastern NSW. The range extends from Clermont and near Rockhampton in the north to the NSW border in the south. There are two recent records from adjacent NSW in the Texas area, but most records are from the Dalby-Tara area of the Darling Downs of south-east Qld.

Habitat: This snake is found in vegetation communities mainly dominated by Brigalow *Acacia harpophylla*, Cypress pine *Callitris glaucophylla*, Bulloak *Allocasuarina luehmannii*, Belah *Casuarina cristata* and dry sclerophyll forest, usually on black clay and clay loam soils. The majority of records are from between 200 and 500 m in altitude. It is known to exist in several protected estates within Qld including Tarong State Forest, Lake Broadwater Conservation Park, Erringibba and Expedition Range NPs.

Ecology: A secretive and nocturnal species, Dunmall's Snake is very rarely encountered and appears to be genuinely scarce. During daylight hours, it shelters under rocks, logs and in soil cracks, avoiding damp or wet areas. Detectability of the species appears to be increased after periods of heavy rain when soil cracks are filled with water. Nothing is known of its prey species except that the stomach of one individual contained a Tree Skink *Egernia striolata*. It is assumed that skinks and some geckos form a large proportion of its diet. Similarly, nothing is known regarding reproduction in Dunmall's Snake, but other species of *Furina* are known to be oviparous, with clutch sizes six to 10 eggs and a breeding season between October to February. Dunmall's Snake is aggressive when disturbed; it is not necessarily inclined to bite but regarded as potentially dangerous because the only recorded bite is said to have caused a severe reaction.

Reasons for decline and current threats: Due to the secrecy and rarity of this snake, it is not known if this species has actually declined in numbers, although records suggest a decline at least in the eastern part of its range. This snake's distribution, however, is confined to the Brigalow Belt: a bioregion that has been highly modified for agriculture, including clearing, pasture improvement and broad-acre cropping, as well as other industries such as timber extraction, natural gas and coal exploration and extraction, and urban development. Much of the habitat of Dunmall's Snake has been highly fragmented and modified especially in core areas such as the Darling Downs. Other threats identified for the species include predation by feral animals (foxes, cats and pigs) and inappropriate roadside management, because much of its core habitat only exists as linear fragments along roadsides and stock routes.

CURRENT RECOVERY EFFORTS

Recovery plan: Richardson R (2006) 'Queensland Brigalow Belt reptile recovery plan 2008–2012'. Report to the Department of the Environment, Water, Heritage and the Arts, Canberra. WWF-Australia, Brisbane.

Other plans: Cogger HG, Cameron EE, Sadlier RA and Eggler P (1993) 'The action plan for Australian reptiles'. Australian Nature Conservation Agency, Canberra.

FUTURE RECOVERY EFFORTS

What you can do:

- Become an advocate for reptile conservation and protection by joining an interest group that has a direct influence on reptile conservation such as a herpetological or natural history society, or an indirect influence such as a Landcare group.

- Do not collect timber from roadsides or reserves because this destroys shelter for wildlife including Dunmall's Snake.

Conservation needs:
- Develop guidelines and encourage landowners to reduce impacts on the habitat of this species outside the protected areas in line with recommendations in the Brigalow Belt Reptile Recovery Plan.
- Promote public awareness through community groups, schools, local media and websites.
- Encourage land managers to retain regrowth vegetation to connect patches of remnant woodland and consider entering into conservation agreements for remnant vegetation.
- Encourage land managers to implement sustainable grazing and fire management practices and to control feral predators.

Research directions:
- Survey known and potential habitat within the known range of the species.
- Research actions to determine if the population is declining and to determine the reasons for decline.
- Implement a long-term monitoring program to determine any changes in population size, habitat use and geographic range.
- Conduct scientific research into the basic biology and ecology of the species.

Sources and further reading

Queensland Murray-Darling Committee (2008) *Reptiles of the Brigalow Belt: An Information Kit. Dunmall's Snake.* WWF-Australia and Queensland Murray-Darling Committee, <http://www.qmdc.org.au/publications/download/39/fact-sheets-case-studies/reptile-recovery/dunmall's-snake.pdf>.

Shine R (1981) Ecology of Australian elapid snakes of the genera *Furina* and *Glyphodon*. *Journal of Herpetology* **15**, 219–224.

Stephenson G and Schmida G (2008) A second record of the elapid snake *Furina dunmalli* from New South Wales. *Herpetofauna* **38**, 22–23.

Worrell E (1963) A new elapine snake from Queensland. *Proceedings of the Royal Zoological Society of New South Wales* **1953–54**, 41–43.

Author: Rod Hobson

GREY SNAKE

Hemiaspis damelii (Gunther, 1876)

Photo: Stewart McDonald-UgMedia.com.au

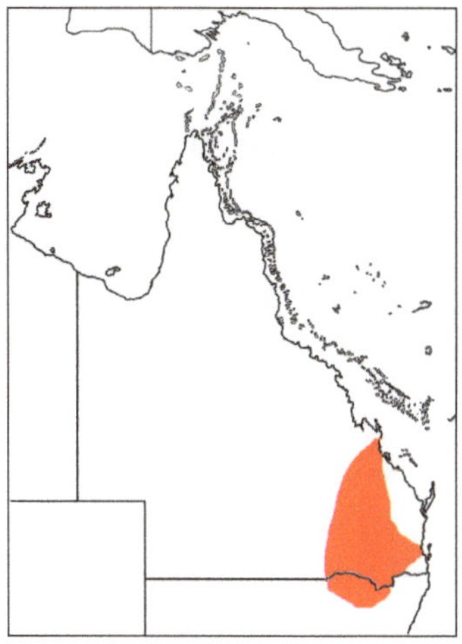

Family: Elapidae

Conservation status: Qld: Endangered

Description: A small snake with a snout–vent length to 700 mm. The body is grey to olive. The scales, especially along sides, may be tipped with black anteriorly. A dark band is present on the neck and base of head but the face is unmarked. The belly is yellowish, cream or white with occasional darker flecks. The juvenile has a dark neck band extended to top of head, which may disappear entirely in the adult. This snake has large eyes with round pupils and dark brown iris with pale upper edge. The body scales are smooth and the mid-body scales number 17. The ventral scales range from 140–170 and the subcaudal scales from 35–50, all single. The anal scale is divided.

Distribution: The Grey Snake occurs in central inland NSW, from the Macquarie Marshes area, with an isolated population on the Lachlan River, through south-east Qld and coastwards to Marburg in the Brisbane Valley and north to Rockhampton in central Qld. Largely confined to the Brigalow Belt, the snake is very sparsely distributed, but can be locally common especially in the Tara, Meandarra and Glenmorgan areas. It also known from the Lockyer Valley and Marburg area in south-east Qld and from protected estate in Qld at Lake Broadwater Conservation Park, Barakula State Forest, Southwood, Erringibba and Bendidee NPs.

Habitat: The Grey Snake has been found in moist areas along the eastern coast and seasonally inundated areas of the drier eastern interior on clay floodplains vegetated by grassland and open woodlands of Brigalow *Acacia harpophylla*, Belah *Casuarina cristata* and dry open eucalypt forest such as Poplar Box *Eucalyptus populnea*. It prefers cracking, flood-prone clay or loam soils and areas with gilgais (or melon holes) and shows some tolerance to cleared and altered habitats, but most records from cleared land have some degree of regrowth vegetation.

Ecology: This snake is usually found under fallen or embedded logs, flood debris or in abandoned burrows during the day, generally in the vicinity of water bodies. It is primarily nocturnal, with the highest activity around dusk, and feeds on frogs and occasionally lizards. Reproduction is viviparous, with females producing from four to 16 young annually between January and March. The neonates of *Hemiaspis* species are among the smallest of Australian elapids at 1.2 g. Females mature in a year and males approximately the same or slightly less. The snake hides its head under the body coils when confronted by a predator. The Grey Snake is not inclined to bite unless provoked or roughly handled and is not considered dangerous, though large individuals should be treated with caution.

Reasons for decline and current threats: The Grey Snake is threatened by destruction of habitat due to urban and rural development. Changes in hydrology, such as draining of wetlands, installation of bores, damming of watercourses and the levelling and cultivation of gilgais, have a negative impact on snake and prey species (frogs). Inappropriate burning that removes ground cover and destroys habitat such as logs and flood debris is also detrimental, as is grazing, which destroys ground cover and alters the soil structure through compaction and erosion. Watering stock, and feeding and wallowing by feral pigs, destroys wetlands through erosion and eutrophication. Pigs destroy the wetland habitat of the snake and its prey and directly compete for its food source (frogs). The spread of the Cane Toad has serious impact on frog-eating snakes, including likely sub-lethal effects that may make them more vulnerable to predation or other stresses.

CURRENT RECOVERY EFFORTS

Recovery plan: Richardson R (2006) 'Queensland Brigalow Belt reptile recovery plan 2008–2012'. Report to the Department of the Environment, Water, Heritage and the Arts, Canberra. WWF-Australia, Brisbane.

This plan is applied to an assemblage of Brigalow Belt bioregional reptiles of conservation significance including the Grey Snake. However, it does not cover the entire range of the species (i.e. in NSW).

FUTURE RECOVERY EFFORTS

What you can do:

- Become an advocate for reptile conservation and protection.
- Become active in, and support, wetland protection and rehabilitation.
- Join an interest group that has a direct influence on reptile conservation such as a herpetological or natural history society, or an indirect influence such as a landcare group.
- Do not collect timber from roadsides or reserves because this destroys shelter for wildlife, including Grey Snake.

Conservation needs:

- Encourage land managers to retain regrowth vegetation to connect patches of remnant woodland and consider entering into conservation agreements for remnant vegetation.
- Encourage land managers to implement sustainable grazing and fire management practices and actively control feral animals, especially pigs.

Research directions:

- Survey known and potential habitat within the known range of the species.
- Research actions to determine if the population is declining and, if so, to determine the reasons for decline.
- Implement a long-term monitoring program to determine any changes in population size, habitat use and geographic range.
- Conduct scientific research into the basic biology and ecology of the species.

Sources and further reading

Queensland Murray-Darling Committee (2008) *Reptiles of the Brigalow Belt: An Information Kit. Grey Snake.* WWF-Australia and Queensland Murray-Darling Committee, <http://www.qmdc.org.au/publications/download/39/fact-sheets-case-studies/reptile-recovery/grey-snake.pdf>.

Shine R (1987) Food habits and reproductive biology of Australian snakes of the genus *Hemiaspis* (Elapidae). *Journal of Herpetology* **21**, 71–74.

Swan G, Shea G and Sadlier R (2004) *A Field Guide to the Reptiles of New South Wales.* 2nd edn. Reed New Holland, Sydney.

Wilson S (2003) *Reptiles of the Southern Brigalow Belt.* WWF-Australia, Brisbane.

Author: Rod Hobson

Birds (Class Aves)

INTRODUCTION TO THREATENED BIRDS

David Parker and Dean Ingwersen

Given the fact that Queensland is the second-largest state in Australia, it is perhaps no coincidence that it is rich in avifauna. There are over 630 bird species considered as either resident, migratory or vagrant within Queensland – more than 70% of Australia's total. Only New South Wales and Western Australia come close in comparison, both of which boast around 550 species. This high diversity of avian species is a reflection of the wide variety of habitats, from the deserts and mulga country to the rainforests and coastal mangroves, as well as the physical location of the state. The tip of Cape York is only 150 kilometres from the megadiverse Papua New Guinea and, as a result, no other state list in the country is so heavily influenced by the trans-Torres Strait migrants such as the Red-bellied Pitta *Pitta erythrogaster.*

The Wet Tropics region is the richest spot within Australia for birds, with well over 400 species found there, including a number of species endemic to the region, such as the Mountain Thornbill *Acanthiza katherina* and Fernwren *Oreoscopus gutturalis*. Significant subspecies are also present in the north of the state, including the eastern population of the 'black-bellied' Crimson Finch *Neochmia phaeton*, and the 'white-bellied' Crimson Finch *Neochmia phaeton evangelinae* restricted to the Cape York region. Queensland is home to some of the most charismatic birds in the country, such as the Southern Cassowary *Casuarius casuarius johnsonii*, Golden Bowerbird *Amblyornis newtonianus* and Golden-shouldered Parrot *Psephotus chrysopterygius*. However, as a result of past management of the landscape, many of these species are now threatened or declining.

Under the EPBC Act, two species or subspecies occurring within Queensland are listed as Critically Endangered, 12 Endangered and 12 Vulnerable. These figures do not include the albatrosses, which are only rare visitors or vagrants to Queensland waters; however, the majority of albatross taxa visiting Queensland waters are threatened on the EPBC Act and the NCA. Eleven species or subspecies listed under Commonwealth legislation are not listed under state legislation. At the time of writing, Queensland's NCA listed 48 bird species or subspecies as threatened – about 7% of species occurring in the state. As well as the extinct Paradise Parrot *Psephotus pulcherrimus*, 25 species were listed as Endangered and 23 as Vulnerable. It should be noted, though, that a review of the Rare status (which is being phased out) of the NCA is occurring and any species listed as such are currently undergoing reassessment to determine which of the new categories most adequately describe their status.

Unlike the case with mammals in Australia, few birds have become extinct, but we may be on the verge of an extinction wave that could include several parrots (Night Parrot *Pezoporus occidentalis*, Orange-bellied Parrot *Neophema chrysogaster*, Western Ground Parrot *Pezoporus flaviventris*, Golden-shouldered Parrot, Carnaby's Black-Cockatoo *Calyptorhynchus latirostris* and possibly even the Swift Parrot *Lathamus discolor*), among other species. The 'Action Plan for Australian Birds 2000' lists eight species as extinct, but only one of these occurred on the mainland – the Paradise Parrot. The distribution of this species once extended from southern-central Queensland to south-east Queensland and, according to some accounts, northern New South Wales. Across this range, the species inhabited open eucalypt woodland and open forest with a grassy understorey. The presence of termite mounds, used for nest sites, was also a crucial habitat requirement. A combination of land clearing, grazing by livestock, changed fire patterns and the early settlers' penchant for tennis courts made of crushed termitaria, all contributed to the loss of the Paradise Parrot. Trapping for aviaries, egg collection, the spread of invasive introduced plant species and drought further contributed to the eventual demise of the species. The last confirmed sighting of this species existing was back in 1927. Searches since have failed to provide any form of conclusive evidence to suggest that the species still hangs on in an isolated, human-free pocket of habitat. As if to hold on to a glimmer of hope, though, the Paradise Parrot was only listed as Extinct in the Wild in Queensland under the NCA in 2005. Only one subspecies, the Southern Star Finch *Neochmia ruficauda ruficauda*, has been lost from Queensland to date, and some species lost from other states persist in Queensland today. Some species, such as the Critically Endangered Night Parrot, are only just hanging on, though.

Threats

Queensland's threatened birds continue to face ongoing pressures and demands on their survival, with the main threats to both terrestrial and wetland birds being land clearing and ongoing insidious land degradation. Despite these facts having been long acknowledged, Queensland cleared more land in the second half of the 20th century than it had in the preceding 150 years, including three million hectares cleared under the Australian and state government's Brigalow Scheme. Despite stricter clearing regulations in recent years, land clearing has continued on a smaller, yet still devastating, scale, especially in the south-east corner of Queensland. These incremental losses continue to place stresses on the bird fauna.

Habitat loss, degradation and fragmentation have been, and remain, a major threat to the survival of species of woodland bird such as the Regent Honeyeater *Anthochaera phrygia* and Swift Parrot. The distribution of the Regent Honeyeater has contracted greatly since European settlement, and it has been estimated that around 75% of the species' habitat has been cleared to make way for agriculture and urban development. Today, this species is thinly dispersed throughout its present range, with an estimated total national population of between 350 and 400 birds. Protecting remnant habitat on both public and private land, and undertaking habitat restoration activities across its range, are essential management actions required to protect this species, and many others like it. For the Red Goshawk *Erythrotriorchis radiatus*, the loss of coastal and subcoastal forests, to make way for urban development and agriculture, has seen the species almost disappear from northern New South Wales and the coastal lowlands of eastern Queensland. Although population estimates are difficult to determine, owing to its sparse and discontinuous distribution, it is estimated that as few as 1400 birds may remain in the wild across the whole of northern Australia.

The influence of grazing regimes has been shown to affect granivorous birds such as the Endangered Plains-wanderer *Pedionomus torquatus* and Gouldian Finch *Erythrura gouldiae*. The overgrazing of native grasslands by domestic livestock and rabbits, especially during drought conditions, can result in the temporary displacement of the Plains-wanderer from its preferred habitat on a local or regional scale. An increase in the rate of mortality may also be experienced during these times, as a result of loss of habitat and also the reduction of food availability. Conversely, the absence or lack of grazing may also have a negative effect on the species, particularly when followed by widespread rainfall and prolific grass growth, though this impact is less significant, particularly in arid regions. Wet season grass species such as Cockatoo Grass *Alloteropsis semialata* and Golden Beard Grass *Chrysopogin aciculatus* – species favoured by the Gouldian Finch – are selectively grazed by cattle and horses, leading to a reduced incidence and extent of these species. Reduced seed production has an impact on both the food resources of the finch and the recruitment of these grasses.

A suite of ground-dwelling birds, such as the Eastern Bristlebird *Dasyornis brachypterus*, are considered to have declined primarily as a result of a change in fire regimes, exacerbated by land clearing and, possibly, predation. For the Eastern Bristlebird, open woodlands with dense, grassy undergrowth near rainforests are its preferred habitat. These woodlands contain native plants such as tussock grasses, small shrubs, ferns and vines. Fire is necessary to maintain this habitat and, in the absence of fire, trees become very dense and shade out the grass. Conversely, too-frequent fires can result in vegetation that is not dense enough for the Eastern Bristlebird to nest in and feed beneath. A fire during the breeding season (July to October) can have devastating impacts on the species. Although the adults may take refuge in the nearby rainforest, eggs and nestlings can be destroyed. Finding the right balance of fire is crucial to the future of the species.

Predation by foxes and feral cats, competition with native species and land degradation by feral rabbits, pigs and goats, are all listed as Key Threatening Processes under the Australian Government's EPBC Act. Introduced predators such as the fox and feral cat are a threat to ground-nesting species such as the Plains-wanderer *Pedionomus torquatus*. Having not evolved with these predators, these ground-dwelling species have no strategies for evasion and as a result make easy prey. The Night Parrot may have declined to near extinction as a result of, among other things, nest predation by cats and foxes. In the early 1990s, the feral cat population in the Diamantina region of south-west Queensland grew rapidly after good rains. This region supports the Night Parrot and the Plains-wanderer. Over a 2-week period, more than 500 feral cats were shot on Diamantina Downs.

The feral pig is another introduced species that can have a negative impact on our native birds. The damage caused to natural habitats through their wallowing and rooting for food in watercourses and swamps destroys the native plants that provide food and nesting sites for waders and other waterbirds (e.g. Australian Painted Snipe *Rostratula austra-*

lis). Likewise, feral goats have had a major impact on native plants and the habitats of our native fauna.

Birds that require tree hollows for nesting, such as the Glossy Black-Cockatoo *Calyptorhynchus lathami*, have suffered greatly from loss of hollows through land clearing and timber production. The loss is now being increasingly exacerbated as a result of hollow invasion by feral honeybees *Apis mellifera*, which use hollows as hives. This prevents access by nesting birds. Competition for food resources by honeybees may also be affecting native birds such as the Swift Parrot, and requires further investigation.

Finally, an even greater threat now looms over these already declining species. Human-induced climate change brings an uncertain future, with predicted temperature increases and alterations in rainfall patterns across Australia. In Queensland, it is predicted that climate change will affect the timing of the wet season. Such changes may influence the frequency and intensity of wildfires, or may favour some plant species, such as grasses, over others that may be more palatable to dependent species. Availability in surface water during the dry season may also affect the persistence of bird populations in some areas. There is also a concern that the shift in mean temperatures will literally push some species, such as the Mountain Thornbill, off the cliff. Predicted increases in temperature will shift upland rainforest species higher up their mountain ranges until such time as the temperatures become suboptimal and persistence of the birds is impossible. Any rise in sea level with increased global temperature will also have an impact on coastlines and offshore islands, and will likely disrupt or ruin breeding attempts in a variety of nesting colonies such as those that occur on Michaelmas Cay or Raine Island.

Is there hope for our threatened bird species?

Having survived for thousands of years and adapted to dramatic changes to the Australia climate, our threatened species continue to battle numerous threats. For some, such as the southern race of the Star Finch, their decline and subsequent disappearance from areas commenced before the process of decline could be described. Thankfully, we understand many of the reasons for decline now. The overwhelming issue now is one of funding of recovery, enforcement of legislation and compliance monitoring. Given the situation we're currently in, these issues have not been adequately dealt with, but hopefully policy shifts in the future will see recovery actions undertaken and species recovering their former status.

CASSOWARIES AND EMUS (ORDER: CASUARIIFORMES)

SOUTHERN CASSOWARY

Casuarius casuarius johnsonii F. Müller, 1866

Photo: www.ataglance.com.au

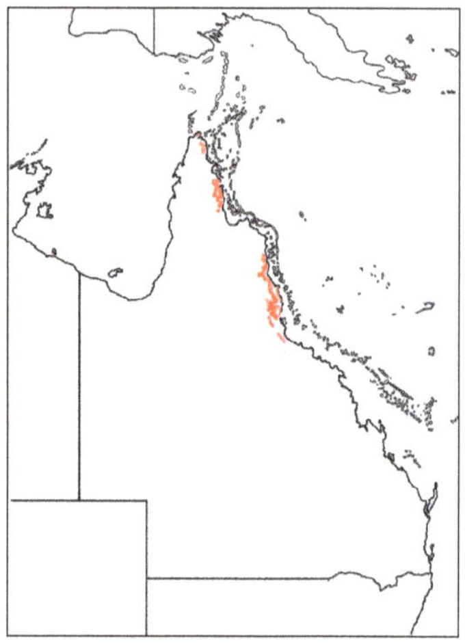

Family: Casuariidae

Other common name: Double-wattled Cassowary

Conservation Status: IUCN: Vulnerable; Aust.: Endangered; Qld: Endangered (Wet Tropics population); Vulnerable (northern population)

Description: This tall, striking, flightless bird has a total length of up to 200 cm and weighs 35–75 kg, with females being larger than males. Its pale blue head is crowned by a brown casque, the featherless neck is blue on the sides and front and orange to purple on the back. Bright red wattles hang from the lower part of the bare neck and the body is covered in glossy black feathers. Its thick legs look scaly and each of three toes on its large feet has a strong nail, the inner being up to 12 cm long and dagger-like. Chicks have longitudinal stripes and subadults are uniform brown.

Distribution: Occurs in Papua New Guinea, neighbouring islands and in northern Australia where it has three more or less isolated populations including far northern Cape York Peninsula on the McHenry Uplands (and previously Lockerbie Scrub), in the McIlwraith and Iron Ranges, and the Wet Tropics from Big Tableland near Cooktown to Paluma near Townsville. The two Cape York populations are listed collectively as the northern population. The population within the Wet Tropics shows a degree of natural fragmentation, with subpopulations being partially separated by natural barriers such as river gorges and habitat discontinuities. This natural fragmentation has been increased by human disturbance.

Habitat: In Australia, it occurs in and near rainforest from montane rainforests >1000 m altitude to lowland tropical vine forest, but it is by no means restricted to rainforest, using the full habitat mosaic available in any area. This may include *Melaleuca*, palm and mangrove swamps, beaches, savannahs and grasslands and, frequently, production landscapes and suburban gardens situated near forest.

Ecology: The Southern Cassowary is solitary, but maintains social contact with neighbours through vocalisations and visits. Adults may feed together when fruits are abundant or during the mating season, which occurs from June to September. Males incubate the eggs and raise the young. Females may mate and lay eggs with several males each season. Cassowaries eat mostly fruit – recorded from 240 species so far – but they also consume invertebrates, flowers, fungi and carrion. They are excellent seed dispersers: one of only a handful of vectors that can handle large-seeded fruits and provide them with occasional long-distance dispersal. Although frequently characterised as a keystone species, recent work with the whole community of seed dispersers has demonstrated they are not.

Reason for decline and current threats: Although no satisfactorily rigorous method of estimating population size has yet been developed or implemented, the Southern Cassowary has been listed on the basis that it is believed to have declined and it could decline further. A rapid decline over only a few generations can be inferred from the extensive clearing of forests on the coastal lowlands and Atherton Tablelands in the Wet tropics, which probably constituted habitats with the highest carrying capacity for the species. However, this rate of clearing has virtually ceased and ~90% of remaining habitat is now protected. It is currently uncertain whether populations are still in decline or stabilised. Measures to tackle the remaining threats – the effects of fragmented habitat; habitat degradation such as by the invasion of weeds and feral animals; vehicle strikes; dog attacks; human interactions that alter Cassowary behaviour and increase risks to these birds; hunting; and possibly disease, such as aspergillosis and avian tuberculosis – require further implementation to continue recovery of the species.

CURRENT RECOVERY EFFORTS

Recovery plan: Latch P (2007) 'National recovery plan for the Southern Cassowary *Casuarius casuarius johnsonii*'. Report to Department of the Environment, Water, Heritage and the Arts, Canberra. Environmental Protection Agency.

Other plans: Department of the Environment, Water, Heritage and the Arts (2010) Significant impact guidelines for the Endangered Southern Cassowary (*Casuarius casuarius johnsonii*). EPBC Act Policy Statement 3.15.

Garnett ST and Crowley GM (2000) 'The action plan for Australian birds 2000'. Environment Australia, Canberra.

Garnett ST, Szabo JK and Dutson G (2011) *The Action Plan for Australian Birds 2010.* CSIRO Publishing, Melbourne.

Conservation actions: A large investment has been made into research on the species, including into its ecology and ecological role, capture, transport, husbandry, disease and veterinary care, identification of individuals, estimation of population size, cultural and social impacts and interactions, and its negative impacts as a disperser of weeds. Numerous organisations and individuals, government and private, have devoted and continue to devote considerable effort and resources to:

- acquiring and protecting habitat;
- establishing legislative and planning tools;
- controlling threats, restoring habitat and creating corridors;
- providing information for and on the Southern Cassowary;
- integrating habitat into suburban and agricultural landscapes;
- controlling the detrimental aspects of human–cassowary interactions and mitigating the effects of urban expansion.

FUTURE RECOVERY EFFORTS

What you can do: You can help by getting involved in restoration and conservation activities already established in your local area (resources and direction can be obtained from the many government and NGO websites that include cassowaries).

Conservation needs:

Measures to conserve the Southern Cassowary are extremely advanced. Future measures should:

- Continue to focus on the wide range of strategies and directions already underway, many of which require long-term investment.
- Protect and restore habitat, connect habitat fragments and create corridors, as well as those that minimise human-induced mortality.

Research directions:

- Develop sound survey methods to estimate population sizes and trends.
- Estimate recruitment and mortality rates and the age structure of populations.

Sources and further reading

Bradford MG and Westcott DA (2010) The consequences of southern cassowary (*Casuarius casuarius*, L.) gut passage and deposition pattern on the germination of rainforest seeds. *Austral Ecology* **35**, 325–333.

Bradford MG, Dennis AJ and Westcott DA (2008) Diet and dietary preferences of the Southern Cassowary (*Casuarius casuarius*) in North Queensland, Australia. *Biotropica* **40**, 338–343.

Crome FHJ and Moore LA (1990) Cassowaries in north-eastern Queensland: report of a survey and a review and assessment of their status and conservation and management needs. *Australian Wildlife Research*. **17**, 369–385.

Dennis AJ and Westcott DA (2007) Estimating dispersal kernels produced by a diverse community of vertebrates. In *Seed Dispersal: Theory and Its Application in a Changing World.* (Eds AJ Dennis, EW Schupp, PJ Green and DA Westcott) pp. 201–228. CAB International, Wallingford, UK.

Westcott DA, Bentrupperbäumer J, Bradford MG and McKeown A (2005) Incorporating patterns of disperser behaviour into models of seed dispersal and its effects on estimated dispersal curves. *Oecologia* **146**, 57–67.

Author: Andrew J Dennis

TROPICBIRDS (ORDER PHAETHONTIFORMES)

RED-TAILED TROPICBIRD

Phaethon rubricauda Boddaert, 1783

Photo: Ian Gynther

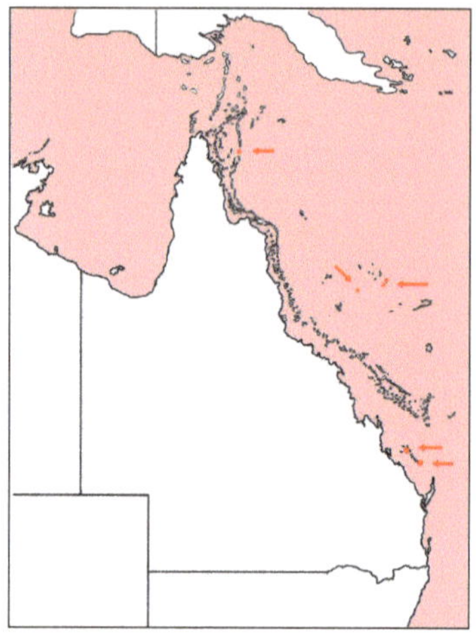

Family: Phaethontidae

Other common names: Red-tailed Bosunbird, Silver Bosunbird

Conservation status: IUCN: Least Concern; Qld: Vulnerable; NSW: Vulnerable

Description: The Red-tailed Tropicbird is the largest of the tropicbirds, measuring 95–104 cm in total length, including the long red tail-streamers (about 35 cm), with a wingspan of up to 119 cm. Adults are white with a black comma-like facial stripe through the eye, black shaft-streaks on the outer primaries, black chevrons on the tertials and dark greyish streaks on the flanks and axillaries. The bill is red. The body feathers have a varying degree of pink shading. Juvenile and immature birds have a black or grey bill, no central tail projections and are heavily barred and scaled black on the upper surface.

Distribution: The Red-tailed Tropicbird is wide-ranging in tropical waters of the Indian and Pacific Oceans. In Australia, it occurs from south-west WA across northern Australia, including through Qld waters to NSW, with occasional records from southern states. Large offshore breeding colonies include Christmas, Lord Howe and Norfolk Islands. Off Qld, the largest breeding colony is on North-East Cay and South-West Cay of the Herald Group in the Coral Sea, with smaller colonies on other cays of the Herald-Coringa Group, and Raine and Lady Elliot Islands in the Great Barrier Reef system.

Habitat: The Red-tailed Tropicbird is a pelagic species with a preference for tropical oceans with sea-surface temperatures of 24–30°C. It breeds largely on islands, including atolls and cays.

Ecology: The breeding season is highly variable between locations, and breeding may take place solitarily, as pairs, or in loose colonies. Nests are shallow scrapes under bushes. North-East Cay, followed by Raine Island (where breeding has been recorded during all months) support the largest colonies off Qld, with Lady Elliot Island supporting a much smaller colony. Considerable transoceanic movements have been documented for the species, with a bird banded as a chick in south-west Australia recovered almost 6000 km away on Réunion Island in the western Indian Ocean as a non-breeding bird. At sea, the Red-tailed Tropicbird is usually solitary, feeding primarily on fish and cephalopods.

Reasons for decline and current threats: A number of threatening processes affect Red-tailed Tropicbirds, including breeding habitat loss due to clearing and mining activities, hunting and predation by invasive species.

The Australian population of the species is estimated at 4000 breeding birds, with a likely continuing decline (as has been seen in south-west Australia, Christmas Island and Raine Island). Nest failure and breeding colony desertion have been attributed to egg and nestling predation by introduced species (rats and cats), predation by unnaturally high populations of Silver Gulls *Chroicephalus novaehollandiae*, which result from human habitation, human interference of nest sites and mining disturbance. Not all of these threats operate at colonies off Qld and the main colonies are relatively free from human disturbance. However, a decline of close to 40% in the population on Raine Island has been noted between 1979–1993 and 1994–2003. Over the same period, there was no evidence of habitat loss or deterioration, or of significant human disturbance, suggesting that broader environmental conditions may be at play. Possible explanations include the depletion of marine food resources linked to changing climatic and oceanographic conditions and/or fishing activities. Nesting sites off Qld may also be threatened by predicted sea level rise.

CURRENT RECOVERY EFFORTS

Recovery plan: None

Other plans: Garnett ST and Crowley GM (2000) 'The action plan for Australian birds 2000'. Environment Australia, Canberra.

Conservation actions:

- Raine Island is protected in the Raine Island NP (Scientific). A Restricted Access – Special Management Area extends out to 500 m from the high water mark. Surrounding waters are within a marine national park (green) zone of the Great Barrier Reef Marine Park.
- North-East Cay is protected in the Commonwealth-managed Coringa-Herald National Nature Reserve.
- Access to Raine Island is by permit only and is restricted to authorised persons undertaking management, scientific or monitoring activities.
- Visitor numbers to Coringa-Herald National Nature Reserve are managed through a limit on the number of commercial tour operators permitted to operate within the reserve.
- The Great Barrier Reef Marine Park Authority and DERM have a joint position statement on managing access to Raine Island.
- A banding program was initiated on Lady Elliot Island in 1993.

FUTURE RECOVERY EFFORTS

What you can do:

- Stay at least 10 m away from any nesting seabirds to minimise human disturbance.
- Keep away from flight corridors used by the adult birds to come and go from the nest.
- Adhere to regulations managing access to seabird colonies in the Great Barrier Reef region.
- If visiting the Great Barrier Reef, choose to stay at resorts that support conservation and research programmes.

Conservation needs:

- Continue to protect nesting sites from human disturbance by controlling visitor access to nesting areas.
- Maintain strict quarantine arrangements at nesting islands to reduce the threat from exotic pests.
- Continue monitoring breeding colonies at Raine Island, Lady Elliot Island and North-East Cay.

Research directions:

- Examine the linkages between changing climatic and oceanographic conditions and breeding in Red-tailed Tropicbirds.
- Examine the movements of Red-tailed Tropicbirds and their ecology at sea.

Sources and further reading

Batianoff GN and Cornelius NJ (2005) Birds of Raine Island: population trends, breeding behaviour and nesting habitats. *Proceedings of the Royal Society of Queensland* **112**, 1–29.

Congdon BC, Erwin CA, Peck DR, Baker GB, Double MC and O'Neill P (2007) Vulnerability of seabirds on the Great Barrier Reef to climate change. In *Climate Change and the Great Barrier Reef. A Vulnerability Assessment.* (Eds JE Johnson and PA Marshall) pp. 427–463. Great Barrier Reef Marine Park Authority, Townsville and Australian Greenhouse Office, Canberra.

Le Corre M, Salamolard M and Portier MC (2003) Transoceanic dispersion of the Red-tailed Tropicbird in the Indian Ocean. *Emu* **103**, 183–184.

Marchant S and Higgins PJ (Eds) (1990) *Handbook of Australian, New Zealand and Antarctic Birds. Volume 1: Ratites to Ducks.* Oxford University Press, Melbourne.

Author: Peter M Kyne and Lee K Curtis

PIGEONS AND DOVES (ORDER: COLUMBIFORMES)

SQUATTER PIGEON (SOUTHERN SUBSPECIES)

Geophaps scripta scripta (Temminck, 1821)

Photo: Craig Eddie

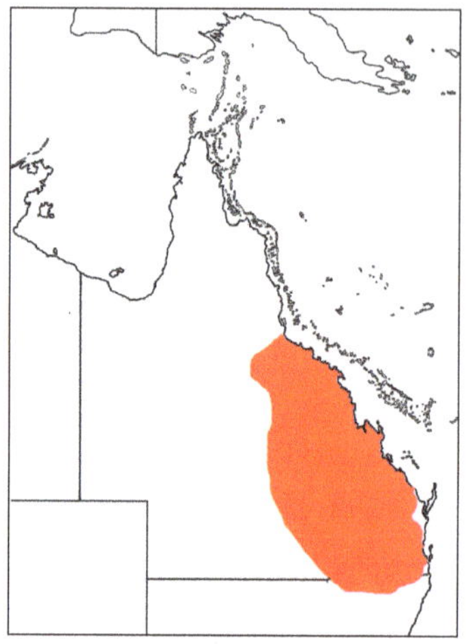

Family: Columbidae

Other common names: Partridge Bronzewing, Blue-eyed Partridge Bronzewing

Aboriginal name: Manga

Conservation status: IUCN: species (*G. scripta*) is Least Concern; Aust.: Vulnerable; Qld: Vulnerable; NSW: Endangered

Description: This is a medium-sized, heavily built, long-legged pigeon, with a total length of 26–32 cm including a short tail, a wingspan of approximately 45 cm, and weighing 200–225 g. It is mostly brown, with a short erectile crest, bold black-and-white face pattern, green or violet iridescent patches in the upperwings, a blue-grey breast and white flanks. The eye-ring is blue-grey.

Distribution: The Squatter Pigeon is largely, if not now wholly, restricted to Qld, south to the NSW border. The southern subspecies occurs north to the Burdekin River and east to Townsville and Proserpine. In the west it reaches Longreach, Barcaldine and Charleville. It is now very localised in south-east Qld, though still recorded in low numbers around Inglewood, Warwick and Esk. The total population size is estimated at 40 000 breeding birds (with low reliability). In 2000, the extent of occurrence of the subspecies was estimated as 440 000 km^2 and stable. The area of occupancy was estimated (with low reliability) as 10 000 km^2. The decline has slowed and the Squatter Pigeon remains locally abundant (e.g. groups of up to 30 are still seen in central Qld).

Habitat: The southern subspecies occurs mainly in dry grassy eucalypt woodlands and open forests, mostly in sandy sites near permanent water. It forages along roads and railway lines, occurs around homesteads and cattle yards, and will use highly modified areas such as bitumen carparks and the lawns of resorts. Although it remains common in heavily grazed country in tropical Qld, it is more common in ungrazed land.

Ecology: The Squatter Pigeon is terrestrial: feeding, resting and nesting on the ground but generally roosting in trees. It is largely granivorous, though invertebrates are seasonally important in its diet. It is tame and unobtrusive and usually found in small flocks, pairs or family groups. In southern Qld, breeding is known from July to May, but breeding may be more influenced by rainfall rather than season. The nest is a shallow depression lined with sparse dead grass and usually sheltered by a tussock, bush or log. There appear to be no large-scale seasonal movements, and the subspecies is probably locally nomadic, or perhaps sedentary.

Reason for decline and current threats: Historically, great numbers were shot for food, and illegal hunting is still suspected. However, changes to the ground cover caused by cattle, sheep and rabbits, and consequent replacement of perennial bushes, herbs and grasses with ephemeral herbs and annual grasses, are the likely cause of the pigeon's decline. The impacts of cattle, sheep and rabbits were exacerbated by overstocking during drought in the 19th century. Where cattle and sheep are common, Squatter Pigeons no longer occur.

In Qld, cattle grazing and drought are probably the most important factors in the decline of the species, with the southern subspecies declining before the land-clearing era. Collapse of traditional Indigenous burning regimes may have affected the species through altered vegetation structure and composition. Only in the south-east corner of Qld is large-scale habitat destruction the likely major cause of the great reduction in numbers.

Currently the southern subspecies is threatened by loss and fragmentation of woodlands, predation by cats and foxes, illegal shooting and the impacts of livestock and rabbits. Some pasture improvement is also a threat. Although the pigeon remains common in much of central Qld in both cleared and relatively intact landscapes, it becomes less common where Buffel Grass *Cenchrus ciliaris* becomes the dominant pasture species.

CURRENT RECOVERY EFFORTS

Recovery plan: None

Other plans: Garnett ST and Crowley GM (2000) 'The action plan for Australian birds 2000'. Environment Australia, Canberra.

FUTURE RECOVERY EFFORTS

What you can do:

- Join/donate to Birds Australia, Birds Qld, Bird Observation and Conservation Australia.
- Report sightings to DERM.
- Donate to privately funded organisations that purchase and manage properties for conservation.

Conservation needs:

- Protect grassy woodlands from clearing, fragmentation or development.
- Revegetate areas linking existing patches of suitable habitat.
- Implement appropriate fire regimes in suitable habitat to maintain cover of native grasses.
- Implement stocking rates to maintain a grazing intensity supporting suitable grass cover and exclude livestock during the breeding season where breeding is known.
- Control rabbits, cats and foxes.
- Implement public awareness campaigns for conservation.
- Develop community-based tree-planting schemes to enhance or rehabilitate suitable habitats.
- Establish population monitoring sites.

Research directions:

- Undertake diet studies to identify important non-native food species for the southern subspecies (the northern subspecies is known to feed on the introduced pasture legume *Stylosanthes humilis*)
- Investigate the effects of Buffel Grass and Parthenium weed *Parthenium hysterophorus* invasion.
- Determine the population size and distribution, as well as the relationships between Squatter Pigeons and tree density, fire frequency and stocking rates.

Sources and further reading

Crome F and Shields J (1992) *Parrots and Pigeons of Australia.* Angus and Robertson, Sydney.

Department of the Environment, Water, Heritage and the Arts (2009) *Geophaps scripta scripta*. In *Species Profile and Threats Database.* <http://www.environment.gov.au/cgi-bin/sprat/public/publicspecies.pl?taxon_id=64440>.

Frith HJ (1982) *Pigeons and Doves of Australia*. Rigby, Adelaide.

Garnett S, Crowley G, Delaney M and Shephard P (1998) Fluctuations in the population of the Squatter Pigeon *Geophaps scripta* at Coen, Cape York Peninsula. *Sunbird* **28,** 53–54.

Higgins PJ and Davies SJJF (Eds) (1996) *Handbook of Australian, New Zealand and Antarctic Birds. Volume 3: Snipe to Pigeons.* Oxford University Press, Melbourne.

Author: Terry Reis

FROGMOUTHS AND NIGHTJARS (ORDER: CAPRIMULGIFORMES)

MARBLED FROGMOUTH (SOUTHERN SUBSPECIES)

Podargus ocellatus plumiferus Gould, 1845

Photo: Mark Clayton

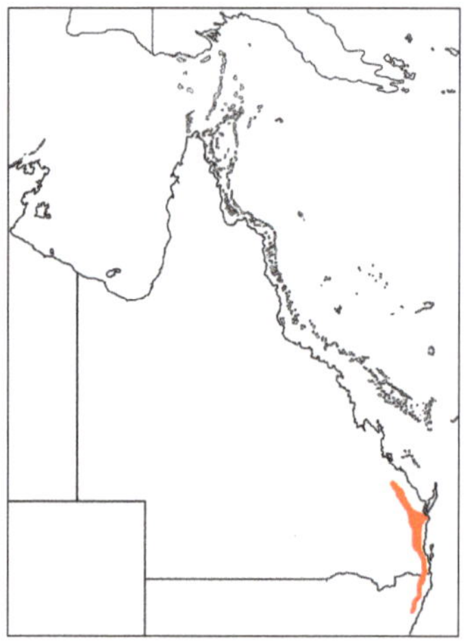

Family: Podargidae

Other common name: Plumed Frogmouth (for southern subspecies)

Conservation status: IUCN: species (*P. ocellatus*) is Least Concern; Qld: Vulnerable; NSW: Vulnerable

Description : This is a small, slender frogmouth, cryptically mottled, with a total length of 40–45 cm, a wingspan of 62–75 cm and weighing 200–320 g. Males are grey or brown and marbled underneath; females are rufous and plainer; both have a banded forehead tuft. The eyes are bright orange. Identification based on physical appearance is problematic, with calls the most reliable characteristic for field identification: a nocturnal series of *kooloo* calls, or downscale gobbles finishing with bill claps.

Distribution: The species is restricted within its Australian range to two isolated populations, representing two separate subspecies; the population in northern Qld (*P. o. marmoratus*) is not considered threatened. The other subspecies occurs in south-eastern Qld and north-eastern NSW (*P. o. plumiferus*), where it is distributed through coastal ranges of the Mount Warning caldera, northern NSW to the Many Peaks Range, south-east Qld. Strongholds for the southern Marbled Frogmouth in Qld appear to be in the Conondale and Border Ranges. Approximately 755 pairs are thought to inhabit the Conondale Ranges: a significant component of the population.

Habitat: Preferred habitat of the southern Marbled Frogmouth is subtropical rainforest, including complex and simple notophyll vine forest and sometimes Araucarian notophyll vine forest. Marbled Frogmouths often roost close to streams in dense vegetation. Southern Marbled Frogmouths inhabit rainforests and wet sclerophyll forest associated with drainage lines in the Conondale Ranges, where they occupy combined territories from 12 to 19 ha in size and individual home ranges from 5 to 18 ha. The home ranges of the male and female of a pair overlap.

Ecology: Contact is maintained through at least five call types, of which three are heard commonly in response to call playback. The typical year consists of a peak in response to call playback in August (owing to territorial behaviour associated with the onset of breeding), a lull in September (associated with egg-laying and brooding), peak calling between November and March (coinciding with juvenile fledging and dispersal) and then a further lull in response during the cooler months of April to July (during the non-breeding season). Intra-pair duets occur primarily between November and February. The nest is a flimsy 'saucer' of vegetation (e.g. moss, ferns or orchids) in the fork of a rainforest tree, and a single young is raised per year after incubation and nestling periods of a month each. Food is mainly nocturnal insects and other arthropods, taken from the rainforest floor and foliage by perch-pouncing.

Reason for decline and current threats: Habitat alteration (i.e. forestry, weed infestation, grazing by introduced animals and die-back), habitat clearing (i.e. for agriculture and forestry), habitat destruction (i.e. for industrial development, mining, housing) and change in fire regimes (intensity, size, frequency and season) are the primary threats to this subspecies.

CURRENT RECOVERY EFFORTS

Recovery plan: None

Other plans: Garnett ST and Crowley GM (2000) 'The action plan for Australian birds 2000'. Environment Australia, Canberra.

Conservation actions:

- A Conondale Range land-use study begun in 1990, in response to a Qld Government decision to convert forestry estate to national park, culminated in the release of recommendations and strategies to enhance conservation of the Marbled Frogmouth in the Conondale Ranges (Conondale Range Consultative Committee 1991; Conondale Range Strategic Plan 1993).
- A Species Management Profile was developed to guide forestry activities in areas of known occurrence.
- In 1999, a South-East Qld Forest Agreement led to the declaration of wider areas of national park across the Conondale Ranges.

FUTURE RECOVERY EFFORTS

Conservation needs:

- Conserve rainforest remnants and undertake supplementary plantings to link patches and increase the size of remnants.
- Regenerate habitat, change fire regimes, manage and conserve habitat, and carry out scientific monitoring using call playback surveys.
- Monitor and assess the population using call playback.

Research directions:

- Undertake survey and monitoring, in the Conondale Ranges and elsewhere in south-east Qld, for comparison with previously known distribution and population estimates, and to determine the population size in poorly known areas; funding will be needed.
- Further investigate the significance of calls (being the key survey technique).
- Use data on population trends and distribution (published information and DERM's WildNet database) to assess whether recovery actions will be required for the subspecies, or whether with relaxation of timber harvesting there has been recovery, or stabilisation, of the population level.
- Use the historical and current data to review the frogmouth's conservation status as possibly Endangered. If the subspecies so qualifies, it will be necessary to identify threats and management actions to ameliorate those threats, to engage stakeholders and to formulate a recovery plan for the subspecies.

Sources and further reading

Beruldsen GR (2003) Breeding biology of the Plumed Frogmouth *Podargus ocellatus plumiferus* and Marbled Frogmouth *P.o. marmoratus*. *Australian Field Ornithology* **20,** 126–138.

Corben C and Roberts G (1993) Some recent records of the Plumed Frogmouth *Podargus ocellatus plumiferus*. *Sunbird* **23**, 61–72.

Higgins PJ (Ed.) (1999) *Handbook of Australian, New Zealand and Antarctic Birds. Volume 4: Parrots to Dollarbirds.* Oxford University Press, Melbourne.

Smith GC and Jones DN (1997) Vocalisations of the Marbled Frogmouth I: descriptions and an analysis of sex differences. *Emu* **97**, 290–295.

Smith GC, Hamley BJ and Lees N (1998) An estimate of the Plumed Frogmouth *Podargus ocellatus plumiferus* population size in the Conondale Ranges. *Pacific Conservation Biology* **4**, 215–226.

Authors: Geoffrey C Smith and Michael Mathieson

TUBE-NOSED SEABIRDS (ORDER: PROCELLARIIFORMES)

ALBATROSSES

Wandering Albatross *Diomedea exulans* Linnaeus, 1758; Royal Albatross *D. epomophora* Lesson, 1825; Black-browed Albatross *Thalassarche melanophris* (Temminck, 1828); Shy Albatross *T. cauta* (Gould, 1841); Grey-headed Albatross *T. chrysostoma* (J.R. Forster, 1785); Yellow-nosed Albatross *T. chlororhynchos* (J.F. Gmelin, 1789); Buller's Albatross *T. bulleri* (Rothschild, 1893); Sooty Albatross *Phoebetria fusca* (Hilsenberg, 1822); Light-mantled Sooty Albatross *P. palpebrata* (J.R. Forster, 1785)

Black-browed Albatross. Photo: Micha V Jackson

Family: Diomedeidae

Conservation status: IUCN: Near Threatened to Critically Endangered; Aust.: Vulnerable or Endangered; Qld: nine species or subspecies listed as Vulnerable: Antipodean Albatross *Diomedea antipodensis antipodensis*, Gibson's Albatross *D. a. gibsoni*, Wandering Albatross *D. exulans*, Sooty Albatross *Phoebetria fusca*, Buller's Albatross *Thalassarche bulleri*, Grey-headed Albatross *T. chrysostoma*, Indian Yellow-nosed Albatross *T. carteri*, Shy Albatross *T. cauta*, White-capped Albatross *T. steadi* (note: some of these names do not match those given above due to differing approaches to albatross taxonomy; see below)

Description: Albatrosses are very large, bulky seabirds, with total lengths of ~65–135 cm and wingspans ranging from ~175 to 350 cm (the Wandering and Royal Albatrosses have the largest wingspans of any bird). Albatrosses can be divided into four major groups: great albatrosses (genus *Diomedea*; two species occurring in Australian territories), North Pacific Albatrosses (*Phoebastria*; one species vagrant to Australia), mollymawks (*Thalassarche*; five Australian species) and sooty albatrosses (*Phoebetria*; two Australian species). Plumage is highly variable with age; see the Sources and further reading list for details on plumage, moult and identification. It should be noted that the systematics and taxonomy of albatrosses are highly debated. Conservative species level treatment recognises 13 species with several subspecies (followed here), while other sources recognise up to 23 species.

Distribution: The great albatrosses, mollymawks and sooty albatrosses are birds of the Southern Ocean, with some species ranging into subtropical, and sometimes tropical, waters mainly in winter and spring. They breed on cold temperate and sub-Antarctic islands, which, in Australian waters, include islands off Tasmania as well as Macquarie, Heard and McDonald Islands. However, not all species listed above breed on Australian territory, with some occurring in Australian waters only as foraging birds. Albatrosses generally only visit Qld waters in winter and spring, and then only in small numbers; four species are rare visitors (Wandering, Black-browed, Shy and Yellow-nosed Albatrosses) and five are vagrants (Royal, Grey-headed, Buller's, Sooty and Light-mantled Sooty Albatrosses) to Qld waters.

Habitat: Albatrosses occurring in Australia are marine and pelagic birds of the Southern Ocean. Breeding habitat is generally among short vegetation on cold temperate and sub-Antarctic islands.

Ecology: Albatrosses take several years to reach maturity, have small clutch sizes (one egg with no replacement laying if the egg or chick is lost), long incubation (2+ months) and nestling (up to 12 months) periods, and low natural mortality. Albatrosses breed in colonies and pair bonds are monogamous and long lasting. Adult pairs may not breed annually. Immature birds and adults between breeding spend their life foraging at sea.

Reason for decline and current threats: Habitat degradation and predation by feral animals, disturbance, intentional shooting, marine pollution, avian parasites and diseases, and significantly, the incidental capture in fisheries have reduced the Australian and global populations of most, if not all albatrosses.

At sea, mortality from the incidental capture in longline and, to a lesser extent, trawl fisheries has severely affected many albatross populations. Entanglement or collision with fishing gear can also result in mortality. Although much has been done to deal with this issue among fishing nations (particularly those member countries of the Agreement on the Conservation of Albatrosses and Petrels; ACAP), bycatch in illegal and high-seas fisheries (which are generally unmonitored, unregulated and unmanaged) is an ongoing issue. The extent to which human-induced climate change will affect albatross breeding and foraging is poorly understood, although population declines have been predicted due to the future warming of the Southern Ocean.

CURRENT RECOVERY EFFORTS

Recovery plan: Department of Sustainability, Environment, Water, Population and Communities (2011) 'National recovery plan for threatened albatrosses and giant petrels 2011–2016'. Commonwealth of Australia, Hobart.

Other plans: Department of the Environment, Water, Heritage and the Arts (2009) 'Threat abatement plan to reduce the impacts of exotic rodents on biodiversity on Australian offshore islands of less than 100 000 hectares'. Department of the Environment, Water, Heritage and the Arts, Canberra.

Department of the Environment, Water, Heritage and the Arts (2008) 'Threat abatement plan for predation by feral cats'. Commonwealth of Australia, Canberra.

Australian Antarctic Division (2006) 'Threat abatement plan for the incidental catch (or bycatch) of seabirds during oceanic longline fishing operations'. Australian Antarctic Division, Kingston.

Garnett ST and Crowley GM (2000) 'The action plan for Australian birds 2000'. Environment Australia, Canberra.

Garnett ST, Szabo JK and Dutson G (2011) *The Action Plan for Australian Birds 2010.* CSIRO Publishing, Melbourne.

Conservation actions:

- Australia is a member country of ACAP, which aims 'to achieve and maintain a favourable conservation status for albatrosses and petrels' through national and international measures.
- The eradication of feral animals on breeding islands is ongoing.
- Australian longline fisheries employ seabird bycatch mitigation devices and strategies.

FUTURE RECOVERY EFFORTS

What you can do:

- Pick up plastic garbage on the beach; at sea, such items can be mistaken by albatrosses and other seabirds for food.
- Join or support the Southern Oceans Seabird Study Association (SOSSA; <http://www.sossa-international.org>).
- Obtain the Australian Marine Conservation Society's *Australia's Sustainable Seafood Guide* and buy only seafood that is harvested from sustainable sources with limited bycatch (see <http://www.amcs.org.au>).

Conservation needs:

The conservation of albatrosses is largely dependent on actions outside of Qld, these include:

- Continue to monitor all Australian breeding populations.
- Ensure integrated pest eradication strategies are implemented on breeding islands.
- Continue to mitigate the impacts of Australian fisheries on all seabirds, and continue to be active in the international community regarding this issue.

Sources and further reading

Barbraud C, Rivalan P, Inchausti P, Nevoux M, Rolland V and Weimerskirch H (2011) Contrasted demographic responses facing future climate change in Southern Ocean seabirds. *Journal of Animal Ecology* **80**, 89–100.

Comben P, Stewart D and Walbridge P (2001) Records of seabirds (Order: Procellariiformes) in south-east Queensland waters. *Sunbird* **31**, 33–72.

Lindsey T (2008) *Albatrosses.* CSIRO Publishing, Melbourne.

Marchant S and Higgins PJ (Eds) (1990) *Handbook of Australian, New Zealand and Antarctic Birds. Volume 1: Ratites to Ducks.* Oxford University Press, Melbourne.

Onley D and Scofield P (2007) *Albatrosses, Petrels and Shearwaters of the World.* Princeton University Press, Princeton.

Author: Peter M Kyne

SOUTHERN GIANT-PETREL

Macronectes giganteus (J.F. Gmelin, 1789)

Photo: Paul Walbridge

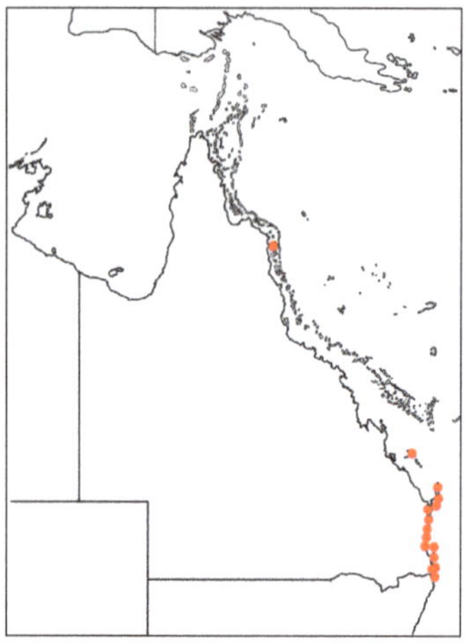

Family: Procellariidae

Other common names: Nellie, Stinker, Giant Fulmar, Glutton, Mollymawk, Mother Carey's or Sea Goose, Stinkpot, Vulture of the Seas, Bone-shaker

Conservation status: IUCN: Least Concern; Aust.: Endangered; Qld: Endangered; NSW: Endangered; SA: Vulnerable; Tas.: Vulnerable; Vic.: Vulnerable; WA: Endangered

Description: The Southern Giant-Petrel is a very large (almost albatross-sized), bulky fulmarine petrel, measuring 85–100 cm in total length, with a wingspan of 150–210 cm. In juvenile birds, the plumage is dark-grey brown overall. The head, neck and upper breast of dark-morph adults are white. Up to 10% of the population is nearly all white, with some dark mottling (white morph). A very small percentage of the population is purely leucistic. The bill is enormous, with the large nasal tube extending 60% or more along the length of the bill. The horn is yellowish-green coloured with usually greenish-tipped ungues (bill tips). The Southern Giant-Petrel is slightly larger than the Northern Giant-Petrel *Macronectes halli*, which usually has dusky brownish to reddish-pink tips to both ungues. The Southern Giant-Petrel's eye is brown to brownish grey, becoming much paler with age.

Distribution: This species is circumpolar in the Southern Ocean, with mostly juvenile birds ranging in winter months into subtropical, and sometimes tropical, waters. It generally breeds further south than the Northern Giant-Petrel on sub-Antarctic and Antarctic islands, and Antarctica itself. In Australian territorial waters, it breeds on Macquarie, Heard and McDonald Islands, as well as in the Australian Antarctic Territory. It visits Qld waters generally in winter, but has become extremely rare in recent years.

Habitat: This is a marine and pelagic bird of the Southern Ocean. The breeding habitat includes exposed coastal areas, rocky bluffs, outcrops, ridges, slopes and flats.

Ecology: This is an opportunistic scavenger and predator, primarily feeding on penguin carcasses around breeding grounds. At sea, it feeds on cephalopods and other birds up to the size of albatrosses. Larger birds are approached on the water where the neck is seized and the head plunged under water thus drowning the bird before devouring it. The clutch size is one egg.

Reason for decline and current threats: Habitat degradation and predation by feral animals, disturbance, and the incidental capture in fisheries have affected the Australian and global populations of the Southern Giant-Petrel. The breeding population on all Australian breeding islands has decreased. Following the successful eradication of feral cats on Macquarie Island in 2000, rabbits caused devastation to vegetation, with severe erosion occurring in places, affecting seabird colonies. The Southern Giant-Petrel will come to vessels, feeding on bycatch and any other offal offered, although it is a little more cautious and less likely to do so than Northern Giant-Petrels. Such behaviour makes it vulnerable to capture and entanglement in fishing gear, and bycatch in oceanic longline fisheries has had an impact upon this species. Human-related waste products, such as plastics, may affect the bird at sea (plastic waste can be mistaken for food).

CURRENT RECOVERY EFFORTS

Recovery plan: Department of Sustainability, Environment, Water, Population and Communities (2011) 'National recovery plan for threatened albatrosses and giant petrels 2011–2016'. Commonwealth of Australia, Hobart.

Other plans: Department of the Environment, Water, Heritage and the Arts (2009) 'Threat abatement plan to reduce the impacts of exotic rodents on biodiversity on Australian offshore Islands of less than 100 000 hectares'. Department of the Environment, Water, Heritage and the Arts, Canberra.

Department of the Environment, Water, Heritage and the Arts (2008) 'Threat abatement plan for predation by feral cats'. Commonwealth of Australia, Canberra.

Australian Antarctic Division (2006) 'Threat abatement plan for the incidental catch (or bycatch) of seabirds during oceanic longline fishing operations'. Australian Antarctic Division, Kingston.

Garnett ST and Crowley GM (2000) 'The action plan for Australian birds 2000'. Environment Australia, Canberra.

Conservation actions:

- The eradication of exotic pests on Macquarie Island is ongoing.
- Australian longline fisheries employ seabird bycatch mitigation devices and strategies.

FUTURE RECOVERY EFFORTS

What you can do:

- If you locate a live Giant-petrel on the shore (an occasional event, particularly during adverse weather conditions or in the case of young inexperienced birds), closely monitor the situation; if, after a while, the bird flies off then it should be safe. If, however, the bird appears to be distressed or is threatened by, for example people or dogs, then cautiously intervene. Giant-petrels have formidable bills that can easily inflict damage. First the bill, then the powerful wings need to be immobilised before housing into a large, dry box or enclosure. Do not keep it in a heated area because Giant-petrels are not particularly tolerant of warm conditions. Take the bird to the local vet or wildlife carer. If you are on the Sunshine Coast, contact Australia Zoo; if you are on the Gold Coast contact either Currumbin Sanctuary or Sea World. If the bird is dead, but in reasonable condition, freeze the specimen in a sealed plastic bag and contact the Qld Museum.
- Join/donate to the Southern Oceans Seabird Study Association, Birds Australia, Birds Qld, Bird Observation and Conservation Australia.
- Anglers should ensure they dispose of plastic bags thoughtfully onshore.

Conservation needs:

- Anglers should develop a substitute method for storing bait (reusable plastic containers, rather than plastic bags).

The conservation of the Southern Giant-Petrel is largely dependent on actions outside of Qld, including at breeding colonies; these include:

- Continue to monitor all Australian breeding populations.
- Ensure that integrated pest eradication strategies are employed to avoid a repeat of the Macquarie Island situation.
- Continue to mitigate the impacts of Australian fisheries on all seabirds, and continue to be active in the international community regarding this issue.

Sources and further reading

Comben P, Stewart D and Walbridge P (2001) Records of seabirds (Order: Procellariiformes) in south-east Queensland waters. *Sunbird* **31**, 33–72.

Harrison P (1985) *Seabirds: An Identification Guide.* Rev. edn. Christopher Helm, London.

Marchant S and Higgins PJ (Eds) (1990) *Handbook of Australian, New Zealand and Antarctic Birds. Volume 1: Ratites to Ducks.* Oxford University Press, Melbourne.

Serventy DL, Serventy V and Warham J (1971) *The Handbook of Australian Sea-birds.* Reed, Sydney.

Author: Paul Walbridge

NORTHERN GIANT-PETREL

Macronectes halli Mathews, 1912

Photo: Paul Walbridge

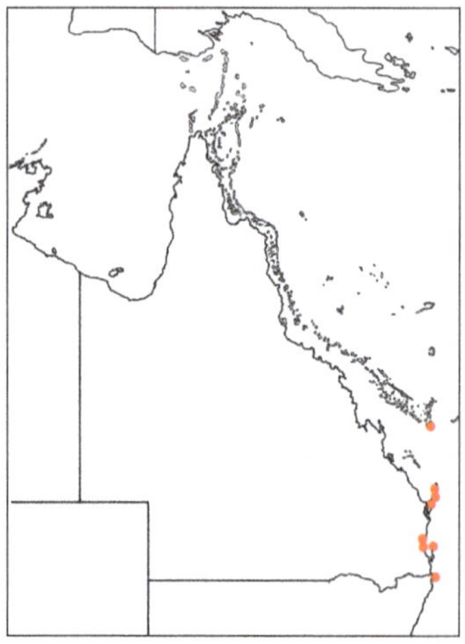

Family: Procellariidae

Conservation status: IUCN: Least Concern; Aust: Vulnerable; Qld: Vulnerable; NSW: Vulnerable; Tas.: Rare; Vic.: Near Threatened

Description: The Northern Giant-Petrel is a very large (almost albatross-sized), bulky fulmarine petrel, measuring 80–95 cm in total length, with a wingspan of 150–210 cm. In juvenile birds, the plumage is sooty-brown overall. In adults, the underparts are pale brownish grey and paler than the upperparts. The upperparts become greyer with age, owing to paler edges to mantle feathers and wing coverts. The upperwings have silvery primary bases, extending onto the secondaries. With age, the adult head becomes a paler mottled grey with some white feathering, but often retaining some dark feathering on the crown. The bill is enormous, with a very long nasal tube, pale yellowish to horn coloured with usually, dusky brownish to reddish-pink tipped ungues (bill tips) (the similar Southern Giant-Petrel *Macronectes giganteus* usually has greenish-tipped ungues). The iris is pale brown, becoming paler with age to almost whitish-grey.

Distribution: This species is circumpolar in the Southern Ocean, with mostly juvenile birds ranging in winter months into subtropical, and sometimes tropical, waters. This species breeds on a few island groups in the Southern Ocean. In Australian territorial waters, it breeds on Macquarie Island. It visits Qld in autumn–spring and in southern Qld waters it was considered to be extremely rare up until the mid-1990s. Sightings have since become more regular. Recent population increases of Northern Giant-Petrels have been recorded at several breeding localities, although decreases have been recorded elsewhere.

Habitat: This is a marine and pelagic bird of the Southern Ocean. Breeding habitat includes sheltered coastal areas, slopes, headlands, among vegetation and under overhangs.

Ecology: This opportunistic scavenger and predator feeds on dead whales, seals and penguins and will also prey on young penguins. At sea, it feeds on fish, cephalopods and other seabirds up to the size of small albatrosses. It can plunge to depths of up to 2 m. The clutch size is one egg.

Reason for decline and current threats: The incidental capture in fisheries is cited as the most significant threat to the Northern Giant-Petrel.

The Northern Giant-Petrel will come to vessels, feeding on bycatch and any other offal offered. Such behaviour makes it vulnerable to capture and entanglement in fishing gear, and bycatch in oceanic longline fisheries has had an impact upon this species. Following the successful eradication of feral cats on Macquarie Island in 2000, rabbits caused devastation to vegetation, with severe erosion occurring in places, and thus altering the landscape. However, monitoring has shown an increase in Northern Giant-Petrel numbers at Macquarie Island. Human disturbance around the breeding grounds has been shown to be a problem, owing to the timidness of the birds, and human-related waste products, such as plastic bags, may affect the bird at sea (plastic waste can be mistaken for food).

CURRENT RECOVERY EFFORTS

Recovery plan: Department of Sustainability, Environment, Water, Population and Communities (2011) 'National recovery plan for threatened albatrosses and giant petrels 2011–2016'. Commonwealth of Australia, Hobart.

Other plans: Department of the Environment, Water, Heritage and the Arts (2009) 'Threat abatement plan to reduce the impacts of exotic rodents on biodiversity on Australian offshore islands of less than 100 000 hectares'. Department of the Environment, Water, Heritage and the Arts, Canberra.

Australian Antarctic Division (2006) 'Threat abatement plan for the incidental catch (or bycatch) of seabirds during oceanic longline fishing operations'. Australian Antarctic Division, Kingston.

Garnett ST and Crowley GM (2000) 'The action plan for Australian birds 2000'. Environment Australia, Canberra.

Conservation actions:

- The eradication of exotic pests on Macquarie Island is ongoing.
- Australian longline fisheries employ seabird bycatch mitigation devices and strategies.

FUTURE RECOVERY EFFORTS

What you can do:

- If you locate a live giant-petrel on the shore (an occasional event, particularly during adverse weather conditions or in the case of young inexperienced birds), closely monitor the situation; if, after a while, the bird flies off then it should be safe. If, however, the bird appears to be distressed or is threatened by, for example people or dogs, then cautiously intervene. Giant-petrels have formidable bills that can easily inflict damage. First the bill, then the powerful wings need to be immobilised before housing into a large, dry box or enclosure. Do not keep it in a heated area because giant-petrels are not particularly tolerant of warm conditions. Take the bird to the local vet or wildlife carer. If you are on the Sunshine Coast, contact Australia Zoo; if you are on the Gold Coast contact either Currumbin Sanctuary or Sea World. If the bird is dead, but in reasonable condition, freeze the specimen in a sealed plastic bag and contact the Qld Museum.
- Join/donate to the Southern Oceans Seabird Study Association, Birds Australia, Birds Qld, Bird Observation and Conservation Australia.
- Anglers should ensure they dispose of plastic bags thoughtfully onshore.

Conservation needs:

- Anglers should develop a substitute method for storing bait (reusable plastic containers, rather than plastic bags).

The conservation of the Northern Giant-Petrel is largely dependent on actions outside of Qld, including at breeding colonies; these include:

- Continue to monitor the Australian breeding population.
- Ensure that integrated pest eradication strategies are employed to avoid a repeat of the Macquarie Island situation.
- Continue to mitigate the impacts of Australian fisheries on all seabirds, and continue to be active in the international community regarding this issue.

Sources and further reading

Agreement on the Conservation of Albatrosses and Petrels (2010) *ACAP Species assessments: Northern Giant Petrel* Macronectes halli. <http://www.acap.aq>.

Comben P, Stewart D and Walbridge P (2001) Records of seabirds (Order: Procellariiformes) in south-east Queensland waters. *Sunbird* **31**, 33–72.

Harrison P (1985) *Seabirds: An Identification Guide.* Revised Edn. Christopher Helm, London.

Marchant S and Higgins PJ (Eds) (1990) *Handbook of Australian, New Zealand and Antarctic Birds. Volume 1: Ratites to Ducks.* Oxford University Press, Melbourne.

Serventy DL, Serventy V and Warham J (1971) *The Handbook of Australian Sea-birds.* Reed, Sydney.

Author: Paul Walbridge

KERMADEC PETREL (WESTERN SUBSPECIES)

Pterodroma neglecta neglecta (Schlegel, 1863)

Photo: Paul Walbridge

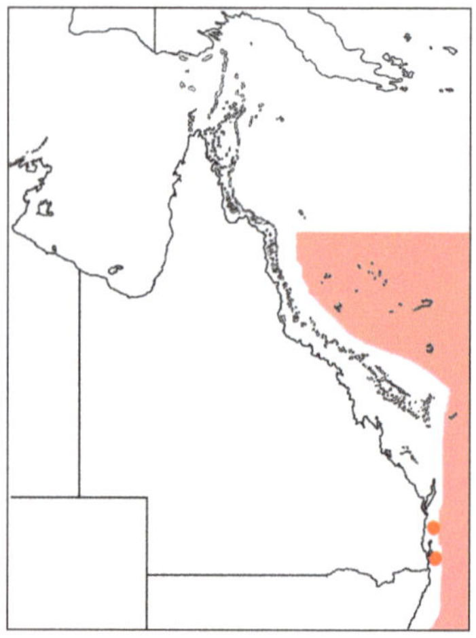

Family: Procellariidae

Conservation status: IUCN: species (*P. neglecta*) is Least Concern; Aust: Vulnerable; NSW: Vulnerable

Description: The Kermadec Petrel is a medium- to large-sized gadfly petrel, stocky in build with a shortish squared off to slightly rounded tail and long, broad wings. It measures 38 cm in total length with a wingspan of 92 cm. The bill is black and moderately robust. It occurs in three different colour morphs. The pale morph has a mostly white head with light greyish barring from forehead to nape. The upperparts (including the wing) are greyish brown except for blackish primaries, which show diagnostic bare shaft bases in 6–8 primaries. The underparts are almost entirely white except for the brownish tail. The underwing is mostly dark except for variable whitish feathering on the inner forewing and a showy white triangular area in the primaries, with a thin white crescent mark in the adjacent coverts. The dark morph is almost all dark brown, apart from the characteristic above-mentioned wing pattern, although in the underwing, the whitish area on the inner forewing is not usually evident. The intermediate morph has highly variable underparts, ranging from mostly white with dark undertail coverts and chin, with variable brownish mottling on the flanks, through to just a small white area on the belly and sometimes the chin and throat. The wing pattern is generally as for pale birds.

Distribution: Widespread in the Pacific, it breeds on tropical and subtropical islands across the south Pacific from Australian territorial waters to islands off the coast of Chile, dispersing northwards when not breeding. In Australian territorial waters, it breeds on Phillip Island off Norfolk Island and on Ball's Pyramid near Lord Howe Island. Once considered a rare visitor to Australian east coast waters, there have been an increasing number of sightings, owing to the increase in organised pelagic seabird-watching trips. Off Southport, south-east Qld, the species has been recorded in most months of the year on such trips; up to eight birds per trip have been recorded, and so it would probably best be rated as uncommon.

Habitat: These marine and pelagic birds live in tropical and subtropical waters. The breeding habitat is mainly on high rocky islands on vegetated coastal slopes, cliffs or inland mountainous terrain.

Ecology: This species is a surface breeder, nesting between rocks and crevices under low vegetation. Its breeding habits are poorly known. The clutch size is one egg. The feeding habits are poorly known; it surface feeds, presumably on small fish, cephalopods and crustaceans.

Reason for decline and current threats: Across the breeding range of the species, habitat degradation and predation by feral animals, as well as human harvesting, have caused a serious decline.

The Kermadec Petrel is extinct on Lord Howe Island, owing to predation by the introduced Black Rat *Rattus rattus*. The Australian breeding population, which is now very small, is restricted to two islets (Ball's Pyramid near Lord Howe Island and Phillip Island off Norfolk Island). Given its small size, this population is vulnerable to stochastic or catastrophic events. The accidental transportation of rats or cats from the main islands to the breeding colonies could be disastrous. Predation by native predators, particularly recently established Nankeen Kestrels *Falco cenchroides* and Purple Swamphens *Porphyrio porphyrio* on Phillip Island, affects the small Australian breeding population. Marine debris (plastic bags, recreational fishing bait bags and helium-filled balloons) poses a risk to various marine life and may be mistaken for food by petrels.

CURRENT RECOVERY EFFORTS

Recovery plan: Director of National Parks (2010) 'Norfolk Island Region threatened species recovery plan'. Department of the Environment, Water, Heritage and the Arts, Canberra. This plan covers the Norfolk Island region only.

Other plans: Department of the Environment, Water, Heritage and the Arts (2009) 'Threat abatement plan to reduce the impacts of exotic rodents on biodiversity on

Australian offshore islands of less than 100 000 hectares'. Department of the Environment, Water, Heritage and the Arts, Canberra.

Garnett ST and Crowley GM (2000) 'The action plan for Australian birds 2000'. Environment Australia, Canberra.

Garnett ST, Szabo JK and Dutson G (2011) *The Action Plan for Australian Birds 2010.* CSIRO Publishing, Melbourne.

Department of the Environment, Water, Heritage and the Arts (2008) 'Threat abatement plan for predation by feral cats'. Commonwealth of Australia, Canberra.

Conservation actions: Cat ownership was ceased on Lord Howe Island and responsible cat ownership is being encouraged on Norfolk Island.

FUTURE RECOVERY EFFORTS

What you can do:

- If you locate a live beach-cast Kermadec Petrel (a rare event, but possible during adverse weather conditions), it must be kept warm in a container. Take the bird to the local vet or wildlife carer. If you are on the Sunshine Coast, contact Australia Zoo; if you are on the Gold Coast, contact either Currumbin Sanctuary or Sea World. If the bird is dead, but in reasonable condition, freeze the specimen in a sealed plastic bag and contact the Qld Museum.
- Contact your respective state and federal MPs and senators and petition for the cessation of mass helium-filled balloon launchings at public events.
- Join/donate to the Southern Oceans Seabird Study Association, Birds Australia, Birds Qld, Bird Observation and Conservation Australia.
- Anglers should ensure they dispose of plastic bags thoughtfully onshore.

Conservation needs:

- Anglers should develop a substitute method for storing bait (reusable plastic containers, rather than plastic bags).

The conservation of the Kermadec Petrel is largely dependent on actions outside of Qld, particularly at breeding colonies, these include:

- Maintain a strict quarantine on vessels visiting Phillip Island and Ball's Pyramid.
- Eradicate Black Rats on Lord Howe and Norfolk Islands by trapping and baiting.
- Gradually cease cat ownership on Norfolk Island after neutering of all cats (with no renewal of breeding stock).

Sources and further reading

Comben P, Stewart D and Walbridge P (2001) Records of seabirds (Order: Procellariiformes) in south-east Queensland waters. *Sunbird* **31**, 33–72.

Harrison P (1985) *Seabirds: An Identification Guide.* Rev. edn. Christopher Helm, London.

Marchant S and Higgins PJ (Eds) (1990) *The Handbook of Australian, New Zealand and Antarctic Birds. Volume 1: Ratites to Ducks.* Oxford University Press, Melbourne.

Serventy DL, Serventy V and Warham J (1971) *The Handbook of Australian Sea-birds.* Reed, Sydney.

Author: Paul Walbridge

HERALD PETREL

Pterodroma heraldica (Salvin, 1888)

Photo: Pablo Caceres Contreras

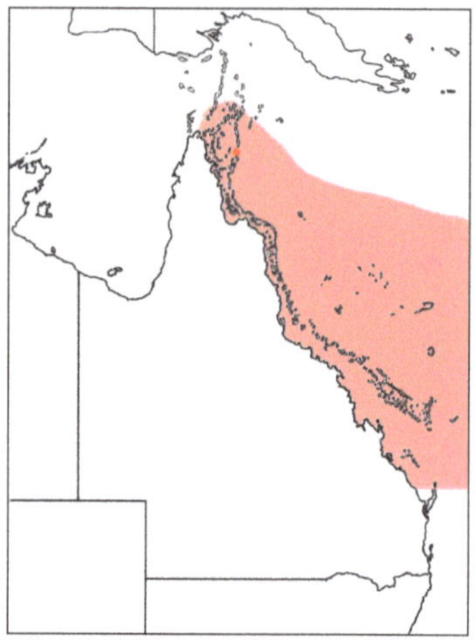

Family: Procellariidae

Conservation status: IUCN: Least Concern; Aust.: Critically Endangered; Qld: Endangered

Description: The Herald Petrel is a medium- to large-sized tropical gadfly petrel, measuring around 37 cm in total length, with a wingspan of around 95 cm. It occurs in three different colour morphs. The pale morph (the majority of Australian records) has a white body, with an ashy-brown head and an indistinct grey band across the chest. The underwing is dark grey with a white leading edge near the body and a white stripe that broadens into a white patch at the base of the primaries, producing the diagnostic dark M-band on the underwing. The upperparts are dark grey-brown with a darker shadowy M-band. The dark morph has a completely grey-brown body and head, and paler underwing markings. The intermediate morph is characterised by plumage between that of the light and dark morphs.

Distribution: The Herald Petrel has a wide range across the tropical and subtropical Pacific Ocean, mostly south of the equator to about 30°S. It breeds on islands and cays between the Coral Sea and Easter Island. Off Qld, it has bred on Raine Island, an outer cay of the northern Great Barrier Reef, and possibly other small cays in the Coral Sea (although surveys have failed to find evidence of this). There are occasional records from the Australian east coast, including south-east Qld and NSW. The Australian population is estimated at <50 mature individuals. The global population is estimated at >10 000 birds.

Habitat: These marine pelagic birds live in tropical and subtropical waters. Breeding habitat on Raine Island is on vegetated sand ridges with nests under dense low vegetation consisting of grasses, creepers and dense shrubs.

Ecology: The breeding habits of the Herald Petrel are poorly known. The nest is a shallow scrape with no lining material. The clutch size is one egg. Records from Raine Island are from February to September, with breeding suspected to occur from July to at least August. There are no records on the island from November to January, and at this time the birds are presumably dispersive at sea. At sea, they fly within 20 m of the water surface and feed on cephalopods (squid) and fish on or just below the surface.

Reason for decline and current threats: The very small Australian population size renders it vulnerable to catastrophic and stochastic events, such as the accidental introduction of predators or adverse weather conditions, which could potentially have a significant impact on a breeding colony.

The species was first recorded on Raine Island in 1959 and was a regular breeding bird in the 1980s, albeit in small numbers. Breeding was irregular (breeding some years and not others); however, only one breeding occurrence was observed on Raine Island from December 1987 to December 2003, and it was last recorded on the island in June 2002 when only one breeding pair was observed. Although conditions are reported to be suitable for breeding on Raine Island, the very low number of recent records suggests a decline in visitation to the island by the birds.

Threats to Herald Petrels include disturbance to nesting birds, erosion of nesting sites, vegetation changes and predation. At sea, pressures may include competition for marine resources with other species and the accidental entanglement in fishing gear (although this is probably not an issue for Herald Petrels because they do not generally follow boats). Their rarity in Australia suggests they may not be able to weather 'bad seasons', as other more common species may. With such a small population size, disease and genetic inbreeding may also pose risks.

CURRENT RECOVERY EFFORTS

Recovery plan: Department of Environment and Heritage (2005) 'Recovery plan for the following seabirds: Round Island Petrel – *Pterodroma arminjoniana*, Herald Petrel – *Pterodroma heraldica*, Antarctic Tern (New Zealand) – *Sterna vittata bethunei*, Antarctic Tern (Indian Ocean) – *Sterna vittata vittata*, Blue Petrel – *Halobaena caerulea*, Fairy Prion (southern) – *Pachyptila tutur subantarctica*, Heard Shag – *Phalacrocorax nivalis*, Macquarie Shag – *Phalacrocorax purpurascens*, Soft-plumaged Petrel – *Pterodroma mollis*, Australian Lesser Noddy – *Anous tenuirostris melanops*. 2005–2010'. Department of the Environment and Heritage, Canberra.

Other plans: Garnett ST and Crowley GM (2000) 'The action plan for Australian birds 2000'. Environment Australia, Canberra.

Garnett ST, Szabo JK and Dutson G (2011) *The Action Plan for Australian Birds 2010*. CSIRO Publishing, Melbourne.

Department of the Environment, Water, Heritage and the Arts (2008) 'Threat abatement plan for predation by feral cats'. Commonwealth of Australia, Canberra.

Conservation actions:
- Raine Island is protected in the Raine Island NP (Scientific). A Restricted Access – Special Management Area extends out to 500 m from the high water mark. Surrounding waters are within a marine national park (green) zone of the Great Barrier Reef Marine Park.
- Access to Raine Island is by permit only and is restricted to authorised persons undertaking management, scientific or monitoring activities.
- The Great Barrier Reef Marine Park Authority and DERM have a joint position statement on managing access to Raine Island.

FUTURE RECOVERY EFFORTS

What you can do:
- Adhere to regulations managing access to seabird colonies in the Great Barrier Reef region.
- Join/donate to the Southern Oceans Seabird Study Association, Birds Australia, Birds Queensland, Bird Observation and Conservation Australia.

Conservation needs:
- Continue to protect breeding habitat from human disturbance by controlling visitor access to Raine Island.
- Maintain strict quarantine arrangements at nesting islands to reduce the threat from exotic pests.
- Continue monitoring the breeding habitat at Raine Island to determine if the bird is present there.

Research directions:
- Determine the size and population trends on Raine Island by timing annual surveys to coincide with the breeding season.
- Survey other Coral Sea cays to establish the size and extent of the Australian population.
- Examine the movements and ecology of this species at sea.

Sources and further reading

Batianoff GN and Cornelius NJ (2005) Birds of Raine Island: population trends, breeding behaviour and nesting habitats. *Proceedings of the Royal Society of Queensland* **112**, 1–29.

Department of Environment and Heritage (2005) 'Issues paper. Population status and threats to ten seabird species listed as threatened under the *Environment Protection and Biodiversity Act 1999*'. Department of the Environment and Heritage, Canberra.

King BR (1984) The Herald Petrel *Pterodroma arminjoniana heraldica* breeding on Raine Island, Qld. *Emu* **84**, 246–247.

Marchant S and Higgins PJ (Eds) (1990) *Handbook of Australian, New Zealand and Antarctic Birds. Volume 1: Ratites to Ducks*. Oxford University Press, Melbourne.

Authors: Peter M Kyne and Lee K Curtis

HAWKS, EAGLES AND ALLIES (ORDER: ACCIPITRIFORMES)

RED GOSHAWK

Erythrotriorchis radiatus (Latham, 1801)

Photo: Chris Field

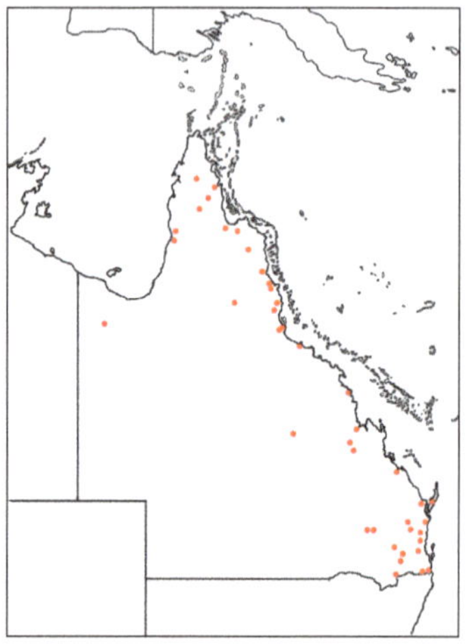

Family: Accipitridae

Conservation status: IUCN: Vulnerable; Aust.: Vulnerable; Qld: Endangered; NSW: Critically Endangered; NT: Vulnerable; WA: Vulnerable

Description: The Red Goshawk is a powerful, medium to large hawk, 48 cm (male) to 60 cm (female) in total length, with a wingspan up to 140 cm. It has broad shoulders and a short spiky crest on the nape. The plumage is reddish and streaky; the wings and tail are strongly barred and the long tail has a squarish tip. The legs are long, thick and yellow. The stance when perched is usually erect, but occasionally inclined; the wing tips do not project beyond the tail tip. The female is more robust and deep-chested than the male. Adults are grey-brown mottled with rufous and black above and rufous below, with whitish face and throat; the face and underparts are streaked black. Females have paler (almost white) underparts, with rufous flanks, 'trousers' and variable breast band. Juveniles are almost entirely rufous with finer streaking, and have cream to pale grey legs. The Red Goshawk soars with its wings extended, primaries spread (looks long-winged) and tail fanned, and glides with the tail closed, wrists pushed forward and outer wings backswept. It flies with rapid, deep and powerful strokes interspersed with glides, or slower, regular flapping with deep strokes for long periods.

Distribution: The Red Goshawk occurs across coastal and subcoastal northern (Kimberley Division of Western Australia to the Top End of the Northern Territory and North West Highlands of Qld) and eastern Australia (Cape York to Border Ranges of south-east Qld). It is apparently absent from the Gulf Plains. It formerly occurred as far south as the Sydney region, but is now suspected to be extinct in New South Wales south of the Border Ranges. It is occasionally recorded (mainly from gorge country) in central Australia and western Qld. The total Australian population is estimated to be about 700 breeding pairs. The historical population in Qld is estimated to have been about 180 pairs, and the present Qld population is around 100–140 pairs.

Habitat: The Red Goshawk is primarily found in extensive areas of coastal and subcoastal open forest and woodland that support a mosaic of vegetation types. Permanent freshwater (watercourses and wetlands) is usually present in close proximity to tall emergent trees that are used for nesting.

Ecology: The Red Goshawk is usually seen singly or in pairs. It soars and flies close to the canopy, or soars high and glides over long distances. It nests high in tall, often emergent, trees (*Eucalyptus*, *Corymbia*, *Angophora* and *Melaleuca* spp.). The nest is a substantial platform of sticks placed on a horizontal limb near the edge of the canopy. It is a very powerful raptor and, ecologically, more like a hawk-eagle than a typical goshawk. It mainly preys on medium to large birds (waterfowl, parrots, pigeons, kookaburras and large passerines), and rarely takes mammals, reptiles and other small animals. Most prey is taken in the air or near the ground; it will snatch prey following a fast or stealthy approach. There is no information concerning the long-term home range and territory dynamics for Red Goshawks. Preliminary telemetry data from the Top End (NT), during the late nestling and early post-fledging periods, showed that a breeding female commonly covered an area within a 5 km radius of the nest (in all directions) and, on one occasion, was recorded at 7 km from the nest, suggesting a home range estimated to be 120 km^2. During the non-breeding season, the male of this pair commonly flew up to 7 km from the nest and was frequently out of signal range (10 km). The male was usually located at distances of 8–8.5 km from the nest, suggesting a home range of 200 km^2.

Reason for decline and current threats: The primary threat is land clearing; that is, the loss of extensive forests and woodlands. It is suspected that the Red Goshawk's decline in eastern Australia may be related to loss of coastal forests and woodlands. The remaining population is now largely restricted to major reserves and inaccessible areas.

Other threats include fragmentation and degradation of remaining forest and woodland, loss of connectivity between remaining suitable habitats, declines in primary prey species (pigeons, declining woodland birds), disturbance, egg collecting and persecution.

CURRENT RECOVERY EFFORTS

Recovery plan: Baker-Gabb DJ (Submitted in 2005) 'Recovery plan for the Red Goshawk *Erythrotriorchis radiatus* 2007–2011'. Elanus Pty Ltd, Melbourne. (Under consideration.)

Other plans: Aumann T and Baker-Gabb DJ (1991) 'The ecology and status of the Red Goshawk in northern Australia'. RAOU Report 75, RAOU, Melbourne.

Garnett ST and Crowley GM (2000) 'The action plan for Australian birds 2000'. Environment Australia, Canberra.

Garnett ST, Szabo JK and Dutson G (2011) *The Action Plan for Australian Birds 2010.* CSIRO Publishing, Melbourne.

Conservation actions: Surveys have been conducted for Red Goshawks, including in north-east Qld, Cape York Peninsula and south-east Qld.

FUTURE RECOVERY EFFORTS

What you can do:

- Report possible sightings to relevant authorities (Qld Museum, DERM, Birds Australia or Birds Qld).
- Do not disturb nests or nesting birds.
- Do not disturb or destroy Red Goshawk habitat.

Conservation needs:

- Monitoring.
- Habitat protection.
- Appropriate fire management.

Research directions: To ensure the survival of the Red Goshawk, studies are needed of the home range and seasonal movements, as well as population biology and recruitment.

Sources and further reading

Czechura GV and Field C (2007) *Raptors of Southern Queensland.* Queensland Museum, Brisbane.

Czechura GV, Hobson RG and Stewart DA (2009) Observations on the biology of the Red Goshawk *Erythrotriorchis radiatus* in Queensland. *Australian Field Ornithology* **26**, 148–156.

Czechura GV, Hobson RG and Stewart DA (2011) Distribution, status and habitat of the Red Goshawk *Erythrotriorchis radiatus* in Queensland. *Corella* **35** (1), 3–10.

Debus S (1998) *The Birds of Prey of Australia: A Field Guide.* Oxford University Press, Melbourne.

Marchant S and Higgins PJ (Eds) (1993) *Handbook of Australian, New Zealand and Antarctic Birds. Volume 2: Raptors to Lapwings.* Oxford University Press, Melbourne.

Author: Greg Czechura

SHOREBIRDS, GULLS, BUTTON-QUAILS AND ALLIES (ORDER: CHARADRIIFORMES)

BEACH STONE-CURLEW

Esacus magnirostris Mathews, 1912

Photo: www.ataglance.com.au

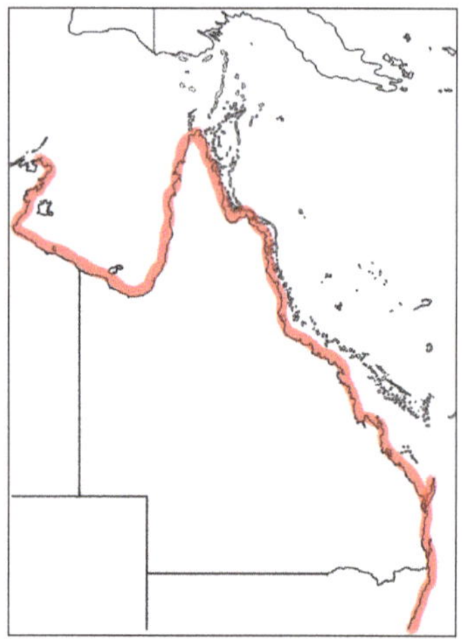

Family: Burhinidae

Other common name: Beach Thick-knee

Aboriginal names: Bilambirri, Kulawitbit, Taniu, Wee-lo

Conservation status: IUCN: Near Threatened; Qld: Vulnerable; NSW: Critically Endangered

Description: The Beach Stone-curlew is a large, thickset, grey-brown shorebird with a massive, yellow-based bill, a black and white face pattern, yellow eyes, a brown and white stripe on the forewings, and bold black and white patches on the wingtips in flight. It is 54–56 cm in total length, with a wingspan of 89–109 cm, and weighs up to 1 kg. It has stout legs with thick 'knees'.

Distribution: The Beach Stone-curlew has a disjunct distribution along the coast and on offshore islands in northern and eastern Australia, including the entire Qld coast. The Australian population is estimated at 5000 birds. However, the density may have declined where there are high levels of disturbance.

Habitat: Exclusively coastal, the Beach Stone-curlew occurs on all types of beaches including muddy, sandy, rocky, sheltered and surf beaches. It does not occur where there are large stretches of mangroves or cliffs. At high tide in the tropics, the birds may forage and roost on adjacent open habitats such as near-coastal airports 2 km inland, and even nest there on occasions.

Ecology: Adults are sedentary and mainly forage in the intertidal zone where they feed on crabs and other marine invertebrates. At high tide they roost in the shade of trees or in mangrove fringes. Beach Stone-curlews nest as single pairs and lay a single egg in a shallow depression on the landward edge of the beach. They may re-lay if nesting fails. Incubation lasts one month. The downy chicks are precocial, can run soon after hatching, and fly well when they are 3 months old. Both adults defend the nest and care for the young that may be dependent for as long as 7–12 months. In Qld, eggs have been recorded from mid-July to mid-November.

Reason for decline and current threats: Sensitivity to human disturbance has led to declines in Beach Stone-curlew densities on beaches in southern Qld and inhabited islands of the Great Barrier Reef and Torres Strait. The species is at risk of further declines in density and local and regional extinctions related to ongoing coastal development, including the loss of nesting and roosting habitat as a result of dune removal and degradation. Disturbance by beach users, boats and vehicles disrupts breeding and may lead to breeding failure. The predation of adults, chicks and eggs by cats, dogs and feral pigs are possible causes of reduced breeding success. Because high tides may wash out nests, rising sea levels resulting from global climate change may also threaten breeding habitat. Illegal egg collectors may occasionally take eggs. The Beach Stone-curlew's food supply is adversely affected by human impacts such as 4WD vehicles on beaches.

CURRENT RECOVERY EFFORTS

Recovery plan: None

Other plans: Garnett ST and Crowley GM (2000) 'The action plan for Australian birds 2000'. Environment Australia, Canberra.

Conservation actions:

- Surveys of Beach Stone-curlew distribution and abundance on sections of the Qld coast and islands have been conducted, usually as part of general wader surveys.

- A database of Beach Stone-curlew records for far north Qld was operated by members of Birds Australia North Qld between 1994 and 2007.
- As part of their coastal bird monitoring strategy, DERM and Great Barrier Reef Marine Park Authority staff collect records of Beach Stone-curlews on Great Barrier Reef islands in the course of routine patrols.

FUTURE RECOVERY EFFORTS

What you can do:

- Avoid driving vehicles on beaches and mudflats.
- Keep dogs under control and well away from feeding, roosting and nesting birds.
- Record sightings of Beach Stone-curlews and contribute records to the *Atlas of Australian Birds* (Birds Australia).
- If you live on the coast, monitor Beach Stone-curlew occurrence and breeding success in your local area.
- Stay informed about coastal development plans and write submissions on development applications, urging controls on beach access and use.

Conservation needs:

- Control coastal development and pollution.
- Limit disturbance, particularly during the breeding season, by controlling use of beaches by people and their dogs and restricting vehicular access.
- Implement control measures for feral animals

Research directions:

- Survey and monitor the distribution, density and breeding success to identify important and/or vulnerable sections of coast and determine long-term trends.
- Investigate the relationship between levels of disturbance and breeding success of the Beach Stone-curlew.

Sources and further reading

Freeman AND (2003) The distribution of Beach Stone-curlews and their response to disturbance on far north Queensland's Wet Tropical Coast. *Emu* **103**, 369–372.

Hole H, Hole B and Mardell C (2001) Observations of nesting Beach Stone-curlews on the mid-north coast of New South Wales. *Australian Bird Watcher* **19**, 49–54.

Maguire GS (2008) *A Practical Guide for Managing Beach-nesting Birds in Australia*. Birds Australia, Melbourne. <http://www.birdsaustralia.com.au/images/stories/current-projects/bnb/bnb_Manual_part1a_sm.pdf>.

Marchant S and Higgins PJ (Eds) (1993) *Handbook of Australian, New Zealand and Antarctic Birds. Volume 2: Raptors to Lapwings*. Oxford University Press, Melbourne.

Milton D (1998) Distribution and abundance of Beach Stone-curlews on northern Great Barrier Reef islands. *Sunbird* **28**, 31–38.

Author: Amanda Freeman

PLAINS-WANDERER

Pedionomus torquatus Gould, 1840

Photo: David Parker

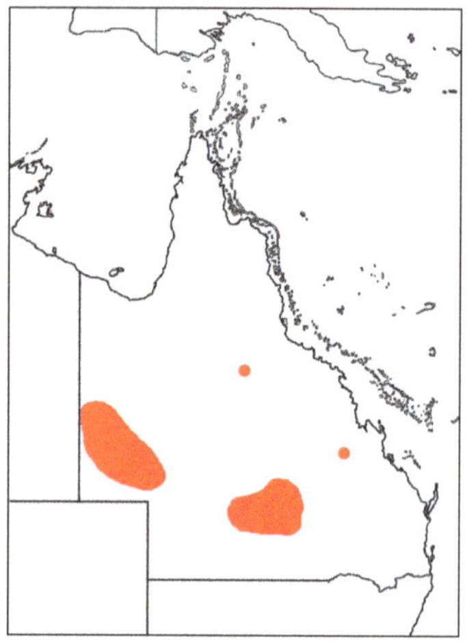

Family: Pedionomidae

Other common names: Plain Wanderer, Collared Plains-wanderer, Turkey Quail

Conservation status: IUCN: Endangered; Aust.: Vulnerable; Qld: Vulnerable; NSW: Endangered; NT: Data Deficient; SA: Endangered; Vic.: Threatened

Description: The Plains-wanderer is a small quail-like bird, with a fawn finely mottled and scalloped black, a pale belly and yellow legs, and a pale wing-bar in flight. It has a total length of 12–15 cm, with a wingspan of 28–36 cm; males weigh 40–80 g and females, 55–95 g. The female has a prominent white-spotted black collar above a rufous breast patch. Juveniles are similar to adult males, but have spotted rather than scalloped underparts.

Distribution: Historically, the Plains-wanderer occurred widely in natural grasslands on the plains of inland eastern Australia, from inland NSW and Vic. to eastern SA, extending north to central Qld and possibly into the southern NT (though that is unconfirmed). Today its stronghold is the NSW Riverina, with populations in north-central Vic., central-western Qld and north-eastern SA. In Qld, it occurs mainly in grasslands of the Channel Country in the far west, particularly in northern Astrebla Downs NP, southern Diamantina Lakes NP and Sandringham Station. Scattered records exist in native grasslands east and south-west from this region. Across this continental range, most Plains-wanderers occur on private land.

Habitat: Plains-wanderers live in sparse, low native grasslands, typically on hard, red-brown soils having about 50% bare ground and widely spaced grass tussocks up to 10–20 cm apart. Most vegetation is less than 5 cm high, with some grass up to 30 cm tall. In arid inland areas, such as Qld, Plains-wanderer habitat commonly occurs in channel country, on hard red-brown soil between the riverine black-soil plains and the dunes. Plains-wanderers occasionally occur in other habitats, including stubble, low cereal crops and grassy chenopod shrub lands.

Ecology: Entirely terrestrial, Plains-wanderers forage during the day on a wide variety of seeds and ground-dwelling insects. They are sedentary in home ranges of 7–21 ha (average 12 ha) per bird, moving only in response to changes in their habitat. Normally solitary, Plains-wanderers form pairs in late winter. In southern Australia, they lay first clutches mainly between late August and early November, with second clutches in January if rains fall during summer. In Qld, Plains-wanderers breed in autumn and early winter. They may not breed during drought, or very poorly in flood years. They lay three to five eggs in a shallow scrape on the ground, sometimes lined or hooded with grass. The male incubates for 23 days and broods and guards the chicks, leaving the female to pair with another male. Chicks are independent about 2 months after hatching and can breed in their first year.

Reason for decline and current threats: Modern agricultural practices – including cultivation, dense pasture growth of fertilised native grasses or introduced ones and overgrazing by livestock and rabbits, as well as the absence of grazing, when followed by widespread rainfall and prolific grass growth, which can render native grasslands too dense – have led to the decline of this species

Current threats include: feral predators (foxes and cats), especially around cultivated lands where mice sustain large predator populations; pesticides sprayed on locust plagues, such as fenitrothion or fipronil, which may kill birds or deplete invertebrate foods; and large-scale, intensive wildfires and drought, which can exacerbate the impacts of overgrazing, deplete food resources (seeding of grasses and forbs, and insects) and prevent the Plains-wanderer from breeding.

CURRENT RECOVERY EFFORTS

Recovery plans: Baker-Gabb D (2002) 'Recovery plan for the Plains-wanderer *Pedionomus torquatus* 2002–2006: Conservation of lowland native grassland dependent fauna'. Environment Australia, Canberra.

NSW Department of Environment and Climate Change (2008) 'NSW revised draft recovery plan for the Plains-wanderer (*Pedionomus torquatus*)'. NSW Department of Environment and Climate Change, Queanbeyan.

Other plans: Garnett ST and Crowley GM (2000) 'The action plan for Australian birds 2000'. Environment Australia, Canberra.

Garnett ST, Szabo JK and Dutson G (2011) *The Action Plan for Australian Birds 2010.* CSIRO Publishing, Melbourne.

Department of the Environment, Water, Heritage and the Arts (2008) 'Threat abatement plan for predation by feral cats'. Commonwealth of Australia, Canberra.

Conservation actions:

- The Australian Plague Locust Commission and NSW DECC conduct research into impacts of locust control chemicals on Plains-wanderer.
- Bush Heritage Australia manages Cravens Peak in far-western Qld: a large habitat reserve (233 000 ha). It also manages two reserves in SA (Boolcoomatta and Bon Bon).
- The Australian Wildlife Conservancy manages the Kalamurina Sanctuary in SA.

FUTURE RECOVERY EFFORTS

What you can do:

- Join/donate to Bush Heritage Australia, The Australian Wildlife Conservancy, Birds Australia, Bird Observation or Conservation Australia.
- Write letters to state and federal environment ministers, urging more funding and the implementation of research recommendations.
- Contact your local NRM officer about best-practice native grassland management.
- Protect remnant grasslands in your community or on your land.
- Support local conservation efforts by joining a local Landcare or Catchment group, field naturalists club, natural history or 'friends of' group, or by volunteering for groups such as Green Corps or the Australian Trust for Conservation Volunteers.

Conservation needs:

- Identify, protect and manage critical habitat, and monitor changes in its status.
- Purchase reserves and establish private refuge areas.

Research directions: Conduct field surveys to determine the population size and distribution in central-western Qld.

Sources and further reading

Baker-Gabb DJ (1998) Native grasslands and the plains-wanderer. Birds Australia Conservation Statement No. 1. *Wingspan* **8**, 1–8.

Maher PN and Baker-Gabb DJ (1993) 'Surveys and conservation of the Plains-wanderer in northern Victoria'. ARI Technical Report No. 132, Department of Conservation and Natural Resources, Melbourne.

Marchant S and Higgins PJ (Eds) (1993) *Handbook of Australian, New Zealand and Antarctic Birds. Volume 2: Raptors to Lapwings.* Oxford University Press, Melbourne.

Parker DG and Oliver DL (2007) *Plains-wanderer Habitat Management Guide: A Guide for Visually Assessing the Grassland Structure of Plains-wanderer Habitat.* 3rd edn. NSW Department of Environment and Climate Change and Murray Catchment Management Authority.

Story P, Oliver DL, Deveson T, McCulloch L, Hamilton G and Baker-Gabb D (2007) Estimating and reducing the amount of Plains-wanderer (*Pedionomus torquatus* Gould) habitat sprayed with pesticides for locust control in the New South Wales Riverina. *Emu* **107**, 308–314.

Author: David G Parker

AUSTRALIAN PAINTED SNIPE

Rostratula australis (Gould, 1838)

Photo: Ian Gynther

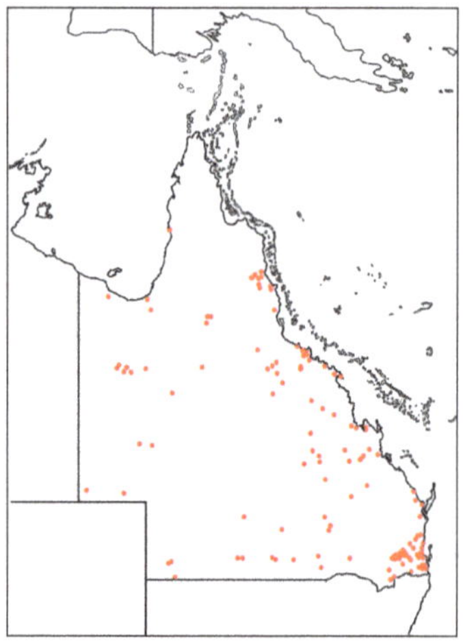

Family: Rostratulidae

Aboriginal name: Puulokor (western Vic.)

Conservation status: IUCN: Endangered; Aust.: Vulnerable; Qld: Vulnerable; NSW: Endangered; NT: Vulnerable; SA: Vulnerable; Vic.: Critically Endangered; WA: Vulnerable

Description: This is a medium-sized shorebird with a long, slightly drooping bill. It has a total length of 24–30 cm with a wingspan of 50–54 cm, and it weighs 100–140 g. It has cryptic plumage, with white underparts, a bold white 'comma' round the eyes, and a unique white 'harness'. Female have rich dark foreparts and their olive-green wing-coverts are finely barred black. Males have duller foreparts and their wing-coverts are spangled buff.

Distribution: The Australian Painted Snipe is scattered throughout Australia, mostly in the eastern half, especially the Murray–Darling Basin. Sightings in Qld include the Channel Country and the Fitzroy Basin and, recently, from floodwater plains of coastal central and north Qld where the birds are suspected to be regular migrants occurring in autumn and winter. Individuals are nomadic, remaining at sites temporarily. Population estimates vary from 1250 to 5000.

Habitat: Specialised when breeding, these birds nest in temporary freshwater or brackish wetlands with low vegetation (avoiding tall dense reeds) during the flush of productivity that follows recent flooding by fresh water. Recently flooded temporary wetlands are their preferred habitats during the non-breeding period, but various other fresh to slightly brackish wetlands are used if necessary, especially if there are shallow muddy shores and nearby low dense vegetation for cover. Artificial habitats that are occasionally used include reservoirs, farm dams, sewage ponds, inundated grassland and leaking irrigation channels. The birds spend most daylight hours at roosts in thick, usually low, vegetation near water.

Ecology: The species feeds from dusk to early morning, mostly in shallow water or adjacent wet mud, on small items including insects, freshwater snails and worms, and on seeds of wetland plants. In southern Australia, it breeds mostly between October and early February. In northern Australia, nesting occurs during, or soon after, the wet-season rains, mainly from March to May, but potentially until at least August, depending on flood levels. Nests are simple cups on the ground, typically set on tiny islands surrounded by water. The birds are suspected to be polyandrous (females mating with two or more males) as suggested by the larger, brighter females sometimes having long-term associations with multiple males, and the males performing the parental care.

Reason for decline and current threats: The decline is attributed to the loss and fragmentation of breeding habitat (i.e. shallow, ephemeral wetlands) through intensive water management and agricultural development, especially in the Murray–Darling Basin. Reduced flood frequency, and loss of much fresh water to irrigation and other diversions, an increase and stabilisation of water levels, creating permanent wetlands with deep water and dense reed-beds, and where islands and complex shallow margins are removed, and cropping on ephemeral wetlands all contribute to the loss and fragmentation of habitat.

Other potentially negative factors include: reduced water quality (e.g. lack of flushing, increased nutrient run-off, saline discharge, pesticide and herbicide run-off or spray drift, and vegetation removal leading to erosion and turbidity) and its effect on the vegetation and invertebrates of temporary wetlands; invasive introduced plants (e.g. Lippia *Phyla canescens* in the Murray–Darling Basin, *Parkinsonia* in the tropics); and grazing (e.g. pugging of wetland edges, damage to vegetation and increased sedimentation and turbidity, especially by cattle and feral pigs).

CURRENT RECOVERY EFFORTS

Recovery plan: Currently in production.

Other plans: Garnett ST and Crowley GM (2000) 'The action plan for Australian birds 2000'. Environment Australia, Canberra.

Garnett ST, Szabo JK and Dutson G (2011) *The Action Plan for Australian Birds 2010.* CSIRO Publishing, Melbourne.

Conservation actions: Birds Australia's Threatened Bird Network conducts biannual Australian Painted Snipe surveys.

FUTURE RECOVERY EFFORTS

What you can do:

- Join volunteer surveys.
- Report confirmed sightings to DERM and Birds Australia.

Conservation needs:

- Protect critical habitat and monitor any changes in its status.
- Establish and maintain periodic environmental flows to wetlands.
- Generate GIS-based habitat models and maps (breeding and non-breeding habitat) to identify and predict critical habitat.
- Conduct extension activities, such as educational material (brochures/DVDs) and regional workshops.
- Trial appropriate hydrological regimes through environmental flows at strategic wetlands and investigate the effects of grazing management on wetlands.
- Assess whether minor modifications to irrigated agricultural practices could be used to provide breeding habitat (e.g. the related Greater Painted Snipe *R. benghalensis* breeds in rice fields in Asia).
- Lobby governments and their agencies, highlighting the need for active management of wetlands, particularly environmental flows.

Research directions:

- Develop survey and monitoring guidelines.
- Undertake extensive field surveys throughout inundated arid and semi-arid lands following floods.
- Conduct detailed research on the Australian Painted Snipe's biology and ecology (e.g. foraging-site attributes, roost-site attributes, breeding-site attributes, water quality, prey sampling, interactions with other species and movement patterns using radiotelemetry).
- Analyse data to identify population trends, distribution, critical habitat, breeding-event patterns and movements.
- Develop a greater knowledge of distribution and movements (e.g. identification of key breeding sites, movement patterns and links to rainfall events).
- Determine habitat requirements (e.g. management of temporary wetlands including water fluctuations and timing, food sources and prey habitat requirements, water quality).
- Continue targeted surveys, linked with other broad-scale wetland survey programs (e.g. ground-truth potentially suitable sites identified from aerial wetland surveys).
- Determine the detectability of the species (e.g. identifying key indicator species, developing wetland survey methods and landscape-scale sampling methods).
- Study demographics (e.g. total population, fecundity, population viability, longevity and mortality).

Sources and further reading

Baker AJ, Pereira SL, Rogers DI, Elbourne R and Hassell CJ (2007) Mitochondrial-DNA evidence shows the Australian Painted Snipe is a full species, *Rostratula australis*. *Emu* **107,** 185–189.

Black R, Houston W and Jaensch R (2010) Evidence of regular seasonal migration by Australian Painted Snipe *Rostratula australis* to the Queensland tropics in autumn and winter. *Stilt* **58,** 1–9.

Oring LW, Rogers D, Oring KE and Tzaros C (2004) Snipes in peril. *Wingspan* **14,** 10–15.

Rogers, D, Hance I, Paton S, Tzaros C, Griffioen P, Herring M, Jaensch R, Oring L, Silcocks A and Weston M (2005) The breeding bottleneck: breeding habitat and population decline in the Australian Painted Snipe. In *Status and Conservation of Shorebirds in the East Asian-Australasian Flyway.* (Ed. P Straw) pp. 15–23. Wetlands International, Sydney.

Thomas J, Purnell C, Ingwersen D, Tzaros C and Rogers D (2010) Meet Australia's newest species. *Wingspan* **20**, 20–25.

Authors: Chris Tzaros, Dean Ingwersen and Danny Rogers

BLACK-BREASTED BUTTON-QUAIL

Turnix melanogaster (Gould, 1837)

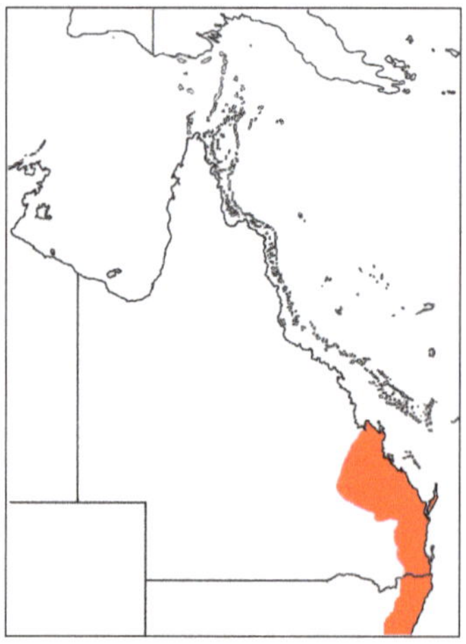

Photo: Micha V Jackson

Family: Turnicidae

Conservation status: IUCN: Vulnerable; Aust.: Vulnerable; Qld: Vulnerable; NSW: Critically Endangered

Description: Males are approximately 18 cm in total length, females are 19 cm, both with wingspans of 32–35 cm. Males weigh approximately 65 g, females up to 100 g. Males have finely patterned backs and wings with brown, black, grey and white mottling; the face and throat are whitish and the breast is black with numerous white crescents. The female colouring is similar in all respects except for having a black face and throat, and a larger dark area over the upper and lower breast with heavier white crescents. The bill of both sexes is grey, the eyes cream-white and the legs are pale yellow. Juveniles resemble males but are duller.

Distribution: The Black-breasted Button-quail is scattered in eastern Qld and NSW from near Byfield to north-eastern NSW, in the Northern Rivers and Tablelands areas, with only 10 reliable recent records from six areas in far north-eastern NSW. Reports from farther south in the Dorrigo and Walcha–Yarrowitch areas – the latter in the 1970s – cannot be verified owing to habitat removal. Most extant populations occur in Qld, with 14 known populations. These generally occur east of the Great Dividing Range, with a few records on its western slopes, up to 300 km inland at locations such as Palmgrove NP. The two main populations occur in south-east Qld, in the semi-evergreen vine thickets of the Yarraman–Blackbutt region and in coastal dune scrubs of the Great Sandy region. There were two reports of Black-breasted Button-quail from the Atherton and Cooktown regions, well north of the accepted range of the species, in the late 1800s to early 1900s, but none since. A crude estimate for the Yarraman–Blackbutt population is 2000 adults.

Habitat: Key habitats of the species include: vine thickets and rainforests that are periodically water-stressed, such as semi-evergreen vine thicket, low microphyll vine forest, Araucarian microphyll vine forest, Araucarian notophyll vine forest and *Brachychiton* scrubs having Bottle Trees *Brachychiton rupestre*, Brigalow *Acacia harpophylla* and Belah *Casuarina cristata*; low thickets or woodlands with a dense understorey but with little ground cover, typically dominated by *Acacia* spp.; and in littoral situations, dry vine scrubs, acacia thickets and areas densely covered in shrubs, particularly Midgen Berry *Austromyrtus dulcis*.

Ecology: This is a serially polyandrous (female breeds with several males), ground-dwelling and ground-nesting species, usually producing three to four young per clutch, which are incubated and tended solely by the male. Black-breasted Button-quails are commonly seen in pairs or occasionally in small groups, particularly males with young. Being territorial, females are occasionally seen singly. Its dispersion patterns are poorly known. It is generally considered to be sedentary, although it can appear intermittently or as a transient in marginal habitats. In a study of 13 rainforest patches in an agricultural landscape, foraging occurred in the three largest patches only (17.4, 40.0 and 63.8 ha), with highest occurrence in the largest patch for periods greater than 100 days, suggesting that patch size is critical to persistence. The preferred habitats of the species typically have thick leaf litter 3–10 cm deep. Foraging occurs in leaf litter by 'pivot-feeding'. The birds scratch at the leaf litter with one leg while pivoting the body on the other, thus displacing leaves and soil, exposing invertebrate prey. This scratching produces distinctive circular feeding depressions 15–25 cm in diameter. Some other button-quail species (e.g. Painted Button-quail *Turnix varia*) feed in a similar fashion and it can be difficult to distinguish species from their feeding scrapes. The diet consists mainly of small, ground-dwelling invertebrates and seeds.

Reason for decline and current threats: These include: the loss and fragmentation of habitat by clearing for a range of purposes (timber-harvesting and other forestry-related practices, agriculture, infrastructure construction and urban development); habitat degradation by domestic stock and feral pigs using habitat, and invasion of habitat by weeds; habitat loss or degradation by fire, including prescribed burning; and predation by feral animals.

CURRENT RECOVERY EFFORTS

Recovery plan: Mathieson MT and Smith GC (2009) 'National recovery plan for the black-breasted button-quail *Turnix melanogaster*'. Report to the Department of the Environment, Water, Heritage and the Arts, Canberra. Department of Environment and Resource Management, Brisbane.

Other plans: Garnett ST and Crowley GM (2000) 'The action plan for Australian birds 2000'. Environment Australia, Canberra.

Garnett ST, Szabo JK and Dutson G (2011) *The Action Plan for Australian Birds 2010*. CSIRO Publishing, Melbourne.

Department of the Environment, Water, Heritage and the Arts (2008) 'Threat abatement plan for predation by feral cats'. Commonwealth of Australia, Canberra.

Conservation actions:
- There is a management plan for Mount Blandy in response to the construction of the Burnett Dam.
- Local councils and land developers are mitigating impacts of housing developments by reducing urban footprint and subdivisions, retaining habitat and placing covenants on land
- Species management is undertaken in state forests (species management profiles).

FUTURE RECOVERY EFFORTS

What you can do:
- Report confirmed sightings to DERM.

Conservation needs:
- Implement the recovery plan.
- Map the habitat of the species and conduct searches for new populations in mapped habitat.
- Develop management practices and protocols for human activities in Black-breasted Button-quail habitat, and regulate land use through state and local authorities.
- Protect areas of habitat and rehabilitate degraded areas, including strategic weed removal coinciding with native habitat replacement by bush-regeneration techniques.
- Establish extension activities with landholders and the development of community networks.
- Investigate and implement feral animal control programs.
- Involve traditional owners in the recovery effort.

Research directions:
- Undertake population estimates.
- Examine the impact of population isolation on genetics.

Sources and further reading

Hamley T, Flower P and Smith GC (1997) Present and past distribution of the Black-breasted Button-quail *Turnix melanogaster* (Gould) in Queensland. *Sunbird* **27**, 1–21.

Lees N and Smith GC (2000) Use of mature hoop pine plantation by the vulnerable Black-breasted Button-quail (*Turnix melanogaster*). *Australian Forestry* **62**, 330–335.

Marchant S and Higgins PJ (Eds) (1993) *Handbook of Australian, New Zealand and Antarctic Birds. Volume 2: Raptors to Lapwings*. Oxford University Press, Melbourne.

Smith GC, Aridis J and Lees N (1998) Radio-tracking revealed home ranges of Black-breasted Button-quail *Turnix melanogaster* in remnant vine scrub between hoop pine plantation and agriculture. *Emu* **98,** 171–177.

Smyth AK and Pavey CR (2001) Foraging by the Black-breasted Button-quail (*Turnix melanogaster*) within fragmented rainforest of an agricultural landscape. *Biological Conservation* **98,** 149–157.

Authors: Geoffrey C Smith and Michael Mathieson

BUFF-BREASTED BUTTON-QUAIL

Turnix olivii Robinson, 1900

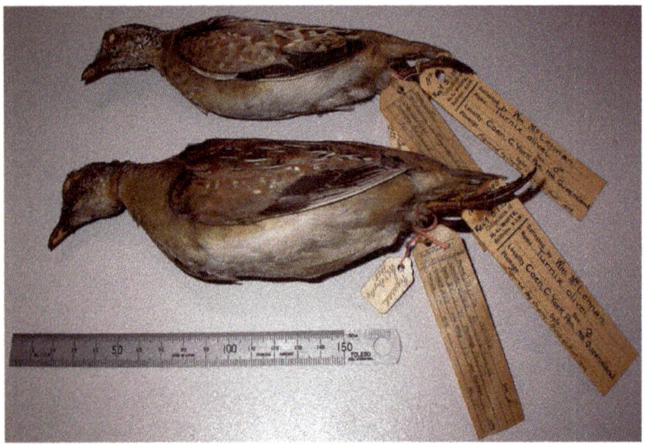

Photo: Geoffrey C Smith

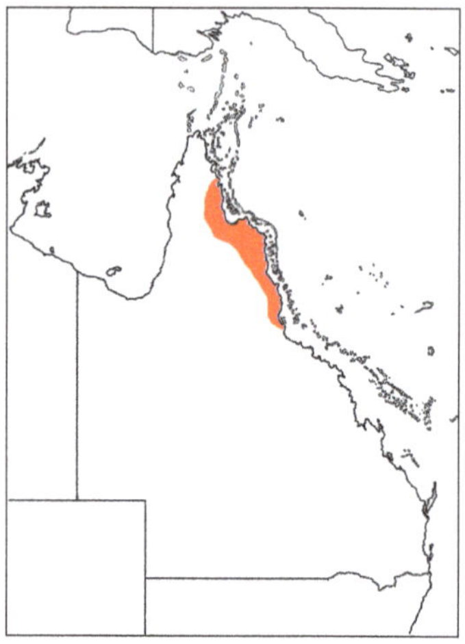

Family: Turnicidae

Other common names: Buff-backed Quail, Buff-breasted Quail, Olive's Quail

Conservation status: IUCN: Endangered; Aust.: Endangered; Qld: Vulnerable

Description: Males are about 18 cm in total length, with females slightly larger (19–22 cm). The species is large and plump, similar in build to Black-breasted Button-quails (65–100 g). The face is grey to pale grey and the eyes are yellow. The dorsal plumage is largely pale brown to chestnut with scattered black markings and white flecks. The flight feathers are dark grey-black. The ventral plumage is largely buff to pale cream. The sexes differ only slightly in plumage, with females having a darker face. The bill is grey and the legs are pale yellow. Juveniles are unknown.

Distribution: This species is endemic to north-east Qld, with scattered records from the eastern parts of Cape York Peninsula, the Wet Tropics and the eastern fringe of the Einasleigh Uplands, and unconfirmed dry season records from the Karumba area. Owing to lack of detailed information, it is not known whether discrete populations exist. The apparently high mobility of the species suggests that there is only a single population throughout its entire known range. Therefore, any location supporting Buff-breasted Button-quail is probably important to the long-term survival of the species.

Habitat: Buff-breasted Button-quail are most frequently reported from stony and/or grassy woodlands and forests. Broad-leaved Paperbark *Melaleuca viridiflora* and/or Small Leafed Paperbark *M. minutifolia* are often common midstorey species. The sparsely wooded, well-drained, slightly sloping bases of hills that support this vegetation appear critical during the breeding season. This species has also been reported from grassy clearings within and adjacent to rainforest patches.

Ecology: The Buff-breasted Button-quail is one of our least-known bird species. There are only seven specimens in museum collections, and few sightings at few localities. It was reported only four times for *The New Atlas of Australian Birds* between 1998 and 2002 and not at all during the survey period for the first *Atlas of Australian Birds* between 1977 and 1981. It is a ground-dwelling and ground-nesting species, usually producing three to four eggs per clutch during the wet season, in mid-January to late March. The nest is a dome of woven dry grasses in a shallow depression in the ground, often at the base of a grass clump or shrub for camouflage. It is assumed (though not yet confirmed) to be polyandrous, as with all other button-quail species. Buff-breasted Button-quails are usually seen in pairs, or occasionally in small groups of up to five birds. The dispersion patterns of this species are not known. Reportedly, birds appear in the Mareeba area in the late dry season and remain until the grass becomes very dense. If the wet season arrives late or is weak, the grasses do not become so dense and the birds seem to remain and breed. The gut contents of four birds consisted of insects and seeds, with coarse sand as gizzard grit. The calls of the Buff-breasted Button-quail apparently include repeated booms, like those made by females of other button-quail species.

Reason for decline and current threats: The reason for the decline of this species and its rarity are largely unknown, but are strongly suspected to be related to habitat degradation and clearing and predation by native animals such as goannas *Varanus* spp. and by feral pigs.

The habitat of the species is prone to degradation from grazing and burning regimes that encourage extensive fires and promote woody-weed invasion. Extensive dry season fires may eliminate large areas of habitat. Some areas in the south of the Buff-breasted Button-quail's range are in danger of being cleared for agriculture. Although the impact of predation by feral animals on the species is not well known, it is likely that the incidence of predation is linked to loss and degradation of habitat. Animals under stress are likely to be subject to increased levels of predation, accentuated by being forced to disperse through unsuitable habitat.

CURRENT RECOVERY EFFORTS

Recovery plan: Mathieson MT and Smith GC (2009) 'National recovery plan for the buff-breasted button-quail *Turnix olivii*'. Report to Department of the Environment, Water, Heritage and the Arts, Canberra. Department of Environment and Resource Management, Brisbane.

Other plans: Garnett ST and Crowley GM (2000) 'The action plan for Australian birds 2000'. Environment Australia, Canberra.

Garnett ST, Szabo JK and Dutson G (2011) *The Action Plan for Australian Birds 2010.* CSIRO Publishing, Melbourne.

Conservation actions: DERM scientists (the authors) are currently developing survey and capture methods for the species, to facilitate future large-scale surveys and radio-tracking studies to determine the extent of occurrence and habitat preferences.

FUTURE RECOVERY EFFORTS

What you can do: Report sighting of this little-known, infrequently observed species to DERM.

Research directions:

- Gather basic information regarding where these birds occur throughout the year and their habitat preferences.
- Develop useful methods for surveying Buff-breasted Button-quail, and implement them.

Sources and further reading

Barrett G, Silcocks A, Barry S, Cunningham R and Poulter R (2003) *The New Atlas of Australian Birds.* Royal Australasian Ornithologists Union, Hawthorn East.

Macdonald JD (1971) Validity of the Buff-breasted Quail. *Sunbird* **2,** 1–5.

Marchant S and Higgins PJ (Eds) (1993) *Handbook of Australian, New Zealand and Antarctic Birds. Volume 2: Raptors to Lapwings.* Oxford University Press, Melbourne.

Squire JE (1990) Some southern records and other observations of the Buff-breasted button-quail *Turnix olivei. Australian Bird Watcher* **13,** 149–152.

White HL (1922) Description of nest and eggs of *Turnix olivii* (Robinson). *Emu* **22,** 2–3.

Authors: Michael Mathieson and Geoffrey C Smith

LITTLE TERN (WESTERN PACIFIC SUBSPECIES)

Sternula albifrons sinensis (Gmelin, 1789)

Photo: Blesbok Enterprises

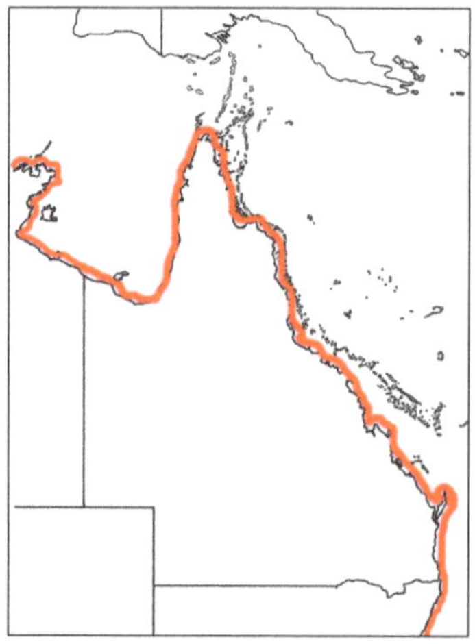

Family: Laridae

Conservation status: IUCN: species (*S. albifrons*) is Least Concern; Qld: Endangered; NSW: Endangered; SA: Endangered; Tas.: Endangered; Vic.: Vulnerable

Description: The Little Tern is a very small and slender tern measuring 20–28 cm in total length with a wingspan of 45–55 cm. It has a small head with a sloping forehead, a bill about as long as the head, very narrow wings and a fairly long and deeply forked tail (the tail tip is much shorter than the wing tips when resting). It is similar in size and appearance to the Fairy Tern *Sternula nereis*, but the Little Tern has dark outer primaries and dark lores in breeding birds. In breeding plumage, the cap is black, the forehead is white and extends back over the eye in a short eyebrow, the bill is pale yellow and usually has a black tip, and the legs are yellow. In non-breeding plumage, there is a black band from in front of the eye extending back to the nape, the forehead and crown are white, the bill and legs are black, and there is a dark grey shoulder bar. Young birds are like non-breeding adults, but have less black in front of the eyes, scaled feathers on the wings and yellowish-black legs.

Distribution: The Little Tern has a widespread distribution throughout large parts of Europe, Africa, Asia and Australia, with three recognised races. Race *sinensis* occurs in South-East and East Asia, New Guinea and northern and eastern Australia. There are three populations of this race that occur in Australia: a northern population, a south-eastern population and a non-breeding migratory population from Asia. In Qld, the northern population breeds north of Mackay into the Gulf of Carpentaria and the southern population breeds south of Bundaberg. Non-breeding Asian migrants also visit Qld.

Habitat: The Little Tern is mainly found along sheltered coastal areas, including beaches, sheltered inlets, estuaries, lakes, sewage farms, lagoons, river mouths and deltas.

Ecology: Little Terns nest on undisturbed and unvegetated sites near estuaries and adjacent freshwater lakes, on islands and on coral cays, with nests between the high tide mark and shore vegetation. Nesting sites are inherently vulnerable to natural disturbance in the form of adverse weather and tidal conditions, and from encroachment of dune vegetation. Birds usually attempt to re-nest if the first clutch is lost. Most south-eastern Australian birds appear to move north outside the breeding season and at least some migrate to Asia. It is possible that the northern Australian population does not migrate. Little Terns feed by plunging into shallow water of channels and estuaries and in surf off beaches, feeding primarily on small fish.

Reason for decline and current threats: An increase in human disturbance, principally at nesting sites, combined with a high natural rate of breeding failure (loss of eggs or chicks owing to the ephemeral nature of breeding sites) threatens the Little Tern within Australia.

Little Terns nest on open ground, leaving their eggs and chicks exposed and vulnerable to disturbance. These are sometimes crushed and nests abandoned as a direct result of human disturbance, including 4WD vehicles, trail bikes and people walking. Unleashed dogs also cause disturbance to nesting birds. Furthermore, nests and adults sitting on them are exposed to predation by introduced predators (foxes, domestic and feral cats and dogs, wild pigs and rats have all been recorded as predators of the Little Tern). Additionally, the presence of weeds can degrade the habitat to the extent where it is no longer functional for nesting. Industrial, residential and recreational development has encroached on Little Tern breeding areas and has resulted in a loss of nesting habitat. Many human activities (pollution, over-fishing, loss of important estuarine habitats and changes to estuarine morphology and hydrology) can affect the food resources available to the Little Tern.

CURRENT RECOVERY EFFORTS

Recovery plan: NSW National Parks and Wildlife Service (2003) 'Little Tern *Sterna albifrons* Recovery Plan'. NSW National Parks and Wildlife Service, Hurstville. This plan covers NSW only.

Other plans: Garnett ST and Crowley GM (2000) 'The action plan for Australian birds 2000'. Environment Australia, Canberra.

Conservation actions:

- There is active management (fencing, signage, education and compliance) and monitoring of some Qld breeding colonies (e.g. Barubbra Island near Bundaberg and South Stradbroke Island).
- Some Qld colonies are covered within the reserve system.

FUTURE RECOVERY EFFORTS

What you can do:

- If you live near a Little Tern roosting or nesting site, contact your local bird group to become involved in monitoring, protecting and maintaining the site.
- Join/donate to the Southern Oceans Seabird Study Association (SOSSA), Birds Australia, Birds Qld, Bird Observation and Conservation Australia.
- If you own a dog, abide by local regulations and avoid walking your dog on beaches where terns are present. Always walk your dog on a lead.
- If you own a 4WD, reconsider your need to drive on beaches.

Conservation needs:

- Continue the monitoring and management of nest sites in areas with high rates of disturbance.
- Protect and maintain known and potential habitat to minimise human disturbance and increase breeding success (e.g. implement protection zones, fence sensitive nesting areas, erect signage and undertake warden patrols).
- Prepare a recovery plan for Qld.
- Control introduced predators around nesting sites.
- Raise public awareness and use local community groups as site wardens.
- Protect and manage significant non-breeding sites (e.g. Caloundra and Noosa Sandbanks).

Research directions:

- Determine the extent of interbreeding with the Fairy Tern.
- Determine the extent of interchange between colonies and philopatry of breeding sites.

Sources and further reading

Chan K and Dening J (2007) Use of sandbanks by terns in Queensland, Australia: a priority for conservation in a popular recreational waterway. *Biodiversity and Conservation* **16**, 447–464.

Egan K and Webb G (2000) A useful strategy for the management of eggs dislodged from the nests of Little Terns *Sterna albifrons*. *Corella* **24**, 13–14.

Higgins PJ and Davies SJJF (Eds) (1996) *The Handbook of Australian, New Zealand and Antarctic Birds. Volume 3: Snipe to Pigeons*. Oxford University Press, Melbourne.

Rose AB (2001) Predation of nesting Little Terns *Sterna albifrons* at Forster, New South Wales. *Australian Bird Watcher* **19**, 79–87.

Taylor IR and Roe EL (2004) Feeding ecology of little terns *Sterna albifrons sinensis* in south-eastern Australia and the effects of pilchard mass mortality on breeding success and population size. *Marine and Freshwater Research* **55**, 799–808.

Author: Tim Holmes

PARROTS AND COCKATOOS (ORDER: PSITTACIFORMES)

GLOSSY BLACK-COCKATOO

Calyptorhynchus lathami (Temminck, 1807)

Photo: Blesbok Enterprises

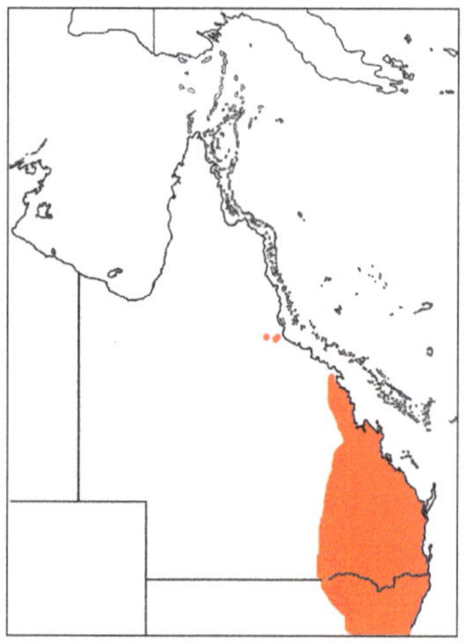

Family: Cacatuidae

Other common names: Casuarina, Casuarine, Glossy or Latham's Cockatoo, Leach's Black-Cockatoo, Leach's Red-tailed Cockatoo, Nutcracker

Aboriginal name: Garadi (Dharawal people of NSW)

Conservation status: IUCN: Least Concern; Qld: Vulnerable; NSW: Vulnerable; SA: Endangered; Vic.: Threatened

Description: This is the smallest of the black-cockatoos, with a total length of 48 cm and a wingspan of 90 cm, and weighing 420 g. It has a broad, bulbous bill and an inconspicuous crest. The body is dull black, with a brownish-black head, neck and underparts, and red tail panels. Adult females have yellow patches on the head and neck, and barred tail panels. Young birds have yellow spots or bars on the breast, belly and flanks, some yellow spots on the head, and barred tail panels. The species is generally more cryptic than, and not as raucous as, the other species of black-cockatoos. It is often confused with the Red-tailed Black-Cockatoo *Calyptorhynchus banksii.*

Distribution: This cockatoo occurs from central-coastal Qld to eastern Vic., with an isolated population on Kangaroo Island; it apparently became extinct in Tasmania and mainland South Australia in historic times. In Qld, the south-eastern subspecies *C. l. lathami* occurs from the NSW border north to Gympie–Chinchilla and west to Mitchell, and the apparently isolated northern subspecies *C. l. erebus* occurs from Bundaberg north to the Paluma Range and west to the Expedition Range. It occurs sparsely throughout, with highest densities east of the Great Dividing Range and scattered occurrences inland.

Habitat: Prefers woodland dominated by she-oaks (Casuarinaceae) or open woodland with a middle stratum of she-oaks, often now as remnant patches in hills and gullies surrounded by cleared agricultural land. In south-east Qld, it is also found in she-oak/Brigalow woodlands, coastal lowlands and offshore islands, and isolated trees in some residential areas close to more extensive habitat.

Ecology: This cockatoo feed almost exclusively on seeds from the cones of a few species of she-oaks, particularly Black She-oak *Allocasuarina littoralis* and Mountain or Forest She-oak *A. torulosa*. It breeds in tree hollows, usually in eucalypts, and rears a single young at a time.

Reason for decline and current threats: The clearing of habitat (especially for agriculture, plantations and timber) has resulted in local extinctions and range contractions. In coastal areas, the clearing of habitat for residential development, resulting in the loss of food and nest trees, is a continuing threat. It nests in hollows in old trees, which take many years to be replaced.

More frequent and intense fires since European settlement have reduced suitable habitat. Burning fire-sensitive she-oaks can leave feeding habitat unsuitable for several years. Once shed, seeds of she-oaks are not viable for long and are susceptible to fire. Frequent fire also prevents seedlings from reaching maturity, and grazing suppresses she-oak regeneration.

Recent introductions and range expansions of other cockatoos have increased competition for a reduced number of nest hollows (owing to land-clearing), from the Long-billed Corella *Cacatua tenuirostris* (introduced to Qld), Galah *Eolophus roseicapillus* (expanded range after clearing for agriculture) and Little Corella *C. sanguinea* (expanded range and new populations from aviary escapes).

Chicks and eggs are collected from the wild for aviculture, but there are few data on the current level of impact. The current rate of decline may be masked by the longevity and slow reproduction of the species.

CURRENT RECOVERY EFFORTS

Recovery plan: None

Other plans: Department of the Environment and Heritage (2005) 'Threat abatement plan for beak and feather disease affecting endangered psittacine species'. Commonwealth of Australia, Canberra.

Garnett ST and Crowley GM (2000) 'The action plan for Australian birds 2000'. Environment Australia, Canberra.

Garnett ST, Szabo JK and Dutson G (2011) *The Action Plan for Australian Birds 2010*. CSIRO Publishing, Melbourne.

Conservation actions:

- Griffith University is leading a Glossy Black-Cockatoo project, which aims to assess the current distribution and habitat use of the species in the Gold Coast area in order to better protect habitat, and thus guarantee the birds' long-term survival.
- A community-based census at sites across Gold Coast City was undertaken in 2009 and similar events are planned for future years.
- A trial radio-tracking study is underway to monitor movements, and a genetic study is being planned to determine regional variation between local populations.

The Glossy Black Conservancy, a coalition of government, non-government, commercial and academic bodies:

- coordinates the assessment of food, nest and water resources across local government areas in south-east Qld, engaging the community to collect sighting records, raising awareness by engaging the general community, schools and environmental groups; provides information, including development of a website; and manages a central database of Glossy Black-Cockatoo records;
- continues research to increase knowledge of the species' use of pioneer vegetation, its fidelity to individual feed trees, requirement for nesting hollows, and diurnal and seasonal range and movement patterns, in order to map its vital resources in south-east Qld;
- monitors key threats and helps to refine management practices, such as the use of fire, thinning and clearing in Glossy Black-Cockatoo habitat;
- leads the development of a Regional Management Plan for Glossy Black-Cockatoos in south-east Qld.

FUTURE RECOVERY EFFORTS

What you can do:

- Report sightings and accurate locations of Glossy Black-Cockatoos, and their nests, drinking and feeding sites (look for chewed cones), to The Glossy Black Conservancy. Participate in community-based censuses of the Gold Coast or other locations coordinated by The Glossy Black Conservancy. Report sightings online: <http://www.glossyblack.org.au/form.html> or by email: GBC@biodiversity.tv
- Join/donate to Birds Australia, Birds Qld and Bird Observation and Conservation Australia.

Sources and further reading

Cameron M (2007) *Cockatoos*. CSIRO Publishing, Melbourne.

Cameron M and Cunningham RB (2006) Habitat selection at multiple spatial scales by foraging Glossy Black-Cockatoos. *Austral Ecology* **31**, 597–607.

Higgins PJ (Ed.) (1999) *Handbook of Australian, New Zealand and Antarctic Birds. Volume 4: Parrots to Dollarbirds*. Oxford University Press, Melbourne.

Robinson T and Paull D (2009) Comparative evaluation of suburban bushland as foraging habitat for the Glossy Black-Cockatoo. *Corella* **33**, 7–12.

Stock MJ and Wild CH (2005) Seasonal variation in Glossy Black-Cockatoo *Calyptorhynchus lathami* sightings on the Gold Coast, Queensland. *Corella* **29**, 88–90.

Author: Tim Holmes

MAJOR MITCHELL'S COCKATOO

Lophochroa leadbeateri (Vigors, 1831)

Photo: www.ataglance.com.au

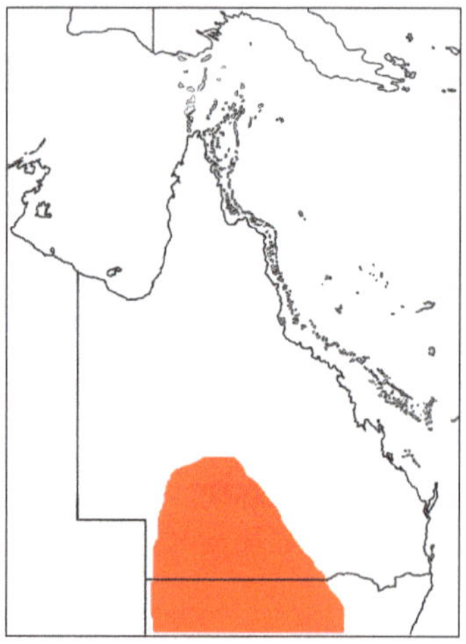

Family: Cacatuidae

Other common names: Leadbeater's, Desert, Pink Cockatoo, Cocklerina, Chockalott, Wee Juggler

Conservation status: IUCN: Least Concern; Qld: Vulnerable; NSW: Vulnerable; SA: Rare; Vic.: Vulnerable, WA: Specially Protected

Description: Major Mitchell's Cockatoo is a small pink and white cockatoo, with a total length of 39 cm, and a wingspan of 80 cm. It has broad rounded wings, a rather short tail and a long, narrow forward-curving crest with a distinct broad peach-red band bisected by a single yellow band, only visible when the crest is erect. It is unmistakable, but broadly resembles a small Sulphur-crested Cockatoo *Cacatua galerita*.

Distribution: It is endemic to Australia, found mainly in inland arid regions. In Qld, it is restricted to southern sections of the south-west and south-central regions, extending from west of Eromanga, north along the Barcoo River to Isisford and east to Roma and St George. Occasionally it extends east to Goondiwindi and the Darling Downs near Warra. Records around Brisbane are aviary escapees. Although the overall population appears to have declined in Qld, there has been a range expansion to the east, with these cockatoos arriving in St George in the 1930s.

Habitat: The preferred habitat is semi-arid and arid regions, in dry woodlands, particularly mallee *Eucalyptus–Callitris–Casuarina* vegetation communities.

Ecology: The species has been studied only in the mallee Wheatbelt of WA, and in Wyperfeld NP in the mallee of Vic. This cockatoo is a cavity-nesting specialist, requiring a large hollow, preferably in the vertical or nearly vertical trunk or limb of a live or dead tree. In the Victorian mallee it prefers to nest in Slender Cypress-pine *Callitris gracilis*, with hollows averaging 53 cm deep and 20 cm wide. These nest trees average 74 cm in diameter at breast height, and are upwards of 170 years old. The cockatoo breeds in spring, from August to December, in Vic. It is entirely or mostly sedentary in areas of permanent water, but is usually nomadic in the arid zone where movements and fluctuations in numbers are dictated by availability of food and water. The diet includes the seeds of grasses and herbaceous plants, fruits, roots, bulbs and unidentified galls and insect larvae.

Reason for decline and current threats: In most parts of Australia its range has declined, mostly owing to fragmentation of habitat, trapping and nest robbing for aviculture, and, in some cases, persecution. In addition, cereal cropping, to which most suitable habitat has been converted, does not provide an adequate variety or continuous supply of food to sustain viable populations.

Current threats include inadequate extent, quality and connectivity of remaining habitat, and continual clearing of habitat and loss of large old, hollow-bearing trees. The total grazing pressure from stock, rabbits and goats further degrades habitat, removes seedling grasses (thus preventing regeneration of food plants) and prevents the successful regeneration of nest-tree species. Inappropriate fire regimes also kill nest trees and can inhibit their regeneration. Weed invasion can also impede the recruitment of nest-tree species. The taking of adults, young and eggs for the domestic and international trade, although illegal throughout Australia, is still considered to be widespread and affecting local populations. Birds at nests are very sensitive to nearby human disturbances. It is known to be persecuted for damage to cereal and almond crops.

CURRENT RECOVERY EFFORTS

Recovery plans: None

Other plans: Department of Sustainability and Environment (1999) 'Flora and Fauna Guarantee Action Statement no. 87: Major Mitchell's Cockatoo'. Victoria Department of Sustainability and Environment, Melbourne.

Department of the Environment and Heritage (2005) 'Threat abatement plan for beak and feather disease affecting endangered psittacine species'. Commonwealth of Australia, Canberra.

Garnett ST and Crowley GM (2000) 'The action plan for Australian birds 2000'. Environment Australia, Canberra.

Garnett ST, Szabo JK and Dutson G (2011) *The Action Plan for Australian Birds 2010*. CSIRO Publishing, Melbourne.

Qld Environment Protection Agency (2007) 'Conservation management profile: Major Mitchell's Cockatoo'.

Conservation actions: The NSW Department of Environment and Conservation has identified eight priority actions to help recover the Major Mitchell's Cockatoo in NSW. <http://www.threatenedspecies.environment.nsw.gov.au/tsprofile/pas_profile.aspx?id=10116>.

FUTURE RECOVERY EFFORTS

What you can do:

- Landholders are encouraged to enter voluntary conservation agreements to protect and manage habitat and other site management agreements.
- Keep Major Mitchell's Cockatoo nest site localities confidential, to minimise illegal collection.
- Notify DERM if you suspect any illegal activities.

Conservation needs:

- Map and assess suitable habitat.
- Identify the remaining populations through targeted survey and community-based monitoring.
- Protect the remaining populations and their habitat from human threats.
- Regenerate and rehabilitate their habitat to improve its extent, quality and connectivity.
- Identify at least two target populations (per year over an initial three years); focus the recovery actions and adaptive management at these sites.
- Identify, protect and monitor nest trees and reproduction.
- Control stock and feral grazing species (especially goats, rabbits and camels) in known foraging areas.
- Tackle the threat of illegal collection (particularly nestlings).
- Conduct regular targeted audits (in late spring) of avicultural collections of this species.

Research directions:

- Further investigate the distribution and ecology of the species in Qld.
- Conduct nest-site surveys to monitor extant populations.
- Investigate the health and age structure (via size classes) of preferred nest tree species to determine projected recruitment rates for future hollow bearing trees.
- Study the dietary preferences and home-range size.
- Conduct research on the impact of feral bees on nest-site availability.
- Determine the level of competition with more common cockatoo species (such as Galahs *Eolophus roseicapillus* and Little Corellas *Cacatua sanguinea*) for nest hollows.

Sources and further reading

Cameron M (2007) *Cockatoos*. CSIRO Publishing, Melbourne.

Gibson M, Florentine S and Hurley VG (2008) 'Age distribution of Slender-Cypress pine (*Callitris gracilis*) within Pine Plains, Wyperfeld National Park'. Unpublished report, prepared for the Department of Sustainability and Environment by the Centre for Environmental Management, University of Ballarat, Victoria.

Higgins PJ (Ed.) (1999) *Handbook of Australian, New Zealand and Antarctic Birds. Volume 4: Parrots to Dollabirds*. Oxford University Press, Melbourne.

Hurley VG (2006) 'Physical characteristics and thermal properties of Major Mitchell's Cockatoo (*Cacatua leadbeateri leadbeateri*) nest hollows, Wyperfeld NP 2006'. Unpublished Report, Department of Sustainability and Environment, Mildura, Victoria.

Hurley VG (2006) 'Survey of Major Mitchell's Cockatoo in Pine Plains, Wyperfeld NP 2006'. Unpublished Report, Department of Sustainability and Environment, Mildura, Victoria.

Hurley VG (2010) 'Results from the 2010 breeding survey of Major Mitchell's Cockatoo (*Lophochroa l. leadbeateri*) in Pine Plains, Wyperfeld NP'. Unpublished report, Department of Sustainability and Environment, Mildura, Victoria.

Author: Victor G Hurley

COXEN'S FIG-PARROT

Cyclopsitta diophthalma coxeni Gould, 1867

Artwork: Sally Elmer

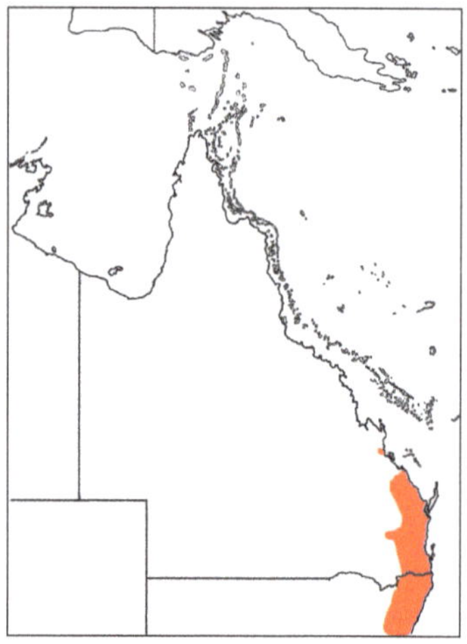

Family: Psittacidae

Other common names: Blue-browed, Red-faced or Southern Fig-Parrot or Lorilet

Conservation status: IUCN: species (*C. diophthalma*) is Least Concern; Aust.: Endangered; Qld: Endangered; NSW: Critically Endangered

Description: The Coxen's Fig-parrot is a small, mostly green parrot with a 'dumpy' appearance, disproportionately large head and bill and a short tail; it has a total length of 15–16 cm. It has a yellow line along the flanks, the outer wings are edged deep blue and there are red spots on the rear inner wings near the body. The face is orange-red, bordered by a blue forehead and mauve-blue lower cheeks (these colours being duller and less extensive in the female). The bill is grey.

Distribution: It is found from Gympie in south-east Qld west to the Bunya Mountains and south to the Richmond River in north-east NSW. Recent credible reports extend the Qld range north to Maryborough, Bundaberg, Gin Gin, Miriam Vale and Rockhampton. The population is very small, with perhaps only 100 mature individuals remaining, in four small subpopulations.

Habitat: The preferred habitat was probably lowland dry and subtropical rainforest. Recent reports come from sea level to 900 m rainforests, including littoral rainforest, gallery forest, dry rainforest, and warm and cool subtropical rainforests. Recent records come from coastal open forest of paperbarks *Melaleuca*, bloodwoods *Corymbia* and cabbage palms *Livistona* north of Bundaberg, and riparian corridors through woodland, open woodland or cleared land, but only where fig *Ficus* spp. or other food species occur. The parrot also visits fruiting trees in gardens and farmland.

Ecology: The diet includes the seeds of ripe or near-ripe figs, the seeds or flesh of other fruits, nectar, insect larvae and lichen. The nest is a chamber excavated by the birds in the dead wood of a vertical trunk or branch or the underside of a sloping limb or trunk in rainforest or nearby open forest, and proximity to fruiting trees may be important. Excavation commences in August, with breeding continuing to December or January. The parrot may be a seasonal altitudinal migrant, moving from highland areas after the breeding season to the lowlands in autumn and winter, possibly in response to the local availability of food. At some lowland sites in south-east Qld, where food is present year-round, birds have been reported in all seasons. In the non-breeding season, the parrots may form small flocks of up to seven.

Reason for decline and current threats: The decline was probably caused by clearing and fragmentation of lowland subtropical and dry rainforest, and logging of rainforest.

Remnant populations may largely be restricted to the forests of the foothills and ranges, where a lack of year-round fruit may place them under additional pressure. Ongoing threats include: inadequate extent, quality and connectivity of habitat (making the search for food more difficult and increasing predation risk when crossing open areas); seasonal gaps in food availability; exotic invasive trees, shrubs and vines, particularly in gallery rainforests and other lowland riparian corridors; and disturbance to breeding areas from inappropriate burning, logging or clearing.

CURRENT RECOVERY EFFORTS

Recovery plans: Coxen's Fig-Parrot Recovery Team (2001) 'Coxen's Fig-Parrot *Cyclopsitta diophthalma coxeni* recovery plan 2001–2005'. Report to Environment Australia, Canberra. Queensland Parks and Wildlife Service, Brisbane.

NSW National Parks and Wildlife Service (2002) 'Approved recovery plan for the Coxen's Fig-Parrot *Cyclopsitta diophthalma coxeni* (Gould)'. NSW National Parks and Wildlife Service, Sydney.

Other plans: Garnett ST and Crowley GM (2000) 'The action plan for Australian birds 2000'. Environment Australia, Canberra.

Garnett ST, Szabo JK and Dutson G (2011) *The Action Plan for Australian Birds 2010.* CSIRO Publishing, Melbourne.

Department of the Environment and Heritage (2005) 'Threat abatement plan for beak and feather disease affecting endangered psittacine species'. Commonwealth of Australia, Canberra.

Conservation actions:
The following research, monitoring, planning, extension activities and captive management are in place:

- Deploying an autonomous call-detection system at a Coxen's Fig-Parrot site; searching known and potential habitat for nesting activity; surveying fruiting fig trees; compiling sighting records, including reports from the community; and maintaining a records database.
- A preliminary genetics project is under way, using museum and fresh DNA of Australian and other subspecies, such as *C. diophthalma.*
- Identification/survey workshops are taking place to enhance community skills via information sessions.
- Liaising with state and local government planning authorities to protect and expand remnant habitat.
- Currumbin Wildlife Sanctuary maintains an analogue captive population of Macleay's Fig-Parrot *C. d. macleayana.*

FUTURE RECOVERY EFFORTS

What you can do:

- Photograph any Coxen's Fig-Parrots seen and record their calls; urgently report all fig-parrot sightings and call records in central and southern Qld to DERM; if you find a dead fig-parrot *anywhere* in Qld, freeze it and immediately contact the Qld Museum.
- Plant known or potential fig-parrot food trees on your property.
- Participate in local rainforest rehabilitation and replanting projects.
- Donate to Currumbin Wildlife Sanctuary's Coxen's Fig-Parrot recovery project.
- Write letters to state and federal environment ministers, urging that funding be made available for implementation of the recovery plan.

Conservation needs:

- Identify the remaining populations through targeted survey and community-based monitoring.
- Protect the remaining populations and their habitat from human threats.
- Assess and map any suitable habitat.
- Regenerate and rehabilitate suitable habitat to improve its extent, quality and connectivity.
- Initiate captive breeding for a release program.
- Increase community participation in the recovery program.

Research directions:

- Further investigate the distribution and ecology of the subspecies.
- Investigate new sighting reports.
- Deploy autonomous call-recording devices.
- Conduct nest-site surveys to locate extant populations.
- Investigate dietary preferences, home-range size, roosting behaviour, activity schedules and movement patterns of other Australian subspecies.

Sources and further reading

Beruldsen GR (2002) Erroneous clutch of Coxen's Fig-Parrot. *Australian Bird Watcher* **19**, 196–198.

Gynther I, Martindale J and O'Reilly P (1998) 'Final report on the Coxen's Fig-Parrot survey. Project number 585'. Unpublished report to the Threatened Species and Communities Section, Environment Australia, Canberra.

Higgins PJ (Ed.) (1999) *Handbook of Australian, New Zealand and Antarctic Birds. Volume 4: Parrots to Dollarbirds.* Oxford University Press, Melbourne.

Holmes G (1990) 'The biology and ecology of Coxen's Fig-Parrot'. RAOU Report No. 65. RAOU, Melbourne.

Martindale J (1986) 'A review of literature and the results of a search for Coxen's Fig-Parrot in South-East Queensland and North-East New South Wales during 1985'. RAOU Report No. 21, RAOU, Melbourne.

Author: Ian Gynther

MACLEAY'S FIG-PARROT

Cyclopsitta diophthalma macleayana Ramsay, 1874

Photo: Michael Cermak

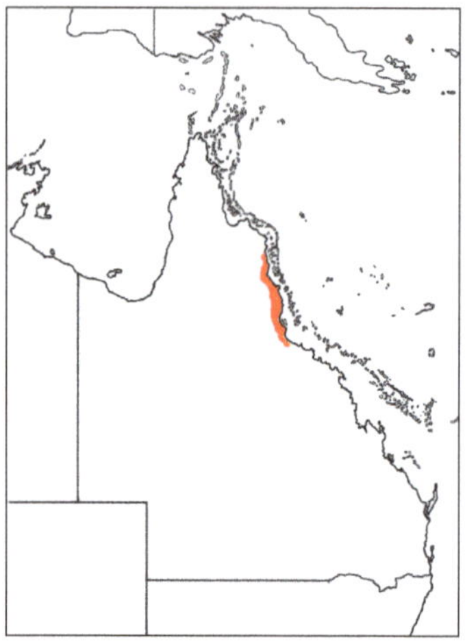

Family: Psittacidae

Other common names: McCoy's, Leadbeater's, Blue-faced or Red-browed Fig-Parrot or Lorilet

Conservation status: IUCN: species (*C. diophthalma*) is Least Concern; Qld: Vulnerable

Description: These fig-parrots are small, with a total length of 13–15 cm, wingspan of 25–28 cm, and weigh about 35 g, with plump bodies, short tails and large heads. They are mostly bright green with a prominent yellow line along the flanks and have blue edges to the outer wings and red spots on the rear inner wings near the body. Male Macleay's Fig-Parrots have a blue face with a red forehead cheek stripe; females have a blue face (paler on the cheeks) with red only on the forehead.

Distribution: Macleay's Fig-Parrot is endemic to the Wet Tropics region of north Qld, from south of Cooktown to Paluma. There are believed to be about 5000 pairs. It is isolated from the other subspecies: Marshall's Fig-Parrot *C.d. marshalli* on Cape York Peninsula, and Coxen's Fig-Parrot *C.d. coxeni* in subtropical coastal south-east Qld and northern NSW.

Habitat: Macleay's Fig-Parrot prefers lowland coastal rainforests, but also occurs up to 1200 m. Its habitat includes tropical semi-deciduous vine forests and gallery forests that include fig trees *Ficus* spp. It also occurs at the edge between rainforest and sclerophyll woodland, and more rarely in cleared or fragmented forests. It sometimes occurs, and even breeds, in parks and gardens in north Qld towns.

Ecology: Fig-parrots breed in self-excavated hollows in the underside of dead branches in living or dead trees, in rainforest and adjoining *Eucalyptus* and *Melaleuca* woodland. They prefer the seeds of figs, but also eat other fruits and nectar. They discard the fig pulp, so the sounds of falling debris, and their soft chattering, can aid in detection. Fig-parrots also tear bark from trees in search of insect larvae. They occur in pairs and small groups during the breeding season (September to January), or in small flocks outside the breeding season. The female does the excavation and most of the parental care, but the male may assist. The clutch of usually three eggs is incubated for 20–24 days, and nestlings fledge after 36–42 days. Fig-parrots form communal roosts, sometimes in hundreds, in tall trees in the rainforest canopy or in parks and gardens.

Reason for decline and current threats: Likely causes of Macleay's Fig-parrot's possible decline include loss of habitat to logging or clearing for farmland, and remnants are small and fragmented, causing difficulties for foraging or dispersing birds through loss of connectivity between foraging areas, increased predation when crossing open areas, and intermittent gaps in food availability. The remaining habitat is also vulnerable to weed invasion and loss of fig trees. Similar habitat degradation has occurred in north Qld: much of the original lowland rainforest has been cleared for agriculture, and weed invasion is a serious threat to many remaining patches. Macleay's Fig-Parrot appears to have similar ecological requirements and conservation issues as Coxen's Fig-Parrot. However, the small size and camouflaged colour of fig-parrots make them difficult to survey and census. Illegal collection for aviculture may continue. Psittacine Beak and Feather Disease may also pose a threat. The two north Qld subspecies are subject to tropical cyclones, including loss of nesting trees and disruption of food supplies, which may increase in frequency with climate change.

CURRENT RECOVERY EFFORTS

Recovery plan: None

Other plans: Garnett ST and Crowley GM (2000) 'The action plan for Australian birds 2000'. Environment Australia, Canberra.

Conservation actions:

- Captive Macleay's Fig-Parrots are held at Currumbin Wildlife Sanctuary, and elsewhere in Australia, to investigate the subspecies' reproductive biology and refine husbandry techniques.

- Personnel from the Coxen's Fig-Parrot Recovery Team locate and monitor wild Macleay's Fig-Parrots, and techniques for attachment of radio-tracking devices are being trialled. This will enable the full investigation of the conservation status of wild Macleay's Fig-Parrots.

FUTURE CONSERVATION EFFORTS

What you can do:

- Document sightings both in natural habitat and around towns, and especially note breeding events.
- Join a Landcare group that control weeds on the edges of rainforests, to stop habitat degradation, and that also implements revegetation programs to restore degraded or former areas of rainforest and to establish corridors between them.

Conservation needs:

- Conserve remaining lowland rainforest in north Qld as a high priority.
- Expand existing community-based tree-planting programs, to include figs and other fruiting trees.
- Community awareness and involvement may be essential for arresting the likely decline in Macleay's Fig-Parrot numbers, in particular to avoid the parlous state of Coxen's Fig-Parrot.

Research directions:

- Undertake a major study of movements, ecology and breeding biology, including the fate of young birds, seasonal movements and the effect of seasonal bottlenecks in food availability over at least 3 years, including the establishment of sites for longer-term monitoring.
- Investigate predation levels and relative population densities by examining the remains in, and near, the nests of likely predators, including goshawks and owls.
- Establish survey protocols to identify the best habitat, and to understand why some areas are better than others.
- Undertake detailed mapping of such 'hotspots' so that conservation efforts can be more targeted.
- Planned genetic research on the taxonomic status of Coxen's Fig-Parrot will also elucidate the taxonomic status of Macleay's Fig-Parrot.

Sources and further reading

Charley D (2000) Predator-avoidance behaviour in the Double-eyed Fig-Parrot. *Australian Bird Watcher* **18**, 203.

Courtney J (2010) An observation of nesting behaviour in Marshall's Fig-Parrot *Cyclopsitta diophthalma marshalli*. *Australian Field Ornithology* **27**, 149–152.

Forshaw JM and Cooper WT (2002) *Australian Parrots*. 3rd edn. Alexander Editions, Robina.

Higgins PJ (Ed.) (1999) *Handbook of Australian, New Zealand and Antarctic Birds. Volume 4: Parrots to Dollarbirds*. Oxford University Press, Melbourne.

Juniper T and Parr M (1998) *Parrots: A Guide to the Parrots of the World*. Pica Press, Sussex, UK.

Author: Robert Heinsohn

ECLECTUS PARROT (CAPE YORK PENINSULA SUBSPECIES)

Eclectus roratus macgillivrayi Mathews, 1913

Photo: www.ataglance.com.au

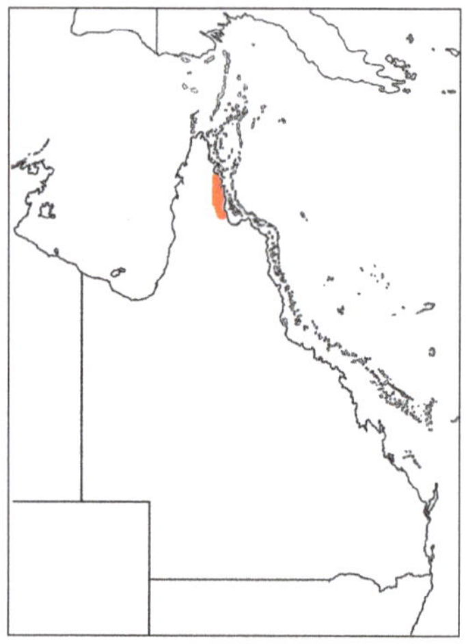

Family: Psittacidae

Aboriginal name: Tinta

Conservation status: IUCN: species (*E. roratus*) is Least Concern; Qld: Vulnerable

Description: This is a large, short-tailed parrot with a total length of 40–48 cm and a wingspan of 83–90 cm, and weighing about 500–630 g. Males are mostly green, with red on the sides of the abdomen extending on to the otherwise dark underwings, and blue outer wing feathers. The beak is orange with a paler tip and the eyes are orange-red. Females are red, darker on the wings, back and tail, with blue outer wings, eye-rings, collar and belly band extending to the underwings. The eyes are yellow and the beak is black.

Distribution: The Australian subspecies is endemic to far north Qld, and is restricted to the Iron and McIlwraith Ranges of eastern Cape York Peninsula. Although suitable habitat is also available in other areas, such as Lockerbie Scrub and Olive River, these rainforest patches are too small to support viable populations. The Australian population is estimated at 2500–3500 birds.

Habitat: Eclectus Parrots mainly inhabit the interior and edges of lowland rainforest, but also higher rainforest regions. They need emergent trees with large hollows in the trunk or larger branches for nesting, and on Cape York Peninsula favour three key species or genera as nest trees: Black Bean *Castanospermum australe*, Milky Pine *Alstonia* spp. and Green Figs *Ficus albipila*.

Ecology: Useable hollows are rare: only one per square kilometre of rainforest. Many nest hollows also flood in heavy rain, causing even large chicks to drown. A good hollow is one that stays dry for at least 4 months over the breeding period: 1 month for incubation plus 3 months from hatching to fledging. The threat of losing the hollow forces the females to stay at and defend their nest trees for up to 11 months each year, avoiding starvation because they are fed by the males. Each female is attended by up to five males, which are unrelated to each other or the female. Males outnumber females by about two to one, and often feed and mate with more than one female at nests up to 10 km apart. All hollows are in bright light, and females sit at the entrance with their bright red head and chest showing as a clear signal to other females. However, such conspicuousness leads to high mortality in young females. Old females with good hollows use them to hide from predators, but young females without hollows are twice as likely to be preyed upon. Their diet consists of fruit, nuts, seeds, flowers and nectar, obtained from the tree canopy over very large ranges.

Reasons for decline and current threats: Habitat degradation, drought, nest failure, extreme weather events as well as illegal trapping and nest-robbing have all contributed to the decline of the Eclectus Parrot population.

The limited area of habitat means that habitat degradation is a serious threat. Also, there are high rates of nest failure: 40% of females fail to produce any fledglings, owing to predation and flooding of nest hollows, and 32% produce fewer than one fledgling per year on average. Owing to unusually severe *El Niño*-induced drought, reproduction was poor in the 2002–03 and 2003–04 breeding seasons, with reproductive rates less than 30% of 'normal' (average of other years in study) in 2002 and less than 50% in 2003. A severe cyclone in early 2006 caused the destruction of 15% of nest trees in the study area, and also severely depleted fruiting sources. Breeding was over 3 months late in 2006 and reproductive success very low (less than 20% of normal). These events suggest that any increase in drought or cyclone activity, associated with global climate change, could further affect the long-term viability of Eclectus Parrots in Qld.

The area of the McIlwraith, Tozer and Iron Ranges contains about 2000 km^2 of suitable habitat, with about 1500 breeding females, of which at most only 900 breed successfully: the minimum number for long-term viability of the population. Removal of nest trees or habitat increases the extinction risk. Thus, logging of rainforests in the Iron-McIlwraith region is not an option if Eclectus Parrots are to be sustained in Australia.

CURRENT RECOVERY EFFORTS

Recover plan: None

Other plans: Department of the Environment and Heritage (2005) 'Threat abatement plan for beak and feather disease affecting endangered psittacine species'. Commonwealth of Australia, Canberra.

Garnett ST and Crowley GM (2000) 'The action plan for Australian birds 2000'. Environment Australia, Canberra.

Garnett ST, Szabo JK and Dutson G (2011) *The Action Plan for Australian Birds 2010.* CSIRO Publishing, Melbourne.

Conservation actions: An intensive research and monitoring program was conducted between 1997 and 2005 on the reproductive success, genetics, habitat preferences, reproductive rates and mating/social systems. The research program completed its intensive phase in 2005. The demographic data will be compared with ongoing surveys and monitoring of population size. Researchers will monitor the reproductive success, count adult birds and collect genetic samples from feathers in nests to establish whether the same females are still occupying the hollows. This will allow estimates of female lifespan: a key component for understanding of population structure and long-term reproductive success.

FUTURE RECOVERY EFFORTS

What you can do: Be vigilant in case there are any plans for logging or major developments in Eclectus Parrot habitat. The remote habitat is difficult to monitor, making it imperative to support conservation groups (e.g. The Wilderness Society, WWF, ACF, Birds Australia) actively involved in those areas.

Conservation needs:

- Keep all of the rainforest in the Iron-McIlwraith region intact and in good condition.
- Discourage commercial logging, even on a small scale, which could seriously harm the population.
- Eclectus Parrots are also threatened by climate change, so any community action against this threat will also help.

Research directions: Conduct ongoing population surveying and monitoring.

Sources and further reading

Heinsohn R (2008) The ecological basis of unusual sex roles in reverse-dichromatic Eclectus Parrots. *Animal Behaviour* **76**, 97–103.

Heinsohn R and Legge S (2003) Breeding biology of the reverse-dichromatic parrot, *Eclectus roratus. Journal of Zoology* **259**, 197–208.

Heinsohn R, Murphy S and Legge S (2003) Overlap and competition for nest holes among eclectus parrots, palm cockatoos and sulphur-crested cockatoos. *Australian Journal of Zoology* **51**, 81–94.

Higgins PJ (Ed.) (1999) *Handbook of Australian, New Zealand and Antarctic Birds. Volume 4: Parrots to Dollarbirds.* Oxford University Press, Melbourne.

Legge S, Heinsohn R and Garnett S (2004) Nest hollow availability and breeding population size of Eclectus Parrots, *Eclectus roratus*, on Cape York Peninsula, Australia. *Wildlife Research* **31**, 149–161.

Author: Robert Heinsohn

SUPERB PARROT

Polytelis swainsonii (Desmarest, 1826)

Photo: Graeme Chapman

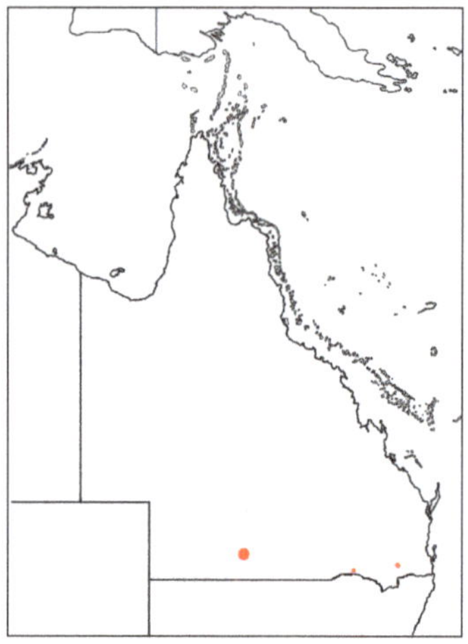

Family: Psittacidae

Other common names: Barraband, Scarlet-breasted Parrot, Barraband Parakeet, Green Leek

Conservation status: IUCN: Vulnerable; Aust.: Vulnerable; ACT: Vulnerable; Qld: Vulnerable; NSW: Vulnerable; Vic.: Endangered

Description: The slender Superb Parrot is 36–42 cm long and weighs 130–160 g. It has a long graduated tail, and pointed backswept wings with a wingspan of 53–57 cm. The adult male is bright green, with a pink bill, extensive yellow face bordered by a red band on the throat, and dull blue outer wings. The female has a dull blue (not yellow) face, pink-red thighs and a pink-marked undertail.

Distribution: The Superb Parrot occurs mainly inland of the Great Dividing Range on the slopes and plains, from north-central Vic. to northern NSW. It is a rare winter migrant as far north as southern Qld, recently found near Cunnamulla and between Goondiwindi and Warwick. Records from other parts of Qld are aviary escapes. Across its range, this species has an estimated population of about 6500 birds.

Habitat: Superb Parrots use grassy eucalypt forest and woodland habitats, especially River Red Gum *Eucalyptus camaldulensis* and box (e.g. Yellow Box *E. melliodora* or Grey Box *E. macrocarpa*), and also box-pine Callitris and Boree *Acacia pendula* woodlands. Their habitats often have an open midstorey of wattles *Acacia* and ballart *Exocarpus*. Superb Parrots usually move along vegetated corridors, seldom crossing extensive open areas.

Ecology: Superb Parrots feed in trees and understorey shrubs, and on the ground. In trees, they feed most often in the outer canopy, on buds, flowers and seed-pods. They also feed in mistletoe, including box mistletoe (parasitic on *Eucalypt* spp.) and grey mistletoe (parasitic on *Acacia* spp.), and in midstorey seeding or fruiting shrubs including wattles, Cherry Ballart *Exocarpos cupressiformis* and other shrubs. Superb Parrots feed mainly on the ground, on native grasses, prostrate shrubs or weeds, or on spilt grain along roadsides. Insects such as lerps, which have a sugary coating, are also gleaned from leaves. Crops of wheat or oats are also used. Superb Parrots require corridors of trees linking feeding areas with nearby woodlands, and are reluctant to fly over open country. There are no known breeding records for the Superb Parrot in Qld; its breeding range is confined to the Riverina and the inland slopes of southern NSW.

Reason for decline and current threats: Current and historical threats facing the species include loss of foraging and nesting habitat and flight paths, degradation of remaining habitat, heavy grazing, fire, road-kills, poisoning and the illegal bird trade.

The main threat to the Superb Parrot is the loss, degradation and fragmentation of box woodlands, especially breeding and foraging habitats, and corridors used for regular movement. Food availability also limits the parrot's population. Clearing of vegetation corridors reduces the ability of the species to forage and to move between breeding and non-breeding habitats. Heavy grazing can reduce food availability for Superb Parrots, especially at critical times. In dry years, grass cover within forests or woodlands becomes patchy, shrubs are browsed by stock, in turn, reducing food availability. River regulation and water exploitation throughout the parrot's range affect the health of its breeding and foraging habitats. Highly regulated flows threaten trees used by the Superb Parrot. River Red Gums can only withstand continuous flooding for about 3 years, after which trees may be killed. Superb Parrots were formerly killed by poisoned grain intended for Galahs *Eolophus roseicappillus* and rabbits. Other threats include trapping, and collisions with vehicles while foraging on roads, especially during harvesting when birds congregate beside the road to feed on spilt grain.

CURRENT RECOVERY EFFORTS

Recovery plan: Baker-Gabb D (2004) 'Draft recovery plan for the Superb Parrot *Polytelis swainsonii*'. Environment Australia, Canberra.

Other plans: Garnett ST and Crowley GM (2000) 'The action plan for Australian birds 2000'. Environment Australia, Canberra.

Department of the Environment and Heritage (2005) 'Threat abatement plan for beak and feather disease affecting endangered psittacine species'. Commonwealth of Australia, Canberra.

Conservation actions:

- Bush Heritage Australia provides habitat for this threatened species.
- The Superb Parrot Project of north-central Vic. and the Southern Riverina Field Naturalists Club in southern NSW undertake an annual census in November to monitor changes in the breeding population in the Riverina.

FUTURE RECOVERY EFFORTS

What you can do:

- Report sightings to DERM and Birds Australia's Threatened Bird Network.
- Join/donate to Bush Heritage Australia, Australian Wildlife Conservancy, Birds Australia, Bird Observation or Conservation Australia.
- Write to state and federal environment ministers, urging greater funding allocations and the implementation of research recommendations.
- Contact your local NRM officer to find out about best practices native grassland management.
- Protect the remnant forests and woodlands in your community or on your land to help provide habitat for native wildlife.
- Support local efforts to conserve threatened species in your area by joining a local organisation such as Landcare or Catchment group, field naturalists club, natural history or a 'friends of' group.
- Volunteer with Birds Australia, Green Corps or Conservation Volunteers Australia.

Conservation needs:

- Identify and protect forests and woodlands used by the Superb Parrot.
- Encourage sympathetic and sustainable grazing management practices in forests and woodlands, with the aim to maintain a shrubby understorey and healthy ground-cover.
- Provide incentives to private landowners who wish to protect and manage areas of their property for conservation.

Research directions:

- Research the occurrence and population size in Qld.
- Study the bird's ecological requirements.

Sources and further reading

Forshaw JM and Cooper WT (2002) *Australian Parrots.* 3rd edn. Alexander Editions, Robina.

Higgins PJ (Ed.) (1999) *Handbook of Australian, New Zealand and Antarctic Birds. Volume 4: Parrots to Dollarbirds.* Oxford University Press, Melbourne.

Leslie D (2005) Is the Superb Parrot *Polytelis swainsonii* population in Cuba State Forest limited by hollow or food availability? *Corella* **29**, 77–87.

Manning AD, Lindenmayer DB and Barry SC (2004) The conservation implications of bird reproduction in the agricultural 'matrix': a case study of the vulnerable Superb Parrot of south-eastern Australia. *Biological Conservation* **120**, 363–374.

Manning AD, Lindenmayer DB, Nix HA and Barry SC (2005) A bioclimatic analysis for the highly mobile Superb Parrot of south-eastern Australia. *Emu* **105**, 193–201.

Authors: David G Parker and Rick Webster

SWIFT PARROT

Lathamus discolor (Shaw, 1790)

Photo: Graeme Chapman

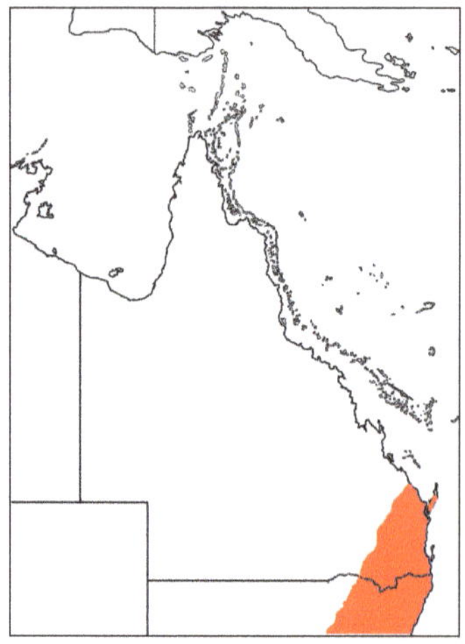

Family: Psittacidae

Other common names: Red-faced or Red-shouldered Parrot, Swift Lorikeet, Clink, Keet

Aboriginal name: Gorail or Wang-u-wang

Conservation status: IUCN: Endangered; Aust.: Endangered; Qld: Endangered; NSW: Endangered; SA: Endangered; Tas.: Endangered; Vic.: Threatened

Description: This parrot has a total length about 25 cm long and weighs 65–95 g. The plumage is bright green with a red face bordered by yellow, blue crown, cheeks and outer wings, red shoulders, underwings and (in adult males) undertail, a red or yellow spot on the rear inner wings and a narrow, dull crimson tail. Its clinking calls are unlike lorikeet screeches.

Distribution: The Swift Parrot occurs in Tasmania and in mainland south-eastern Australia, from south-east SA (rarely) to south-east Qld (formerly north to the Dawson River). Small numbers regularly reach northern coastal NSW and south-east Qld. The total population was estimated at 1320 pairs in 1988, 940 pairs in 1995 and 1000 pairs in 2004, with a continuing decline likely.

Habitat: Swift Parrots occur in eucalypt forests and woodlands that provide food sources during their migratory cycle. The nesting habitat is mature or old-growth forest with hollows, near forage trees.

Ecology: The Swift Parrot breeds only in Tasmania, in summer, then migrates to mainland Australia in winter. It feeds mainly on nectar and lerp in the canopy of *Eucalyptus* and *Corymbia* trees, notably box-ironbark and, on the coast, Forest Red Gum *E. tereticornis*, Spotted Gum *C. maculata* and Swamp Mahogany *E. robusta*. Mature trees provide the best sources of food and, during drought, coastal refuges are essential. Swift Parrots use the same sites repeatedly, though not necessarily every year.

Reason for decline and current threats: The major threat is habitat loss. Other threats include habitat alteration, collision mortality, competition, Psittacine Beak and Feather Disease, frequent fire, illegal capture and reduced rainfall under climate change.

Swift Parrots collide with wire mesh fences, windows and cars, especially in breeding and coastal drought-refuge habitats. Larger or more aggressive species, and introduced birds and bees, compete for food and nest sites. Beak and Feather Disease may increase under stress from loss of habitat and food resources, and the infection may spread via the rehabilitation and release of injured parrots. Climate change may affect the parrot's breeding success, migration patterns, physiology and food resources. Where fire is too frequent, flowering and maturation of nectar-rich plants may be reduced, and nesting hollows and nest-tree recruits may be destroyed.

CURRENT RECOVERY EFFORTS

Recovery plans: Saunders D and Tzaros C (2010) 'National recovery plan for the Swift Parrot *Lathamus discolor*'. NSW Department of Environment and Climate Change and Water, Queanbeyan, and Birds Australia, Melbourne.

Swift Parrot Recovery Team (2001) 'National Swift Parrot recovery plan 2001–2005'. Department of Primary Industries, Water and Environment, Hobart.

Other plans: Department of the Environment and Heritage (2005) 'Threat abatement plan for beak and feather disease affecting endangered psittacine species'. Commonwealth of Australia, Canberra.

Garnett ST and Crowley GM (2000) 'The action plan for Australian birds 2000'. Environment Australia, Canberra.

Garnett ST, Szabo JK and Dutson G (2011) *The Action Plan for Australian Birds 2010.* CSIRO Publishing, Melbourne.

Threatened Species Scientific Committee (2001) 'Threat abatement plan for dieback caused by the Root-rot

Fungus *Phytophthora cinnamomi*'. Commonwealth of Australia, Canberra.

Threatened Species Scientific Committee (2001) 'Key threatening process: Loss of terrestrial climatic habitat caused by anthropogenic emissions of greenhouse gases. Advice to the Minister for the Environment and Heritage on a public nomination of a Key Threatening Process under the Environment Protection and Biodiversity Conservation Act 1999'. Commonwealth of Australia, Canberra.

Department of the Environment, Water, Heritage and the Arts (2008) 'Threat abatement plan for predation by feral cats'. Commonwealth of Australia, Canberra.

Conservation actions:

- Universities conduct and support ecological research.
- Private landowners manage habitat on their land, and many community organisations support the recovery program.
- Over 300 volunteers participate in the national Swift Parrot survey twice a year, supported by Birds Australia, and provide incidental records.
- About 260 Swift Parrots are held in captivity in Australia and breed successfully (although there is no official captive breeding program).

FUTURE RECOVERY EFFORTS

What you can do:

- Volunteer for Swift Parrot surveys, or data entry at Birds Australia.
- Report sightings to Birds Australia/Swift Parrot Recovery Team (mail@birdsaustralia.com.au)
- Become a 'Friend of the Swift Parrot' (email swiftparrots@yahoo.com.au)
- Protect/fence off remnant habitat and promote habitat regeneration.
- Plant suitable forage trees, and encourage community and school involvement.
- Install and promote collision prevention measures in areas used by Swift Parrots.
- Publicise the Swift Parrot video, now available on YouTube, or provide a copy of the DVD (available from Birds Australia at swiftparrots@yahoo.com.au) to local libraries and schools.
- Donate to support research and conservation efforts by the Swift Parrot Recovery Program through Birds Australia.

Conservation needs:

- Identify, map and protect nesting, foraging and roosting habitat, and migration pathways, throughout the range of the species.
- Monitor and manage collisions, via a database of parrot injuries and deaths, public education and improved building standards.
- Instigate Beak and Feather Disease monitoring and management by fauna rescue and state conservation organisations, including quarantine for rescued birds and testing for disease before release.
- Extend the volunteer program to monitor the population trends and habitat use more comprehensively throughout the range of the species.
- Refine the methods for effective population monitoring.
- Conduct an analysis of population viability.
- Lobby governments for more effective threatened species protection.

Research directions:

- Determine the best way to monitor the parrot population.
- Determine the extent of remaining habitats and threats to these.
- Identify migration strategies and migratory pathways.
- Study the ecology, breeding biology and success, population regulation, genetic structure and prevalence of disease in the wild population.
- Investigate the potential impact of climate change on the Swift Parrot and its habitat.

Sources and further reading

Pfennigwerth S (2008) 'Minimising the Swift Parrot collision threat. Guidelines and recommendations for parrot-safe building design'. WWF-Australia, Sydney.

Saunders D, Brereton R, Tzaros C, Holdsworth M and Price B (2007) Conservation of the Swift Parrot *Lathamus discolor* – management lessons for a threatened migratory species. *Pacific Conservation Biology* **13**, 111–119.

Saunders DL (2008) Ecology and conservation of the Swift Parrot *Lathamus discolor* – an endangered austral migrant. PhD thesis. Australian National University, Canberra.

Saunders DL and Heinsohn R (2008) Winter habitat use by the endangered, migratory Swift Parrot (*Lathamus discolor*) in New South Wales. *Emu* **108**, 81–89.

Tzaros C, Ingwersen D, Saunders D and Webb M (2009) Swift Parrots – dynamic migrants within a drought-prone environment. *Wingspan* **19**, 14–15.

Author: Debbie Saunders

GOLDEN-SHOULDERED PARROT

Psephotus chrysopterygius Gould, 1857

Photo: Graeme Chapman

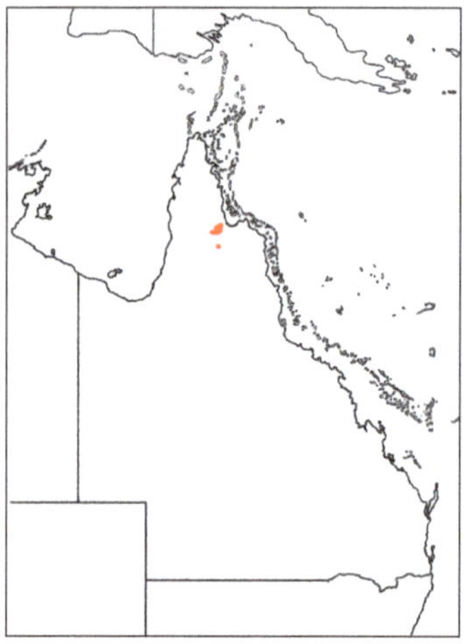

Family: Psittacidae

Other common names: Antbed, Anthill, Chestnut-crowned or Golden-winged Parrot or Parakeet

Conservation status: IUCN: Endangered; Aust.: Endangered; Qld: Endangered

Description: This is a small seed-eating parrot that nests in termite mounds; it has a total length of 23–28 cm and weighs about 55 g. The male is turquoise with a black crown, bright yellow on the wings and forehead, and a salmon-pink belly. Females and immature birds are mostly green with a turquoise rump.

Distribution: The Golden-shouldered Parrot formerly occurred across most of Cape York Peninsula, between Croydon, Chillagoe, Laura, Coen and Weipa, but is now restricted to two populations in central Cape York Peninsula. The range of the population in the headwaters of the Morehead River catchment is about 1380 km^2 and contracting. The population primarily in Staaten River NP occupies at least 300 km^2.

Habitat: Golden-shouldered Parrots live where night-time temperatures rarely fall below 10°C in the breeding season. The parrots nest in grasslands and grassy woodlands on hills and drainage depressions where there are suitable conical termite mounds. They spend most of the dry season in eucalypt woodland on sand ridges. At the start of the wet season, they move to gravel slopes, rocky hills and glimmer grass flats, where fallen seeds persist the longest. Through the wet season they feed primarily in the hills, but also use storm-burnt areas in the lowlands.

Ecology: The life cycle of Golden-shouldered Parrots is controlled by seasonal changes in seed availability. Throughout the dry season, the parrots feed on the fallen seeds of annual grasses, particularly Fire Grass *Schizachyrium* spp. Abrupt declines in food availability occur in the early wet season, when rain causes most seeds to germinate. The parrots must feed for extended periods at this time, so are particularly vulnerable to predation. They gather beneath the nests of Black-faced Woodswallows *Artamus cinereus*, because these birds alert the parrots to the presence of predators. Food remains scarce until Cockatoo Grass *Alloteropsis semialata* seeds about 6 weeks after the first rains. As seed supplies again become reliable towards the end of the wet season, the parrots begin nesting: the female excavating a chamber in a termite mound and laying four to six eggs. Most nests are built in the conical mounds of *Amitermes scopulus*, although other mounds are also used. The parrots remain in family groups until late in the dry season, when unmated juvenile birds flock together.

Reasons for decline and current threats: Current threats include: predation (particularly by butcherbirds); vegetation thickening; inappropriate fire regimes; wet season food shortage (loss of perennial grasses), grazing by feral pigs; overgrazing by cattle; and loss of woodswallows.

Possible threatening processes include shortage of nest sites (loss of termite mounds), trapping, disease and habitat clearance. The introduction of pastoralism led to a decrease in hot fires late in the year. As a result, grasslands contracted and grassy woodlands thickened. Because attacks by Pied Butcherbirds *Cracticus nigrogularis* and other ambush predators are more successful in dense vegetation than in open vegetation, this has reduced both survival and nesting success of the parrots. Vegetation thickening appears to disadvantage Black-faced Woodswallows, as well as parrots. Pigs and cattle reduce fuel loads, feed on important food plants (including Cockatoo Grass) and damage termite mounds.

CURRENT RECOVERY EFFORTS

Recovery plan: Garnett ST and Crowley GM (2003) 'Recovery plan for the Golden-shouldered Parrot *Psephotus chrysopterygius* 2003–2007'. Report to Environment Australia, Canberra by Queensland Department of Environment, Brisbane.

Other plans: Department of the Environment and Heritage (2005) 'Threat abatement plan for beak and feather disease affecting endangered psittacine species'. Commonwealth of Australia, Canberra.

Garnett ST and Crowley GM (2000) 'The action plan for Australian birds 2000'. Environment Australia, Canberra.

Garnett ST, Szabo JK and Dutson G (2011) *The Action Plan for Australian Birds 2010.* CSIRO Publishing, Melbourne.

Conservation actions:

- Since 1994, the open vegetation structure in Golden-shouldered Parrot habitat has been maintained using storm-burning.
- Supplementary feeding has been provided where appropriate.
- Methods to minimise habitat damage by pigs and cattle are being developed in cooperation with landholders.

FUTURE RECOVERY EFFORTS

Conservations needs:

- Continue to use storm-burning as a means of maintaining the open structure of Golden-shouldered Parrot habitat.
- Provide seed for parrots during the wet season in key areas.
- Monitor populations at selected sites once every five years.
- Monitor and assess cattle and pig impacts on the parrot's habitat, notably on Cockatoo Grass and termite mounds, and develop and implement harm-minimisation strategies.
- Initiate a captive-breeding program for re-establishing a population in Mungkan Kandju NP.
- Restore habitat in the eastern section of Mungkan Kandju NP in preparation for reintroduction of the parrot, and undertake procedural preparation for reintroduction.
- Work in co-operation with all landholders and traditional owners with an interest in Golden-shouldered Parrots and their habitat.

Research directions:

- Determine the impact of Pied Butcherbirds, and develop control measures if required.
- Examine the dependence of the parrot on Black-faced Woodswallows, and develop conservation actions to maintain Black-faced Woodswallows.

Sources and further reading

Crowley G, Garnett S and Shephard S (2009) Impact of storm-burning on *Melaleuca viridiflora* invasion of grasslands and grassy woodlands on Cape York Peninsula, Australia. *Austral Ecology* **34**, 196–209.

Crowley GM, Garnett ST and Shephard S (2004) Management guidelines for golden-shouldered parrot conservation. Queensland Parks and Wildlife Service, Brisbane.

Forshaw JM and Cooper WT (2002) *Australian Parrots.* 3rd edn. Alexander Editions, Robina.

Franklin DC (1999) Evidence for disarray among granivorous bird assemblages in the savannas of northern Australia, a region of sparse human settlement. *Biological Conservation* **90**, 53–68.

Higgins PJ (Ed.) (1999) *Handbook of Australian, New Zealand and Antarctic Birds. Volume 4: Parrots to Dollarbirds.* Oxford University Press, Melbourne.

Author: Gabriel Crowley

GROUND PARROT

Pezoporus wallicus (Kerr, 1792)

Photo: Graeme Chapman

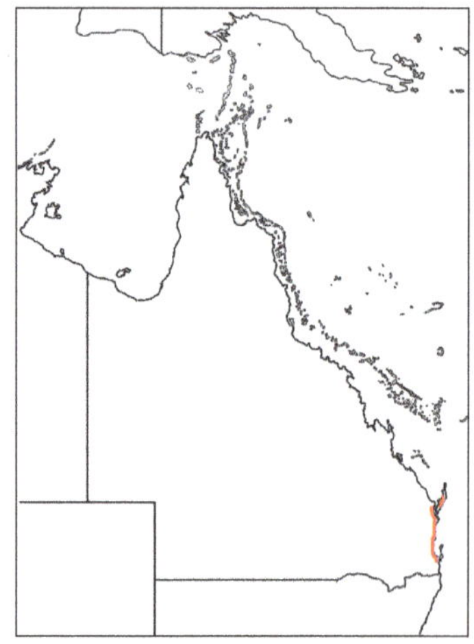

Family: Psittacidae

Other common name: Swamp Parrot

Conservation status: IUCN: Least Concern; Qld: Vulnerable; NSW: Vulnerable; SA: Extinct; Vic.: Vulnerable

Description: This is a medium-sized parrot with a total length of 28–33 cm, and weighing 70–80 g. Adults are mainly grass-green with black and yellow markings, alternate fine bands of yellow and black on the tail, and a bright red band above the beak. Juveniles have duller plumage. These parrots often stand upright, lacking the waddling gait of other parrots.

Distribution: This distribution of the species is patchy, with records from south-east Qld through coastal NSW, Vic., western Tas. and (historically) SA west to Adelaide. In Qld, the parrot was formerly found from Woodgate to the northern suburbs of Brisbane; its current stronghold is the Great Sandy region (Fraser Island, Wide Bay Training Area and Cooloola NP) and Sunshine Coast (Noosa NP). In 1989, its population in these areas was estimated at 2900–4900 birds. Recent censuses in Noosa NP (1997–2007) and Cooloola NP (2001–2004) indicate similar densities today.

Habitat: Ground Parrots occur mainly in diverse heathland and sedgeland with low, dense vegetation and few trees. The floristics and structure vary across the parrot's geographic range, and with local topography and time since fire.

Ecology: These solitary birds roost and nest on the ground, and feed on the ground or in low vegetation. They eat small seeds from a wide variety of shrubs, sedges and rushes. They are active by day but rarely fly except around dawn and dusk. Being cryptic, they are usually only seen if disturbed or when they call, which is most frequent around dawn and dusk. Females lay one clutch of usually three or four eggs per year; males provide all the food for the brooding female and young.

Reason for decline and current threats: Ground Parrots live in fire-prone habitats, including where fire regimes have been altered, which has contributed to their decline since European settlement. Optimal fire regimes for maintaining Ground Parrot habitat are debated; burning at fixed intervals is no longer considered appropriate. If controlled burning or fuel-load reduction is not carried out, local populations may be eliminated by wildfires or by back-burning needed to contain such fires. Controlled burns in Qld currently try to take the Ground Parrot's needs into account, though with difficulty near human habitation.

Feral predators, such as cats and foxes, prey on Ground Parrots. Cats may have contributed to Ground Parrot extinction in SA, and to the recent decline of the western taxon (now a full species on genetic grounds: the Critically Endangered Western Ground Parrot *P. flaviventris*). The impact of predation is likely to be greater near urban areas and in smaller habitat patches.

Habitat modification and destruction, including drainage of wetlands, urban development and stock grazing, is a major cause of Ground Parrot decline. In Qld, most Ground Parrots now occur in national parks and on Commonwealth land where active management attempts to minimise habitat loss. Outside protected areas, destruction and/or degradation of heathlands and sedgelands is ongoing. Continuing habitat fragmentation may cause further decline, through inbreeding or the inability of Ground Parrots to disperse and recolonise patches after fire. The Qld population has low genetic diversity equivalent to that found in threatened species worldwide: a key finding for future management.

CURRENT RECOVERY EFFORTS

Recovery plan: None

Other plans: Department of the Environment, Water, Heritage and the Arts (2008) 'Threat abatement plan for predation by feral cats'. Commonwealth of Australia, Canberra.

Garnett ST and Crowley GM (2000) 'The action plan for Australian birds 2000'. Environment Australia, Canberra.

Garnett ST, Szabo JK and Dutson G (2011) *The Action Plan for Australian Birds 2010*. CSIRO Publishing, Melbourne.

Conservation actions:

- Population monitoring occurs annually in the Noosa, Mount Coolum and Mooloolah River NPs on the Sunshine Coast, irregularly in the Poona and Great Sandy NPs (DERM).
- Other sightings are recorded; for example, through the *Atlas of Australian Birds* (Birds Australia).
- Controlled burns are carried out in Noosa, throughout the Great Sandy region national parks including areas in Cooloola, Fraser Island as well as Mount Coolum and Mooloolah River NPs, with the timing and frequency taking Ground Parrot needs into account.

FUTURE RECOVERY EFFORTS

What you can do:

- Join or donate to Birds Australia.
- Be aware of the distribution of the species and its sensitivity to cat predation. If you live close to known or potential Ground Parrot habitat, ensure your cats and dogs are controlled, particularly during the breeding season.
- If you own property that includes, or is close to, Ground Parrot habitat, see information on the DERM website for advice on how to manage your land to minimise the impacts on heathland and sedgeland habitat.

Conservation needs:

- The species requires a patchwork of connected habitats, where regular, long-term monitoring of Ground Parrot numbers and distribution can be carried out.
- Implement adaptive fire management, based on actual Ground Parrot numbers rather than fixed intervals. Preserve all known and potential habitat.
- Explore the feasibility of a reintroduction program.
- Plan, monitor and manage the populations at a landscape level.
- Submit records of Ground Parrot observations to the *Atlas of Australian Birds* (Birds Australia).

Research directions:

- Research the parrot's social behaviour.
- Study the seasonal and age-related dietary requirements, movement and dispersal patterns.
- Carry out studies of population genetic structure using microsatellite, mitochondrial and nuclear DNA.
- Study the responses of habitat to fire, to enable management for its recovery at a landscape scale. Monitor the subpopulations not currently being monitored.

Sources and further reading

Baker J, Whelan DJ, Evans L, Moore S and Norton M (2010) Managing the Ground Parrot in its fiery habitat in south-eastern Australia. *Emu* **110**, 279–284.

Chan K, Glover DR, Ramage CM and Harrison DK (2008) Low genetic diversity in the Ground Parrot (*Pezoporus wallicus*) revealed by randomly amplified DNA fingerprinting. *Annales Zoologici Fennici* **45**, 211–216.

Higgins PJ (Ed.) (1999) *Handbook of Australian, New Zealand and Antarctic Birds. Volume 4: Parrots to Dollarbirds.* Oxford University Press, Melbourne.

Spearritt A and Krieger G (2007) 'A review of a long term monitoring program of the Ground Parrot (*Pezoporus wallicus wallicus*) population within the national parks of the Sunshine Coast, South-East Queensland'. Queensland Parks and Wildlife Service, Moggill.

Tasker E and Baker J (2005) Fire and feathers: managing fire-sensitive birds on a flammable continent. *Fire and Birds: Fire Management for Biodiversity*, Supplement to *Wingspan* **15**(3), 18–20.

Author: Elen Shute

NIGHT PARROT

Pezoporus occidentalis (Gould, 1861)

Artwork: WT Cooper

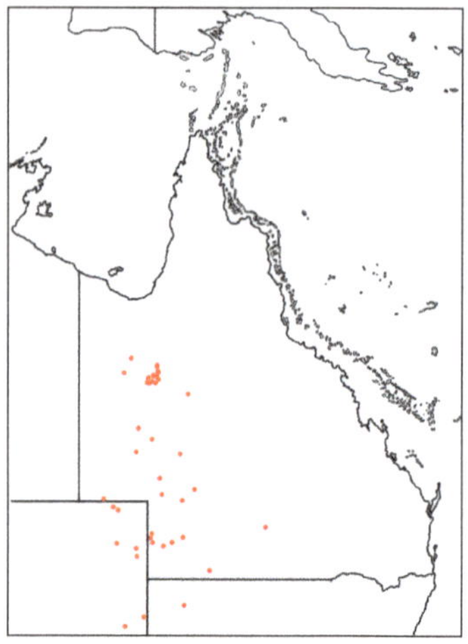

Family: Psittacidae

Other common name: Spinifex Parrot

Aboriginal name: Tnokkapaltara (Arunta, Central Australia); Myrrlumbing (ethnography uncertain)

Conservation status: IUCN: Critically Endangered; Aust.: Endangered; Qld: Endangered; WA: Critically Endangered; NT: Critically Endangered; SA: Endangered; NSW: Presumed Extinct; Vic.: Regionally Extinct.

Description: This is a medium-sized parrot resembling a Ground Parrot *Pezoporus wallicus* with a short tail; it has a total length of 22–25 cm and a wingspan of 44–46 cm. The plumage is mostly green, streaked and barred black and yellow, with yellower underparts. It has been likened to a 'fat budgie'; it is quail-like when flushed by day; it is larger than the partly nocturnal Bourke's Parrot *Neopsephotus bourkii*.

Distribution: The Night Parrot is widespread across the arid zone. The parrot was formerly recorded from the Gawler Ranges and Cooper Creek areas, SA, to the Pilbara, WA. Recent reliable records come from western/north-western Qld (the Channel Country and Selwyn Range) and the Pilbara. Two dead birds and numerous reports have come from western Qld since 1990, which suggests a focus for field searches. One recent specimen was a juvenile female, which suggests a breeding population.

Habitat: Records are associated both with spinifex *Triodia* hummock grasslands, especially on rocky ranges, or breakaways in WA, and with open herbaceous flats around salt lakes (samphire), and occasionally downs and flats with Mitchell Grass *Astrebla* and low herbaceous cover.

Ecology: This parrot is virtually unknown, beyond its habitat associations. It is thought to move into spinifex areas following rain, when spinifex seeds. It drinks at water sources during the night.

Reason for decline and current threats: The primary threats to the species appear to be competition for food and degradation of habitat by grazing of domestic or feral animals (especially near artificial water sources by stock or rabbits), reduced availability of natural water sources (especially as a result of over-use by feral camels), predation by foxes and cats, and altered fire regimes, particularly causing the loss of mature spinifex stands.

CURRENT RECOVERY EFFORTS

Recovery plan: Blyth J (1997) 'Night Parrot (*Pezoporus occidentalis*) interim recovery plan for Western Australia 1996–1998'. In 'Interim recovery plans 4–16 for Western Australian critically endangered plants and animals' (Eds J Pryde, A Brown and AA Burbidge). Department of Conservation and Land Management, Perth.

Other plans: Bamford MJ and Metcalf BM (2005) 'Fortescue Metals Group management plan for the Night Parrot *Pezoporus occidentalis* in the vicinity of the Fortescue Marshes'. Unpublished report by Bamford Consulting Ecologists to Fortescue Metals Group, Perth.

Department of the Environment, Water, Heritage and the Arts (2008) 'Threat abatement plan for competition and land degradation by unmanaged goats'. Commonwealth of Australia, Canberra.

Department of the Environment, Water, Heritage and the Arts (2008) 'Threat abatement plan for competition and land degradation by feral rabbits'. Commonwealth of Australia, Canberra.

Department of the Environment, Water, Heritage and the Arts (2008) 'Threat abatement plan for predation by feral cats'. Commonwealth of Australia, Canberra.

Department of the Environment, Water, Heritage and the Arts (2008) 'Threat abatement plan for predation by feral foxes'. Commonwealth of Australia, Canberra.

Garnett ST and Crowley GM (2000) 'The action plan for Australian birds 2000'. Environment Australia, Canberra.

Garnett ST, Szabo JK and Dutson G (2011) *The Action Plan for Australian Birds 2010*. CSIRO Publishing, Melbourne.

Conservation actions: Current efforts are aimed at locating a population that can be studied.

FUTURE RECOVERY EFFORTS

What you can do:

- Search for Night Parrots in the arid zone by spotlighting along tracks or at waterholes, in varied terrain after good rains.
- Write letters to governments and corporations, urging funding and in-kind support (e.g. transport, personnel and equipment) for conservation projects.
- Report sightings to DERM and/or to the Qld Museum.
- Monitor the effects your land management is having on the natural environment, such as through photographic reference points over time, for future adaptive management.

Conservation needs:

- Practise sustainable land management, including biodiversity management.
- Promote the value of ungrazed or lightly grazed lands for biodiversity conservation. Keep artificial water points well spaced to minimise concentrated grazing, by ensuring that some land is much farther than 5 km from water in sheep country and 10 km in cattle country.
- Minimise the impacts on natural ecosystems from adjacent land use; e.g. fence to prevent stock from trampling natural water sources (with wildlife-friendly fencing, not barbed wire). Manage burning regimes to prevent wildfires damaging vegetation, particularly mature spinifex.
- Eradicate feral animals such as foxes, cats and rabbits, and control environmental weeds such as Buffel Grass *Cenchrus ciliaris*.
- Avoid fire regimes that spread weeds and practise wash-down hygiene on earthmoving equipment to minimise the spread of weeds. Reduce goat and camel numbers to a level that does not threaten the vegetation.
- Raise awareness of the value of regional conservation reserves through community education.
- There is a need for threat abatement, habitat regeneration, change of fire regimes, translocation and management techniques, possibly including captive breeding.

Research directions:

- Carry out scientific research, surveys and monitoring of this species.
- When a population has been located, a program of research and monitoring should acquire the basic natural history and ecological data on food, reproductive biology, home range and movements, estimates of population size and, eventually, estimates of genetic diversity.
- Model the possible relationships between occurrence of the birds and rainfall, fire, seeding events of spinifex and other habitat variables, which should increase the likelihood of finding the birds in new locations.

Sources and further reading

Cupitt R and Cupitt S (2008) Another recent specimen of the Night Parrot *Pezoporus occidentalis* from western Queensland. *Australian Field Ornithology* **25**, 69–75.

Davis RA and Metcalf BM (2008) The Night Parrot (*Pezoporus occidentalis*) in northern Western Australia: a recent sighting from the Pilbara region. *Emu* **108**, 233–236.

Higgins PJ (Ed.) (1999) *Handbook of Australian, New Zealand and Antarctic Birds. Volume 4: Parrots to Dollarbirds.* Oxford University Press, Melbourne.

McDougall A, Porter G, Mostert M, Cupitt R, Cupitt S, Joseph L, Murphy S, Janetzki H, Gallagher A and Burbidge A (2009) Another piece in an Australian ornithological puzzle – a second Night Parrot is found dead in Queensland. *Emu* **109**, 198–203.

Olsen PD (2008) Night Parrots: fugitives of the inland. In *Boom and Bust.* (Eds L Robin, R Heinsohn and L Joseph) pp. 121–146. CSIRO Publishing, Melbourne.

Authors: Leo Joseph, Allan Burbidge and Steve Murphy

OWLS (ORDER: STRIGIFORMES)

POWERFUL OWL

Ninox strenua (Gould, 1838)

Photo: David Hollands

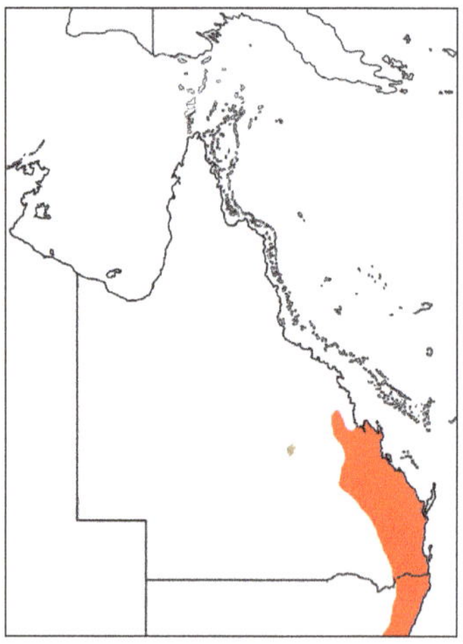

Family: Strigidae

Aboriginal names: Nangaw, Wirimul, Yuuitch Peen, Yuuitch Peetch, Yuitch Pilap

Conservation status: IUCN: Least Concern; Qld: Vulnerable; NSW: Vulnerable; SA: Endangered; Vic.: Vulnerable

Description: This is a very large owl with a total length of 50–60 cm and a wingspan of 110–140 cm; males average 1.45 kg and females 1.25 kg. The adult is brown, narrowly barred, on the upperparts and white, coarsely marked with V-shaped bars on the underparts. It has prominent orange-yellow eyes and fully feathered legs with large yellow feet. The juvenile is white and downy on the head and on the underparts, with dark eye-patches and fine streaks on the breast. The owl's characteristic call is a loud, slow double hoot *woo-hoo*, mournful but 'big', with a far-carrying, ringing quality. Sometimes a single hooting note is given.

Distribution: The Powerful Owl occurs from central-coastal Qld to south-east SA. In south-east Qld, it is found in coastal and upland areas from the Clarke Range (north of Eungella) to the NSW border, inland to Carnarvon Gorge, Blackdown Tableland and Durikai State Forest (west of Warwick). Populations occur in extensive native forest. The total national population has been estimated at 7000 breeding birds, of which there may be 10–15% or 700–1000 in Qld.

Habitat: The Powerful Owl is found in wet sclerophyll forest, including that with rainforest trees in the midstorey, and dry sclerophyll forest and woodland, extending inland in riverine woodland. Roost and nest sites are usually in gullies. Roost sites are sheltered groves of midstorey trees, or sometimes pine plantations, and paperbark *Melaleuca* thickets on inland rivers. Blackwood *Acacia melanoxylon*, Red Turpentine *Syncarpia glomulifera*, Coachwood *Ceratopetalum apetalum*, Lilly-pilly *Acmena smithii*, Sassafras *Doryphora sassafras* or groves of she-oaks *Allocasuarina* may indicate potential roosting gullies. This owl also occurs in large areas of urban bushland and large botanic gardens in cities, and at night ranges into adjacent suburban gardens.

Ecology: This owls uses secluded roosting sites on bare branches 2–20 m above the ground under a dense leafy canopy, to avoid discovery and mobbing by other birds. It preys mainly on mammals, particularly medium-sized (often hollow-dependent) arboreal marsupials such as possums and gliders, but also on flying-foxes and other species. It also takes birds and some large insects. Powerful Owls are resident in large, permanent home-ranges of about 600 ha in highly productive, moist coastal forest to 3000 ha in dry inland forest. They lay eggs in autumn and winter, usually in May–June. The nest is a large hollow 45 cm to 100 cm wide and 45 cm to 200 cm deep in a big old tree, usually live, but sometimes dead, 8–45 m above the ground. The clutch is usually two eggs. Incubation takes 35–40 days and the nestling period 8–9 weeks. Breeding success is often high in coastal forest, with broods of two frequently raised. Young are dependent on their parents for several months after fledging.

Reasons for decline and current threats: This owl is threatened by clearance and fragmentation of habitat, by intensive forestry practices that remove hollow trees over wide areas, and by fire (which may destroy hollow trees and reduce the owl's food supply). Foxes prey on some owlets. The species persists in native wood-production forests, breeding in unlogged areas such as gully reserves and foraging in logged or regenerating areas, but is disappearing from some fragmented inland forests. Local threats include road-kill and human disturbance to nest sites.

CURRENT RECOVERY EFFORTS

Recovery plan: NSW Department of Environment and Conservation (2006) 'Recovery plan for the large forest owls'. NSW Department of Environment and Conservation, Sydney.

Other plans: Garnett ST and Crowley GM (2000) 'The action plan for Australian birds 2000'. Environment Australia, Canberra.

Conservation actions: Limited surveys for owls have been conducted.

FUTURE RECOVERY EFFORTS

What you can do:

- Report sightings of the owls and their nests to the *Atlas of Australian Birds* (Birds Australia) and to DERM.
- Join/donate to Birds Australia, Bird Observation and Conservation Australia, Birds Qld and the Australasian Raptor Association.

Conservation needs:

- Conserve adequate areas of suitable nesting habitat via reserves, native forest management plans, conservation agreements, clearance controls and development mitigation strategies.
- Develop and implement appropriate forestry practices, especially the preservation of suitable nesting trees and protection of riparian vegetation.
- Develop appropriate wildfire management strategies on public and private land (avoid destroying old hollow-bearing trees).
- Enhance community awareness of Powerful Owls, their environmental significance and their conservation requirements.
- Undertake a refined population estimate.
- Re-establish and reconnect riparian forest through cleared landscapes.
- Publish significant observations on the owls in bird journals; use recent literature as a guide.
- Erect suitable nest boxes in suburban bushland with high prey densities but few hollows.
- Control foxes.

Research directions:

- Conduct ecological research, comparable to that done in relation to forestry and owl population genetics in NSW and Vic.
- Conduct systematic surveys for this owl across its range, using established call-playback methods, and report (publish) results.
- Develop techniques for identifying individual owls, including by analysis of calls.

Sources and further reading

Bilney RJ, Cooke R and White J (2006) Change in diet of Sooty Owls (*Tyto tenebricosa*) since European settlement: from terrestrial to arboreal prey and increased overlap with Powerful Owls. *Wildlife Research* **33**, 17–24.

Cooke R, Wallis R, Hogan F, White J and Webster A (2006) The diet of Powerful Owls (*Ninox strenua*) and prey availability in a continuum of habitats from disturbed urban fringe to protected forest environments in south-eastern Australia. *Wildlife Research* **33**, 199–206.

Hollands D (2008) *Owls, Frogmouths and Nightjars of Australia*. Bloomings Books, Melbourne.

Kavanagh RP (2004) Conserving owls in Sydney's urban bushland: current status and requirements. In *Urban Wildlife: More than Meets the Eye*. (Eds D Lunney and S Burgin) pp. 93–108. Royal Zoological Society of NSW, Sydney.

Soderquist T and Gibbons D (2007) Home-range of the Powerful Owl (*Ninox strenua*) in dry sclerophyll forest. *Emu* **107**, 177–184.

Author: Stephen JS Debus

RUFOUS OWL (EASTERN SUBSPECIES)

Ninox rufa queenslandica Mathews, 1911

Photo: David Hollands

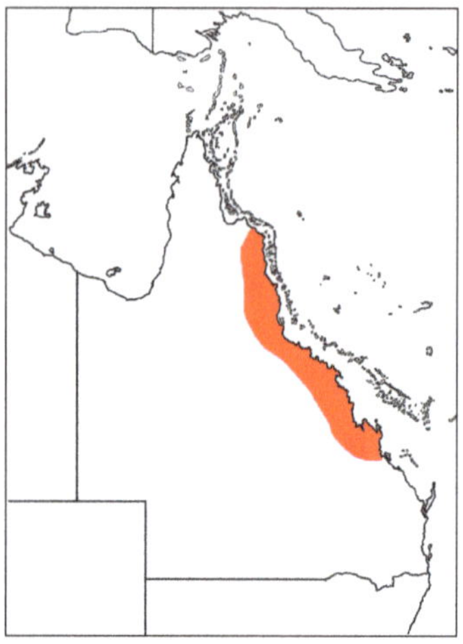

Family: Strigidae

Conservation status: IUCN: species (*N. rufa*) is Least Concern; Qld: Vulnerable

Description: This is a large owl with a total length of 46–57 cm and a wingspan of about 115 cm; males average 950 g and females 660 g. The adult is rufous-brown with fine bars on the upperparts and white finely barred rufous on the underparts. It has prominent yellow eyes and fully feathered legs with large yellow feet. The juvenile is white and downy on the head and underparts, with dark eye-patches. The characteristic call is a slow double hoot *woo-hoo*, more nasal and subdued than that of the Powerful Owl *Ninox strenua*, and lacking the ringing, far-carrying quality. Sometimes a single hooting note is given.

Distribution: This subspecies is found in coastal and upland north-east Qld, from Cooktown to Rockhampton and inland to Mareeba and the Blackdown Tableland. The southern limit is uncertain, owing to the similarity of its calls to those of the Powerful Owl; calling birds in the zone of overlap should be sighted if possible. In dry forest or woodland types, populations are restricted to the vicinity of moist refuges such as gallery forest or swamp forest, but in extensive rainforest the owls are more widespread. The total population across the range of the subspecies (i.e. Qld) has been estimated at 3000 breeding birds. The only (localised) population census recorded 19 birds in an area of about 5000 km^2 on the Atherton Tableland. A survey for Powerful Owls located a few Rufous Owls near the latter's southern limit.

Habitat: The Rufous Owl is found in rainforest, or sclerophyll forest and woodland near rainforest. Roosting sites are located in sheltered groves of dense, shady trees. Rainforest trees, paperbarks *Melaleuca* or gallery rainforest on watercourses through eucalypt forest or woodland are potential roosting areas. This owl also occurs in swamps, mangroves and large botanic gardens in cities, and at night ranges into suburban gardens.

Ecology: The Rufous Owl preys mainly on mammals, including various arboreal and terrestrial species, but particularly on arboreal (often hollow-dependent) marsupials such as medium-sized possums and gliders, and on flying-foxes. It also takes birds, including rainforest species, and some large insects. The Rufous Owl lays eggs in winter and spring, from June to October. The nest is a large hollow 45–60 cm wide and 30 cm to 200 cm deep in a big old tree, often a eucalypt or melaleuca, but also rainforest trees, 3–40 m above the ground. The clutch is usually two eggs. Incubation takes 36–38 days, and the nestling period 6–8 weeks. Broods of two owlets often fledge. The causes of nesting failure include invasion of the nest hollow by green ants and predation on owlets by goannas. Young owls are dependent on their parents for several months.

Reasons for decline and current threats: The Rufous Owl is threatened by clearance and fragmentation of habitat, by fire (especially late dry season fires that destroy hollow trees), and possibly by intensive forestry practices that remove hollow trees over wide areas. Locally, it may have been threatened by brodifacoum-based rodenticides until their use in canefields was curtailed.

CURRENT RECOVERY EFFORTS

Recovery plan: None

Other plans: Garnett ST and Crowley GM (2000) 'The action plan for Australian birds 2000'. Environment Australia, Canberra.

Conservation actions:

- The use of brodifacoum-based rodenticides has been curtailed.
- Some owl nest boxes have been erected around canefields in north Qld.
- Limited surveys for owls have been conducted.

FUTURE RECOVERY EFFORTS

What you can do:

- Report sightings of the owls and their nests to the *Atlas of Australian Birds* (Birds Australia) and to DERM.
- Join/donate to Birds Australia, Bird Observation and Conservation Australia, Birds Qld and the Australasian Raptor Association.

Conservation needs:

- Protect riparian habitat within areas proposed for clearing, with buffers of at least 500 m around known owl nest sites.
- Encourage non-toxic control of pests such as rodents, particularly by planting trees along the edges of sugar cane and the erection of owl nest boxes.
- Publish significant observations on the owls in bird journals; use recent literature as a guide.
- Erect suitable nest boxes in tropical suburban bushland with high prey densities but few hollows.
- Re-establish and reconnect riparian forest through cleared landscapes.
- Modify fire regimes to avoid destroying old hollow trees.
- Identify any other threats.

Research directions:

- Ecological research, comparable to that done for the Powerful Owl in relation to forestry and owl population genetics in NSW and Vic., is required.
- Study the diet and foraging ecology, the home-range size and habitat use (by radiotelemetry, particularly in relation to logging history and to densities of prey species and hollow trees), breeding biology and population parameters (dispersal, survival, recruitment, longevity and genetic structure).
- Conduct a refined population survey and other systematic surveys for owls, particularly in relation to forestry and to determine its southern range limit. Apply established call-playback methods and report (publish) results.

Sources and further reading

Debus S (2009) *The Owls of Australia: A Field Guide to Australian Night Birds*. Envirobook, Sydney.

Harrington GN and Debus SJS (2000) Dietary items of the Rufous Owl *Ninox rufa* on the Atherton Tableland, north Queensland. *Australian Bird Watcher* **18**, 251–252.

Higgins PJ (Ed.) (1999) *Handbook of Australian, New Zealand and Antarctic Birds. Volume 4: Parrots to Dollarbirds*. Oxford University Press, Melbourne.

Hollands D (2008) *Owls, Frogmouths and Nightjars of Australia*. Bloomings Books, Melbourne.

Legge S, Heinsohn R, Blackman C and Murphy S (2003) Predation by Rufous Owls on Eclectus Parrots and other animals at Iron Range NP, Cape York. *Corella* **27**, 45–46.

Author: Stephen JS Debus

MASKED OWL (NORTHERN SUBSPECIES)

Tyto novaehollandiae kimberli Mathews, 1912

Photo: David Hollands

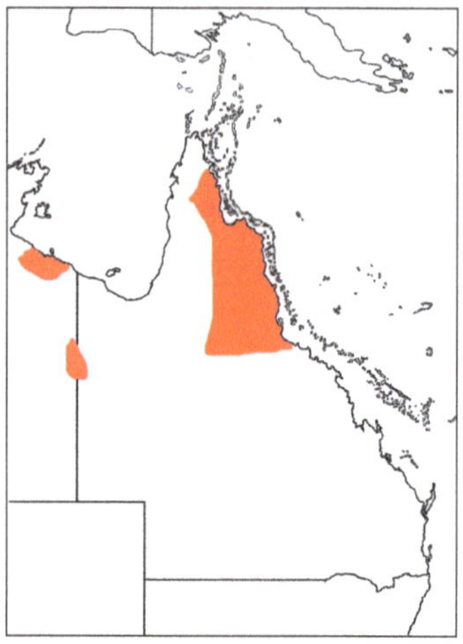

Family: Tytonidae

Aboriginal names: Atanokadi, Weerong-arra, Wiaratja

Conservation status: IUCN: species (*T. novaehollandiae*) is Least Concern; Aust.: Vulnerable; Qld: Vulnerable; NT: Vulnerable

Description: This is a medium-sized owl with a total length of 33–45 cm and a wingspan of about 90–115 cm. One male weighed 455 g and two females 680–730 g. The adult has large eyes, surrounded by a brown patch and set in a rounded, heavily rimmed white face. The feet are large and powerful, with legs fully feathered down to the toes. The upperparts are grey with buff mottling and fine pale spots; the wings and tail are strongly barred. The underparts are white to buff with variable dark spotting. The owl's characteristic call is a loud, harsh screech.

Distribution: This owl occurs across tropical northern Australia, west to the Kimberley (WA). In Qld, it occurs in the northern third from near Townsville north and west to the NT border. It lives mostly in coastal and upland areas, possibly inland to the Barkly Tableland. The total national population has been estimated at 12 000 breeding birds, of which there may be about half or 6000 in Qld.

Habitat: Sclerophyll forest and woodland, often near ecotones with open areas, such as grassland, heath or canefields, and typically grassy or with a mosaic of sparse and dense ground-cover. Roosting sites are tree-hollows, caves or dense foliage (3–8 m above the ground) and, rarely, buildings.

Ecology: The Masked Owl preys mainly on terrestrial mammals, particularly rodents, up to the size of bandicoots. It also takes arboreal mammals (e.g. small gliders), and some birds, frogs, lizards and large insects. Masked Owls are resident in large home-ranges of 400–1500 ha or sometimes more than 3000 ha. This owl lays eggs from February to October, commonly March-July, with normally one brood per year. The nest is a large hollow in a big old eucalypt, usually live, but sometimes dead; the hole is 10-45 m above the ground, with the entrance more than 20 cm wide, and the chamber at least 45 cm wide and up to 500 cm deep. The clutch is one to four eggs, usually two or three. Incubation takes 33-35 days and the nestling period 8-9 weeks. Fledged broods usually number one or two, rarely three. Young are dependent on their parents for 1-3 months after fledging.

Reasons for decline and current threats: The northern Masked Owl is threatened by clearance and fragmentation of habitat, by intensive forestry practices that remove hollow trees over wide areas, by loss of old hollow trees on private land, and by fire (especially late dry season fires that destroy hollow trees). Some owls are killed by collisions with barbed-wire fences and vehicles. Locally, the owl may have been threatened by brodifacoum-based rodenticides until the use of these chemicals in canefields was curtailed. The owl persists in native wood-production forests, but is less abundant in logged than in unlogged areas, probably because dense post-logging regeneration may be unsuitable for foraging. The owl may be affected by a decline in its food supply (small and medium-sized native mammals) across the northern Australian woodlands.

CURRENT RECOVERY EFFORTS

Recovery plan: Woinarski JCZ (2004) 'National multi-species recovery plan for the Partridge Pigeon [eastern subspecies] *Geophaps smithii smithii*; Crested Shrike-tit [northern (sub)species] *Falcunculus (frontatus) whitei*; Masked Owl [northern Australian mainland subspecies] *Tyto novaehollandiae kimberli*; and Masked Owl [Tiwi Islands subspecies] *Tyto novaehollandiae melvillensis*, 2004–2008'. NT Department of Infrastructure, Planning and Environment, Darwin.

Other plans: Garnett ST and Crowley GM (2000) 'The action plan for Australian birds 2000'. Environment Australia, Canberra.

Garnett ST, Szabo JK and Dutson G (2011) *The Action Plan for Australian Birds 2010.* CSIRO Publishing, Melbourne.

Conservation actions:
- The use of brodifacoum-based rodenticides has been curtailed.
- Some owl nest boxes have been erected around canefields in north Qld.
- DERM requires protective measures for forestry operations in state forests, which include buffers around known nests, retention of habitat trees and recruits, and removal of bushfire fuel loads around nest trees.

FUTURE RECOVERY EFFORTS

What you can do:
- Report sightings of the owls and their nests to the *Atlas of Australian Birds* (Birds Australia) and to DERM.
- Join/donate to Birds Australia, Bird Observation and Conservation Australia, Birds Qld and the Australasian Raptor Association.

Conservation needs:
- Publish significant observations on the owls in bird journals; use recent literature as a guide.
- Erect suitable nest boxes in tropical suburban bushland with high prey densities but few hollows.
- Re-establish and reconnect riparian forest through cleared landscapes.
- Modify fire regimes to avoid destroying old hollow trees.

Research directions:
- Conduct ecological research, comparable to that done in NSW and Vic. Gaps in our knowledge include: diet and foraging ecology, home-range size and habitat use, breeding biology, population parameters and genetic status of alleged subspecies *T. n. galei* on Cape York Peninsula.
- Conduct a refined population survey and other systematic surveys for owls, particularly in relation to forestry and to determine its southern range limit. Apply established call-playback methods and report (publish) results.

Sources and further reading

Debus SJS and Wood C (2007) Growth of a captive nestling Masked Owl *Tyto novaehollandiae. Australian Field Ornithology* **24**, 49–53.

Fitzsimons JA and Rose AB (2008) Notes on the diet of the northern Masked Owl *Tyto novaehollandiae kimberli* in north Queensland. *Memoirs of the Queensland Museum* **52**, 148.

McNabb EG, McNabb J and Barker K (2003) Post-nesting home range, habitat use and diet of a female Masked Owl *Tyto novaehollandiae* in western Victoria. *Corella* **27**, 109–117.

Thomson CN (2006) A trial of the use of artificial nest-boxes for the Masked Owl *Tyto novaehollandiae* near Newcastle, New South Wales. *Australian Field Ornithology* **23**, 192–197.

Todd M (2006) Prey partitioning and behaviour of breeding Masked Owls *Tyto novaehollandiae* on the Central Coast of New South Wales. *Australian Field Ornithology* **23**, 186–191.

Author: Stephen JS Debus

PERCHING BIRDS OR SONGBIRDS (ORDER: PASSERIFORMES)

RUFOUS SCRUB-BIRD (NORTHERN SUBSPECIES)

Atrichornis rufescens rufescens (Ramsay, 1866)

Photo: Roger Brown-AUSCAPE

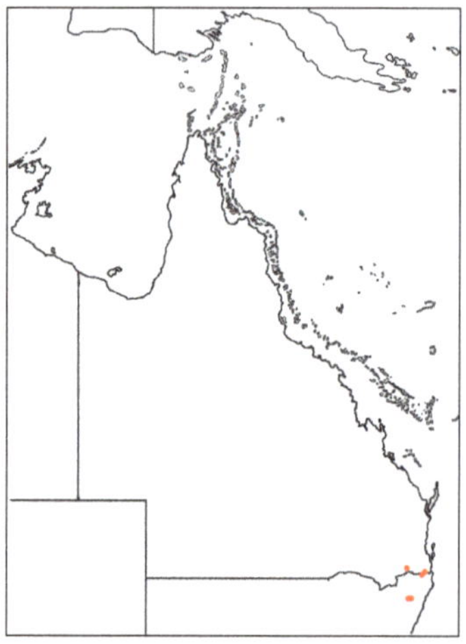

Family: Atrichornithidae

Other common names: Eastern Scrub-bird, Mousebird, Mysterybird

Conservation status: IUCN: species (*A. rufescens*) is Vulnerable; Qld: Vulnerable; NSW: Vulnerable

Description: This is a small ground-dwelling, almost flightless bird (males 19 cm, females 16.5 cm in total length) with short wings, a long tail and strong legs. The upperparts are rufous-brown with fine black barring. The throat is white, becoming buff on the breast and abdomen (yellower in females). The breast and sides are mottled black, and the flanks and undertail are rufous. It is almost invisible in its habits and habitat, though its loud song is distinctive.

Distribution: The Rufous Scrub-bird now occurs discontinuously along the Great Dividing Range escarpment from south-east Qld south to Barrington Tops (NSW), mostly above 600 m elevation. The northern subspecies *A. r. rufescens* occurs from the Mount Mistake area, south-east Qld, south to Gibraltar Range (NSW), and the southern subspecies *A. r. ferrieri* occurs from the Dorrigo Plateau (NSW) south to Barrington Tops. In Qld, the Rufous Scrub-bird occurs only in scattered localities near the NSW border, from Lamington NP west to Mount Barney NP and north to the upper reaches of Dalrymple Creek and Mount Cordeaux in Main Range NP. It is still present at Mount Cordeaux, although it has disappeared from many areas such as Lionsville–Washpool, Mount Warning, Cunningham's Gap and Spicer's Gap. The subspecies has declined substantially, particularly in northern NSW were it once occurred in lowland habitat that is now cleared. In the 1800s, the subspecies may have numbered more than 9600 individuals. Even in suitable habitat, its territories are widely spaced with a maximum of six pairs/km^2. Surveys, mostly in the Border Ranges to Gibraltar Range between 1979 and 1983, estimated about 730 territories. From recent limited surveys, the subspecies appears to have declined further in Qld and northern NSW although has been stable at other territories.

Habitat: The Rufous Scrub-bird occurs mainly in upland closed forest and wet open forest, particularly in areas of beech (*Nothofagus*) forest and thick undergrowth. Its habitat typically has ground cover a metre high, a moist microclimate and deep leaf litter, as occurs in rainforest and adjacent wet eucalypt forest. The rainforest habitat usually has canopy openings, such as along watercourses, around natural tree-falls or where selectively logged, that promote dense undergrowth. Eucalypt forest is only suitable habitat if it is well buffered from fire by nearby patches of rainforest.

Ecology: The Rufous Scrub-bird is secretive, staying within dense vegetation in poorly lit areas of rainforest gullies and beech forests. Incapable of sustained flight, it creeps around and over logs and rocks and through dense vegetation. The male has a wide variety of loud penetrating calls, including a song of repeated, swelling 'chip' notes and mimicry of other birds (e.g. cuckoos, parrots, robins, honeyeaters and scrubwrens). Rufous Scrub-birds breed during spring and early summer, with nests usually built close to the ground, either within grass tussocks or among ferns. The clutch is of two eggs, incubated for 38 days, and chicks fledge after about 28 days. Rufous Scrub-birds eat a wide variety of insects and worms found among the leaf litter on the forest floor.

Reason for decline and current threats: Clearing of rainforest in south-east Qld ceased many years ago, though the

Rufous Scrub-bird population has continued to decline, which suggests that other (unidentified) threatening processes are now important. Rufous Scrub-birds may require disturbance (e.g. selective logging) to open up the canopy, allowing scrubby growth on the rainforest floor.

Climate change may eliminate the Rufous Scrub-bird in Qld in the long term, with modelling suggesting that beech forests will contract southwards, thus reducing the limited areas of habitat.

CURRENT RECOVERY EFFORTS

Recovery plan: None

Other plans: Garnett ST and Crowley GM (2000) 'The action plan for Australia birds 2000'. Environment Australia, Canberra.

Garnett ST, Szabo JK and Dutson G (2011) *The Action Plan for Australian Birds 2010.* CSIRO Publishing, Melbourne.

Conservation actions:

- Government-funded volunteer surveys, managed by Ekerlogic Consultancy Services, have been conducted in Qld and NSW.
- Since 2006, DERM staff have also conducted surveys.
- In Qld, most remaining populations are secure in protected lands (national parks or state forests), at least in the short to medium term, although their continued existence may be threatened by inappropriate forest management or accidental burning.

FUTURE RECOVERY EFFORTS

What you can do:

- Report sightings to DERM.
- Volunteer with DERM to help conduct Rufous Scrub-bird surveys.

Conservation needs:

- Ensure that each subpopulation remains viable by protecting its habitat, especially from fire.
- Translocate individuals to maintain genetic diversity and/or to supplement small subpopulations in isolated habitats.
- Determine the Rufous Scrub-bird's current distribution and monitor for changes, using recent habitat models.

Research directions: Investigate the interaction between climate change, area of beech forests and Rufous Scrub-bird populations.

Sources and further reading

Chisholm AH (1951) The story of the scrub-birds. *Emu* **51**, 88–112; 258–297.

Higgins PJ, Peter JM and Steele W (Eds) (2001) *Handbook of Australian, New Zealand and Antarctic Birds. Volume 5: Tyrant-flycatchers to Chats.* Oxford University Press, Melbourne.

Portelli DJ (2004) The singing behaviour of the noisy scrub-bird, *Atrichornis clamosus*: congeneric comparisons and the feasibility of using individual variation in song as a census tool. *Emu* **104**, 273–281.

Schodde R and Mason IJ (1999) *The Directory of Australian Birds: Passerines.* CSIRO Publishing, Melbourne.

White HL (1920) The Rufous Scrub-bird (*Atrichornis rufescens*) in Queensland. A new sub-species. *Emu* **19**, 257–258.

Author: David Stewart

PURPLE-CROWNED FAIRY-WREN

Malurus coronatus Gould, 1857

Photo: Graeme Chapman

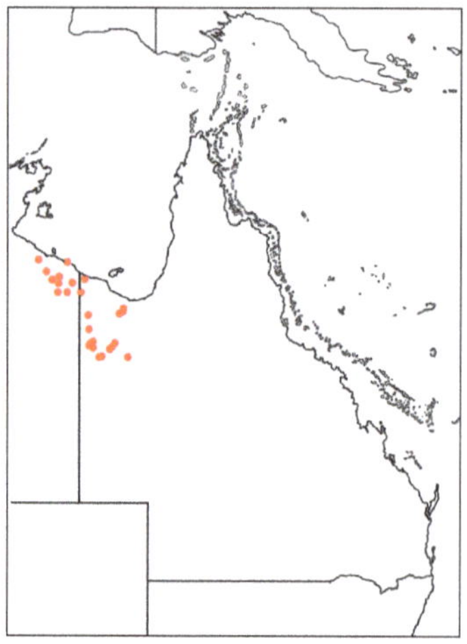

Family: Maluridae

Conservation status: IUCN: species (*M. coronatus*) is Least Concern; Aust.: Vulnerable (western subspecies *Malurus coronatus coronatus*); Qld: Vulnerable; NT: Vulnerable; WA: declared threatened fauna

Description: The Purple-crowned Fairy-wren is Australia's largest fairy-wren, with a total length of 14 cm (including a tail of 6.5–8.3 cm) and a wingspan of 5–5.5 cm, and weighs 9–13 g. All birds have warm brown to grey backs, white fronts with a buff wash, upright long blue tails and a dark bill. Adult breeding males develop a purple crown with a black patch in the centre and a black face mask. Non-breeding males and adult females are distinguished by their ear coverts, which are black or dark grey in males, and chestnut in females. Juveniles are similar to non-breeding adults, except for their pale gape. The chestnut ear coverts of female Purple-crowned Fairy-wrens distinguish them from females of other similar species.

Distribution: The species occurs on a subset of rivers from the Kimberley in WA to the western edge of the Gulf of Carpentaria in Qld. Two subspecies are recognised: the western (*M. c. coronatus*) occurs in the Kimberley and Victoria River District; and the eastern (*M. c. macgillivrayi*) occurs along the Gulf of Carpentaria. The species occurs on Lawn Hill Creek, the Nicholson and O'Shannassy Rivers, and the Gregory River catchment in Qld. Wild populations are currently estimated to contain 18 000 birds in the Gulf District, 2500–4500 birds in the Kimberley and 5500–7500 birds in the Victoria River District.

Habitat: The Purple-crowned Fairy-wren is largely confined to the river-fringing vegetation, characterised by a dense midstorey and high canopy of trees that grow beside permanent freshwater pools along rivers and billabongs. Based on midstorey structure, the habitat type is classified as dominated either by River Pandanus *Pandanus aquaticus* or by Rivergrass *Chionachne cyanthopoda.*

Ecology: The species is a territorial cooperative breeder, living in sedentary groups composed of a breeding pair that is helped by one to six offspring from previous broods. Helpers may stay with parents for several years before attempting to breed themselves. Territories are maintained year round and generally occupy between 20 and 300 m of river length. Breeding can occur at any time of year, with peaks in the early and late dry season (March–May and August–November). Nests are built mainly of fine rootlets, grass, leaves and strips of bark. Incubation of clutches of two to three eggs takes 14 days and chicks fledge after 10 days. The species is mainly insectivorous and consumes a range of small invertebrates. Most dispersal occurs when helpers abandon their natal territories in search of a breeding territory of their own. Males generally disperse shorter distances than females. The species has not been observed at great distances from riparian habitat, suggesting that dispersal predominantly occurs along watercourses.

Reasons for decline and current threats: The species has declined following degradation and loss of its habitat through overgrazing by introduced herbivores, weed invasion, erosion and changed fire conditions. These threats have caused major change to the understorey composition and structure of the wren's habitat in parts of its range with the western subspecies thought to have been most affected. During the early 20th century, the loss of riparian habitat due to pastoral activity led to declines, in both range and number, along vast stretches of river in the Kimberley. More recently, declines have also been detected on the Victoria and Ord Rivers. In contrast, the eastern subspecies may be in better shape. The only declines that have been noted were on the Leichardt River in Qld following damming of the river. However, the eastern populations are still probably threatened by the combined effects of grazing, fire and weed incursion.

CURRENT RECOVERY EFFORTS

Recovery plan: None

Other plans: Commonwealth conservation advice on *Malurus coronatus coronatus* (Purple-crowned Fairy-wren (western)) (Threatened Species Scientific Committee, 2008fr) [Conservation Advice].

Garnett ST and Crowley GM (2000) 'The action plan for Australian birds 2000'. Environment Australia, Canberra.

Garnett ST, Szabo JK and Dutson G (2011) *The Action Plan for Australian Birds 2010*. CSIRO Publishing, Melbourne.

Conservation actions:

- Scientific research projects have focused on populations of the Purple-crowned Fairy-wren at Riversleigh (Qld), on the Victoria River (NT) and within the Kimberley (WA). Research has investigated the breeding ecology, population structure and sensitivity of the species to grazing of riparian habitat.
- The NT DNREAS is studying population trends and ecology of the species on the Victoria River.
- The Max-Planck Institute for Ornithology is studying behaviour, ecology and dispersal.
- The conservation biology of the western species is being studied by the Australian Wildlife Conservancy and the Australian National University, with additional funding from the Threatened Species Network (WWF-Australia), Department of Environment and Conservation, Western Australia and Birds Australia.

FUTURE RECOVERY EFFORTS

What you can do: Advocate reduced grazing of river frontages and the production of guidelines for managers to promote conservation of riparian habitat and the Purple-crowned Fairy-wren.

Conservation needs:

- Map the habitat, improve the distributional knowledge of the species and monitor population trends.
- Protect Purple-crowned Fairy-wren habitat by reducing stock access to river frontage.

Research directions:

- Information on the status of the eastern subspecies of the Purple-crowned Fairy-wren is required to direct conservation.
- To help prioritise actions, research into how habitat arrangement affects the population structure and influences the dispersal of the species across the landscape is required.

Sources and further reading

Rowley I (1993) The Purple-crowned Fairy-wren *Malurus coronatus*. I. History, distribution and present status. *Emu* **93**, 220–234.

Rowley I and Russell E (1993) The Purple-crowned Fairy-wren *Malurus coronatus* II. Breeding biology, social organisation, demography and management. *Emu* **93**, 235–250.

Skroblin A and Legge S (2010) The distribution and status of the western subspecies of the Purple-crowned Fairy-wren (*Malurus coronatus coronatus*). *Emu* **110**, 339–347.

Smith LA and Johnstone RE (1977) Status of the Purple-crowned Wren (*Malurus coronatus*) and Buff-sided Robin (*Poecilodryas superciliosa*) in Western Australia. *Western Australian Naturalist* **13**, 185–188.

Van Doorn A (2008) Ecology and conservation of the Purple-crowned Fairy-wren (*Malurus coronatus coronatus*) in the Northern Territory, Australia. PhD thesis. University of Florida, Gainsville.

Author: Anja Skroblin

GREY GRASSWREN (BULLOO SUBSPECIES)

Amytornis barbatus barbatus Favaloro & McEvey, 1968

Photo: Graeme Chapman

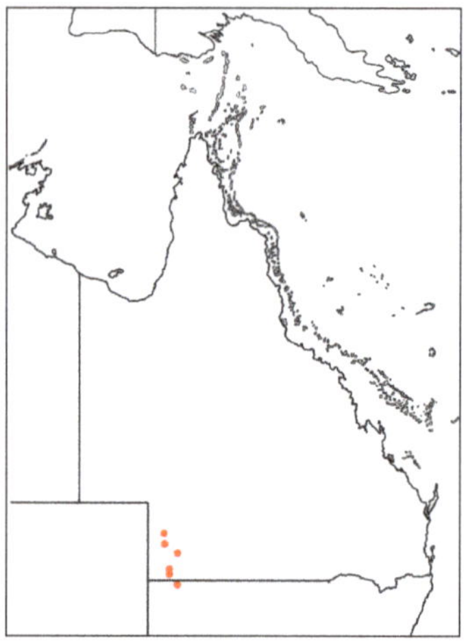

Family: Maluridae

Conservation status: IUCN: species (*A. barbatus)* is Least Concern; Aust.: Vulnerable; Qld: Vulnerable; NSW: Endangered

Description: The Grey Grasswren is 18–22 cm in total length, including a long tail, with a wingspan of up to 21 cm, and weighs 15–23 g. The upperparts are primarily grey and cinnamon streaked black and white, and the underparts are white, streaked black on the breast and becoming pale rusty on the flanks. The face is boldly striped with black and white, including a black stripe through the eyes and white eyebrows. Females have narrower black facial markings and fainter breast streaks, and juveniles are duller.

Distribution: The Bulloo subspecies of the Grey Grasswren is found in the Bulloo River drainage in south-western Qld and north-western NSW, including the Bulloo Overflow and the Caryapundy and Jerrira Swamps. Its entire population may be restricted to an area of less than 100 km^2, now only 80 km^2 in NSW, although new populations may yet be found in unsurveyed potential habitat in Qld. The population is thought to fluctuate depending on the level of flooding, the extent of drought and the intensity of cattle grazing. Recent sightings of Grey Grasswrens on the Cooper Creek floodplain were closer to the Bulloo population than to the Diamantina River population (subspecies *A. b. diamantina*), and were of undetermined subspecies, though probably *A. b. diamantina*; unsuitable or discontinuous habitat exists between the Cooper and Bulloo. The size of the Bulloo population was estimated, with low reliability, as 10 000 breeding birds; with 193 banded at one site of 11 ha (Caryapundy Swamp) in Qld. Since then, the NSW population and its range have declined by 90% to 1600 birds.

Habitat: The Bulloo subspecies of the Grey Grasswren lives on swampy floodplains covered in two key plant species: dense, tall clumps of Lignum *Muehlenbeckia florulenta* and tussocks of Swamp Canegrass *Eragrostis australasica*, which are seasonally inundated. It also occurs in adjacent Oldman Saltbush *Atriplex nummularia* and samphire flats. High-quality lignum is an important drought refuge. The Diamantina subspecies, at least, also uses other floodplain vegetation types.

Ecology: This well-camouflaged ground-dwelling bird lives among the lignum and canegrass, including when flooded, where it nests and feeds on insects, larvae and seeds and occasionally on water snails. The semi-domed grass nest is well concealed low in the centre of a clump of lignum or canegrass, and the clutch is of two or three eggs. Its social organisation is poorly known. During the breeding season, the species appears to occur in pairs or small family groups, with larger congregations being observed in suitable habitat after breeding. The birds also congregate in habitat refuges during droughts.

Reasons for decline and current threats: The existence of this subspecies is precarious, because its small range is vulnerable to catastrophic events and its small population fluctuates with seasonal conditions. Its prospects appear to be entirely related to habitat alteration or destruction by burning or clearing of lignum for cattle grazing, and grazing and other habitat destruction by feral animals such as rabbits, goats, horses and pigs. Potential threats include: changes in flooding patterns, should water be diverted from the Bulloo River; predation by cats and foxes; and the possible spread of invasive and noxious weeds (such as *Parkinsonia aculeata* and *Prosopis*).

CURRENT RECOVERY EFFORTS

Recovery plan: NSW National Parks and Wildlife Service (2003) 'Grey Grasswren (*Amytornis barbatus*) draft recovery plan'. NSW National Parks and Wildlife Service, Sydney.

Other plans: Garnett ST and Crowley GM (2000) 'The action plan for Australia birds 2000'. Environment Australia, Canberra.

Garnett ST, Szabo JK and Dutson G (2011) *The Action Plan for Australian Birds 2010*. CSIRO Publishing, Melbourne.

Conservation actions:
NSW DECCW has identified the following priority actions:

- Control vertebrate pest populations (e.g. foxes, cats and rabbits) that either prey on or compete for resources.
- Implement appropriate fire regimes to avoid burning all habitat and food resources within a region at one time, and to promote natural succession.
- Reduce stock intensity (or exclude/control grazing) to allow regeneration of vegetation for habitat, such as food sources or nest sites.
- Apply appropriate water regulation in wetland habitats.
- Retain understorey shrubs and allow them to complete their life cycle (i.e. seed set, germination, establishment, growth to maturity).
- Prevent clearing of habitat such as nesting sites and food sources.
- Protect and conserve wetland habitats.

FUTURE RECOVERY EFFORTS

Conservation needs:

- DERM recovery objectives are to retain and monitor a viable population, implement appropriate grazing regimes and monitor populations.
- A recent study concluded that the only workable solution for conservation of the Bulloo population is for NSW DECCW to acquire Narriearra Station as a reserve.

NSW DECCW, in its recovery strategy and draft recovery plan for the NSW Grey Grasswren (Bulloo), identifies the following required actions:

- Establish a community and landholder liaison and heighten public awareness and/or education.
- Habitat management must include feral pest control.
- Establish a fire management plan for the Bulloo region.
- Carry out ongoing environmental impact assessment advice to consent and planning authorities.
- Implement site protection (e.g. fencing/signage).
- Consider the acquisition of appropriate habitat in the Bulloorine (possibly as an addition to Sturt NP or as a separate reserve).
- Encourage the protection of habitat on private land through voluntary conservation agreements or the development of management plans for the species (which would include fire, pest and grazing management and monitoring).

Research directions: Undertake population monitoring, survey/mapping and habitat assessment.

Sources and further reading

Carpenter G (2002) The Grey Grasswren on Cooper Creek south-west Queensland. *Sunbird* **32**, 52–55.

Hardy J (2002) A banding study of the Grey Grasswren *Amytornis barbatus barbatus* in the Caryapundy Swamp of south-western Queensland. *Corella* **26**, 106–109.

Hardy JW (2010) Distribution, status and options for the future management of the Grey Grasswren *Amytornis barbatus barbatus* in New South Wales. *Corella* **34**, 25–35.

Higgins PJ, Peter JM and Steele WK (Eds) (2001) *Handbook of Australian, New Zealand and Antarctic Birds Volume 5: Tyrant-flycatchers to Chats.* Oxford University Press, Melbourne.

Jaensch R and McFarland D (2002) A population of Grey Grasswren *Amytornis barbatus* in the Diamantina Channel Country, Queensland. *Sunbird* **32**, 56–61.

Author: Lee K Curtis

EASTERN BRISTLEBIRD (NORTHERN SUBSPECIES)

Dasyornis brachypterus monoides Schodde & Mason, 1999

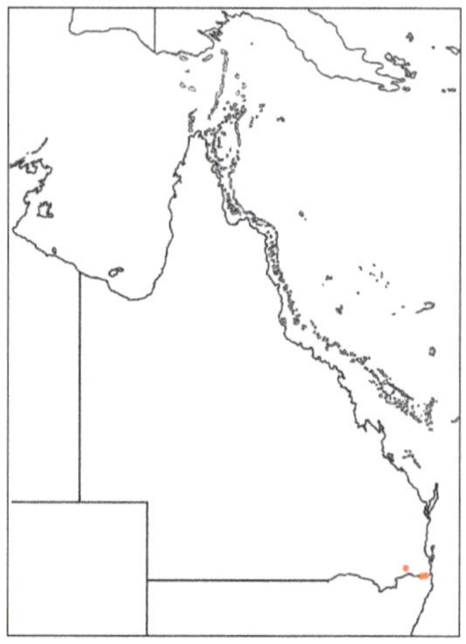

Photo: Zan Gubler

Family: Dasyornithidae

Other common name: Brown Bristlebird

Conservation status: IUCN: species (*D. brachypterus*) is Endangered; Aust.: Endangered; Qld: Endangered; NSW: Endangered

Description: The Eastern Bristlebird is a small, long-tailed bird, with a total length of about 21 cm, and weighing about 42 g, that lives in dense undergrowth. It has short wings and weak flight. It is brown with rufous flight feathers, a paler and greyer underside, a pale face with red eyes, and robust legs and feet. The tail appears 'frayed' and is sometimes raised and fanned. The territorial song is loud and melodic.

Distribution: The southern subspecies *D. b. brachypterus* occurs patchily in southern NSW and far eastern Vic. The northern subspecies *D. b. monoides* now occurs between the Conondale Range, Qld and the Tweed Range, NSW. In Qld, it occurs patchily in the Conondale, Main Range, Lamington and Mount Barney NPs. It is thought to have declined by about 80% over the last three generations or 15 years. A survey for Eastern Bristlebird in 2007 found only 15 birds, distributed between Lamington, Main Range and Conondale NPs with 25–30 birds at 10–15 sites in NSW.

Habitat: The northern subspecies occurs mostly in upland open forest with dense grassy undergrowth, but also in montane heath. Colonies occur either adjacent to, or within 750 m of, rainforest, which acts as a fire and drought refuge, and often along intermittent watercourses. The diverse ground layer includes tussock grasses, particularly mature Wild Sorghum *Sorghum leiocladum*, Kangaroo Grass *Themeda australis* and Blady Grass *Imperata cylindrica*, with scattered shrubs, logs, tall ferns and/or tangled vines. Bristlebirds require dense foliage near the ground, the structure and floristics being determined by the fire regime, and typically a mosaic of age classes.

Ecology: The northern subspecies is less well known than the southern species, and most information derives from the latter. Bristlebirds are sedentary in territories of about 1–4 ha, which are advertised and defended by singing all year, with a peak in spring. Birds sing from dense cover, in early morning or during cool and calm weather, in bouts of up to 15–20 minutes, with antiphonal duets between male and female. Bristlebirds build a globular nest low in dense vegetation, such as a grass tussock (particularly Wild Sorghum), or occasionally in low branches of shrubs among herbage. The clutch is usually two eggs, with one successful brood (of only one chick) raised per year. Eggs are laid in August and September, incubated for 21 days, and the chicks fledge after 15 days. Nests disturbed during incubation are readily abandoned. Bristlebirds forage on the ground and also glean foliage or snatch flying insects. Food includes insects (e.g. ants, bugs, beetles and orthopterans), seeds, small fruits and earthworms.

Reason for decline and current threats: An inappropriate fire regime (i.e. frequent high-intensity fires) appears to be a major cause of the decline or extinction of Eastern Bristlebird colonies in south-east Qld and north-east NSW. Other potential threatening processes include weed invasion, predation or habitat damage by feral pigs, predation by foxes or cats, habitat damage by recreational 4WDs, inappropriate grazing, including weeds spread by deer and cattle, human disturbance to active nests and illegal collection of eggs.

CURRENT RECOVERY EFFORTS

Recovery plans: Stewart D (2002) 'Recovery plan for the northern population of the Eastern Bristlebird *Dasyornis brachypterus monoides* 2001–2005' (unpublished). Queensland Parks and Wildlife Service, Brisbane.

A new draft recovery plan is in preparation and available for comment:

Thomas V (2010) 'National recovery plan for the Eastern Bristlebird *Dasyornis brachypterus*'. Draft 1 – May 2010 (unpublished).

Other plans: Department of the Environment, Water, Heritage and the Arts (2008) 'Threat abatement plan for predation by feral cats'. Commonwealth of Australia, Canberra.

Garnett ST and Crowley GM (2000) 'The action plan for Australia birds 2000'. Environment Australia, Canberra.

Garnett ST, Szabo JK and Dutson G (2011) *The Action Plan for Australian Birds 2010.* CSIRO Publishing, Melbourne.

Holmes G (1989) 'Eastern Bristlebird: Species management plan for northern populations' (draft). Queensland National Parks and Wildlife Service, Brisbane/NSW National Parks and Wildlife Service, Sydney.

Stewart D (2006) 'Threat management plan for Eastern Bristlebird populations in the Lamington National Park'. Unpublished report for Queensland Parks and Wildlife Service, Brisbane.

Stewart D (2006) 'Threat management plan for the Eastern Bristlebird population in the Conondale Range'. Unpublished report for Queensland Parks and Wildlife Service, Brisbane.

Conservation actions:

- Surveys and monitoring over the last decade, by Birds Qld volunteers, NSW National Parks and Wildlife Service and the QPWS, are continuing. Habitat on private and protected lands has been protected by a network of firebreaks.
- A pilot captive-breeding program at the David Fleay Wildlife Park (DERM) on the Gold Coast has successfully bred Eastern Bristlebirds, with 13 birds raised over four years. In 2008–09, four were released in Qld and four in NSW.
- A volunteer project run by Birds Qld searches for new Eastern Bristlebird colonies and potential habitat in montane areas of south-east Qld. Two associated long-term studies are:
  - Analysis of vegetation regeneration after fire in areas occupied by Eastern Bristlebirds. In unburnt areas, this project investigates how Eastern Bristlebirds cope with successional stages in the forest structure.
  - Long-term monitoring of the released captive-bred birds.

FUTURE RECOVERY EFFORTS

What you can do: Volunteer for the Birds Australia bristlebird surveys.

Conservation needs:

- Implement threat abatement: change of fire regimes, fire management, control domestic cattle stocking numbers, limit human activities in bristlebird habitat, especially during the breeding season.
- Develop captive breeding and translocation.
- Regenerate suitable habitat.

Research directions:

- Conduct scientific research into the biology and genetics of the species.
- Conduct ongoing surveys and continuously monitor this species.
- Study food (invertebrate) availability in bristlebird habitat.

Sources and further reading

Hartley S and Kikkawa J (1994) 'The population management of the Eastern Bristlebird *(Dasyornis brachypterus)*: Findings on the biology, threats and management of the Eastern Bristlebird in Queensland and Northern New South Wales'. Zoology Department, University of Queensland, St Lucia.

Higgins PJ and Peter JM (Eds) (2002) *Handbook of Australian, New Zealand and Antarctic Birds. Volume 6: Pardalotes to Shrike-thrushes.* Oxford University Press, Melbourne.

Holmes G (1997) 'Conservation status of Eastern Bristlebird in south-east Queensland and north-east New South Wales'. Unpublished report to Department of Environment, Brisbane.

Pyke G, Saillard R and Smith J (1995) Abundance of Eastern Bristlebirds in relation to habitat and fire history. *Emu* **95**, 106–110.

Stewart DA, Gillman S and Rounsevell D (2004) The recovery process for the Eastern Bristlebird *Dasyornis brachypterus* in Queensland. *Sunbird* **34**, 66–79.

Author: David Stewart

REGENT HONEYEATER

Anthochaera phrygia (Shaw, 1794)

Photo: Blesbok Enterprises

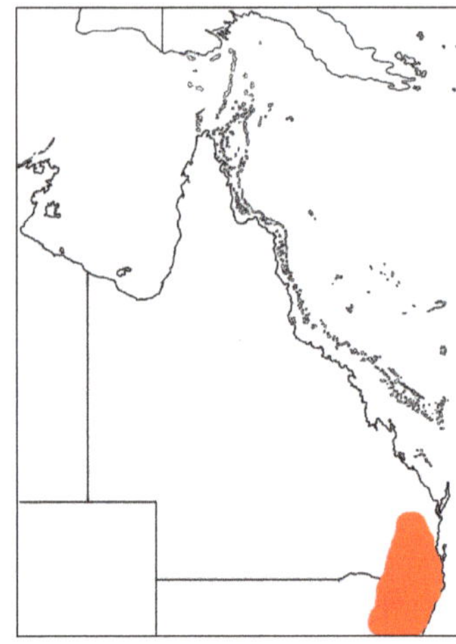

Family: Meliphagidae

Conservation status: IUCN: Endangered; Aust.: Endangered; ACT: Endangered; Qld: Endangered; NSW: Critically Endangered; SA: Endangered; Vic.: Critically Endangered

Description: The Regent Honeyeater has a total length of 20–23 cm and weighs 31–50 g. It is mostly black with yellow edges to the tail and wing feathers. The body feathers, except for the head and neck, are broadly edged pale yellow or white. A patch of yellow to pink, bare, warty skin surrounds each eye. The overall impression is of a blackish bird boldly embroidered yellow and white, with yellow flashes in the wings and tail.

Distribution: Formerly distributed in south-eastern Australia from Port Pirie (SA) to 100 km north of Brisbane (Qld), the honeyeater's southerly range has contracted to Bendigo (Vic.). On the western edge of its range it occurs as far inland as Dalby, Narrabri, Warrumbungle NP, Dubbo, Parkes and Finley. Its distribution is now extremely patchy, with a small number of known breeding sites. In Qld, breeding occurs regularly west of Warwick, though only by small numbers of birds. The size of its national population is unknown, but has been estimated at between 350 and 400. A highly mobile species thinly spread over large areas, it is difficult to obtain reliable data on densities.

Habitat: Regent Honeyeaters occur mainly in box-ironbark woodlands, particularly those dominated by Mugga Ironbark *Eucalyptus sideroxylon*, Yellow Box *E. melliodora* or White Box *E. albens*. They prefer wetter, more fertile sites with high nectar flows, such as creek flats, river valleys and lower slopes. Other habitat types include River She-oak *Casuarina cunninghamiana* gallery forests supporting Needle-leaved Mistletoe *Amyema cambagei*. Remnant stands of timber, roadside reserves, travelling stock routes and street trees also provide important habitat for Regent Honeyeaters at times. Regent Honeyeaters also use wet lowland coastal forest dominated by Swamp Mahogany *E. robusta* or Spotted Gum *Corymbia maculata*, perhaps mainly as a drought refuge.

Ecology: The honeyeaters select individual trees or stands with abundant nectar and prefer larger trees, which are thought to produce the most nectar. The preferred 'key' tree species have fairly predictable flowering periods and produce heavy nectar flows. Regent Honeyeaters may use different areas in different years, depending on food resources, and may move large distances to do so. Breeding has been recorded in all months between June and February, with a core period from early August to late November. The annual cycle at key Regent Honeyeater sites includes: arrival in winter or early spring and occurrence in loose flocks, with communal roosting; breeding through spring and early summer, often as aggregations of pairs; and dispersal in mid- to late summer after young have fledged. Regent Honeyeaters feed on nectar and arthropods, and occasionally fruit. They also eat other sugary plant or insect exudates including lerp, honeydew and manna. Insects taken, especially during breeding, are gleaned from foliage and bark and taken in flight.

Reasons for decline and current threats: The major cause of the decline has been the clearing and fragmentation of woodland and forest containing the birds' preferred eucalypt species.

Continuing threats include: the further degradation of habitat, the loss of large mature trees (through senescence, cutting for fence posts or firewood, clearing for grazing or from the impact of eucalypt dieback) which threatens remnants and the lack of regeneration of eucalypt woodland and riverine casuarinas. Noisy Miners *Manorina melanocephala*, through their aggressive behaviour, exclude Regent Honeyeaters from resources and even from some habitat remnants. Miners prefer open areas adjoining woodland, and are abundant in fragmented and degraded habitat. Honeybees may also compete with Regent Honeyeaters for nectar, but the extent is unknown.

CURRENT RECOVERY EFFORTS

Recovery plan: Menkhorst P, Schedvin N and Geering D (1999) 'Regent Honeyeater (*Xanthomyza phrygia*) recovery plan 1999–2003'. Department of Natural Resources and Environment, Melbourne. (New plan to 2012 in preparation.)

Other plans: Garnett ST and Crowley GM (2000) 'The action plan for Australia birds 2000'. Environment Australia, Canberra.

Garnett ST, Szabo JK and Dutson G (2011) *The Action Plan for Australian Birds 2010*. CSIRO Publishing, Melbourne.

Conservation actions:

- A full-time Recovery Coordinator is employed and hosted by Birds Australia.
- Volunteer regional operations groups are operating at key locations implementing on-ground actions to protect and enhance areas containing 'significant habitat'.
- Research, including the monitoring of populations and changes in distribution; determining movements by colour-banding and determining post-breeding dispersal by radio-telemetry are all being coordinated by the recovery team.
- Maintaining a genetically viable captive population, and supplementing regional 'at risk' subpopulations with captive-bred birds, coordinated by Taronga Conservation Society.

FUTURE RECOVERY EFFORTS

What you can do:

- Report all sightings immediately to the national Regent Honeyeater Recovery Coordinator on Freecall 1800 621 056.
- Record, and report, the colour-band combination on any birds seen (i.e. the position and colour of each band on each leg).
- Landholders can also covenant all or parts of their properties if they contain habitat of significant quality or type. Areas containing 'key' species are thus protected in perpetuity: a key action for a species that has so much important habitat on private property.

Conservation needs:

- Restore box-ironbark woodlands.
- Reduce grazing pressure to allow the natural regeneration of eucalypts and riparian vegetation. Funding through Catchment Management Authorities and/or NRM agencies may assist with fencing, revegetation and weed control.

Sources and further reading

Geering D and French K (1998) Breeding biology of the Regent Honeyeater *Xanthomyza phrygia* in the Capertee Valley, New South Wales. *Emu* **98**, 104–116.

Higgins PJ, Peter JM and Steele WK (Eds) (2001) *Handbook of Australian, New Zealand and Antarctic Birds. Volume 5: Tyrant-flycatchers to Chats.* Oxford University Press, Melbourne.

Oliver DL (2000) Foraging behaviour and resource selection of the Regent Honeyeater *Xanthomyza phrygia* in New South Wales. *Emu* **100**, 12–30.

Oliver DL (2001) Activity budget of the Regent Honeyeater, *Xanthomyza phrygia*, in northern New South Wales. *Australian Journal of Zoology* **49**, 695–712.

Powys V (2010) Regent Honeyeaters – mapping their movements through song. *Corella* **34**, 92–102.

Author: David Geering

YELLOW CHAT (GULF SUBSPECIES)

Epthianura crocea crocea Castelnau & Ramsay, 1877

Photo: Graeme Chapman

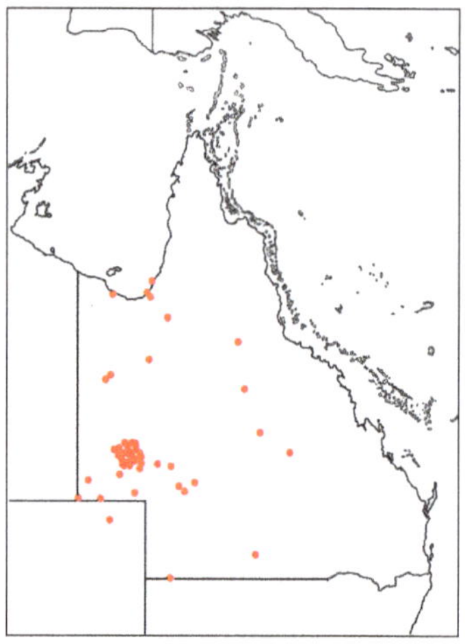

Family: Meliphagidae

Other common name: Inland Yellow Chat

Conservation status: IUCN: species (*E. crocea*) is Least Concern; Qld: Vulnerable

Description: The Yellow Chat is a small bird about 11 cm in length and weighing 9 g, with distinctive long legs. Adult males and females are olive brown on the back, with a yellow belly, chest and rump, pale eyes and white edges to the outer wings. Breeding males are golden with a brighter golden-yellow head and a diagnostic, usually crescent-shaped, black band on the chest. Females are more lemon yellow, and lack the breast bar and golden crown. Newly fledged juveniles are pale, with yellow only on the lower belly and rump; they rapidly develop browner plumage with varying amounts of yellow on the chest.

Distribution: The Yellow Chat occurs patchily across northern and eastern inland Australia, from the Kimberley (WA) and Top End (NT) to southern Qld. The Gulf subspecies of Yellow Chat is widely, but locally, distributed across inland Qld on the Lake Eyre Basin rivers, particularly the Georgina, Diamantina and Cooper, and occasionally on the internally draining Hay and Bulloo rivers in south-west Qld. It also occurs on upper and coastal sections of rivers draining into the Gulf of Carpentaria, including the Nicholson, Leichardt, Flinders, Norman and Gilbert. Many Lake Eyre Basin sightings were around bores, many of which have been recently closed down. The population trend of this subspecies is unknown.

Habitat: This subspecies occurs in the Lake Eyre and Gulf of Carpentaria drainage basins, where it inhabits low vegetation around the margins of ephemeral wetlands, especially floodplains, swamps and bore drains, but also vegetated dams, salt lakes, clay pans, lagoons, sewage ponds or reservoirs. It forages in adjacent grasslands, flooded samphire or on drier ground dominated by chenopods and, on the Norman River, in tall grassy swamps. Most records come from bores, where typically the vegetation includes tall cover such as sedges, Bulrush *Typha*, reeds *Phragmites* and rank grasses (e.g. canegrass) frequently interspersed with sparser vegetation of low shrubs, usually chenopods. In wetlands of the Georgina and Diamantina basins, Yellow Chats occur mainly in sparsely vegetated lignum swamps with Bluebush, *Sesbania* and *Eleocharis*.

Ecology: Chats are predominantly insectivorous, taking mostly flies, with some beetles, bugs, small damselflies, lacewing larvae, ants and nematode worms. They forage on low vegetation and along the bases of sedges and bulrushes, on exposed root mats, and in mud and shallow water between plants. Most breeding occurs in spring–summer, but occasionally in autumn or early winter. The nests are cups of fine, dry grass, well concealed in emergent wetland vegetation, or in tall grass near the shore, in thistles or in flooded lignum. The clutch size is usually three eggs. Apparently, this chat can disperse and colonise new habitat, as suggested by its colonisation of bores created in the 1890s. However, it may have been present in nearby natural wetlands. Local movements between breeding habitat and dry-season habitat seem likely.

Reasons for decline and current threats: Threats include habitat loss from the closure of artesian bores, although there may be compensatory benefits. Threats to natural artesian springs include feral animals, particularly pigs, but also camels, goats and horses, and weeds. Threats to the chat's preferred habitat (ephemeral swamps and lakes) include habitat degradation by reduction in water flows (by human constructions), weed invasion, feral animals, fire regimes and possibly overgrazing. Reduction or alteration of flows into ephemeral swamps and lakes is the primary threat to these wetlands, by reducing the productivity upon which the breeding of chats is thought to depend.

In inland Qld and the NT, invasion by *Parkinsonia aculeata* has the potential to alter chat breeding habitat. This shrub forms very dense thickets along watercourses and around watering points, smothering the wetland habitat. Overgrazing by cattle can affect wetlands by removing important vegetative cover, such as sedges, rushes, reeds and rank grasses, which are used by chats for breeding.

CURRENT RECOVERY EFFORTS

Recovery plan: None

FUTURE RECOVERY EFFORTS

What you can do:

- Report sightings to DERM.
- Volunteer with DERM to help with on-ground management actions at Astrebla Downs NP (e.g. fencing or pest management).
- Because most chat habitat is under freehold tenure, donations to DERM to aid landholders manage lands placed under Nature Refuge tenure may be the way to improve conservation prospects.
- Graziers can help by taking an interest in sedges around wetlands and channels in the same way they take an interest in promoting grass growth. Reducing trampling of samphire vegetation and spelling of wetlands during the wet season will also improve chat habitat.
- Landholders can help by fencing bores, applying to NRM groups for advice on support for conservation benefits (e.g. fencing significant chat habitat and spelling paddocks for mutual benefits) and biodiversity incentives schemes and avoiding overgrazing of known chat habitat.

Research directions:

- Assess the implications of bore closures.
- Carry out surveys and monitoring to establish baseline data on population levels, particularly of the recently identified natural wetlands in the core area in the Georgina and Diamantina catchments.
- Map the extent of regional ecosystems known to support Gulf Yellow Chats.

Sources and further reading

Black AB, Duggan G, Pedler JA and Pedler LP (1983) The Yellow Chat *Epthianura crocea* at Pandiburra Bore, north-eastern South Australia. *South Australian Ornithologist* **29**, 42–45.

Higgins PJ, Peter JM and Steele WK (Eds) (2001) *Handbook of Australian, New Zealand and Antarctic Birds. Volume 5: Tyrant-flycatchers to Chats.* Oxford University Press, Melbourne.

Jaensch R (2004) Yellow Chat *Epthianura crocea* in natural wetlands of the Georgina and Diamantina Channel Country. *Sunbird* **34**, 58–65.

Reynolds IS, Walter JC and Woodall PF (1982) Observations on Yellow Chats (*Epthianura crocea*) in western Queensland. *Sunbird* **12**, 21–29.

Strong BW and Fleming MR (1987) Recent observations of the distribution and habitat of the Yellow Chat *Epthianura crocea* in the Northern Territory. *South Australian Ornithologist* **30**, 98–102.

Author: Wayne Houston

YELLOW CHAT (CAPRICORN SUBSPECIES)

Epthianura crocea macgregori Keast, 1958

Photo: Robert Black

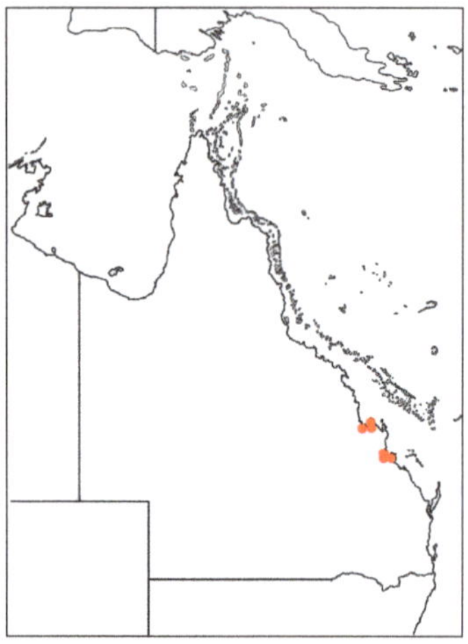

Family: Meliphagidae

Other common name: Capricorn Yellow Chat

Conservation status: IUCN: species (*E. crocea*) is Least Concern; Aust.: Critically Endangered; Qld: Endangered

Description: The Yellow Chat is a small bird about 11 cm in total length and weighing 9 g, with distinctive long legs. Adult males and females are olive brown on the back, with a yellow belly, chest and rump, pale eyes and white edges to the outer wings. Breeding males are golden, with a brighter golden-yellow head and a diagnostic, usually crescent-shaped, black band on the chest. Females are more lemon yellow, and lack the breast bar and golden crown. Newly fledged juveniles are pale, with yellow only on the lower belly and rump; they rapidly develop browner plumage with varying amounts of yellow on the chest.

Distribution: The Yellow Chat occurs patchily across northern and eastern inland Australia, from the Kimberley (WA) and Top End (NT) to southern Qld. The isolated Capricorn subspecies of the Yellow Chat, on the Qld coast, occurred at the Fitzroy River Delta in 1859 and Torilla Plain north of Rockhampton in 1917, but went unrecorded until found at Curtis Island near Gladstone in 1992. Surveys have confirmed its historical distribution, and located further sites in the Broad Sound area (encompassing Torilla Plain) and Fitzroy Delta. The Curtis Island population has declined during drought, although the birds persist. The population numbers about 250 birds, with most at Torilla Plain.

Habitat: The subspecies of this chat occurs mainly on coastal plains, which are treeless other than littoral mangroves, with many shallow areas that flood during the wet season (December–March). Most sites have a mosaic of microhabitats of salt-tolerant and freshwater plants: wetlands supporting stands of sedges and tall grasses, and adjacent more open habitats of sparser grasslands and samphire, with mangroves at tidal sites.

Ecology: This chat forages mostly on wetland muddy edges, the flowers and seed heads of sedges, forbs of damp pastures, in low shrubs such as samphire and *Sesbania*, and occasionally on nearby grasses and mangrove shrubs. During the dry season, they forage on grassy verges of dried wetlands, on cracked dry soils in swales and depressions, and in samphire. They take spiders, moths, caterpillars, damselflies, flying ants, beetles, grasshoppers and cockroaches. These chats breed mostly in summer and autumn, but also in spring or winter. The nests are cups of loosely woven plant fibres, placed low within grass, sedge or samphire clumps, often surrounded by shallow water. The clutch size is usually two or three eggs. Chats occur annually at the same breeding locations and undertake local movements during the dry season.

Reasons for decline and current threats: This chat's population trend is unknown. Its small population and restricted distribution render it vulnerable to extinction.

Foraging pigs damage sedges in chat habitat. Other threats include: habitat loss to industry and mining; loss or reduction of water flows to breeding habitat; overgrazing of tall sedges and grasses; removal of grazing from introduced pastures, leading to smothering of sedges and thickening of open grasslands; wildfire; siltation of wetlands following soil disturbance and lack of conservation tenure.

CURRENT RECOVERY EFFORTS

Recovery plan: Houston W and Melzer A (2008) 'Yellow Chat (Dawson subspecies) *Epthianura crocea macgregori* recovery plan 2007–2011'. Report to Department of Environment and Heritage, Canberra by Queensland National Parks and Wildlife Service, Brisbane.

Other plans: Hall C (2005) 'Curtis Island vertebrate pest management plan'. Report to Queensland Parks and Wildlife Service by Central Queensland University.

Garnett ST and Crowley GM (2000) 'The action plan for Australian birds 2000'. Environment Australia, Canberra.

Garnett ST, Szabo JK and Dutson G (2011) *The Action Plan for Australian Birds 2010.* CSIRO Publishing, Melbourne.

Conservation actions:

- Surveys have been undertaken on the Torilla Plain, Fitzroy Delta and Curtis Island, with potential habitat searched and further sites identified.
- Regular monitoring of chats has occurred at Torilla Plain and Fitzroy Delta, and the link between wetland inundation, food availability and breeding success researched.
- Ecological research has been carried out on the subspecies.
- Trials of techniques for monitoring and controlling vertebrate pests, pig control at Curtis Island, and trials on the effect of pigs and cattle on habitat at Curtis Island NP have taken place.
- A fire management plan for Curtis Island has been prepared.
- The preliminary fine-scale mapping of chat habitat for Torilla Plains and Curtis Island has been undertaken.
- *Harrisia* cactus mealy bug has been distributed to control *Harrisia* cactus populations.

FUTURE CONSERVATION EFFORTS

What you can do:

- Report any sightings to CQU or DERM (07) 4936 0511.
- Birdwatchers can assist with surveys of potential habitat (contact Wayne Houston at CQU).
- Graziers can help by reducing the trampling of samphire vegetation, and by spelling of marine plains during the wet season.
- Volunteer with DERM to aid on-ground management actions at Curtis Island. Donations to DERM will help landholders manage lands placed under Nature Refuge tenure.
- Landholders can help by developing management plans for pests such as pigs and ponded pasture weeds, jointly with neighbours for these widespread pests; they can seek advice on support for conservation benefits (e.g. fencing chat habitat and spelling paddocks), biodiversity incentives schemes and how to avoid overgrazing known chat habitat.

Conservation needs:

- Increase the area under conservation tenure.
- Implement more general actions to preserve wetlands.

Research directions:

- Conduct surveys and monitoring to establish baseline data on population levels, in order to understand population trends and their relationship with wetland inundation.
- Re-map regional ecosystems occupied by chats at a finer scale.

Sources and further reading

Houston W, Elder R, Black R and McCabe J (2006) Conservation significance of coastal wetland habitats for birds at Twelve Mile Creek, Fitzroy River, Central Queensland. *Sunbird* **36**, 20–36.

Houston W, Jaensch R, Black R, Elder R and Black L (2009) Further discoveries extend the range of Capricorn Yellow Chat in coastal central Queensland. *Sunbird* **39**, 29–38.

Houston W, Porter G, Elder R, Black R and Sheaves M (2004) Rediscovery of Yellow Chats (Capricorn subspecies) on the Fitzroy River Delta, Central Queensland. *Sunbird* **34**, 36–42.

Houston W, Porter G, O'Neill P and Elder R (2004) The ecology of the critically endangered Yellow Chat *Epthianura crocea macgregori* on Curtis Island. *Sunbird* **34**, 10–23.

Jaensch R, Houston W, Black R, Campbell L, Elder R and McCabe J (2004) Rediscovery of the Capricorn subspecies of Yellow Chat *Epthianura crocea macgregori* at Torilla Plain, on the mainland coast of central Queensland. *Sunbird* **34**, 24–35.

Author: Wayne Houston

PAINTED HONEYEATER

Grantiella picta (Gould, 1838)

Photo: Graeme Chapman

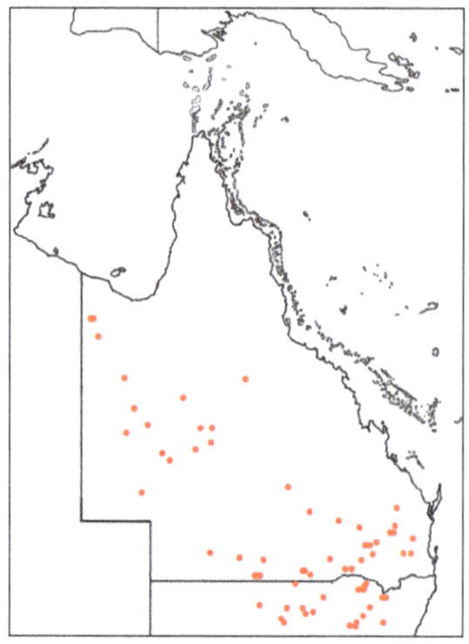

Family: Meliphagidae

Conservation status: IUCN: Vulnerable; ACT: Vulnerable; Qld: Vulnerable; NSW: Vulnerable; Vic.: Vulnerable

Description: The Painted Honeyeater is 14–15 cm in total length with a wingspan of 16 cm, and weighs 22 g. It has a striking white underside, glossy black back and bold yellow edging to the tail and wing feathers, a small white marking beneath the ear and black speckles from the throat to the flanks. The bill is pink and the eyes deep red. Females are noticeably browner and duller than adult males.

Distribution: Formerly widespread in eastern Australia from Mount Gambier (SA) to Katherine (NT) and occasionally to the Kimberley (WA), it is now more sparsely distributed and rarely recorded east of the Great Dividing Range. Its breeding range is now largely restricted to woodlands in inland NSW and southern Qld (south of Roma). Breeding populations in Vic. and the ACT have decreased. Records from the arid zone and regions north of 26°S are non-breeding birds. The current population size is unknown, but has been estimated at less than 10 000. It is a highly mobile species, thinly spread over large areas; the non-breeding densities are unknown, but breeding densities can be as high as five pairs per hectare.

Habitat: Painted Honeyeaters occur mainly in habitats on deep, productive soils and rely on abundant mistletoes for food. Acacia-dominated woodlands, particularly those dominated by Brigalow *Acacia harpophylla*, Yarran *A. homalophylla*, Boree or Weeping Myall *A. pendula* and Mulga *A. aneura* are favoured, with stands of Womal *A. maranoensis*, River Cooba *A. stenophylla* and Silver Wattle *A. dealbata* also used. Woodlands of Belah *Casuarina cristata* and Bulloak *Allocasuarina luehmanni* are often used, as are riparian woodlands of Black Box *Eucalyptus largiflorens* and River Red Gum *E. camaldulensis*, especially those with associated acacia shrubs. In the east, box-ironbark woodlands and White Cypress-pine *Callitris glaucophylla* can support Painted Honeyeaters if many mistletoes are present. River She-oak *Casuarina cunninghamiana* gallery forests with abundant mistletoes are also used, and may serve as a drought refuge. Many of these habitats occur in roadside stands, travelling stock routes and small remnants within agricultural land. Breeding Painted Honeyeaters depend primarily on two species: Needle-leaved Mistletoe *Amyema cambagei* and Grey Mistletoe *A. quandang*.

Ecology: Painted Honeyeaters have a highly specialised diet, depending almost entirely on mistletoe fruit, although they occasionally take other fruit of the same size, including Peppercorn Tree *Schinus molle* and grapes *Vitis vinifera*. Ingested mistletoe seeds remain sticky, so birds often use their bill to remove the strands from their vent. Nesting sites are located near abundant fruiting mistletoes and, although nests are occasionally placed within mistletoe clumps, they are usually located in drooping branches of trees. Fledging typically coincides with peaks in mistletoe fruit abundance. As well as mistletoe fruit, older nestlings are fed spiders. Non-breeding birds in northern and inland Australia are poorly known, but are closely associated with flowering mistletoes. They occasionally use other flowering plants, such as nectar from eucalypt blossom. Breeding birds often form aggregations, where they are highly vocal, giving their distinctive call from a prominent perch. In contrast, birds outside the breeding areas are easily overlooked, rarely calling and moving rapidly. Painted Honeyeaters arrive at key sites in early spring, with the males arriving first and establishing territories in areas of abundant mistletoes. Breeding occurs in late spring to early summer and, once chicks have fledged, all birds depart in late summer.

Reason for decline and current threats: The major cause of decline has been the clearing and fragmentation of woodland and forest containing the birds' preferred mistletoe species. This has been severe, given that the Painted Honeyeater's preference for habitats on deep productive soil – land that was preferentially cleared initially and continues to be threatened by agricultural intensification.

Current threats include: incremental clearing on private land; further habitat degradation on public land; frequent fires that reduce mistletoe densities, degrading habitats and removing critical breeding habitat; lack of regeneration of acacia and casuarina woodland, owing to inappropriate grazing regimes; and the removal of mistletoe from roadside trees and remnant woodlands.

CURRENT RECOVERY EFFORTS

Recovery plan: None

Other plans: Garnett ST and Crowley GM (2000) 'The action plan for Australian birds 2000'. Environment Australia, Canberra.

Garnett ST, Szabo JK and Dutson G (2011) *The Action Plan for Australian Birds 2010.* CSIRO Publishing, Melbourne.

Department of the Environment, Water, Heritage and the Arts (2008) 'Threat abatement plan for predation by feral cats'. Commonwealth of Australia, Canberra.

FUTURE RECOVERY EFFORTS

What you can do:

- Learn the bird's distinctive call and locate new breeding areas through strategic surveying.
- Landholders can covenant all or parts of their properties that contain habitat of significant quality or type. Areas containing 'key' species are thus protected in perpetuity: a key action for a species that has so much important habitat on private property.
- Do not remove mistletoe from your property; it is not harmful to healthy trees and provides homes and food to a multitude of wildlife.

Conservation needs:

- Restore acacia-dominated woodlands, especially Brigalow, Boree and Yarran.
- Reduce grazing pressure to allow natural regeneration of acacias and riparian vegetation.
- Obtain funding through Catchment Management Authorities and/or NRM agencies to assist with fencing, revegetation and weed control.

Research directions:

- Gather further information about the distribution and ecology of Painted Honeyeaters outside the breeding season.
- Conduct targeted research to establish population estimates – there may be as few as 4000 birds throughout their entire range.

Sources and further reading

Barea LP (2008) Nest-site selection by the Painted Honeyeater (*Grantiella picta*), a mistletoe specialist. *Emu* **108**, 213–220.

Barea LP and Watson DM (2007) Temporal variation in food resources determines onset of breeding in an Australian mistletoe specialist. *Emu* **107**, 203–209.

Higgins PJ, Peter JM and Steele WK (Eds) (2001) *Handbook of Australian, New Zealand and Antarctic Birds. Volume 5: Tyrant-flycatchers to Chats.* Oxford University Press, Melbourne.

Oliver DL, Chambers MA and Parker DG (2003) Habitat and resource selection of the Painted Honeyeater (*Grantiella picta*) on the northern floodplains region of New South Wales. *Emu* **103**, 171–176.

Watson DM (2011) *Mistletoes of Southern Australia.* CSIRO Publishing, Melbourne.

Author: David M Watson

BLACK-THROATED FINCH (WHITE-RUMPED SUBSPECIES)

Poephila cincta cincta (Gould, 1837)

Photo: Greg Calvert

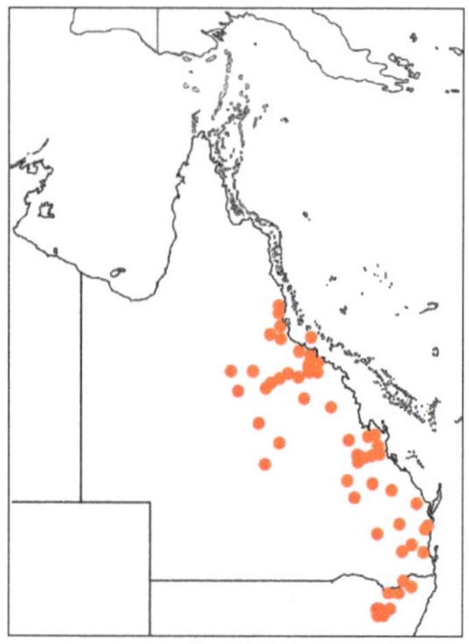

Family: Estrildidae

Other common names: Diggles Finch, Parson Finch

Conservation status: IUCN: species (*P. cincta*) is Near Threatened; Aust.: Endangered; Qld: Endangered; NSW: Endangered

Description: This is a stocky grass-finch with a total length of about 10 cm and weight of 13–17 g. The plumage is mainly shades of brown with a grey head, prominent black 'bib' and white undertail. The eyes, short conical bill, short tail and patch on the rear flanks are also black. The northern subspecies has a black rump, whereas the southern subspecies has a white rump. Hybrids, with both black and white feathers in the rump, are known. The legs and feet are bright pink. The black bill is unique among adult *Poephila* finches and the tail is shorter than in the Masked *P. personata* and Long-tailed Finches *P. acuticauda*.

Distribution: The northern subspecies *P. c. atropygialis* occurs on Cape York Peninsula south to about Mareeba and the Gulf plains west to the NT border; the southern subspecies *P. c. cincta* formerly occurred in eastern Qld south of about Mareeba, as far south as north-eastern NSW. Very few have been reported in Qld south of Rockhampton since the late 1970s. The wild population, past or present, is unknown. The maximum number of birds recorded during annual surveys at waterholes in the Townsville region is 461, but these indicate, at best, *relative* abundance and not population size. No southern Black-throated Finches are known to exist on conservation reserves.

Habitat: The Black-throated Finch occurs in grassy, open woodlands and forests dominated by *Eucalyptus*, *Corymbia* and *Melaleuca*. It is usually found within a few kilometres of water. It sometimes forages in modified habitats such as grassy roadsides, rail corridors and beneath powerlines, where these occur near more-or-less intact woodlands. It is rarely found in highly modified peri-urban environments and does not use urban habitats. In Qld, it has been recorded from a variety of vegetation types, including open forest on ridges and grassy hillsides. The main surviving coastal population in the Townsville coastal plain uses intact and modified environments but occurs mainly near more-or-less intact woodlands.

Ecology: The Black-throated Finch requires access to water, grass seeds and nest trees. It must drink daily and, though it can use various kinds of watering points, including natural waterholes, farm dams and stock troughs, the suitability of these sources is affected by the characteristics of the surrounding vegetation. The Black-throated Finch is apparently sedentary, moving locally as it exploits patchily distributed resources. It is likely to be restricted to the close proximity of its nest site during the breeding season, but becomes more mobile during the non-breeding season.

Reasons for decline and current threats: The Black-throated Finch has undergone a major contraction in range that is related to land clearing, livestock grazing, changed fire regimes, drought, urbanisation, feral predators (cats) and collecting for aviculture. Most of these factors are associated with agricultural and pastoral development. This subspecies requires intact woodlands, a water source and a year-round supply of grass seeds. Clearing removes intact woodland and fragments the habitat. Overgrazing reduces grass seed production and probably exacerbates the 'bottleneck' in seed availability that typically occurs in the early wet season. Small fragmented populations are likely to be more vulnerable to drought-induced food shortages, predation and, historically at least, collection for aviculture. Currently, urbanisation is a threat to some surviving populations.

CURRENT RECOVERY EFFORTS

Recovery plan: Black-throated Finch Recovery Team (2007) 'National recovery plan for the Black-throated Finch southern subspecies *Poephila cincta cincta*'. Report to the Department of the Environment and Water Resources, Canberra. Department of Environment and Climate Change (NSW), Sydney and Queensland Parks and Wildlife Service, Brisbane.

Other plans: Garnett ST and Crowley GM (2000) 'The action plan for Australian birds 2000'. Environment Australia, Canberra.

Garnett ST, Szabo JK and Dutson G (2011) *The Action Plan for Australian Birds 2010*. CSIRO Publishing, Melbourne.

Conservation actions: Since 2002, annual counts of birds watering at selected waterholes have been conducted in the Townsville region.

FUTURE RECOVERY EFFORTS

What you can do:
- Report sightings to the Black-throated Finch Recovery Team (<http://www.blackthroatedfinch.com>).
- Managers of pastoral land supporting Black-throated Finches can help conserve quality habitat by allowing perennial grasses to set seed and by retaining extensive areas of intact woodland. Resting areas from grazing during the wet season may be one means of doing this.

Conservation needs:
- Identify and quantify the threats to this species.
- Locate any additional populations and manage the data.
- Protect and enhance finch habitat.
- Coordinate a recovery program.
- In the southern part of its range, the recovery of the species may depend upon reintroduction, given the fragmented nature of the landscape, so there is a need to develop a capacity for reintroduction.

Research directions:
- Expand the knowledge of current distribution and abundance.
- Measure the genetic variation within the species.
- Quantify the habitat requirements of the species, particularly its foraging requirements during the different seasonal and annual conditions, as well as the bird's responses to different land management practices, particularly grazing and burning.
- Study the local movement patterns of the species and what drives them.

Sources and further reading

Franklin DC (1999) Evidence of disarray amongst granivorous bird assemblages in the savannas of northern Australia, a region of sparse human settlement. *Biological Conservation* **90**, 53–68.

Higgins PJ, Peter JM and Cowling SJ (Eds) (2006) *Handbook of Australian, New Zealand and Antarctic Birds. Volume 7: Boatbills to Starlings*. Oxford University Press, Melbourne.

Ley AJ and Cook SM (2001) The Black-throated Finch *Poephila cincta* in New South Wales. *Australian Bird Watcher* **19**, 115–120.

Mitchell DF (1996) Foraging ecology of the Black-throated finch *Poephila cincta cincta*. MSc thesis. James Cook University of North Queensland, Townsville.

Zann R (1976) Distribution, status and breeding of Black-throated Finches *Poephila cincta* in northern Queensland. *Emu* **76**, 201–206.

Author: Anthony Grice

CRIMSON FINCH (WHITE-BELLIED SUBSPECIES)

Neochmia phaeton evangelinae d'Albertis & Salvadori, 1879

Photo: www.ataglance.com.au (species indicative)

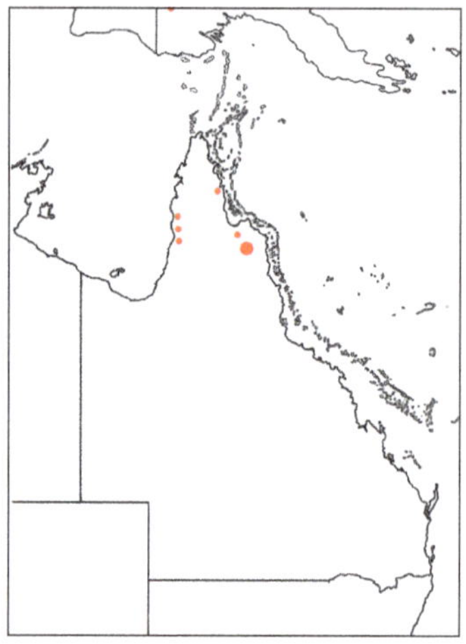

Family: Estrildidae

Other common names: Blood Finch, Cape York Crimson Finch, Pheasant Finch, Pale Crimson Finch, Red Finch, White-bellied Finch, Australian Firefinch

Conservation status: IUCN: species (*N. phaeton*) is Least Concern; Aust.: Vulnerable; Qld: Endangered

Description: The white-bellied subspecies of the Crimson Finch has a total length of 12–13 cm and weighs 7.5–12 g. The adult male is mostly red, including the long tail, with a blue-grey cap, red-tinged brown upperparts, white spots on the flanks and a white belly. The female is mostly brown, with a red face and tail, and white spots on the flanks. Juveniles are brown, with a black bill and a red wash to the wings and tail.

Distribution: Five populations occurred on Cape York Peninsula: on the Claudie River; on watercourses in Lakefield NP; on the Archer River near Aurukun; on the Edward and Chapman Rivers near Pormpuraaw; and on Magnificent Creek near Kowanyama. The finch has not been seen north of Princess Charlotte Bay since 1913 and is thought to have declined on the Archer River in the 1930s or '40s. Populations in Lakefield NP and those near Pormpuraaw and Kowanyama are thought to be stable. The wild population is estimated at 2000 breeding birds, with low reliability, owing to the difficulty of surveying in remote areas.

Habitat: These finches occur in paperbark woodlands, pandanus-fringed rivers and wetlands, and floodplains containing thick understorey of tall grasses such as Canegrass *Chionachne cyathopoda*. They often occur near *Pandanus*, tall grass stands or Cabbage Palm *Corypha utan*, and travel up to a kilometre to reach suitable grass seed food resources.

Ecology: Crimson Finches nest during the wet season, producing two or more broods of five to six young per season. Pairs are monogamous, often nesting in loose colonies. The dome-shaped nests are built in the tops of pandanus, palms or tall grasses. The finches typically live in small flocks of 6–12 birds, although large groups of 30–80 juveniles occur after the breeding season and into the dry season, until October when they disperse to become breeding adults. The finches take some small insects, nectar and lerps, but feed mainly on grass seeds, relying on a wide variety of grasses throughout the year.

Reasons for decline and current threats: Grazing, changed fire regimes and exotic plants have caused declines. Grazing removes tall grasses used as nesting and roosting sites, as well as reducing the food supply of the finches. The decline of the Aurukun population has been linked to overgrazing and trampling of riparian habitat by cattle. Destruction of wetland habitat by pigs and cattle in other areas of the Cape may be a limiting factor.

Hot dry season fires destroy large areas of grassy woodland, thus concentrating finches in wet or riparian areas protected from fire. Grazing, low-intensity fires and/or fire suppression have led to woody plants, such as Vivid Flower Paperbark *Melaleuca viridiflora*, invading grasslands and open woodlands, thus reducing the food supply of seed-eating birds.

Exotic weeds such as Rubber Vine *Cryptostegia grandifolia* and Grader Grass *Themeda quadrivalvis* are invading large areas of Cape York Peninsula, eliminating the finch's habitat, such as on the Laura and Mitchell Rivers. Illegal trapping for the pet trade may still continue, but is rare.

CURRENT RECOVERY EFFORTS

Recovery plan: Dorricott KE and Garnett ST (2007) 'National recovery plan for the white-bellied subspecies of the Crimson Finch *Neochmia phaeton evangelinae* and the northern subspecies of the Star Finch *Neochmia ruficauda clarescens*'. Report to Environment Australia, Canberra, by Queensland Parks and Wildlife Service, Brisbane.

Other plans: Garnett ST and Crowley GM (2000) 'The action plan for Australian birds 2000'. Environment Australia, Canberra.

Garnett ST, Szabo JK and Dutson G (2011) *The Action Plan for Australian Birds 2010*. CSIRO Publishing, Melbourne.

Conservation actions:

- There has been reduction of cattle grazing in Lakefield NP.
- Mapping of fire history of the Laura Basin has been undertaken, and ecological research on the Star Finch *Neochmia ruficauda clarescens*, which occupies the same habitats and has similar ecology, has taken place.
- The University of Wollongong, Australian Wildlife Conservancy, Charles Darwin University, NT DNREAS, and Birds Australia are studying Crimson Finch health in response to grazing, fire and exotic plants.
- Finch populations are being monitored by the Kowanyama Land and Sea Management Centre.
- Relevant lands are managed by DERM, the Kowanyama Land and Sea Management Centre, the Pormpuraaw Council and the Pormpuraaw Land and Sea Management Centre.

FUTURE RECOVERY EFFORTS

What you can do:

- Report sightings to DERM.
- Write letters to state and federal environment ministers and agencies, urging the approval and funding of recovery plan actions.
- Practice appropriate fire and weed management on properties and while camping (e.g. wash down vehicles to remove weed seeds).
- Support the work of Australian conservation organisations.

Conservation needs:

- Review the conservation status of the subspecies.
- Manage exotic weeds in finch habitat within Lakefield NP and near Kowanyama.
- Erect fencing to exclude pigs from finch breeding areas and wetlands.
- Incorporate the protection of finch habitat within regional and national park fire management plans. Develop monitoring programs for finch populations in Lakefield NP, Pormpuraaw and Aurukun.

Research directions:

- Conduct surveys of the current distribution of Crimson Finches.
- Carry out further research into the effect of exotic plants on the finch to assess the threat from Rubber Vine and exotic grasses.
- Determine the feasibility of reintroduction in Aurukun and Lockhart River, using habitat quality surveys and consultation with traditional owners.

Sources and further reading

Higgins PJ, Peter JM and Cowling SJ (Eds) (2006) *Handbook of Australian, New Zealand and Antarctic Birds. Volume 7: Boatbills to Starlings*. Oxford University Press, Melbourne.

Todd MK (2002) Nest-site and breeding-season data for the Crimson Finch *Neochmia phaeton* in Australia. *Australian Bird Watcher* **19**, 161–171.

Todd MK, Felton A and Garnett ST (2003) Morphological and dietary differences between common and uncommon subspecies of Crimson Finch, *Neochmia phaeton*, and Star Finch, *Neochmia ruficauda*, in northern Australia. *Emu* **103**, 141–148.

Authors: Kim Maute and Sarah Legge

STAR FINCH (SOUTHERN SUBSPECIES)

Neochmia ruficauda ruficauda (Gould, 1837)

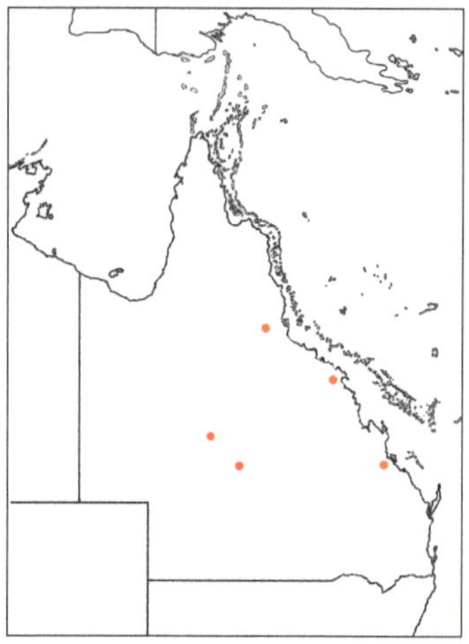

Photo: Michael Todd (species indicative)

Family: Estrildidae

Other common names: Red-tailed finch, Firetail, Rufous-tailed Finch, Red-faced Finch

Conservation status: IUCN: species (*N. ruficauda*) is Near Threatened; Aust.: Endangered; Qld: Endangered; NSW: Extinct in the Wild

Description: This is a small finch with a total length of 10–12 cm and weighing 9–11 g. It is sleek, with a red bill, face, rump and tail, an olive body, pale belly and yellow legs. All the face, breast, flank and rump feathers are spotted with white. The sexes are likely to be similar, but pale relative to the other subspecies. Although information is lacking, juveniles are likely to look similar to the young of other subspecies: a black bill and light brown plumage, a pale to white underside and a red wash to the brown rump and tail.

Distribution: The former range of the southern subspecies of the Star Finch is not well known. It is thought to have been restricted to inland Qld, from Bowen in the north to Winton in the west and Wowan in the south. The current wild population is estimated at 20–50 individuals. The reliability of this estimate is low, owing to the fact that no confirmed sightings have been made since 1994, despite targeted surveys in 1993–94 and again in 1996–97, and the former range is large and remote. Declines are thought to have begun in the late 19th century, and this subspecies may now be extinct in the wild. It formerly occurred in northern inland NSW, but is extinct there.

Habitat: The Star Finch occurs in grasslands and grassy woodlands, near permanent water, and often in or near cleared suburban areas. One report suggested that they also inhabited river banks dominated by native grasses and sedges.

Ecology: There is little information on the ecology of the subspecies, but, based on studies of other Star Finch subspecies, it is likely that it nested during the wet season, producing one or more broods of four to six young per season. Pairs are monogamous and often nest in loose colonies. The dome-shaped nests are built in low dense shrubs or grass tussocks. These finches typically feed and drink in small flocks of 6–12 birds. Star Finches are strict granivores, adapted to feeding on small seeds and consequently have a more restricted diet than some other finches residing in northern Australia's grasslands.

Reasons for decline and current threats: Because the decline of southern Star Finches occurred before research could link this decline with a threatening process, not much is known about the cause of their disappearance. However, the likely cause of this decline is the degradation of habitat by livestock grazing. Cattle and sheep would have reduced the quality of Star Finch breeding and foraging habitat by removing grasses used as nesting and roosting sites, as well as reducing the productivity of grasses that provide seed.

CURRENT RECOVERY EFFORTS

Recovery plan: None

Other plans: Garnett ST and Crowley GM (2000) 'The action plan for Australian birds 2000'. Environment Australia, Canberra.

Garnett ST, Szabo JK and Dutson G (2011) *The Action Plan for Australian Birds 2010.* CSIRO Publishing, Melbourne.

FUTURE RECOVERY EFFORTS

What you can do:

- Report any sightings to DERM.
- To assist in finding remnant populations, write letters to state and federal environment ministers and agencies, urging approval and funding of recovery actions.
- Ensure that fires started in potential habitat are done so with the ecological consequences on native grasses in mind.
- Control the spread of exotic weeds by using appropriate vehicle hygiene.

Conservation needs:

- The recovery of the southern Star Finch is dependent on the discovery and protection of a wild population. Although it is estimated that over 20 000 Star Finches exist in captive populations in Australia and all over the world, it is unlikely that *N. r. ruficauda* has persisted in captivity, owing to interbreeding with the much more common *N. r. clarescens* and *N. r. subclarescens*.
- Immediately follow up any sighting of the southern Star Finch with surveys that attempt to locate the core habitat of the population. Once inhabited areas are located, make efforts to coordinate with the landowner(s) to maintain and improve finch habitat in that area.

Research directions: Identify requirements of key food plants and other plant resources.

Sources and further reading

Garnett ST, Clarkson JR, Felton A, Harrington, GN and Freeman AND (2005) Habitat and diet of the Star Finch (*Neochmia ruficauda clarescens*) in the early wet season at Princess Charlotte Bay, Cape York Peninsula, Australia. *Emu* **105**, 81–85.

Higgins PJ, Peter JM and Cowling SJ (Eds) (2006) *Handbook of Australian, New Zealand and Antarctic Birds. Volume 7: Boatbills to Starlings*. Oxford University Press, Melbourne.

Holmes G (1996) Distribution and status of the southern Star Finch. *Sunbird* **26**, 49–59.

Neldner VJ, Fensham, RJ, Clarkson JR and Stanton JP (1997) The natural grasslands of Cape York Peninsula, Australia: Description, distribution and conservation status. *Biological Conservation* **81**, 121–136.

Todd MK, Felton A and Garnett ST (2003) Morphological and dietary differences between common and uncommon subspecies of Crimson Finch, *Neochmia phaeton*, and Star Finch, *Neochmia ruficauda*, in northern Australia. *Emu* **103**, 141–148.

Authors: Kim Maute and Sarah Legge

GOULDIAN FINCH

Erythrura gouldiae (Gould, 1844)

Photo: www.ataglance.com.au

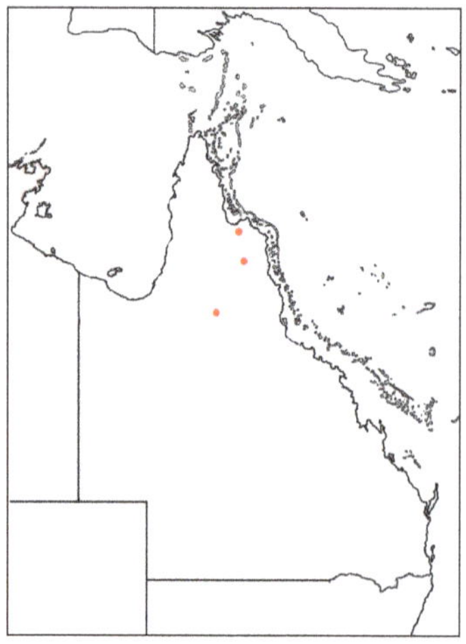

Family: Estrildidae

Other common names: Gouldian Grass-finch, Painted Finch, Purple-breasted Finch, Purple-chested Finch, Rainbow Finch, Scarlet-headed Finch

Conservation status: IUCN: Endangered; Aust.: Endangered; Qld: Endangered; NT: Endangered; WA: Endangered

Description: This is a small finch 13–14 cm in length and weighing 10–14 g. Adult males have a black, bright red or yellow-orange face, purple breast, yellow belly and blue rump, and are otherwise green dorsally. The tail is grey-black with a short streamer. The female is a pastel version of the male, with a shorter tail streamer. Juveniles have a pale grey-green head, buff underside and green upperparts. The bill is pale and the legs are pink.

Distribution: Historically, the Gouldian Finch ranged throughout Australia's tropical savannas, from the north-west coast of WA south-east to Chillagoe, Qld. In Qld, it was found in good numbers near Laura up until the 1950s. Since then, the species has undergone a sharp decline and large range contractions across northern Australia. Current breeding populations are now small: in the order of 2500 individuals scattered throughout the Northern Territory and Kimberley regions. Breeding populations in Qld have not been observed in recent decades, though small groups and individual birds are still occasionally sighted near Mount Isa and Georgetown and on Cape York Peninsula.

Habitat: The Gouldian Finch is found in open tropical woodland with a grassy understorey. It typically breeds in the hollow limbs of Snappy Gum *Eucalyptus brevifolia* and Ringing Gum *E. tintinnans,* often in rocky hills or low escarpment country. In the dry season, it feeds in habitat dominated by annual *Sarga* spp. grasses. In the wet season, it feeds in sites ranging from Spinifex *Triodia* spp. dominated landscapes to wet lowlands characterised by diverse perennial grasses.

Ecology: The Gouldian Finch starts nesting during the early wet season, producing two or more broods of five to six young per season. Pairs are monogamous and build nests inside tree hollows. They typically feed and drink in groups of 10–200 individuals, depending on the local population size and time of year. Compared with other grass-finches, the Gouldian Finch is more strictly granivorous, rarely supplementing its diet with insects or other food. This reliance on seed may explain why it is opportunistically nomadic.

Reasons for decline and current threats: Trapping for the pet trade was widespread until wild capture was banned in 1986. Improper fire management and the effects of grazing are the main threats to the survival of this species.

Historic and continued grazing by exotic herbivores, frequent fires and introduced weeds have likely reduced seed yields and the abundance of grasses used by the finch. Perennial grasses that are important to finches in the wet season are especially sensitive to any form of defoliation. In the early wet season, birds begin to breed and defend nest hollows for long periods. If seed is not locally abundant, nest defence may become unsustainable, owing to the energy expended seeking seed, and the birds may either loose condition or be forced to minimise breeding activity.

CURRENT RECOVERY EFFORTS

Recovery plan: O'Malley C (2006) 'Recovery plan for the Gouldian Finch (*Erythrura gouldiae*)'. WWF-Australia, Sydney and Parks and Wildlife, Northern Territory Department of Natural Resources, Environment and the Arts, Northern Territory Government, Palmerston.

Other plans: Garnett ST and Crowley GM (2000) 'The action plan for Australian birds 2000'. Environment Australia, Canberra.

Garnett ST, Szabo JK and Dutson G (2011) *The Action Plan for Australian Birds 2010.* CSIRO Publishing, Melbourne.

Conservation actions:

- In the mid-1990s, DERM conducted surveys to locate populations and Birds Australia North Qld has actively promoted searches.
- Current research is focused on better understanding the impacts of fire and grazing on the Gouldian Finch and how to better manage land for finches.
- The Wildlife Conservancy of Tropical Qld has attempted several reintroductions at Mareeba Wetlands, with JCU studying factors affecting survival after reintroduction. No viable wild populations have yet resulted and further attempts are ongoing. However, before reintroduction can be successful, suitable habitat in Qld needs to be identified and rehabilitated, involving the coordination of landowners and conservation agencies in core areas of the former range of the species, to increase the productivity of food grasses and other key finch habitat characteristics.

FUTURE RECOVERY EFFORTS

What you can do: Join/donate to the Australian Wildlife Conservancy, the Wildlife Conservancy of Tropical Qld and Birds Australia.

Conservation needs:

- The survival and recovery of Gouldian Finch populations in Qld will depend on the protection and rehabilitation of remaining suitable habitat from the effects of overgrazing, fire and exotic plants. Therefore, the most important action is for property managers to implement best-practice fire and grazing management (see http://savanna.cdu.edu.au/information/land_manager_website.html).
- NRM groups should treat the Gouldian Finch as a high priority and seek support from DERM and advice from the recovery team.

Research directions: The environmental conditions causing large-scale regional movements of Gouldian Finch populations are unknown and warrant investigation.

Sources and further reading

Brazill-Boast J, Pryke SR and Griffith SC (2010) Nest-site utilisation and niche overlap in two sympatric, cavity-nesting finches. *Emu* **110**, 170–177.

Dostine PL, Johnson GC, Franklin DC, Zhang Y and Hempel C (2001) Seasonal use of savanna landscapes by the Gouldian Finch, *Erythrura gouldiae*, in the Yinberrie Hills area, Northern Territory. *Wildlife Research* **28**, 445–458.

Higgins PJ, Peter JM and Cowling SJ (Eds) (2006) *Handbook of Australian, New Zealand and Antarctic Birds. Volume 7: Boatbills to Starlings*. Oxford University Press, Melbourne.

Holmes G (1995) 'Survey of Gouldian Finch in Queensland during dry season of 1995 – with review of distribution and status'. Unpublished report to Conservation Commission of the Northern Territory, Darwin.

Tidemann SC (1996) Causes of the decline of the Gouldian Finch, *Erythrura gouldiae*. *Bird Conservation International* **6**, 49–61.

Authors: Kim Maute and Sarah Legge

Mammals (Class: Mammalia)

INTRODUCTION TO THREATENED MAMMALS

Andrew J. Dennis

Approximately half of Australia's mammals are marsupials and monotremes, the rest are placental mammals. Although the placental mammals on land comprise only mice, rats, bats and the Dingo *Canis lupus dingo*, the marsupials and monotremes are more diverse. They range from tiny carnivores, such as ningaui and planigale, to larger ones, such as the quolls. In the Australian seas, the mammals include whales, dolphins and the Dugong *Dugong dugon*.

Representatives from many of these groups are either extinct, such as the Thylacine *Thylacinus cynocephalus*, which was wiped out by people more than 70 years ago, or are currently at risk of extinction. Most species that are already extinct were small to medium-sized and occurred in the dry inland areas of southern Australia. As discussed previously (see Extinct in Queensland pp. 10–14), they suffered numerous impacts, creating significant stresses on populations, as Australia was transformed by European settlement, finally succumbing to predation by the Red Fox *Vulpes vulpes* and feral Cat *Felis catus*. There are still many other threatened animals that fit the pattern and characteristics of those that are already extinct: some are only just hanging on, others are now found only on offshore islands. In Queensland, several such species occur in the south-west corner of the state, primarily in the Channel Country but extending to the north, west and east. Examples include the Greater Bilby *Macrotis lagotis* the Brush-tailed Mulgara *Dasycercus blythi* and the Kowari *Dasyuroides byrnei*.

Australia is in the embarrassing position of having the world's worst record of mammal extinctions. Although much can be blamed on the introduction of foxes and cats, there are many other factors influencing the high probability of extinction for many species. Several of our threatened mammals have suffered incredible destruction directly at our hands. These include some that have been hunted and killed in very large numbers, such as many of our whales, the Brush-tailed Rock-wallaby *Petrogale penicillata* and the Spectacled Flying-fox *Pteropus conspicillatus*. Many others have suffered from the huge loss of habitat associated with transforming the Australian landscape for primary production and urban development. In some places, such as the Brigalow Belt, well over 90% of a community was cleared, leaving the species that relied on it – such as the South-eastern Long-eared Bat *Nyctophilus timoriensis* – with only tiny fragmented remnants. In other cases, such as for the Mahogany Glider *Petaurus gracilis*, continuous small-scale clearing and fragmentation, and steady habitat alteration through weed invasion and fire suppression may result in extinction – 'death by 1000 cuts'.

A few of the threatened mammals in Queensland occur primarily outside the state. These can be divided into species that have distributions beyond Australian borders, such as the whales and some of the bats, and those that occur primarily in other states but just make it into Queensland such as the Greater Bilby, Brush-tailed Mulgara and Brush-tailed Rock-wallaby. In some of these cases, it would be easy to dismiss the state's role in the recovery of the species. However, there are clear examples to indicate that every part of a species' range can be critical; again, the Brush-tailed Rock-wallaby provides a case in point, where populations at its northern limit in southern Queensland seem to have declined the least. The population of Greater Bilby in Queensland is now isolated from those in other states, clearly indicating that the Channel Country provides a more secure environment than in much of the rest of its previous range.

The whales require a different mind set. Most of them are widely distributed and require the cooperation of numerous nations and states to affect their recovery. Although opportunity for 'on ground' efforts in Queensland may be minimal for some species that only just reach our waters, involvement in the broader effort is still warranted and, for some species, Queensland waters (5 km off shore), or Australian waters off the Queensland coast (200 nm offshore), are critical birthing areas. Our state legislation reflects this. The '(Queensland) Nature Conservation (Whales and Dolphins) Conservation Plan 1997' covers activities including whale watching and dealing with stranded and entangled cetaceans. Within the Great Barrier Reef Marine Park, activities interfering with cetaceans are regulated through the 'Great Barrier Reef Marine Park Zoning Plan 2003'. In addition, the Australian Whale Sanctuary provides high levels of protection to whales in Australian waters. The latter comprise the Australian Exclusive Economic Zone (EEZ), generally extending 200 nm from the coast, as well as waters around the Australian Antarctic Territory, Macquarie, Heard and McDonald Islands. Measures such as these are frequently beyond those afforded to species in other parts of their global range.

Some of our bats occur in neighbouring countries, such as Papua New Guinea, and frequently their status in those countries is unknown. In some cases, they may be worse off elsewhere than in Australia. However, the bats raise a set of issues that are common to a few other taxonomic groups. Although most mammals are reasonably well known, bats, as a whole, are not. Many of the bats listed as threatened and appearing in this book have been listed on the basis of very little data. For many species, basic survey work to determine distribution and abundance is required to have greater certainty of their risk of extinction. This could be assisted with a readily available library of echolocation calls and an updatable Atlas of Australian Bats.

Although most bats are best described as Data Deficient, there are a range of measures that can be undertaken that will benefit threatened bats, and many other species besides. Protecting their roosts is a critical one. Readers might be surprised to learn that many bats roost in tree hollows, limbs and trunks. Therefore, protecting hollow bearing trees – even in urban back yards and especially in logged forests – and providing artificial hollows with varying entrance styles and sizes can be highly beneficial to bat conservation. Of course, protecting and minimising disturbance to cave roosts is important to other species. Perhaps more critical is maintaining sufficient natural vegetation in a healthy state for foraging areas. Without the broader healthy and well-connected natural landscape in which to find all the resources it requires, no species is likely to remain resilient enough to persist long into the future.

Of the 45 mammals described in the following profiles, 20 are listed as Endangered: at very high risk of extinction without intervention from us. Some of these are among Australia's most threatened; for example, the Northern Hairy-nosed Wombat *Lasiorhinus krefftii* numbers a little over 100 individuals and the Bramble Cay Melomys *Melomys rubicola* less than 100. Yet they hang on; and, given the right conditions and help, may continue to do so and even recover their numbers. All of these species can be assisted by us as individuals or as members of groups in various ways, so please take advantage of the opportunities outlined in the following profiles and in other parts of this book and help them out.

DASYURIDS (ORDER: DASYUROMORPHIA)

BRUSH-TAILED MULGARA

Dasycercus blythi (Waite, 1904)

Photo: www.ataglance.com.au

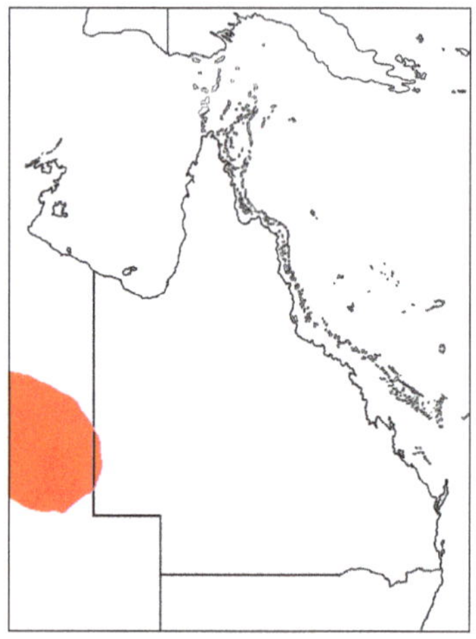

Family: Dasyuridae

Aboriginal names: Not differentiated from Ampurta in Qld

Conservation status: IUCN: Least Concern; Aust.: Vulnerable; Qld: Vulnerable

Description: This rat-sized carnivorous marsupial has a head and body length of 130–165 mm (males) or 120–140 mm (females), and a tail that reaches 75–100 mm (males) or 60–90 mm (females). It weighs 60–130 g (males) or 45–80 g (females). Its fat tail, sturdy build, and pointy nose distinguish it from all similar-sized mammals except the Crest-tailed Mulgara *Dasycercus cristicauda*, with which it had been synonymised until recently. The Brush-tailed Mulgara is generally smaller, with sandy grey fur above and greyish white below and its conspicuous, fat tail has black hairs on the last two thirds that do not form a dorsal crest like the latter species.

Distribution: The distribution of the Brush-tailed Mulgara is uncertain because of its taxonomic confusion. However, it appears to have occupied much of arid Australia, including the Simpson Desert in south-west Qld. Currently, it occurs in only a small area in south-west Qld, but is still widespread in spinifex grasslands in WA and NT. Range contraction has occurred mostly from the south, with no animals recorded in SA since 1966 despite substantial surveys occurring in potential habitat in the last 20 years.

Habitat: They occur on sand plains, dunes and laterite rises dominated by hummock grasslands, with or without mulga woodlands. They appear to associate with hummocky spinifexes such as *Triodia basedowii* and *T. pungens*. However, they have also been found in tussock grasslands burrowing under *Eromophila* spp.

Ecology: The Brush-tailed Mulgara produces a litter of up to six young, once a year between June and October, despite having a polyoestrous reproductive cycle. Most individuals survive for two breeding seasons, but few survive a third. In captivity they can live for 5 years. The population declines during the breeding season then increases after the influx of juveniles in spring. On occasion, adult males may be completely absent by the time pouch young are evident in September. Populations can also vary between years, as well as seasons, in many areas, although at Uluru Kata Tjuta NP little change was measured from 1986 to 1994 despite substantial changes in climatic conditions. However, the same population went through an increase and rapid decline in 2002 after large wildfires and predator baiting. Work in Qld and WA has also shown measurable effects of fire and rainfall, though they are not consistent between locations and at different scales. The Brush-tailed Mulgara is solitary with high site fidelity. Males and females maintain home ranges of 1.4–14 ha, which generally overlap less than 20%. Sub-adult males are more mobile than females, moving away from their natal area before the breeding season, whereas females tend to stay within or near to the natal area. This species eats invertebrates, reptiles and small mammals, with some differences in composition between locations and seasons.

Reasons for decline and current threats: Chenopod shrublands once supported populations of mulgara but no longer do so because of the reduction in cover and change in structure resulting from grazing by domestic stock and rabbits. This mulgara is still widely distributed in spinifex grasslands, where domestic stock and rabbits are few or non-existent. Hence, there is no certainty that remaining populations are still in decline, but two other lines of evidence suggest the species is still vulnerable. Cats and foxes may periodically exert excessive predation and decimate, and possibly extirpate, mulgara populations. This occurs when rodent populations erupt a couple of years after a heavy summer rainfall. All predators increase in number (including mulgara) but when rodents begin to crash, introduced predators switch to eating mulgaras.

Brush-tailed Mulgara prefer areas with 15–60% spinifex cover. Areas with low cover rarely support them. Fires remove cover, so changes in fire regimes could have an impact on the security of the species, especially widespread fires with even destruction of spinifex.

CURRENT RECOVERY EFFORTS

Recovery plan: None

FUTURE RECOVERY EFFORTS

What you can do:

- Write letters to the state and federal environment ministers to urge greater and long-term funding for the implementation of research recommendations.
- Contact the Desert Ecology Research Group, University of Sydney for their research directions and volunteer opportunities in the Simpson Desert.

Conservation needs: Most actions currently centre on clarifying uncertainties. However, land management practices should focus on ensuring patches of good spinifex cover are maintained during controlled burns and the impacts of introduced herbivores and predators are minimised.

Research directions:

- Clarify the population size and trends and the impacts and possible thresholds of threats mentioned above, particularly the introduced predators in Qld.
- Differentiate the distribution and habitat requirements of the Brush-tailed Mulgara and Crest-tailed Mulgara to best understand how to manage threats.
- Confirm the habitat preferences and determine if the species occur predictably in these habitats
- Monitor the distribution and relative abundance over large areas using track counts.

Sources and further reading

Letnic M and Dickman CR (2005) The responses of small mammals to patches regenerating after fire and rainfall in the Simpson Desert, central Australia. *Austral Ecology* **30**, 24–39.

Masters P (1993) The effects of fire-driven succession and rainfall on small mammals in spinifex grasslands at Uluru National Park, Northern Territory. *Wildlife Research* **20**, 803–813.

Masters P (1998) The Mulgara *Dasycercus cristicauda* (Marsupialia: Dasyuridae), at Uluru National Park, Northern Territory. *Australian Mammalogy* **20**, 403–407.

Masters P (2003) Movement patterns and spatial organisation of the Mulgara, *Dasycercus cristicauda* (Marsupialia: Dasyuridae), in central Australia. *Wildlife Research* **30**, 339–344.

Woolley PA (2005) The species of *Dasycercus* Peters, 1875 (Marsupialia: Dasyuridae). *Memoirs of Museum Victoria* **62**, 213–221.

Author: Pip Masters

CREST-TAILED MULGARA

Dasycercus cristicauda (Krefft, 1867)

Photo: Jean-Paul Ferrero-AUSCAPE

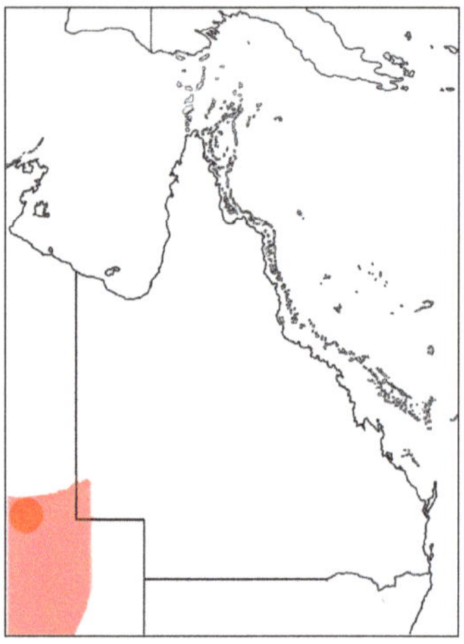

Family: Dasyuridae

Aboriginal name: Ampurta (Arrernte/Ilyowra), Mudagoora (Wankangurru)

Conservation status: IUCN: Least Concern; Aust.: Vulnerable, (Endangered under *D. hillieri*); Qld: Vulnerable; NT: Vulnerable; WA: Vulnerable

Description: This rat-sized carnivorous marsupial has a head and body length of 130–230 mm (males) or 120–170 mm (females), and a tail that reaches 80–125 mm (males) or 80–110 mm (females). It weighs 110–185 g (males) or 65–120 g (females). It has a black-tipped, often-fattened tail and pointed nose. It is similar to the Brush-tailed Mulgara *Dasycercus blythi*, but the Crest-tailed Mulgara is generally bigger with a ginger tinge to the fur above and creamy white below. It has a conspicuous, fin-like crest of black hairs on the dorsal surface of the tail. It has eight nipples in the pouch (six in the Brush-tailed Mulgara) and three upper premolars (the third very small and sometimes present on one side only) compared with two in the Brush-tailed Mulgara.

Distribution: This mulgara's previous distribution included Sandringham Station in south-west Qld, most of arid SA including the Musgrave Ranges and the Lake Eyre Basin, the Canning Stock Route in the Great Sandy Desert in WA, and south-eastern NT. At Sandringham Station, captures were made up until February 1968. Subsequent trapping in 1971 and in the 1990s was unsuccessful. It is currently known from the southern Simpson Desert, slightly west of where the borders of NT, Qld and SA converge, and in the Tirari and Strzelecki Deserts of SA.

Habitat: This mulgara's known habitat is primarily dunes with a low cover of Sandhill Canegrass *Zygochloa paradoxa*, or areas around salt lakes with Nitre Bush *Nitraria billardierei*; unlike the Brush-tailed Mulgara, which occurs in spinifex grasslands with medium to high cover. Most burrows are at the base of large canegrass clumps or Nitre Bush hummocks. Latrines are commonly located near the burrow entrance, with scats being up to 90 mm long and 6 mm wide. Crest-tailed Mulgara densities are high in the central part of the Simpson Desert, where vegetation cover is low and surface water is scarce, compared with surrounding pastoral properties on the fringes of the Simpson Desert.

Ecology: Little ecological work has been undertaken on this mulgara, partly because of the taxonomic confusion and remoteness of occupied habitat. Its lifespan in captivity is around 5 years, but under natural conditions they are likely to live for only 2–3 years. Based on knowledge of the Brush-tailed Mulgara, females probably mature in their first year but produce litters of up to eight rather than a maximum of six for the Brush-tailed Mulgara. The Crest-tailed Mulgara seems to be a seasonal breeder, birthing in winter/early spring with juveniles emerging in spring and possibly early summer. They occupy complex burrow systems with multiple entrances. The Crest-tailed Mulgara is an opportunistic carnivore eating a range of invertebrates, lizards and small mammals.

Reasons for decline and current threats: The extent and reasons for its decline are unclear. As with Brush-tailed Mulgara, cats and foxes may periodically exert excessive predation and decimate, and possibly extirpate, mulgara populations. This occurs when rodent populations erupt a couple of years after a heavy summer rainfall. All predators increase in number (including mulgara), but when rodents begin to crash, introduced predators switch to eating mulgaras. Some evidence suggests that some introduced herbivores, possibly including rabbits, cattle, sheep and camels, may cause declines at certain thresholds. Researchers have found that the abundance of the Crest-tailed Mulgara declines as that of camels increases, but not so with rabbits. Grazing by cattle damages dunes and the vegetation, which could cause detrimental habitat change. Climate change could have catastrophic effects in the next 50 years. It is expected to create higher mean temperatures, decrease the effective rainfall and increase the occurrence of extreme climatic conditions. Combined, these factors are likely to have a significant impact on natural desert systems.

CURRENT RECOVERY EFFORTS

Recovery plan: None

Conservation actions: Research and monitoring efforts to date have focused on genetics, past and present distribution and habitat requirements.

FUTURE RECOVERY EFFORTS

What you can do:

- Write letters to the state and federal environment ministers to urge greater and long-term funding for the implementation of research recommendations.
- Contact the Diversity, Ecology and Evolution Research Group, University of Sydney for their research directions in the Simpson Desert.

Conservation needs: Land management practices need to be based around minimising the impacts of introduced herbivores and predators.

Research directions:

- Conduct a comparative study of the distribution and habitat separation of the two species of mulgara.
- Study the drivers of population fluctuations and their ecology.
- Conduct a study examining the impacts of introduced herbivores and predators on mulgara populations, with particular reference to possible threshold densities, which will allow the broader management of the landscape.

Sources and further reading

Adams M, Cooper N and Armstrong J (2000) 'Revision of *Dasycercus* systematics'. Unpublished report to the South Australian Department of Environment and Heritage, Adelaide.

Baker L (1996) The distribution, habitat requirements and status of the Mulgara (*Dasycercus cristicauda*, Krefft). MSc thesis. University of New England, Armidale.

Foulkes J (1999) 'The status and distribution of a new species of Mulgara (*Dasycercus* spp.) in South Australia: results of 1996 survey'. Unpublished report to the South Australian Reserves Advisory Committee under the Wildlife Conservation Fund grant scheme. South Australian Department for Environment, Heritage and Aboriginal Affairs.

Masters P (2008) Crest-tailed Mulgara *Dasycercus cristicauda*. In *The Mammals of Australia*. (Eds S Van Dyck and R Strahan) pp. 49–50. Reed New Holland, Sydney.

Woolley PA (2005) The species of *Dasycercus* Peters, 1875 (Marsupialia: Dasyuridae). *Memoirs of Museum of Victoria* **62**, 213–221.

Author: Pip Masters

KOWARI

Dasyuroides byrnei Spencer, 1896

Photo: Peter Canty

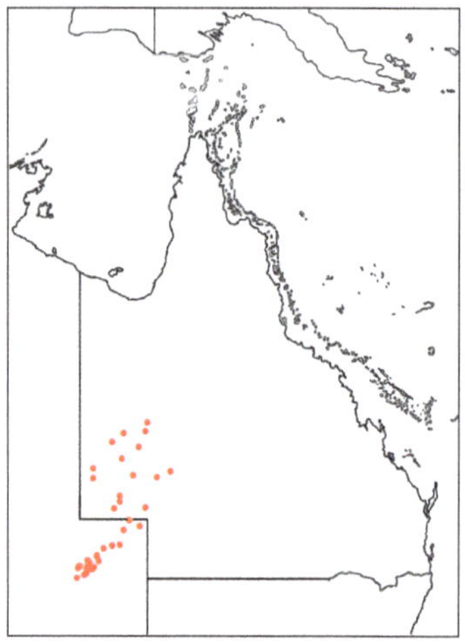

Family: Dasyuridae

Other common names: Byrne's Crest-tailed Marsupial Rat, Brush-tailed Marsupial Rat, Bushy-tailed Marsupial Rat

Aboriginal name: Kowari

Conservation status: IUCN Vulnerable; Aust.: Vulnerable; Qld: Vulnerable; SA: Vulnerable; NT: Data Deficient

Description: This rat-sized marsupial carnivore is most notable for its large brush of dense black hairs on its graceful tail, its prominent pale eye ring and only four toes on the rear foot. Its head and body length is 140–180 mm (males) or 135–160 mm (females), with a tail measuring 110–160 mm. It weighs 85–175 g (males) or 70–140 g (females). It has sandy-grey fur and dark guard hairs above, with pale grey to white fur below.

Distribution: The Kowari occurs in Sturt's Stony Desert, which extends from south-west Qld into north-east SA. Historic records include stony desert areas along the SA–NT border on the western edge of the Simpson Desert, a population now considered extinct. Sub-fossil remains from old owl roosts suggest Kowaris once occurred around the northern Flinders Ranges. Modelling suggested a total population ranging from 1000 to 10 000, the variation reflecting the typical desert-dwelling species dependence on seasonal conditions.

Habitat: The Kowari occurs in stony desert landscapes and sparsely vegetated gibber pavements. Within these pavements of very flat, tightly packed gibber stones, the Kowari occurs where the landscape features a range of other nearby or interspersed habitats such as small ephemeral swamps, isolated dunes and broader areas of soil cover over gibber. These habitat 'islands', particularly sand mounds, are generally the only places where Kowaris can readily dig their refuge and breeding burrows and find much of their prey.

Ecology: The Kowari is an opportunistic hunter, eating a range of invertebrates, small mammals, reptiles, ground-dwelling and nesting birds and probably carrion. It is solitary and has a home range that can be in the order of >20 km^2, which is extraordinarily large for a small mammal. This may reflect the effort required to obtain sufficient prey in barren gibber pavement. The Kowari can breed twice a year and can have up to six young at a time when prey is abundant. This allows them to respond rapidly when rainfall brings abundant growth. Kowaris make use of a number of burrows scattered throughout their home range. These vary from simple, single entrance tunnels to multiple entrance, multiple chamber complexes, presumed to be used by females with dependent young.

Reasons for decline and current threats: Historical records of Kowaris are scarce and any assessment of causes of decline cannot be based on direct data. However, several factors, such as high levels of cattle and sheep grazing and periods of extreme drought early last century, combined with the incursions of rabbits and foxes, caused declines in many small mammals and may be associated with their decline. In addition, Kowaris occur in greater numbers with increasing distance from stock watering points, suggesting grazing may have a negative impact.

Other potential threats could include: fences across gibber pavement, which provide vantage points for owls, and encourage incursions into habitat otherwise avoided by foxes and dingoes; spraying insecticides over stony deserts to combat locust plagues; baiting for dingoes; and direct harvesting of gibbers for road maintenance.

Some Kowari habitat is protected in Qld in national parks (Astrebla and Diamantina Lakes). However, the Kowari requires appropriate management outside reserves to prosper. Close and evenly spaced stock water points potentially threaten sand mound habitat patches. The long-term conservation of the species may require that new water points are not established near preferred habitat patches.

CURRENT RECOVERY EFFORTS

***Recovery plan*:** The original plan from the 1990s is no longer active: Lim L (1992) 'Recovery plan for the Kowari, *Dasyuroides byrnei* Spencer, 1896 *(Marsupialia, Dasyuridae)*'. A report submitted to Australian National Parks and Wildlife Service Endangered Species Programme.

Conservation actions: During the 1990s, surveys were conducted that demonstrated the species could be down listed from Endangered to Vulnerable. Subsequent research focused on ecology and management issues.

FUTURE RECOVERY EFFORTS

What you can do: Animals found dead (e.g. road-kills) can be frozen or preserved in 75% alcohol or methylated spirits, along with sound location data, and submitted to DERM or Qld Museum.

Conservation needs:

- Protect enough critical habitat to maintain or increase the current distribution.
- Increase public awareness of stony desert wildlife using the Kowari as an iconic species to assist in achieving sensitive and sustainable use of stony deserts.
- Establish a recovery team for threatened arid zone mammal fauna for south-west Qld.

Research directions:

- Determine the entire current distribution and key habitat features.
- Regularly monitor the key locations throughout the range of the species
- Monitor the health of essential habitats.
- Complete research on the Kowari's diet to help determine the source habitats and the importance of local habitat/landform diversity and integrity (numerous scats and prey species samples are available for this from SA).
- Carry out detailed research into Kowari reproduction in the wild, including local population genetics and factors that limit effective recruitment and range expansion.
- Clearly identify the threats to current populations and determine how those threats can be managed or reduced.
- A genetic study should be carried out, starting with approximately 50 samples of ear tissue and blood that are currently available from one site.

Sources and further reading

Canty P and Brandle R (2008) 'Kowari *Dasycercus byrnei* distribution monitoring in Sturts Stony Desert, South Australia, Spring 2007'. Science and Conservation, SA Dept Environment and Heritage, 2008 for SA Arid Lands Natural Resources Management Board.

Lim L (1998) 'Kowari – technical background information, final report research phase (National Recovery Plan)'. Unpublished report for Department Environment and Heritage, Queensland and Environment Australia.

Lim L (2008) Kowari *Dasyuroides byrnei*. In *The Mammals of Australia*. 3rd edn. (Eds S Van Dyck and R Strahan) pp. 52–54. Reed New Holland, Sydney.

Woolley PA (2005) The species of *Dasycercus* Peters, 1875 (Marsupialia: Dasyuridae). *Memoirs of Museum of Victoria* **62**, 213–221.

Author: Peter Canty

NORTHERN QUOLL

Dasyurus hallucatus Gould, 1842

Photo: www.ataglance.com.au

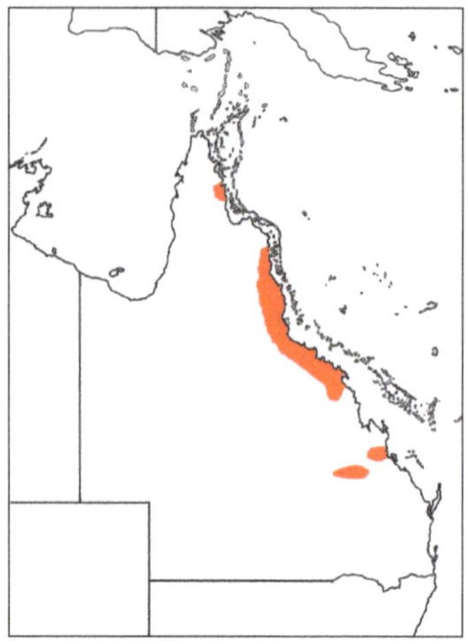

Family: Dasyuridae

Other common name: Northern Native Cat

Aboriginal names: Ankata (Angkamuthi), Atamuoi (Atampaya), Diguy (Mbabaram), Gya (Mpakwithi), Jaquol (Guugu Yimidhirr), Uyaba (Yadhaykenu).

Conservation status: Aust.: Endangered; Qld: Least Concern; NT: Critically Endangered; WA: Endangered

Description: Males of this small cat-sized marsupial carnivore are larger than females, with a head and body length of 270–370 mm (males) or 249–310 mm (females) and a tail length of 222–345 mm (males) or 202–300 mm (females). It weighs 240–1120 g. Its coat has white spots over a grey-brown background. Spotting may extend onto base of tail, but never extensively. It is distinguished from other Qld quolls by its short hind foot (<57 mm), striated pads on its sole and five toes.

Distribution: This quoll once occurred from Maroochydore in the east to Broome in the West. It has declined from vast areas and is now only known from the most rugged and remote parts of its former range. In Qld, the Northern Quoll is now known from about Rockhampton to the Wenlock River on Cape York Peninsula, though within this range its populations are disjunct, centred around the Expedition Range, Rockhampton, Mackay, Townsville, the Atherton Tablelands to Cooktown, Coen, Portland Roads and along the Wenlock River.

Habitat: Northern Quolls occupy a range of dry sclerophyll habitats and vine thickets and tend to incorporate rocky areas into their ranges. Females prefer rocky areas during the non-breeding season and non-rocky areas during the lactation phase of their life cycle.

Ecology: The Northern Quoll is an opportunistic predator and scavenger, eating invertebrates, mammals, fruit, reptiles and birds. The Northern Quoll is territorial and solitary and is capable of long-range movements in relatively short periods. In Qld, mating and births occur in July and August and up to eight young usually become independent and begin dispersing during November and December. The timing of reproduction appears to vary between sites and years. This quoll is the world's largest semelparous mammal, a condition in which most males die at the age of about 12 months after a short, synchronised breeding period. This life history pattern renders populations of Northern Quoll extremely susceptible to extinction from relatively minor increases in mortality, such as may occur from consumption of poisoned baits, toads or increased predation.

Reasons for decline and current threats: The large decline in extent of occurrence of the Northern Quoll in Qld is at least partly a result of poisoning from the introduced Cane Toad as it expanded its range. Nevertheless, the Northern Quoll and the toad co-exist in some areas of north Qld. This is encouraging for the recovery of decimated populations such as those on Cape York Peninsula and elsewhere. Other factors, including intensive fire regimes and intensive pastoralism, which lead to loss of ground cover, appear to result in declines in quoll populations as a result of increased predation by wild dogs and dingoes. Poison baiting for wild dogs is also a possible threat.

CURRENT RECOVERY EFFORTS

Recovery plan: Hill BM and Ward SJ (2010) 'National Recovery Plan for the Northern Quoll *Dasyurus hallucatus*'. Department of Natural Resources, Environment, The Arts and Sport, Darwin.

Conservation actions:

- The 'Protecting quolls in Qld landscapes' project was run by the Wildlife Preservation Society of Qld (WPSQ). Its purpose was to raise awareness through 'quoll discovery days', targeted mail outs, media releases and via a grant to landholders to quoll proof their poultry yards. It also seeks sighting records.
- The Land Manager website has information specific to management of Northern Quoll habitats.
- Scientists from the University of the Sunshine Coast are conducting a range of research projects looking

at survey and monitoring techniques, distribution and population sizes, and responses to Cane Toads, in Qld.

- A collaborative research project is underway between University of the Sunshine Coast and What About the Animals (Saddlers Springs, via Injune), to test the effectiveness of trained sniffer dogs for locating quolls in the wild.
- The Quoll Seekers Network provides a repository for sightings and information to manage quoll habitats and quoll–human interactions.
- The South Cape York Catchments Group received a Caring for Our Country grant in June 2009 for 'Community Solutions for Managing Natural Resource Challenges in South Cape York' and carried out a community project to search for Northern Quolls in the Lakeland Downs and Laura areas of Cape York Peninsula.

FUTURE RECOVERY EFFORTS

What you can do:

- Quoll-proof your poultry pen (see <http://www.wildlife.org.au>).
- Don't use strychnine baits and bury 1080 poison baits at least 10 cm below ground.
- Join the Quoll Seekers Network and report sightings to their webpage (above) and be kept up to date with quoll conservation, survey tips and volunteer opportunities.
- Keep domestic animals restrained at night.
- Join/donate to the Quoll Seekers Network and/or your local environmental organisation(s).
- Inspect and quarantine equipment and vehicles that are taken to toad-free islands that have quolls.

Research directions:

- Conduct studies on the distribution and population sizes.
- Determine the most effective quoll census techniques.
- Assess the role and nature of the pet industry in threatened species conservation
- Study the impacts of 1080 baiting on quoll populations, as well as the impacts of foxes, wild dogs and cats.
- Study the impact of burning and grazing on quoll populations.
- Conduct research on how Qld's Northern Quolls may provide insights into the responses and resilience of Australian predator populations to Cane Toads. We don't know why some populations survived Cane Toad colonisation while others perished, but those that did survive are those that occupy the most productive, wettest and remote areas of the former range. Animals from these populations may provide a resilient founder stock for reintroductions.

Sources and further reading

Burnett SE (1997) Colonising cane toads cause population declines in native predators; reliable anecdotal information and management implications. *Pacific Conservation Biology* **3**, 65–72.

King DR (1989) An assessment of the hazard posed to northern quolls (*Dasyurus hallucatus*) by aerial baiting with 1080 to control dingoes. *Australian Wildlife Research* **16**, 569–574.

Oakwood M (2002) Spatial and social organization of a carnivorous marsupial *Dasyurus hallucatus* (Marsupialia: Dasyuridae). *Journal of Zoology* **257**, 237–248.

Oakwood M and Foster P (2008) 'Monitoring extinction of the northern quoll'. Report to the Australian Academy of Science, Canberra.

Woinarski J, Oakwood M, Winter J, Burnett S, Myles H, Foster P and Holmes B (2008) 'Surviving the cane toads: patterns of persistence of the northern quoll *Dasyurus hallucatus* in Queensland'. Report to the Natural Heritage Trust, Canberra.

Author: Scott Burnett

SPOTTED-TAILED QUOLL (NORTH QLD SUBSPECIES)

Dasyurus maculatus gracilis (Ramsay, 1888)

Photo: www.ataglance.com.au

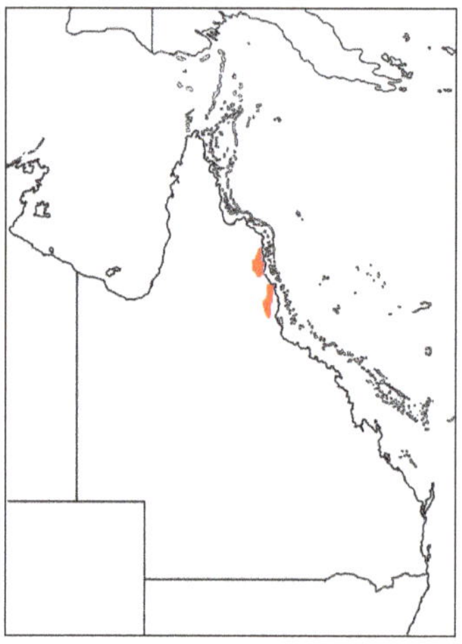

Family: Dasyuridae

Other common names: Tiger Quoll, Tiger Cat, Northern Tiger Cat

Aboriginal names: Yarri (Djirbalngan), Burrumbul (Yidinjdji)

Conservation status: ICUN: species (*D. maculatus*) is Near Threatened; Aust.: Endangered; Qld: Endangered

Description: This is the small race of the largest quoll species, with a head and body length of 310–510 mm and a tail length of 310–440 mm. It weighs 800–2540 g, with males being 50% heavier than females. The coat has distinctive white or cream spots over a grey, reddish, chocolate or black background with spotting extending onto the tail. It is easily distinguished from other quolls by its hind foot length of >57 mm.

Distribution: It is estimated that 540 individual Northern Spotted-tailed Quolls occur in the Wet Tropics. Historically, it was known to have occurred from Paluma in the south to the Big Tableland, south of Cooktown, in the north. It appears to have occurred in three broad populations centred on the Daintree Uplands (subpopulations on Finnegan, Thornton, Windsor and Carbine uplands), Atherton Uplands (subpopulations Lamb Range, Bellenden Ker, Atherton Tablelands and southern Cardwell Range) and Mount Spec Uplands. There have been no confirmed sightings of the species in the southern Wet Tropics since about the 1950s.

Habitat: The Northern Spotted-tailed Quoll is confined to continuous tracts of structurally complex mountain vine-forests and adjacent tall eucalypt forests that cloak the ranges. Although occurring from sea-level to the highest mountain tops, it is most abundant above 600 m altitude, like its prey. It uses boulder piles, fallen logs and hollows in standing trees as shelter sites. Maternal den sites are constructed within boulder piles and standing trees, in hollows with naturally occurring long, narrow access tunnels.

Ecology: Scat analyses reveal that small- and medium-sized mammals (70–2000 g), including the Fawn-footed Melomys *Melomys cervinipes* (found in 22% of scats), the Daintree and Herbert River Ringtail Possums, *Pseudochirulus cinereus* and *P. herbertensis* (13%), the Musky Rat-kangaroo, *Hypsiprymnodon moschatus* (12%), the Long-nosed Bandicoot, *Perameles nasuta* (12%), and Bush Rat *Rattus fuscipes* (10%), form the bulk of the diet of Northern Spotted-tailed Quolls. It also preys on invertebrates, reptiles and birds, and scavenges, feeding on anything of animal origin in any state of decomposition. Females can raise a single litter of up to six young per year, but more often wean four to five. Most births occur in June or July. Females reach maturity and breed the following year and do not breed beyond their second year. They produce two litters in their lifetime. The Northern Spotted-tailed Quoll is solitary and wide ranging, with a home range in excess of 200 ha for females and a much greater, though unquantified, home range for males. Males have been recorded moving 6 km (straight line distance) and females 1.8 km in 24 hours.

Reasons for decline and current threats: The Northern Spotted-tailed Quoll has declined on the Atherton Tablelands due to habitat loss. The decline from the Mount Spec uplands (Paluma area) is attributed to persecution at poultry yards in the first half of the 1900s. Although many populations are exposed to this threat, the small size of the Mount Spec population made it very susceptible. Poisoning from consuming toads, and baits laid for pests such as wild dogs and rodents, have probably also contributed to declines.

Most of its habitat is now protected in the Wet Tropics World Heritage Area and national parks. However, because it ranges widely, most individuals can traverse private lands that surround the protected areas. Current threats on private land include killing at poultry yards, direct and secondary poisoning from baits laid for wild dogs, pigs and rodents, poisoning by toads, road-kill and predation by domestic and feral carnivores.

Climate change could have a significant impact on this species (and its prey). Modelling suggests the climatic and habitat envelope it currently occupies may reduce by up to 50% with an increase of 1°C.

CURRENT RECOVERY EFFORTS

Recovery plan: Submitted and under review.

Conservation actions:

- See 'Protecting quolls in Qld landscapes' and 'Quoll Seekers Network' under Northern Quoll.
- Researchers at the University of the Sunshine Coast are working to determine effective monitoring techniques and to collect data on populations of Northern Quoll throughout Queensland.

FUTURE RECOVERY EFFORTS

- ***What you can do:*** Join/donate to the Quoll Seekers Network and/or your local environmental organisation(s).

Conservation needs: See actions under Northern Quoll and apply to the Wet Tropics distribution of the Northern Spotted-tailed Quolls.

Research directions:

- Gather data on the distribution and population sizes of quolls.
- Investigate the most effective quoll census techniques.
- Further assess the impacts of 1080 baiting on quoll populations in Qld.
- Assess the impacts of wild dogs and cats on quolls.
- Assess the feasibility and impact of Cane Toad control in significant areas (e.g. Windsor and Carbine Tablelands, and Lamb Range).

Sources and further reading

Burnett S (2001) Ecology and conservation status of the northern spot-tailed quoll, *Dasyurus maculatus* with reference to the past, present and future of Australian carnivorous marsupials. PhD thesis. James Cook University of North Queensland, Australia.

Burnett S and Marsh H (2004) Conservation of the spotted-tailed quoll, *Dasyurus maculatus*: a conceptual and applied model with particular reference to populations of the endangered *D. m. gracilis*. In *Conservation of Australia's Forest Fauna*. (Ed. D Lunney) pp. 624–638. Royal Zoological Society New South Wales, Mosman.

Krockenberger AK, Kitching RL and Turton SM (2003) 'Environmental crisis: climate change and terrestrial biodiversity in Queensland'. Cooperative Research Centre for Tropical Rainforest Ecology and Management. Rainforest CRC, Cairns.

Author: Scott Burnett

SPOTTED-TAILED QUOLL (SOUTHERN SUBSPECIES)

Dasyurus maculatus maculatus (Kerr, 1792)

Photo: Bruce Thomson

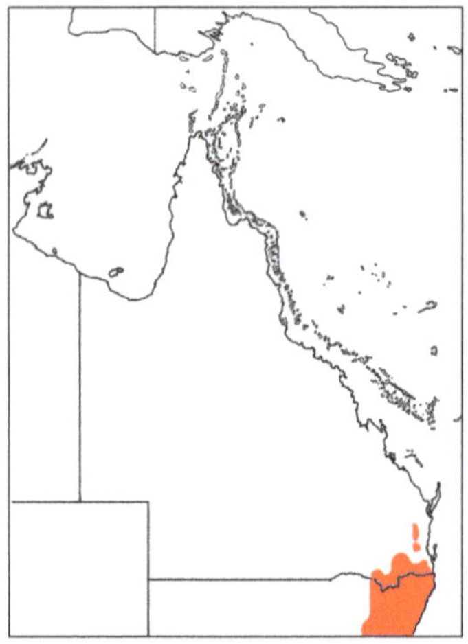

Family: Dasyuridae

Other common names: Tiger Quoll, Tiger Cat

Aboriginal name: Mumdjin (Bundjalung)

Conservation status: Aust.: Endangered; Qld: Vulnerable

Description: This is the largest of the quolls with a head and body length of 235–425 mm and a tail length of 305–475 mm. It weighs 1.2–4.3 kg, with males being 78% heavier than females. The coat has distinctive white or cream spots over a grey, reddish, chocolate or black background with spotting extending onto the tail. It is easily distinguished from other quolls by its hind foot length of >57 mm.

Distribution: Historically, this subspecies ranged throughout coastal and sub-coastal regions from south-east SA to the Rockhampton area. In Qld, this included the coast and ranges from the NSW border north to the Berserker Range at Rockhampton. It no longer occurs in the northern half of its Qld range nor across most of Vic. or SA. Currently, the Granite Belt and Border Ranges are the only regions where the species is still regularly recorded in Qld, although it probably still occurs in very low numbers in the mountain country from Gympie south to the Qld border.

Habitat: In Qld, the southern Spotted-tailed Quoll occupies large tracts of vine-forest, wet and dry sclerophyll forests and woodlands. It uses boulder piles, fallen logs and standing tree hollows as shelter sites.

Ecology: This species (including the Northern Spotted-tailed Quoll) is the only marsupial hypercarnivore on mainland Australia, with mammals constituting ~80% of its diet; the rest comprises invertebrates, reptiles and birds. The diet composition is flexible. Medium-sized mammals such as the Brushtail Possum *Trichosurus vulpecula*, European Rabbits *Oryctolagus cuniculus* and Greater Glider *Petauroides volans* form most of the diet in the Granite Belt, though they will consume more reptiles or large mammals in response to low availability of smaller mammals. Like its northern relative, the southern Spotted-tailed Quoll is an inveterate scavenger. Between 77 to 95% of females give birth to a single litter during the months of July and August, usually raising four to six young successfully. Females mature and breed after 1 year and do not usually breed beyond their second year. The life expectancy in the wild is 3 years although a very few individuals survive into their fourth year. Little is known of their movements. Three studies (the first from Qld's Granite Belt) found homes ranges in the order of 783–1202 ha (for males over 5 days), 621–2561 ha for males and 88–65 ha for females (southern NSW), and 108–757 ha for males and 76–175 ha for females (south-east NSW).

Reasons for decline and current threats: The primary reason for their decline is habitat loss due to clearing for primary production and urbanisation, most of which occurred in the early and mid-1990s. Other reasons contributing to their decline include: persecution at poultry yards; poisoning by baits (particularly strychnine) and Cane Toads *Rhinella marinus*; competition and predation by domestic and feral dogs, cats and foxes; and death from vehicle strikes. Recent studies suggest that 1080 baits, which have largely replaced strychnine baits for dog and fox control, are less lethal to quolls than strychnine.

Broad-scale clearing still threatens populations in the southern Granite Belt, though much of its remaining habitat is protected within national parks and forest reserves. The other most serious current threats include drought, climate change and persecution at poultry yards. Road-kill is probably only a significant threat in areas where heavily used roads bisect quoll habitat, such as Cunningham's Gap.

CURRENT RECOVERY EFFORTS

***Recovery plan*:** Submitted and under review.

Conservation actions:

- See 'Protecting quolls in Qld landscapes' and 'Quoll Seekers Network' under Northern Quoll.
- DERM provides advice on managing quoll habitats via a Spotted-tailed Quoll species management

profile available for download at <http://www.derm.qld.gov.au/register/p02353aa.pdf>.

- Recent projects include:
  - Stephanie Meyer–Gleaves PhD project, The Threat of Extinction: Ecology and Conservation of the Spotted-tailed Quoll (*Dasyurus maculatus maculatus*) in Southern Qld.
  - Qld Department of Primary Industries and Fisheries. The effects of 1080 baiting on a population of the Spotted-tailed Quoll, *Dasyurus maculatus,* in the southern Granite Belt found that these quolls appear not to be negatively impacted by 1080 baiting.
  - A number of studies have been undertaken in collaboration between the Quoll Seekers Network and the University of the Sunshine Coast including: The 'Quolls in Qld's Border Ranges' study conducted in 2008 and again in 2010 indicated a decline in quoll sightings in the area over the past 10–17 years; The 'Quolls in the Beaudesert area' a study conducted in 2006 confirmed the presence of quolls and raised awareness of them; The 'Quolls in the Mary River headwaters' study conducted in 2009 confirmed the presence of quolls and raised awareness of them, and revealed that they have declined and are very rare in the area.

FUTURE RECOVERY EFFORTS

What you can do: Join/donate to the Quoll Seekers Network and/or your local environmental organisation(s).

Conservation needs: The protection of quoll habitats not currently within the conservation estate (especially in south-east Qld's Granite Belt).

Research directions:

- Gather data on the distribution and population sizes.
- Investigate the most effective quoll census techniques.
- Further assess the impacts of 1080 baiting on quoll populations in Qld.
- Assess the impacts of foxes, wild dogs and cats on quolls.
- Assess the feasibility and impact of Cane Toad control in significant areas.
- Create models to assess the impacts of climate change.

Sources and further reading

Burnett S and Marsh H (2004) Conservation of the spotted-tailed quoll, *Dasyurus maculatus*: a conceptual and applied model with particular reference to populations of the endangered *D. m. gracilis*. In *Conservation of Australia's Forest Fauna*. (Ed. D Lunney) pp. 624–638. Royal Zoological Society New South Wales, Mosman.

Claridge AW, Paull D, Dawson J, Mifsud G, Murray A, Poore R and Saxon MJ (2005) Home range of the spotted-tailed quoll (*Dasyurus maculatus*), a marsupial carnivore, in a rainshadow woodland. *Wildlife Research* **32**, 7–14.

Glen AS and Dickman CR (2008) Niche overlap between marsupial and eutherian carnivores: does competition threaten the endangered spotted-tailed quoll?' *Journal of Applied Ecology* **45**, 700–707.

Kortner G (2007) 1080 aerial baiting for the control of wild dogs and its impact on spotted-tailed quoll (*Dasyurus maculatus*) populations in eastern Australia. *Wildlife Research* **34**, 48–53.

Meyer-Gleaves S (2008) The threat of extinction: ecology and conservation of the spotted-tailed quoll (*Dasyurus maculatus maculatus*) in southern Queensland. PhD thesis. Griffith University, Australia.

Author: Scott Burnett and Stephanie Meyer-Gleaves

CARPENTARIAN PSEUDANTECHINUS

Pseudantechinus mimulus (Thomas, 1906)

Family: Dasyuridae

Other common name: Carpentarian Antechinus

Conservation status: IUCN: Endangered; Aust.: Vulnerable; NT: Endangered

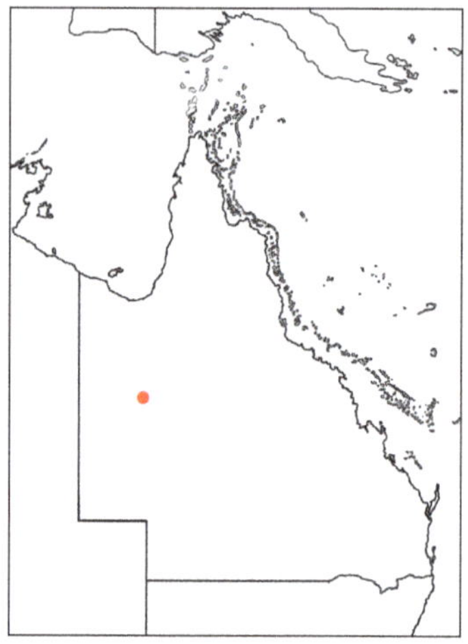

Description: This very small marsupial insectivore has a head and body length of 70–90 mm and a tail length of 60–75 mm. It weighs 15–25 g. It is buff-brown above with distinct light eye-rings and cinnamon patches behind the ears and grey to white on the underside. Its tail is relatively short and can enlarge to become quite fat in good conditions.

Distribution: The Carpentarian Pseudantechinus is known from the Mount Isa region in Qld, Alexandria Station in the NT, and the Sir Edward Pellew Group of islands plus nearby coastal mainland in the Gulf of Carpentaria. First collected in 1905 on Alexandria Station, it has not been recorded in that region since. Records from each location are few, so it is difficult to comment on the total size of the population. It possibly occurs in an area of as little as about 400 km^2 across these three widely separated locations.

Habitat: Known records of this species are from rocky areas, or in woodlands close to rocky areas. They include the sides of sandstone ridges supporting open woodland of *Eucalyptus leucophloia, E. normantonensis, Corymbia terminalis, Atalaya hemiglauca* and *Acacia* spp. over hummock grassland (near Mount Isa) and sandstone boulders, pavements and outcrops with hummock grasses (*Triodia* spp.) in woodland or shrubland, including *Eucalyptus aspera, E. kombolgiensis, Acacia latifolia, A. multisiliqua, Bossiaea bossiaeoides* and *Calytrix* spp. (on the Pellew Islands).

Ecology: It is similar to other pseudantechinus species, being nocturnal and a voracious predator of invertebrates and, probably, small vertebrates. It spends the day among rocks, and hunts with great agility across the rocks at night. The species is a seasonal breeder, mating probably in June–July, giving birth August–September and weaning young in October–November. One female was carrying six pouch young, but the typical litter size is probably four to six. It is likely that some males and females survive to breed in a second year.

Reasons for decline and current threats: No threatening processes have been demonstrated to cause the decline of the species, though cats and fire are likely influences. Carpentarian Pseudantechinuses are of a size that they would be eaten by cats, though their rugged rocky habitats should provide them some protection and they persist in areas that have cats, albeit relatively recently arrived in the case of the Pellew Islands. Changed fire regimes, particularly the increase in hot extensive late dry season fires, probably change the abundance and availability of their prey and the structure of their habitat.

Although the impact of the Cane Toad on pseudantechinus is unknown, it now exists throughout their range. The bufotoxins produced by toads are likely to be lethal to pseudantechinuses, but other small dasyurids have learnt to avoid Cane Toads.

The Carpentarian Pseudantechinus was first recorded from Qld during a study of the impacts upon biota of sulphur dioxide emissions from the Mount Isa copper smelter. Although captures were too few for statistical certainty, Carpentarian Pseudantechinus occurred only in two sites with the lowest levels of sulphur dioxide emissions (>20 km from source), and not in similar environments at four sites with higher emission levels.

CURRENT RECOVERY EFFORTS

Recovery plan: Woinarski JCZ (2004) 'National multi-species recovery plan for the Carpentarian Antechinus *Pseudantechinus mimulus*, Butler's Dunnart *Sminthopsis butleri* and Northern Hopping-mouse *Notomys aquilo*, 2004–2009'. Northern Territory Department of Infrastructure Planning and Environment, Darwin.

Conservation actions:

- NT NRETAS has mounted irregular surveys in the Sir Edward Pellew Islands, in conjunction with the li-Anthawirriyarra Sea Rangers.
- General fauna surveys monitoring the impacts of mining and processing operations of the Mount Isa mines have found a few animals.
- The Mabunji Aboriginal Resource Association's li-Anthawirriyarra Sea Ranger Unit and Desert Wildlife Services are running trials to control cats on West Island (in the Sir Edward Pellew group), with the intention to expand control to other islands in the group.
- Threatened Species Network (TSN; WWF and Australian Government) has funded past surveys and has supported local Indigenous ranger units.

FUTURE RECOVERY EFFORTS

What you can do:

- Join/donate to WWF-Australia.

- Write letters to state and federal environment ministers to urge much greater funding allocations to threatened species research and recovery.

Conservation needs:
- Manage fire regimes in known habitat of Carpentarian Pseudantechinus to increase habitat suitability.
- Control feral cats.

Research directions:
- Develop a quantitative monitoring technique and apply it to assess the impacts of different fire regimes, levels of predation, and emission levels from the Mount Isa smelter.
- Develop and trial management practices to minimise the negative effects of these factors on pseudantechinus populations.
- Survey more areas to determine the geographic distribution of the species.
- Study the basic ecology of the species.

Sources and further reading

Griffiths AD (1998) 'Impact of sulphur dioxide emissions on savanna biodiversity at Mt Isa, Queensland'. Report to Mt Isa Mines Ltd. Tropical Savannas Cooperative Research Centre, Darwin.

Johnson KA, Woinarski JCZ and Langford DG (2008) Carpentarian Pseudantechinus *Pseudantechinus mimulus*. In *The Mammals of Australia*. 3rd edn. (Eds S Van Dyck and R Strahan) pp. 71–72. Reed New Holland, Sydney.

Woinarski J (2007) Carpentarian antechinus *Pseudantechinus mimulus*. In *Lost from Our Landscape: Threatened Species of the Northern Territory*. (Eds S Woinarski, C Pavey, R Kerrigan, I Cowie and S Ward) p. 221. Northern Territory Government, Darwin.

Woolley PA (2011) *Pseudantechinus mimulus*: a little known dasyurid marsupial. *Australian Mammalogy* **33**, 57–67.

Author: Simon Ward

JULIA CREEK DUNNART

Sminthopsis douglasi Archer, 1979

Photo: Greg Calvert

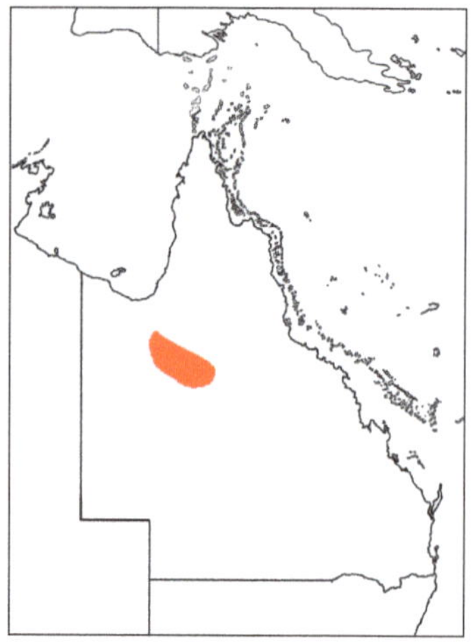

Family: Dasyuridae

Conservation status: IUCN: Near Threatened; Aust.: Endangered, Qld: Endangered

Description: This small marsupial carnivore is the largest of the *Sminthopsis* group, with a head and body length of 110–135 mm and a tail length of 110–130 mm. It weighs 40–70 g, with males being larger than females. Its body is brown, speckled with grey above and buff-white below. It has a prominent facial stripe, but is distinguished by dark hairs in rings around the eyes and on the outer edge of the ears. It has rufous hairs on the cheeks and at the base of the ears and dark hairs towards the tip of its long tapering tail that is fattened at the base.

Distribution: The Julia Creek Dunnart is restricted to Mitchell Grass Downs in north-west Qld. Until 1991, only four specimens, collected between Richmond and Julia Creek, were known; it was thought to be extinct. A chance rediscovery in 1991 and subsequent searches have found 25 locations across the Mitchell Grass Downs and adjacent Desert Uplands bioregion. However, its occurrence remains patchy and abundance low.

Habitat: The Mitchell Grass Downs, its primary habitat, are grass-covered cracking clay soils of two types: ashy and stony. The species occurs on both in areas of sparse and dense ground cover, and in a variety of soil cracking conditions, though they prefer sites with high density of cracks and holes and dense vegetation.

Ecology: The Julia Creek Dunnart is nocturnal, sheltering during the day in cracking clay soil cavities or in vegetation. It is insectivorous, feeding on crickets, cockroaches, silverfish, arachnids, slaters and small reptiles. A female can produce two litters of up to eight young between September and February. A female matures rapidly and may be able to breed in its first year and for up to 3 years, though generally two. A male can also breed for two consecutive years.

Reasons for decline and current threats: The reasons for decline are not known, but climatic factors, cats and stock grazing were suggested initially. Subsequent investigations have confirmed that cats are a significant and novel predator, and may be responsible for the disappearance of Julia Creek Dunnarts on the Lyrian property.

Introduced weeds have also been suggested. Prickly Acacia *Acacia nilotica* is thought to have a negative impact on the quality of dunnart habitat by eliminating native ground vegetation and changing the composition of native grasses. Although grazing by sheep and cattle has been implicated, the Julia Creek Dunnart persists on active grazing properties, particularly those with modest stocking rates.

CURRENT RECOVERY EFFORTS

Recovery plan: Department of Environment and Resource Management (2009) 'National recovery plan for the Julia Creek Dunnart (*Sminthopsis douglasi*)'. Report to the Department of the Environment, Water, Heritage and the Arts, Canberra. Queensland Parks and Wildlife Service, Brisbane.

Other plans: Maxwell S, Burbidge AA and Morris K (Eds) (1996) 'The 1996 action plan for Australian marsupials and monotremes'. Wildlife Australia, Canberra.

Conservation actions:

- The Julia Creek Dunnart research and recovery program has been successful so far. In 20 years, the distribution of the species has been more thoroughly documented, ecological studies have been completed, a captive breeding program has been established and on-ground actions are being implemented by the community to protect habitat for the species.
- In 2005, a memorandum of understanding was signed between the Qld Government and McKinlay Shire outlining cooperative arrangements to progress the recovery of the Julia Creek Dunnart.
- Biological control agents are now being investigated by Biosecurity Qld to reduce the impacts of Prickly Acacia.
- Twenty captive bred individuals were re-introduced in 2007. Because some were eaten by cats, a predator-proof fence was erected in 2008 and more releases made.

- Because most habitat is not protected, effort has focused on promoting and encouraging land management practices that support the persistence of dunnarts in production systems. These include control of introduced predators and Prickly Acacia and conservative management of grazing to maintain grass cover.
- Surveys continue to detect new populations, better define its distribution and inform management actions.
- The establishment of voluntary nature refuge agreements over areas known to support populations and habitat of the Julia Creek Dunnart is a current focus of the recovery program. Examples currently being negotiated include the Julia Creek Airport Sanctuary and Toorak Research Station.
- The recovery team encourages and supports landholders to manage Mitchell Grass habitats sustainably to support conservation of the dunnart. This can include weed management, careful management of fire, conservative stocking rates and control of feral predators.
- Information brochures have been produced and McKinlay Shire is currently developing an information video to promote the Julia Creek Dunnart as a regional icon for broader biodiversity conservation in the Mitchell Grass Downs within the context of sustainable production.

FUTURE RECOVERY EFFORTS

Conservation needs:
The long-term conservation of suitable habitat and the sustainable management of grazing lands are essential for the survival of the species. Achieving this includes:

- controlling Prickly Acacia, *Acacia nilotica*. Landholders are also advised to clear mustering lanes, fence off dense infestations and clear around watering points.
- encouraging private and commercial gardeners and horticulturalists to prevent the trade in, and distribution of Prickly Acacia throughout Australia.

Research directions:

- Investigate the interactions between predators, water sources and grazing management and their impacts on the Julia Creek Dunnart.
- Examine the interactions between this species and sympatric small mammals to better manage the entire system.
- Extend research on initial studies that indicated that the timing and extent of burning are critical considerations to avoid impacting on resident dunnarts.

Sources and further reading

Kutt AS (2003) New records of the Julia Creek dunnart *Sminthopsis douglasi* in central-north Qld. *Australian Zoologist* **32**, 257–260.

Mifsud G (1999) Ecology of the Julia Creek dunnart *Sminthopsis douglasi* (Marsupialia: Dasyuridae). MSc thesis. La Trobe University.

Woolley PA (1992) New records of the Julia Creek Dunnart, *Sminthopsis douglasi* (Marsupialia: Dasyuridae). *Wildlife Research* **19**, 779–783.

Woolley PA (2000) 'Monitoring the effectiveness of predator control on populations of the Julia Creek Dunnart on Toorak Research Station, and surveys to locate other populations'. Unpublished report to the Queensland Environmental Protection Agency, Brisbane.

Woolley PA (2008) Julia Creek Dunnart *Sminthopsis douglasi*. In *The Mammals of Australia*. 3rd edn. (Eds S Van Dyck and R Strahan) pp. 136–137. Reed New Holland, Sydney.

Author: Geoff Lundie-Jenkins

BANDICOOTS AND BILBIES (ORDER: PERAMELEMORPHIA)

GREATER BILBY

Macrotis lagotis (Reid, 1837)

Photo: www.ataglance.com.au

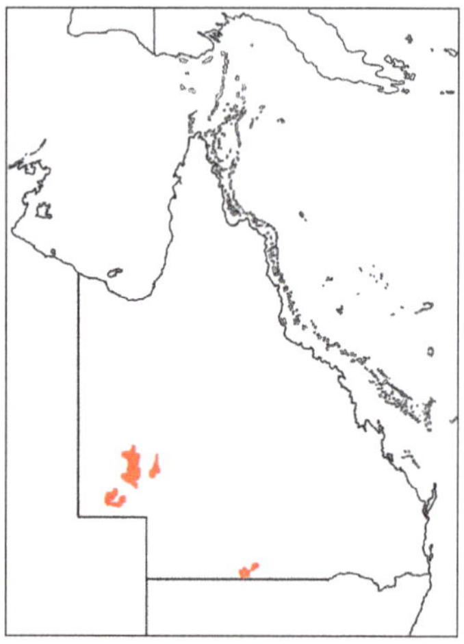

Family: Peramelidae

Other common names: Rabbit-eared Bandicoot, Bilby

Aboriginal names: Thalka (Qld), Ninu, Thulka, Pinkie (WA and NT)

Conservation status: IUCN: Vulnerable; Aust.: Vulnerable; Qld: Endangered; NSW: Presumed Extinct; NT: Threatened; SA: Vulnerable; WA: Rare

Description: The Greater Bilby is a burrowing bandicoot with large naked ears, a long pointed snout, and a striking crested black and white tail. The head and body length is 290–550 mm and its tail length is 200–290 mm. It weighs 1000–2500 g (males) or 800–1100 g (females).

Distribution: Once widespread throughout arid and semi-arid Australia, the Greater Bilby now occurs in arid north-western Australia (parts of WA and NT) and in south-western Qld. In Qld, it occurs in a number of locations on the stony and clay plains of the Diamantina region.

Habitat: This bilby occupies primarily the clay areas, but also some stony plains, of the Channel Country, mainly west of the Diamantina River. These vast plains are flat to gently undulating and support a diverse range of annual and perennial grasses and forbs. Burrows are generally associated with shallow drainage depressions and the slopes of stony ridges that are scattered throughout the plains.

Ecology: The Greater Bilby is a burrowing marsupial omnivore. It is strictly nocturnal and retreats during the day to one of a number of burrows in its home range. Burrows are deep and may extend for 2 m underground. It digs for most of its food, leaving behind many characteristic shallow conical excavations or 'feed scrapes'. Greater Bilbies eat a wide range of invertebrates and may also take small reptiles and mammals. Seeds, bulbs and fungi are also consumed. They do not require access to free water. Females breed all year after 6 months of age and generally have two young at a time, following a remarkably short gestation period of 12–14 days.

Reasons for decline and current threats: The species is extinct in over 90% of its former range, and nearly 99% of its previous range in Qld. It no longer occurs in NSW or SA and is still declining in the deserts of the NT and WA. The decline started with loss and modification of its habitat during agricultural expansion in the early 1800s. During the mid-1800s, the introduction of the rabbit and the fox, and possibly cats, caused further declines. Sheep and cattle may also have had an impact.

Many arid zone species, including the Greater Bilby, are now threatened by the spread of water points for stock, particularly in Qld. Formerly arid and semi-arid landscapes with permanent drinking water enable domestic livestock and exotic predators to occupy areas that were previously only available for short periods during good seasons and therefore extend their competitive and predatory impacts beyond the ability of native species to cope.

CURRENT RECOVERY EFFORTS

Recovery plan: Pavey C (2006) 'National recovery plan for the Greater Bilby *Macrotis lagotis*'. Northern Territory Department of Natural Resources, Environment and the Arts.

Other plans: Department of the Environment, Water, Heritage and the Arts (2008) 'Threat abatement plan for competition and land degradation by rabbits'. Department of the Environment, Water, Heritage and the Arts, Canberra.

Department of the Environment, Water, Heritage and the Arts (2008) 'Threat abatement plan for predation by the European Red Fox'. Department of the Environment, Water, Heritage and the Arts, Canberra.

Department of the Environment, Water, Heritage and the Arts (2008) 'Threat abatement plan for predation by feral cats'. Department of the Environment, Water, Heritage and the Arts, Canberra.

Conservation actions:
- Ongoing support for bilby research and conservation is provided by DERM – e.g. monitoring the population at Astrebla Downs NP annually and broad-scale aerial burrow surveys every 5–8 years.
- Captive breeding at Charleville (DERM) and Dreamworld have provided founders for re-introductions to a 29 km^2 feral animal free enclosure at Currawinya NP since late 2005. Funding was provided by public donations to the Save the Bilby Fund (SBF) and state and federal environment agencies.
- Additional sites for predator controlled re-introductions are currently being selected in Qld.
- SBF has raised about $1 000 000 and much awareness of the bilby throughout Australia and internationally.

FUTURE RECOVERY EFFORTS

Conservation needs:
- Support the Save the Bilby Fund.
- Lobby local, state and federal governments to regulate and control cats.
- Maintain public awareness of the species.
- Implement effective long-term control programs for exotic species, including foxes, cats, rabbits and feral goats, to protect existing populations.
- Encourage effective conservation agreements with measurable outcomes for known populations on private land.
- Assist with re-introduction programs to establish new populations.

Research directions:
- Continue monitoring known populations to measure trends and identify emerging threats.
- Consider the management of water in arid and semi-arid areas and its effect on native arid zone animals and vegetation.
- Study the interaction of water points with introduced competitors and predators and their flow on effects to arid zone animals.

Sources and further reading

Abbott I (2001) The bilby *Macrotis lagotis* (Marsupialia: Peramelidae) in south western Australia: original range limits, subsequent decline and presumed regional extinction. *Records of the Western Australian Museum* **20**, 271–305.

Gibson LA (2001) Seasonal changes in the diet, food availability and food preference of the Greater Bilby (*Macrotis lagotis*) in south-western Queensland. *Wildlife Research* **28**, 121–134.

McRae PD (2004) Aspects of the ecology of the Greater Bilby *Macrotis lagotis* in Queensland. University of Sydney.

Southgate RI (1990) The distribution and abundance of the Greater Bilby *Macrotis lagotis* Reid (Marsupialia: Peramelidae). In *Bandicoots and Bilbies*. (Eds JH Seebeck, RL Wallis and CM Kemper) pp. 293–302. Surrey Beatty & Sons, Sydney.

Author: Peter McRae

KOALAS, WOMBATS, POSSUMS AND MACROPODS (ORDER: DIPROTODONTIA)

KOALA (SOUTH-EAST QLD BIOREGION)

Phascolarctos cinereus (Goldfuss, 1817)

Photo: www.ataglance.com.au

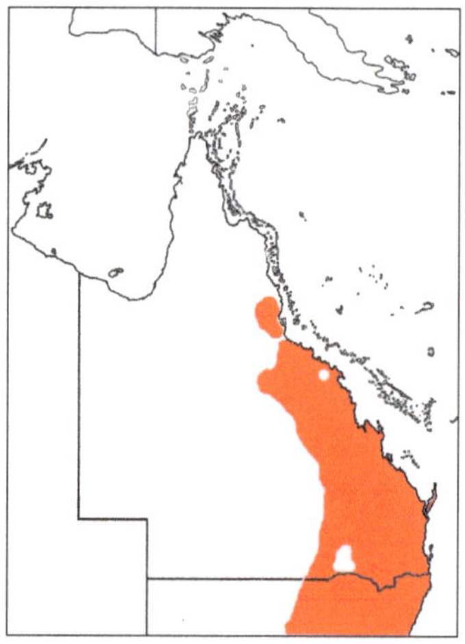

Family: Phascolarctidae

Aboriginal names: Many are recorded. Qld: Dumbirbi (Yagara people)

Conservation status: IUCN: species (*P. cinereus*) is Least Concern; Qld: Vulnerable (south-east Qld bioregion); NSW: Vulnerable (+ two Endangered populations)

Description: The Koala has a short compact body and large head, large ears and no functional tail. It has a head and body length of 674–736 mm and weighs 4.2–9.1 kg (in Qld). It has dense fur, which is white below and brown to grey above with white patches, a prominent black nose and large claws. Males are up to 50% larger than females and have a larger head, more prominent nose and a more obvious sternal (chest) scent gland. Southern Koalas are larger (up to 60% heavier) and darker than northern Koalas. Three subspecies have been described: *P. c. adustus* (Qld), *P. c. cinereus* (NSW) and *P. c. victor* (Vic.).

Distribution: The Koala was common and widespread in forests and woodlands along the east coast of Australia from north Qld to south-eastern SA. Its distribution is strongly linked to the presence and abundance of eucalypt (and some non-eucalypt) food plants. Its range has now contracted due to clearing of habitat, by about 30% within Qld.

Habitat: The species of eucalypt and the extent of tree cover are the most important habitat characteristics for the Koala. Soil fertility and water regime may also be important. It eats about 50 species of eucalypts across its range, but preferences vary regionally and locally. Koala food choice, and thus habitat suitability, is determined not just by tree species, but by chemotypes within species. A complex group of phenolic chemicals called formyl phloroglucinol compounds (FPCs) act as deterrents. These compounds vary in concentration and activity between eucalypt species, and within species among individual trees, so that quality of Koala habitat can vary on a small scale.

Ecology: A diet of eucalypt leaves, with their low digestibility and chemical defences, places constraints on the energy budget of the Koala. In consequence, the Koala is notoriously inactive, spending around 19 h per day sleeping, and has the lowest known input of energy to reproduction of any mammal measured, with corresponding slow growth rates. Koalas can live >12–15 years in the wild and females give birth to a single young between October and May, which occupies the pouch for 6 to 8 months but continues to depend on its mother for 12 months.

Reasons for decline and current threats: In south-east Qld, Koala populations have declined due to habitat clearing and urbanisation and associated issues such as dog and road-kills, habitat fragmentation, inappropriate fire management and disease. Wildfire and drought can have devastating effects on Koala populations. Historically, Indigenous hunting seemed to limit Koala density and distribution. After European settlement, Indigenous populations (and hunting) declined, allowing Koala populations to grow. By the 1860s to '70s the Koala was so abundant that a substantial trade in Koala fur began. In Qld, 450 000 to nearly 1 000 000 Koalas were shot each season, in six open seasons from 1906 to 1927. Despite the massive harvest, which probably caused some short-term declines, there was no evidence that hunters caused permanent population declines, although the public had perceived a decline and opinion turned against the harvest. In some areas, overpopulation results in over-browsing of food plants, particularly in small habitat islands, which then results in population crashes, often preceded by an increase in the incidence of disease. The most significant permanent conservation problem remains loss of habitat. The most serious (and almost intractable) current threat for Koalas is local and regional decline due to urban expansion at several locations, such as is occurring in south-east Qld.

CURRENT RECOVERY EFFORTS

Recovery plan: None

Other plans: Natural Resource Management Ministerial Council November 2009. 'National Koala Conservation and Management Strategy 2009–2014'. Department of Environment, Water, Heritage and the Arts.

Conservation (Koala) 'Conservation Plan 2006 and Management Program 2006–2016.' Queensland Department of the Environment and Resource Management.

Conservation actions:

- Numerous organisations and individuals, government and private, are devoted to the care and protection of the Koala.
- Many resources are devoted to controlling threats to Koalas, integrating habitat into urban, suburban and agricultural footprints and, in some places, controlling Koalas themselves so that they do not overpopulate and destroy habitat for themselves and other wildlife.
- Intensive habitat management is being carried out in south-east Qld in an attempt to mitigate the effects of urban expansion.
- A large investment has been made into research on the species, to the point where we begin to understand impacts of the chemical composition of individual trees on population processes.

FUTURE RECOVERY EFFORTS

Conservation needs: The conservation of the Koala in areas where it is threatened will require the integration of wildlife habitat of sufficient quality and connectedness into the design of urban and production areas, habitat rehabilitation and restoration in parts of its former range, and management of the impact of climate change.

Research directions:

- Identify and protect suitable habitat and identify the factors determining distribution.
- Continue to monitor key populations.
- Carry out research into managing the impact of urbanisation and into the impact of climate change, as well as mitigation of that impact.

Sources and further reading

Gordon G, Hrdina F and Patterson R (2006) Decline in the distribution of the koala *Phascolarctos cinereus* in Queensland. *Australian Zoologist* **33**, 345–358.

Krockenberger AK (2003) Meeting the energy demands of reproduction in female koalas, *Phascolarctos cinereus*: evidence of energetic compensation. *Journal of Comparative Physiology B* **173**, 531–540.

Lee KE, Seddon JM, Corley SW, Ellis WAH, Johnston SD, De Villiers DL, Preece HJ and Carrick FN (2009) Genetic variation and structuring in the threatened koala populations of Southeast Queensland. *Conservation Genetics* **11**, 2091–2103.

McAlpine CA, Rhodes JR, Callaghan JG, Bowen ME, Lunney D, Mitchell DL, Pullar DV and Possingham HP (2006) The importance of forest area and configuration relative to local habitat factors for conserving forest mammals: a case study of koalas in Queensland, Australia. *Biological Conservation* **132**, 153–165.

Moore BD and Foley WJ (2005) Tree use by koalas in a chemically complex landscape. *Nature* **435**, 488–490.

Authors: Andrew Krockenberger, Greg Gordon and Andrew J Dennis

NORTHERN HAIRY-NOSED WOMBAT

Lasiorhinus krefftii (Owen, 1872)

Photo: David Watts

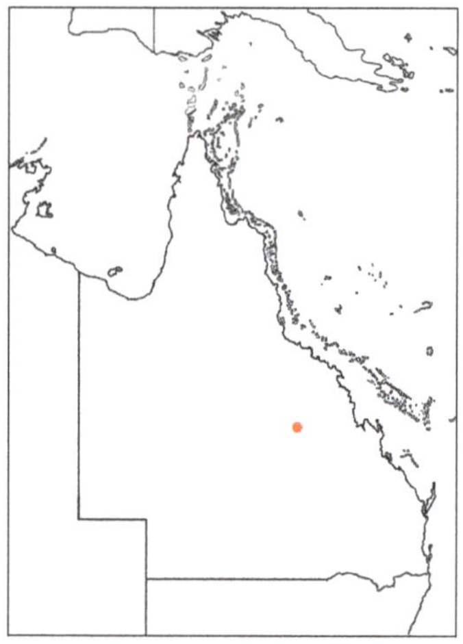

Family: Vombatidae

Aboriginal name: Yaminon (St George area)

Conservation status: IUCN: Critically Endangered; Aust.: Endangered; Qld: Endangered.

Description: The Northern Hairy-nosed Wombat is the largest burrowing herbivore in the world, with a head and body length of about 1150 mm and an insignificant tail of about 52 mm. It weighs about 32 kg, with females slightly larger than males. This stocky, rounded animal has dark to ashen, silky fur and short, powerful legs with soft padded feet and long, broad nails.

Distribution: Fossil records indicate a historical range in the semi-arid zone from Charters Towers in north Qld into Vic., probably in areas with deep, sandy soils. Live animals have been recorded from north-west of Clermont in Qld, the St George area in southern Qld and at Deniliquin in southern NSW. The populations near St George and Deniliquin were extinct by 1908, leaving the species restricted to a single, very small population in a 300 ha area at Epping Forest NP (Scientific) near Clermont in central Qld. In the 1980s, the population had reduced to ~35. With subsequent intensive management, this increased to ~115 by 2000. In 2007, it had climbed to ~138 individuals. During 2009 and 2010, a small number of wombats were translocated from Epping Forest NP to the Richard Underwood Nature Refuge at Yarran Downs near St George, reintroducing them to that area after an absence of >100 years.

Habitat: Epping Forest NP is a flat, semi-arid environment with grasslands, woodlands and open forests, each with varying degrees of tree density. Temperatures range from <0°C to 40°C and the area receives an average of 450 mm per year, which falls predominantly in the summer. Droughts lasting up to 6 years are common. There are also occasional floods. Deep, alluvial, sandy soils suitable for digging burrows, are a key habitat feature. Most burrows are generally in the vicinity of an ancient creek bed that no longer flows. Buffel Grass *Cenchrus ciliaris* has rapidly dominated the understorey, almost to complete exclusion of native grass species in most areas.

Ecology: The Northern Hairy-nosed Wombat survives on grass (90% of diet) which fluctuates seasonally in both quality and abundance. A highly efficient digestive system enables it to assimilate maximum nutrition value from scarce, poor quality food during extended drought periods. It has extremely low water requirements and low energy expenditures. Northern Hairy-nosed Wombats spend most of their time deep underground in burrows they construct to provide insulation from extreme temperatures and to provide an optimal humidity level to conserve water and energy. Individuals in the wild have been recorded living >26 years.

Reasons for decline and current threats: The population declined rapidly after the introduction of agriculture, particularly pastoral activities. Competition from sheep and cattle in combination with frequent and extended drought probably caused the species decline. Predation by wild dogs killed at least 10 wombats at Epping in 2001 – 10% of the total population at the time. Predation is considered a significant ongoing threat.

At Epping, competition from kangaroos and wallabies inside the fenced wombat habitat protection area (see below) poses a threat that needs to be managed, particularly during droughts.

CURRENT RECOVERY EFFORTS

Recovery plan: Horsup A (2004) 'Recovery plan for the northern hairy-nosed wombat *Lasiorhinus krefftii* 2004–2008'. Report to the Department of Environment and Heritage, Canberra. Environmental Protection Agency/Queensland Parks and Wildlife Service, Brisbane.

Conservation actions:

- A partnership project between the Qld Government, Xstrata and private landholders has established a population of Northern Hairy-nosed Wombats at the Richard Underwood Nature Refuge at Yarran Downs near St George.

- A predator- and cattle-proof fence at Epping Forest NP excludes wild dogs and cattle to prevent predation and competition. It also prevents wombats, particularly adult females, dispersing to unsuitable habitat and perishing. A similar fence surrounds the Richard Underwood Nature Refuge.
- Food and water stations have been provided to supplement their natural diet, particularly during times of drought at Epping Forest NP and Yarran Downs.
- A fire management plan exists for Epping Forest NP.
- Strips through grassland areas have been slashed to manipulate pasture and stimulate growth of native grasses.
- When competition appears to be a problem, macropods within the fenced wombat habitat area at Epping Forest NP are moved to adjacent sections of the Park.
- Sites for translocations have been assessed within the former range of the species in Qld. The Wombat Foundation provides funding to facilitate research, and assist with the other parts of the recovery program.
- The Foundation for Australia's Most Endangered Mammals has provided financial assistance for censuses.
- Project Kial, Australian Animals Care and Education (AACE) is currently working in conjunction with UQ on perfecting artificial insemination in Southern Hairy-nosed Wombats. Once developed, these techniques could be applied to Northern Hairy-nosed Wombats.

FUTURE RECOVERY EFFORTS

What you can do: Join/donate to the Wombat Foundation, the Wombat Survival Fund and Project Kial (AACE).

Conservation needs:

- Continue the current and intensive management.
- Prepare a new recovery plan.
- Introduce new populations to additional, carefully selected and managed sites to increase the number of populations where and when appropriate.

Research directions:

- Assess the requirements of recovery, such as habitat management and manipulation to improve resilience and reproductive output.
- Carry out ongoing monitoring to determine the effectiveness or otherwise of management actions.
- Study the animal's basic ecology.
- Undertake research on the impact of introduced pasture grasses, particularly Buffel Grass, on Northern Hairy-nosed Wombats.
- Undertake a detailed risk assessment with subsequent research into the most plausible risks, such as the spread of disease and parasites.

Sources and further reading

Banks SC, Hoyle SD, Horsup A, Sunnucks P and Taylor AC (2003) Demographic monitoring of an entire species (the northern hairy-nosed wombat, *Lasiorhinus krefftii*) by genetic analysis of non-invasively collected material. *Animal Conservation* **6**, 101–107.

DERM: <http://www.derm.qldgov.au/wildlife-ecosystems/wildlife/threatened_plants_and_animals/endangered/northern_hairynosed_wombat/index.html>.

Gordon G, Riney T, Toop J, Lawrie B and Godwin MD (1985) Observations on the Queensland Hairy-nosed Wombat *Lasiorhinus krefftii* (Owen). *Biological Conservation* **33**, 165–95.

Johnson CN (1991) Utilisation of habitat by the northern hairy-nosed wombat, *Lasiorhinus krefftii*. *Journal of Zoology* **225**, 495–507.

Taylor AC, Sherwin WB and Wayne RK (1994) Genetic variation of microsatellite loci in bottlenecked species: the northern hairy-nosed wombat *(Lasiorhinus krefftii)*. *Molecular Ecology* **3**, 277–290.

Author: Andrew Dinwoodie

BRIDLED NAILTAIL WALLABY

Onychogalea fraenata (Gould, 1841)

Photo: www.ataglance.com.au

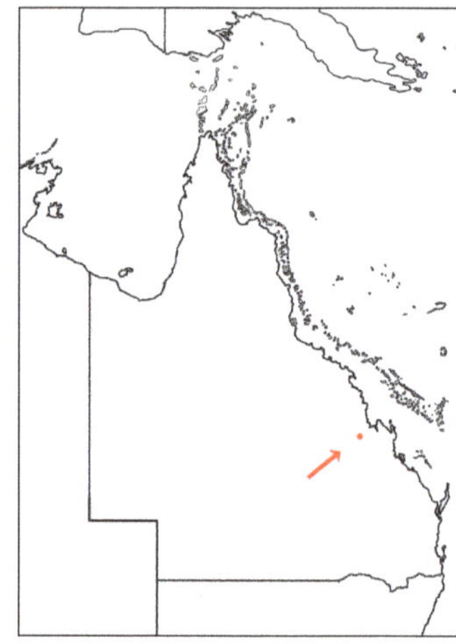

Family: Macropodidae

Other common name: Flashjack, Waistcoat Wallaby, Pademelon

Aboriginal name: Merrin

Conservation status: IUCN: Endangered; Aust.: Endangered; Qld: Endangered; NSW: Extinct in the Wild; Vic.: Extinct

Description: This is a small wallaby with a head and body length of 470–590 mm and a 470–540 mm tail with a black pointed 'finger' nail at the tip. It weighs 2–8 kg, with males being larger than females. The body fur is grey with some white and yellowish on the belly, legs and tail and a rufous wash to the neck. The 'bridle' consists of distinctive white stripes with dark borders between the ears, down across the shoulders and under the forearms.

Distribution: Before 1788, these wallabies occurred west of the Great Dividing Range from near Ravenswood in north Qld south to northern Vic. Today, the wild population is limited to a tiny area of Taunton NP and was recently estimated at around 70 individuals. They have been re-introduced in Avocet Nature Refuge west of Rockhampton, and in 1994 at Idalia NP south of Longreach, as insurance against the potential loss of the wild population at Taunton brought on by wildfire, habitat loss, drought and disease. The Idalia population was estimated at 63 to 109 individuals at different times in 2008 and at 96 individuals in 2009.

Habitat: This wallaby originally occupied acacia shrubland and grassy woodland in the semi-arid regions. Today the species inhabits remnant Brigalow *Acacia harpophylla* areas and large alluvial flats. The remaining population can be found on ecotones on relatively fertile soils, where complex Brigalow woodland meets open eucalypt woodland.

Ecology: This species is capable of breeding rapidly, providing seasons are good and predation is controlled. Gestation is 22 days, with joeys becoming independent at 120 days. Young females become mature at 136 days and males at 270 days. They are nocturnal, spending the day in shaded scrapes under low shrubs or fallen branches or in hollow logs, and graze and browse on non-woody broad leaf plants, chenopods and succulents, including pigweed, during the night. They tend to be solitary except when females are in oestrus and are followed by small groups of males or when small groups of animals feed together.

Reasons for decline and current threats: The Bridled Nailtail Wallaby disappeared soon after European settlement for three main reasons: loss of habitat, changes in the structure and floristics of remaining habitat, and predation by cats, dogs and foxes. Of these, the change in structure and floristics of habitats is the most pervasive and complex. These changes occurred due to cattle grazing and changes in fire regimes and patterns (often exclusion). The species was considered extinct by the late 1930s, but was rediscovered in 1973 in central Qld. Current threats include changes to its habitat and feral predators, such as cats and dogs, but the current role of foxes is not known. Buffel Grass *Cenchrus ciliaris* is causing structural and floristic change to the preferred habitat by altering fire-regimes and outcompeting forbs.

The loss of genetic diversity in the translocated populations is possibly a long-term threat. Finally, as with all species that occur as few populations in limited areas of occupancy, stochastic events such as disease outbreak, drought and wildfire could threaten their survival.

CURRENT RECOVERY EFFORTS

Recovery plan: Lundie-Jenkins G and Lowry J (2005) 'Recovery plan for the Bridled Nailtail Wallaby (*Onychogalea fraenata*) 2005–2009'. Report to the Department of Environment and Heritage (DEH), Canberra. Environmental Protection Agency/Queensland Parks and Wildlife Service, Brisbane.

Conservation actions:

- Australian Animals Care and Education (AACE's Project Kial) and Scotia Sanctuary (NSW) hold captive colonies and breed for release and research.
- Idalia NP serves as a release site for captively bred offspring and is continuously managed for feral animals, fire and weed.

FUTURE RECOVERY EFFORTS

What you can do:

- Join/donate/volunteer to Australian Animals Care and Education (AACE) Project Kial: A Bridled Nailtail Recovery Project in Central Qld and the Bridled Nailtail Wallaby Trust.
- Volunteer with Bush Heritage Australia.

Conservation needs:

- Locate and evaluate potential release sites where ongoing management is guaranteed.
- Carry out ongoing control of feral cats, rabbits, foxes and dingos.
- Eradicate Buffell Grass in Taunton Scientific NP.

Research directions: Carry out immunology research and further studies on the diseases and parasites affecting this animal.

Sources and further reading

Evans M and Gordon G (2008) The Bridled Nailtail Wallaby *Onychogalea fraenata*. In *The Mammals of Australia*. 3rd edn. (Eds S Van Dyck and R Strahan) pp. 355–357. Reed New Holland, Sydney.

Fisher D (1998) Endangered bridled nailtail wallaby: a strategy for survival. *Proceedings of the Royal Society of Queensland* **107**, 115–118.

Gordon G and Lawrie BC (1980) The rediscovery of the bridled nail-tailed wallaby, *Onychogalea fraenata*, (Gould) (Marsupialia: Macropodidae) in Queensland. *Australian Wildlife Research* **7**, 339–345.

Horsup A and Evans M (1992) Predation by feral cats, *Felis catus*, on an endangered marsupial, the bridled nailtail wallaby, *Onychogalea fraenata*. *Australian Mammalogy* **16**, 85–86.

Johnson P (2003) *Kangaroos of Queensland*. Queensland Museum, Brisbane.

Author: Lee K Curtis

BRUSH-TAILED ROCK-WALLABY

Petrogale penicillata (Gray, 1825)

Photo: www.ataglance.com.au

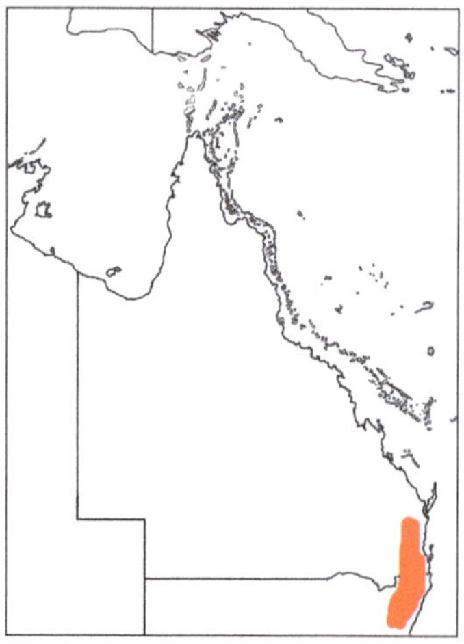

Family: Macropodidae

Conservation status: IUCN: Near Threatened; Aust.: Vulnerable; Qld: Vulnerable; NSW: Endangered; Vic.: Critically Endangered

Description: This medium-sized wallaby has a head and body length of 510–585 mm with a 500–700 mm dark brown or black cylindrical tail with a prominent brushed end. It weighs 4.7–10.9 kg, with males being larger than females. It is dark grey to dark brown with rufous on the rump and grey on neck and shoulders. It has a conspicuous white or cream marking on the chest and a cream to buff cheek stripe.

Distribution: The Brush-tailed Rock-wallaby is patchily distributed along the Great Dividing Range, from Yarraman, 200 km north of the Qld border, to the upper Snowy River in eastern Vic. Genetic analysis has identified distinct southern, central and northern populations that should be treated as separate units or species (evolutionary significant units). A Qld survey in the late 1990s found evidence of the Brush-tailed Rock-wallaby at 131 sites. Half were on private land and 48 (74%) of those were occupied during the survey. Fifty-four sites (41%) were in conservation reserves and most of these (46–85%) were occupied. None of the 12 (9%) sites in state forest were shown to be occupied.

Habitat: The Brush-tailed Rock-wallaby inhabits rock faces with large tumbled boulders, ledges and caves in a range of vegetation types from rainforest, through tall eucalypt to open woodland. Populations inhabiting cliffs or piles that are not structurally complex tend to die out, presumably because they lack refuges from the introduced fox.

Ecology: Similar to other rock-wallabies, the Brush-tailed Rock-wallaby is gregarious and shelters by day in caves and deep fissures. It is active by night, but does bask in the sun on rocky platforms in winter. It eats mostly grasses, forbs and browse, but also takes flowers, fruits and seeds. Females mature at 18 months and, after 31 days gestation, young clamber into the pouch where they remain for 204 days. This rock-wallaby occupies a range of 2–30 ha, the size probably relating to habitat quality. In the north, a narrow hybrid zone sees interbreeding with the neighbouring Herbert's Rock-wallaby.

Reasons for decline and current threats: In the late 1800s and early 1900s, the species was decimated in the south of its range by hunting for skins and for bounties because it was deemed an agricultural pest. Over 500 000 bounties were paid and over 90 000 skins were marketed through a single company in just one year. Since then, habitat degradation, predation by the fox, and competition with goats have continued the decline and these remain significant threats. Because habitat is naturally dispersed in small patches, natural recovery may not be possible.

CURRENT RECOVERY EFFORTS

Recovery plan: State of NSW and Department of Environment and Climate Change NSW (2008) 'Recovery plan for the brush-tailed rock-wallaby (*Petrogale penicillata*)'.

Other plans: Environment Australia (1999) 'Threat abatement plan for competition and land degradation by feral goats'. Biodiversity Group, Environment Australia.

Environment Australia (1999) 'Threat abatement plan for competition and land degradation by feral rabbits'. Biodiversity Group, Environment Australia.

Maxwell S, Burbidge AA and Morris K (Eds) (1996) 'The 1996 action plan for Australian marsupials and monotremes'. Wildlife Australia, Canberra.

Conservation actions:

- Extensive surveys have been carried out throughout the rock-wallaby's range and its current distribution and the location of most extant populations is well known.
- Multiple captive populations have been established from several source populations. These have been used to develop husbandry techniques and rapid reproduction through cross fostering using the Tammar Wallaby *Macropus eugenii* and the Yellow-footed Rock-wallaby *Petrogale xanthopus*.

- Fox control programs have been established at a number of colony sites in NSW. The success of these is still uncertain, though the signs are promising.
- Goat control programs have been established in some areas specifically to benefit the Brush-tailed Rock-wallaby. Examples include Warrumbungles NP and Oxley Wild Rivers NP.
- A significant national recovery program is established for this species. Both NSW and Vic. have their own recovery teams and actions coordinating through the national team.

FUTURE RECOVERY EFFORTS

What you can do:

- Protect native vegetation on private land in and near colonies.
- Report sightings to the national recovery team via website links <http://btrw.org/recovery/vic/>.
- Visit the national recovery team website to join or establish a support group <http://btrw.org/recovery/vic/>.
- Report harmful activities at colonies to DSEWPC - Compliance and Enforcement Branch (freecall: 1800 110 395).

Conservation needs:

- Establish a Qld branch of the recovery team.
- Protect native vegetation on private land in and near colonies.
- Control foxes, goats, rabbits and domestic animals at, and near, colonies.

Research directions:

- Review and improve the current threat abatement measures.
- Establish management plans for individual colonies.
- Enhance our understanding of the rock-wallaby's ecology.
- Better define the boundaries and degree of genetic separation between the three distinct groups.
- Develop a standardised survey technique.
- Evaluate sites and establish protocols and monitoring for re-introduction programs.

Sources and further reading

Eldridge MDB and Close RL (2008) Allied rock wallaby *Petrogale assimilis*. In *The Mammals of Australia*. 3rd edn. (Eds S Van Dyck and R Strahan) pp. 363–364. Reed New Holland, Sydney.

Hazlitt SL, Eldridge MDB and Goldizen AW (2004) Fine-scale spatial genetic correlation analyses reveal strong female philopatry within a brush-tailed rock-wallaby colony in southeast Queensland. *Molecular Ecology* **13**, 3621–3632.

Johnson P (2003) *Kangaroos of Queensland*. Queensland Museum, Brisbane.

Laws RJ and Goldizen AW (2003) Nocturnal home-ranges and social interactions of the brush-tailed rock-wallaby *Petrogale penicillata* at Hurdle Creek, Queensland. *Australian Mammalogy* **25**, 169–176.

Murray JV, Low Choy S, McAlpine CA, Possingham HP and Goldizena AW (2008) The importance of ecological scale for wildlife conservation in naturally fragmented environments: a case study of the brush-tailed rock-wallaby (*Petrogale penicillata*). *Biological Conservation* **141**, 7–22.

Author: Geoff Lundie-Jenkins

PROSERPINE ROCK-WALLABY

Petrogale persephone Maynes, 1982

Photo: Barry Nolan

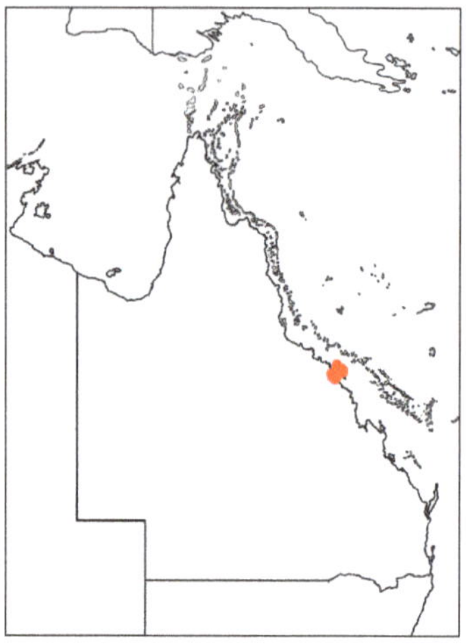

Family: Macropodidae

Other common name: Tree-kangaroo

Conservation status: IUCN: Endangered; Aust.: Endangered; Qld: Endangered

Description: This large rock-wallaby has a head and body length of 490–630 mm and a long, often white-tipped tail of 515–710 mm. It weighs 3.5–10 kg, with males being larger than females. It has prominent ears, a long narrow muzzle and is predominately grey with a mauve tint, darker points and a pale underside. It has a dark patch behind the armpit and light reddish brown on its legs, forearms and at the base of its tail.

Distribution: This species is restricted to 14 500 ha of patchy habitat in the Whitsunday Shire. Populations are known from only 24 sites in four separate areas: Gloucester Island, Clarke Range, Conway Range and Dryander. In addition, it was recently introduced to Hayman Island (1998–2002). Its distribution is within that of the Unadorned Rock-wallaby *Petrogale inornata*.

Habitat: On the mainland, it occupies rocky outcrops, boulder piles, rocky slopes and gullies in semi-deciduous vine-thickets and occasionally the outer margins of rainforest. It also includes rocky *Acacia* woodland on Gloucester Island.

Ecology: This rock-wallaby shelters within rock piles, caves, crevices or dense vegetation during the day, emerging to feed on grasses, forbs, tree and vine leaves, fruits, flowers and seedlings at night. It often moves from vine-thickets into adjacent grassy woodlands or residential gardens to feed. It regularly sunbathes in the early morning. A home range averages 20 ha, but is larger for males than females. Young are born throughout the year and mature at 20–24 months after a 33–34 day gestation, 7 months in the pouch and 4–5 months at foot before weaning. Oestrus and mating follow within 36 hours of a birth.

Reasons for decline and current threats: The population has declined as a result of clearing for grazing, farming, urban and tourist development and deaths due to road traffic, as well as domestic and feral dogs and cats. Its limited and fragmented distribution makes it vulnerable to further habitat loss or degradation. The same threats that caused the initial decline continue and have been extended to include exotic diseases, such as toxoplasmosis and hydatidosis (which are spread by domestic and feral cats/dogs), as well as the potential consumption of toxic introduced plants both in gardens and environmental weeds (e.g. Pink Periwinkle *Catharanthus roseus* and Rubber Vine *Cryptostegia grandiflora*). In addition, natural dispersal resulting in recolonisations and gene flow among populations is retarded by fragmentation of the landscape, resulting in higher probabilities of multiple and sustained local extinctions.

CURRENT RECOVERY EFFORTS

Recovery plan: Nolan BJ and Johnson PM (2001) 'Recovery plan for the Proserpine rock-wallaby *Petrogale persephone* 2000–2004'. Queensland Parks and Wildlife Service, Brisbane, about to be superseded by Department of Environment and Resource Management (2008) 'National recovery plan for the Proserpine rock-wallaby *Petrogale persephone*'. Report to Department of the Environment, Water, Heritage and the Arts, Canberra. Department of Environment and Resource Management, Brisbane.

Other plans: Maxwell S, Burbidge AA and Morris K (Eds) (1996) 'The 1996 action plan for Australian marsupials and monotremes'. Wildlife Australia, Canberra.

Conservation actions:

- Known populations are monitored annually.
- An area of 1100 ha has been added to Dryander NP (1996) and 650 ha as Environmental Reserve in the Clarke Range.
- The are seven Nature Refuge agreements (plus another being prepared) protecting Proserpine Rock-wallaby populations on private land. Additional areas of habitat have been protected through Land for Wildlife agreements.
- Special roadside reflectors and removal of palatable grass from verges have reduced road-kills in identified hot-spots.

- Public education and community awareness has increased to assist species recovery and reduce the incidence of toxic introduced plants in gardens in Proserpine Rock-wallaby areas.
- Research has examined: habitat and extant population distributions (JCU); reproduction and ageing of pouch young (DERM); parasites (UQ, MU); disease (JCU); genetics (AM); ecology and diet of a wild population (JCU); habitat preference throughout distribution (JCU).
- Numerous individuals and organisations have become involved in their recovery, including Hayman Island Resort and Whitsunday Regional Council. Private individuals/companies manage land containing over 10 wild populations.
- Between 1998 and 2002, 27 captive bred animals were released on Hayman Island with the population subsequently increasing and spreading to most patches of suitable habitat on the island.
- A captive colony at David Fleay Wildlife Park was wound up in 2008.

FUTURE RECOVERY EFFORTS

What you can do: Drive carefully in areas with Proserpine Rock-wallaby populations and be an environmentally responsible pet owner and gardener.

Conservation needs:

- Increase the area of Proserpine Rock-wallaby habitat protected through acquisitions, declaration of Wildlife Refuges, voluntary conservation agreements, etc.
- Lobby governments to continue threat abatement including tight regulation of urban/tourist development in and near Proserpine Rock-wallaby habitat, as well as greater control of domestic and feral dogs/cats near populations.
- Continue the implementation of strategies to reduce road-kills.
- Eradicate Pink Periwinkle and Rubber Vine on Gloucester Island.
- Further educate land managers and residents in Whitsunday Shire about species presence, habitat requirements and ways to minimise human impacts.
- Lobby governments for increased resources for population monitoring, and research into the biology and ecology of threatened species.

Research directions:

- Survey potential habitat to identify and map all the extant populations.
- Investigate the genetic differentiation among the four major clusters of habitat patches.
- Investigate ways to ameliorate the impact of habitat fragmentation.
- Conduct detailed studies of habitat use around refuge sites.
- Determine the appropriate size of buffer zones around developments.
- Test the effectiveness of harm minimisation strategies associated with urban development.
- Test the effectiveness of strategies to reduce road-kills.
- Quantify the impact of domestic/feral cats/dogs (predation and disease).
- Compare the ecology of the Proserpine and the Unadorned Rock-wallaby to determine the potential for competition or exclusion of one by the other.

Sources and further reading

Eldridge MDB (1997) Rock-wallaby conservation: essential data and management priorities. Proceedings of the 1994 National Rock-wallaby Symposium Workshops. *Australian Mammalogy* **19**, 325–330.

Eldridge MDB, Johnston PG and Close RL (1992) Chromosomal rearrangements in rock wallabies, *Petrogale* (Marsupialia: Macropodidae). VI. Determination of the plesiomorphic karyotype: G- banding comparison of *Thylogale* with *Petrogale persephone*, *P. xanthopus*, and *P. l. lateralis*. *Cytogenetics and Cell Genetics* **61**, 29–33.

Johnson PM (2003) *Kangaroos of Queensland*. Queensland Museum, Brisbane.

Johnson PM, Nolan BJ and Schaper DN (2003) Introduction of the Proserpine rock-wallaby *Petrogale persephone* from mainland Queensland to nearby Hayman Island. *Australian Mammalogy* **25**, 61–71.

Author: Mark Eldridge

PURPLE-NECKED ROCK-WALLABY

Petrogale purpureicollis Le Souef, 1924

Photo: www.ataglance.com.au

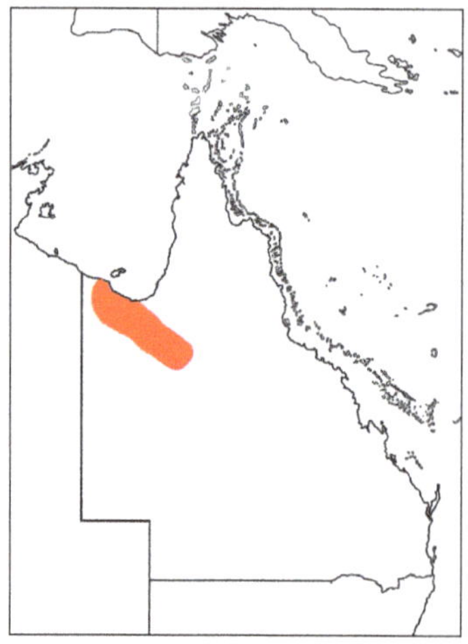

Family: Macropodidae

Conservation status: IUCN: Least Concern; Qld: Vulnerable

Description: This medium-sized rock-wallaby has a head and body length of 390–610 mm and a tail length of 445–580 mm. It weighs 3-7 kg, with males being larger than females. It is generally light grizzled brown (greyer in winter), with darker points and a paler underside and has distinct cheek- and mid-dorsal head-stripes. It has characteristic purple pigmentation over the face, head, neck and shoulders that varies in intensity, from a faint pink wash to bright red/purple. The pigmentation also varies in coverage, from being confined to the head to extending well down the shoulders. The pigment is removable and water soluble.

Distribution: The populations of this rock-wallaby are scattered from near Winton, north-west to Mount Isa, Cloncurry and beyond Lawn Hill almost to the NT border. It is common in the centre of its distribution (Mount Isa, Cloncurry and Dajarra) but rare to the north-west and has declined in the south-east, with some population extinctions (e.g. Manhole waterhole and Bladensburg NP ~1975). No populations remain in the ~250 km between Winton and McKinley.

Habitat: This rock-wallaby occurs on limestone, sandstone and quartzite outcrops, boulder piles, cliffs and gorges, as well as rocky slopes and gullies within dry eucalypt and acacia woodlands with Spinifex in the understorey.

Ecology: This species shelters in cool rocky refuges during the day, emerging around dusk or later to feed on grass and browse in the surrounding woodland. In cool weather, it will sunbathe in the morning and late afternoon. Being social, it can be found in colonies of over 20 individuals, which breed throughout the year. It matures at 18–22 months (female–male), after a 33–35 day gestation, 6 months in the pouch and 3–6 months at foot before weaning.

Reasons for decline and current threats: The cause of the decline is unclear, but competition with, and habitat degradation by, feral goats appears likely in the south-east of the range around Winton. Other current threats may include habitat degradation by cattle or feral goats and horses, changes in fire regimes and predation of young by cats.

CURRENT RECOVERY EFFORTS

Recovery plan: None

Conservation actions:

- Surveys conducted by DERM and AM from 1991–2000 identified new populations and expanded the known distribution.
- AM research used genetics to clarify taxonomy and investigate relationships with other species of rock-wallaby.
- DERM, AM, UQ and UM have conducted other research into its ecology and threat management, reproductive biology and parasites.
- A captive colony established at DERM's Northern Regional Centre, Pallarenda in 1993, is now disbanded.

FUTURE RECOVERY EFFORTS

Conservation needs:

- Increase and sustain the control of feral herbivores around known populations: goats in the south-east and horses in the north-west.

- Educate land managers throughout distribution about the presence of the species, its habitat requirements and the effects of feral herbivores.
- Lobby governments for increased resources for biodiversity surveys to document species distribution, population monitoring, feral animal control and research into biology and ecology of threatened species.

Research directions:

- Conduct targeted surveys to locate all extant populations in the known range, especially in the south-east where species has declined, and to clarify limits of distribution (e.g. does species occur in the NT?).
- Monitor populations throughout its range to detect decline or recovery.
- Investigate the impact of potential threatening processes, especially cattle grazing, feral herbivores, altered fire regimes and feral predators.
- Gain a better understanding of the basic biology and ecology of the species to better inform management.

Sources and further reading

Eldridge MDB (1997) Rock-wallaby conservation: essential data and management priorities. Proceedings of the 1994 National Rock-wallaby Symposium Workshops. *Australian Mammalogy* **19**, 325–330.

Eldridge MDB and Close RL (1997) Chromosomes and evolution in rock-wallabies, *Petrogale* (Marsupialia: Macropodidae). *Australian Mammalogy* **19**, 123–135.

Eldridge MDB, Wilson AC, Metcalfe CJ, Dollin AE, Bell JN, Johnson PM, Johnston PG and Close RL (2001) Taxonomy of rock-wallabies, *Petrogale* (Marsupialia: Macropodidae). III. Molecular data confirms the species status of the purple-necked rock-wallaby *Petrogale purpureicollis* Le Souef 1924. *Australian Journal of Zoology* **49**, 323–343.

Johnson PM and Delean S (2002) Reproduction in the purple-necked rock-wallaby, *Petrogale purpureicollis* Le Souef (Marsupialia: Macropodidae) in captivity, with age estimation and development of the pouch young. *Wildlife Research* **29**, 463–468.

Sharman GB, Close RL and Maynes GM (1990) Chromosome evolution, phylogeny and speciation of rock wallabies (*Petrogale*: Macropodidae). *Australian Journal of Zoology* **37**, 351–363.

Author: Mark Eldridge

SHARMAN'S ROCK-WALLABY

Petrogale sharmani Eldridge & Close, 1992

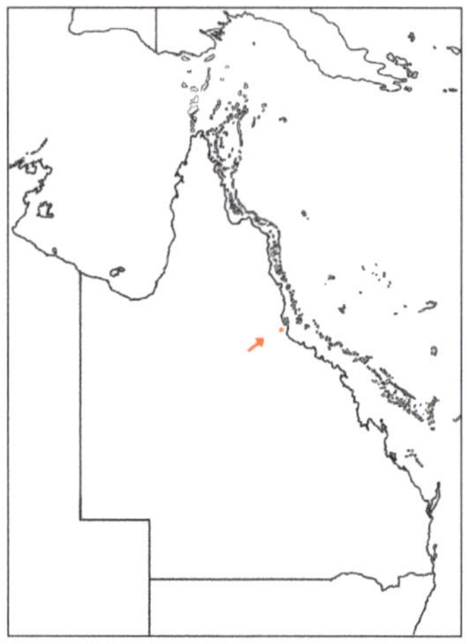

Family: Macropodidae

Other common name: Mount Claro Rock-wallaby

Conservation status: IUCN: Near Threatened; Qld: Endangered

Description: This medium-sized rock-wallaby has a head and body length of 455–530 mm and a long, dark-tipped, slightly brushy tail 435–532 mm long. It weighs 3–5 kg, with males being only slightly larger than females. It is grey-brown with paler underparts and occasionally has pale cheek-stripes or a pale mid-dorsal head-stripe. It appears more grey after the autumn moult, then gradually becomes more brown as the year progresses.

Distribution: A highly localised species, it occurs within 200 000 ha of the Seaview and Coane Ranges west of Ingham. Only 10 scattered populations are currently known, which probably number less than 800 individuals in total. The closely related Mareeba Rock-wallaby *Petrogale mareeba* occurs to the north and the Allied Rock-wallaby *Petrogale assimilis* to the west and south.

Habitat: This rock-wallaby occurs in or near boulder piles, rocky slopes and gullies, rocky outcrops, cliffs and gorges within tropical woodland or open forest with a grassy understorey.

Ecology: It shelters in rocky refuges or dense vegetation during the day, emerging around dusk to feed in the surrounding woodland. It is a social species that can be found in colonies of over 40 individuals that frequently bask in the sun during cooler weather, especially in the early morning and late afternoon. It appears to breed throughout the year. Other aspects of its ecology are unknown, but are likely to be similar to other north-east Qld rock-wallabies.

Reasons for decline and current threats: This species is not known to have declined, but its limited distribution and small number of populations makes this species so vulnerable to any perturbation that it qualifies for Endangered status. Possible threats could include land clearing for cattle grazing, changes in fire regimes, habitat degradation by cattle or goats, predation by cats and competition with other more common rock-wallaby species should they invade its range.

CURRENT RECOVERY EFFORTS

Recovery plan: None

Other plans: Maxwell S, Burbidge AA and Morris K (Eds) (1996) 'The 1996 action plan for Australian marsupials and monotremes'. Wildlife Australia, Canberra.

Conservation actions:

- The Australian Wildlife Conservancy (AWC) acquired 'Taravale' and 'Mt Zero' properties in 2002 to be managed for wildlife conservation. Surveys of these properties by AWC, DERM and AM staff discovered three new populations of Sharman's Rock-wallaby and confirmed the continued existence of a fourth population first identified in 1981.
- DERM manages national parks containing wild populations.
- The Australian Museum provides research and expertise on taxonomy, distribution and genetics.

FUTURE RECOVERY EFFORTS

What you can do:

- Donate or volunteer to assist with the management of AWC's Mount Zero-Taravale Wildlife Sanctuary.
- Help raise species awareness with landholders and managers in the area.

Conservation needs: Lobby state and federal governments for increased resources for biodiversity surveys to document species distribution, research into biology and ecology of threatened species and population monitoring.

Research directions:

- Conduct surveys to locate all extant populations and clarify the extent of the rock-wallaby's distribution, especially to the west and north of current known limits.
- Conduct surveys to determine whether or not this species is present on Hinchinbrook Island. Monitor the known colonies, given the tiny total population of the species.
- Research its biology and ecology and the potential impacts of possible threatening processes such as cattle grazing, altered fire regimes and feral predators.

Sources and further reading

Eldridge MDB and Close RL (1992) Taxonomy of rock wallabies, *Petrogale* (Marsupialia: Macropodidae). I. A revision of the eastern *Petrogale* with the description of three new species. *Australian Journal of Zoology* **40**, 605–625.

Eldridge MDB and Close RL (2008) Sharman's rock-wallaby *Petrogale sharmani*. In *The Mammals of Australia*. 3rd edn. (Eds S Van Dyck and R Strahan) pp. 391-392. Reed New Holland, Sydney.

Eldridge MDB, Johnson PM, Hensler P, Holden JK and Close RL (2008) The distribution of three parapatric, cryptic species of rock-wallaby (*Petrogale*) in north-east Queensland: *P. assimilis, P. mareeba* and *P. sharmani*. *Australian Mammalogy* **30**, 37–42.

Sharman GB, Close RL and Maynes GM (1990) Chromosome evolution, phylogeny and speciation of rock wallabies (*Petrogale*: Macropodidae). *Australian Journal of Zoology* **37**, 351–363.

Author: Mark Eldridge

NORTHERN BETTONG

Bettongia tropica (Wakefield, 1967)

Photo: Andrew J Dennis

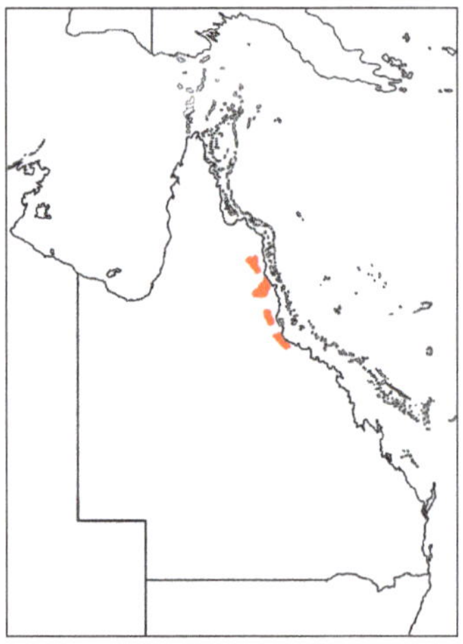

Family: Potoroidae

Other common names: Brush-tailed Bettong, Brush-tailed Rat-kangaroo, Rat-kangaroo

Conservation status: IUCN: Endangered; Aust.: Endangered; Qld: Endangered

Description: This small potoroid has a head and body length of 300–430 mm and a 270–350 mm tail with a short black crest on the last third. It weighs 900–1400 g. Its fur is grizzled grey above and pale grey to white below. The short fore-limbs have hands that sport long nails for digging.

Distribution: The Northern Bettong recently occurred at four locations on the western edge of the Wet Tropics of Qld and historically in the Ravenshoe area of the Wet Tropics (until 1920s) and the Dawson Valley near Rockhampton (1884). In the Wet Tropics, it occurs in small pockets on the Mount Windsor Tablelands in the north (last seen in 2003), the western edge of the Mount Carbine Tableland (~1 km diameter area), the Lamb Range (largest population with southern and northern genotypes) and the Coane Range in the south (last confirmed in 2003, but possible sightings since).

Habitat: The Northern Bettong occurs in eucalypt forests, generally with grassy understoreys, along the western edge of rainforests. The habitats vary from tall eucalypt forests to medium height and drier forests and change in floristic and structural composition with decreasing rainfall and increasing seasonality with distance from rainforest. The feature limiting their distribution is probably the presence of truffles (hypogeal sporocarps of ectomycorrhizal fungi) in sufficient diversity and abundance and with low seasonality.

Ecology: This nocturnal and solitary animal relies on truffles for food (~36 spp.), although at different times other items such as roots, tubers, grass stems (e.g. *Alloteropsis semialata*), herbs, invertebrates and seeds may contribute 30–70% of the diet. It breeds continuously when conditions allow and carries its single young in the pouch for 102–112 days, with weaning occurring 60–80 days later. It rests by day in a hidden, grassy nest and by night ranges over 50–75 ha, with males ranging further than females, occasionally up to 120 ha. This bettong tolerates fire well and maintains its pre-fire range, but switches to sheltering by day in what cover it can find.

Reasons for decline and current threats: The Northern Bettong has disappeared from the Dawson Valley and the greater Ravenshoe area. Up until 2003, it was restricted to four, very small, disjunct populations. In 2010, it is confirmed to still occur at only two of those, the largest being an area of 75 km^2, the other being >1 km^2. Its listing is based on both the declines and the tiny area of occupancy and the small number and size of populations. The largest population to disappear was around Ravenshoe. The reasons for the declines are thought to relate to changes in habitat structure and suitability for both the bettong and the fungi it consumes. Changes were probably influenced to varying degrees by grazing and logging, weed invasion, changed fire regimes and the invasion of rainforest into tall eucalypt forest. The greatest future threat to the Northern Bettong is the potential expansion of fox into its range. Pigs have been identified as a possible competitor or as something that may detrimentally restructure habitat, but evidence of their impact is still scant and unconvincing. Future climate change may benefit this species by expanding the area in which current habitats occur.

CURRENT RECOVERY EFFORTS

Recovery plan: Dennis AJ (2001) 'Recovery plan for the northern bettong, *Bettongia tropica* 2000–2004'. Report to Environment Australia, Canberra. Queensland Parks and Wildlife Service, Brisbane.

Other plans: Maxwell S, Burbidge AA and Morris K (Eds) (1996) 'The 1996 action plan for Australian marsupials and monotremes'. Wildlife Australia, Canberra.

Conservation actions:

- Populations have been surveyed and monitored by WWF, DERM and CSIRO, and by conservation organisations (e.g. Kuranda Envirocare and AWC).
- Significant research has been undertaken examining its biogeography and habitat use, general ecology and reproduction, captive husbandry, fire ecology,

population genetics and mating systems, diet and resource distribution and abundance, and modelling the impacts of climate change (JCU, CSIRO, DERM, UQ), all of which have informed its management and conservation needs.

- Most of its habitat has now been incorporated into the protected area estate, both government and private (AWC), and is being managed for the bettong's conservation by de-stocking, ceasing logging, managing fire and weeds and monitoring the vegetation and the bettong.
- Surveys of fox distribution and abundance and the development of a fox control plan have been initiated.
- Conservation groups and the public have been engaged and provide information in bettong areas.
- A trial re-introduction was attempted near Ravenshoe, but failed, and the captive colony from which the re-introduction was initiated no longer exists.

FUTURE RECOVERY EFFORTS

What you can do: Join/donate to Kuranda Envirocare, the Tree-kangaroo and Mammal Group and Australian Wildlife Conservancy.

Conservation needs:

- Protect and manage the bettong's habitat.
- Identify its ecological requirements. These are likely to focus on the importance of truffles, and the role of fire, weed and exotic animal management to maintain optimal habitat conditions.
- Impose buffer areas between known Northern Bettong habitat and future developments, monitoring for, and finalising control strategies for, the fox.
- Identify, restore and prepare suitable sites for re-introduction programs.

Research directions:

- Conduct research to clarify the responses of the suite of truffles, grasses and eucalypts in core habitat to fire regimes and climate variability.
- Research habitat restoration based on the effects of cattle and weed and rainforest invasion.
- Develop protocols and infrastructure to safely manage fire during dry periods.

Sources and further reading

Johnson CN and McIllwee AP (1997) Ecology of the Northern Bettong *Bettongia tropica*, a tropical mycophagist. *Wildlife Research* **24**, 549–559.

Johnson PM (2003) *Kangaroos of Queensland*. Queensland Museum, Brisbane.

Pope LC, Estoup A and Moritz C (2000) Phylogeography and population structure of an ecotonal marsupial, *Bettongia tropica*, determined using mtDNA and microsatellites. *Molecular Ecology* **9**, 2041–2053.

Vernes K and Pope LC (2001) Stability of nest range, home range and movement of the northern bettong (*Bettongia tropica*) following moderate-intensity fire in a tropical woodland, north-eastern Queensland. *Wildlife Research* **28**, 141–150.

Author: Andrew J Dennis

LONG-NOSED POTOROO (SOUTH-EAST MAINLAND SUBSPECIES)

Potorous tridactylus tridactylus (Kerr, 1792)

Photo: Lochman Transparencies

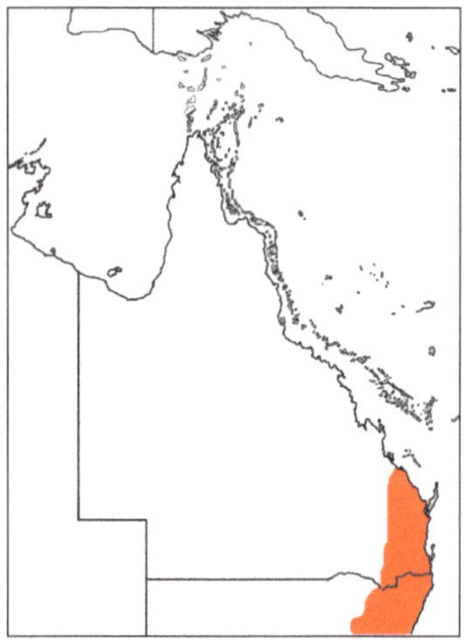

Family: Potoroidae

Other common names: Long-nosed Rat-kangaroo, Common Rat-kangaroo, Three-toed Potoroo

Aboriginal name: Potoroo

Conservation status: IUCN: species (*P. tridactylus*) is Least Concern; Aust.: Vulnerable; Qld: Vulnerable; NSW: Vulnerable; Vic.: Endangered; SA: Endangered

Description: This small potoroid has a head and body length of 340–380 mm and a 150–240 mm, lightly haired tail, which is commonly white-tipped in the south of its range. It weighs 740–1640 g, with males being slightly larger than females. It is brown to grey above, sometimes with a reddish-brown or sandy tint, and is light grey to cream below. It has coarse outer hairs with a grey underfur layer. The snout is long and pointed, with bare skin extending onto the muzzle from the nose. The snout on individuals from the north of the range is relatively shorter than in the south. The hind foot is shorter than the head, a feature distinguishing them from the Long-footed Potoroo *Potorous longipes*.

Distribution: The subspecies *P. t. tridactylus* occurs along the eastern seaboard from south-east Qld to south-east SA, while *P. t. apicalis* is in northern and eastern Tas. and some Bass Strait islands and is not considered threatened. The subspecies is generally found within 50 km of the coast, between sea level and 800 m (or sometimes higher in the north), in areas with rainfall exceeding 760 mm. Within this range, its distribution is patchy.

Habitat: The species uses habitats with a dense understorey, including dense coastal scrubs, wet heaths, woodlands with a heath understorey and inland moist forests and rainforest, but not treeless heath communities in Qld. It uses dense, structurally complex patches for shelter and predator avoidance and relatively open, floristically diverse patches for foraging.

Ecology: Long-nosed Potoroos breed throughout the year. After a gestation period of 38 days, a single young is born and has a pouch life of 17–18 weeks. Females mature at one year of age and can produce 2.7 young per year. Sub-adults of both sexes appear to disperse from natal home ranges. The species is solitary, with male and female home ranges overlapping considerably. Home range size varies from 2–5 ha, often influenced by habitat quality. It can be active day and night, but shelters in a shallow depression under dense vegetation when not active. It eats mostly fungi, primarily the underground fruiting bodies of mycorrhizal fungi that associate with myrtaceous plants, particularly *Eucalyptus* trees. Fungus makes up 30–90% of its diet, depending on the location and season; this is supplemented with invertebrates, fruits, flowers, seeds and plant tissue.

Reasons for decline and current threats: The introduction of foxes (and possibly cats and dogs), habitat loss and fragmentation caused by clearing for agriculture and human settlement, and habitat degradation (with the loss of dense understorey cover through changed fire regimes and overgrazing by stock) are the four key drivers of its decline. Each is still a major threat today.

High levels of fox predation were observed in a recent study, despite the presence of dense vegetative cover. Habitat degradation and the frequent use of prescribed fire are considered significant threats to the species and should be avoided. Fire management should aim to perpetuate a mosaic of habitat types with variable floristic and structural diversity, in particular ensuring some patches are left unburnt for 20 years and longer. To avoid further habitat degradation, efforts need to be made to minimise damage to lower stratum vegetation and the litter layer in grazed areas and during forestry operations. Excluding forestry activities and grazing from refuge areas used in times of stress (e.g. fires and drought) would benefit the long-term persistence of the species.

CURRENT RECOVERY EFFORTS

Recovery plan: Martin D and Temple-Smith P (2010) 'National recovery plan for the Long-nosed Potoroo *Potorous tridactylus tridactylus*. Department of Sustainability and Environment, East Melbourne. (Submitted for comment).

Other plans: Department of the Environment, Water, Heritage and the Arts (2008) 'Threat abatement plan for predation by the European Red Fox'. Department of the Environment, Water, Heritage and the Arts, Canberra.

NSW National Parks and Wildlife Service (2010) 'Threat abatement plan for predation by the Red Fox (*Vulpes vulpes*)'. Department of Climate Change and Water, Hurstville.

Conservation actions:

- A number of research projects are currently underway on the Long-nosed Potoroo, including examining its genetic variation across its range, the effects of prescribed burning, phylogeography and population dynamics, the impact of cat toxicants and baiting methods on the species and its distribution and abundance across some of its range.
- In addition, small numbers of captive individual are held in a number of zoos, sanctuaries and research facilities, which could form a captive breeding stock if any should become necessary.

FUTURE RECOVERY EFFORTS

What you can do:

- Implement fox, feral dog and cat control on your property.
- Keep pet cats and dogs out of key habitat.
- Control weeds, particularly those affecting the understorey layer.
- Exclude livestock from key habitat.
- Protect suitable habitat from inappropriate fire-regimes.
- Revegetate areas with suitable habitat.
- Consider the formal protection of known habitat under a voluntary conservation agreement.

Conservation needs:

- Implement appropriate fire regimes, excluding fire from moist habitats such as gullies, wet forests and rainforests and managing fire in other habitat such that there is a mosaic burn, with sufficient long-unburnt areas remaining for habitat, including fungal food availability.
- Manage forestry operations to minimise their impacts.
- Protect and restore habitat, targeting links between existing populations.
- Control weeds in and adjacent to habitat and conduct landscape-scale predator control.
- Ensure that landholders and managers are aware of the species and its requirements for survival.
- Develop policies and processes to avoid further clearing and fragmentation of habitat.

Research directions:

- Research appropriate fire regimes.
- Study the impact of predation.
- Determine any additional actions required during times of stress (e.g. fire and drought).
- Study the genetics of the species across its range.
- Survey and monitor the distribution and abundance of the species across its range.

Sources and further reading

Bennett AF (1987) Conservation of mammals within a fragmented forest environment: the contributions of insular biogeography and autecology. In *Nature Conservation: The Role of Remnants of Native Vegetation* (Eds DA Saunders, GW Arnold, AA Burbidge and AJ Hopkins) pp. 41–52. Surrey Beatty & Sons, Sydney in association with CSIRO and CALM.

Claridge A, Seebeck J and Rose R (2007) *Bettongs, Potoroos and the Musky Rat-kangaroo.* CSIRO Publishing, Melbourne.

Johnson PM (2003) *Kangaroos of Queensland.* Queensland Museum, Brisbane.

Norton MA (2009) Habitat associations of the Long-nosed Potoroo (*Potorous tridactylus*) at multiple spatial scales. MSc thesis. University of Wollongong, Wollongong.

Seebeck JH, Bennett AF and Scotts DJ (1989) Kangaroos, wallabies and rat-kangaroos. In *Ecology of the Potoroidae – A Review.* (Eds G Grigg, P Jarman and I Hume) pp. 67–88. Surrey Beatty & Sons, Sydney.

Author: Melinda Norton

YELLOW-BELLIED GLIDER

Petaurus australis Shaw, 1791 (unnamed subspecies)

Photo: Andrew J Dennis

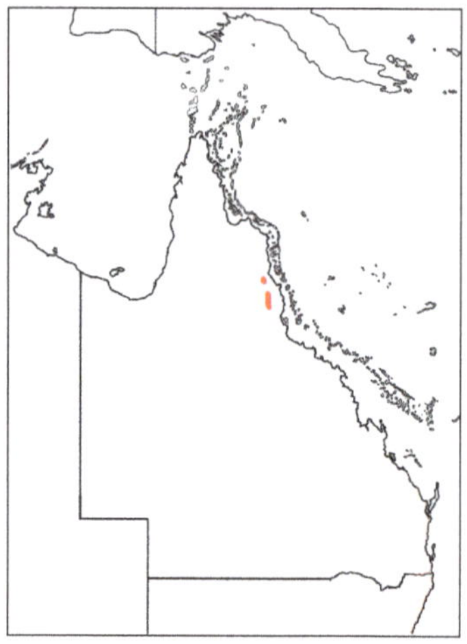

Family: Petauridae

Other common names: Fluffy Glider

Conservation status: IUCN: Least Concern; Aust.: Vulnerable (unnamed subspecies in Wet Tropics); Qld: Vulnerable (unnamed subspecies in Wet Tropics)

Description: This mid-sized gliding possum has a head and body length of 240–310 mm, with a 380–470 mm long fluffy tail. Males weigh 470–725 g and females 435–660 g. Although males are heavier than females, their head and body length is only marginally longer and females have a longer tail than males. The fur is grey to brown with a black stripe along the back and down each thigh. The face, arms legs, tail and the edge of the gliding membrane are all dark to black and the nose pink. This glider is whitish to yellowish below and has large relatively bare ears.

Distribution: The Yellow-bellied Glider listed here is considered an unnamed subspecies, which is restricted to three populations along the western edge of the Great Dividing Range in the Wet tropics of Queensland, from Kirrima in the south to the Windsor Tablelands in the north. These three populations are further subdivided into over a dozen subpopulations. The species occurs to the Vic./SA border and is also listed as Vulnerable in NSW.

Habitat: The Yellow-bellied Glider in the Wet Tropics occurs in tall eucalypt forest above 600 m altitude along the western margin of the Wet Tropics rainforests. From Kirrima to Atherton, this occurs as a broken narrow strip, often less than 1km wide, and is steadily being invaded by rainforest. The northern populations occupy patches of the same habitat up to 5 km wide and generally with less rainforest invasion. Their habitat always includes Flooded Gum *Eucalyptus grandis* and Red Mahogany *E. resinifera* and generally Turpentine *Syncarpiaglomulifera* and Coast Banksia *Banksia integrifolia.*

Ecology: This highly vocal glider lives in family groups of one adult male, one or more females and their offspring, resulting in groups of up to six individuals. Each group defends a home range of 25–120 ha, over which they can travel rapidly by gliding 40–100 m or more from tree to tree. By day, groups occupy hollows in *E. grandis*, rarely using hollows in any other species. In the Wet Tropics, they feed primarily on sap harvested from incisions made into the trunks or major branches of *E. resinifera.* Incisions can be V shaped, but are more often heart-shaped, oval or circular. They also consume nectar and pollen from a range of plants (with *Banksia* being particularly important), invertebrates, the occasional small vertebrate and the honeydew and manna produced by a range of insects.

Reasons for decline and current threats: Two lines of evidence indicate that the distribution of this subspecies has contracted in recent times. Firstly, some areas contain trees with only very old, inactive feeding cuts, indicating the species was once present but is no longer. These instances may have been due to the loss of den and feed trees during logging operations in the past. Secondly, a large proportion (up to 70%) of tall eucalypt forests have become dominated, and eventually replaced by, rainforest, rendering the habitat unsuitable to the glider. Rainforest expansion is due to long-term climate trends providing suitable conditions for rainforest plant growth, combined with few hot fires penetrating tall eucalypt forest due to human infrastructure, grazing and weed invasion. Lack of fire to maintain tall eucalypt forest remains a primary threat. Other threats include: continued fragmentation of habitat by developments and infrastructure; continuing degradation of habitat by incremental loss of feed and den trees; small isolated populations dying out with no opportunity for recolonisation because of the fragmented nature of suitable habitats; and inbreeding depression. Barbed wire fences pose a threat in some areas.

CURRENT RECOVERY EFFORTS

Recovery plan: None

Other plans: Maxwell S, Burbidge AA and Morris K (Eds) (1996) 'The 1996 action plan for Australian marsupials and monotremes'. Wildlife Australia, Canberra.

'Wet Tropics Management Plan 1998' (Queensland)

Wet Tropics Management Authority (2004) 'Wet Tropics Conservation Strategy: the conservation, rehabilitation and transmission to future generations of the Wet Tropics World Heritage Area'. Wet Tropics Management Authority, Cairns.

Conservation actions:

- Most Yellow-bellied Glider habitat is now protected within national parks and World Heritage areas.
- The Tree Kangaroo and Mammal Group have mapped glider habitat, restoration priorities and developed a management strategy for the Herberton and Ravenshoe areas.
- DERM continues to monitor populations and manage habitats with periodic fires far from towns, particularly in the northern end of the glider's range.

FUTURE CONSERVATION EFFORTS

What you can do:

- Replace top strand barbed with plain wire (support The Wildlife Friendly Fencing Project <http://www.wildlifefriendlyfencing.com>).
- Control woody weeds and assist and promote the use of fire for managing tall eucalypt forest.
- Contact TKMG, WPSQ, the Qld Glider Network and/or Conservation Volunteers Australia to assist in hands-on work with this species.

Conservation needs:

- Manage habitat to reduce or reverse rainforest and weed invasion.
- Map and protect feed and den trees in areas susceptible to disturbance or those being managed with fire.
- Ensure that landowners and managers and linear infrastructure managers and their contractors protect individual feed and den trees in areas suitable for Yellow-bellied Gliders.
- Ensure that populations do not become further fragmented through developments of housing and infrastructure.

Research directions:

- Re-assess the condition and trajectory of glider habitat with a view to implementing or adjusting management.
- Continue to monitor populations.
- Examine the genetic structure of the subspecies and its populations.
- Assess the feasibility of re-introducing the Yellow-bellied Glider to areas of vacant habitat.
- Design and implement safe methods for fire management in tall eucalypt forest with the aim of reducing rainforest invasion in glider habitat.

Further reading

Goldingay RL and Quin D (2004) Components of the habitat of the yellow-bellied glider in north Queensland. In *The Biology of Australian Possums and Gliders* (Eds RL Goldingay and SM Jackson) pp. 369–375. Surrey Beatty & Sons, Sydney.

Quin D, Goldingay R, Churchill S and Engel D (1996) Feeding behaviour and food availability of the Yellow-bellied Glider in North Queensland. *Wildlife Research* **23**, 637–646.

Russell R (1984) Social behaviour of the Yellow-bellied Glider, *Petaurus australis reginae* in North Queensland. In *Possums and Gliders*. (Eds AP Smith and ID Hume) pp. 343–353. Australian Mammal Society, Sydney.

Winter JW, Dillewaard HA, Williams SE and Bolitho EE (2004) Possums and gliders of north Queensland: distribution and conservation status. In *The Biology of Australian Possums and Gliders* (Eds RL Goldingay and SM Jackson) pp. 26–50. Surrey Beatty & Sons, Sydney.

Author: Andrew J Dennis

MAHOGANY GLIDER

Petaurus gracilis (de Vis, 1883)

Photo: www.ataglance.com.au

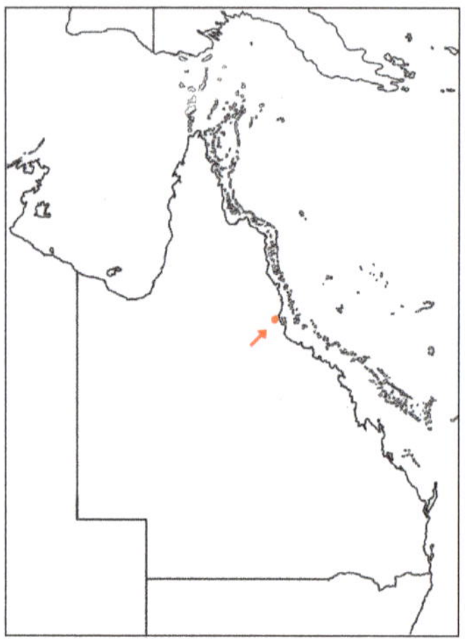

Family: Petauridae

Aboriginal name: Mungurru

Conservation status: IUCN: Endangered; Aust.: Endangered; Qld: Endangered

Description: This small gliding possum has a head and body length of 225–275 mm with a 335–405 mm long and relatively short-haired tail (cf. Squirrel Glider *P. norfolcensis*). It weighs 310–500 g, with males being larger than females. Males have a relatively bigger head, while females have a relatively longer tail. The fur along its gliding membrane can be buff, rich orange or deep mahogany and it has a dark stripe extending from the eyes to the rump.

Distribution: The Mahogany Glider is restricted to coastal lowlands in the southern Wet Tropics of north Qld. Its distribution extends in a narrow and highly fragmented band of open woodland and forests between the Hull River near Tully and Ollera Creek ~40 km south of Ingham: only 122 km north to south. There are five large habitat areas and three smaller fragmented habitats within that distribution. Modelling predicts that the five larger areas should be adequate for the long-term survival of populations, if kept resilient.

Habitat: The Mahogany Glider occurs in mixed open forest or woodland below 120 m elevation. These forests are a mix of different vegetation associations in small patches and provide complex structure and floristic diversity. More than 20 tree and shrub species, including eucalypts, bloodwoods, melaleucas, acacia, *Albizia procera* and *Xanthorrhoea* form essential habitat. Fire maintains the necessary habitat structure; without it, the vegetation becomes more dense.

Ecology: The Mahogany Glider is nocturnal and can glide up to 60 m between trees, requiring an open forest structure to do so. It eats mostly nectar but also takes pollen, insects, honeydew, sap and wattle exudates. It will forage down as low as 1 m on Grass Tree *Xanthorrhoea johnsonii* flower spikes. Siris, Forest *Albizia procera* sap is favoured in summer when few species flower. Dens occur in hollows in Poplar Gum *Eucalyptus platyphylla*, Forest Red Gum *E. tereticornis*, Pink Bloodwood *Corymbia intermedia* and Clarkson's Bloodwood *C. clarksoniana*. Each den is lined with eucalypt leaves and individuals, pairs or family groups may occupy up to 13 dens in their normal range of movements. Male and female Mahogany Gliders pair bond and share most of their range (86%), which overlaps little with neighbours (12%). Individuals and pairs will range over 19–23 ha in continuous forest and half of that in fragmented patches of habitat.

Reasons for decline and current threats: Around 80% of their habitat has been cleared and the remainder is severely fragmented. About 45% of the remaining essential habitat is protected, but, because the remaining habitat is fragmented, 'colonies' are at risk of becoming isolated from one another, facilitating local extinctions. Currently, implementing fire management is compromised by pioneer rainforest plants and weed invasion in long unburnt habitat. These changes render the habitat less suitable by creating a denser forest, which limits the ecological benefits of gliding. Lantana *Lantana camara*, Sicklepod *Senna* spp., Knobweed *Hyptis capitata*, Urena Burr *Urena lobata*, Bluetop *Ageratum houstonianum*, Calapo *Calopogonium mucunoides* and Centro *Centrosema pubescens* are weeds that suppress fire in essential habitat. Individuals are also being killed crossing highways or getting tangled in barbed-wire.

CURRENT RECOVERY EFFORTS

Recovery plan: Parsons M and Latch P (2007) 'Recovery plan for the mahogany glider *Petaurus gracilis*'. Report to the Department of the Environment, Water, Heritage and the Arts, Canberra. Environmental Protection Agency, Brisbane.

Other plans: National Lantana Management Group (2009) 'Draft plan to protect environmental assets from Lantana'. Department of Employment, Economic Development and Innovation, Yeerongpilly, Queensland.

Conservation actions:

- Old mature Lantana thickets have been controlled across Girringun NP since 2008.
- DERM protects, manages fire and weeds, helps to restore habitat and communicates information on the Mahogany Glider.
- Government funding is directed towards habitat recovery on the DERM estate: replacing top strand barbed wire with high tensile plain wire under Good

Neighbour agreements; and maintaining habitat mapping.

- The Thorsborne Trust has provided funding for recovery plan actions.
- BMD Constructions has sponsored glider crossing poles at Corduroy Creek as part of the Murray Flats highway upgrade.
- WPSQ promotes provision of artificial dens and the rescue, care and release of injured Mahogany Gliders.
- Powerlink has sponsored student research into how gliders cross power easements at Mullers Creek.

FUTURE CONSERVATION EFFORTS

What you can do:

- Support Wildlife Friendly Fencing (<http://www.wildlifefriendlyfencing.com>).
- Replace top strand barbed with plain wire.
- Assist in planting corridors with native eucalypt species.
- Control woody weeds.
- Install den boxes.
- Contact WPSQ, the Qld Glider Network and or Conservation Volunteers Australia to assist in hands on work with this species.

Research directions:

- Study the condition of the essential habitat, including fire regimes (particularly restoring habitat from long unburnt intervals), coastal forest transition, weed control and ecosystem restoration.
- Research ways to augment the habitat, such as glider crossing poles and other ways of enhancing and connecting corridors and fragments.
- Study the genetic structure of the species and potentially subpopulations.
- Explore points of sympatry with squirrel glider populations on the northern, southern and western geographic boundary.
- Integrate fire and pest management to retain resilient landscapes.

Sources and further reading

Jackson SM (1999) Preliminary predictions of the impacts of habitat area and catastrophes on the viability of mahogany glider *Petaurus gracilis* populations. *Pacific Conservation Biology* **5**, 56–62.

Jackson SM (2000) Population dynamics and life history of the mahogany glider *Petaurus gracilis* and sugar glider *Petaurus breviceps* in North Queensland. *Wildlife Research* **27**, 21–37.

Jackson SM (2001) Foraging behaviour and food availability of the mahogany glider *Petaurus gracilis* (Petauridae: Marsupialia). *Journal of Zoology* **253**, 1–13.

Jackson SM and Johnson CN (2002) Time allocation to foraging in the mahogany glider *Petaurus gracilis* (Marsupialia, Petauridae) and a comparison of activity times in exudivorous and folivorous possums and gliders. *Journal of Zoology* **256**, 271–277.

Van Dyck S (1993) The taxonomy and distribution of *Petaurus gracilis* (Marsupialia: Petauridae), with notes on its ecology and conservation status. *Memoirs of the Queensland Museum* **33**, 77–122.

Author: Mark Parsons

BATS (ORDER: CHIROPTERA)

LARGE-EARED PIED BAT

Chalinolobus dwyeri Ryan, 1966

Photo: Terry Reardon

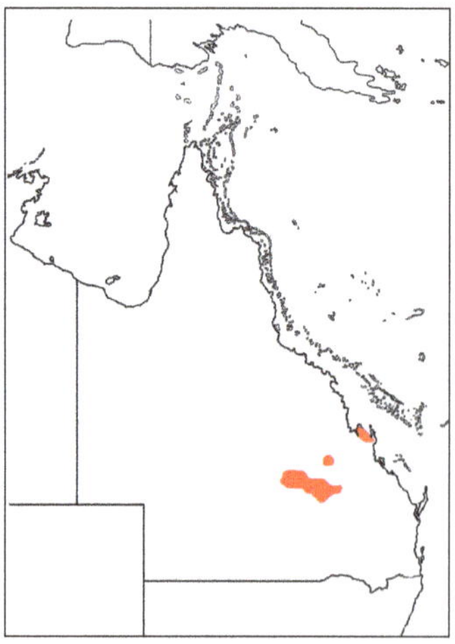

Family: Vespertilionidae

Other common name: Large Pied Bat

Conservation status: IUCN: Near Threatened; Aust.: Vulnerable; Qld: Vulnerable; NSW: Vulnerable

Description: This small bat has a head and body length of around 46 mm, a tail length of 41 mm, forearm of 40 mm and a wingspan of ~277 mm. It weighs around 7.7 g. The Large-eared Pied Bat has features typical of a *Chalinolobus,* but exaggerated compared with its relatives. It has large ears, curly wattles and uniform glossy black fur on the back. The belly is washed with brown, usually with a band of bright white fur along the sides of the body where the wing membranes join. These white lateral stripes converge in the pubic region to form a V-shape. The ears are very large and easily touch when pressed together above the head. The lobe (wattle) at the corner of the mouth is well developed, as is a second lobe of skin along the side of the lower lip.

Distribution: This species occurs in areas with extensive cliffs and caves, primarily in the central Qld sandstone belt – Carnarvon Ranges, Blackdown Tableland and Cania Gorge. It is also known from further south, although records are scant. An old mine known to contain a small population was flooded during construction of the Glenlyon Dam near Texas in south-east Qld.

Habitat: This bat has been recorded from sandstone gorges in tall open eucalypt forest, dry sclerophyll forests and woodlands, as well as at the edge of rainforest, wet sclerophyll forest and *Callitris* dominated forest.

Ecology: This bat is known to roost in caves, crevices and pock-holes in cliffs, Fairy Martin *Petrochelidon ariel* nests and mines, where they cluster in groups of three to 40 in indentations in the ceiling. It is also possible that tree roosts are used. When roosting in shallow caves, crevices and pock-holes, these bats occupy twilight areas not far from the entrance, but they are also known to roost deep inside caves. Males have been found alone in deep torpor during winter, mostly in abandoned mines but also in an abandoned Fairy Martin nest. This bat feeds on insects, but beyond that nothing is known. They fly slowly with rapid, shallow wing beats and are quite manoeuvrable. They usually fly at mid-canopy level 6–10 m above the ground, though have also been seen flying low along a creek bed. The time of mating, autumn or spring, is not confirmed. However, males have enlarged testes in autumn, which makes it likely that they conform to the pattern seen in other *Chalinolobus*, and many vespertilionids, where mating occurs in late autumn and winter, with females then storing sperm over winter. Pregnancy then commences when the females become active at the beginning of spring. Groups of 20–40 females have been recorded raising young from November through to February. They give birth, commonly to twins, in November, and the young are independent by late February. The young leave the cave soon after, while the females remain until late March and return to the same cave each summer over many years.

Reasons for decline and current threats: As with many species of bat, the nature and extent of declines remain poorly documented, partly due to a lack of baseline information on their distribution and ecology. Most knowledge of this species comes from observations at Copeton mine and this colony was drowned with the formation of Copeton Dam. In addition, at least one other known roost site appears to be abandoned. With so few known, these losses and the general threats faced by many bats have resulted in a Vulnerable status.

CURRENT RECOVERY EFFORTS

Recovery plan: None

Other plans: Duncan A, Baker GB and Montgomery N (1999) 'The action plan for Australian bats'. Natural Heritage Trust, Canberra.

Conservation actions: To date, recovery objectives and conservation actions have been listed in the action plan and by NSW ECCW. Although many actions have been identified, little progress has been reported. The greatest need is for basic ecological research that identifies key aspects of resource use, behaviour and also the genetic structuring of populations. This is required to understand habitat use and the constraints that may influence population distribution and abundance.

FUTURE RECOVERY EFFORTS

Conservation needs:

- Protect known and potential roost sites and/or manage the impacts of mining and recreational caving.
- Exclude feral goats from overhang roost sites along cliff faces.
- Identify the maternity roost requirements and protect known maternity sites.
- Protect, manage and/or rehabilitate native vegetation around known roost sites.
- Report sightings to DERM for inclusion in the Wildnet database.

Research directions:

- Clarify the current distribution, abundance and genetic structure of the species.
- Determine its habitat requirements, foraging patterns and diet details.
- Determine its susceptibility to pesticides.
- Investigate and ameliorate the impacts of longwall mining.
- Incorporate into fire management studies and strategies.

Sources and further reading

Churchill S (2008) *Australian Bats.* 2nd edn. Allen and Unwin, Crows Nest, NSW.

Department of Environment and Climate Change, NSW: Large-eared pied bat profile: <http://www.threatenedspecies.environment.nsw.gov.au/tsprofile/profile.aspx?id=10157>.

Hoye GA and Schulz M (2008) Large-eared Pied Bat *Chalinolobus dwyeri*. In *The Mammals of Australia*. 3rd edn. (Eds S Van Dyck and R Strahan) pp. 531–532. Reed New Holland, Sydney.

Author: Andrew J Dennis

FAWN LEAF-NOSED BAT

Hipposideros cervinus Gould, 1854

Photo: Bruce Thomson

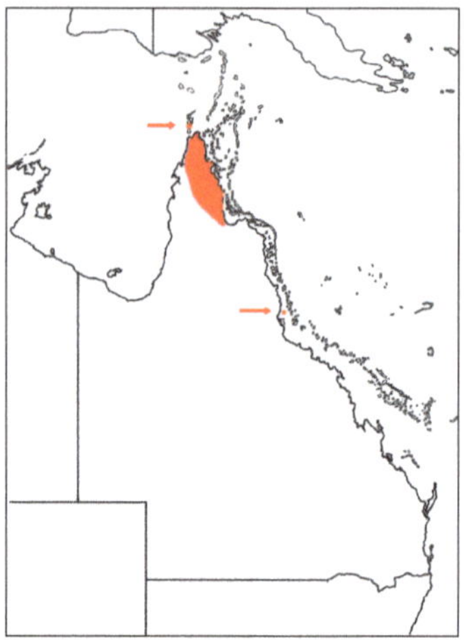

Family: Hipposideridae

Other common names: Fawn Round-leafed Bat, Fawn Horseshoe Bat

Conservation status: IUCN: Least Concern; Qld: Vulnerable

Description: This small bat has a head and body length of 46 mm with a 27 mm tail and a 47 mm forearm. It weighs 7 g. It has broad triangular ears and a distinctive nose-leaf, which is small and narrows below the nostrils to expose two well-developed supplementary leaflets. It ranges in colour from grey to orange. As with many bats, the orange hues may be due to bleaching by high concentrations of ammonia and high humidity in particular roosts.

Distribution: This species is distributed through Malaysia, Indonesia, the Philippines, New Guinea, Solomon Islands and Vanuatu. In Australia, it is found on Cape York Peninsula, north of Coen, and in the Torres Strait, with an isolated population on Dunk Island in North Qld.

Habitat: This bat forages in a range of habitats including rainforest, riparian rainforest and open woodland. It may also forage through human habitation including around lights, in buildings and through parkland.

Ecology: This species is most often observed roosting in caves and mines in colonies of 21 to 900 individuals. Roosts seem to be of high humidity and, although once thought to be always in caves, recent observations in Australia and South-East Asia suggest it may roost in the bases of large trees, under earth banks along creeks and in other sheltered, semi-dark situations, including old sheds and buildings when humidity is high. When roosting, it hangs free from the ceiling by its toes, often from one foot, with individuals spaced 10–15 cm apart. It is always alert and flies readily when approached. The Fawn Leaf-nosed Bat does not enter torpor and is ectothermic between 31–35°C. It feeds on a range of insects including moths, beetles, bugs, cockroaches, spiders, flies and parasitic wasps. It catches them on the wing or by gleaning while flying close to vegetation, the ground or water using a slow fluttering flight. It also hunts from perches, a technique that uses less energy than aerial hunting and which may account for its relatively low metabolic rate – 58% of that expected for a mammal of its weight. Its flight can be so slow and manoeuvrable that it is one of the only bats capable of hovering and rising out of a conventional harp trap. When commuting from roosting to foraging areas, generally along established pathways, flight is fast and direct. A colony exiting a roost site may be seen to subdivide as individuals and small groups commute to separate foraging sites. The Fawn Leaf-nosed Bat produces a single young at the start of the wet season, in November or December. This species does not form maternity colonies separate to its normal roosts, but rather occupies the same roost with its young. Echolocation calls are between 120–145 kHz, which is at the upper end of the detection range of the most commonly used bat detector system in Australia (AnaBat).

Reasons for decline and current threats: This bat's restricted distribution in Qld and the loss of a colony in the abandoned Stewart River Mine, which was destroyed when mining recommenced, has led to its listing. No other concrete data are available, but roost disturbance and habitat alteration may also be detrimental.

CURRENT RECOVERY EFFORTS

Recovery plan: None

FUTURE RECOVERY EFFORTS

Conservation needs: Protect roosts from disturbance and destruction.

Research directions: Conduct surveys of potential roost sites to better determine distribution, abundance and susceptibility to disturbance.

Sources and further reading

Churchill S (2008) *Australian Bats*. 2nd edn. Allen and Unwin, Crows Nest, NSW.

Pavey CR and Burwell CJ (2008) Fawn Leaf-nosed Bat, *Hipposideros cervinus*. In *The Mammals of Australia*. 3rd edn. (Eds S Van Dyck and R Strahan) pp. 459–461. Reed New Holland, Sydney.

Video link: *Hipposideros cervinus* filmed by Tigga Kingston: <http://www.vimeo.com/groups/2343/videos/3035544>.

Author: Andrew J Dennis

SEMON'S LEAF-NOSED BAT

Hipposideros semoni Matschie, 1903

Photo: Bruce Thomson

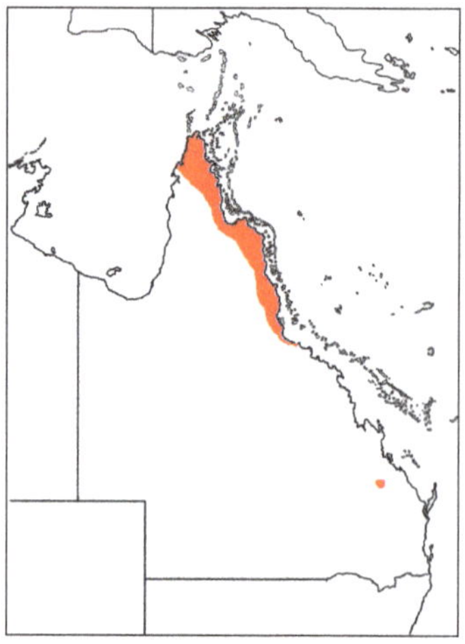

Family: Hipposideridae

Conservation status: IUCN: Data Deficient; Aust.: Endangered; Qld: Vulnerable

Description: This small bat has a head and body length of 40–50 mm, a 22–26 mm tail, a 42–50 mm forearm and, a wingspan up to 320 mm. It weighs 6–10 g. Its ears are long and narrow with a narrow point. Its long fur is dark brown or smoky grey, paler on the belly and where it extends on to the base of the wing membranes. Its roughly square nose-leaf extends over most of the face and its two wart-like protuberances help to distinguish it from close relatives: the lower being longer than the upper and longer than in other species (6–8 mm).

Distribution: This bat's full distribution is as yet unclear. It is known to occur from New Guinea along eastern Cape York Peninsula to Townsville. Beyond this, there is possibly an isolated population at Kroombit Tops south of Gladstone and other possible records, based on echolocation calls, suggest they may occur as far south as Maryborough.

Habitat: This species has been caught in rainforest in both wet and seasonally wet environments. Its broad wings and slow flight suggest it is well adapted for foraging among dense vegetation, which it appears to do within a metre or two of the ground. However, it is not restricted to rainforest and has been recorded in tall eucalypt forest and open woodland.

Ecology: Semon's Leaf-nosed Bat has usually been found roosting alone, although small groups have also been noted both roosting and foraging. It seems able to use any small nook or cranny, at least as temporary shelter, but more often roosts in tree hollows, rock fissures and caves, often in twilight areas where it will abandon the roost rapidly on disturbance. It is known to eat moths and probably also gleans spiders and beetles, though little is known about its diet. Its slow, low and manoeuvrable flight supports the suggestion it may glean, as well as catch prey, on the wing. Pregnant females have been caught in late October and a male with scrotal testes, which generally indicates sexual activity, was caught at the same time. Nothing else is known about its reproduction. Semon's Leaf-nosed Bat is another species with unusual echolocation calls, with males calling at 94 kHz and females at 74 kHz.

Reasons for decline and current threats: A decline has been inferred based on a relative reduction in the number of recent versus historical records, despite more sophisticated survey techniques being employed in recent times. No data are available on why a decline may have occurred. Guesses include loss or disturbance of roosts, though it is unlikely that either of these would have changed significantly within their distribution during the timeframe of the decline. Habitat loss has also been mentioned, but the species is distributed in one of the least cleared areas of Qld. Changes in vegetation structure and floristics brought about by introduced herbivores and changed fire regimes could be a more likely cause.

CURRENT RECOVERY EFFORTS

Recovery plan: Thomson B, Pavey C and Reardon T (2001) 'National recovery plan for cave-dwelling bats, *Rhinolophus philippinensis, Hipposideros semoni* and *Taphozous troughtoni* 2001–2005'. Department of the Environment, Water, Heritage and the Arts.

Other plans: Duncan A, Baker GB and Montgomery N (1999) 'The action plan for Australian bats'. Natural Heritage Trust, Canberra.

Conservation actions:

- A few sites in the southern part of their confirmed range have been resurveyed to compare with historical surveys, but more are needed throughout the greater part of their range.

- A mine was gated at Iron Range to minimise disturbance.

FUTURE RECOVERY EFFORTS

Conservation needs: Protect a significant sample of roost sites throughout their range.

Research directions:

- Determine its current distribution, particularly in the south.
- Conduct further surveys (based on historical ones) to better determine population trends and better understand its conservation status.
- Conduct further research into all its ecological requirements. Of particular interest is the understanding of its roost site requirements. Observations of similar species in South-East Asia suggest they are capable of using a wide range of cave-like and hollow-like structures, which may indicate that disturbance at mine roosts may be a distraction from the real cause of decline.

Sources and further reading

Churchill S (2008) *Australian Bats*. 2nd edn. Allen and Unwin, Crows Nest, NSW.

de Oliveira MC and Schulz M (1997) Echolocation and roost selection in Semon's Leaf-nosed Bat *Hipposideros semoni*. *Memoirs of the Queensland Museum* **41**, 158.

Author: Andrew J Dennis

NORTHERN LEAF-NOSED BAT

Hipposideros stenotis Thomas, 1913

Photo: Bruce Thomson

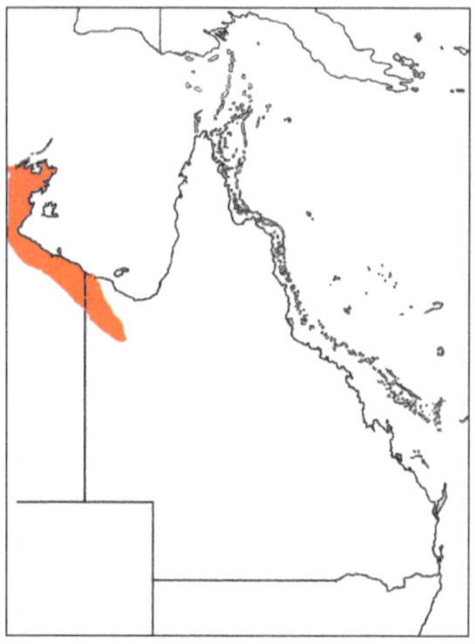

Family: Hipposideridae

Conservation status: IUCN: Least Concern; Qld: Vulnerable; NT: Near Threatened

Description: This small bat is similar in many respects to Semon's Leaf-nosed Bat, but smaller and without the long lower wart-like projection on the nose-leaf. It has a head and body length of 40–45 mm, a 22–28 mm tail, a 42–46 mm forearm and a wingspan up to 275 mm. It weighs 4.5–7.5 g. Like Semon's Leaf-nosed Bat, its long, pointed ears are noticeable when the bat is roosting and its long fur has a slightly shaggy appearance and is paler on the belly than the back.

Distribution: Found only in Australia, it appears to occur as isolated populations in suitable habitat dotted across northern Australia, from the escarpments of the Kimberley across the Top End into the Gulf of Carpentaria (e.g. the Nicholson River and Lawn Hill areas) and as far south as Mount Isa, but no further east.

Habitat: This bat is almost always associated with escarpments, rocky ranges or boulder piles, suggesting its distribution is determined by roost sites. Habitats adjacent to roosts have been known to include tussock and spinifex grasslands, open woodland vine thickets and riparian forests.

Ecology: This species seems to prefer to roost alone in shallow crevices and blow-outs along the edge of sandstone escarpments where it is protected from direct sunlight, but nevertheless in quite open and well lit sites. It also uses road culverts, boulder piles and old mines, but seems to avoid limestone caves. It is easy to disturb from its roost and often flies out of a cave rather than deeper into one. It has a slow fluttering flight and generally hunts close to the ground, navigating through the vegetation to catch insects, mostly moths and beetles. It appears not to commute far from its roost to feed which may account for its perceived rarity, requiring appropriate roost sites and feeding grounds in close proximity. Its birthing season is long, predominantly from October to January, though pregnant females caught in July were likely to have given birth to their single young well before October.

Reasons for decline and current threats: Why this species is listed is unclear. A couple of roost sites have been destroyed, though others nearby remained occupied (Koolan Island) and there is a perception of fewer individuals in the Pine Creek area in the NT. This does not demonstrate a decline, but with so few records of roost sites for the species and only a small part of its distribution being in Qld, concern is warranted. There is also an unquantified perception that it is sensitive to disturbance, perhaps due to its tendency to rapidly vacate a roost when disturbed, a behaviour that may make it more tolerant to disturbance than less.

CURRENT RECOVERY EFFORTS

Recovery plan: None

Other plans: Duncan A, Baker GB and Montgomery N (1999) 'The action plan for Australian bats'. Natural Heritage Trust, Canberra.

FUTURE RECOVERY EFFORTS

Conservation needs: Review its conservation status.

Research directions:

- Survey this bat's distribution and abundance to establish a baseline for measuring population trends.

- Conduct ecological research into its diet and foraging, roost and maternity site selection, which will assist long-term management.

Sources and further reading

Churchill S (2008) *Australian Bats.* 2nd edn. Allen and Unwin, Crows Nest, NSW.

Pavey CR and Burwell CJ (2008) Fawn Leaf-nosed Bat, *Hipposideros cervinus.* In *The Mammals of Australia.* 3rd edn. (Eds S Van Dyck and R Strahan) pp. 459–461. Reed New Holland, Sydney.

Schulz M and Menkhorst K (1985) Notes on the Lesser Warty-nosed Horseshoe Bat (*Hipposideros stenotis*). *Australian Bat Research Newsletter* **20**, 14–16.

Author: Andrew J Dennis

GHOST BAT

Macroderma gigas (Dobson, 1880)

Photo: Bruce Thomson

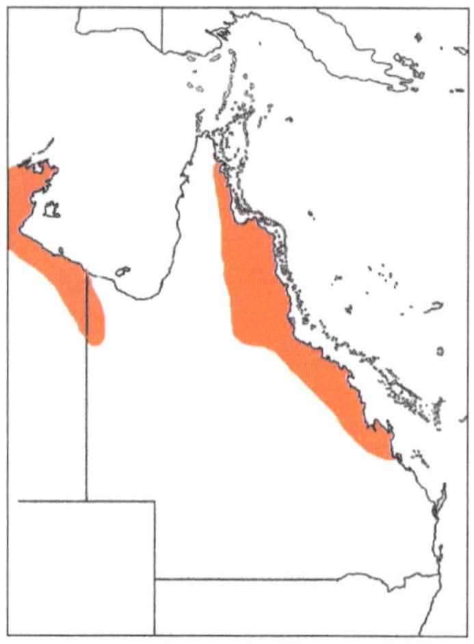

Family: Megadermatidae

Other common name: Australian False Vampire Bat

Conservation status: IUCN: Vulnerable; Qld: Vulnerable

Description: This large carnivorous bat has a head and body length of 100–130 mm, a tail membrane, but no tail, and a 96–113 mm forearm. Weighing 140–165 g, it is the largest microchiropteran in Australia and third largest in the world. Its ears are long, oval ended and join together above the head. Its fur is light to dark grey above and paler below. It has large eyes and a simple nose-leaf.

Distribution: Endemic to Australia, its disjunct colonies are currently distributed across tropical and subtropical Australia to 27°S in Qld and 28°S in WA, though it was once more widespread. Based entirely on known colonies, without projections for unknown colonies, counts for Australia range from 4000–6000 individuals, based on 750–850 in Qld, 2500–3500 in the NT and around 1500 in WA.

Habitat: The Ghost Bat occurs in many habitats, from hot and dry mulga country to the wet tropical forests of far north Qld. While using a wide variety of habitats for temporary and/or feeding sites, permanent roosts, especially maternity sites, are in deep natural cave systems or large disused mines and have particular structural and microclimatic features. Only 14 maternity colonies are currently known. Individuals may disperse well away from maternity roosts when not breeding and can move between several roost sites seasonally.

Ecology: The Ghost Bat is Australia's only carnivorous bat. During the summer or wet season, large insects, such as grasshoppers and beetles, dominate its diet. However, during winter or the dry season, it eats vertebrate prey including frogs, lizards, birds and mammals, including small bats. It gleans its prey from the ground or other surfaces rather than catching it in flight. It does not continuously echolocate during flight but listens and looks to locate prey. Once located, it will land on its prey, envelope it in its wings and kill it with powerful bites. Ghost Bats generally forage within 1–2 km of their daytime roost site. Mating probably occurs in July and August, with females giving birth to single pups between September and November. Mothers form nursery colonies, which include males. Juveniles hunt with their mothers until they become fully independent. Generation length is probably around 4 years and recruitment is very low. A major population genetic analysis has shown that Ghost Bat populations are highly structured, being genetically distinct at both regional and local scales, implying that there is virtually no movement of individuals among sites. Southern populations are separated by a minimum distance of 300 km from the others. There are even significant genetic differences between the east and west of the Pilbara and in different parts of the Kimberley region. Although populations within the NT and far north Qld are distinct from each other, there is greater movement of male bats between populations than in the west and south.

Reasons for decline and current threats: Fossil and subfossil evidence suggests the range of the species contracted markedly both before and after European settlement.

Historical contractions are speculated to coincide with continental drying about 10 000 years ago. The species is highly sensitive to human disturbance. Cave tourism is a problem, especially during breeding, but the most serious threats are from quarrying and the reworking and collapse of old mines. In the late 1980s, a campaign to prevent mining activities destroying caves used by this bat at Mount Etna failed and this population has subsequently declined. Habitat modification may intensify competition for prey with foxes and cats.

For some populations, particularly in the south where only a few suitable roost sites are separated by large geographic distances, genetic studies have shown that colonies are isolated and unlikely to be recolonised by immigration following local extinctions.

CURRENT RECOVERY EFFORTS

Recovery plan: None

Other plans: Duncan A, Baker GB and Montgomery N (1999) 'The action plan for Australian bats'. Natural Heritage Trust, Canberra.

Conservation actions:

- There are breeding populations of Ghost Bats in nocturnal houses at Sydney's Featherdale Wildlife Park, the Territory Wildlife Park at Berry Springs, NT and the Perth Zoo.
- There have been long-term population studies and monitoring in Qld.

FUTURE RECOVERY EFFORTS

What you can do:

- Report any reliable sightings to the relevant Parks authority or government agency responsible for their management.
- Clear Lantana *Lantana camara* from private properties and public lands (multiple benefits).
- Install or insert non-barbed wire fencing where practical and critical. Support Wildlife Friendly Fencing <http://www.wildlifefriendlyfencing.com>.
- Join/donate to the Australasian Bat Society.
- Be mindful if entering tourist caves that they are homes to many organisms, some of which are vulnerable to disturbance.

Conservation needs:

- Identify and protect other unprotected maternity colonies.
- Although visitation to caves and monitoring of disturbance in national parks is well managed, there is a general need for careful regulation of cave tourism and visitation for the conservation of bats.

Research directions:

- Conduct more surveys for maternity sites and collect samples for genetic work, particularly from the Kimberley, WA.
- Conduct additional ecological research to help inform the management of landscapes in which the bats occur.

Sources and further reading

Armstrong KN and Anstee SD (2000) The ghost bat in the Pilbara, 100 years on. *Australian Mammalogy* **22**, 93–101.

Richards GC, Hand S, Armstrong KN and Hall LS (2008) Ghost bat *Macroderma gigas*. In *The Mammals of Australia*. 3rd edn. (Eds S Van Dyck and R Strahan) pp. 449–450. Reed New Holland, Sydney.

Tidemann CR, Priddel D, Nelson JE and Pettigrew JD (1985) Foraging behaviour of the Australian ghost bat, *Macroderma gigas* (Microchiroptera, Megadermatidae). *Australian Journal of Zoology* **33**, 705–713.

Toop J (1985) Habitat requirements, survival strategies and ecology of the ghost bat, *Macroderma gigas* Dobson (Microchiroptera, Megadermatidae) in central coastal Queensland. *Macroderma* **1**, 37–41.

Worthington Wilmer J, Hall L, Barratt E and Moritz C (1999) Genetic structure and male-mediated gene flow in the Ghost Bat (*Macroderma gigas*). *Evolution* **53**, 1582–1591.

Author: Jessica Worthington Wilmer

FLUTE-NOSED INSECTIVOROUS BAT

Murina florium Thomas, 1908

Photo: Bruce Thomson

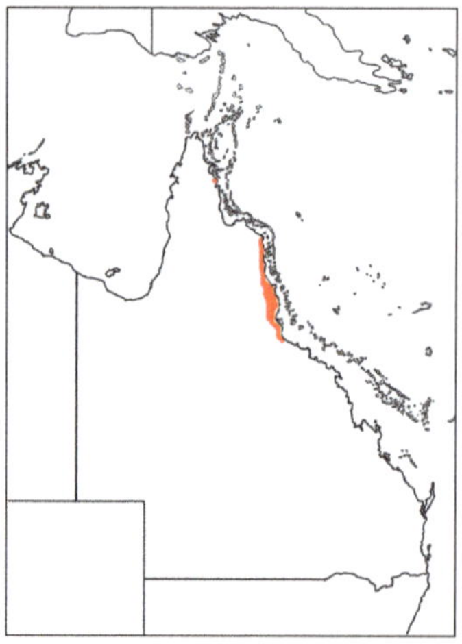

Family: Vespertilionidae

Other common name: Tube-nosed Insectivorous Bat

Conservation status: IUCN: Least Concern; Qld: Vulnerable

Description: This tiny bat has a head and body length of 45–60 mm, a 31–37 mm tail with a membrane covered in long hairs and a 32–36 mm forearm. It weighs 6–11 g. Its ears are small and rounded (10–14 mm). The grizzled, woolly body fur ranges from rufous-brown to dark grey-brown and is slightly paler on the belly. Its most distinctive features are its tubular nostrils and the presence of thick hair on the skin over the bones on the wing.

Distribution: There are two populations known for the species in Australia: a northern population based around Iron Range and a southern population in the Wet Tropics. It also occurs in eastern Indonesia and New Guinea.

Habitat: This bat occurs in rainforest and tall eucalypt forests and has been recorded from near sea-level on the Alexandra and Marlborough-Thomson Ranges to elevations of over 1100 m.

Ecology: This species forages in the forest canopy, but is most often caught during adverse weather (e.g. rain) when it forages in the sub-canopy and understorey. It eats a range of invertebrates including beetles, spiders, caddis flies, caterpillars, moths, flies and mayflies, which it gleans from surfaces, occasionally hovering to do so. Roosts have been located in the disused nests of the Yellow-throated Scrubwren *Sericornis citreogularis*, the Fernwren *Oreoscopus gutturalis* and a kingfisher nest in a suspended termite mound, in tree hollows and the curled up sheath of palm leaves held aloft by vines. As well as an echolocation call, it produces an audible whistle-like call, which appears to be social. This call sounds similar to the calls of the Tube-nosed Blossom Bat *Nyctimene robinsoni*, but is higher in frequency (pitch). Captured individuals seem to attract free-flying individuals by these vocalisations, resulting in this species often being captured in groups rather than individually.

Reasons for decline and current threats: The Flute-nosed Insectivorous Bat was first recorded in Australia in 1981 and was first thought to be restricted to mountain tops. Its relatively recent discovery led some to believe it was one of Australia's rarest mammals. Changes in the techniques and technology used to capture and survey bats in the early '80s, have redefined this species from being considered a rare mountain-dwelling one to being more widespread but difficult to catch. There is no threat known to have caused a decline and because most of its distribution is within protected areas, its conservation outlook is probably good. However, habitat loss and fragmentation could have significant impacts outside protected areas and climate change may have impacts across its range if the climate becomes drier.

CURRENT RECOVERY EFFORTS

Recovery plan: None

Other plans: Duncan A, Baker GB and Montgomery N (1999) 'The action plan for Australian bats'. Natural Heritage Trust, Canberra.

Conservation actions: The Wet Tropics Management Authority sponsored a survey of the Wet Tropics region for this species, resulting in numerous new records and the identification of its social and echolocation calls. This survey suggests the species could be down-graded to Near Threatened. This is an excellent example of a relatively cryptic species being considered to be threatened due to a lack of data rather than a real restricted distribution or low number of individuals.

FUTURE RECOVERY EFFORTS

What you can do: Volunteer with Trees for the Evelyn and Atherton Tableland (TREAT), which actively plants trees to reduce the impacts of fragmentation and habitat clearing in this species' range.

Conservation needs: Revise the conservation status on the basis of recent surveys.

Research directions: Conduct further surveys and ecological studies to determine the requirements of the species, particularly the function and importance of social structures.

Sources and further reading

Clague CI (2000) The tube nosed insectivorous bat. *Nature Australia* **26**, 22–23.

Clague CI, Coles RB, Whybird OJ, Spencer HJ and Flemons P (1999) The occurrence and distribution of the Tube-nosed Insectivorous Bat (*Murina florium*) in Australia. *The Proceedings of the Linnean Society of New South Wales* **121**, 175–191.

Richards GC, Hall LS, Helman PM and Churchill SK (1981) First discovery of a species of the rare tube-nosed insectivorous bat (*Murina*) in Australia. *Australian Mammalogy* **5**, 149–151.

Schulz M and Hannah D (1998) Relative abundance, diet and roost selection of the tube-nosed insect bat, *Murina florium*, on the Atherton Tablelands, Australia. *Wildlife Research* **25**, 261–271.

Author: Chris Clague

SOUTH-EASTERN LONG-EARED BAT

Nyctophilus corbeni Parnaby, 2009

Photo: Bruce Thomson

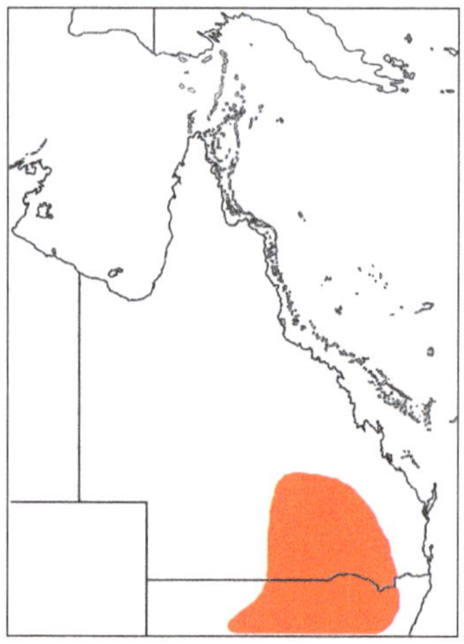

Family: Vespertilionidae

Other common names: Eastern Long-eared Bat, Greater Long-eared Bat (south-eastern form) *Nyctophilus* species 2

Conservation status: Aust.: Vulnerable; Qld: Vulnerable; NSW: Vulnerable; SA: Vulnerable; Vic.: Vulnerable. Originally listed as *Nyctophilus timoriensis*, south-eastern form.

Description: This is the largest and most robust of the long-eared bats with a head and body length of 50–75 mm, a forearm length of 40–50 mm and a tail length of 35–50 mm. It weighs 11–21 g, with males being slightly smaller than females. It has long ears (21–28 mm) that are erect when the bat is alert, but fold back when at rest. Its fur is light brown to grey-brown on the back and noticeably lighter on the underside. This species is readily distinguished from other Qld species of long-eared bats by forearm size, outer canine width (5–5.8 mm) and phallus morphology.

Distribution: There are surprisingly few records of this species from Qld given that published maps show that the state accounts for around 30% of the total distribution. Records are from <30 localities, mainly in the Brigalow Belt South bioregion which suggests that the species is not common within the known extent of its Qld distribution. There are no data on population size.

Habitat: Individuals are capable of flying large distances (>7 km in one night) between roosts and foraging grounds and what little is known of their habitat in Qld is based solely on the limited number of capture records. Captures have been made in open dry woodland and forest, which may include *Corymbia citriodora, C. bloxsomei, Eucalyptus crebra, E. melanophloia, E. populnea, E. major, E. pilligaensis, E. chloroclada, E. fibrosa, Angophora leiocarpa, Allocasuarina leuhmanii, Dodonaea viscosa, Callitris glaucophylla, Acacia harpophylla, Ac. leiocalyx, Ac. conferta, Casuarina cristata* and *Geijera parviflora*.

Ecology: Most of what is known of this species comes from research and observation from outside Qld. Individuals or small colonies roost in tree-hollows and cracks in tree limbs and occasionally under shedding bark, or even within the foliage. Individuals may move regularly between roosts, every 1.3 days and over long distances in one study. They are reported to catch insects in flight and by gleaning from the vegetation or the ground. They mostly forage just under the tree canopy. Like many microchiropterans, this species will drink from dams, open tanks or any natural open water. Mating is thought to occur in autumn and winter, with the females storing sperm until early spring when fertilisation occurs and gestation commences. Twin young are usually born in late spring and are not fully weaned until mid summer. The lifespan for small bats is typically 7–8 years, although over 30 years has been recorded for a few species.

Reasons for decline and current threats: This species is considered threatened because of a population decline inferred from major habitat loss over a large proportion of its distribution, primarily with clearing of Brigalow *Acacia harpophylla* for agriculture. This is compounded because much of the remaining vegetation is under pressure from grazing, timber harvesting and fires (wild and prescribed). Recent research in the NSW Brigalow Belt has shown the 84% of trees that collapsed after fuel reduction burns were mature hollow-bearing trees. Flow-on effects from loss of habitat include increased competition for hollows from other animals and increased exposure to predators. A significant, but often overlooked, threat is the lack of general appreciation of microbats and therefore, the lower priority they receive for recovery action.

CURRENT RECOVERY EFFORTS

Recovery plan: Schulz M and Lumsden L (2010) '(Draft) National recovery plan for the South-eastern Long-eared Bat *Nyctophilus corbeni*'. Victorian Department of Sustainability and Environment.

Other plans: Duncan A, Baker GB and Montgomery N (1999) 'The action plan for Australian bats'. Natural Heritage Trust, Canberra.

Conservation actions: A number of surveys have been conducted, which included records of this species in Qld.

FUTURE RECOVERY EFFORTS

Conservation needs:

- Prevent native vegetation clearing.
- Encourage the protection and enhancement of understorey vegetation.
- Reduce grazing pressure in native forests.
- Promote the protection of native vegetation and hollow-bearing trees on private lands.
- Retain hollow-bearing trees during logging and timber removal operations.
- Review fire management practices to retain hollow-bearing trees.
- Develop and promote state-wide bat awareness programs.
- Restore degraded habitat.
- Install escape nets in water tanks so that bats falling in while drinking can climb out again.

Research directions:

- Conduct further surveys to determine the full extent of the distribution.
- Determine the key attributes for roost requirements, including maternity sites and roost fidelity.
- Determine the risks to roosts (fire, clearing?).
- Determine the diet, foraging ecology and foraging habitat characteristics.
- Set up population monitoring sites to assess long-term changes.
- Investigate breeding success and survival rates.
- Investigate the effects of fragmentation.
- Determine the effectiveness of artificial roosts.
- Quantify the benefits of local bat populations to reducing the impact of insect pests on commercial crops and maintaining the health of remnant vegetation.

Sources and further reading

Parnaby H (2009) A taxonomic review of Australian greater long-eared bats previously known as *Nyctophilus timoriensis* (Chiroptera: Vespertilionidae) and some associated taxa. *Australian Zoologist* **35**, 39–81.

Parnaby H, Lunney D, Shannon I and Fleming M. (2010) Collapse rates of hollow-bearing trees following low intensity prescription burns in the Pilliga forests, New South Wales. *Pacific Conservation Biology* **16**, 209–220.

Pennay M (2002) Greater Long-eared Bat *Nyctophilus timoriensis*. In 'Appendix 2, Brigalow Belt South Stage Two, Vertebrate Fauna Survey, Analysis and Modelling'. Unpublished report. pp. 30–31. Resource and Conservation Division, Planning NSW, Sydney.

Turbill C and Ellis M (2006) Distribution and abundance of the south eastern form of the Greater Long-eared Bat *Nyctophilus timoriensis*. *Australian Mammalogy* **28**, 1–7.

Venz M, Mathieson M and Schulz M (2002) 'Fauna of the Lower Dawson River Floodplain – an assessment of fauna downstream of the proposed Nathan Dam'. Forest Ecosystem Research and Assessment, Queensland Parks and Wildlife Service, Indooroopilly.

Author: Terry Reardon

SPECTACLED FLYING-FOX

Pteropus conspicillatus Gould, 1850

Photo: www.ataglance.com.au

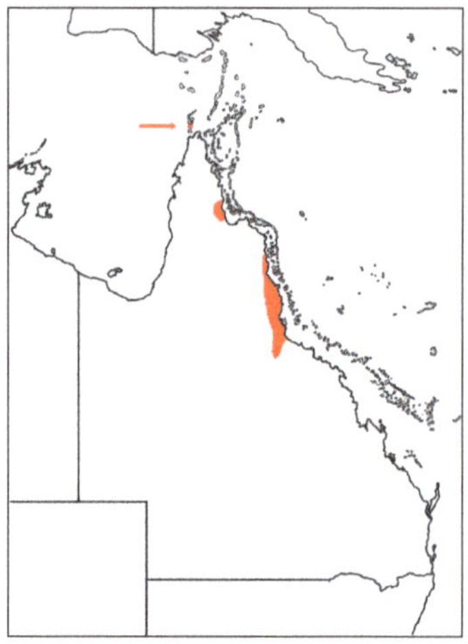

Family: Pteropodidae

Conservation status: IUCN: Least Concern; Aust.: Vulnerable

Description: This large bat has a head and body length of 220–240 mm and a 150–189 mm forearm. It weighs 450–1000 g, with males being larger than females. It has black, naked wings, black body fur, often with silvery guard hairs and yellowish or straw-coloured eye-rings and muzzle fur and a similar coloured neck ruff, sometimes red-stained and of variable extent.

Distribution: The Spectacled Flying-fox occurs in New Guinea, some of its offshore islands, and north-east Qld. The Australian populations are currently considered subspecific (*P. c. conspicillatus*). Australian populations occur in the vicinity of tropical rainforest in the Iron Range and Wet Tropics regions.

Habitat: In Australia, the Spectacled Flying-fox is associated with, but not restricted to, tropical rainforests. It uses resources in eucalypt forests, melaleuca swamps, littoral and coastal mixed forests and mangroves, farmlands and urban and suburban gardens, as well as rainforest. Although 'camps' tend to be within or near rainforest, the bats forage widely, easily covering 50–100 km each night.

Ecology: This social bat occupies camps of varying size and permanence. Individuals can move between camps often, even though a camp may seem permanent. Although females may mature at 1–2 years, successful mothering skills may not be learnt until their fourth year, despite most giving birth by their third year. Males also mature at 3–4 years. It breeds seasonally, producing a single young between October and December, which is weaned at 5 months but cared for in crèches for many more months while it learns foraging, navigation and social skills. This bat disperses seeds, mosses, micro-organisms and pollen. Its role as a pollinator is not yet well understood. It disperses small seeds (<5 mm) over long distances and large seeds over short distances.

Reasons for decline and current threats: The population size and trend is still uncertain due to difficulties in measuring it, but several studies have indicated the sensitivity of the species to population decline. However, a decline is inferred due to considerable habitat loss within its range. In addition, the bats have faced and, in some cases, still face, a suite of new mortality factors including: persecution at camps, particularly in and near towns; persecution at fruit orchards; tick paralysis from the Paralysis Tick *Ixodes holocyclus*; collisions with powerlines and barbed-wire fences; and possibly a cleft palate syndrome, the origins of which are unknown but it may be chemically induced. Combined with a low reproductive rate at the population level and its natural death rate, these new mortality factors could cause a long-term decline, which is why they are considered Vulnerable. As with many species, climate change may be a future threat.

CURRENT RECOVERY EFFORTS

Recovery plan: Queensland Department of Environment and Resource Management (2010) 'National recovery plan for the spectacled flying-fox *Pteropus conspicillatus*'. Report to the Department of Sustainability, Environment, Water, Population and Communities, Canberra.

Other plans: Duncan A, Baker GB and Montgomery N (1999) 'The action plan for Australian bats'. Natural Heritage Trust, Canberra.

Department of the Environment and Heritage (2003) 'Administrative guidelines on significance – supplement for the Spectacled Flying-fox, for the 2003–2004 fruit-growing season'.

Conservation actions:

- Currently, legislation is in place that stops the killing of this species at fruit orchards, which was recently a source of mortality that was used to model the population's long-term decline. Most fruit growers now protect these bats and use non-lethal deterrents at orchards, particularly nets that do not entangle them.
- A number of individuals and organisations play significant roles in the rescue and rehabilitation of tick victims and their babies and bats orphaned or injured by other means, as well as educating people about bats and issues surrounding them (e.g. see www.tolgabathospital.org).
- Significant research programs are also underway by a range of research organisations and individuals, many of them focused on understanding aspects of the bat's ecology and management, disease profiles and methods to measure population size and trends (e.g. see <http://www.csiro.au/science/Spectacled-Flying-Fox--ci_pageNo-1.html).

FUTURE RECOVERY EFFORTS

What you can do:

- Support Wildlife Friendly Fencing (<http://www.wildlifefriendlyfencing.com>).
- Join/donate to a bat hospital and the Australian Bat Society.

Conservation needs:

- Ensure the protection of sufficient foraging habitat and resources at a landscape scale appropriate to the movements of and use by the Spectacled Flying-fox.
- Assist fruit growers to maintain or install non-lethal protection for their crops.
- Provide accurately quantified data on the most successful and cost effective methods to deter bats at fruit crops.
- Ensure compliance with non-lethal control measures at orchards.
- Educate people about the ecological roles of flying-foxes and risks associated with urban and suburban camps and assist them in finding ways to tolerate their presence.
- Investigate options for buffer zones around camps near human habitation and the possibility of providing long-term camp sites on islands in decorative lakes and using rehabilitated bats and feeding stations to establish these.

Research directions:

- Research and monitor natural food resources to be able to predict years when fruit orchards may be targeted by bats.
- Develop methods of measuring and monitoring population size and trends at appropriate scales, then implement them.
- Further clarify how the Spectacled Flying-fox comes into contact with the Paralysis Tick and develop methods to prevent this.
- Continue commercial research into the most successful and cost effective methods to deter bats at fruit crops.
- Investigate the incidence of disease, its transmission to humans and other wildlife and potential movement to and from New Guinea populations.

Sources and further reading

Fox S, Luly B, Mitchell C, Maclean J and Westcott DA (2008) Demographic indications of decline in the spectacled flying fox (*Pteropus conspicillatus*) on the Atherton Tablelands of northern Qld. *Wildlife Research* **35**, 417–424.

McIlwee A and Martin L (2002) On the intrinsic capacity for increase of Australian flying-foxes (*Pteropus* spp Megachiroptera). *Australian Zoologist* **32**, 76–100.

Parsons J, Cairns A, Johnson C, Robson S, Shilton LA and Westcott DA (2006) Diet variation in Spectacled Flying Foxes (*Pteropus conspicillatus*) in the Wet Tropics of Australia. *Australian Journal of Zoology* **54**, 417–428.

Westcott DA, Dennis AJ, McKeown A, Bradford MG and Margules CR (2001) 'The spectacled flying-fox, *Pteropus conspicillatus*, in the context of the World Heritage values of the Wet Tropics World Heritage Area'. A peer-reviewed report to Environment Australia, June 2001.

Author: Andrew J Dennis

GREY-HEADED FLYING-FOX

Pteropus poliocephalus Temminck, 1825

Photo: Bruce Thomson

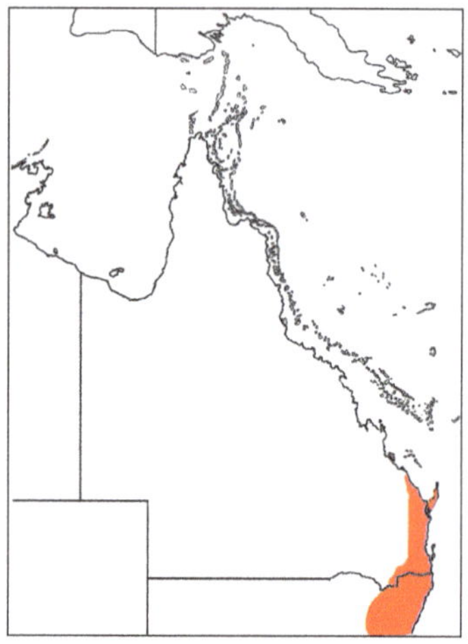

Family: Pteropodidae

Aboriginal names: Girraman (Sunshine Coast), Bilin (Gold Coast), Graymon (Boonah)

Conservation status: IUCN: Vulnerable; Aust.: Vulnerable; NSW: Vulnerable; Vic.: Vulnerable

Description: This is a large bat with a head to body length of 220–280 mm and a forearm length of 155–175 mm. It weighs 750–1000 g (males) or 620–800 g (females). It has a medium to dark grey body with many light-tipped hairs, a black to light grey head, an orange or russet collar encircling the neck and black wing membranes. The leg fur extends to the ankle.

Distribution: This species is found throughout coastal south-eastern Australia from Melbourne to Mackay. During the past few years, the Grey-headed Flying-fox has been sighted in Adelaide as well. It extends inland to at least the Great Diving Range and as far as the tablelands and western slopes in northern NSW and southern Qld. Although, in the past 20 years it may have become more consistent in urban settings, its patterns of occurrence and relative abundance vary widely between regions, seasons and years and are determined primarily by the availability of food.

Habitat: This bat feeds in a variety of forest and woodland communities and urban and production landscapes. A diverse range of vegetation communities is required to access year round food supplies. It typically camps in closed forests >8 m high, >1 ha in area and dominated by rainforest, Broad-leaved Paperbark *Melaleuca quinquenervia*, mangrove or *Casuarina* spp. Camps are generally within 50 km of the coast or <65 m altitude, near water and on level ground or gentle slopes.

Ecology: This bat is nomadic and, although a few individuals may remain resident in a camp, most move distances >500 km and even >1000 km in response to changing food availability. It disperses pollen and seeds and can provide long distance dispersal. Females give birth to single young in October/November and lactate for approximately 6 months. It can mature at 2 years but few <3 years successfully reproduce. The mean age of reproductive adults is 6 years and individuals can reach >16 years in the wild.

Reasons for decline and current threats: A decline of at least 30% between 1989 and 2004 has been estimated. The loss of feeding habitat is thought to have been the main cause of the decline and is probably still the primary threat to the species. Clearing of forest reduces resources, but forestry practices, eucalypt dieback, drought and fire can also reduce floristic diversity, interrupt flowering and fruiting patterns and simplify the age structure of forests, reducing resource availability. Urbanisation on coastal floodplains in northern NSW and southern Qld threatens reliable winter foraging areas. Food shortages in winter and spring can reduce reproduction by 30–75%. Spring or summer food shortages also bring the bat in to feed at fruit orchards where historically they were killed in large numbers, contributing to their decline. Although killing is now illegal in Qld and Vic., licensed killing continues in NSW. Other threats are thought to include possible competition with the Black Flying-fox *P. alecto*, accidental injury and death on powerlines, entanglement in loose netting and barbed-wire fencing and the impacts of climate change on food availability and heat-related mortality.

CURRENT RECOVERY EFFORTS

Recovery plan: Department of Environment, Climate Change and Water NSW (2009) 'Draft national recovery plan for the grey-headed flying-fox *Pteropus poliocepha-*

lus'. Department of Environment, Climate Change and Water NSW, Sydney.

Other plans: Department of Environment and Climate Change NSW (2007) 'Flying-fox camp management policy'. Department of Environment and Climate Change NSW, Sydney.

Conservation actions:
- Large-scale movements and habitat use are being studied.
- Injured flying-foxes are rehabilitated and released.
- Revegetation and habitat protection programmes are under way at various camps (e.g. Woodend, Ipswich; Tooan Tooan, Hervey Bay; and Cascade Gardens, Gold Coast).
- Community groups are educating people about wildlife friendly fencing and backyard netting.
- DERM and DSEWPC are educating people to reduce human/flying-fox conflict, reduce deaths in fruit orchards and improve stewardship.
- DSEWPC and DECCW NSW are ranking the significance to Grey-headed Flying-foxes of feeding habitat in their range and informing land managers.
- Databases of camps have been created (state-wide: DERM; range-wide: DECCW NSW).
- DERM is monitoring the populations of camps at monthly intervals.

FUTURE RECOVERY EFFORTS

What you can do:
- Join your local wildlife rehabilitation groups.
- Volunteer with DERM to participate in monitoring camp populations.
- Join your local bush regeneration groups to become involved with replanting.
- Plant bat food trees in your backyard.
- Ensure bat-safe netting on your backyard fruit trees and encourage others to do the same.
- Support wildlife friendly fencing by removing or covering unnecessary barbed wire and encouraging others to do the same.

Conservation needs:
- Identify, protect and enhance key foraging and roosting habitat.
- Substantially reduce killing in commercial fruit crops.
- Reduce negative public attitudes and conflict with humans.
- Tackle the impact of powerlines, loose netting and barbed wire fences.
- Develop accurate methods for monitoring population size.
- Improve knowledge of demographics and population structure.

Research directions:
- Clarify the western boundary of the range.
- Develop accurate methods for monitoring population size and monitor trends.
- Further define the selection criteria for camp sites.
- Develop methods to monitor nectar availability at a landscape scale.
- Determine how flying-foxes track food over large areas.
- Describe dispersal behaviour, habitat requirements and mortality in juveniles.
- Examine the impacts of relocating camps.
- Trial the capacity of artificial roost structures to enhance degraded camp sites.
- Investigate the use of camps north of Bundaberg.
- Determine if different flying-fox species compete.

Sources and further reading

Eby P and Lunney D (Eds) (2002) *Managing the Grey-headed Flying-fox as a Threatened Species in NSW.* Royal Zoological Society of NSW, Mosman.

McDonald-Madden E, Schreiber ESG, Forsyth DM, Choquenot D and Clancy TF (2005) Factors affecting Grey-headed Flying-fox (*Pteropus poliocephalus*: Pteropodidae) foraging in the Melbourne metropolitan area, Australia. *Austral Ecology* **30**, 600–608.

Roberts B, Eby P, Catterall C, Kanowski J and Bennett G (2011) Analysis of the costs and outcomes of attempts to relocate an urban flying-fox camp at Maclean, NSW. In *Biology and Conservation of Australasian Bats.* Royal Zoological Society of NSW, Sydney.

Tidemann CR and Nelson J (2004) Long-distance movements of the grey-headed flying fox (*Pteropus poliocephalus*). *Journal of Zoology* **263**, 141–146.

Welbergen JA, Klose SM, Markus N and Eby P (2007) Climate change and the effects of temperature extremes on Australian flying-foxes. *Proceedings of the Royal Society B* **275**, 419–425.

Authors: Peggy Eby and Billie J Roberts

LARGE-EARED HORSESHOE BAT (LARGE FORM)

Rhinolophus philippinensis Waterhouse, 1843

Photo: Bruce Thomson

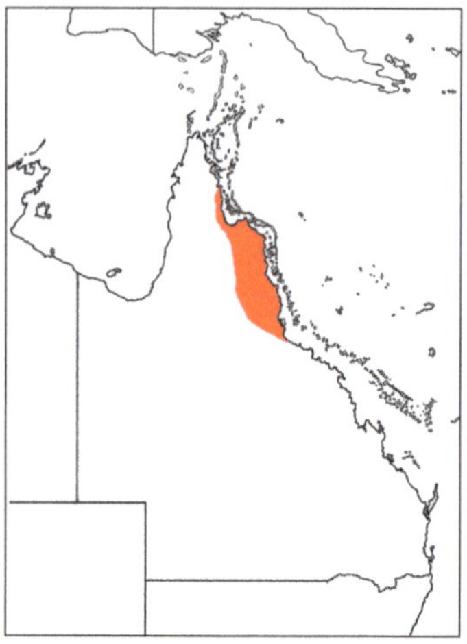

Family: Rhinolophidae

Note: The taxonomy of this bat remains unclear: a genetic study failed to differentiate small and large forms in Australia, and six subspecies of *R. philippinensis* are recognised across their range, one of which incorporates *R. robertsi* Tate, 1952. For the purposes of this book, we use *R. philippinensis* to match listings in legislation but recognise that the taxonomy is yet to be finalised.

Conservation status: IUCN: species (*R. philippinensis*) is Least Concern; Aust.: Endangered; Qld: Endangered

Description: This small bat has a head and body length of 50–60 mm, a 30–35 mm tail, a 52–59 mm forearm and a wingspan up to 365 mm. It weighs 10–16 g. Its ears are very large and pointed at the tip. Its long fur is fine and grizzled, grey-brown and paler on the belly. Its broad and long noseleaf is horse-shoe shaped and displays an equally long and broad projection (sella). Its large ears and nose make it very distinctive. The large form discussed here shows a cline in the size getting smaller in the south.

Distribution: This bat occurs from Iron Range on Cape York Peninsula to Townsville in the south and west to Chillagoe. The form discussed here occurs only in Australia, though a number of other forms of the species occur all the way north into the Philippines.

Habitat: This species is known to occupy a range of coastal and near coastal habitat types including lowland rainforest, riparian forests, open woodland, *Melaleuca* forest and the edge of grassy clearings in rainforest.

Ecology: This bat tends to roost individually or in small scattered groups, sometimes in association with other species of bat. It is known to use caves, mines and culverts, but probably also uses tree hollows and cavities, boulder piles and other suitable sites. It feeds mostly on moths and beetles, but also takes lacewings and Orthoptera (grasshoppers and crickets). It catches prey during slow, fluttery flight, often foraging close to the ground (0.5 to 2 m) and sometimes rapidly changing direction and height in pursuit of potential prey. Heavily pregnant females have been found in October and November and the single young are thought to be born in those months. Its echolocation calls are considered very low, ranging between 28 and 34 kHz.

Reasons for decline and current threats: The listing of this bat is based on a decline inferred from the loss of roost sites, including five natural caves from one site, and two mines, one of which has been recolonised. In addition, fears were held for individuals at another mine that was closed with a gate, causing the Large-eared Horseshoe Bat to use a ventilation shaft for access. Whether this fear has eventuated has not yet been reported. Current threats may include further loss and disturbance of roosts by closures of mines and mining of limestone. However, given it is likely that it is not reliant on caves or mines, the impact of these losses is far from certain. Because it appears to be naturally rare, any losses are of concern. Perhaps a greater threat is the disturbance, fragmentation and clearing of foraging habitat and the cumulative loss of all potential roosts sites brought about by significant habitat alteration.

CURRENT RECOVERY EFFORTS

Recovery plan: Thomson B, Pavey C and Reardon T (2001) 'National recovery plan for cave-dwelling bats, *Rhinolophus philippinensis, Hipposideros semoni* and *Taphozous troughtoni* 2001–2005'. Department of the Environment, Water, Heritage and the Arts, Canberra.

Other plans: Duncan A, Baker GB and Montgomery N (1999) 'The action plan for Australian bats'. Natural Heritage Trust, Canberra.

Conservation actions: A range of roost caves has been protected in the protected area estate, particularly in the Wet Tropics.

FUTURE CONSERVATION EFFORTS

Research directions:

- Resolve the taxonomic status of small and large forms in Australia.
- Examine the use of caves and other roost sites to determine the relative importance of caves and mines.
- Conduct further surveys to determine the population size and trends and better understand its conservation status.
- Review this bat's conservation status.
- Analyse its foraging habitat selection and use and determine the impact of vegetation changes at broad scales.

Sources and further reading

Cooper SJB, Reardon TB and Skilin J (1998) Molecular systematics of Australian rhinolophid bats (Chiroptera: Rhinolophidae). *Australian Journal of Zoology* **46**, 203–220.

Kutt AS (2005) Clarification of the distribution of the Long-eared Horseshoe bat *Rhinolophus philippinensis* complex in Australia. *Australian Zoologist* **32**, 629–631.

Pavey CR and Kutt AS (2008) Large-eared horseshoe bat *Rhinolophus philippinensis*. In *The Mammals of Australia*. 3rd edn. (Eds S Van Dyck and R Strahan). Reed New Holland, Sydney.

Author: Andrew J Dennis

ORANGE LEAF-NOSED BAT

Rhinonicteris aurantia (Gray, 1845)

Photo: Bruce Thomson

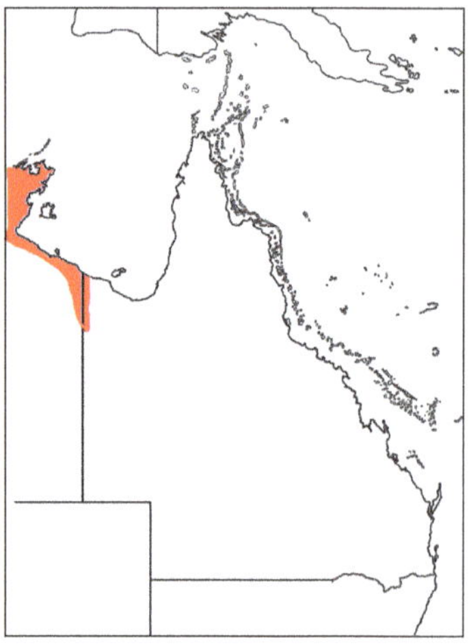

Family: Hipposideridae

Conservation status: IUCN: Least Concern; Aust.: Vulnerable; Qld: Vulnerable; WA: Rare or likely to become extinct.

Description: This small bat has a head and body length of 45–55 mm, a 23–29 mm tail completely enclosed in a membrane and a 45–50 mm forearm. It weighs 6.5–11.2 g. Its ears are 12–14 mm long and pointed at the tip. It has distinctive bright orange fur, with the dorsal surface appearing darker because of brownish tips to each hair. It can also occasionally be fawn, lemon or very pale.

Distribution: The Orange Leaf-nosed Bat occurs in the Kimberley (WA), the Top End of the NT and in far north-west Qld. An isolated Pilbara (WA) population is currently designated as a separate form because of genetic, acoustic and subtle morphological differences. In Qld, it is known from the very deep Kalkadoon Cave in the Camooweal Caves NP (~70 bats) and a relatively shallow cave in Lawn Hill Gorge NP (~20 bats). However, targeted surveys for the species have been too few to be certain of its distribution and population size in Qld. It is likely that other caves are used in the known areas, although it has not been found in caves and mines of the hard rock areas around Mount Isa and Cloncurry, despite extensive searches. Bones indistinguishable from this species occur among the 3 million year old fossils at Riversleigh north of Camooweal. An extinct species of *Rhinonicteris* and other closely related genera have also been described from this area.

Habitat: Both known roosts in Qld are in limestone caves. Elsewhere, it roosts in sandstone and silcrete caves, Devonian limestone, banded ironstone and disused mines in igneous greenstone terrain. All caves it roosts in are warm and humid. Outside Qld, colonies appear to be restricted to the deepest and most humid caves in the driest periods, suggesting that movement between distant roosts is probably only done during widespread very wet periods. It forages along steep-sided rocky gorges containing pools, open watercourses containing ephemeral pools lined with eucalypts or tall melaleuca and over spinifex (*Triodia*) or other grasslands with open eucalypt woodland.

Ecology: This bat is an acrobatic, high energy flier that intercepts insect prey while in flight, feeding in open spaces adjacent to vegetation. Its high frequency echolocation call is designed as a 'flutter detection' system. A wide range of flying insects is taken, but primarily moths and beetles. Its known roosts contain from 20 to 20 000 individuals, but numbers fewer than 100 are most likely to be encountered. In the monsoonal NT, mating occurs in July when males and females congregate together in the larger colonies. In December and January, after the birth of young, females disperse to unknown locations. Babies suckle over the early wet season and wean by late February.

Reasons for decline and current threats: The population trend in Qld is unknown, but assumed to be stable. The current listing of this species as Vulnerable in Qld is based on the few known records and colonies, and small known population size. The Pilbara form is listed as threatened because of the possibility of decline from the extensive mining activity in that region, and known concentrations in a few disused mines. Known roosts in Qld occur in natural caves, and the threat from mining activity is not likely to be as great as elsewhere. However, in Qld the species is susceptible to disturbance at roosts, and perhaps inappropriate fire regimes. It is highly prone to dehydration and loss of body heat and few caves have microclimates that allow it to maintain water and heat balance. In addition, it will easily abandon a roost after disturbance by people, sometimes for months or years. This species may be susceptible to changes in its foraging habitat that result from large scale burning and long dry periods, particularly when they coincide. It prefers gorges and gullies with water sources in arid areas. Its ability to disperse may be limited and survivorship may decline if quality of foraging habitat is reduced by fire and/or drought, and a suitable alternative roost is not within nightly flight range. This bat is susceptible to collisions with motor vehicles due to its low foraging height and curiosity for lights.

CURRENT RECOVERY EFFORTS

Recovery plan: None

Other plans: Duncan A, Baker GB and Montgomery N (1999) 'The action plan for Australian bats'. National Heritage Trust, Canberra.

Department of the Environment, Water, Heritage and the Arts (2010) 'Survey guidelines for Australia's threatened bats'. Department of the Environment, Water, Heritage and the Arts, Canberra.

Conservation actions: Known major roosts are protected and managed in both Qld and the NT.

FUTURE CONSERVATION EFFORTS

What you can do:

- Avoid entering caves where there are colonies of bats.
- Send road-kill samples to Qld Museum, but bone and guano deposits in caves should be left undisturbed because of their *in situ* scientific value.

Research directions:

- Conduct non-invasive surveys based only on echolocation detection at cave entrances, with consultation of appropriate survey guidelines before field surveys, particularly during dry periods.
- Conduct exploratory surveys to locate additional roosts and refine the knowledge of its distribution and habitat associations.
- Undertake a small-scale genetic study to inform future management as to whether the Qld population might be isolated from the one in the NT.

Sources and further reading

Armstrong KN (2006) Phylogeographic structure in *Rhinonicteris aurantia* (Chiroptera: Hipposideridae): implications for conservation. *Acta Chiropterologica* **8**, 63–81.

Armstrong KN and Coles RB (2007) Echolocation call frequency differences between geographic isolates of *Rhinonicteris aurantia* (Chiroptera: Hipposideridae): implications of nasal chamber size. *Journal of Mammalogy* **88**, 94–104.

Baudinette RV, Churchill SK, Christian KA, Nelson JE and Hudson PJ (2000) Energy, water balance and the roost microenvironment in three Australian cave-dwelling bats (Microchiroptera). *Journal of Comparative Physiology B* **170**, 439–446.

Author: Kyle Armstrong

BARE-RUMPED SHEATHTAIL BAT

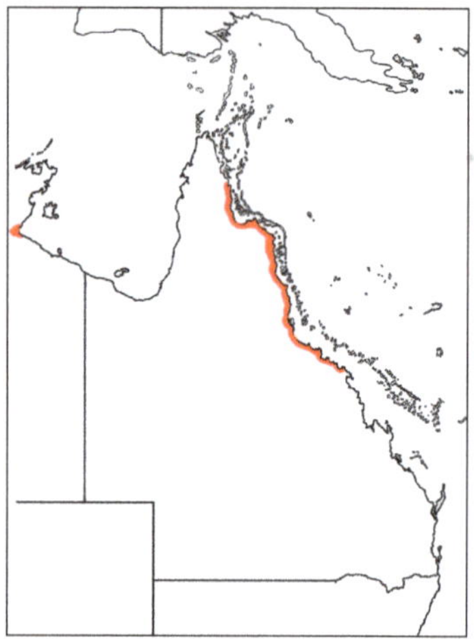

Saccolaimus saccolaimus nudicluniatus (Temminck, 1838)

Family: Emballonuridae

Conservation status: IUCN: species (*S. saccolaimus*) is Least Concern; Aust.: Critically Endangered; Qld: Endangered

Description: This large insectivorous bat has a head and body length of 81–99 mm, a 23–40 mm tail mostly enclosed in a membrane and a 72–80 mm forearm. It weighs 31–61 g. Its ears are small (16–22 mm) and narrow towards the blunt apex. Its fur ranges from brown to dark reddish brown or black and is slightly paler on the belly in Qld specimens, but white on the belly in the NT. It usually has white flecks in the fur and is generally hairless from the hips down. Males have a throat pouch, which is rudimentary in females.

Distribution: This bat occurs in north-east Qld in coastal areas from Bowen to Iron Range on Cape York Peninsula, in the NT on the western edge of the Gulf of Carpentaria and on the north-west coast and extensively in Australasia from the Solomon Islands in the east to India in the west.

Habitat: In Australia, the Bare-rumped Sheathtail Bat occupies tropical woodland and tall open forests where it roosts in long, wide hollows in the trunks of various eucalypts in groups of 4–40. In Qld it appears to prefer coastal eucalypt forests with high rainfall. Although not found in caves in Australia, it does roost in caves in other parts of its range, so may do so here as well.

Ecology: This species forages in the open air above the forest canopy. Hence, it has long narrow wings and a rapid, straight flight, although it is also known to be reasonably manoeuvrable in flight. The reproductive timing seems to vary with location. The few observations available suggest that, in Qld, females give birth to their single young in December and January, whereas in the NT birth is probably over a much wider period, if not all year.

Reasons for decline and current threats: So little is known about this species that threats to its populations are mostly unknown. Its listing is based on the lack of records of the species over a 16-year period and only a few since then. One of the reasons for this may be that collectors no longer use shotguns to sample fast, high-flying bats but rely on recording echolocation calls. This species cannot be differentiated from a more common species by its echolocation call. However, one clear threat is the clearing of coastal eucalypt forests for urban expansion, other infrastructure and farming. Several roost trees have been destroyed and some of the few specimens collected have been taken from trees being felled. Other possible threats listed for the species include increased competition for hollows from native and exotic birds and bees, the selective removal of hollow-bearing trees, such as those considered unsafe in parks and along roads and tracks and, because the bats seem to occupy a narrow climatic range, climate change. Disease is also listed, though apart from a related species being known to carry Australian Bat Lyssavirus, no explanation as to how it might be a threat is offered.

CURRENT RECOVERY EFFORTS

Recovery plan: Schulz M and Thomson B (2007) 'National recovery plan for the bare-rumped sheathtail bat *Saccolaimus saccolaimus nudicluniatus*'. Report to Department of the Environment and Water Resources, Canberra. Queensland Parks and Wildlife Service, Brisbane.

Other plans: Duncan A, Baker GB and Montgomery N (1999) 'The action plan for Australian bats'. Natural Heritage Trust, Canberra.

FUTURE CONSERVATION EFFORTS

What you can do: Anyone finding bats that may be this species should notify state environment authorities, a museum or a local wildlife carer who handles bats and record relevant sighting information. Live animals should not be handled by people who are not vaccinated against rabies, and dead animals should be handled only by people experienced taking precautions against fatal infectious diseases.

Conservation needs:

- Protect potential roosts – trees with deep wide hollows in their trunks.
- It is essential that any records of its existence are made known.

Research directions:

- Develop a technique to reliably detect and record the presence of this species.
- Conduct a targeted survey program to assess the distribution of the species, its population size and status and habitat preferences.
- Further clarify its taxonomic status.

Sources and further reading

Churchill S (2008) *Australian Bats*. 2nd edn. Allen and Unwin, Crows Nest, NSW.

Hall LS, Thomson BG and Milne DJ (2008) Bare-rumped Sheathtailed Bat, *Saccolaimus saccolaimus*. In *The Mammals of Australia*. 3rd edn. (Eds S Van Dyck and R Strahan) pp. 475–475. Reed New Holland, Sydney.

Author: Andrew J Dennis

COASTAL SHEATHTAIL BAT

Taphozous australis Gould, 1854

Photo: Bruce Thomson

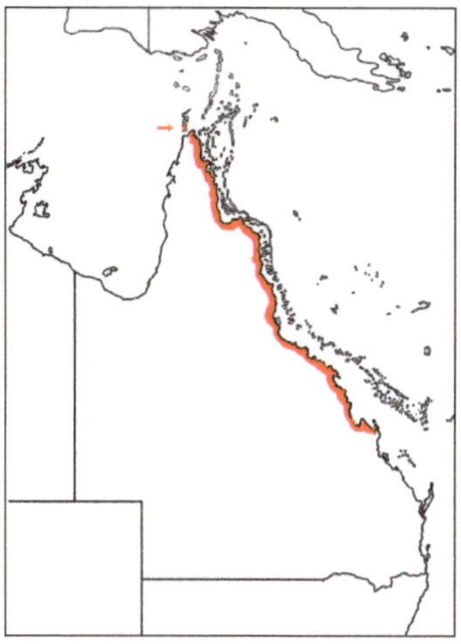

Family: Emballonuridae

Other common names: Little Sheath-tailed Bat, North-eastern Sheath-tailed Bat

Conservation status: IUCN: Near Threatened; Qld: Vulnerable

Description: This large insectivorous bat has a head and body length of 60–90 mm, a 20–30 mm tail, mostly enclosed in a membrane, and a 61–68 mm forearm. It weighs 20–50 g. Its ears are small (19–25 mm) and roughly triangular. Its fur ranges from light brown or fawn to dark grey-brown and it is slightly paler on the belly. Males have a functional throat pouch, which is rudimentary in females.

Distribution: This species occurs along the Qld coast from Shoalwater Bay to the southern coast of New Guinea, and appears to remain within a few kilometres of the sea, including on offshore islands and through the Torres Strait.

Habitat: The Coastal Sheathtail Bat occupies habitats close to the sea including coastal dune communities, melaleuca swamps, mangroves and rainforests and any other habitats within foraging range of its roost sites. It roosts in the twilight zone of coastal rock shelters including caves, overhangs, boulder fields and anything offering a similar level of shelter.

Ecology: This species has long slender wings and flies fast in the open air above the forest canopy where it catches beetles and other insects. These it brings back to a foraging roost, which may be the same as its daytime roosts, to consume. It is known to roost on islands and commute to the mainland to forage. The testes of males descend to the scrotum in April and retract in September. Young are born from September to November and young are large by January.

Reasons for decline and current threats: Declines in numbers in roost sites near Cape Hillsborough and one near Townsville, which have both been attributed to disturbance by people, first suggested a threat to the species. However, there is no evidence that this translates to a population decline; bats may simply have moved to alternative sites to roost. More important is the very narrow distribution of the species, within a few kilometres of the coast: an area of constant human disturbance and interference. In addition, it appears to occupy a higher percentage of sea caves in the southern part of its range where disturbance is more likely to increase.

CURRENT RECOVERY EFFORTS

Recovery plan: None

Other plans: Duncan A, Baker GB and Montgomery N (1999) 'The action plan for Australian bats'. Natural Heritage Trust, Canberra.

FUTURE RECOVERY EFFORTS

Conservation needs:

- Raise awareness of the species and its susceptibility to disturbance.
- Protect roosts and maternity caves.
- Carefully record and report data on confirmed sightings.

Research directions:

- Undertake a survey program, perhaps in combination with surveys for the Bare-rumped Sheathtail Bat *Saccolaimus saccolaimus nudicluniatus*, to assess the population size and status of the species and identify roost and maternity sites in danger of disturbance.
- Once sufficient data are collected and collated, its conservation status should be reassessed and further actions determined.

Sources and further reading

Churchill S (2008) *Australian Bats*. 2nd edn. Allen and Unwin, Crows Nest, NSW.

Hoye G (1985) Observations on bats of Cape Hillsborough national Park, Queensland. *Macroderma* **1**, 48–51.

Richards GC (2008) Coastal Sheath-tailed Bat, *Taphozous australis*. In *The Mammals of Australia*. 3rd edn. (Eds S Van Dyck and R Strahan) pp. 476–477. Reed New Holland, Sydney.

Author: Andrew J Dennis

RODENTS (ORDER: RODENTIA)

BRAMBLE CAY MELOMYS

Melomys rubicola Thomas, 1924

Photo: Andrew J Dennis

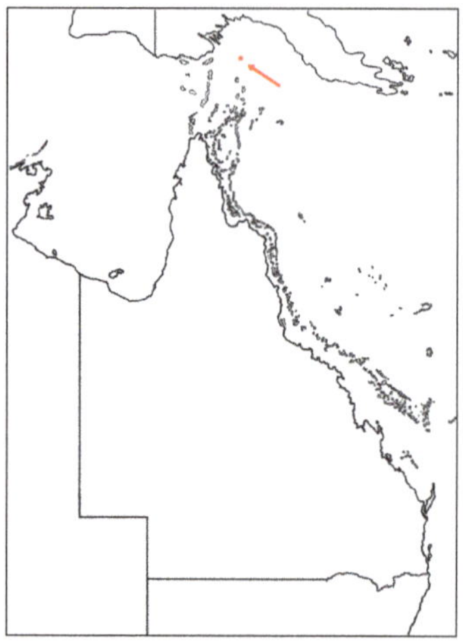

Family: Muridae

Conservation status: IUCN: Critically Endangered; Aust.: Endangered; Qld: Endangered

Description: This large mouse has a head and body length of 148–165 mm and a long tail of 145–185 mm, but short ears and large feet. It weighs 78–164 g. The tail is roughly scaled and prehensile at the tip. This melomys has red-brown fur with long black guard hairs above and grey-brown below. Like most melomys, it has a slightly 'roman' nose.

Distribution: This mouse occurs only on Bramble Cay, a small (4–5 ha) cay made of the skeletons of marine diatoms and perched on a coral reef, which is itself attached to a rock outcrop. It is in the far north-east of the Torres Strait, ~50 km from the mouth of the Fly River in Papua New Guinea. Bramble Cay Melomys do not occupy the entire area of the cay, instead they occur with the vegetated part: an area of ~2 ha.

Habitat: The Bramble Cay Melomys occurs in the vegetation growing on the cay, which includes a few herbs and grasses forming a low cover up to 400 mm high. Up to 11 species of plant have now been recorded on the cay but only three to five are generally present or abundant at any one time. Common species include *Boerhavia albiflora*, *Portulaca oleracea* (a succulent herb), *Cenchrus echinatus* (a grass) and *Amaranthus viridis* (a leafy herb). Bare patches of guano also occur on the cay, though they are probably not used by the melomys. Perhaps most distinctively, their habitat is characterised by huge numbers of seabirds and by the seasonal disturbance of nesting Green Turtles *Chelonia mydas*.

Ecology: Trap data demonstrated that the Bramble Cay Melomys has a strong female-biased sex ratio. Pregnant and lactating females, juveniles and subadults have all been captured in July, suggesting an extended breeding season at least over winter months. Most observations of the species, and all captures, are from the vegetated area of the cay, although some observations were made on the foreshore. Melomys avoid areas of high seabird density, evidenced by a strong negative correlation between sea birds and trap success, preferring instead to forage in dense vegetation where they have been observed feeding on *P. oleracea*. They have also been recorded eating turtle eggs and probably use all available resources at different times.

Reasons for decline and current threats: There is no decline measured for the Bramble Cay Melomys. It is listed as Endangered on the basis of its tiny distribution, population and area of occupancy. Early reports in 1845 noted 'large numbers'; later, in 1978, hundreds were seen. The only population estimate available (~90) was based on 42 captures in 1998, while subsequent trapping studies in 2002 and 2004 only yielded 10 and 12 individuals. It is very likely that actual numbers, but also trap success, would fluctuate considerably with seasons, rainfall and disturbance. Current threats centre on three areas: the fact that only a single population exists; the high potential for disturbance to the cay; and the fact that the species appears to be inbred. Disturbance to the cay could take a range of forms including excessive erosion and/or salt water inundation caused by storms and tides killing vegetation and reducing habitat area or the introduction of detrimental weeds, animals or disease, any of which could quickly push this tiny population to extinction.

CURRENT RECOVERY EFFORTS

***Recovery plan*:** Latch P (2008) 'Recovery plan for the Bramble Cay Melomys *Melomys rubicola*'. Report to Department of the Environment, Water, Heritage and the Arts, Canberra. Environmental Protection Agency, Brisbane.

Other plans: Lee AK (1995) 'The action plan for Australian rodents.' Australian Nature Conservation Agency, Canberra.

Conservation actions:

- The species' taxonomic status has been confirmed and it is distinct from all Australian and all described New Guinean melomys on both morphological and genetic grounds.
- Baselines for monitoring the area and status of the cay and its melomys population were established in 1998 and re-measured in 2002 and 2004.
- The traditional owners of the cay (Erubam Le) and the Australian Coast Guard have been informed about and engaged in work on the melomys since 1998 and are integral to its recovery planning and the execution of actions.

FUTURE CONSERVATION EFFORTS

Conservation needs:

- Educate visitors to the cay about the melomys and the threats to its survival.
- Enforce strict quarantine measures on visitors by restricting visitation to the cay and possibly using engineering techniques to stabilise or enlarge the cay.
- Establish one or more other populations either in captivity or on another carefully selected island as a safeguard.

Research directions:

- Continue monitoring the cay and its melomys.
- Conduct targeted surveys on Torresian islands. Although not described from New Guinea, it remains to be determined if it has gone undetected in the Fly River catchment and, despite significant surveys in the Torres Strait, there is still good justification for further targeted surveys.
- Study its ecology and biology, including its diet, water balance, physiological tolerances, social behaviour and reproductive cues, rates and longevity.
- Conduct more extensive investigations into the likely impacts of sea level rise in the Torres Strait: whether the area rises out of the sea due to isostatic rebound or whether it will be inundated.

Sources and further reading

Dennis A and Storch D (1998) 'Conservation and taxonomic status of the Bramble Cay melomys, *Melomys rubicola*'. Unpublished report to Environment Australia Endangered Species Program Project No. 598. Queensland Department of Environment.

Dennis AJ (2008) Bramble Cay Melomys *Melomys rubicola*. In *The Mammals of Australia*. 3rd edn. (Eds S Van Dyck and R Strahan) pp. 449–450. Reed New Holland, Sydney.

Limpus CJ, Parmenter CJ and Watts CHS (1983) *Melomys rubicola*, an endangered murid rodent endemic to the Great Barrier Reef of Queensland. *Australian Mammalogy* **6**, 77–79.

Menzies JI (1996) A systematic revision of Melomys (Rodentia: Muridae) of New Guinea. Australian *Journal of Zoology* **44**, 367–426.

Author: Andrew J Dennis

NORTHERN HOPPING-MOUSE

Notomys aquilo Thomas, 1921

Photo: Frank Woerle-AUSCAPE

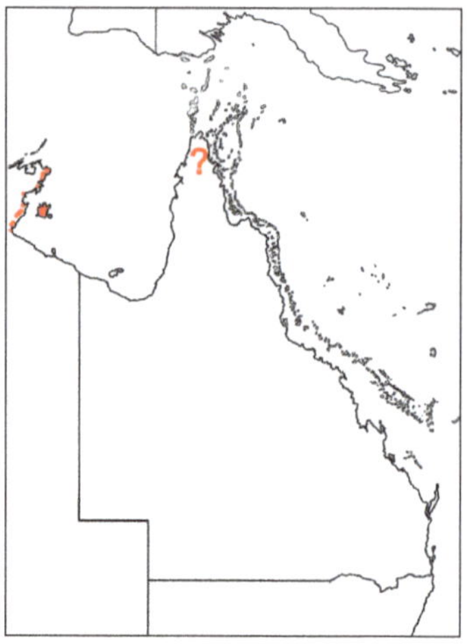

Family: Rodentia

Aboriginal names: Woorrentinta, Djirrkin

Conservation status: IUCN: Endangered; Aust.: Vulnerable; Qld: Vulnerable; NT: Vulnerable

Description: This small mouse has a body length of 90–115 mm and a long tail (140–175 mm) with a distal tuft of dark hairs. It weighs 30–50 g. It has large ears and eyes and elongated hind feet. It has thin, sandy-brown fur above with a pure white belly. Both sexes have a small throat pouch.

Distribution: First collected before 1867 on Cape York Peninsula, this species has not been seen in Qld since. The original collection locality remains uncertain, though it may have been near Somerset at the tip of the peninsula because this was where the specimen's collector, John Cockerell, was based. However, Cockerell also spent much time sailing the Gulf of Carpentaria and the Arafura Sea and he appears to have been an untrustworthy record keeper who misrepresented the source of many of the specimens that he obtained. If he was in error here, and collected this specimen from the western side of the Gulf of Carpentaria where the Northern Hopping-mouse still occurs, it is possible that this species has never occurred in Qld at all. The species occurs patchily in coastal areas of the western Gulf region, in parts of Arnhem Land and more abundantly on Groote Eylandt. There are early records from the Roper River.

Habitat: The Northern Hopping-mouse occurs in a wide variety of habitats with a sandy substrate, from coastal dunes with grass or heath, to shrubland, open forest and the edges of dense rainforest thickets. It occurs in a distribution with 1000–1400 mm of rain a year: substantially wetter and more productive than habitats occupied by other species of hopping-mice.

Ecology: This species is very social and lives communally in extensive and elaborate burrow systems. These comprise a vertical entrance shaft that descends to 45 cm or more, several smaller 'pop-holes' and a spoil heap of excavated sand, all within an area of several square metres. Burrows reach 2 m or more in length, and animals occupy a slightly enlarged nest chamber lined with scraps of vegetation. Breeding has been recorded from January to June on Groote Eylandt, but extends into the second half of the year in captivity. Females have four teats and rear one to five young at a time. Gestation may take as long as 7 weeks and animals mature at 3–4 months. It is nocturnal and elusive, detected most reliably by its distinctive tracks and burrows. It does not climb and feeds mostly on the seeds of grasses, low herbs and shrubs, as well as on green vegetation and occasional insects.

Reasons for decline and current threats: The listing of this species is based on its small area of occurrence, few populations and the possibility of a decline. Its disappearance from Cape York Peninsula is uncertain because it may not have ever occurred there. In Arnhem Land, the existence or extent of a decline is uncertain, though one is suspected. Despite the uncertainty of a decline, clear threats are evident. These include: habitat change from grazing by livestock, buffalo and feral pigs; invasion of unpalatable weeds; and a change in fire regimes towards more frequent and intense, late dry season wildfires. Predation from cats likely has further impacts. Cats may exploit the social behaviour of the hopping-mice, waiting at the entrance of burrows for animals to emerge and eating entire colonies over short periods: a strategy made easier after fire.

CURRENT RECOVERY EFFORTS

Recovery plan: Woinarski JCZ (2004) 'National multi-species recovery plan for the Carpentarian Antechinus *Pseudantechinus mimulus*, Butler's Dunnart *Sminthopsis butleri*, and Northern Hopping-mouse *Notomys aquilo*, 2004–2009'. Northern Territory Department of Infrastructure Planning and Environment, Darwin.

Other plans: Department of the Environment, Water, Heritage and the Arts (2008) 'Threat abatement plan for predation by feral cats'. Department of the Environment, Water, Heritage and the Arts, Canberra.

Conservation actions:

- Three Indigenous Protected Areas (IPAs), Dhimurru, Anindilyakwa and Laynhapuy, totalling ~3.9 million ha, are being managed for fire, weeds and feral animals, thus helping to reduce the threats to the Northern Hopping-mouse.

- The NT Department of Natural Resources, Environment and the Arts and rangers from the Anindilyakwa Land Council have collaborated since 2006 to find that Northern Hopping-mice prefer moderately open and floristically diverse heathland, and that mice are most easily detected by spotlighting.

FUTURE RECOVERY EFFORTS

What you can do:

- Do not own a cat.
- Practice good ecological hygiene – do not bring invasive species such as Cane Toads, weeds or their seeds with you (e.g. regularly wash down under cars).
- Report sightings of hopping mice on Cape York Peninsula (to DERM) and in Arnhem Land (to NTDNREA) – it is very likely to be a Northern Hopping-mouse.

Conservation needs:

- Support Indigenous Protected Areas (IPAs) - all current records are from IPAs.
- Help to increase its profile in the NT and Qld – information about the hopping-mouse needs to be communicated among the general public and interested stakeholders.
- Support the establishment of a recovery team – recovery actions need to be planned, implemented and monitored.
- Develop an effective survey technique.

Research directions:

- Conduct standardised surveys of animal's tracks or apply new more cost-effective techniques using cameras.
- Carry out surveys to identify where the species occurs, its population status, population trends, habitat, other resource needs and key threats.

Sources and further reading

Threatened Species Network (2007) Northern Hopping-mouse *Notomys aquilo*. <http://www.environment.gov.au/biodiversity/threatened/publications/pubs/tsd07-n-hopping-mouse.pdf>.

Watts CHS and Aslin HJ (1981) *The Rodents of Australia*. Angus & Robertson, Sydney.

Woinarski JCZ and Flannery TF (2008) Northern Hopping-mouse *Notomys aquilo*. In *The Mammals of Australia*. 3rd edn. (Eds S Van Dyck and R Strahan) pp. 599–600. Reed New Holland, Sydney.

Woinarski JCZ, Gambold N, Wurst D, Flannery TF, Smith AP, Chatto R and Fisher A (1999) Distribution and habitat of the Northern Hopping-mouse, *Notomys aquilo*. *Wildlife Research* **26**, 495–511.

Author: Chris R Dickman

DUSKY HOPPING-MOUSE

Notomys fuscus (Jones, 1925)

Photo: Peter Canty

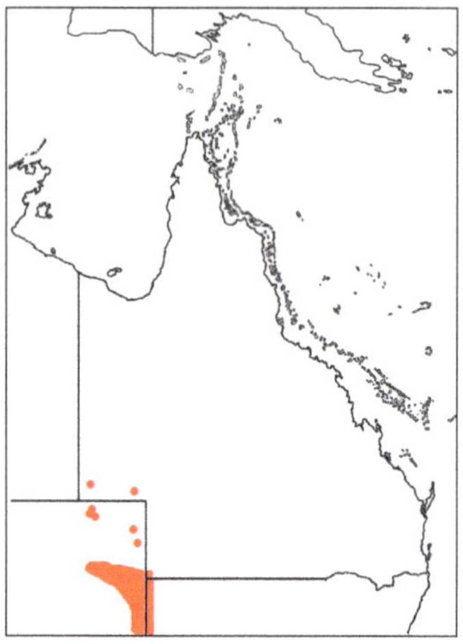

Family: Muridae

Other common names: Birdsville Hopping-mouse, Wood Jones' Hopping-mouse

Aboriginal name: Wilkinti

Conservation status: Aust.: Vulnerable; Qld: Vulnerable; NSW: Endangered; SA: Vulnerable; NT: Probably Extinct

Description: This small mouse has a head and body length of 76–115 mm and a long tail of 115–155 mm with a distal tuft of longer hairs. It weighs 26–55 g. It has large ears and eyes, elongated hind feet. The fur is orange to fawn above and white below. While similar to the sympatric Fawn Hopping-mouse *Notomys cervinus*, it has a longer arched snout and a well developed throat-pouch, which is covered by a distinct ball-like swirl of stiff white hair.

Distribution: This mouse once occurred on the north-western and eastern margins of the Simpson Desert in NT and Qld and from the north of Coopers Creek to the Southern Strzelecki Desert in SA. Subfossil remains and early specimens found in the 1930s indicate an earlier more southerly distribution into central SA. Current records in Qld are all from an area near Beetoota, with one record from the Simpson Desert in 1984. Populations are known to 'boom' following good rainfall events, which have been recorded for the southern Strzelecki Desert in 1845, during the early 1990s and from 2005 to 2008. This last 'boom' period has enabled the species to be regularly detected in north-western NSW after an absence of more than 150 years.

Habitat: Arid dune fields with perennial Sandhill Canegrass *Zygochloa paradoxa*, or Nitre Bush *Nitraria bilardierei*, with scattered Umbrella Wattles *Acacia ligulata*, support current known populations of this mouse during dry seasons. When populations are high, the mice can often be found in a larger variety of sandy habitats. Recently they were found on a gibber plain with chenopod shrublands and a sandy creek line at the Beverly uranium mine. Their temporary presence at these monitored sites was attributed to good rainfall.

Ecology: Dusky Hopping-mice excavate deep burrow systems on the stable crests and slopes of dunes. These consist of one or more horizontal tunnels at 1–2 m deep, connected to the surface by vertical shafts, 2–3 cm in diameter, and terminating in a flat hole at the surface. Litters comprise up to five young, which are weaned after 30 days. Breeding has been recorded throughout the year and is triggered by significant rainfall events. The oldest recorded wild animal was a little over 3 years. The Dusky Hopping-mouse appears to eat seeds, small amounts of green vegetation including flowers, leaves and stems, insects and even small lizards.

Reasons for decline and current threats: The reasons for the decline of the Dusky Hopping-mouse from the eastern and western margins of the Simpson Desert have not been determined. It is likely that habitat degradation and competition from rabbits and other introduced herbivores may have led to localised extinctions during droughts. Dingo control may also have had an impact because their reduction is often accompanied by more foxes (the fox suppresses the mouse populations more effectively than the dingo).

CURRENT RECOVERY EFFORTS

Recovery plan: None

Other plans: Lee AK (1995) 'The action plan for Australian rodents.' Australian Nature Conservation Agency, Canberra.

Conservation actions:

- The South Australian Department for Environment and Heritage monitored two populations, one in the Strzelecki Desert SA and near Betoota Qld, from 1992 to 2002. Continuing work in SA is planned by the SA Government.
- There is no program in place for the Qld population.

FUTURE CONSERVATION EFFORTS

What you can do: Collect dead hopping-mice (e.g. road-kills) and drop them in to your nearest QPWS office or the relevant state museums.

Conservation needs:

- Find out if this species still persists in the Simpson Desert.
- Researchers conducting fauna trapping should accurately identify all potential *Notomys* species. Specimens with good location data should be

collected for the Qld Museum in accordance with Qld wildlife legislation.

- Establish a recovery team for threatened arid zone fauna for south-west Qld, who could have close links with a similar group in SA and NSW covering the same suite of species. The recovery team should provide advice on the design of research projects and oversee the implementation of the recovery plan. Other species could include the Fawn Hopping-mouse, Kowari *Dasyuroides byrnei*, Kultarr *Antechinomys laniger* and Plains Mouse *Pseudomys australis.*
- Use agreements to protect critical habitat on public or private land and manage impacts around watering points and dingo control in and around the core habitat.
- Adequately protect enough critical habitat to maintain or increase current distribution.

Research directions:

- Develop detection and monitoring techniques; tracks and signs are distinct, but trapping may be required where multiple *Notomys* species co-occur.
- Determine the current distribution and occupancy levels (broadly mapped in SA, but unknown in Qld).
- Identify any potential habitat that can be occupied during dry, non-boom periods.
- Clearly identify the threats to the current populations and determine how those threats can be managed or reduced. The initial focus should be on exotic herbivores and the impacts of predators and interactions between predators.

Sources and further reading

Dickman CR, Pressey RL, Lim L and Parnaby HE (1993) Mammals of particular conservation significance in the western division of NSW. *Biological Conservation* **65**, 219–248.

Letnic M, Crowther MS and Koch F (2009) Does a top-predator provide an endangered rodent with refuge from predation by an invasive mesopredator? *Animal Conservation* **12,** 302–312.

Letnic M, Koch F, Gordon C, Crowther MS and Dickman CR (2009) Keystone effects of an alien top-predator stem extinctions of native mammals. *Proceedings of the Royal Society of London B* **276,** 3249–3256

Moseby KE, Owens H, Brandle R, Bice JK and Gates J (2006) Variation in population dynamics and movement patterns between two geographic populations of the dusky hopping-mouse (*Notomys fuscus*). *Wildlife Research* **33**, 223–232.

Waudby HP and How T (2008) An additional record of the dusky hopping-mouse *Notomys fuscus* in South Australia. *Australian Mammalogy* **30**, 47–49.

Author: Robert Brandle

PLAINS MOUSE

Pseudomys australis Grey, 1832

Photo: Peter Canty

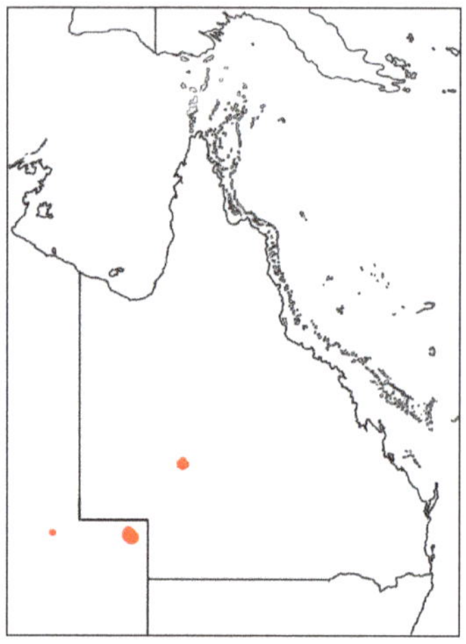

Family: Muridae

Other common names: Plains Rat, River Rat, Eastern Mouse, Eastern Rat

Aboriginal name: Palyoora

Conservation status: IUCN: Vulnerable; Aust.: Vulnerable; Qld: Endangered; NSW: Extinct in the Wild; SA: Vulnerable; NT: Vulnerable

Description: This large mouse has a head and body length of 90–145 mm and a tail length of 85–125 mm. It weighs 30–65 g. It has large ears and the tail is covered with fine fur that is grey on top and white below, becoming lighter towards the tip. It has grey to grey-brown fur above and white or cream below, which can tend to grey on some animals.

Distribution: The Plains Mouse was distributed from the edge of the Great Dividing Range in central Qld and northern NSW, through much of SA to the southern coast of WA. Today it appears to occur in central northern SA and southern NT and has recently expanded into sandy habitats within a predator and herbivore controlled area (Arid Recovery Project) north of Roxby Downs, SA. Qld specimens come from Barcarolle on the Diamantina River (pre 1936). Recently, the closest specimen to Qld is from 1968 on Cordillo Downs Station in north-east SA.

Habitat: This mouse originally appeared to occupy a wide variety of habitats, ranging from mallee communities, vegetated swamps and floodplains, dune fields supporting Mulga *Acacia aneura* or Sandhill Canegrass *Zygochloa paradoxa* to stony plains supporting only ephemeral plants. More recently, it has been detected primarily in cracking clay habitats, particularly more productive depressions and minor drainage lines on stony plains with chenopod low open shrublands and ephemeral grasslands/herblands. It is also found in sandy habitats at Roxby Downs.

Ecology: When populations expand after above average rainfall, animals burrow in friable or stable soils around the roots of perennial shrubs and temporarily inhabit a wide variety of habitats including dune fields. Burrows can be shallow warrens to 60 cm deep with numerous tunnels and several chambers up to 2 m^2 in area. During exceptional population booms, up to 20 individuals can share a burrow, though more usually only one animal and its offspring share a warren. The Plains Mouse breeds continuously in captivity and in the wild breeds opportunistically in response to rainfall events. The oestrus cycle averages 8.5 days, gestation lasts 30–31 days and the average litter size is 3.6. The Plains Mouse has four teats and females give birth at 12 weeks of age. The young attach themselves to the mother's teats and may be dragged around during lactation. Analysis of faecal pellets indicates the species feeds on seeds, together with some green plant material and invertebrates.

Reasons for decline and current threats: There are no clear causes of its severe decline and range contraction. However, habitat loss through extensive clearing and habitat degradation from introduced herbivores are likely past and current threats, particularly in gilgai habitats. Stock can remove vegetation cover, trample burrows and silt up cracks in gilgai. Rabbits may have been involved in the disappearance of the Plains Mouse from habitats other than gilgai, by consuming all the food and raising fox numbers. Predation by cats and foxes may suppress subpopulation irruptions and prevent large-scale dispersal events. Local subpopulation irruptions and resultant dispersal are likely to be important mechanisms for maintaining area of occupancy, extent of occurrence and genetic diversity. Predation by cats and foxes during drought conditions may also threaten populations. Because their shallow burrows are easy to excavate, populations are more vulnerable to local extinction during drought. The recent increase in the baiting of dingoes north of the dingo exclusion fence may threaten the remaining populations. Dingoes can suppress fox populations and possibly also cats.

CURRENT RECOVERY EFFORTS

Recovery plan: Moseby KM (2006) 'Recovery plan for *Pseudomys australis*'. Report for the South Australian Department for Environment and Heritage.

Other plans: Lee AK (1995) 'The action plan for Australian rodents.' Australian Nature Conservation Agency, Canberra.

FUTURE RECOVERY EFFORTS

Conservation needs:

- Report location records for specimens or burrows thought to be this species.
- Adequately protect enough critical habitat to maintain or increase the current distribution. This can be done by ensuring that at least 30% of known habitat is remote from high grazing pressure around water sources.
- Establish a recovery team for threatened arid zone fauna across all states and ensure the specific needs of this species are incorporated into actions and research.
- Ensure there is a latitudinal spread of currently and previously occupied habitat throughout its range as a buffer against climate change.
- Identify appropriately managed and available sites for re-introductions in Qld from wild populations during the next irruption.
- Ensure that the Moon Plain near Coober Pedy remains remote from watering points because it is the largest area of gypseous cracking clay known to be inhabited by the Plains Mouse and is likely to be an important source population.

Research directions:

- Develop methods for detecting the Plains Mouse at low densities when trapping becomes ineffective.
- Conduct surveys in previously occupied areas to try to detect remnant populations.
- Conduct surveys to identify and map known and potential critical habitat used during both wet and dry periods in all states.
- Clearly identify threats to current populations including, determining the impact of stock grazing and predation.
- Determine the effects of watering points on the species. For example, monitor persistence, shelter preference and breeding of colonies close to and remote from water.
- Determine the effects of watering points on the threats to the species, including the impact of cat and fox predation and herbivore impacts near and far from water.

Sources and further reading

Brandle R and Moseby KE (1999) Comparative ecology of two populations of *Pseudomys australis* in northern South Australia. *Wildlife Research* **26**, 541–564.

Brandle R, Moseby KE and Adams M (1999) The distribution, habitat requirements and conservation status of the plains rat, *Pseudomys australis* (Rodentia: Muridae).*Wildlife Research* **26**, 463–477.

Corbett LK, Newsome AE and Jones MA (1975) *Pseudomys australis minnie* – a new record for the Northern Territory. *Australian Mammalogy* **1**, 392–393.

Letnic M, Koch F, Gordon C, Crowther MS and Dickman CR (2009) Keystone effects of an alien top-predator stem extinctions of native mammals. *Proceedings of the Royal Society of London B* **276,** 3249–3256.

Author: Robert Brandle and Chris R Pavey

HASTINGS RIVER MOUSE

Pseudomys oralis Thomas, 1921

Photo: Paul Meek

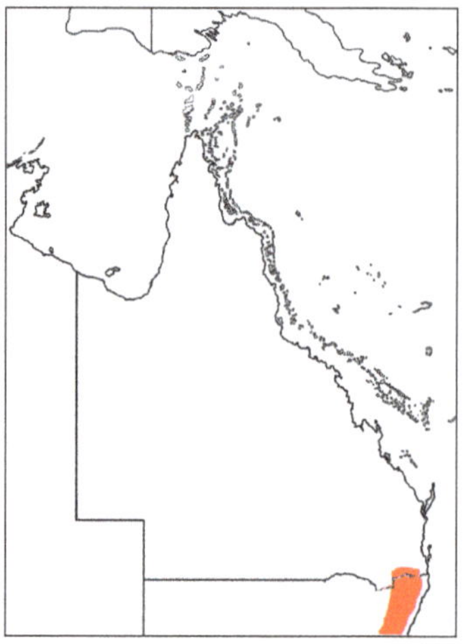

Family: Muridae

Other common name: Hastings River Rat

Aboriginal name: Koontoo

Conservation status: IUCN: Vulnerable; Aust.: Endangered; Qld: Vulnerable; NSW: Endangered

Description: This medium-sized mouse has a head and body length of 120–155 mm and a tail length of 110–150 mm. It weighs 80–100 g, with males being slightly larger than females. It has large protuberant eyes surrounded by black eye rings and a distinctly rounded 'roman nose'. Its slender tail is dark above and pale below. It has soft grey-brown fur above with greyish-white below. Its white feet have distinctive hairs that protrude over the toe nails.

Distribution: Subfossil evidence demonstrates that the Hastings River Mouse occurred from Mapleton in southern Qld to Buchan in north-east Vic. Living animals occur in ~40 subpopulations from south-east Qld to central NSW at elevations from 300 to 1250 m. In Qld, it is known from the Great Dividing and McPherson Ranges, with populations present in and adjacent to Main Range NP and Lamington NP. The total population is unknown, although the NSW NPWS Atlas reports over 700 individuals captured at approximately 222 sites between 1969 and 2007 across both NSW and Qld. Recent surveys in NSW have increased the known locations and total records of this mouse.

Habitat: The species prefers eucalypt open forests with variable understoreys that include good cover of ferns, grasses, sedges, heath, herbs and forbs. It spends most of its time beneath this ground cover, which also includes fallen logs. Refuge and nest habitat such as fallen logs, rock piles and root holes are critical to its persistence. In Qld, the Hastings River Mouse is mostly found on volcanic soils in association with New England Blackbutt *Eucalyptus campanulata* on the Great Dividing Range and other tree associations elsewhere. In NSW, it also occurs with New England Blackbutt forest types (*E. campanulata*, *E. microcorys*, *E. pilularis* and *E. saligna*), primarily on granodiorite, siltstone and siltstone-combinations and rainfall, temperature and solar radiation also correlate with its distribution.

Ecology: The Hastings River Mouse appears adapted to periodic disturbance of its habitat to maintain plant associations in a mid-successional phase. The species declines when the understorey complexity diminishes due to over-frequent fire or when a mesic shrub layer develops in late succession communities. It is able to cope well with logging in managed state forests using fallen trees, as well as fallen head and butt residue from logging operations. Individuals range over 0.53 to 0.8 ha and rarely up to 2 ha. Juveniles can disperse over a kilometre at 4–5 month of age. The Hastings River Mouse can breed up to three times per year, starting in August and ending as late as April, with generally two young, but up to four in rare cases. The young nipple-cling for a few weeks before being left in the nest. They usually mature in their second year, but early born young occasionally breed in their first year and live for 2–3 years. This mouse eats leaves, stems, seeds, flowers, pollen, fern sporangia, insects and fungi, with a preference for leaf and seed in summer and leaf in winter. The main food plants include *Taraxacum, Echinopogon, Entolasia, Imperata, Lomandra, Poa, Stipa, Themeda, Juncus, Cyperus, Gahnia, Carex* and *Restio*.

Reasons for decline and current threats: Several factors combine to suggest this species has declined in recent times and is threatened: subfossil evidence suggests a historical decline; reasonably fresh owl pellets contain its bones in areas north of the current distribution; recent monitoring in Qld has demonstrated declines at some sites; its populations appear naturally small and scattered; and it seems susceptible to some significant threatening processes. Threats may include clearing and removal of rock, wood and dead trees from the habitat, inappropriate fire regimes and predation by foxes and cats.

CURRENT RECOVERY EFFORTS

Recovery plan: NSW Department of Environment and Climate Change (2005) 'Recovery plan for the Hastings River Mouse (*Pseudomys oralis*)'. Department of Environment and Climate Change (NSW), Hurstville.

Other plans: Department of the Environment, Water, Heritage and the Arts (2008) 'Threat abatement plan for predation by feral cats'. Department of the Environment, Water, Heritage and the Arts, Canberra.

Conservation actions:

- Monitoring of some populations and surveys for others are conducted in south-east Qld and several sites in NSW. Monitoring occurs at seven populations including Main Range NP and Lamington NP in Qld and Marengo State Forest, Hyland Nature Reserve, Carrai Plateau NP, Border Ranges and Mount Royal NP in NSW.
- Investigations into the role of fire in the ecology of the species is underway (UNE).
- Habitat models, both descriptive and GIS-based, have been produced to predict the distribution.
- The importance of fungi in the diet of small murids is being investigated (UNE).
- The use of remote cameras for surveying Hastings River Mouse is being evaluated (NSW DPI).

FUTURE RECOVERY EFFORTS

What you can do:

- Do not remove woody debris or rocks from its habitat.
- Report sightings/findings to DERM or NSW DECCW.

Conservation needs:

- Manage fire and protect woody debris.
- Exclude cattle, control weeds and predators from known population areas.
- Communicate with DERM or DECCW for assistance in managing freehold land with this species.

Research directions:

- Investigate the mouse's requirement for a particular successional stage in its habitat.
- Gain knowledge about, and how to control, threats.
- Research its responses to fire, grazing and weeds.
- Study predation rates and control of foxes and cats.

Sources and further reading

Fox BJ, Read DG, Jeffreys E and Luo J (1994) Diet of the Hastings River Mouse (*Pseudomys oralis*). *Wildlife Research* **21**, 491–505.

Meek PD and Shields J (2005) *Hastings River Mouse Forum: Proceedings of a Forum Hosted by Forests NSW, Cumberland State Forest, Sydney 4–5 March 2004*. NSW Department of Primary Industries, Orange.

Meek PD, Radford SL and Tolhurst BL (2006) Summer–Autumn home range and habitat use of the Hastings River Mouse *Pseudomys oralis*. *Australian Mammalogy* **28**, 39–50.

Read DG (1993) Body size in Hastings River Mouse *Pseudomys oralis* from old and new locations. *Australian Zoologist* **29**, 117–23.

Townley SJ (2000) The ecology of the Hastings River Mouse *Pseudomys oralis* (Rodentia: Muridae) in northeastern New South Wales and southeastern Queensland. PhD thesis. Southern Cross University, Lismore.

Author: Paul D Meek

WATER MOUSE

Xeromys myoides Thomas, 1889

Photo: C Hans and Judy Beste/Lochman Transparencies

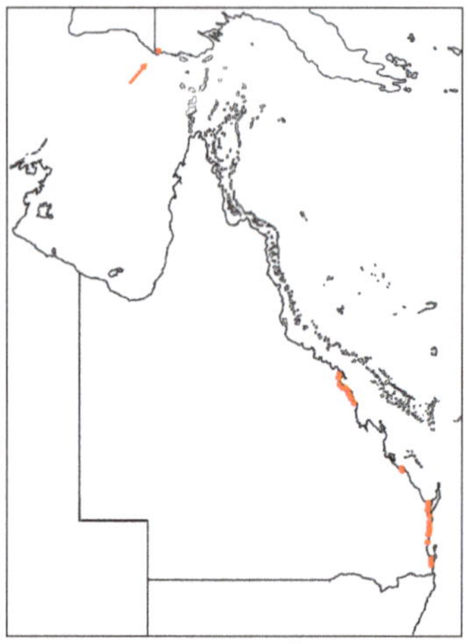

Family: Muridae

Other common name: False Water-rat

Aboriginal name(s): Yirrkoo

Conservation status: IUCN: Vulnerable; Aust.: Vulnerable; Qld: Vulnerable

Description: This small rodent has a head to body length of 72–126 mm and a 62–99 mm tail. It weighs 32–64 g. It is terrestrial and nocturnal, with small eyes, short, dense, silky fur and short, rounded ears. Its smooth tail and the top of its feet are covered with fine white hairs. Young animals have steel-grey backs and pure white bellies, but as they age the dorsal fur can become grey-brown and spotted with small white flecks.

Distribution: The Water Mouse has a patchy distribution from southern Papua New Guinea on the Bensbach River floodplain, coastal NT, coastal areas near Proserpine, Mackay and Rodds Peninsula to Eurimbula Creek south of Gladstone, then from Hervey Bay to the Coomera River. Populations are also known from Fraser, Bribie, Russell, and North and South Stradbroke Islands. No records are available from coastal mangroves east of Brisbane between Donnybrook and Steiglitz and from northern NSW.

Habitat: The Water Mouse is associated with near-coastal freshwater and saline habitats. In central Qld, these include permanent freshwater swamps densely covered with tall grass, shrubs and pandanus (as described for locality near Mackay, probably Orphanage Swamp), and mangrove forests of *Ceriops tagal* with some *Bruguiera* spp. in the high intertidal zone. In south-east Qld, it is found in sedgeland, chenopod shrubland, *Sporobolus* grassland, and mangroves, with isolated captures from freshwater swamps and brackish lagoons.

Ecology: The Water Mouse is nocturnal, often nesting and sheltering in mounds constructed of mud, peat and vegetation. Its nesting structures may be free-standing, built on small 'islands' amid sedgeland, incorporated into living or dead trees and their roots, within an earth bank or associated with human-made spoil heaps. A leaf-filled chamber inside the nest is above high water. Up to eight animals may share a mound, generally with one sexually active male. A nest mound may be used by successive generations over many years. In tidal areas, the species forages for invertebrate prey including grapsid crabs, other intertidal crustaceans, molluscs and marine polyclad flatworms. The ecology of this species in freshwater environments is unknown. Capture records suggest it is capable of reproducing throughout the year and four young were recorded from one nest.

Reason for decline and current threats: Two recent declines have been noted from Qld. A local extinction occurred adjacent to an urban real estate development at Coomera Waters on the Gold Coast over a period during which the following processes were documented: damage to nests, mangrove vegetation and intertidal substrate by vehicles and cattle; acid sulfate contamination; and regular spraying of mangroves and saltmarshes with new generation mosquito control agents. Around Mackay, declines were concurrent with diminishing grapsid crab populations, which were attributed to changes in salinity and sediment loads caused by discharge of stormwater from adjacent urbanisation. Elsewhere, other forms of pollution including thermal effluent, sewage, industrial wastes and oil spills are also clear threats.

The loss, degradation and fragmentation of freshwater and intertidal wetland communities are threats that are associated with a wide range of coastal developments and extractive industries. However, water quality and productivity in mangrove ecosystems can be reduced with little immediate or visible impact on the vegetation, but significant impact on animals. This can include activities affecting the quality and quantity of water in the watertable and catchment, patterns of water flow, the integrity of the supralittoral bank, and the stability of associated reed and freshwater zones. Fires in adjacent terrestrial communities can burn into nests and may remove sedges and grasses in the upper intertidal zone, exposing individuals to a greater risk of predation. Populations are also thought to be at high risk of local extinction through fox, cat, dog and pig predation. Sea level rise could affect many current nesting sites.

CURRENT RECOVERY EFFORTS

Recovery plan: Department of Environment and Resource Management (2011) 'National recovery plan for the Water Mouse (false water rat) *Xeromys myoides*'. Report to Department of Sustainability, Environment, Water, Population and Communities , Canberra. Department of Environment and Resource Management, Brisbane. This was preceded by a community-led interim plan: Caneris A and Costello G (2002) 'Draft interim regional recovery plan for the Water Mouse *Xeromys myoides* 2003–2008. New South Wales/Queensland Border to Hervey Bay'. Australian Wildlife Management and Services for the WPSQ, Unpublished.

Conservation actions:

- In the past 10 years, researchers have located populations, investigated nesting, feeding and social behaviour, identified reasons for declines, trialled artificial nests and raised awareness of the species throughout its range.
- Potential re-introduction sites have been identified from Deception Bay to Scarborough, including Kippa-Ring and Redcliffe airport, and a tissue library is held in the Qld Museum.

FUTURE RECOVERY EFFORTS

What you can do:

- Volunteer for *MangroveWatch* at Nudgee Beach, which is sponsored by WPSQ.
- Search for nesting mounds in mangroves and supralittoral habitats (being mindful of crocodiles in north and central Qld); report findings to the Qld Museum.
- Monitor the health of your local mangroves, including monitoring inflows from drain pipes and recognising declines in crab numbers or the build up of leaf litter inside the littoral zone.
- Look for fox, dog, pig and cat tracks (and scats) in mangrove mud. Notify local wildlife managers regarding your findings.
- Encourage schools close to mangrove habitats to incorporate the Water Mouse into natural science and conservation studies. Contact Qld Museum Learning Resources about *Mangrove Challenge*.

Conservation needs:

- Establish a formal recovery process and implement the recovery plan.
- Rehabilitate the habitat at priority sites.
- Engage local communities to protect existing populations.
- Use voluntary agreements to manage the threats on private land.

Research directions:

- Study metapopulation dynamics and genetic variation of the species across its distribution.
- Study water quality and runoff issues, the use and impacts of pesticide and herbicidal sprays in natural environments, and the impacts of habitat loss (including sea level rise) on the quality and extent of mangrove and other intertidal habitats.

Sources and further reading

Ball D (2004) Distribution and habitat of the false water rat, *Xeromys myoides* Thomas, 1889 (Rodentia: Muridae) in intertidal areas of central eastern Queensland. *Memoirs of the Queensland Museum* **49**, 487–494.

Gynther IC and Janetzki H (2008) Water mouse *Xeromys myoides*. In *The Mammals of Australia*. 3rd edn. (Eds S Van Dyck and R Strahan) pp. 664–666. Reed New Holland, Sydney.

Van Dyck S (1997) *Xeromys myoides* Thomas, 1889 (Rodentia: Muridae) in mangrove communities of North Stradbroke Island, south-east Queensland. *Memoirs of the Queensland Museum* **42**, 337–366.

Van Dyck S and Gynther I (2003) Nesting strategies of the Water Mouse *Xeromys myoides* in south-east Queensland. *Memoirs of the Queensland Museum* **49**, 453–479.

Van Dyck S, Janetzki H and Gynther I (2003) Artificial nesting mounds for the Water Mouse *Xeromys myoides*. *Memoirs of the Queensland Museum* **49**, 480.

Authors: Steve Van Dyck and Ian Gynther

DUGONGS (ORDER: SIRENIA)

DUGONG

Dugong dugon (Müller, 1776)

Photo: Leonie Mynott

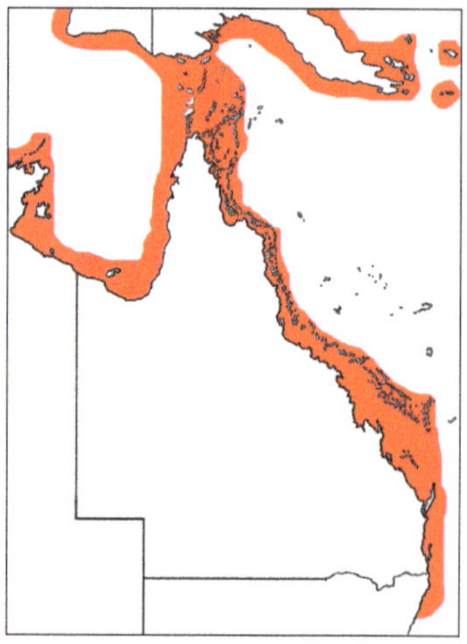

Family: Dugongidae

Aboriginal names: Numerous, including: Girribithi (Guugu Yimmithirr); Dangal (Torres Straits)

Other common name: Sea Cow

Conservation status: IUCN: Vulnerable; Qld: Vulnerable; NSW: Vulnerable; NT: Protected wildlife; WA: Specially Protected Fauna

Description: The Dugong has a head and body length of 3 m and looks rather like a slow-moving, rotund dolphin with the head of a Walrus and without a dorsal fin. Adults are grey, but often appear brown from a distance. The head is distinctive, with the mouth opening ventrally beneath a broad, flat muzzle. The tusks of mature males, and some old females, erupt on either side of the upper jaw. The eyes are small and the ears are small bilateral openings. There are two mammary glands, each opening via a single teat situated in the 'armpit' or axilla. Hind limbs are absent. The tail of the Dugong is triangular, like that of a whale.

Distribution: The Dugong's range extends along 140 000 km of coastline across 40 countries in tropical and subtropical coastal and island waters of the Indo-West Pacific, between about 26° and 27° north and south of the equator. Extensive aerial surveys indicate that most of the world's remaining Dugongs are in northern Australian waters, from Moreton Bay in the east to Shark Bay in the west. These are the most abundant marine mammal in our northern inshore waters. Although some areas of suitable habitat have not been surveyed, population estimates of areas surveyed since 2005 add to 68 000 Dugongs. Survey data are likely to underestimate the actual populations.

Habitat: Dugongs are most often found in inshore waters supporting sizable seagrass beds, but may be seen in deeper protected waters that support deep-water seagrasses. Evidence from depth recorders indicates that Dugongs spend more than 70% of their time less than 3 m from the sea surface. Shallow waters, such as tidal sandbanks and estuaries, have been reported as sites for calving.

Ecology: Growth layers in their tusks indicate that Dugongs can live for up to 70 years and bear their first calf when they are between 6 and 17 years old, after approximately 13 months gestation. The single calf suckles for about 18 months. The period between successive births is variable; estimates range from 3 to 7 years. Dugongs eat seagrasses, preferring those that are relatively low in fibre and high in available nutrients such as nitrogen and starch. Satellite-tracking indicates that both male and female Dugongs exhibit individualistic and apparently directed movements of up to 600 km from their capture sites. Solitary individuals or small groups are occasionally recorded in the waters off isolated oceanic islands after years of apparent absence, indicating that Dugongs can cross ocean trenches.

Reasons for decline and current threats: Dugongs are vulnerable to human impacts because of their life history and their dependence on coastal seagrass communities. Because Dugongs breed slowly, the sustainable level of anthropogenic mortality is about 2% of the female population per year. This rate is even lower in areas where the reproductive rate has been reduced by food shortage, such as can be caused by habitat loss, seagrass dieback, decline in the nutrient quality of available seagrass, or a reduction in the time available for feeding due to disturbance from boat traffic. Accidental entangling in fishing and shark control nets were historically major sources of Dugong mortality in Qld that continue to a lesser degree; between 1962 and 1999, shark nets caught over 800 Dugongs. Hunting Dugongs for food and oil used to be widespread throughout their range, and still continues in some regions. Research suggests that the level of legal hunting by Indigenous people currently occurring in Torres Strait and the adjacent remote coastal waters of Cape York may not be sustainable.

CURRENT RECOVERY EFFORTS

Recovery plan: None

Conservation actions:

- The establishment and rezoning of an extensive series of large-scale multiple use marine protected areas, especially the 348 000 km^2 Great Barrier Reef World Heritage Area, has helped to protect Dugong habitat.
- There have been interventions to improve water quality.
- Partnerships have been established between management agency staff and Indigenous Traditional Owners to develop and implement mutually acceptable, legal frameworks for sustainable Dugong hunting at culturally relevant scales.
- The replacement of shark control nets by baited hooks has occurred at most locations.
- Vessel lanes and/or speed restrictions have been established to protect Dugongs from vessel strikes in several major habitats.
- The use of the herbicide diuron, detected in Dugong tissues and the sediments in seagrass beds has been reviewed.
- A marine wildlife carcass salvage program has been implemented.
- A Dugong research strategy is under way, which has included a monitoring program based on aerial surveys. The results of 20 years of aerial surveys suggest that numbers are now stable in eastern Qld.
- A National Wildlife Conservation Plan for Dugongs, which will include research priorities, is being negotiated.

FUTURE CONSERVATION EFFORTS

What you can do:

- Observe speed limits and designated vessel lanes when boating.
- Report Dugong carcasses to the Qld Marine Stranding Hotline (1300 130 372).
- Reduce terrestrial runoff into the marine environment, particularly herbicides and pesticides.
- Support Indigenous efforts to manage Sea Country.

Research directions:

- Determine the distribution and relative abundance of Dugong.
- Obtain data on the genetic structure, gene flow and age structure of Dugong populations.
- Establish baseline health parameters and major drivers of disease.
- Investigate spatial and temporal variation in life history parameters, especially survivorship.
- Study Dugong social structure and communications and the potential for change in response to disturbance.
- Determine the spatial distribution of cumulative risks to Dugongs, including quantifying all causes of mortality.
- Develop criteria for assessing the cumulative impacts of the risks outlined above at various spatial scales.
- Gather information on status and trends in seagrass and Dugong populations to detect a decline, other than a steep one, within a timeframe suitable to management responses.
- Determine the relationship between the distribution of species of seagrass and the causes, timing and patterns of Dugong movements.
- Identify the environmental features that determine habitat 'quality' and food choice, and determine the impact of grazing.
- Determine the carrying capacity of seagrass habitats.
- Model the impacts of climate change on habitat, distribution and population density.
- Set up appropriate performance indicators for assessing the effectiveness of actions at various spatial scales.

Sources and further reading

Daley BP, Griggs P and Marsh H (2008) Exploiting marine wildlife in Queensland: the commercial Dugong and marine turtle fisheries, 1847–1969. *Australian Economic History Review* **48**, 227–265.

Grech A and Marsh H (2008) Rapid assessment of risks to a mobile marine mammal in an ecosystem-scale marine protected area. *Conservation Biology* **22**, 711–720.

Marsh H (2008) Dugong dugon. In *IUCN 2008. 2008 IUCN Red List of Threatened Species*, Version 2010.4. IUCN, Gland, Switzerland and Cambridge, UK, <http://www.iucnredlist.org>.

Marsh H, De'Ath G, Gribble N and Lane (2005) Historical marine population estimates: Triggers or targets for conservation? The Dugong case study. *Ecological Applications* **15**, 481–492.

Smyth D, Kennett R and NAILSMA (Eds) (2006) *Dugong and Marine Turtle Knowledge Handbook: Indigenous and Scientific Knowledge of Dugong and Marine Turtles in Northern Australia.* North Australian Indigenous Land and Sea Management Alliance (NAILSMA), Darwin.

Authors: Helene Marsh, Kristen Weiss and Alana Grech

WHALES (ORDER: CETACEA)

SEI WHALE

Balaenoptera borealis Lesson, 1828

Photo: © Doug Perrine / SeaPics.com

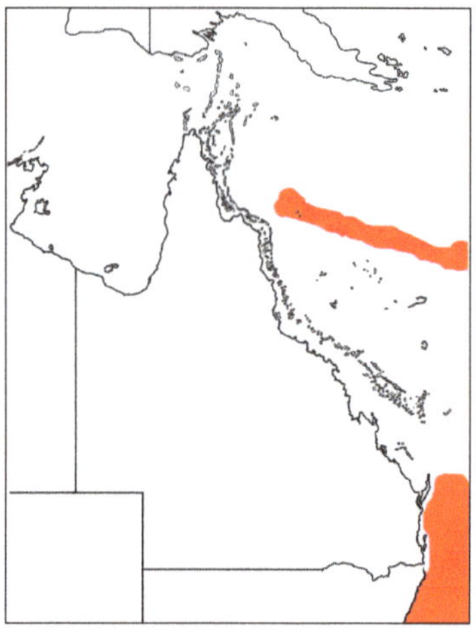

Family: Balaenopteridae

Conservation status: IUCN: Endangered; Aust.: Vulnerable

Description: This is a large slender streamlined whale that may reach a head and body length of 18 m and weigh up to 45 tonnes. The rostrum is pointed, with a single central ridge leading back to the blowholes (whereas Bryde's Whale *Balaenoptera edeni* – a similar species more commonly sighted in Qld waters – has three rostral ridges). The steep and sharply hooked dorsal fin is situated about two-thirds back along the body. The skin colour is dark grey or brown, almost black in some individuals, while the belly skin may be almost white. The body may be pocked with numerous small whitish scars, probably from cookie cutter shark (*Isistius* spp.) attacks. The blow is columnar and may exceed 3 m. The baleen plates are black with grey-white tips and white almost silky inner fringes, and are less than 80 cm long. This whale rarely raise its flukes when diving and generally behaves inconspicuously.

Distribution: Sei Whales occur worldwide, but are most numerous in the Southern Hemisphere. They have been seen at high latitudes in the Antarctic but are most common in the southern polar frontal waters of the deep Southern Ocean. They are presumed to migrate to tropical waters during winter, and are occasionally seen off Australia's coasts, but the wintering areas are unknown and calves are rarely seen. The size of the southern hemisphere population may once have been around 100 000, but was reduced by whaling to fewer than 10 000, though there is no indication of how many genetically distinct populations exist.

Habitat: Sei Whales are often sighted along the oceanic frontal systems of the Southern Ocean, such as the Sub-Tropical Front where subtropical and sub-Antarctic water masses meet south of Australia, and the Polar Front further south, where sub-Antarctic and Antarctic waters meet. These regions are rich in plankton prey such as copepods. The whales have been sighted near the south of the Ross Sea: as far south as whales can go. They are occasionally seen feeding in the Bonney Upwelling Blue Whale feeding area: the only area in Australian waters where they may be expected regularly. They presumably migrate to warm tropical waters for breeding.

Ecology: In general, little is known of the ecology of Sei Whales because they are rarely sighted and very difficult to study. Unlike their larger cousins, Blue Whales *Balaenoptera musculus*, which feed almost entirely on krill, Sei Whales are more generalist feeders that prey on a range of food sizes including copepods, krill and schooling fish such as anchovies. Sei Whales often skim slowly through swarms of copepods like right whales, though they may also feed on krill or fish using sharp lunges, which bring the head well clear of the water. They are fast swimmers, probably the fastest of all cetaceans. Calving occurs in wintering areas and calves are weaned at 6–8 months on summer feeding grounds. They are presumed to fast during winter migrations, but are more irregular in their movements than other large baleen whales. They are usually found in small groups of two to five, though larger numbers may aggregate where food is abundant.

Reasons for decline and current threats: When their larger cousins Fin Whales *Balaenoptera physalus* and Blue Whales had been sufficiently decimated by whaling, the whalers turned to Sei Whales and about 200 000 were taken in the Southern Ocean, mainly between the 1950s and 1970s. Currently threats are not well defined, although they are probably similar to the range of threats facing Blue Whales, including ship strike, acoustic disturbance from a range of sources, entanglement, pollution and the effects of climate change on their feeding ecology.

CURRENT RECOVERY EFFORTS

Recovery plan: Anonymous (2005) 'Blue, Fin and Sei Whale recovery plan 2005–2010'. Department of the Environment and Heritage, Canberra.

Other plans: '(Queensland) Nature conservation (whales and dolphins) conservation plan 1997'. *Nature Conservation Act 1992.*

Bannister JL, Kemper CM and Warnecke RM (1996) 'The action plan for Australian cetaceans'. Australian Nature Conservation Agency, Canberra.

Conservation actions: No specific research effort is currently directed at Sei Whales in Australian waters because they are so rarely sighted. They are sighted incidentally during Blue Whale surveys in the Bonney Upwelling, where some Sei Whales feed at times. In Antarctic surveys, they are sighted rarely, but this is not surprising given their rather more northerly distribution.

FUTURE RECOVERY EFFORTS

Research directions: As with other baleen whales, future research needs to consider the possibility that changes in sea surface temperatures may affect the abundance and distribution of prey species.

Sources and further reading

Bannister J (2008) *Great Whales.* CSIRO Publishing, Melbourne.

Baumgartner MF and Fratantoni DM (2008) Diel periodicity in both sei whale vocalisation rates and the vertical migration of their copepod prey observed from ocean gliders. *Limnology and Oceanography* **53**, 2197–2209.

Department of the Environment, Water, Heritage and the Arts (2009) 'Species profiles and threats database – sei whales'. Department of the Environment, Water, Heritage and the Arts, Canberra. <http://environment.gov.au/cgi-bin/sprat/public/publicspecies.pl?taxon_id=34>.

Horwood J (2002) Sei whale. In *Encyclopedia of Marine Mammals.* (Eds WF Perrin, B Würsig and JGM Thewissen) pp. 1069–1071. Academic Press, San Diego.

Kawamura A (1974) Food and feeding ecology of the southern sei whale. *Scientific Reports of the Whales Research Institute* **26**, 25–144.

Author: Peter Gill

BLUE WHALE

Balaenoptera musculus Linnaeus, 1758

Photo: Blue Whale Study

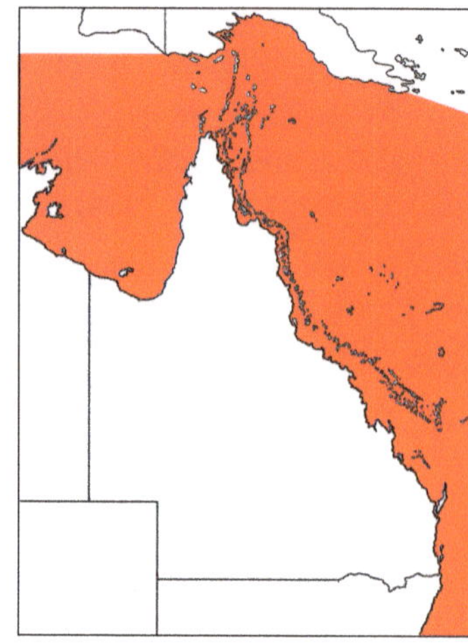

Family: Balaenopteridae

Conservation status: IUCN: Endangered; Aust.: Endangered; NSW: Endangered; SA: Endangered; Vic.: Critically Endangered

Description: Blue Whales are the largest living animal, sometimes exceeding a head and body length of 30 m and weighing 150 tonnes. Slender and streamlined, they are variable grey, although underwater they appear luminous blue or aqua. The mottled flank patterns allow individuals to be identified. The rear end of the body is often pocked with numerous small whitish scars from Cookie-cutter Sharks (*Isistius* spp.). There is a small dorsal fin about three-quarters along the very long back which becomes visible when it surfaces. Tail flukes may be raised when diving. The blow is typically tall and straight, reaching 10 m or more. A prominent rostral ridge and 'splashguard' lie in front of the blowholes. The baleen plates are uniformly black, and there are 60–88 throat pleats.

Distribution: Blue Whales occur worldwide, with Antarctic seas previously their greatest stronghold. Around Australia, their summer–autumn feeding areas are the Perth Canyon, WA, and the Bonney Upwelling and adjacent waters from the Great Australian Bight to Bass Strait. They also pass close to shore off Eden, NSW and Geographe Bay, WA, possibly en route to Antarctic waters. There are two southern hemisphere subspecies, the Antarctic Blue *B. m. intermedia* and the Pygmy Blue *B. m. brevicauda*. The former inhabit Antarctic seas; the latter are usually north of 55°S in the Indian and Pacific Oceans. Although Blue Whales are not known to use Qld waters for ecologically important activities, several strandings have occurred, and both subspecies may transit Qld waters during their migration to tropical breeding areas. The current estimate for Antarctic Blue Whales is around 2000 (less than 1% of pre-whaling numbers), while there are no reliable estimates for Pygmy Blues.

Habitat: Blue Whales are strongly associated with the sea-ice zone, which is prime habitat for Antarctic Krill *Euphausia superba*; in temperate regions, they are typically found at productive upwellings and oceanic frontal regions where krill are abundant.

Ecology: Blue Whales feed almost exclusively on krill, in depths between the surface and 500 m. To meet their enormous energy needs, Blue Whales move rapidly between widely separated feeding areas. Productivity in these areas is usually variable in space and time, so Blue Whales must be flexible in feeding habitats and methods. During daylight, they forage for scattered hard-to-find krill swarms and lunge into them, gulping enormous mouthfuls of krill and water. Unlike Humpback *Megaptera novaeangliae* and Southern Right Whales *Eubalaena australis*, they do not fast during winter so require abundant food year round. They are usually found alone or in pairs, though larger numbers aggregate where prey are abundant. They may maintain acoustic contact over great distances. The tropical mating and calving areas are largely unknown, although satellite tagging has recently confirmed migration between WA and Indonesia: a likely winter breeding area. Calves are 6–7 m long when born in winter and are weaned after 7 months on summer feeding grounds, when they are 16 m long.

Reasons for decline and current threats: Blue Whales were the major target of 20th century whaling, with more than 360 000 taken in the Southern Hemisphere. Antarctic Blue Whales were hunted to the edge of extinction between 1904 and 1972, while Pygmy Blues were only 'discovered' in the Indian Ocean in the late 1950s and targeted until 1972, despite global protection in 1965. The Soviet Union illegally killed about 9000 Blue Whales between 1947 and 1972, many just south of Australia. Current threats include: ship strike; pollution; ocean noise including military operations, seismic surveys and shipping; disturbance by large and small vessels; entanglement in fishing gear; and the effects of climate change, including changes in ocean circulation and chemistry, reduction of sea-ice cover, etc. There have been calls for Blue Whales to be included in Japan's 'scientific' whaling program.

CURRENT RECOVERY EFFORTS

Recovery plan: Anonymous (2005) 'Blue, Fin and Sei Whale recovery plan 2005–2010'. Department of the Environment and Heritage, Canberra.

Other plans: '(Queensland) Nature conservation (whales and dolphins) conservation plan 1997'. *Nature Conservation Act 1992.*

Bannister JL, Kemper CM and Warnecke RM (1996) 'The action plan for Australian cetaceans'. Australian Nature Conservation Agency, Canberra.

Conservation actions:

- Current actions include monitoring population abundance and trends; defining critical feeding and breeding habitat; mitigating threats where they occur; and determining migration routes and threats in Qld waters.
- Current management actions include: countering arguments for whaling; protection of critical habitat; monitoring prey over-harvesting; and monitoring the effects of climate change.
- Research is being undertaken in south-east Australia, WA, Indonesia and Antarctica by Curtin University (WA), Western Whale Research (WA), the Centre for Whale Research (WA), the Blue Whale Study (Vic./SA), Deakin University (Vic.), Macquarie University (NSW), Apex Environmental (Indonesia) and the Australian Antarctic Division (Antarctica/Australia).

FUTURE RECOVERY EFFORTS

What you can do:

- Report sightings to the National Marine Mammal Database.
- Support anti-whaling campaigns where appropriate.

Research directions: Determine the importance of Qld waters as a migratory pathway between winter breeding and summer feeding locations.

Sources and further reading

Branch TA, Stafford KM, Palacios DM, Allison C, Bannister JL, Burton CLK *et al.* (2007) Past and present distribution, densities and movements of blue whales *Balaenoptera musculus* in the Southern Hemisphere and northern Indian Ocean. *Mammal Review* **37**, 116–175.

Calambokidis J and Steiger G (1997) *Blue Whales.* Voyageur Press, Stillwater, Minnesota.

Clapham PJ, Young SB and Brownell Jr RL (1999) Baleen whales: conservation issues and the status of the most endangered populations. *Mammal Review* **29**, 35–60.

Gill PC, Morrice MG, Page B, Pirzl R, Levings AH and Coyne M (2011) Blue whale habitat selection and within-season distribution in a regional upwelling system off southern Australia. *Marine Ecology Progress Series* **421**, 243–263.

Yochem PK and Leatherwood S (1985) Blue whale *Balaenoptera musculus* (Linnaeus, 1758). In *Handbook of Marine Mammals. Volume 3: The Sirenians and Baleen Whales.* (Eds SH Ridgeway and RJ Harrison) pp. 193–240. Academic Press, London.

Author: Peter Gill and Margie Morrice

FIN WHALE

Balaenoptera physalus (Linnaeus, 1758)

Photo: Mark Carwardine-AUSCAPE

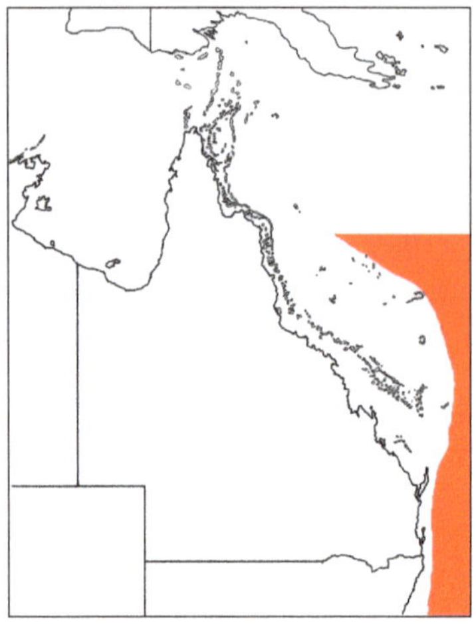

Family: Balaenopteridae

Aboriginal name: Galbol ('whale'- Torres Strait)

Conservation status: IUCN: Endangered; Aust.: Vulnerable

Description: This is a very large whale with a head and body length to 25 m (male) or 27 m (female) long and weighing up to 90 tonnes. It has a long slender body and slender head with a distinct medial ridge, short narrow flippers, a tall hooked dorsal fin, a caudal peduncle that is distinctly ridged and compressed laterally, and a triangular tail. Numerous ventral grooves reach back to the navel. It is dark grey above and on the sides and white below, including on the undersides of flippers and tail. It often has a greyish-white chevron, with the apex orientated forward, just behind the head. It has an asymmetrical lower jaw, which is white on the right and dark on the left, mirrored by a group of white baleen plates on the right side of the mouth, while the remainder (over 400 each side) are black with coarse yellowish-white fringes.

Distribution: The species occurs worldwide, with separate northern and southern hemisphere populations, which are further subdivided into possibly eight separate breeding populations; one – the 'east Australian' – spends winter in the Coral Sea and occasionally in the Great Barrier Reef waters off Qld. They are highly migratory, undertaking extensive journeys between warm water breeding grounds and cold water feeding grounds. In the southern hemisphere, they travel to around 65°S south. There are no population estimates specifically for the Australian region, though relatively large numbers have been sighted in the Antarctic south of the continent recently, including a report of more than 100 animals off the Ross Sea.

Habitat: Off Australia, Fin Whales are generally oceanic, infrequently recorded near the coast, and rarely strand themselves.

Ecology: Fin Whales first reproduce at 6–10 years, calving at 2–3 year intervals after a gestation period of around 11 months. The breeding grounds are assumed to be in tropical oceanic waters, but no exact locations have been identified. Some may calve in the Coral Sea off Qld. In the southern hemisphere, they feed almost exclusively on Antarctic Krill *Euphausia superba* during summer, although elsewhere they can feed on small schooling fish. They generally dive between 100 and 200 m to feed. They are usually found in small groups of up to 10 animals, in pairs or singly, although larger aggregations can occur on the feeding grounds. There is distinct segregation on migration, with pregnant females travelling first, followed by adult males; lactating females and juveniles bringing up the rear.

Reasons for decline and current threats: Fin Whales were second only to Blue Whales *Balaenoptera musculus* in importance to the Antarctic whaling industry, with over 700 000 killed between 1905 and 1976. Most were taken from south of the Atlantic and western Indian Oceans, but substantial numbers were also taken from south of Australia, including from our east coast population. The result was a massive decline to as few as 15 000 by the 1980s from an original southern hemisphere population of around 500 000.

With no commercial whaling, the major threat is currently in abeyance, with some recent signs of increase in numbers. However, the Japanese scientific whaling program (JARPA II) has proposed killing up to 50 whales per year in alternate years in Antarctic 'Area V' (between 130°E and 170°W, i.e. on the 'east Australian' population). So far this has not been implemented, but, if it is, it could have an effect at low population levels.

The most significant potential threat is the likely long-term effect of climate change on their food, Antarctic Krill.

CURRENT RECOVERY EFFORTS

Recovery plan: Anonymous (2005) 'Blue, Fin and Sei Whale recovery plan, 2005–2010'. Department of the Environment and Heritage, Canberra.

Other plans: '(Queensland) Nature conservation (whales and dolphins) conservation plan 1997'. *Nature Conservation Act 1992.*

Conservation actions: Apart from passive acoustic studies, such as off the Western Australian Coast (Perth

Canyon area), there are currently no studies directed at Fin Whales in Australian coastal waters. Acoustic studies south of the continent, such as in and along the Polar Front at 55°S, can monitor the presence of cetaceans including Fin Whales. They have also been detected recently south of eastern Australia by international sightings cruises conducted under the auspices of the International Whaling Commission (IWC-SOWER Programme), and the Japanese Antarctic Research Programme (JARPA).

FUTURE RECOVERY EFFORTS

What you can do: Join/donate to one or more of the many cetacean conservation organisations, internationally (including IUCN, WWF, Greenpeace and IFAW) or nationally. For a listing of Australian Cetacean organisations see the Oceania Project, which includes the Whale and Dolphin Conservation Society (WDCS), the Australian Conservation Foundation (ACF) and ORRCA (marine mammal rescue and research).

Sources and further reading

Bannister JL (2008) *Great Whales.* CSIRO Publishing, Melbourne.

Ensor P, Minami K, Morse L, Olson P and Sekiguchi K (2008) 2007–2008 International Whaling Commission–Southern Ocean Whale and Ecosystem Research (IWC–SOWER) cruise. Paper SC/60/IA1 presented to the Scientific Committee of the International Whaling Commission, Santiago, Chile, May 2008 (unpublished, available from the International Whaling Commission, 135 Station Road, Impington, Cambridge CB24 P UK).

Gedamke J (compiler) (2008) 'Australia, Progress Report on cetacean research, January 2007 to December 2007, with statistical data for the calendar year 2007'. Document SC/60/Progrep Australia presented to the International Whaling Commission Scientific Committee, Santiago, Chile, May 2008 (unpublished, available from the International Whaling Commission, 135 Station Road, Impington, Cambridge CB24 P UK).

Matsuoka K, Hakamada T, Kiwada H, Murase H and Nishiwaki S (2006) Distribution and standardised abundance estimates for humpback, fin and blue whales in the Antarctic Areas IIIE, IV, V, and VIW (35°E–145°W), south of 60°S. Document SC/D06/J7 presented to the IWC JARPA Review Meeting, December 2006 (unpublished, available from the International Whaling Commission, 135 Station Road, Impington, Cambridge CB24 P UK).

Perrin WF, Wursig B and Thewissen JGM (Eds) (2002) *Encyclopedia of Marine Mammals.* Academic Press, San Diego.

Author: John Bannister

SOUTHERN RIGHT WHALE

Eubalaena australis (Desmoulins, 1822)

Photo: Doug Allan-OSF-AUSCAPE

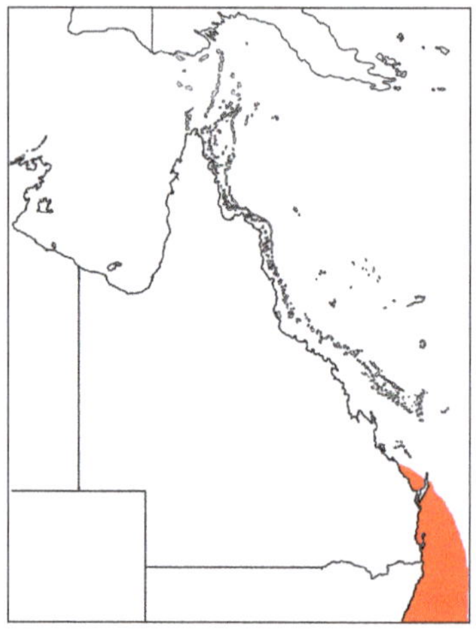

Family: Balaenidae

Aboriginal name: Galbol ('whale'-Torres Strait)

Conservation status: IUCN: Least Concern; Aust.: Endangered

Description: This is a large whale with a head and body length to 17 m and weighing to ~80 tonnes. It has a stocky body with a large head with prominent pale callosities (hard thick patches) and strongly arched upper jaws, a distinctive 'neck', broad paddle shaped flippers, triangular tail flukes with a smooth trailing edge, no dorsal fin and a V-shaped blow. The basic body colour is black with a white chin and belly splashes and occasional white or grey dorsal marks. Some 3% of calves are born white with a black collar and black spots and dashes, darkening to grey-brown within a year and then darkening further though the pattern of dots and splashes persists. The baleen plates are very long, mainly black with lighter coloured, fine inner fringes.

Distribution: Southern Right Whales are migratory, breeding in warm waters to as far north as around 25°S and feeding in cold waters, mainly to around 55°S but they can be found further south. Up to 11 subpopulations have been described, but only three persist in any numbers – off the coasts of eastern South America, South Africa, and Southern Australia and sub-Antarctic New Zealand. Off Australia, they occur close to the coast in winter, particularly off southern WA and SA as far east as Ceduna; smaller numbers occur further east, to as far up the east coast as Hervey Bay. The Australian population has recently been estimated at around 3000 animals, from an original population size of probably more than 10 000. Two other species of Right Whale occur in the North Atlantic, *E. glacialis*, and the North Pacific, *E. japonica*.

Habitat: Adult females come in very close to the coast in shallow waters, 5 m deep, in winter/early spring to give birth and nurse their calves. Adult males and juveniles of both sexes can also be found there at such times. Their summer habitat is less well known, but ranges from around 45°S to 65°S.

Ecology: Southern Right Whale females become sexually mature at around 8–9 years and calve in coastal waters, mainly at 3-year intervals after gestation of around 12 months. On the breeding grounds, males compete for females in interactive but unstable small 'surface active groups' (SAGs). Southern Right Whales generally fast during the breeding season, but in summer they migrate south to the Subtropical Convergence (~45°S) to feed on small plankton such as copepods, but they can also feed further south, on Antarctic krill *Euphausia superba*.

Reasons for decline and current threats: 'Traditional, open boat', whaling reduced the worldwide population of Southern Right Whales from around 65 000 to perhaps as low as 300 in the first half of the 19th century. Once protected, there was some recovery, although illegal whaling by the Soviet Union interrupted that recovery in the 1960s. From the mid-1970s, the south-east American, South African and western Australian populations have been increasing at around 7% per year.

With no commercial whaling, the major threat is currently in abeyance. There have been no proposals for catching Southern Right Whales under the Japanese Antarctic scientific whaling programs (JARPA I or II). Potential threats include disturbance from seismic operations, collision with large vessels, entanglement in fishing gear and pollution. More threatening at the population level is the likely long-term effect of climate change on their food. Breeding success in this species, feeding on krill off South Georgia in the South Atlantic, has been adversely influenced by rises in sea surface temperature, and a similar effect has recently been reported for 'Australian' Southern Right Whales.

CURRENT RECOVERY EFFORTS

Recovery plan: Anonymous (2005) 'Southern Right Whale recovery plan 2005–2010'. Department of the Environment and Heritage, Canberra.

Other plans: '(Queensland) Nature conservation (whales and dolphins) conservation plan 1997'. *Nature Conservation Act 1992.*

Bannister JL, Kemper CM and Warnecke RM (1996) 'The action plan for Australian cetaceans'. Australian Nature Conservation Agency, Canberra.

Director of National Parks (2005) 'The Great Australian Bight Marine Park Management Plan 2005–2012'. Australian Government, Canberra.

Conservation actions:
Current efforts focus on protection of the species and its habitats to allow for continued recovery at the optimum biological rate (7% per year). To achieve this, the Australian Whale Sanctuary provides high levels of protection, including within 200 nm from the coast and waters around the Australian Antarctic Territory, Macquarie, Heard and McDonald Islands. In addition, there are a number of studies being conducted on:

- the population status of Australian Southern Right Whales, which undertakes annual aerial surveys, since 1993, between Cape Leeuwin (WA) and Ceduna (SA). It includes counts and photo-identification and covers most of the current population;
- the life history and population biology of Southern Right Whales, Head of Bight, SA, including counts and photo-identification, annually since 1990;
- the systematic aerial line transect survey off the eastern Tas. coast, 2007;
- photo-identification studies off Tas., southern NSW and southern Australia;
- incidental sightings and strandings information, SA;
- sightings information in Vic., including from whale watching platforms at Logans Beach, and Warrnambool, which are available from the Dept of Sustainability and Environment, Vic.

FUTURE RECOVERY EFFORTS

What you can do:
- Join/donate to or volunteer for a cetacean conservation organisation or project (e.g. IUCN, WWF, Greenpeace, IFAW and see The Oceania Project).
- Volunteer with one or more of the following: the Oceania Project, UQ Cetacean Ecology and Acoustics Laboratory, the Southern Cross University Whale Research Centre, the Pacific Whale Foundation, the Australian Antarctic Division and the Australian Marine Mammal Centre.

Research directions:
- Monitor the population size and abundance to measure recovery.
- Study the impacts of increasing sea surface temperatures on the distribution, abundance and reproductive rate of prey species.
- Study the impacts of increasing sea surface temperatures on the reproductive success of Southern Right Whales.

Sources and further reading

Bannister JL (2001) Status of southern right whales (*Eubalaena australis*) off Australia. *Journal of Cetacean Research and Management* **2**, 103–110.

Bannister JL (2009) 'Southern right whale aerial survey, southern Australian coast, August 2009'. Final report to Island Foundation, Massachusetts, USA (unpublished).

Burnell SR (2001) Aspects of the reproductive biology, movements and site fidelity of right whales off Australia. *Journal of Cetacean Research and Management* **2**, 89–102.

Burnell SR (2008) Estimates of demographic parameters of southern right whales off Australia. Document BRG16 presented to the Scientific Committee of the International Whaling Commission, Santiago, Chile, June 2008.

Pirzl R, Patenaude N, Burnell S and Bannister J (2009) Movement of southern right whales (*Eubalaena australis*) between Australian and sub-Antarctic New Zealand populations. *Marine Mammal Science* **25**, 455–461.

Author: John Bannister

HUMPBACK WHALE

Megaptera novaeangliae (Borowski, 1781)

Photo: Stewart Macdonald-UgMedia

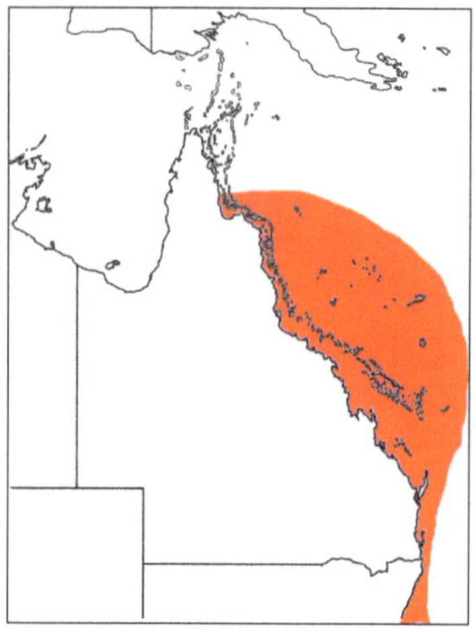

Family: Balaenopteridae

Aboriginal name: Galbol ('whale'-Torres Strait)

Conservation status: IUCN: Vulnerable (worldwide; Oceania subpopulation Endangered); Aust.: Vulnerable; Qld: Vulnerable

Description: This is a large whale with a head and body length to 16 m (male) or 18 m (female) and weighing up to 45 tonnes. It has a short, stocky body, very long (oar-like) and mobile flippers and broad tail flukes that are scalloped on the trailing edge. Its dorsal fin varies from a small rounded hump to a well defined hook, often with a fatty step or ledge at the base. A relatively few wide ventral grooves extend back to the navel. The basic body colour is black above and white below; its flippers and tail are white underneath. The baleen plates are greyish black with lighter coloured, moderately coarse inner fringes.

Distribution: This species is distributed worldwide in two separate northern and southern hemisphere populations. Seven separate southern breeding stocks are currently recognised: one each side of the three continents and another in mid-Pacific (Oceania). The breeding stock off the east coast of Australia migrates close to the coast and is closely related to two other subpopulations – breeding further east, around New Caledonia and Fiji/Tonga, migrating through the Tasman Sea and around New Zealand. The east Australian subpopulation was estimated at approximately 9600 animals in 2007.

Habitat: Humpback Whales are a familiar sight on migration close to the Australian coast. In Qld waters, they swim north within the 100 m depth contour until past Hervey Bay, where they are assumed to spread out over the Great Barrier Reef. Off Stradbroke Island, they pass in a narrow band only about 10 km wide.

Ecology: Humpback Whales first reproduce at 8–10 years of age, calving every 2–3 years and rarely in successive years, following a 10–11 month gestation. Births off the east coast are believed to occur mainly in the Great Barrier Reef area. Southern hemisphere humpbacks feed largely on Antarctic Krill *Euphausia superba* during summer, although elsewhere they can feed on small schooling fish or localised plankton swarms. Males compete aggressively for females, with six to ten animals often observed in tight groups on the breeding grounds. On migration, they travel mainly as pairs or singles although sometimes in groups of up to six to seven, at 3–6 km/h. Immature animals and females accompanied by yearling calves generally migrate north first, followed by males and non-pregnant females; pregnant females bring up the rear. Going south, the cows with their new calves travel last.

Reasons for decline and current threats: Humpback Whale numbers were drastically reduced worldwide by commercial whaling. In the southern hemisphere, they were the first species targeted by 'modern' whalers in the early 1900s. Substantial whaling off the east and west coasts from 1949 and in the Antarctic caused a drastic decline to only a few hundred animals by 1963. Eight thousand animals were killed from Byron Bay and Moreton Island between 1949 and 1963, when Australian Humpback whaling ceased. Since the 1970s, the east coast stock has increased substantially at around 10% per annum (mirrored on the west coast). A current threat is the likely long-term effect of climate change on their food, Antarctic Krill. Rises in sea surface temperature are known to suppress breeding success in Southern Right Whales *Eubalaena australis*, which also feed on krill. Other potential threats include disturbance from seismic operations, collision with large vessels, entanglement in fishing gear and pollution, though none have yet been demonstrated to be significant.

CURRENT RECOVERY EFFORTS

Recovery plan: Anonymous (2005) 'Humpback Whale recovery plan, 2005–2010'. Department of the Environment and Heritage, Canberra.

Other plans: '(Queensland) Nature conservation (whales and dolphins) conservation plan 1997'. *Nature Conservation Act 1992.*

Bannister JL, Kemper CM and Warnecke RM (1996) 'The action plan for Australian cetaceans'. Australian Nature Conservation Agency, Canberra.

Conservation actions: The Australian Whale Sanctuary provides high levels of protection to whales in Australian waters, including within 200 nm from the coast and waters around the Australian Antarctic Territory, Macquarie, Heard and McDonald Islands. In addition, there are a number of studies being conducted:

- East coast Humpback Whale population abundance – UQ Cetacean Ecology and Acoustics Laboratory.
- Further investigation of abundance estimates of migrating Australian Humpback Whales – UQ.
- The use of unmanned vehicles for surveys of marine mammals including Humpback Whales – UQ Cetacean Ecology and Acoustics Laboratory.
- The final development of new computerised fluke matching system and creation of fluke database for east coast Humpback Whales 1999–2005 – SCU.
- Controlled exposure experiments on Humpback Whales (effects of noise) – UQ Cetacean Ecology and Acoustics Laboratory.
- Population structure and sub-structure of Australian Humpback Whales – Australian Antarctic Division.
- Review of existing datasets to assess spatial and temporal distribution of Humpback Whales within the Great Barrier Reef to identify potential mating/calving grounds – UQ.
- International Whaling Commission – SOWER Programme – annual cruises in the Antarctic to provide sightings and acoustic information on all cetaceans.
- The Japanese Antarctic Research Programme (JARPA) – provides sighting and density information on large whales.
- Sightings and photo-ID, off Byron Bay (and Ballina) NSW. SCU Whale Research Centre.
- Photo-ID and sightings, in Hervey Bay – Pacific Whale Foundation.

FUTURE RECOVERY EFFORTS

What you can do: Join/donate to and/or volunteer with IUCN, WWF, Greenpeace, IFAW, the Oceania Project, UQ Cetacean Ecology and Acoustics Laboratory, the SCU Whale Research Centre, the Pacific Whale Foundation, the Australian Antarctic Division and the Australian Marine Mammal Centre.

Sources and further reading

Chittleborough RG (1965) Dynamics of two populations of the humpback whale, *Megaptera novaeangliae* (Borowski). *Australian Journal of Marine and Freshwater Research* **16**, 33–128.

Dawbin WH (1966) The seasonal migratory cycle of humpback whales. In *Whales, Dolphins and Porpoises.* (Ed. KR Norris) pp. 145–170. University of California Press, Berkeley and Los Angeles.

Dawbin WH (1996) Temporal segregation of humpback whales during migration in southern hemisphere waters. *Memoirs of the Queensland Museum* **42**, 105–138.

Noad MJ, Dunlop R, Paton D and Cato D (2008) An update of the east Australian humpback whale population (E1) rate of increase. Document SC/60/SH31 presented to the International Whaling Commission Scientific Committee, Santiago, Chile, May 2008 (unpublished, available from the International Whaling Commission, 135 Station Road, Impington, Cambridge CB24 9NP UK).

Paterson R, Paterson P and Cato DH (2004) Continued increase in east Australian humpback whales in 2001, 2002. *Memoirs of the Queensland Museum* **49**, 712.

Author: John Bannister

Glossary (including abbreviations and acronyms)

Acid frogs: frog species associated primarily with acidic wallum swamps and lakes.

ACT: Australian Capital Territory.

Action plan: a list of specific actions, associated costs, responsibilities, timeframes, etc.

Adelphophagous: a reproductive specialisation in the Grey Nurse Shark where embryos, once they have depleted their own yolk supply, feed on their uterine siblings (once these are depleted, they feed on eggs produced by the mother; see oophagous).

Adit: a horizontal tunnel in a mine.

Aestivate: undergo dormancy during the dry season.

Alluvial: pertaining to sediment – clay, silt, gravel, and so on – deposited by flowing water such as a river or stream.

AM: Australia Museum.

Amphibian: a cold-blooded vertebrate, such as a frog, toad, or salamander, sharing many characteristics with fish and reptiles at various stages of their development.

Amplexus: the manner in which male frogs clasp the female from behind during mating.

Anterior: the forward end of the animal.

Aperture: the opening of a shell.

Aquatic: living in water.

Arboreal: living in trees.

Arthropod: an invertebrate with jointed limbs, segmented body and an exoskeleton made of tough semitransparent protective substance (chitin).

Avifauna: the bird species inhabiting a region.

Axillary amplexus: where an amplexing male frog grasps a female's upper body (in the armpits).

Baleen: the fibrous matted plates that hang from the upper jaw of a baleen whale, which are used for filtering prey from mouthfuls of seawater.

Barbel: the slender, whisker-like tactile organ extending from the chin or head of fish or turtles.

Bifid: cleft or divided into two parts; forked.

Biodiversity: the variety of living organisms in a given place at a given time.

Biodiversity action plan: a plan containing proposed actions for the regional group to consider (or encourage others to consider, in partnership with the regional group) for focusing on-ground action, planning, monitoring and research and community engagement where it will be most effective in working to achieve species conservation.

Biofilm: a thin film of micro-organisms on the surface of stagnant ponds, submerged rocks, woody debris and reeds.

Bioindicator species: a species that responds to environmental changes.

Bioregion: a geographically distinct area with related or similar climate, geology, landform and/or ecosystems throughout. There are 85 major bioregions in Australia and 18 overlapping Queensland.

Biota: the species of plants and animals found in a given region at a given time.

Blastocyst: an early embryonic stage comprising fewer than 100 cells.

Blow: the visible condensed vapour cloud of a whale's exhaled breath.

Blowholes: the nostrils of a whale, which lie on top of the head.

Brigalow: the species *Acacia harpophylla* and/or the vegetation community it dominates.

Bursae: in reptiles, a pouch or sack.

Bycatch: the incidental catch of non-target species in directed fishing, which are often discarded, either dead or alive.

CAMBA: China Australia Migratory Bird Agreement.

Canines: the longest and most distinctive pointed teeth, often larger in carnivores.

Canopy: a distinct layer of vegetation, such as branches and leaves, occurring above the ground.

Captive breeding: the breeding of animals in captivity.

Carapace: a hard bony outer covering. A crustacean's exoskeleton or the fused plates of a turtle.

Carnivore: an animal that subsists primarily on vertebrate prey, such as mammals, birds, reptiles and amphibians (usually does not include fish; see piscivore).

Caudal fin: the tail fin of a fish.

Cephalopods: a class of molluscs with grasping tentacles and an outer or inner shell that occur only in marine habitats, such as nautiloids, cuttlefish, squids and octopuses.

Cerci: paired sensory appendages located on the posterior segments of certain insects and arachnids.

Cetacean: an order of large aquatic mammals, including whales and dolphins.

Chenopod: plants in the family Chenopodiaceae. Examples include saltbush and bluebush.

Chevron: V-shaped pigmentation (when seen from above) on the upper back of some whale species.

Chytridiomycosis: a disease caused by the chytrid fungus *Batrachochytrium dendrobatidis*, which only affects frogs in Australia.

Circum-umbilical: surrounding the umbilicus.

CITES: Convention of International Trade in Endangered Species of Wild Fauna and Flora. An international agreement that aims to ensure that international trade in wildlife does not threaten the survival of species.

Clear-felling: cutting down every tree in a given area.

Cloaca: in many animals, a chamber and passageway where the intestinal, urinary and reproductive canals open.

CMS: Convention on the Conservation of Migratory Species of Wild Animals (also referred to as the Bonn Convention). An intergovernmental treaty that aims to conserve terrestrial, marine and avian migratory species throughout their range.

Congener: belonging to the same genus.

Conservation: the preservation and careful management of natural resources and the natural environment.

Conservation covenant: a voluntary agreement between a landholder and an authorised body, such as the Qld Government.

Conservation dependent (CD): 'a native species that is the focus of a specific conservation program, the cessation of which would result in the species becoming vulnerable, endangered or critically endangered.' (EPBC definition)

Conservation status: a system of rating the level of risk of extinction for a species.

Continental shelf: the gently sloping, shelf-like part of the seabed adjacent to the coast, generally extending to a depth of about 200 m.

Continental slope: the often steep, sloping part of the seabed extending from the edge of the continental shelf to a depth of about 2000 m.

Controlled burn: a fire ignited intentionally to remove flammable undergrowth in order to reduce the chance of wildfires or to clear land for agriculture.

Cookie cutter sharks: small sharks of the genus *Isistius* that often feed by biting ice-cream-scoop-sized chunks of flesh out of whales.

Copepods: small swarming crustaceans that are the preferred food of Right and Sei Whales.

Corridor: a strip of habitat that facilitates the movement of animals between separated or isolated habitats.

Costa: insect wing margin; in dragonflies, the first longitudinal vein (C) of the wing.

Crenulated: having a notched or scalloped edge.

Crepuscular: active at dusk and/or dawn.

Critical habitat: the habitat vital to the survival of a species; it may be used for feeding, breeding or nursing young.

Critically Endangered (CR: IUCN; CE: EPBC): 'a taxon that is considered to be facing an extremely high risk of extinction in the wild.' (IUCN 2001 definition) 'When a native species is facing an extremely high risk of extinction in the wild in the immediate future, as determined in accordance with the prescribed criteria.' (EPBC definition)

Cryptic: features or behaviours that make a species difficult to find or observe, such as camouflage or staying close to cover.

CSIRO: (Australian) Commonwealth Scientific and Industrial Research Organisation.

Cubital vein: one of the main butterfly wing veins.

Data Deficient (DD): 'when there is inadequate information to make a direct, or indirect, assessment of its risk of extinction based on its distribution and/or population status.' (IUCN 2001 definition)

DECCW: NSW Department of Environment, Climate Change and Water.

DEEDI: Department of Employment, Economic Development and Innovation Qld.; formerly Department of Primary Industries and Fisheries (DPI&F).

Deforestation: the act of stripping land of trees or forests.

DEH: Department of Environment and Heritage (Commonwealth).

Demersal: occurring or living near or on the bottom of the ocean.

DENR: Department of the Environment and Natural Resources (South Australia).

DERM: Department of Environment and Resource Management (Qld); formerly the Environmental Protection Agency (EPA).

DEWHA: Department of the Environment, Water, Heritage and the Arts; formerly Department of Environment and Heritage (DEH) (Commonwealth).

Dextral: on the right-hand side of the body.

Diapause: the period during which living organisms suspend development in response to adverse conditions.

Distribution: the known limits or geographic area in which a given species is known to occur.

Diurnal: active by day.

Domestic animal: an animal that is fed and housed by humans and dependent on them for survival.

Dorsal fin: the fin on the midline of a whale's back or on the dorsal surface of a fish; it may be highly variable in location, shape and size.

DPI&F: Department of Primary Industries and Fisheries; now Department of Employment, Economic Development and Innovation (DEEDI) Qld.

DSEWPC: Department of Sustainability, Environment, Water, Population and Communities (Commonwealth); formerly Department of the Environment, Water, Heritage and the Arts (DEWHA).

Dystrophic: containing brown acidic water with high levels of humus, low oxygen content and unable to sustain much plant or animal life.

Ecological community: a collection of species populations inhabiting the same area and potentially interacting.

Ecosystem: the ecological community and physical environment in which members interact.

Ecotone: a transitional zone between two vegetation communities containing the characteristic species of each; or an ecological zone or boundary where two or more ecosystems meet, often accompanied by a change in soil type.

Elasmobranch: any species of the subclass Elasmobranchii – the sharks and rays.

Endangered (E: IUCN, EPBC, NCA): 'a taxon that is considered to be facing a very high risk of extinction in the wild.' (IUCN 2001 definition); 'When a native species is not critically endangered and is facing a very high risk of extinction in the wild in the near future, as determined in accordance with the prescribed criteria.' (EPBC definition); 'A species is endangered if there have not been thorough searches conducted for the wildlife and the wildlife has not been seen in the wild over a period that is appropriate for the life cycle or form of the wildlife; or the habitat or distribution of the wildlife has been reduced to an extent that the wildlife may be in danger of extinction; or the population size of the wildlife has declined, or is likely to decline, to an extent that the wildlife may be in danger of extinction; or the survival of the wildlife in the wild is unlikely if a threatening process continues.' (NCA definition)

Endemic: when the natural occurrence of a given species is restricted to a specific region.

Environment: the living and non-living surroundings of a plant or animal.

EPBC: The Commonwealth of Australia's *Environment Protection Biodiversity Conservation Act 1999.*

Epigastric crest: a ridge towards the middle of the top of the shell above the stomach of many decapod crustaceans.

Epipelagic: the zone of the open ocean from the water surface to a depth of 200 m.

Epiphragm: a membranaceous or calcareous covering of the aperture produced by some pulmonate snails during aestivation to reduce moisture loss.

Epiproct: dorsal part of the eleventh abdominal segment in insects.

ETBF: Eastern Tuna and Billfish Fishery.

Eutrophication: runoffs of excessive nutrients from the land into bodies of water that cause prolific plant growth, which in turn deplete the water's oxygen supply upon decomposition thereby killing all animal life.

Exotic species: a species not native to Australia.

Extant: living.

Extinct (EX): 'a taxon is Extinct when there is no reasonable doubt that the last individual has died.' (IUCN 2001 and EPBC definitions)

Extinct in the Wild (EW: IUCN, EPBC; EiW: NCA): 'when no confirmed sightings of an animal in the wild have been made within 50 years.' (IUCN definition); 'When a taxon is known only to survive in cultivation, in captivity or as a naturalised population well outside its past range; or it has not been recorded in its known and/or expected habitat, at appropriate seasons, anywhere in its past range, despite exhaustive surveys over a time frame appropriate to its life cycle and form.' (EPBC definition); 'A species is extinct in the wild if there have been thorough searches conducted for the wildlife; and the wildlife has not been seen in the wild over a period that is appropriate for the life cycle or form of the wildlife.' (NCA definition)

Extralimital: used to describe a species or group of organisms found outside a given area, such as a subspecies occurring outside of Australia.

Exuviae: the cast-off skin or shell coverings of various organisms such as insect nymphs and crabs.

Family: one of the taxonomic groups of organisms, containing related genera; related families are grouped into orders.

Fauna: the animals of a given geographic region.

Fecundity: a measure of the capacity of the maternal adult to produce young.

Feral animal: an animal established in the wild after escaping human captivity.

Fire management: the controlled burning of grasses, brush, undergrowth and trees carried out to ensure a balanced biodiversity and the prevention of wildfires.

Fire regime: the sequence of fires in any given area. Differences in fire regimes are characterised by differences in fire intensity, fire type, fire frequency and the season in which they occur.

Flagellum: a narrow whip-like extension of a tadpole's tail.

Flagship species: a charismatic species that attracts public attention, funding and conservation efforts.

Flora: the plants of a given region.

Flukes: the two lobes of a whale's tail, often separated by a small notch.

Focal species: a small group of species, the distributions and populations of which are well known, used in conservation planning.

Forb: a wide-leafed herb – not a grass.

Forest: land where tree canopy coverage exceeds 30%.

Fossorial: adapted for digging or burrowing.

Fragment: a patch of remaining habitat formed when a landscape is altered.

Frons: in dragonflies, the upper anterior portion of the head capsule, usually a distinct sclerite between the epicranium and clypeus.

Frugivore: an animal that subsists primarily on fruit.

Gallery forest: forest vegetation occurring adjacent to waterways (i.e. riparian forest).

GBR: Great Barrier Reef.

GBRMPA: Great Barrier Reef Marine Park Authority.

Genetic diversity: refers to the amount of variability in the genome of a species.

Genus (pl: genera): one of the taxonomic groups of organisms, containing related species; related genera are grouped into families.

Gestation: the period between conception and birth in live-bearing (viviparous) animals.

Gilgai: a small shallow pond or pool of water.

Global warming: a medium- to long-term rising trend in the temperature of the Earth's atmosphere.

Granivore: an animal that subsists primarily on seed.

Greenhouse effect: the resulting increase in temperature caused by the trapping of gases such as carbon dioxide, methane and water vapour in the Earth's atmosphere.

Greenhouse gases: gases found in the atmosphere that contribute to the greenhouse effect.

Gregarious: Living in loosely organised groups; social.

Habitat: the physical environment in which an animal survives and reproduces.

Heath: vegetation consisting primarily of shrubs, often growing on sandy soils.

Herbaceous: relating to herbs or plants.

Herbivore: an animal that subsists primarily on plants.

Home range: the area in which an animal habitually moves around.

Hotspot: a particularly important location; it may have high biodiversity of particularly large populations of a particular species of interest.

Hypercarnivore: a carnivore that feeds on the flesh of other vertebrates for >70% of its intake, as opposed to carnivores that may feed primarily on insects for example.

IFAW: International Fund for Animal Welfare.

Inbreeding: breeding between individuals more closely related than randomly selected individuals.

Indigenous species: belonging to, or found naturally in, a particular environment and forms part of the natural biodiversity of a place.

Infralabial scale: one of the row of scales bordering the lower lip.

Inguinal amplexus: where an amplexing male frog grasps a female in the groin.

Insectivore: an animal that subsists primarily on insects.

Interorbital: between the eyes.

Introduced: a plant, animal or organism brought in from outside a given area.

Invertebrate: an animal without a backbone such as a worm, insect or shellfish.

Ischium: The second of the seven segments of a decapod crustacean's leg.

IUCN: International Union for the Conservation of Nature.

JAMBA: Japan Australia Migratory Bird Agreement.

JCU: James Cook University.

Keel: a raised ridge, projection, such as keeled breastbones on birds or keeled scales on reptiles or fish.

Keratinisation: the process that occurs when epithelial cells convert to horny tissue, such as nails, horns, feathers, hooves, claws, etc.

Krill: schooling crustaceans of the family Euphausiidae, which are common prey of baleen whales.

Labial tooth row formula: the configuration of labial tooth rows surrounding the mouth of tadpoles. Here, tooth row formulae are described using the following notation: X(X–X)/Y(Y–Y), where X represents the number of upper labial tooth rows above the mouth opening and Y the number of tooth rows below. Tooth rows with a medial gap or break are listed in brackets (). The numbering of tooth rows is from front to back.

Lateral: on the side.

Laterodorsal: on the upper half of the body and out towards the sides (i.e. away from the midline).

Least Concern (LC; IUCN; NCA): 'a taxon that does not qualify for Critically Endangered, Endangered, Vulnerable or Near Threatened. Widespread and abundant taxa are included in this category.' (IUCN 2001 definition); 'An animal is least concern if the wildlife is common or abundant and is likely to survive in the wild.' (NCA definition)

Lerp: sugary substances produced by sap sucking insects (psyllids) on leaves.

Leucism: animals having white feathers, fur or skin due to lack of pigment.

Life history: the traits that define the biology of a species, specifically in relation to its age, growth and reproduction.

Lignum: a dense shrub that has long tangled stems and forms dense thickets in wetlands of the inland.

Ligula: collective name for glossae and paraglossae in all insects.

Loreal scale: the scale between a snake's nasal scales (bordering the nostril) and the preocular scales (immediately in front of eye).

Lores: the area between the eyes and the base of the bill of a bird.

Mabi Forest: a tall nutrient-rich, vine forest with a dense shrub zone below. Complex Notophyll Vine Forest type 5b and Queensland Regional Ecosystem 7.8.3.

Macropod: animals belonging to the family Macropodidae, including kangaroos, wallaroos, wallabies, pademelons, rock-wallabies and hare-wallabies.

Mallee: small, multi-stemmed mallee scrub trees (e.g. *Eucalyptus dumosa* and *Eucalyptus oleosa*) growing to 5–8 m over a variety of shrubs as well as grasses and herbs.

Management plan: an official government plan for the management of any given threatened species, feral animals, weeds and/or ecosystems.

Mantle (avian): the back, scapular (shoulder feathers), and inner wing plumage of a bird, especially when of the same colour and different from the rest of the body.

Mantle (marine): the tissue surrounding the visceral mass of a mollusc.

Marsupial: mammals that give birth to undeveloped young, which then attach themselves to their mother's teat, usually located in a pouch, to continue their growth.

Maxilliped: in crustaceans, three pairs of leg-like appendages specially modified for manipulating food into the mouth.

Merus: The third of the seven segments of a decapod crustacean's leg; often the longest segment closest to the body.

Metapopulation: population groups of the same species that live in separate places but interact at some level.

Microhabitat specialist: a species that selects a certain type of habitat on a small-scale and is usually restricted to that habitat.

Migratory: the systematic (as opposed to random) movement of individuals of a species from one location to another.

Mist net: mesh net, usually nylon, stretched between two poles and used to catch birds or bats.

Monoculture: a single species of plant having survived numerous burns and/or herbicides and pesticide treatments that now dominates a large area of land.

Myrmecophilous: in butterflies, referring to the association that larvae have with ants.

Native vegetation: plants that have always grown in a specific area.

NCA: Queensland's *Nature Conservation Act 1992.*

Near Threatened (NT: IUCN, NCA): 'a taxon which does not qualify for Critically Endangered, Endangered or Vulnerable now, but is close to qualifying for or is likely to qualify for a threatened category in the near future.' (IUCN 2001 definition); 'An animal is near threatened if the population size or distribution of the wildlife is small and may become smaller; or the population size of the wildlife has declined, or is likely to decline, at a rate higher than the usual rate for population changes for the wildlife; or the survival of the wildlife in the wild is affected to an extent that the wildlife is in danger of becoming vulnerable.' (NCA definition)

Nectarivore: an animal that subsists primarily off of nectar from flowers.

Nematodes: free-living or parasitic roundworms, which are elongated, smooth and cylindrical.

Neonate: a new-born individual.

NHT: Natural Heritage Trust.

Nocturnal: active by night.

Not Evaluated (NE): 'a taxon is Not Evaluated when it is has not yet been evaluated against the criteria.' (IUCN 2001 definition)

NP: National Park.

NRETAS: Natural Resources, Environment, The Arts and Sport (Northern Territory).

NRM: Natural Resource Management.

NSW: New South Wales.

NT: Northern Territory.

Nuptial pads: a hard pad, patch or clump of spines, usually dark, located on the thumbs of breeding male frogs.

Occiput: the lower part of the back of the skull.

Ocellus: the simple, single lens eye, with a few sensory cells, of certain invertebrates.

Old growth forest: a general concept with different interpretations around the world. In Australia it refers to forests with a canopy made primarily of mature, 'over-mature' and senescent trees, often with complex habitat structural characteristics such as many hollows and dead wood.

Oligotrophic: containing low levels of nutrients such as phosphorous and nitrogenous compounds.

Oophagous: a reproductive specialisation in some sharks where embryos, once they have depleted their own yolk supply, feed (in the uterus) on eggs produced by the mother.

Order: one of the taxonomic groups of organisms, containing related families.

Oviparous: a reproductive mode where the adult female lays eggs (as opposed to viviparity).

Oviposition: the act of egg-laying.

Parasite: an animal that grows, feeds and lives on another animal without contributing anything to its host.

Pelagic: occurring or living in open waters or near the water surface and having little contact with or dependency on the ocean floor.

Periostracal: referring to the periostracum.

Periostracum: the skin-like, proteinaceous covering of the shell surface in molluscs.

Phenology: the relationship between a periodic biological phenomenon and climatic conditions.

Philopatry: a behaviour, such as that in sea turtles, wherein the animal returns to its birthplace.

Photo ID: photo identification of individuals using natural markings.

Piscivore: an animal that subsists primarily on fish.

Pitfall trap: a buried container with its opening level with the ground into which mobile animals fall.

Plankton: small drifting aquatic organisms that occur in great numbers in productive regions.

Plastron: the lower section of the turtle shell made of bone and scutes.

Polyandrous: describes when a female breeds with several males in any given season, with the males assuming all parental responsibilities.

Portunid: in crabs, any member of the Portunidae family.

Postclypeus: in dragonflies, the upper of 2 parts into which the clypeus is divided.

Postmental scale: a reptile scale immediately behind the mental (chin) scale.

Postorbital crest: a ridge located behind the orbit or eye socket.

Precautionary Principle: a principle that states that lack of full scientific certainty should not be used as a reason for postponing measures to prevent environmental damage to species or habitats when there is a threat of serious or irreversible environmental degradation or loss of biodiversity.

Precocial: requiring little or no parental care at birth.

Predator: an animal that hunts, catches and eats other animals.

Prementum: the distal part of the dragonfly's labium (the lower lip of the mouth structure).

Prepollex: a hard lump on the base of a frog's thumb.

Protected fauna: animals that are protected by local, state and/or federal government laws.

Protected flora: plants that are protected by local, state and/or federal government laws.

Prothorax: the section of an insect's thorax nearest the head.

Protogynous hermaphrodite: a species that matures first as a female and later changes sex to become a male.

Qld: Queensland.

QPWS: Queensland Parks and Wildlife Service.

Quantification: the determination of the quantity, or amount, of something.

Quaternary: the most recent of the Cenozoic Era's three periods, and which includes the Pleistocene and Holocene epochs and spans the last 2 million years.

Rainforest: a humid forest with a continuous canopy or generally >70% canopy cover. Rainforests occur in wet areas in temperate and subtropical zones but are most widespread in tropical areas.

RAOU: Royal Australasian Ornithologists Union (also known as Birds Australia).

Raptor: a carnivorous bird that hunts and kills animals; a bird of prey.

Recovery: the return of a population of a species to its former favourable condition.

Recovery plan: 'Recovery plans set out the research and management actions necessary to stop the decline of, and support the recovery of, listed threatened species or threatened ecological communities. The aim of a recovery plan is to maximise the long-term survival in the wild of a threatened species or ecological community.' (DSEWPC definition)

Reflected (snail): bent backwards, usually in reference to the lip.

Regrowth: plants that regrow when the original population of plants have been cleared.

Remnant vegetation: the remains of native vegetation that have remained uncleared since European settlement.

Reproductive periodicity: the frequency at which a species breeds; i.e. biannually, annually, biennially, etc.

Reptile: a cold blooded vertebrate such as a lizard, turtle, snake, skink or crocodile.

Ridglet: a short, elongate sculptural element on a fish.

Riparian: referring to the banks and vegetation community adjacent to waterways.

Rookery: a turtle or colonial bird nesting site.

Rostral ridge: a raised ridge on the rostrum of a whale.

Rostrum: the upper surface of a whale's head, in front of the blowholes.

Rugose: a wrinkled or uneven surface.

SA: South Australia.

Samphire: low shrubs that are commonly found growing on tidal flats and saline lakes.

Scat: animal excrement.

SCU: Southern Cross University.

Sculpture: the ornamentation on the shell surface.

Scute: the bony external plate or scale on a turtle's shell or crocodile skin.

Sedges: tufted erect plants that sometimes resemble grasses and grow commonly in damp locations.

Semelparity: the condition of an organism that has only one reproductive cycle during its lifetime.

Sinistral: on the left-hand side of the body.

Site fidelity: describes the relationship between an animal and a particular location. May be strong or weak depending on the frequency or regularity of visits.

SOI: Southern Oscillation Index.

Spatial: relating to how things are distributed across the landscape.

Species: individuals of the same type able to interbreed and produce offspring, under natural conditions, which contribute to a common gene pool.

Spiracle: an opening out of which water exits the gill chamber/s of a tadpole.

Spire: the coils of a shell above the body whorl.

Subglobose: having a shape that is not quite spherical.

Subspecies: a specific geographic or independent breeding group of a species consistently characterised by given features.

Sustainability: 'the ability to conduct an activity without depleting resources beyond their regenerative capacity.' (DERM).

Sympatry/sympatric: where two or more species occur together.

Synthorax: (= pterothorax) in dragonflies, the closely fused meso- and metathorax.

Tas: Tasmania.

Taxon: refers to a species, a particular form of a species or a closely related group of species.

TED: turtle exclusion device – a rigid grid placed in a trawl net, which directs large bycatch species (such as turtles, large sharks and rays, and sponges) out an escape opening in the net. TEDs are mandatory in Queensland trawl fisheries.

Telson: a separated projection of the last true segment of a crustacean's body; the central part of the tail fan of shrimps and lobsters.

Temporal: relating to things over time.

Temporal scale: in reptiles, the scale behind the postocular scales (immediately behind eye), below the parietal scale (last enlarged head scales dorsally) and above the supralabial scales (bordering the upper lip).

Termen: the insect wing's edge furthest from the body.

Terrestrial: ground dwelling.

Territory: an area inhabited by one or more animals, which is defended against other members of their species.

Tertials: the third (innermost) row of feathers nearest the bird's body on the rear edge of its wing.

Threat abatement plan: 'a plan made or adopted under section 270B. Section 271(1): A threat abatement plan must provide for the research, management and other actions necessary to reduce the key threatening process concerned to an acceptable level in order to maximise the chances of the long-term survival in nature of native species and ecological communities affected by the process.' (DSEWPC)

Threatened species: a plant or animal assigned a conservation status (e.g. Vulnerable, Endangered or Critically Endangered).

Threatening process: 'a process that threatens, or may threaten, the survival, abundance or evolutionary development of a native species or ecological community.' (DSEWPC)

Throat pleats: the parallel pleats on the underside of a baleen whale's head, which expand to greatly enlarge the size of the mouth when feeding.

Tornal spots: in butterflies, spots located at the posterior corner of the wing.

Tornus: in butterflies, the posterior corner of the wing.

UM: University of Melbourne.

Umbilicus: the hollow axis at the base of a snail's shell.

UNE: University of New England.

Univoltine: having one generation in a year.

UQ: University of Queensland.

USC: University of the Sunshine Coast.

Vagrant: an animal only occasionally found in a specific area, usually there by accident.

Vent tube: a short tube out which faeces and urine exit in tadpoles.

Venter: in reptiles, the under surface of the animal.

Ventral: on the lower half or underside of the body.

Ventrolateral: on the lower half of the body and out towards the sides (i.e. away from the midline).

Vic.: Victoria.

Viviparity: a reproductive mode where the adult female gives birth to live young (as opposed to oviparity).

Vulnerable (VU: IUCN; V: EPBC, NCA): 'a taxon which is considered to be facing a high risk of extinction in the wild.' (IUCN 2001 definition); 'A native species that is not critically endangered or endangered and is facing a high risk of extinction in the wild in the medium term future, as determined in accordance with the prescribed criteria.' (EPBC definition); 'A species is vulnerable if its population is decreasing because of threatening processes, or its population has been seriously depleted and its protection is not secured, or its population, while abundant, is at risk because of threatening processes, or its population is low or localised or depends on limited habitat that is at risk because of threatening processes.' (NCA definition)

WA: Western Australia.

Wallum: the sandy heathland country of coastal Qld and NSW in which grows the small shrubby tree, *Banksia aemula*.

Wetlands: lands dominated by water saturation either on a permanent or seasonal basis.

Wilderness: a natural environment not yet significantly modified by humans.

Wildfire: a fire resulting from natural causes (i.e. lightning) or arson.

Woodland: one of four broad types of woodlands in Australia characterised by 10–30% tree coverage.

World Heritage site: a natural or human-made site of particular significance as outlined by the international World Heritage Programme administered by the UNESCO World Heritage Committee.

WTBF: Western Tuna and Billfish Fishery.

Appendix I: Organisations involved in Queensland threatened animals recovery and management

(Anyone actively involved in recovery and management of Qld threatened species who would like to be added to this list in future editions, please email lee@ataglance.com.au)

Australasian Bat Society

PO Box 481, Lindfield, NSW 2070
Tel: 02 6229 7099
Email: president@ausbats.org.au
Web site: www.ausbats.org.au

The Australasian Bat Society is an active advocate for the conservation and sound management of bats and their habitat. We provide input and advice to relevant government and non-government agencies, industry, landholders and to the general public; and actively raise positive public awareness of bats through the preparation and circulation of position papers, information brochures, books, leaflets, electronic information and through lectures and encourage scientific research on bats.

Australian Animals Care and Education (AACE) and Project Kial

PO Box 47, Marlborough, Qld 4075
Tel: 07 4935 6076 **Mob:** 0408 613 914
Email: tina_janssen@bigpond.com
Web site: www.aace.org.au

AACE is an environmental organisation dedicated to the rehabilitation and release of native Australian animals. Under the auspices of AACE and DERM, Project Kial: A Bridled Nailtail Wallaby and Northern Hairy-nosed Wombat recovery project currently runs a successful Bridled Nailtail Wallaby captive breeding and release program and is currently working in conjunction with University of Queensland on perfecting artificial insemination in Southern Hairy-nosed Wombats to be able to ensure the survival of the Northern Hairy-nosed Wombats.

Australian Freshwater Turtle Conservation and Research Association (AFTCRA Inc.)

PO Box 963, Cooroy, Qld 4563
Tel: 07 5408 2616
Email: admin@aftcra.org.au
Web site: www.aftcra.org.au

We are a not-for-profit organisation led by a group of volunteers dedicated to ensuring all Australian freshwater turtle populations can remain secure in the wild. This is done by undertaking field studies and captive breeding programs, working with government agencies and other environmental organisations and raising awareness by providing educational workshops for the general public.

Australian Koala Foundation

GPO Box 2659, Level 1, 40 Charlotte Street,
Brisbane, Qld 4001
Tel: 07 3229 7233 **Fax:** 07 3221 0337
Email: akf@savethekoala.com
Web site: www.savethekoala.com

The Australian Koala Foundation (AKF) is the principal not for profit organisation dedicated to the conservation and effective management of the wild Koala and its habitat. The non-government funded scientific organisation is committed to the pursuit of conservation-driven research, world-leading models, and solutions to issues with far-reaching implications.

Australian Mammal Society

Email: amsconservation@australianmammals.org.au
Web site: www.australianmammals.org.au

The Australian Mammal Society (AMS) is an interdisciplinary society of biologists whose common interest is in the biology and conservation of Australian mammals. Research by members of the Society spans a wide variety of biological disciplines including behavioural and community ecology, population genetics, management and conservation.

Australian Marine Conservation Society (AMCS)

PO Box 5136, Level 2/75 Cambridge Pde, Manly, Qld 4179
Tel: 07 3393 5811 **Fax:** 07 3393 5833
Email: amcs@amcs.org.au
Web site: www.amcs.org.au

The Australian Marine Conservation Society played a critical role in improving protection for Queensland's Critically Endangered Grey Nurse Shark and also works to protect Queensland's sea turtles, which are threatened with extinction. AMCS pushed for increased protection zones in the Great Barrier Reef and Moreton Bay Marine Parks, and significantly expanded the green zones where all types of fishing, mining and exploitation are prohibited, and marine wildlife is safe from human harm.

Australian National University

Canberra, ACT 0200
Tel: 1800 620 032
Email: admiss.enq@anu.edu.au
Web site: www.anu.edu.au/studyat

Australian Rivers Institute – Coast & Estuaries

School of Environment, Australian Rivers Institute (ARI) – Coast & Estuaries
Griffith University – Gold Coast campus, Qld 4222
Tel: 07 5552 8638 **Fax:** 07 5552 8067
Email: j.meynecke griffith.edu.au
Web site: www.griffith.edu.au

We investigate the movement of Humpback Whales in relation to physical water properties and coastal urbanisation in the Gold Coast area.

Australian Rivers Institute

School of Environment, Griffith University – Gold Coast Campus
Parklands Drive, Southport, Qld 4215
Fax: 07 5552 8067
Email: ari@griffith.edu.au
Web site: www.griffith.edu.au

We study the fine-scale habitat use and long range movement of Great White Shark in subtropical Queensland.

Australian Tropical Research Foundation

Cape Tribulation Tropical Research Station,
PMB 5, Cape Tribulation, Qld 4873
Tel: 07 4098 0063
Email: hugh@austrop.org.au
Web site: www.austrop.org.au

Austrop is an independent tax-exempt, tax-deductible research and conservation organisation, with particular interests in Spectacled Flying-Fox conservation, rainforest rehabilitation and weed control. Current relevant activities include developing tracking and non-lethal deterrent measures for flying-foxes.

Australian Wildlife Conservancy

PO Box 8070, Subiaco East
Suite 5, 280 Hay St, Subiaco, WA 6008
Tel: 08 9380 9633 **Fax:** 08 9380 9631
Web site: www.australianwildlife.org

AWC establishes sanctuaries throughout Australia on which it implements on-ground conservation programs to manage fire, control feral animals and weeds and reintroduce threatened species. Many of these reintroduced populations have been husbanded to a point where they are now self-sustaining and have provided a source of animals for translocation to additional sites, including national parks. AWC has an Science Program which measures the effectiveness of the land management programs and carries out strategic research to address the key knowledge gaps that limit our ability to conserve wildlife. AWC also has a visitor and public education program at some of its sanctuaries.

Australian Wildlife Health Network

PO Box 20, Mosman, NSW 2088
Tel: 02 9932 4368
Email: awhn@zoo.nsw.gov.au
Web site: www.wildlifehealth.org.au/AWHN/home.aspx

The Australian Wildlife Health Network (AWHN) is a national initiative of the Commonwealth Government and is managed under the Wildlife Exotic Disease Preparedness Program (Australian Department of Agriculture, Fisheries and Forestry). Its mission is to promote and facilitate collaborative links across Australia in the investigation and management of wildlife health in support of human and animal health, biodiversity and trade. Its vision is for a nationally integrated wildlife health system for Australia.

Australia Zoo Wildlife Warriors Worldwide

1638 Steve Irwin Way, Beerwah, Qld 4519
Tel: 07 5436 2331 **Fax:** 07 5436 2066
Email: esu@wildlifewarriors.org.au
Web site: www.wildlifewarriors.org.au

We are involved in threat management through our radio tracking of Koalas residing in potential development sites. We also investigating the prevalence and incidence of disease in wild Koala populations, and conducting disease research in collaboration with universities throughout Qld (e.g. chlamydiosis and Koala retrovirus).

Barung Landcare Association, Inc.

PO Box 1074, Shop 3, Riverside Centre,
8 Maple Street, Maleny, Qld 4552
Tel: 07 5494 3151 **Fax:** 07 5494 3141
Email: info@barunglandcare.org.au
Web site: www.barunglandcare.org.au

Barung Landcare is a leading community-based organisation working to develop ethical solutions for a sustainable future. We support local and regional community empowerment to conserve and enhance the rich diversity of natural resources of the Blackall Range and environs.

Biodiversity Assessment and Management Pty Ltd

38 Middle Street, Cleveland, Qld 4163
Email: baam@biodiversity.tv
Web site: www.biodiversity.tv

Our team of highly skilled ecologists is committed to applying systematic science and our wealth of experience to project and conservation planning.

Birds Australia (National Office and Library)

Suite 2-05, 60 Leicester Street, Carlton, Vic. 3053
Tel: 03 9347 0757 or 1300 730 075
Fax: 03 9347 9323
Email: mail@birdsaustralia.com.au
Web Site: www.birdsaustralia.com.au

A national organisation working for the conservation and protection of Australia's native birds and their habitats.

Queensland Branches

Southern Queensland (BASQ)
PO Box 224, Crows Nest Qld, 4355
Tel: 07 4698 1727
Email: basq@birdsaustralia.com.au
Web site: www.basq.org.au

Capricornia (BA-CAP)
192 Palm Valley Road, Coowonga, Qld 4702
Tel: 07 4935 4645
Email: bacap@birdsaustralia.com.au
Web site: www.irock.com.au/~birdscapricornia/index.htm

North Queensland (BANQ)
PO Box 1012,
Thuringowa Central, Qld 4817
Email: banq@birdsaustralia.com.au
Web site: www.birdsaustralianq.org

Birds Australia Special Interest Groups

Australasian Raptor Society
www.birdsaustralia.com.au/the-organisation/australasian-raptor-association.html

Australasian Seabird Group
www.birdsaustralia.com.au/the-organisation/australasian-seabird-group.html

Australasian Waders Studies Group
www.birdsaustralia.com.au/the-organisation/australasian-wader-studies-group.html

Bird Observation and Conservation Australia (BOCA)
PO Box 185, Nunawading Vic. 3131
Tel: 03 9877 5342 or 1300 305 342 **Fax:** 03 9894 4048
Email: information@boca.org.au
Web site: www.boca.org.au

Bird Observation & Conservation Australia (BOCA) is a major birding organisation with more than 40 Branches and Affiliated Groups in Australia, including five Branches in Queensland. BOCA's objectives can be summarised as appreciation, education and conservation of Australian birds and their habitat.

Birds Queensland

PO Box 2273, Milton, Qld 4064
Tel: 07 3343 6323
Email: secretary@birdsqueensland.org.au
Web site: www.birdsqueensland.org.au

The aim of Birds Queensland is to promote the scientific study and conservation of birds by all means possible, with particular reference to the birds of Queensland.

Bribie Island Environmental Protection Association (BIEPA)

PO Box 350, Bribie Island, Qld 4507
Tel: 07 3410 0757
Email: diane@oxenford.net
Web site: www.biepa.org

BIEPA: coordinates volunteers to carry surveys on threatened frogs; files monthly reports about the Little Tern and Beach Stone-curlew and works to control disturbance and manage habitat; coordinates a survey on the Water Mouse; and monitors Loggerhead Turtles and identifies their seasonal nesting activity on the island.

Brisbane City Council (BCC)

GPO Box 1434, Brisbane, Qld 4001
Tel: 07 3403 8888
Web site: www.brisbane.qld.gov.au

Brisbane City Council's Conservation Action Statements identify and prioritise actions needed to save local threatened species. These actions include protection, management and monitoring. BCC is also a member of the Glossy Black Conservancy: a regional partnership to conserve the Glossy Black-Cockatoo in south-east Queensland.

Burnett Mary Regional Group (NRM)

PO Box 501, 193 Bourbong Street, Bundaberg, Qld 4670
Tel: 07 4181 2999 **Fax:** 07 4154 1427
Email: admin@bmrg.org.au
Web site: www.bmrg.org.au

The BMRG Project Team has developed a multi-species action plan so that future on-ground works will benefit as many species as possible. This plan provides an integrated and cost-effective approach for conservation activities and species recovery.

Bush Heritage Australia

PO Box 329, Flinders Lane, Melbourne, Vic. 3009
Level 5, 395 Collins Street, Melbourne, Vic. 3000
Tel: 1300 NATURE (1300 628 873) **Fax:** 03 8610 9199
Email: info@bushheritage.org.au
Web site: www.bushheritage.org.au

Bush Heritage Australia is a national, independent, non-profit organisation committed to preserving Australia's biodiversity by protecting the bush. It is Australia's most widely supported organisation dedicated to protecting species and habitats through the creation of reserves on private land.

Butterfly and Other Invertebrates Club Inc.

Butterfly and Other Invertebrates Club, Inc.
PO Box 2113, Runcorn, Qld 4113
Web site: www.boic.org.au

BOIC seeks: to increase knowledge of invertebrates; to raise public awareness of threats to biological diversity; to encourage restoration of habitats and the planting of appropriate host plants to facilitate recovery of invertebrate populations.

Cape York Sustainable Futures (NRM)

PO Box 646, N Cairns, Qld 4870
4/275 McCormack Street, Manunda, Qld 4870
Tel: 07 4053 2856 **Fax:** 07 4053 2942
Email: cypda@cypda.com.au
Web site: www.cypda.com.au/node/43

The Cape York Sea Turtle Project, run by Cape York Sustainable Futures, works in collaboration with key organisations and Indigenous communities in addressing turtle nest predation on western Cape York. Flatback, Olive Ridley and Hawksbill Turtles are in serious trouble, with over 90% of nests being destroyed by feral pigs along much of the coastline.

Cape York Turtle Rescue

PO Box 3352, Bangor, NSW 2234
Tel: 07 4069 9978 **Mob:** 0412 756 295 or 0428 283 380
Email: info@capeyorkturtlerescue.com
Web site: www.capeyorkturtlerescue.com

Part of the Cape York Turtle Rescue strategy is to better understand the needs of the turtles from a scientific perspective. Participants in our turtle expeditions work with Rangers to assist with identifying nesting turtles, installing Feral Pig Exclusion Devices (FPEDs) and collecting and compiling data to develop a sustainable management strategy.

Carpentaria Ghost Nets Programme

PO Box 155, Karumba, Qld 4891
Tel: 07 4745 9661
Email: riki.ghostnets@northerngulf.com.au
Web site: www.ghostnets.com.au

The Carpentaria Ghost Net Programme is an alliance of Indigenous communities in northern Australia working together to tackle the 'Ghostnets' – derelict fishing nets – part of vast rafts of pollution arriving from Asia that are fouling the coastline of northern Australia and that present an enormous threat to the threatened populations of marine turtles, Dugongs and sawfish.

Carpentarian Land Council Aboriginal Corporation

Gulf: Lot 87 Musgrave Street, PO Box 71,
Burketown, Qld 4830
Cairns: 2/14A Aplin Street, Cairns, Qld 4870
Tel: Gulf: 07 4745 5132; Cairns: 07 4041 3833
Fax: Gulf: 07 4745 5204; Cairns: 07 4041 3533
Email: receptionist@clcac.com.au
Web site: www.clcac.com.au

CLCAC has three Indigenous Land and Sea Ranger units based at Normanton, Mornington Island and Burketown that undertake management of natural and cultural resources across the lower Gulf of Carpentaria from the Northern Territory/Queensland border around to the Staaten River on western Cape York. CLCAC's Land and Sea management program undertakes various forms of management of native species, including protection and management of threatened species that occur in the area.

Central Queensland University

Bruce Highway, North Rockhampton, Qld 4702
Tel: 07 4930 9777
Web site: www.cqu.edu.au

Charles Darwin University

Ellengowan Drive, Casuarina,
Darwin, NT 0909
Tel: 08 8946 6666 **Fax:** 08 8927 0612
Email: changeyourworld@cdu.edu.au
Web site: www.cdu.edu.au

Charles Sturt University

PO Box 789, 624 Olive Street, Albury, NSW 2640
Tel: 02 6051 6000 **Fax:** 02 6051 6629
Web site: www.csu.edu.au

Community for Cassowary and Coastal Conservation (C4)

PO Box 165, Mission Beach, Qld 4852
Tel: 07 4068 7197 **Fax:** 07 4068 7298
Email: c4@cassowaryconservation.asn.au
Web site: www.cassowaryconservation.asn.au

Through the working groups and the vision statement of 'Knowing Keeping Sharing', C4 undertakes to protect Mission Beach's natural values to ensure a healthy environment, community and economy. We believe Mission Beach can develop and maintain its unique identity, including being a place where you can still see a Southern Cassowary in the wild.

Condamine Alliance (NRM)

3 Clifford St, Toowoomba, Qld 4350
Tel: 07 4620 0100 **Fax:** 07 4620 0100
Email: admin@condaminealliance.com.au
Web site: www.condaminealliance.com.au

Condamine Alliance is the designated natural resource management body of the Condamine Catchment and is responsible for enabling its community to achieve sustainable natural resource management. Condamine Alliance has been involved in threatened species recovery both generally, through the Back on Track Biodiversity Action Planning process in 2008, and specifically through funding of on-ground works to support Murray Cod, Eastern Bristlebird and Bulloak Jewel Butterfly habitat restoration.

Conservation Volunteers Australia

Brisbane
102 McDonald Road, Windsor, Qld 4101
Tel: 07 3357 1124 **Fax:** 07 3357 1129
Email: brisbane@conservationvolunteers.com.au
Web site: www.conservationvolunteers.com.au

Cairns
PO Box 954, Manunda, Qld 4870
38 Hayward St, Mooroobool, Qld 4870
Tel: 07 4032 0844 **Fax:** 07 4032 0833
Email: cairns@conservationvolunteers.com.au

Gladstone
PO Box 8243, 14 Goondoon St, Gladstone, Qld 4680
Tel: 07 4972 4969 **Fax:** 07 4972 4989
Email: gladstone@conservationvolunteers.com.au

Mackay
PO Box 1772, 62 Wellington St, Mackay, Qld 4740
Tel: 07 4951 0933 **Fax:** 07 4951 0922
Email: mackay@conservationvolunteers.com.au

Sunshine Coast
PO Box 982, 455 North Maleny Road, Maleny, Qld 4553
Tel: 0404 828 921

Townsville
PO Box 827, Townsville, Qld 4810
Suite 1, 65 Palmer Street, South Townsville, Qld 4810
Tel: 07 4721 4077 or 07 4721 4351 **Fax:** 07 4772 3077
Email: townsville@conservationvolunteers.com.au

CVA engages the global community in practical conservation projects in support of our threatened species throughout Queensland. CVA managed teams of volunteers can assist with a wide range of on-ground recovery actions including revegetation, weed management, fencing, surveys, etc.

CSIRO Sustainable Ecosystems, Atherton

PO Box 780, Maunds Road, Atherton, Qld 4883
Tel: 07 4091 8827
Web Site: www.csiro.au/places/atherton.html

CSIRO scientists are currently involved in community engagement projects and conducting research on Southern Cassowary genetics, population, ecology, habitat and resource use, movement, and monitoring and surveys.

Currumbin Wildlife Sanctuary

28 Tomewin Street, Currumbin, Qld 4223
Tel: 07 55 34 08 01
Web site: www.cws.org.au

Currumbin Sanctuary is currently involved with the Coxen's Fig-parrot Recovery Team and is running breeding programs for them as well as for Brush-tailed Rock-wallabies and Koalas.

Department of the Environment and Resource Management (Queensland)

PO Box 15155, City East, Qld 4002

EPA Customer Service Centre

160 Ann Street, Brisbane, Qld 4000
Tel: 1300 130 372
Email: info@epa.qld.gov.au.
Web site: www.epa.qld.gov.au

Department of Transport and Main Roads (Queensland)

Road Safety and System Management, PO Box 645, Toowoomba, Qld 4350
Tel: 07 4639 0852 **Mob:** 0428 113 282
Email: www.mainroads.qld.gov.au/contact-us
Web site: www.mainroads.qld.gov.au/en/Environment/Fauna-and-flora/Fauna-management.asp

Transport and Main Roads is committed to minimising the impact of its road operations by applying the most effective fauna mitigation techniques. We also manage and maintain the linear corridors to enable effective fauna and flora conservation.

Desert Channels Queensland (NRM)

PO Box 601, 92 Galah Street, Longreach, Qld 4730
Tel: 07 4658 0600 **Fax:** 07 4658 0122
Web site: www.dcq.org.au

DCQ (Desert Channels Queensland) is the regional NRM group responsible for the Qld portion of the Lake Eyre Basin. Part of our activity is to maintain healthy biodiversity across the region. We do this by supporting our community to protect viable ecosystems, through sustainable land management. We work with community to assist their efforts to maintain healthy landscapes. This could be through training, education and awareness, and direct on ground works, such as fencing off a waterhole or wetland to exclude stock or impacts from feral animals such as feral goats or pigs.

Desert Ecology Research Group

Room 325a, Institute of Wildlife Research,
School of Biological Sciences,
Heydon-Laurence Building A08, University of Sydney,
Sydney, NSW 2006
Tel: 02 9351 8577 **Fax:** 02 9351 4119
Email: bobby.tamayo@sydney.edu.au; aaron.greenville@sydney.edu.au; clbeh@bio.usyd.edu.au
Web site: www.sydney.edu.au/science/biology/sites/dickmanlab/index.shtml

The Desert Ecology Research Group has followed the extraordinary 'boom' and 'bust' cycles that characterise inland Australia over 20 years of continuous research in the Simpson Desert. Predators are also the focus of a study looking at the effects introduced species like the Red Fox have on small mammals, lizards and the broader prey community.

Dreamworld

Dreamworld Parkway, Coomera, Qld 4209
Tel: 07 5588 1111
Web site: www.dreamworld.com.au

Dreamworld has a huge and diverse animal collection, including threatened Southern Cassowary, Bilby and Kroombit Tinkerfrog, which is driven by conservation and education objectives. Dreamworld's dedicated wildlife team has achieved amazing results with education, breeding, habitat and re-introduction programs in external projects.

Fitzroy Basin Association Inc. (NRM)

34 East Street, 4th Floor, Suncorp Metway Building, Rockhampton, Qld 4700
Tel: 07 4999 2809 **Fax:** 07 4921 2860
Email: admin@fba.org.au
Web site: www.fba.org.au

Our biodiversity and vegetation program is focused on protecting a representative range of regional ecosystems, both on and off reserves, and addresses fragmentation and loss of biodiversity. The program assists in sharing the costs of protecting biodiversity across the whole community.

Foundation for Australia's Most Endangered Species (FAME)

PO Box 482, Mitcham, SA 5062
Tel: 08 8131 0066
Email: fame@fame.org.au
Web site: www.fame.org.au

FAME's vision is to prevent any further extinctions of Australian wildlife. We actively support projects that will save endangered wildlife, and we work in partnership with a wide range of like-minded organisations.

Frog Decline Reversal (FDR) Project, Inc.

Cairns Frog Hospital
Post Office Box 949, Edmonton, Qld 4869
Tel: 07 4045 0373
Email: curator@fdrproject.org.au
Web site: www.fdrproject.org.au

The FDR Project is a frog conservation organisation specialising in rehabilitation techniques and the detection and identification of new disease threats to all Queensland amphibian species, including toads. The group has already been involved in uncovering the presence of several new syndromes including immune deficiency in *Litoria infrafrenata* and is currently looking into the involvement of soil diseases in frog decline.

Glossy Black Conservancy

Email: GBC@biodiversity.tv
Web site: www.glossyblack.org.au

GBC is an independent association of local councils, environmental consultancies, birding groups and academic institutes in south-east Queensland. Our aim is to: coordinate the assessment of Glossy Black-Cockatoo resources across local government areas in south-east Queensland; engage with the community to assist in record collection; manage a central database of GBC records which can be accessed by government agencies; continue research to increase current knowledge of the species' use of pioneer vegetation, fidelity to individual feed trees, requirement for nesting hollows and diurnal and seasonal range, with a priority to map vital resources for GBC in south-east Queensland and develop a Regional Management Plan for Glossy Black-Cockatoo.

Gold Coast City Council

135 Bundall Road, PO Box 5042, GCMC,
Surfers Paradise, Qld 9729
Tel: 1300 130 854
Web site: www.goldcoast.qld.gov.au

The Gold Coast City Council: funds Griffith University frog surveys and Glossy Black-Cockatoo research; is currently working with Marine Parks and the Queensland Wader Study Group on a shorebird monitoring project and with Biolink to determine the distribution and range of the Long-nosed Potoroo in Springbrook Reserves; and has undertaken the long-term East Coomera Koala Conservation Project.

Greening Australia, Queensland

GPO Box 9868, Brisbane, Qld 4001
333 Bennetts Road, Norman Park, Qld 4170
Tel: 07 3902 4444 **Fax:** 07 3902 4422
Email: Qldgeneral@greeningaustralia.org.au
Web site: www.greeningaustralia.org.au

Greening Australia is a solutions-driven organisation that is committed to doing something practical about Australia's environmental problems. Greening Australia tackles critical issues like salinity, declining water quality, soil degradation, climate change and biodiversity loss through an innovative blend of practical experience, science and community engagement.

Griffith University, Nathan Campus 170 Kessels Road, Nathan, Qld 4111

Tel: 07 3735 7700
Web site: www.griffith.edu.au

Gympie Regional Council

PO Box 155, 242 Mary Street, Gympie, Qld 4570
Tel: 1300 307 800 **Fax:** 07 5481 0801
Email: council@gympie.qld.gov.au
Web site: www.gympie.qld.gov.au

Gympie Regional Council received funding from *Caring for our Country* Community Coastcare grants and subcontracted to the Mary River Catchment Coordinating Committee (MRCCC) to undertake the 'Implementing the National wallum frog recovery plan in the Great Sandy Strait Ramsar wetland' project. Target species include *Litoria olongburensis* (Wallum Sedgefrog), *Litoria freycineti* (Wallum Rocketfrog) and *Crinia tinnula* (Wallum Froglet).

Infofish Australia

PO Box 9793, Frenchville, Qld 4701
Tel: 07 4928 6133 **Mob:** 0417 075 277

INFOFISH Australia and Qld DEEDI are conducting education campaigns with anglers to emphasise best-practice handling and release techniques for accidental catches of threatened species.

Ipswich City Council

PO Box 191, 37 South Street, Ipswich, Qld 4305
Tel: 07 3810 6666 **Fax:** 07 3810 6206
Email: council@ipswich.qld.gov.au
Web site: www.ipswich.qld.gov.au

Ipswich City Council protects and manages over 6000 hectares of habitat within its conservation estates and bushland reserves. Council also supports a comprehensive private partnership program to assist landowners manage and rehabilitate habitat on their properties. Protection and management of this habitat within Ipswich assists the survival of many threatened species. Council supports the protection and recovery of specific threatened species such as the Glossy Black-Cockatoo, Bilby, Koala, Grey-headed Flying-Fox, Black-breasted Button-quail and Brush-tailed Rock-wallaby.

James Cook University

Townsville
James Cook University, 1 James Cook Drive, Douglas, Qld 4354
James Cook University, Townsville, Qld 4811
Tel: 07 4781 5255 **Fax:** 07 4781 4644

Cairns
James Cook University, McGregor Road, Smithfield, Qld 4878
James Cook University, PO Box 6811, Qld 4870
Tel: 07 4042 1000 **Fax:** 07 4042 1128
Email: EnquiriesTownsville@jcu.edu.au; EnquiriesCairns@jcu.edu.au
Web site: www.jcu.edu.au

Kuranda EnviroCare

PO Box 494, Kuranda, Qld 4881
Tel: 07 4093 8989
Email: secretary@envirocare.org.au
Web site: www.envirocare.org.au

We are involved in population surveys for the Myola Frog, the Common Mistfrog, *Litoria rheccola,* as well as habitat restoration and rehabilitation for a variety of threatened animals including the Southern Cassowary.

Lady Elliot Island Eco Resort

P O Box 348, Runaway Bay, Qld 4216
Tel: 07 5536 3644 **Fax:** 07 5599 5783
Email: reception@ladyelliot.com.au
Web site: www.ladyelliot.com.au

Lady Elliot Eco Resort, in conjunction with DERM, closely monitors nesting Red-tailed Tropicbirds and Green and Loggerhead Turtles. Nestlings and hatchlings are also monitored closely with egg counts carried out to determine hatching success.

Logan and Albert Conservation Association (LACA)

PO Box 557, Beaudesert, Qld 4285
Tel: 07 3297 0624
Email: mkpage@bigpond.com
Web site: www.laca.org.au

LACA receives information from the community about wildlife sightings and forwards this information onto Wildnet (DERM); e.g. of Glossy Black-Cockatoos, quolls, Koala and gliders. We have also been involved in applying for grants to run community education workshops, such as the 2007 Koalas and 2010 Wildlife Workshops. We also network with other groups, such as Wildlife Preservation Society of Queensland, to support actions to help threatened species, such as quolls. We attend local community events to raise the profile about threatened habitats such as *Melaleuca irbyana* and Veresdale Scrub (dry vine forest).

Macquarie University

Balaclava Road, North Ryde, NSW 2109
Tel: 02 9850 7111
Email: mqinfo@mq.edu.au
Web site: www.mq.edu.au

Mary River Catchment Coordinating Committee

PO Box 1027, Gympie, Qld 4570

MRCCC Resource Centre – Tozer Park Rd, Gympie, Qld 4750 (former Cooloola Shire Council depot)
Tel: 07 5482 4766 **Fax:** 07 5482 5642
Email: mrccc@ozwide.net.au
Web site: www.mrccc.org.au

The Mary River Catchment Coordinating Committee has been working with threatened fauna in the Mary and adjacent coastal catchments for the past 7 years, concentrating efforts on freshwater dependant species. Through research into local distributions of species under threat from habitat loss and alteration, feral animals and disease, the MRCCC can contribute to increasing knowledge for better management of natural and impacted areas, and can engage landholders in activities that improve habitat condition and environmental services for a sustainable catchment.

Mitchell River Watershed Management Group

PO Box 646N, Cairns North, Qld 4870
4/275 McCormack St, Manunda, Qld 4870
Tel: 07 4053 3471 **Fax:** 07 4031 3049 **Mob:** 0439 665 202
Email: info@mitchell-river.com.au
Web site: www.mitchell-river.com.au

The Mitchell River Watershed Management Group (MRWMG) is an independent, not-for-profit organisation responsible for grass roots community engagement, capacity building, and sustainable and integrated management of the Mitchell River catchment area. MRWMG works in partnership with local communities to maintain environmental flows in streams, revegetate degraded land, manage local plant and animal pests, run education and youth programs at local schools, raise community awareness of ecologically sustainable development and, in 2008 –2010, operated a small grants scheme to assist landowners with habitat restoration.

North Australian Indigenous Land and Sea Management Alliance (NAILSMA)

Charles Darwin University, Darwin, NT 0909
Tel: 08 8946 7691 **Fax:** 08 8946 6388
Email: nailsma@cdu.edu.au
Web site: www.nailsma.org.au

The North Australian Indigenous Land and Sea Management Alliance (NAILSMA) is an alliance between Kimberley Land Council, Northern Land Council, Carpentaria Land Council Aboriginal Corporation and Balkanu Cape York Development Corporation. NAILSMA supports Aboriginal and Torres Strait Islander land and sea management using strategic approaches to care for country with an emphasis on practical management by Traditional Owners across north Australia.

Northern Gulf Resource Management Group (NRM)

Savannah House, St George Street, Georgetown, Qld 4871
PO Box 63, Georgetown, Qld 4871
Tel: 07 4062 1330 **Fax:** 07 4062 1488
Email: admin@northerngulf.com.au
Web: www.northerngulf.com.au

The well-known ghost nets program works with many Indigenous groups to remove nets and marine debris from the coastline so they not pose a threat to marine life. Research is focused on enhancing baseline information on the region's biodiversity to ensure management and development decisions are based on good science.

North Queensland Dry Tropics (Formerly Burdekin Tropics) (NRM)

PO Box 1466, Townsville, Qld 4810
2 McIlwraith St, Townsville, Qld 4810
Tel: 07 4724 3544 **Fax:** 07 4724 3577
Email: info@bdtnrm.org.au
Web site: www.bdtnrm.org.au

NQ Dry Tropics has produced a number of publications, *Woodland Birds of the Burdekin Dry Tropics*, *Coastal Birds of the Burdekin Dry Tropics* and *Healthy Grazing Country Indicators Native Plants and Animals* to familiarise the community with these species and their habitats.

We have also acted as the Trustee of the Black Throated Finch Trust, which focuses on the protection of habitat for the Black-Throated Finch in the Townsville Region.

Oceania Chondrichthyan Society (OCS)

Email: info@oceaniasharks.org.au
Web: www.oceaniasharks.org.au

The OCS is an international non-profit organisation created 'To promote and facilitate the advancement of the scientific study of chondrichthyan fishes (sharks, skates, rays, and chimaeras), the promotion of education, conservation, and sustainable utilisation of natural resources, and to provide a medium for the exchange of information and ideas among those studying chondrichthyan fishes.'

Oceania Project

PO Box 646, Byron Bay, NSW 2481
Tel: 0418 797 326
Email: wally.trish@oceania.org.au
Web sites: www.oceania.org.au; iwhales.org

The Oceania Project is a not-for-profit research and education organisation affiliated with the Southern Cross University Whale Research Centre and South Pacific Whale Research Consortium. We are dedicated to raising awareness about Cetacea (whales, dolphins and porpoises) and the ocean environment.

Parrot Society of Australia

PO Box 1442, Stafford, Qld 4053
Mob: 0403284665 **Fax:** 07 3388 8493
Email: shaneahancock@gmail.com
Web site: www.parrotsociety.org.au

As the largest avicultural organisation in Australia for parrot enthusiasts, this dynamic club actively supports international and Australian parrot conservation, education and research projects and has established the Parrot Trust of Australia for this purpose.

Pittsworth District Landcare Association Inc.

'Nyleta' 168 Hogg Rd, Mt Tyson Qld 4356
Tel: 07 4693 8186 **Fax:** 07 4693 8205
Email: phalford2@bigpond.com

We are a Landcare group dedicated to promoting the adoption of land use and land management practices that ensure the future livelihood of local people while ensuring a balanced ecosystem is retained. We strongly advocate educating the community about the importance of conserving natural landscapes to support the recovery of our threatened animals as well as fostering cooperation between land users wherever action is needed to control land degradation.

Queensland Frog Society Inc.

PO Box 7017, East Brisbane, Qld 4169
Tel: 07 3366 1868 **Fax:** 07 3366 1868
Email: Qldfrogs@bigpond.net.au
Web site: www.qldfrogs.asn.au

The Queensland Frog Society was established 20 years ago to create awareness of Queensland's special frog fauna and to form an important link between the general public, business and government organisations. As a member of the Threatened Species Network since its inception, we are active in frog conservation. We carry out surveys, lobby for conservation, rehabilitate of sensitive habitat, liaise with other groups, volunteer representatives for recovery teams and offer an annual scholarship that helps researchers continue their work on threatened frogs.

Queensland Murray-Darling Committee (NRM)

PO Box 6243, 127b Campbell Street,
Toowoomba West, Qld 4350
Tel: 07 4637 6200 **Fax:** 07 4632 8062
Email: info@qmdc.org.au
Web site: www.qmdc.org.au

QMDC aims to improve the quality and extent of remnant native vegetation, improve the conservation status of native species and improve the wider community's awareness and understanding of the region's biodiversity. QMDC provides mapping and planning support, financial incentives, and technical advice to private and public land managers across a range of themes including vegetation and biodiversity, water and wetlands, land and soils, weeds and pest animal management.

Queensland Museum Brisbane

PO Box 3300, South Brisbane, Qld 4101
Cnr Grey and Melbourne Streets, South Brisbane, Qld 4101
Tel: 07 3840 7706 **Fax:** 07 3846 1226
Email: inquirycentre@qm.qld.gov.au
Web site: www.intranet.qm.qld.gov.au

Queensland Museum's internationally recognised scientists study the Museum's unique collections and make critical contributions to the protection of Queensland's threatened animals. We produce a wide range of publications on our state's threatened animals and offer a variety of services to the public including workshops, volunteer opportunities, outreach programs, training as well as the taxonomic collection and surveys of Queensland's threatened fauna.

Quoll Seekers Network

c/o Wildlife Preservation Society of Queensland,
95 William St, Brisbane, Qld 4000
Tel: 07 3221 0194 **Fax:** 07 3221 0701
Email: wpsq@wildlife.org.au
Web site: www.wildlife.org.au/projects/quolls/index.

Quoll Seekers Network (QSN) was established to raise community awareness of quolls in Queensland, gather information on quoll populations, and to help people enjoy living alongside quolls. The network aims to be Queensland's central non-government body for collecting and disseminating information about quoll populations in order to achieve good conservation outcomes for quoll species.

Rainforest Rescue

PO Box 40, Mullumbimby, NSW 2482
Tel: 02 6684 4360 or 1300 763 611 **Fax:** 02 6684 6737
Email: info@rainforestrescue.org.au
Web site: www.rainforestrescue.org.au

We purchase property in the Daintree lowland rainforest and protect the land in perpetuity as nature refuges or wildlife refuges and managed as habitat for the Southern Cassowary. We are also involved in rainforest rehabilitation work in the Daintree, clearing weed infested areas and planting tens of thousands of trees.

Reef Catchments Mackay Whitsunday Inc. (NRM)

PO Box 815, Mackay, Qld 4740
172 Victoria Street, Mackay, Qld 4740
Tel: 07 4968 4200 **Fax:** 07 4968 4299
Email: katrina.hanson@reefcatchments.com.au
Web site: www.reefcatchments.com.au

Reef Catchments partners with community groups and partner organisations in projects such as Coasts and Communities to develop, negotiate and implement on-ground works designed to protect key coastal species from ongoing disturbance and coastal development and the Clarke Connors Range Bushfire Consortium aimed at improving the capacity of landholders to manage conservation values through improved fire management.

Regent Honeyeater Project Inc.

PO Box 124, Benalla, Vic. 3672
Tel: 03 5761 1515
Email: ray@regenthoeneyeater.org.au
Web site: www.regenthoneyeater.org.au

A well established (15 years) Victoria-based and landholder supported revegetation project focusing on the three remaining key Regent Honeyeater habitats left in that state with more than 420,000 indigenous plants planted on over 400 individual sites, and a grand total of around 1160 ha of habitat restored. Project also includes activities such as seed collection, plant propagation, fencing, planting, direct seeding, mistletoe removal, environmental weeding, nest box placement, and wildlife monitoring.

Richmond Birdwing Conservation Network

PO Box 855, Kenmore, Qld 4069
Email: birdwing@wildlife.org.au
Web site: www.richmondbirdwing.org.au

The Richmond Birdwing Conservation Network (RBCN) is dedicated to ensuring the survival of the butterfly and propagating the host vines and habitat that it requires for survival. RBCN is also collaborating closely with DERM in a captive breeding project to overcome in-breeding depression in wild populations of the Richmond Birdwing Butterfly.

Save the Bilby Fund

PO Box 155, Charleville, Qld 4470
Tel: 07 4654 1255 **Fax:** 07 4654 1418
Email: bilby@epa.qld.gov.au
Web site: www.savethebilby.icemedia.com.au

The Save the Bilby Fund was founded to support research into the biology and ecology of the species, to identify the reasons for population and distribution decline and to re-establish populations, based on the biological knowledge gained through the research.

School for Field Studies (SFS)

PO Box 141, Yungaburra, Qld 4884
Tel: 07 4095 3656 **Fax:** 07 4095 3633
Web site: www.fieldstudies.org

The Australian branch of the US's oldest and largest environmental study abroad program, which combines hands-on environmental studies with scientific research to develop sustainable solutions to critical environmental problems. SFS students work with communities to discover practical ways to manage their natural resources.

Sea Turtle Foundation

PO Box 1190, Townsville, Qld 4810
Tel/Fax: 07 4721 2699 **Mob:** 0431 259 129
Email: info@seaturtlefoundation.org
Web site: www.seaturtlefoundation.org

Sea Turtle Foundation is conserving and protecting threatened sea turtles by conducting scientific research and conservation activities, campaigning, raising community awareness and creating school education programs to ensure there are turtles for tomorrow.

Southern Oceans Seabird Study Association (SOSSA)

PO Box 142, Unanderra, NSW 2526
Tel: 02 4271 6004 **Fax:** 02 4272 4626 **Mob:** 0418 603 007
Email: albatross@sossa-international.org; sossa@tpg.com.au
Web site: www.sossa-international.org

SOSSA encourages naturalists, both amateur and professional, to research the biodiversity of the Southern Ocean. SOSSA also aims to promote the use of standard methodologies and the publication of all findings. SOSSA will assist study groups or individuals in a standard format. SOSSA is willing to assist other studies and has the resources and trained personnel that may be of benefit to small independent studies or organisations.

South West NRM Ltd

Head Office: 66 Galatea Street, Charleville, Qld 4470
Tel: 07 4656 8500 **Fax:** 07 4654 1600
Email: swnrm@southwestnrm.org.au
Web site: www.southwestnrm.org.au

South West NRM is the natural resource management group for the Warrego, Paroo, Bulloo and Nebine catchments in South West Queensland. In the region, South West NRM engages the community in projects to protect riparian areas, control weeds and pests, recover native pastures and reduce soil erosion.

Sydney Aquarium Conservation Fund (SACF)

Sydney Aquarium, Darling Harbour, NSW 2000
Tel: 02 8251 7875 **Fax:** 02 9262 2583
Email: conservation@sacf.org.au
Web site: www.sacf.org.au

In June 2010, with the support of SACF, a team of researchers and volunteers from the University of Queensland, Sydney Aquarium, Sea World and Taronga Western Plains Zoo worked together to catch Dugongs from Moreton Bay, Queensland, to assess health and reproductive status of this vulnerable species. SACF also contributed funding for a satellite tag, which was deployed on a Hawksbill Turtle in Milman Island to determine movements and behaviours in key breeding areas of the species.

Symbiont Ecology Group Pty Ltd

51 Tristania St, Bangalow, NSW 2479
Tel: 02 6687 2142 **Fax:** 02 6697 2142
Email: symbiontecology@gmail.com
Web site: www.symbiontecology.com

Symbiont Ecology Group conducts research projects on the population and behavioural ecology of a variety of marine mammals, chondrichthyans and cheloniids that provides the scientific background and first-hand experience upon which to base their conservation programs that are delivered to schools, universities, communities and businesses. The aims of these programs are to raise public awareness to the plight of marine threatened species, mitigate threats to their environment and improve legislation.

Taronga Zoo

PO Box 20, Bradleys Head Rd, Mosman, NSW 2088
Tel: 02 9969 2777
Email: tz@zoo.nsw.gov.au
Web site: www.taronga.org.au

Taronga and Taronga Western Plains Zoos supports many Queensland threatened species, including mammals, birds, reptiles and amphibians. The zoos operate diverse programs ranging from animal husbandry, managed conservation breeding programs, breed-for release, recovery programs, field conservation projects, disease management and veterinary expertise.

Terrain (NRM)

Level 1, 88 Rankin Street, Innisfail, Qld 4860
PO Box 1756, Innisfail, Qld 4860
Tel: 07 4043 8000 **Fax:** 07 4061 4677
Email: info@terrain.org.au
Web site: www.terrain.org.au

Terrain uses a focal species approach to work with community groups and partner organisations to mobilise conservation actions that benefit a large range of species. In a unique science partnership with CSIRO, a framework for landscape scale conservation has been developed with a focus on the habitat needs of the Southern Cassowary at Mission Beach. This approach, which mobilises the local community, governments and other stakeholders to tackle complex issues for managing the long-term health of habitats and the threatened species that occupy them.

Threatened Species Network (TSN)

WWF-Australia, PO Box 4010, Wembley, WA 6913
Tel: 08 9387 6444 **Fax:** 08 9387 6180
Web site: www.wwf.org.au/ourwork/species/tsn

TSN works with community groups and other conservation partners to help stop the decline of threatened species. We bring together science with community needs, on-the-ground action with Australia's decision makers, and national issues with local concerns to help protect threatened species.

Tiaro and District Landcare group

PO Box 6, Tiaro, Qld 4650

Tel: 07 4129 6206
Email: tiarolandcare1@bigpond.com
Web site: www.maryriverturtle.com

Tiaro Landcare's Mary River Turtle project incorporates in-situ nest protection, predator control program, establishment of a Mary River Turtle support scholarship ($5000 cash + in kind) for PhD student undertaking new research; creation of a public display in Tiaro Craft Cottage on the Bruce Highway; fund raising through sales of homemade chocolate turtles; publication of the *Mary River Turtle Yesterday Today & Tomorrow*; creation of a website: www.maryriverturtle.com; and various educational activities.

Tolga Bat Hospital and Visitors Centre

PO Box 685, Atherton, Qld 4883
Tel: 07 4091 2683
Email: info@tolgabathospital.org
Web site: www.tolgbathospital.org.au

Tolga Bat Hospital is a community group, Landcare group and charity working specifically for the conservation of bats and their habitat. Our rescue work puts about 550 Spectacled Flying-foxes back into the wild each year, and our education work in schools and the onsite Visitor Centre raises awareness of the threats faced by all bats.

Tree-kangaroo and Mammal Group (TKMG)

PO Box 1409, Atherton, Qld 4883
Email: info@tree-kangaroo.net
Web site: www.tree-kangaroo.net/

The TKMG is an incorporated community group based on the Atherton Tablelands in Far North Queensland. Our aims are to conserve Far North Queensland mammals by promoting the awareness and knowledge of these animals, to undertake and assist with studies into their habitats and to liaise with groups and organisations with similar objectives.

University of New England

University of New England, Armidale, NSW 2351
Tel: 02 6773 3333 **Fax:** 02 6773 3100
Email: www.une.edu.au/askune
Web site: www.une.edu.au

University of Queensland

University of Queensland, Brisbane, Qld 4072
Tel: 07 3365 1111
Email: AdmissionsEnquiries@uq.edu.au
Web site: www.uq.edu.au

University of the Sunshine Coast

University of the Sunshine Coast, Locked Bag 4, Maroochydore DC, Qld 4558
Sippy Downs Drive, Sippy Downs, Qld 4556
Tel: 07 5430 1234
Email: information@usc.edu.au
Web site: www.usc.edu.au

University of Sydney

The University of Sydney, Sydney, NSW 2006
Tel: 02 9351 2222
Email: infocentre@mail.usyd.edu.au
Web site: www.usyd.edu.au

Wildlife Conservancy of Tropical Queensland

PO Box 526, Edge Hill, Qld 4870
Tel: 0408 932 303 or 0418 756 094
Email: info@wildlifequeensland.com.au
Web site: www.wildlifequeensland.com.au; www.mareebawetlands.org

WCTQ is at the forefront of work with the threatened Gouldian Finch, having run a breed and release program for several years from its base at the Mareeba Wetlands. It is also actively involved in researching one of Australia's least known birds, the Buff-breasted Button-Quail.

Wildlife Friendly Fencing Project

Tel: 07 4091 2683
Email: info@wildlifefriendlyfencing.com
Web: www.wildlifefriendlyfencing.com

Each year thousands of animals (more than 75 species) face a cruel death or permanent injury from entanglement on barbed wire, usually on the top strand. The Wildlife Friendly Fencing Project raises awareness about the impact of fencing on Australian wildlife, and develops guidelines for good practice.

Wildlife Preservation Society of Queensland (WPSQ)

95 William St, Brisbane, Qld 4000
Tel: 07 3221 0194 **Fax:** 07 3221 0701
Email: wpsq@wildlife.org.au
Web: www.wildlife.org.au

Wildlife Preservation Society of Queensland is a community environmental organisation with a wonderfully diverse and dedicated team of members, volunteers, supporters and staff united by a deep passion for wildlife and conservation. Quoll Seekers Network is one of the many programs that Wildlife Queensland runs – it is Queensland's central non-government body for collecting and disseminating information about this threatened carnivorous marsupial and campaigns to achieve the best conservation outcomes for quoll species.

Brisbane Branch WPSQ
PO Box 395, Clayfield, Qld 4011
Tel: 07 3221 0194
Email: brisbane@wpsq.org.au
Web site: www.wildlife.org.au/contact/branch/brisbane.html

Brisbane Branch runs a yearly summer season of Batty Boat Cruises up the Brisbane River, to educate the public about Flying-foxes and microbats.

Fraser Coast Branch WPSQ
PO Box 7396, Hervey Bay, Qld 4655
Tel: 07 4129 5979
Email: frasercoast@wildlife.org.au
Web site: www.wildlife.org.au/branches

Fraser Coast Branch carries out threatened frog and Koala surveys, is involved in revegetation and signage projects regarding the Grey-headed Flying-Fox, conducts threatened species workshops and educational activities for school children and is actively involved in lobbying for threatened species.

North West Brisbane Branch
PO Box 134, Albany Creek, Qld 4035
Email: nwbrisbane@wildlife.org.au
Web site: www.wildlife.org.au/branches

The North West Brisbane Branch is working in partnership with SEQ Catchments Glossy Black planting of food trees. This is done in conjunction with the local schools and environmental education centres and local council. It is not directly involved in management and recovery, but supports improving the biodiversity for the Glossy Black Cockatoos.

Wombat Foundation

GPO Box 2188, Sydney, NSW 2001
Email: enquiries@wombatfoundation.com.au
Website: www.wombatfoundation.com.au

The Wombat Foundation is a charitable organisation set up to support activities that aim to bring the Northern Hairy-nosed Wombat back from the brink of extinction. The Northern Hairy-nosed Wombat is one of the world's most threatened species. At last count there were only 138 individuals.

WWF-Australia

Head Office Sydney
Level 13, 235 Jones St, Ultimo, NSW 2007
PO Box 528, Sydney, NSW 2001
Tel: 02 9281 5515 **Fax:** 02 9281 1060

Brisbane
Level 3, 129 Margaret St, Brisbane, Qld 4000
Tel: 07 3003 1480 **Fax:** 07 3229 4431
Email: Form on web site
Web site: wwf.org.au

WWF is addressing major threats to species survival in Australia through broad-ranging campaigns and programs. We work directly with governments to ensure improved policy on species conservation, and our on-ground turtle conservation and shorebirds programs provide a coordinated, national approach to reversing population decline throughout Australia.

Appendix II: How to report a sighting

Each profile will have the names of organisations that are interested in hearing about any sightings of this species. If their web site does not offer a report form to fill in online, email or post them the information below. The correct identification of a species is necessary, otherwise the data is useless. Some organisations will require photographic evidence to accompany species records before they will be accepted for incorporation within corporate systems.

To report sightings to DERM, all public enquiries/reports are to go through the Government Customer Service Centre 13 74 68 or 13 QGOV or info@derm.qld.gov.au.

Observer information

Name:

Address:

Phone number:

Email:

Other observers (Names and contact info):

Sighting information

Date:

Time:

Suburb/Town/Location:

Address of property:

Nearby landmark:

UBD reference or GPS coordinates:

Habitat description (e.g. broad vegetation type, dominant tree species, topography, altitude):

Sighting conditions (time of day, weather, visibility, duration of observation, etc.):

How confident are you of your sighting? (e.g. 75%, 100%):

Who confirmed record (e.g. identified specimen or photographs)?

Species sighted

Number of animals sighted:

Estimated size:

Sex of animal:

Age class (e.g. egg, hatchling, juvenile, sub-adult, adult, tadpole)

Description:

Documentation (photos, tracks, carcass):

Other info

Is this a regular sighting?

Have you reported this sighting to any other organisations?

If yes, which ones?

Additional observations: How long did you watch it, what was it doing, etc.?

Contacts for sightings

Birds

Atlas of Australian Birds
<http://www.birdata.com.au/about_atlas.vm>

Birds Queensland
<http://birdsqueensland.org.au/report_sightings.php>

Glossy Black-cockatoo: Report sightings online: <www.glossyblack.org.au/form.html> or by email: GBC@biodiversity.tv and to Moreton Bay Regional Council: <http://www.moretonbay.qld.gov.au/environment.aspx?id=17197>

Powerful Owl: Report sightings to Moreton Bay Regional Council: <http://www.moretonbay.qld.gov.au/environment.aspx?id=17197>

Threatened Bird Network on (03) 9347 0757, or via email: tbn@birdsaustralia.com.au

Marine Life

Report sick, injured, stranded or dead marine life by calling the DERM Hotline 1300 130 372 or 13 74 68.

Sharks and Rays

Report any live sightings, fishing catches or dead beach-washed specimens to the Queensland Museum: 07 3840 7555 or email inquirycentre@qm.qld.gov.au.

Photographs with accompanying details of location and date can provide valuable records of the species in order to better understand its occurrence and distribution in Queensland. If a specimen is found dead or dies after capture, contact the Queensland Museum, because any specimens may provide important biological information on the species and contribute to collections, which may be used for future scientific study.

Whales

Report sightings to National Marine Mammal Database <http://www.marinemammals.gov.au/ammc-databases-and-applications>

Miscellaneous threatened animals

DERM: reports are to go through the Government Customer Service Centre (13 74 68 or 13 74 68 or email info@derm.qld.gov.au) only if the sighting can be confirmed.

Queensland Museum: 07 3840 7555 or email inquirycentre@qm.qld.gov.au.

Queensland Natural Resource Management Groups: <http://www.nrm.gov.au/nrm/qld.html>

Appendix III: Additional resources

(Anyone actively involved in recovery and management of Qld threatened species who would like to be added to this list in future editions, please email lee@ataglance.com.au)

Today, the recovery and management of our increasingly threatened flora and fauna is not only the responsibility of our governments, but also of each individual citizen. Our federal, state and local governments no longer have the funding necessary to carry out the huge amount of work necessary to ensure the wellbeing of our biodiversity. As a result, NGOs, private organisations, community groups, and even big business entities are pitching in to help recover and manage our threatened animals. Below are additional resources to help readers contribute to the recovery and management of our threatened species.

Web sites

The Land Manager website
www.landmanager.org.au

This website was created by the Tropical Savannas CRC and is supported by the Natural Heritage Trust. It aims to meet the information needs of north Australian land managers including pastoralists, Landcare groups, Indigenous groups and government agencies.

It also provides information specific to management of Northern Quoll habitats.

The Australian Wildlife Health Network
www.wildlifehealth.org.au/AWHN/home.aspx

The Australian Wildlife Health Network (AWHN: the Network) is an initiative of the Australian Government and is managed under the Wildlife Exotic Disease Preparedness Program (Australian Government Department of Agriculture, Fisheries and Forestry). Its mission is to promote and facilitate collaborative links across Australia in the investigation and management of wildlife health in support of human and animal health, biodiversity and trade. Its vision is for a nationally integrated wildlife health system for Australia.

The Australian Wildlife Health Network Bulletin
www.wildlifehealth.org.au/AWHN/Forum/ListForum.aspx

This section of the site summarises current wildlife health activities across Australia under a number of different areas of interest. It allows you to keep the rest of Australia up to date with what you are doing, and lets you and others see what is happening around the country.

Threatened species and ecological communities
www.environment.gov.au/biodiversity/threatened/index.html

DERM land management website
www.derm.qld.gov.au/land/index.html

Volunteer opportunities

Desert Ecology Research Group
Room 325a, Institute of Wildlife Research, School of Biological Sciences,
Heydon-Laurence Building A08, University of Sydney 2006
Tel: 02 9351 8577 **Fax:** 02 9351 4119
Email: bobby.tamayo@sydney.edu.au
aaron.greenville@sydney.edu.au
clbeh@bio.usyd.edu.au
Web site: <http://sydney.edu.au/science/biology/sites/dickmanlab/index.shtml>

DERM
<www.epa.qld.gov.au/about_the_epa/working_at_the_epa/volunteers_work_experience_and_cadetships/volunteers/where_to_start/>

Wildlife Preservation Society of Queensland
www.wildlife.org.au

MangroveWatch at Nudgee Beach, sponsored by WPSQ:
www.mangrovewatch.org.au/index.php?option=com_content&view=section&layout=blog&id=16&Itemid=300182

Queensland Museum Learning Resources Mangrove Challenge www.qm.qldgov.au/education/resources/mangrove

Wildlife Friendly Fencing
<www.wildlifefriendlyfencing.com>

Fish

Fish Atlas of North Australia
www.jcu.edu.au/vhosts/actfr/Projects/FishAtlas/Index.htm

Department of Primary Industries and Fisheries Queensland
www.dpi.qld.gov.au/28.htm

Department of Agriculture, Fisheries and Forestry
www.daff.gov.au/fisheries

Birds

Asia Pacific Shorebird Network
www.shorebirdnetwork.org

Australian Bird Study Association
www.absa.asn.au

Birds Australia (BA)
www.birdsaustralia.com.au

Birds Australia Capricornia
www.irock.com.au/~birdscapricornia

Birds Australia Southern Queensland
www.basq.org.au

Birds Australia North Queensland
http://birdsaustralianq.org

Bird Observation and Conservation Australia (BOCA)
www.birdobservers.org.au

Townsville Region BOCA
www.trboc.org.au

Birds Queensland (Queensland Ornithological Society)
http://birdsqueensland.org.au

Birding Southern Queensland
www.aviceda.org

Nature Refuge Landholders Association Queensland
www.naturerefuges.com

Queensland Conservation
www.qccqld.org.au/

Queensland Naturalists Club
www.qnc.org.au

Toowoomba Birds Observer's
<http://toowoombabirdobservers.org.au

Tweed Bird Observer's
www.bigvolcano.com.au/custom/birdos

Write letters to:

Your local council: www.dip.qld.gov.au/local-government-directory.html

Your regional Natural Resource Management Group: www.regionalnrm.qld.gov.au/my_region/index.html

Qld state government officials: www.qgd.qld.gov.au/cabinet_ministers.html#addresses

DERM: www.derm.qld.gov.au/contactus/index.html

Federal Minister for Sustainability, Environment, Water, Population and Communities: www.environment.gov.au/about/contacts/index.html

Sea Turtles

If you see a sea turtle nesting or at sea, please note where and when you saw it. If possible, also try to identify the species and how many turtles you saw.

If you find a sea turtle with a tag clamped on its flipper, write down:

- the number stamped on the tag
- when, how and where the turtle was caught or seen
- what happened to the turtle
- your name and address.

Don't remove the tag from the turtle unless it has died!

Send this information (with the tag if recovered from a dead turtle) to:

Queensland Turtle Research
Department of Environment & Resource Management
GPO Box 2454
Brisbane, Qld 4001

Please report all sightings of any sick, injured or dead sea turtles, Dugongs, whales or other marine life by calling the Marine Stranding Reporting line: 1300 130 372. **Further information about Qld sea turtles can be obtained from:**

Mon Repos Conservation Park
Queensland Parks and Wildlife Service
PO Box 1735
Bundaberg, Qld 4670
Tel: 07 4159 1652 **Fax:** 07 4159 1630

Suggested reading

Allen GR, Midgley SH and Allen M (2002) *Field Guide to the Freshwater Fishes of Australia*. Western Australian Museum, Perth.

Bannister J (2008) *Great Whales*. CSIRO Publishing, Melbourne.

Braby MF (2000) *Butterflies of Australia*. CSIRO Publishing, Melbourne.

Burbidge A (2004) *Threatened Animals of Western Australia*. Department of Land Conservation and Management, Kensington, WA.

Cary G, Lindenmayer D and Dovers S (2003) *Australia Burning: Fire Ecology, Policy and Management Issues*. CSIRO Publishing, Melbourne

Cavanagh RD, Kyne PM, Fowler SL, Musick JA and Bennett MB (2003) 'The conservation status of Australasian chondrichthyans: report of the IUCN Shark Specialist Group Australia and Oceania Regional Red List Workshop, Queensland, Australia, 7–9 March 2003'. School of Biomedical Sciences, University of Queensland, Brisbane.

Cermak M (2008) *Spectacular Snakes of Australia*. CSIRO Publishing, Melbourne.

Christidis L and Boles WE (2008) *Systematics and Taxonomy of Australian Birds*. CSIRO Publishing, Melbourne.

Churchill S (2009) *Australian Bats*. Allen & Unwin, Sydney.

Cogger HG (2000) *Reptiles and Amphibians of Australia*. Reed New Holland, Sydney.

Collis B (2002) *Fields of Discovery: Australia's CSIRO*. Allen & Unwin, Crows Nest, NSW.

CSIRO Wildlife and Ecology (1999) *Of Beauty Rich and Rare*. CSIRO Publishing, Melbourne.

Ehman H (1992) *Encyclopedia of Australian Animals: Reptiles*. Collins Angus & Robertson, Sydney.

Fox ID, Neldner VJ, Wilson GW and Bannink PJ (2001) *The Vegetation of the Australian Tropical Savannas*. Environmental Protection Agency, Brisbane.

Garnett ST, Szabo JK and Dutson G (2011) *The Action Plan for Australian Birds 2010*. CSIRO Publishing, Melbourne.

Grant EM (2008) *Grant's Guide to Fishes*. 11th edn. EM Grant Pty Limited, Brisbane.

Heinsohn R (2008) *Life in the Cape York Rainforest*. CSIRO Publishing, Melbourne.

Higgins PJ, Marchant S and Davies SJJF (2003) *Handbook of Australian, New Zealand and Antarctic Birds* (HANZAB). Oxford University Press, Melbourne. (Seven Volumes)

Hoskin C and Héro J-M (2008) *Rainforest Frogs of the Wet Tropics, North-East Australia*. Griffith University, Gold Coast.

Hutton D and Connors (1999) *A History of the Australian Environment Movement*. Cambridge University Press, Cambridge.

IUCN (2001) IUCN Red List Categories and Criteria. Version 3.1. IUCN Species Survival Commission, Gland, Switzerland.

IUCN (2010) IUCN Red List of Threatened Species. Version 2010.4. IUCN, Cambridge, UK, <http://www.iucnredlist.org>.

Last PR and Stevens JD (2009) *Sharks and Rays of Australia*. 2nd edn. CSIRO Publishing, Melbourne.

Lindenmayer D (2003) *Wildlife on Farms: How to Conserve Native Animals*. CSIRO Publishing, Melbourne.

Lindenmayer D (2007) *On Borrowed Time: Australia's Environmental Crisis and What We Must Do About It*. Penguin, Melbourne.

Lindenmayer D, Dovers S, Harriss Olson M and Morton S (Eds) (2008) *Ten Commitments: Reshaping the Lucky Country's Environment*. CSIRO Publishing, Melbourne.

Low T (1999) *Feral Future*. Viking/Penguin Books, Australia.

Low T (2002) *The New Nature*. Viking/Penguin Books, Australia.

Markus N (2009) *On Our Watch*. Melbourne University Press, Melbourne.

Marshall AJ (Ed.) (1966) *The Great Extermination*. William Heineman Ltd, Melbourne.

Morritz C and Kikkawa L (2004) *Conservation Biology in Australia and Oceania*. Surrey Beatty & Sons, Sydney.

Orr A and Kitching R (2010) *The Butterflies of Australia*. Allen & Unwin, Sydney.

Pittock B (2003) *Climate Change: An Australian Guide to the Science and Potential Impacts*. Australian Greenhouse Office, Canberra.

Pusey B, Kennard M and Arthington A (2004) *Freshwater Fishes of North-Eastern Australia*. CSIRO Publishing, Melbourne.

Robin L, Heinsohn R and Joseph L (Eds) (2009) *Boom and Bust: Bird Stories for a Dry Country*. CSIRO Publishing, Melbourne.

Sattler PS and Williams RD (Eds) (1999) *The Conservation Status of Queensland's Bioregional Ecosystems*. Environmental Protection Agency, Brisbane.

Shine R (1991) *Australian Snakes: A Natural History*. Reed New Holland, Sydney.

Stanisic J, Shea M, Potter D and Griffiths O (2010) *Australian Land Snails Volume 1*. Queensland Museum, Brisbane.

Theischinger G and Hawking JH (2006) *The Complete Field Guide to Dragonflies of Australia*. CSIRO Publishing, Melbourne.

UNEP, Global Biodiversity Assessment (1995) *Human Influences on Biodiversity*. Cambridge University Press, Cambridge.

Van Dyck S and Strahan R (Eds) (2008) *The Mammals of Australia*. 3rd edn. Reed New Holland, Sydney.

Wilson S (2003) *Reptiles of the Southern Brigalow Belt*. World Wide Fund, Australia.

Wilson S (2005) *A Field Guide to Reptiles of Queensland*. New Holland Publishers, Sydney.

Wilson S and Swan G (2010) *A Complete Field Guide to Reptiles of Australia*. 3rd edn. New Holland Publishers, Sydney.

Woinarski J, Pavey C, Kerrigan R, Cowie I and Ward S (Eds) (2007) *Lost From our Landscape: Threatened Species of the Northern Territory*. Northern Territory Government, Darwin.

Index

www.ingramcontent.com/pod-product-compliance
Lightning Source LLC
LaVergne TN
LVHW071200210926
848482LV00014B/694